Introduction to Numerical Analysis Using MATLAB®

License, Disclaimer Of Liability, And Limited Warranty

The CD-ROM that accompanies this book may only be used on a single PC. This license does not permit its use on the Internet or on a network (of any kind). By purchasing or using this book/CD-ROM package(the "Work"), you agree that this license grants permission to use the products contained herein, but does not give you the right of ownership to any of the textual content in the book or ownership to any of the information or products contained on the CD-ROM. Use of third party software contained herein is limited to and subject to licensing terms for the respective products, and permission must be obtained from the publisher or the owner of the software in order to reproduce or network any portion of the textual material or software (in any media) that is contained in the Work.

Infinity Science Press LLC ("ISP" or "the Publisher") and anyone involved in the creation, writing or production of the accompanying algorithms, code, or computer programs ("the software") or any of the third party software contained on the CD-ROM or any of the textual material in the book, cannot and do not warrant the performance or results that might be obtained by using the software or contents of the book. The authors, developers, and the publisher have used their best efforts to insure the accuracy and functionality of the textual material and programs contained in this package; we, however, make no warranty of any kind, express or implied, regarding the performance of these contents or programs. The Work is sold "as is" without warranty (except for defective materials used in manufacturing the disc or due to faulty workmanship);

The authors, developers, and the publisher of any third party software, and anyone involved in the composition, production, and manufacturing of this work will not be liable for damages of any kind arising out of the use of (or the inability to use) the algorithms, source code, computer programs, or textual material contained in this publication. This includes, but is not limited to, loss of revenue or profit, or other incidental, physical, or consequential damages arising out of the use of this Work.

The sole remedy in the event of a claim of any kind is expressly limited to replacement of the book and/or the CD-ROM, and only at the discretion of the Publisher.

The use of "implied warranty" and certain "exclusions" vary from state to state, and might not apply to the purchaser of this product.

INTRODUCTION
TO
NUMERICAL ANALYSIS
USING MATLAB®

Rizwan Butt, Ph.D.

INFINITY SCIENCE PRESS LLC
Hingham, Massachusetts
New Delhi

Publisher: David Pallai

INFINITY SCIENCE PRESS LLC
11 Leavitt Street
Hingham, MA 02043
Tel. 877-266-5796 (toll free)
Fax 781-740-1677
info@infinitysciencepress.com
www.infinitysciencepress.com

This book is printed on acid-free paper.

Rizwan Butt. *Introduction to Numerical Analysis Using MATLAB®*.
ISBN: 978-1-934015-23-0

Library of Congress Cataloging-in-Publication Data

Butt, Rizwan.
Introduction to numerical analysis using MATLAB / Rizwan Butt.
p. cm.
Includes index.
ISBN-13: 978-1-934015-23-0 (hardcover with cd-rom : alk. paper)
1. MATLAB. 2. Numerical analysis--Data processing. I. Title.
QA297.B88 2007
518--dc22

2007019296

07 08 09 5 4 3 2 1

Our titles are available for adoption, license or bulk purchase by institutions, corporations, etc. For additional information, please contact the Customer Service Dept. at 877-266-5796 (toll free).

Requests for replacement of a defective CD-ROM must be accompanied by the original disc, your mailing address, telephone number, date of purchase and purchase price. Please state the nature of the problem, and send the information to INFINITY SCIENCE PRESS, 11 Leavitt Street, Hingham, MA 02043.

The sole obligation of INFINITY SCIENCE PRESS to the purchaser is to replace the disc, based on defective materials or faulty workmanship, but not based on the operation or functionality of the product.

Printed in Canada

To my parents, who gave their children the
opportunities to develop a love of learning.

Contents

Notations

$\mathbb{R}$	Set of real numbers
$\mathbb{R}^n$	Real n-dimensional space
$\mathbf{x}$	Column vector or element of $\Re^n$
$\mathbf{x}^T$	Transpose of the vector $\mathbf{x}$
A^T	Transpose of a matrix A
$\det(A)$	Determinant of the matrix A
M_{ij}	Minor of a matrix
A^{-1}	Inverse matrix of the matrix A
$[A\ \mathbf{b}]$	Augmented matrix
$\mathbf{I_n}$	$n \times n$ identity matrix
$\rho(A)$	Spectral radius of a matrix A
$K(A)$	The condition number of a matrix A
$\forall$	For all
$\Vert\ \Vert$	An arbitrary norm
$\Vert\Vert_1$	The l_1 norm of vector or matrix
$\Vert\ \Vert_2$	the l_2 norm of vector or matrix
$\Vert\ \Vert_\infty$	The l_∞ norm of vector or matrix
Δ	Forward difference operator
∇	Backward difference operator
δ	Central difference operator
$J(\mathbf{x})$	Jacobian matrix
$f[.]$	Divided difference of the function f
$\begin{pmatrix} n \\ k \end{pmatrix}$	The kth binomial coefficient of order n
$C[a,b]$	Set of all functions continuous on [a,b]
$C^n[a,b]$	Set of all functions having n continuous derivatives on [a,b]
$O(.)$	Order of convergence
$\approx$	Approximately equal to
ϵ	Small positive number
IVP	Initial-value problem
BVP	Boundary-value problem
$(\lambda, \mathbf{X})$	Eigenpair for $n \times n$ matrix A

Preface

I have written this book for introductory and advanced courses in numerical methods and numerical analysis for mathematicians, computer scientists, engineers, and other scientists. Numerical analysis is the branch of mathematics concerned with the theoretical foundations of numerical algorithms for the solution of problems arising in scientific applications. The subject addresses a variety of questions ranging from the approximation of functions and integrals to the approximate solution of algebraic, transcendental, differential and integral equations, with particular emphasis on the stability, accuracy, efficiency and reliability of numerical algorithms.

The intention of this book is to provide a gentle and sympathic introduction to many of the problems of scientific computing, and the wide variety of methods used for their solutions. The presentation of each numerical method is based on the successful teaching methodology of providing examples and geometric motivation for a method, and a concise statement of the steps to carry out the computation, before giving a mathematical derivation of the process or a discussion after more theoretical issues that are relevant to the use and understanding of the topic. Each topic is illustrated by examples that range in complexity from very simple to moderate. Geometrical or graphical illustrations are included whenever they are appropriate.

To provide maximum teaching flexibility, each chapter and each section begins with the basic, elementary material and gradually builds up to the more advanced material. The last section of each chapter gives a brief summary of the methods for solving the kinds of problems covered in the chapter. The level of mathematical justification is determined largely by the desire to keep the mathematical prerequisites to a minimum. Thus, for example, no knowledge of linear algebra is assumed beyond the basic matrix algebra, and analytical results are based on a sound knowledge of the calculus.

In this book, practical justification of the methods is presented through computer examples through the use of MATLAB. In recent years, the number of MATLAB users has dramatically increased and now includes professionals who were trained in other high-level languages, Fortran, C, etc. but are now switching to MATLAB as well as students who are learning MATLAB as their first programming language. The surge of popularity in MATLAB is related to the increasing popularity of UNIX and computer graphics. To what extent numerical computations in the future will be programmed in MATLAB is uncertain. Nonetheless, there is no question that a need exists for a comprehensive text especially geared to the requirements of those who want to learn, or use, numerical methods in MATLAB. This book has been written in response to this need. A short introduction of MATLAB is given in Appendix B and the programs in the text serve as further examples.

The objectives of using MATLAB in this book include: *(1)* to be easily understood by undergraduate students with minimal knowledge of MATLAB, *(2)* to enable students to practice the methods in MATLAB, *(3)* to provide the short programs that can be easily used for scientific applications with or without modifications, and *(4)* to provide software that is easy to understand. Topics covered in this book range from the basics of numerical methods through to intermediate level subjects. Some chapters (first six) can be taught in undergraduate courses with certain omissions, while the chapter seven is recommended for an introductory graduate course in numerical methods. The material in this book has been tested many times at both undergraduate and graduate levels and has been adopted in many universities worldwide. Chapter 1 covers different kinds of errors that are preparatory subjects for numerical computations. To explain the sources of these errors, there are brief discussions of Taylor series and how numbers are computed and saved in computers.

Chapter 2 concerns iterative methods for solving nonlinear equations, including the bisection method, false position method, Newton's method and secant method for simple roots, Scholder's method and modified Newton method for multiple roots, Newton method for system of nonlinear equations, and finally Horner's method, Muller's method, and Bairstow's method for polynomials.

Chapter 3 covers the basic concepts of matrices and determinants and describes the basic computational methods to solve nonhomogeneous linear equations. Cramer rule, Gaussian elimination, Gauss-Jordan and LU decomposition methods are discussed. Ill-conditioned problems are discussed. The Jacobi, Gauss- Seidel, SOR and Conjugate Gradient iterative methods are also explained.

Chapter 4 describes Lagrange and Newton-divided difference interpolations in the start. Then, it covers the interpolation by cubic splines. The causes and behavior of interpolation errors are explained with an intuitive approach. This chapter also describes curve fitting of experimental data based on least square methods.

Chapter 5 covers the basic concepts of the differentiation. It develops a systematic method to derive difference approximations with the truncation error term. This chapter also covers numerical integration methods, starting with simple but fundamental integration methods such as Trapezoidal and Simpson's rules. The Newton-Cotes formulas are then introduced. It also includes the Repeated Trapezoidal rule, Romberg method and Gauss integration method.

Chapter 6 describes numerical methods of ordinary differential equations, including Taylor's method of order n, Runge-Kutta methods and predictor-corrector methods. Also, we discussed the Shooting and finite difference method for the solution of two-point boundary-value problems.

Chapter 7 covers the selected methods of computing matrix eigenvalues. The approach discussed here should help the students to understand the relation of eigenvalues to roots of the characteristic equations. We define eigenvalues and eigenvectors and study several examples. We discuss diagonalization of matrices and the computation of powers of diagonalizable matrices. Various numerical methods are also discussed for the eigenvalues of matrices. Among them are, the power iterative methods, the Jacobi method, the Given's method, the Householder method, the QR iteration method, the LR method and the Singular Value Decomposition method.

In each chapter we discussed a number of examples whose purpose is to enhance the students understanding of the relevant numerical method. Moreover, we have not discussed such topics such as the numerical solution of partial differential equations, or optimization. These topics properly fall in the domain of advanced numerical methods.

In Appendix *A* we discuss some mathematical preliminaries and in Appendix *B* we discuss the basic commands for the software package MATLAB. Finally, in Appendix *C* we give answers to selected odd-numbered exercises for each chapter. In general, the proofs in the theoretical exercises are omitted.

Rizwan Butt
January, 2007

Acknowledgment

First of all I thank God for the health, strength, and perseverance needed to complete this book. This book evolved over a period of several years through many different courses populated by hundreds of undergraduate and graduate students. To all my students and colleagues who have offered suggestions, corrections, criticisms, or just moral support, I offer my heartfelt thanks, and I hope to see as many of you as possible at some point in the future, so that I can convey my feelings to you in person. I express my appreciation to the many talented people whose knowledge and skills have contributed to this book: Kh. Zafar Ellahi, Yacine Benhadid, Esia Al-Said, Omer Hamid, and Salah Hasan. I want to offer particular thanks to Prof. Khalida Noor and Prof. Aslam Noor, two exceptionally fine teachers, for giving me a chance to pursue my dreams. I am particularly indebted to Mr. Sarwar Khan and Mohammed Balfageh for suggestions and helping on writing the manuscripts in Latex format. Last, and certainly not least, neither this book nor anything else I have done in my career would have been possible without the love, help, and unwavering support from Saima, my friend, partner, and my wife. I dedicate this book to Saima and my children, Fatima, Usman, and Fouzan, and to the memory of my parents.

Rizwan Butt
January, 2007

Chapter 1

Number Systems and Errors

1.1 Introduction

This book provides an introduction to numerical analysis and is intended to be used by students of mathematics, engineering, and science. Numerical analysis can be defined as the development and implementation of techniques to find numerical solutions to mathematical problems. It involves engineering and physics in converting a physical phenomenon into a mathematical method; it involves mathematics in developing techniques for the approximate solution of mathematical equations describing the model, and finally it involves computer science for the implementation of these techniques in an optimal fashion for the particular computer available. With the accessibility of computers, it is now possible to get rapid and accurate solutions to many complex problems that create difficulties for the mathematician, engineer, and scientist. In elementary calculus we learned how to differentiate and integrate to get exact answers to a remarkably diverse range of realistic problems that could not be solved by purely algebraic methods. Unfortunately, from a practical point of view, the techniques of elementary (or even advanced) calculus alone are not adequate for solving calculus type problems such as solving polynomial equations of degrees greater than four or even simple equations such as

$$x = \cos x$$

or to evaluate integrals of type

$$\int_a^b e^{x^2}\,dx \quad \text{and} \quad \int_a^b \frac{\sin x}{x}\,dx \qquad \text{etc.}$$

However, it is impossible to get the exact solutions of these problems. Even when an analytical solution can be found, it may be more theoretical than practical. Fortunately, one rarely needs exact answers. Indeed, in the *real world* problems themselves are usually inexact because they are generally possessed in terms of parameters that are measured, hence only approximate. What we are likely to require in a realistic situation is not an exact answer but rather one having a prescribed accuracy. The basic approach used to solve problems in numerical analysis is algorithm, which are used to describe a step-by-step procedure and require a finite number of steps. So a numerical method is an algorithm, which consists of a sequence of arithmetic and logical operations and which produces an approximate solution within any prescribed accuracy. There are different numerical methods for the solution of a problem but the particular method chosen depends on the context from which the problem is taken.

Numerical methods deal with numbers. In the following sections we consider methods for representing numbers on computers and the error introduced by these representations. In addition, we exam the sources of various types of computational errors. In the end of the chapter we also give a brief summary of the methods for solving the problems covered in the chapter.

1.2 Number Representation and Base of Numbers

The number system we use daily is called the *decimal system*. The *base* of the decimal number system is **10**. The familiar decimal notation for numbers employs the digits $0, 1, 2, 3, 4, 5, 6, 7, 8, 9$. When we write down a whole number such as 478325, the individual digits represent coefficients of powers of 10 as follows:

$$\begin{aligned} 478325 &= 5 + 20 + 300 + 8000 + 70000 + 400000 \\ &= 5 \times 10^0 + 2 \times 10^1 + 3 \times 10^2 + 8 \times 10^3 + 7 \times 10^4 + 4 \times 10^5. \end{aligned}$$

Thus, in general, a string of digits represents a number according to the formula

$$a_n a_{n-1} \cdots a_1 a_0 = a_0 \times 10^0 + a_1 \times 10^1 + \cdots + a_{n-1} \times 10^{n-1} + a_n \times 10^n. \quad (1.1)$$

This takes care of the positive whole numbers. A number between 0 and 1 is represented by a string of digits to the right of a decimal point. For example,

$$0.8543 = 8 \times 10^{-1} + 5 \times 10^{-2} + 4 \times 10^{-3} + 3 \times 10^{-4}.$$

Thus, in general, a string of digits represents a number according to the formula

$$0.b_1 b_2 b_3 \cdots = b_1 \times 10^{-1} + b_2 \times 10^{-2} + b_3 \times 10^{-3} + \cdots. \quad (1.2)$$

For a real number of the form

$$a_n a_{n-1} \cdots a_1 a_0 . b_1 b_2 \cdots = \sum_{k=0}^{n} a_k 10^k + \sum_{k=1}^{\infty} b_k 10^{-k} \quad (1.3)$$

the *integer part* is the first summation in the expansion and the *fractional part* is the second. Computers, however, don't use the decimal system in computations and memory but use the *binary system.* The binary system is natural for computers because computer memory consists of a huge number of electronic and magnetic recording devices and each element has only "on" and "off" statues.

In the binary system the base is 2, and the integer coefficients may take the values 0 or 1. The digits 0 and 1 are called *bits,* which is short for binary digits. For example, the number 1110.11 in the binary system represents the number

$$1 \times 2^4 + 1 \times 2^3 + 1 \times 2^2 + 0 \times 2^1 + 1 \times 2^{-1} + 1 \times 2^{-2}$$

in the decimal system.

There are other base systems used in computers, in particular, the *octal* and *hexadecimal* systems. The base for the octal system is **8** and for the hexadecimal system **16**. These two systems are close relatives of the binary system and can be translated to and from binary easily. Expressions in octal or hexadecimal are shorter than in binary, so they are easier for humans to read and understand. Hexadecimal also provides more efficient use of memory space for real numbers. If we use another base, say, β, then numbers represented in the β system look like this:

$$(a_n a_{n-1} \cdots a_1 a_0 . b_1 b_2 \cdots)_\beta = \sum_{k=0}^{n} a_k \beta^k + \sum_{k=1}^{\infty} b_k \beta^{-k}. \quad (1.4)$$

The digits are $0, 1, 2, \ldots, \beta - 1$ in this representation. If $\beta > 10$, it is necessary to introduce symbols for $10, 11, \ldots, \beta - 1$. In this system based on 16, we use A, B, C, D, E, F for $10, 11, 12, 13, 14, 15$, respectively. Thus, for example,

$$(2BED)_{16} = D + E \times 16 + B \times 16^2 + 2 \times 16^3 = 11245.$$

The base of a number system is also called the *radix*. The base of a number is denoted by a subscript, for example, $(4.445)_{10}$ is 4.445 in base 10 (decimal), $(1011.11)_2$ is 1011.11 in base 2 (binary), and $(18C7.90)_{16}$ is $18C7.90$ in base 16 (hexadecimal). The conversion of an integer from one system to another is fairly simple and can probably best be presented in terms of an example. Let $k = 275$ in decimal form; that is, $k = (2 \times 10^2) + (7 \times 10^1) + (5 \times 10^0)$. Now $(k/16^2) > 1$ but $(k/16^3) < 1$, so in hexadecimal form k can be written as $k = (\alpha_2 \times 16^2) + (\alpha_1 \times 16^1) + (\alpha_0 \times 16^0)$. Now, $275 = 1(16^2) + 19 = 1(16^2) + 1(16) + 3$, and so the decimal integer, 275, can be written in hexadecimal form as 113; that is,

$$275 = (275)_{10} = (113)_{16}.$$

The reverse process is even simpler. For example,

$$(5C3)_{16} = 5(16^2) + 12(16) + 3 = 1280 + 192 + 3 = (1475)_{10}.$$

Conversion of a hexadecimal fraction to a decimal is similar. For example,

$$\begin{aligned}(0.2A8)_{16} &= (2/16) + (A/16^2) + (8/16^3) = (2(16^2) + 10(16) + 8)/16^3 \\ &= 680/4096 = 0.166\end{aligned}$$

carries only three digits in decimal form. Conversion of a decimal fraction to hexadecimal (or binary) proceeds as in the following example. Consider the number $r_1 = 1/10 = 0.1$ (decimal form). Then there exist constants $\{\alpha_k\}_{k=1}^{\infty}$ such that

$$r_1 = 0.1 = \alpha_1/16 + \alpha_2/16^2 + \alpha_3/16^3 + \cdots.$$

Now

$$16r_1 = 1.6 = \alpha_1 + \alpha_2/16 + \alpha_3/16^2 + \cdots.$$

Thus, $\alpha_1 = 1$ and

$$r_2 \equiv .6 = \alpha_2/16 + \alpha_3/16^2 + \alpha_4/16^3 + \cdots.$$

Again,

$$16r_2 = 9.6 = \alpha_2 + \alpha_3/16 + \alpha_4/16^2 + \cdots$$

so $\alpha_2 = 9$, and

$$r_2 \equiv .6 = \alpha_3/16 + \alpha_4/16^2 + \cdots.$$

From this stage on we see that the process will repeat itself, and so we have $(0.1)_{10}$ equals the infinitely repeating hexadecimal fraction, $(0.1999\cdots)_{16}$. Since $1 = (0001)_2$ and $9 = (1001)_2$ we also have the infinite binary expansion

$$r_1 = (0.1)_{10} = (0.1999\cdots)_{16} = (0.000110011001\cdots)_2.$$

Example 1.1 *The conversion from one base to another base is*

$$\begin{array}{llllll} (17)_{10} & = & (10001)_2 & = & (21)_8 & = & (11)_{16} \\ (13.25)_{10} & = & (1101.01)_2 & = & (15.2)_8 & = & (D.4)_{16}. \end{array}$$

1.2.1 Normalized Floating-point Representation

Unless numbers are specified as integers, they are stored in the computer in what is known as normalized floating-point form. This form is similar to the scientific notation used as a compact form for writing very small or very large numbers. For example, the number 0.0000123 may be written in scientific notation as 0.123×10^{-4}. In general, every nonzero real number x has a floating-point representation

$$x = \pm\mathbf{M} \times \mathbf{r}^{\mathbf{e}}, \quad \text{where} \quad \frac{1}{\mathbf{r}} \leq \mathbf{M} < 1, \quad \text{or} \quad -\frac{1}{\mathbf{r}} \geq \mathbf{M} > -1,$$

where

$$\mathbf{M} = \sum_{k=1}^{t} \mathbf{d}_k \mathbf{r}^{-k}.$$

Here, $\mathbf{M}$ is called the *mantissa*, $\mathbf{e}$ is an integer called the *exponent*, $\mathbf{r}$ the *base*, $\mathbf{d_k}$ is the value of the *kth* digit, and $\mathbf{t}$ is the maximum number of digits allowed in the number. When $r = 10$, then the nonzero real number x has a normalized floating decimal point representation

$$x = \pm\mathbf{M} \times 10^{\mathbf{e}},$$

where the normalized mantissa $\mathbf{M}$ satisfies $\frac{1}{10} \leq \mathbf{M} < 1$. Normalization consists of finding the exponent $\mathbf{e}$ for which $|x|/10^{\mathbf{e}}$ lies in the interval $[\frac{1}{10}, 1)$, then taking $\mathbf{M} = |x|/10^{\mathbf{e}}$. This corresponds to "floating" the decimal point to the left of the leading significant digit of x's decimal representation, then adjusting $\mathbf{e}$ as needed. For example,

$$\begin{array}{rlrr} -12.75 & \text{has representation} & -0.1275 \times 10^2 & (\mathbf{M} = 0.1275, \mathbf{e} = 2). \\ 0.1 & \text{has representation} & +0.1 \times 10^0 & (\mathbf{M} = 0.1, \mathbf{e} = 0). \\ \frac{1}{15} = 0.066 & \text{has representation} & (\frac{1}{15} \times 10) \times 10^{-1} & (\mathbf{M} = 0.66, \mathbf{e} = -1). \end{array}$$

A *machine number* for a calculator is a real number that it stores exactly in normalized floating-point form. For the calculator storage, a nonzero x is a machine number if and only if its normalized floating decimal point representation is of the form

$$x = \pm\mathbf{M} \times 10^{\mathbf{e}},$$

where

$$\begin{array}{rcl} \mathbf{M} & = & 0.d_1d_2 \cdots d_k (d_k = 0, 1, 2, \ldots, 9), \quad \text{with} \quad d_1 \neq 0 \\ \mathbf{e} & = & -100, -99, \cdots, +99. \end{array}$$

The condition $d_1 \neq 0$ ensures normalization (that is, $\mathbf{M} \geq \frac{1}{10}$).

Computers use a normalized floating-point binary representation for real numbers. The computer stores a binary approximation to x:

$$x = \pm\mathbf{M} \times 2^{\mathbf{e}}.$$

Normalization in this case consists of finding the unique exponent e for which $|x|/2^e$ lies in the interval $(\frac{1}{2}, 1)$, and then taking $|x|/2^e$ as $\mathbf{M}$. For example,

$$\begin{array}{lll} -12.75 = -\frac{51}{4} & \text{can be represented as} & -(\frac{51}{4}.\frac{1}{16}).2^4 \quad (\mathbf{M} = \frac{51}{64}, \mathbf{e} = 4). \\ 0.1 = \frac{1}{10} & \text{can be represented as} & +(\frac{1}{10}.8).2^{-3} \quad (\mathbf{M} = \frac{4}{5}, \mathbf{e} = -3). \\ \frac{1}{15} = 0.06666\cdots & \text{can be representation} & (\frac{1}{15}.8).2^{-3} \quad (\mathbf{M} = \frac{8}{15}, \mathbf{e} = -3). \end{array}$$

Computers have both an integer mode and floating-point mode for representing numbers. The *integer mode* is used for performing calculations that are known to be integer values and have limited usage for numerical analysis. *Floating-point numbers* are used for scientific and engineering applications. It must be understood that any computer implementation of equation $x = \pm\mathbf{M}\times 2^{\mathbf{e}}$ places restrictions on the number of digits used in the mantissa $\mathbf{M}$, and the range of the possible exponent $\mathbf{e}$ must be limited. Computers that use 32 bits to represent single-precision real numbers use eight bits for the exponent and 24 bits for the mantissa. They can represent real numbers whose magnitude is in the range $2.938736E-39$ to $1.701412E+38$ (that is, 2^{-128} to 2^{127}) with six decimal digits of numerical precision (for example, $(2^{-23} = 1.2 \times 10^{-7})$.

Computers that use 48 bits to represent single-precision real numbers might use eight bits for the exponent and 40 bits for the mantissa. They can represent real numbers in the range $2.9387358771E-39$ to $1.701418346E+38$ (that is, 2^{-128} to 2^{127}) with 11 decimal digits of precision (for example, $2^{-39} = 1.8 \times 10^{-12}$). If the computer has 64 bit double-precision real numbers, it might use 11 bits for the exponent and 53 bits for the mantissa. They can represent real numbers in the range $5.56284646268003 \times 10^{-309}$ to $8.988465674311580 \times 10^{307}$ (that is, 2^{-1024} to 2^{1023}) with about 16 decimal digits of precision (for example, $2^{-52} = 2.2 \times 10^{-16}$). There are two commonly used ways of translating a given real number x into the digits floating point number, *rounding* and *chopping*, which we shall discuss in the following section.

1.2.2 Rounding and Chopping

When one gives the number of digits in a numerical value, one should not include zeros in the beginning of the number, as these zeros only help to denote where the decimal point should be. If one is counting the number of decimals, one should of course include leading zeros to the right of the decimal point. For example, the number 0.00123 is given with three digits but has five decimals. The number 11.44 is given with four digits but has two decimals. If the magnitude of the error in approximate number p does not exceed $\frac{1}{2} \times 10^{-k}$, then p is said to have k *correct decimals*. The digits in p, which occupy positions where the unit is greater than or equal to 10^{-k} are called, then, *significant digits* (any initial zeros are not counted). For example, 0.001234 ± 0.000004 has five correct decimals and three significant digits, while, 0.0012342 ± 0.000006 has four correct decimals and two significant digits. The number of correct decimals gives one an idea of the magnitude of the absolute error, while the number of significant digits gives a rough idea of the magnitude of the relative error. There are two ways of rounding off number s to a given number (k) of decimals. In *chopping*, one simply leaves off all the decimals to the right of the *kth*. That way of abridging a number is not recommended since the error has, systematically, the opposite sign of the number itself. Also, the magnitude of the error can be as large as 10^{-k}. A surprising number of computers use chopping on the results of every arithmetical operation. This usually does not do much harm because the number of digits used in the operations is generally far greater than the number of significant digits in the data. In *rounding* (sometimes called "correct rounding"), one chooses from the numbers that are closest to the given number. Thus, if the part of the number, which stands to the right of the *kth* decimal is less than $\frac{1}{2} \times 10^{-k}$ in magnitude, then one should leave the *kth* decimal unchanged. If it is greater than $\frac{1}{2} \times 10^{-k}$, then one raises the *kth* decimal by 1. In the boundary case, when the number that stands to the right of the *kth* decimal is exactly $\frac{1}{2} \times 10^{-k}$, one should raise the *kth* decimal if it is odd or leave it unchanged if it is even. In this way, the error is positive or negative equally often. Most computers, which perform rounding always, as in the boundary case mentioned above, raise the number by $\frac{1}{2} \times 10^{-k}$ (or corresponding operation in a base other than 10), because this is easier to realize technically. Whichever convention one chooses in the boundary case, the error in the rounding will always lie on the interval $[-\frac{1}{2} \times 10^{-k}, \frac{1}{2} \times 10^{-k}]$. For example, shorting to three decimals:

0.2397	rounds to	0.240	(is chopped to 0.239).
−0.2397	rounds to	−0.240	(is chopped to − 0.239).
0.23750	rounds to	0.238	(is chopped to 0.237).
0.23650	rounds to	0.236	(is chopped to 0.236).
0.23652	rounds to	0.237	(is chopped to 0.236).

1.3 Error

An approximate number p is a number that differs slightly from an exact number α. We write

$$p \approx \alpha.$$

By error E of an approximate number p, we mean the difference between the exact number α and its computed approximation p. Thus, we define

$$E = \alpha - p. \tag{1.5}$$

If $\alpha > p$, the error E is positive, and if $\alpha < p$, the error E is negative. In many situations, the sign of the error may not be known and might even be irrelevant. Therefore, we define ***absolute error*** as

$$|E| = |\alpha - p|. \tag{1.6}$$

The *relative error* (RE) of an approximate number p is the ratio of the absolute error of the number to the absolute value of the corresponding exact number α. Thus,

$$RE = \frac{|\alpha - p|}{|\alpha|}, \qquad \alpha \neq 0. \tag{1.7}$$

If we approximate $\frac{1}{3}$ by 0.333, we have

$$E = \frac{1}{3} \times 10^{-3} \qquad \text{and} \qquad RE = 10^{-3}.$$

Note that a relative error is generally a better measure of the extent of the error than the actual error. But one should also note that the relative error is undefined if the exact answer is equal to zero. Generally, we shall be interested in E (or sometimes $|E|$) rather than RE, but when the true value of a quantity is very small or very large, relative errors are more meaningful. For example, if the true value of a quantity is 10^{15}, an error of 10^6 is probably not serious, but this is more meaningfully expressed by saying that $RE = 10^{-9}$. In actual computation of the relative error, we shall often replace the unknown true value by the computed approximate value. Sometimes the quantity

$$\frac{|\alpha - p|}{|\alpha|} \times 100\% \tag{1.8}$$

is defined as *percentage error*. From the above example, we have

$$PE = 0.001 \times 100 = 0.1\%.$$

In investigating the effect of the total error in various methods, we shall often mathematically derive an error, called error bound which is a limit on how large the error can be. We shall have reason to compute error bounds in many situations. This applies to both absolute and relative errors. Note that the error bound can be much larger than the actual error and that this is often the case in practice. Any mathematically derived error bound must account for the worst possible case that can occur and is often based on certain simplifying assumptions about the problem, which in many practical cases cannot be actually tested. For the error bound to be used in any practical way, the user must have a good understanding of how the error bound was derived in order to know how crude it is; that is, how likely it is to overestimate the actual error. Of course, whenever possible, our goal is to eliminate or lessen the effects of errors, rather than trying to estimate them after they occur.

1.4 Sources of Errors

In analyzing the accuracy of numerical results, one should be aware of the possible sources of errors in each stage of the computational process and of the extent to which these errors can affect the final answer. We will consider that there are three types of errors, which occur in a computation. We discuss them step by step as follows.

1.4.1 Human Error

These types of errors arise when the equations of the mathematical model are formed, due to sources such as the idealistic assumptions made to simplify the model, inaccurate measurements of data, miscopying of figures, the inaccurate representation of mathematical constants (for example, if the constant π occurs in an equation, we must replace π by 3.1416 or 3.141593) etc..

1.4.2 Truncation Error

This type of error is caused when we are forced to use mathematical techniques that give approximate, rather than exact, answers. For example, suppose that we use Maclaurin's series expansion to represent $\sin x$, so that

$$\sin x = x - \frac{x^3}{3!} + \frac{x^5}{5!} - \frac{x^7}{7!} + \cdots.$$

If we want a number that approximates $\sin(\frac{\pi}{2})$, we must terminate the expansion in order to obtain

$$\sin(\frac{\pi}{2}) = \frac{\pi}{2} - \frac{(\pi/2)^3}{3!} + \frac{(\pi/2)^5}{5!} - \frac{(\pi/2)^7}{7!} + E.$$

where E is the truncation error introduced in the calculation. Truncation errors in numerical analysis usually occur because many numerical methods are iterative in nature, with the approximations theoretically becoming more accurate as we take more iterations. As a practical matter, we must stop the iteration after a finite number of steps, thus introducing a truncation error. The Taylor series is the most important means used to derive numerical schemes and analysis truncation errors.

1.4.3 Round-off Error

These types of errors are associated with the limited number of digits in the computer. For example, by rounding off 1.32463672 to six decimal places we get 1.324637. A further calculation involving such a number will also contain an error. We round off numbers according to the following rules:

1. If the first discarded digit is less than 5, leave the remaining digits of the number unchanged. For example,

$$48.47263 \approx 48.4726.$$

2. If the first discarded digit exceeds 5, add 1 to the retained digit. For example,

$$48.4726 \approx 48.473.$$

3. If the first discarded digit is exactly 5 and there are nonzeros among those discarded, add 1 to the last retained digit. For example,

$$3.0554 \approx 3.06.$$

4. If the first discarded digit is exactly 5 and all other discarded digits are zero, the last retained digit is left unchanged if it is *even*, otherwise 1 is added to it. For example,

$$\begin{aligned} 3.05500 &\approx 3.06 \\ 3.04500 &\approx 3.04. \end{aligned}$$

With these rules, the error is never larger in magnitude than one-half unit of the place of the *nth* digit in the rounded number. To understand the nature of round-off errors, it is necessary to learn the ways numbers are stored and additions and subtractions are performed in a computer.

1.5 Effect of Round-off Errors in Arithmetic Operations

Here, we discuss the effect of rounding-off errors in calculations in detail. Let a_r be the rounded-off value a_e, which is the exact value of a number that is not necessarily known. Similarly, b_r, b_e, c_r, c_e, etc., are the corresponding values for other numbers. The number $E_A = a_r - a_e$ is called the error in the number a_r. Similarly, E_B is the error in b_r, etc. The error E_A will be positive or negative accordingly as a_r is greater or less than a_e. It is, however, usually impossible to determine the sign of E_A. Therefore, it is normal to consider only the value of $|E_A|$, which is called the absolute error of number a_r. To indicate that a number has been rounded off to two significant figures or four decimal places it is followed by $2S$ or $4dp$ as appropriate.

1.5.1 Rounding-off Errors in Addition and Subtraction

Let a_r and b_r be two approximate numbers and c_r be their sum, which has been rounded off be represented by

$$c_r = a_r + b_r, \tag{1.9}$$

which is an approximation for

$$c_e = a_e + b_e. \tag{1.10}$$

Then by subtracting (1.10) from (1.9), we have

$$\begin{aligned} c_r - c_e &= (a_r - a_e) + (b_r - b_e) \\ E_C &= E_A + E_B. \end{aligned} \tag{1.11}$$

Then

$$|E_C| \leq |E_A| + |E_B|. \tag{1.12}$$

That is, the absolute error to the sum of two numbers is less than or equal to the sum of the absolute error of the two numbers. Note that this can be extended to the sum of any number. One should follow a similar argument for the error involved in the *difference* of two rounded-off numbers; that is

$$c_r = a_r - b_r, \quad \text{with} \quad a_r > b_r \tag{1.13}$$

and one should find that the same result is obtained, which is

$$|E_C| \leq |E_A| + |E_B|. \tag{1.14}$$

This can be also extended to any number of terms. For example, consider the error in the numbers $1015 + 0.3572$ where both numbers have been rounded off. The first number $1.015(a_r)$ has been rounded off to 3dp so that the exact value must lie between 1.0145 and 1.0155, which implies that $-0.0005 \leq E_A \leq 0.0005$. This means that the absolute error is never greater than 0.0005 or $\frac{1}{2} \times 10^{-3}$; that is, $|E_A| \leq \frac{1}{2} \times 10^{-3}$. Note that if a number is rounded off to n decimal places then the absolute error is less than or equal to $\frac{1}{2} \times 10^{-n}$. Similarly, the other given number $0.3572(b_r)$ has been rounded off to 4dp then $E_B \leq \frac{1}{2} \times 10^{-4}$. Since

$$c_r = a_r + b_r$$

then

$$c_r = 1.015 + 0.3572 = 1.3722$$

but

$$\begin{aligned} |E_C| &\leq |E_A| + |E_B| \\ &\leq (\tfrac{1}{2} \times 10^{-3}) + (\tfrac{1}{2} \times 10^{-4}) \\ &\leq (0.5 + 0.05) \times 10^{-3} \\ &\leq 0.55 \times 10^{-3}. \end{aligned}$$

So the exact value of this sum must be in the range

$$1.3722 \pm 0.55 \times 10^{-3}$$

that is, between 1.37165 and 1.37275, so this result may be correctly rounded off to 1.37; that is, to only 2dp.

1.5.2 Rounding-off Errors in Multiplication

Let a_r and b_r be the rounded-off values and c_r be the product of these numbers; that is,

$$c_r = a_r b_r$$

the number that approximates to the exact number

$$c_e = a_e b_e.$$

Then

$$\begin{aligned} c_r - E_C &= (a_r - E_A)(b_r - E_B) \\ &= a_r b_r - E_A b_r - E_B a_r + E_A E_B. \end{aligned}$$

Since $c_r = a_r b_r$, so

$$E_C = E_A b_r + E_B a_r - E_A E_B$$

and

$$\frac{E_C}{c_r} = \frac{E_A b_r + E_B a_r - E_A E_B}{a_r b_r}$$

$$= \frac{E_A}{a_r} + \frac{E_b}{b_r} - \frac{E_A E_B}{a_r b_r}.$$

The last term has its numerator as the product of two very small numbers, both of which will also be small compared to a_r and b_r so we neglect the last term, then we have

$$\frac{E_C}{c_r} = \frac{E_A}{a_r} + \frac{E_b}{b_r}. \tag{1.15}$$

The number E_A/a_r is called the relative error in a_r. Then from (1.15), we have

$$\left|\frac{E_C}{c_r}\right| \leq \left|\frac{E_A}{a_r}\right| + \left|\frac{E_b}{b_r}\right|. \tag{1.16}$$

Hence, the relative error modulus of a product is less than or equal to the sum of the relative error moduli of the factors of the product. Having found the relative error modulus of a product from this result the absolute error is usually then obtained by multiplying the relative error modulus by a_r; that is,

$$RE = |E_A|/|a_r|.$$

This result can be extended to the product of more than two numbers and simply increases the number of terms on the right-hand side of the formula. For example, consider the error in 1.015×0.3573, where both numbers have been rounded off. Then

$$1.015 \times 0.3573 = 0.3626595 \approx 0.363.$$

So the relative error modulus is given by

$$\left|\frac{E_C}{0.363}\right| \leq \frac{\frac{1}{2} \times 10^{-3}}{1.015} + \frac{\frac{1}{2} \times 10^{-4}}{0.3573}$$

$$\leq (0.49 \times 10^{-3}) + (1.4 \times 10^{-4}).$$

Hence,

$$\left|\frac{E_C}{0.363}\right| \leq (0.49 \times 10^{-3}) + (0.14 \times 10^{-3})$$

$$\leq 0.63 \times 10^{-3}.$$

So, we have

$$\begin{aligned} |E_C| &\leq 0.63 \times 0.363 \times 10^{-3} \\ &\leq 0.23 \times 10^{-3}. \end{aligned}$$

Hence, the exact value of this product lies in the range

$$0.3626595 \pm 0.00023$$

that is, between 0.3624295 and 0.3628895, so that this result may be correctly rounded off to 0.36; that is, to 2dp.

1.5.3 Rounding-off Errors in Division

Let a_r and b_r be rounded off values and c_r be the division of these numbers; that is,

$$c_r = a_r/b_r$$

the number that approximates to the exact number

$$c_e = a_e/b_e.$$

Then

$$\begin{aligned} c_r - E_C &= \frac{(a_r - E_A)}{(b_r - E_B)} \\ &= \frac{a_r(1 - E_A/a_r)}{b_r(1 - E_B/b_r)} \\ &= \frac{a_r}{b_r}\left(1 - \frac{E_A}{a_r}\right)\left(1 - \frac{E_B}{b_r}\right)^{-1}. \end{aligned}$$

The number $\left(1 - \dfrac{E_B}{b_r}\right)^{-1}$ is expanded using the binomial series neglecting those terms involving powers of the relative error E_B/b_r. Thus,

$$\begin{aligned} c_r - E_C &= \frac{a_r}{b_r}\left(1 - \frac{E_A}{a_r}\right)\left(1 + \frac{E_B}{b_r} + \cdots\right) \\ &= \frac{a_r}{b_r}\left(1 - \frac{E_A}{a_r} + \frac{E_B}{b_r}\right) \quad \left(\text{neglecting } \frac{E_A E_B}{a_r b_r}\right) \\ &= \frac{a_r}{b_r} - \frac{E_A}{b_r} + \frac{E_B a_r}{b_r^2}, \end{aligned}$$

which implies that

$$E_C = \frac{E_A}{b_r} - \frac{E_B a_r}{b_r^2}$$

and

$$\begin{aligned}\frac{E_C}{c_r} &= \frac{E_A}{b_r} - \frac{E_B a_r}{b_r^2} \div \frac{a_r}{b_r} \\ &= \frac{E_A}{a_r} - \frac{E_B}{b_r}.\end{aligned}$$

Hence,

$$\left|\frac{E_C}{c_r}\right| \leq \left|\frac{E_A}{a_r}\right| + \left|\frac{E_B}{b_r}\right|$$

gives the same result as the product of two numbers. It follows that it is possible to extend this result to quotients with two or more factors in the numerator or denominator by simply increasing the number of terms on the right-hand side. For example, consider the error in $17.28 \div 2.136$ where both numbers have been rounded off. Then

$$17.28/2.136 = 8.0898876 \approx 8.09.$$

Therefore,

$$\begin{aligned}\left|\frac{E_C}{8.09}\right| &\leq \frac{\frac{1}{2}\times 10^{-2}}{17.28} + \frac{\frac{1}{2}\times 10^{-3}}{2.136} \\ &\leq (0.029 \times 10^{-2}) + (0.23 \times 10^{-3}) \\ &\leq (0.29 \times 10^{-3} + 0.23 \times 10^{-3}) \\ &\leq 0.52 \times 10^{-3}\end{aligned}$$

so that

$$\begin{aligned}|E_C| &\leq 4.2 \times 10^{-3} \\ &\leq 0.42 \times 10^{-2}.\end{aligned}$$

Hence, the exact value of this quotient lies in the range

$$8.08989 \pm 0.000432$$

that is, between 8.08569 and 8.09409 so that this result may be correctly rounded off to 8.09; that is, to 2dp. The value of $|E_C|$ suggested this directly. This could be given to 3dp as 8.090 but with a large error of up to five units in the third decimal place.

Example 1.2 *Consider the error in* $5.381 + (5.96 \times 17.89)$ *where all numbers have been rounded off. We first find the absolute error in* $|E_C|$. *So*

$$5.96 \times 17.89 = 106.6244 \approx 106.6.$$

Then

$$\begin{aligned} \left|\frac{E_C}{106.6}\right| &\leq \frac{\frac{1}{2} \times 10^{-2}}{5.96} + \frac{\frac{1}{2} \times 10^{-2}}{17.89} \\ &\leq (0.084 \times 10^{-2}) + (0.028 \times 10^{-2}) \\ &\leq 0.112 \times 10^{-2}, \end{aligned}$$

which gives

$$|E_C| \leq 0.12.$$

The absolute error for 5.381 *is* $\frac{1}{2} \times 10^{-3}$, *so that the maximum absolute error for the sum is*

$$0.12 + 0.0005 = 0.1205.$$

But by the calculator

$$5.381 + (5.96 \times 17.89) = 112.0054$$

so the exact value lies in the range

$$112.0054 \pm 0.1205$$

that is, between 111.8849 *and* 112.1259. *This means that the result may be correctly rounded off to 3S or 0dp as a error of 0.1205 suggested or could be given as 112.0 with an error of up to one unit in the first decimal place.*

1.5.4 Rounding-off Errors in Powers and Roots

Let a_r and b_r be rounded off values and

$$b_r = (a_r)^p,$$

where the power is exact and may be rational. This approximates to the exact number

$$b_e = (a_e)^p.$$

Using $a_e = a_r - E_A$, we have

$$\begin{aligned} b_r - E_B &= (a_r - E_A)^p \\ &= (a_r)^p (1 - E_A/a_r)^p \\ &= (a_r)^p \left(1 - \frac{pE_A}{a_r} + \cdots\right) \end{aligned}$$

using the binomial series and neglecting those terms involving powers of the relative error E_A/a_r. This gives

$$b_r - E_B = (a_r)^p - pE_A(a_r)^{p-1},$$

which implies that

$$E_B = pE_A(a_r)^{p-1}$$

and so

$$\begin{aligned} \frac{E_B}{b_r} &= \frac{pE_A(a_r)^{p-1}}{(a_r)^p} \\ &= p\frac{E_A}{a_r}. \end{aligned}$$

Hence,

$$\left|\frac{E_B}{b_r}\right| = |p|\left|\frac{E_A}{a_r}\right|$$

that is, the relative error modulus of a power of a number is equal to the product of the modulus of the power and the relative error modulus of the number. For example, consider $\sqrt{8.675}$, where 8.675 has been rounded off. Here, $p = \frac{1}{2}$ and by calculator $\sqrt{8.675} = 2.9453$ retaining 4dp. Thus,

$$\begin{aligned} \left|\frac{E_B}{2.945}\right| &\leq \frac{\frac{1}{2} \times 10^{-3}}{8.675} \\ &\leq 0.029 \times 10^{-3} \end{aligned}$$

so that

$$|E_B| \leq 0.85 \times 10^{-4}.$$

This means that $\sqrt{8.675}$ may be correctly rounded off to 2.945; that is, to 3dp or may be given to 4dp with an error of up to one unit in the fourth decimal place.

1.6 Summary

In this chapter we discussed the storage and arithmetic of numbers on a computer. Efficient storage of numbers in computer memory requires allocation of a fixed number of bits to each value. The fixed bit size translates to a limit on the number of decimal digits associated with each number, which limits the range of numbers that can be stored in computer memory. The three number systems most commonly used in computing are binary (base 2), decimal (base 10), and hexadecimal (base 16). Techniques have been developed for transforming back and forth between the number systems. Binary numbers are a natural choice for computers because they correspond directly to the underlying hardware, which features transistors that are switched on and off. The absolute and relative errors were discussed as a measure of difference between exact x and approximate $\hat{x}$. They were applied to the storage mechanisms of chopping and rounding to estimate the maximum error introduced on storing a number. Rounding is somewhat more accurate than chopping (ignoring excess digits), but chopping is typically used because it is simpler to implement in hardware. Round-off error is one of the principal sources of errors in numerical computations. Mathematical operations on floating-point values introduce round-off errors because the results must be stored with a limited number of decimal digits. In numerical calculations involving many operations, round-off errors gradually corrupt the least significant digits of the results. The other main source of error in numerical computations are called truncation errors. Truncation error is the error that arises when approximations to exact mathematical expressions are used, such as the truncation of an infinite series to a finite number of terms. Truncation error is independent of round-off errors, although these two sources of error combine to affect the accuracy of a computed result. Truncation error considerations are important in many procedures and are discussed throughout the book.

1.7 Exercises

1. Convert the following binary numbers to decimal form:

$$(1010)_2, \ (100101)_2, \ (.1100011)_2.$$

2. Convert the following binary numbers to decimal form:

$$(101101)_2, \ (1010)_2, \ (100101)_2, \ (10000001)_2.$$

3. Find the first five binary digits of $(0.1)_{10}$. Obtain values for the absolute and relative errors in yours results.

4. Convert the following:

   (a) decimal numbers to binary numbers form:

   $$165, 3433, 111, 2345, 278.5, 347.45.$$

   (b) decimal numbers to hexadecimal decimal numbers:

   $$1025, 278.5, 14.09375, 1445, 347.45.$$

   (c) hexadecimal numbers to both decimal and binary:

   $$1F.C, \ FFF.118, \ 1A4.C, \ 1023, \ 11.1.$$

5. What is the absolute error in approximating 1/3 by 0.3333? What is the corresponding relative error?

6. Evaluate the absolute error in each of the following calculations and hence give the answer to a suitable degree of accuracy:

   (a) $9.01 + 9.96$, (b) $4.65 - 3.429$, (c) 0.7425×0.7199, (d) $0.7078 \div 0.87$.

7. Find the absolute and relative errors in approximating π by 3.1416. What are the corresponding errors in the approximation $100\pi \approx 314.16$?

8. Calculate the error, relative error, and number of significant digits in the following approximations, with $p \approx x$:

   (a) $x = 25.234, \quad p = 25.255.$
   (b) $x = e, \quad p = 19/7.$
   (c) $x = \sqrt{2}, \quad p = 1.414.$

9. Write each of the following numbers in (decimal) floating-point form, starting with length m and exponent e:

   $$13.2, \quad -12.532, \quad 2/125.$$

10. Find the absolute error in each of the following calculations (all numbers are rounded):

    (a) $187.2 + 93.5$.
    (b) 0.281×3.7148.
    (c) $\sqrt{28.315}$.
    (d) $\sqrt{(6.2342 \times 0.82137)/27.268}$.

11. Express the base of natural logarithm e as a normalized floating-point number, using both chopping and symmetric rounding for each of the following systems:

    (a) base 10 with 4 significant figures.
    (b) base 10 with 7 significant figures.
    (c) base 2 with 10 significant figures.

12. Write down the normalized binary floating-point representations of 1/3, 1/5, 1/7, 1/9, and 1/10. Use enough bits in the mantissa to see the recurring patterns.

Chapter 2

Solutions of Nonlinear Equations

2.1 Introduction

In this chapter we study one of the fundamental problems of numerical analysis, namely the numerical solutions of nonlinear equations. Most equations arising in practice are nonlinear and are rarely of a form that allows the roots to be determined exactly. Consequently, numerical methods are used to solve nonlinear algebraic equations when the equations prove intractable to ordinary mathematical techniques. These numerical methods are all iterative, and they may be used for equations that contain one or several variables. These techniques can be divided into two categories; one-point (needing one initial approximation) and two-point (needing two initial approximations) methods. A nonlinear equation may be considered any one of the following types:

1. An equation may be an *algebraic equation (a polynomial equation of degree n)* expressible in the form:

$$a_n x^n + a_{n-1}x^{n-1} + \cdots + a_1 x + a_0 = 0, \quad a_n \neq 0, \quad n > 1,$$

where $a_n, a_{n-1}, \ldots, a_1$, and a_0 are constants. For example, the following equations are nonlinear:

$$x^2 + 5x + 6 = 0, \quad x^3 = 2x + 1, \quad x^{100} + x^2 + 1 = 0.$$

2. The power of the unknown variable involved in the equation must be difficult to manipulate. For example, the following equations are nonlinear:

$$x^{-1} + 2x = 1, \quad \sqrt{x} + x = 0, \quad x^{2/3} + \frac{2}{x} + 4 = 0.$$

3. An equation may be a *transcendental equation* which involves trigonometric functions, exponential functions, and logarithmic functions. For example, the following equations are nonlinear:

$$x = cos(x), \qquad e^x + x - 10 = 0, \qquad x + \ln x = 10.$$

All the numerical methods described in this chapter are applicable to general nonlinear functions. Special techniques do, however, exist for certain restricted classes of functions, such as polynomials. Some of these are described in the books by [58] and [93]. The iterative methods we will discuss in this chapter are basically of two types: one in which convergence is guaranteed and the other in which convergence depends on the initial approximation.

Definition 2.1 (Root of an Equation)

Assume that $f(x)$ is a continuous function. A number α for which $f(\alpha) = 0$ is called a root of the equation $f(x) = 0$ or a zero of function $f(x)$.

There may be many roots of the given nonlinear equation but we will seek the approximation of only one of its root in the given interval $[a, b]$. This root may be *simple* (not repeating) or *multiple* (repeating). Now, we shall discuss the methods for nonlinear equations in a *single variable.* The problem here can be simply written down as:

$$f(x) = 0. \tag{2.1}$$

We seek values of x called the *roots* of (2.1) or *zeros* of function $f(x)$ such that (2.1) is true. The roots of (2.1) may be real or complex. Here we will look for the approximation of the real root of (2.1). There are many methods, which will give us information about the real roots of (2.1). The methods we will consider in this chapter are iterative methods and they are the bisection method, the false position method, the fixed-point method, the Newton method (also called, the Newton-Raphson method), and the secant method, which gives us the approximation of the single (or simple) root of (2.1). For the multiple roots of (2.1) we will use other iterative methods called the first modified Newton's method (also called Schroeder's method) and the second modified Newton's method. The iterative methods for the approximation of the simple root can be used also for the approximation of the multiple roots but they are very slow. We will use Aitken's method for improving the rate of convergence for a slow convergence sequence of iterates. We will also discuss Newton's method for the system of nonlinear equations. Finally, we will use Horner's method for evaluating a polynomial at a given point, Muller's method to approximate the zeros of the polynomials, and Bairstow's method to find all the possible roots of the polynomials.

2.2 Method of Bisection

This is one of the simplest iterative techniques for determining the roots of (2.1) and it needs two initial approximations to start. It is based on the *Intermediate Value Theorem*. This method is also called the *interval-halving method* because the strategy is to bisect or halve the interval from one endpoint of the interval to the other endpoint and then retain the half interval whose end still brackets the root. It is also referred to as the *bracketing method* or sometimes called *Bolzano's method*. The fact that the function is required to change sign only once gives us a way to determine which half interval to retain; we keep the half on which $f(x)$ changes sign or becomes zero. The basis for this method can be easily illustrated by considering the following function:

$$y = f(x).$$

Our object is to find an x value for which y is zero. Using this method, we begin by supposing $f(x)$ is a continuous function defined on the interval $[a, b]$ and then by evaluating the function at two x values, say, a and b, such that

$$f(a).f(b) < 0.$$

The implication is that one of the values is negative and the other is positive. These conditions can be easily satisfied by sketching the function (see Figure 2.1). Obviously, the function is negative at one endpoint, a, of the interval and positive at other endpoint, b, and is continuous on $a \leq x \leq b$. Therefore, the root must

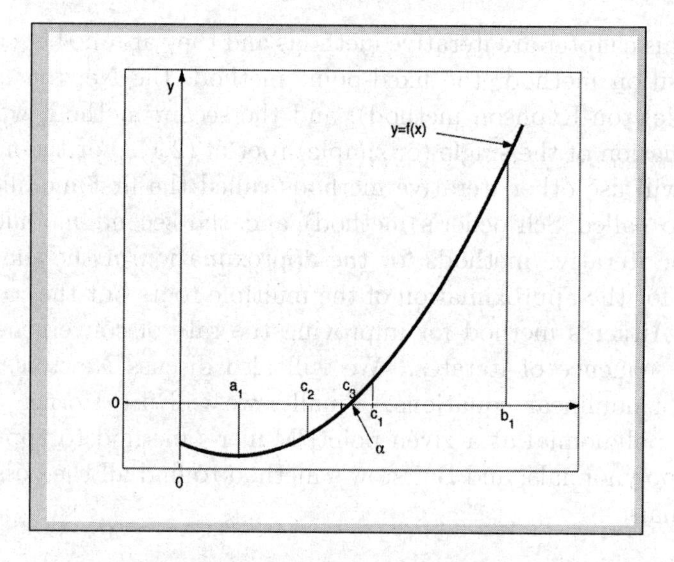

Figure 2.1: Bisection method.

lie between a and b (by *Intermediate Value Theorem*) and a new approximation to the root, α, may be calculated as

$$c = \frac{a+b}{2},$$

and, in general

$$c_n = \frac{a_n + b_n}{2}, \qquad n \geq 1. \tag{2.2}$$

Iterative formula (2.2) is known as the *bisection method.*

If $f(c) \approx 0$, then $c \approx \alpha$ is the desired root, and, if not, then there are two possibilities. Firstly, if $f(a).f(c) < 0$, then $f(x)$ has a zero between point a and point c. The process can then be repeated on the new interval $[a, c]$. Secondly, if $f(a).f(c) > 0$ it follows that $f(b).f(c) < 0$ since it is known that $f(b)$ and $f(c)$ have opposite signs. Hence, $f(x)$ has zero between point c and point b and the process can be repeated with $[c, b]$. We see that after one step of the process, we have found either a zero or a new bracketing interval, which is precisely half the length of the original one. The process continues until the desired accuracy is achieved. We use the bisection process in the following example.

Example 2.1 *Use the bisection method to find the approximation of the root of the equation*

$$x^3 = 2x + 1,$$

that is located on the interval $[1.5, 2.0]$ *accurate to within* 10^{-2}.

Solution. *Since the given function* $f(x) = x^3 - 2x - 1$ *is a polynomial function and so is continuous on* $[1.5, 2.0]$, *starting with* $a_1 = 1.5$ *and* $b_1 = 2$, *we compute:*

$$\begin{array}{lll} a_1 &=& 1.5: \quad f(a_1) = -0.625 \\ b_1 &=& 2.0: \quad f(b_1) = 3.0 \end{array}$$

and since $f(1.5).f(2.0) < 0$, *so that a root of* $f(x) = 0$ *lies in the interval* $[1.5, 2.0]$. *Using formula (2.2) (when* $n = 1$*), we get:*

$$c_1 = \frac{a_1 + b_1}{2} = 1.75; \qquad f(c_1) = 0.859375.$$

Hence, the function changes sign on $[a_1, c_1] = [1.5, 1.75]$. *To continue, we squeeze from right and set* $a_2 = a_1$ *and* $b_2 = c_1$. *Then the midpoint is:*

$$c_2 = \frac{a_2 + b_2}{2} = 1.625; \qquad f(c_2) = 0.041056.$$

Then continuing in this manner we obtain a sequence $\{c_k\}$ *of approximation shown by Table 2.1.*

Table 2.1: Solution of $x^3 = 2x + 1$ by bisection method.

n	Left Endpoint a_n	Midpoint c_n	Right Endpoint b_n	Function Value $f(c_n)$
01	1.500000	1.750000	2.000000	0.8593750
02	1.500000	1.625000	1.750000	0.0410156
03	1.500000	1.562500	1.625000	-0.3103027
04	1.562500	1.593750	1.625000	-0.1393127
05	1.593750	1.609375	1.625000	-0.0503273
06	1.609375	1.617188	1.625000	-0.0049520

We see that the functional values approach zero as the number of iterations increase. The desired approximation of the root of the given equation is $c_6 = 1.617188 \approx \alpha$ *after six iterations with accuracy* $\epsilon = 10^{-2}$.

To use the MATLAB command for the bisection method, first we define a function m-file as fn.m for the equation as follows:

```
function y = fn(x)
y = x.^3 - 2*x - 1;
```

then use the single command:

```
>> s = bisect('fn', 1.5, 2, 1e-2)
s =
    1.617188
```

.

Program 2.1

MATLAB m-file for the Bisection Method

```
function sol=bisect(fn,a,b,tol)
fa = feval(fn,a); fb = feval(fn,b);
if fa*fb > 0; fprintf('Endpoints have same sign') return
end
while abs (b-a) > tol
c = (a+b)/2; fc = feval(fn,c);
if fa*fc < 0; b = c; else a = c; end
end; sol=(a+b)/2;
```

Theorem 2.1 (Bisection Convergence and Error Theorem)

Let $f(x)$ be a continuous function defined on the initial interval $[a_0, b_0] = [a, b]$ and suppose $f(a).f(b) < 0$. Then bisection method (2.2) generates a sequence $\{c_n\}_{n=1}^{\infty}$ approximating $\alpha \in (a, b)$ with the property

$$|\alpha - c_n| \leq \frac{b-a}{2^n}, \quad n \geq 1. \tag{2.3}$$

Moreover, to obtain accuracy of

$$|\alpha - c_n| \leq \epsilon,$$

(for $\epsilon = 10^{-k}$) it suffices to take

$$n \geq \frac{\ln\{10^k(b-a)\}}{\ln 2}, \tag{2.4}$$

where k is nonnegative integer.

Proof.

Since both the root α *and the midpoint* c_1 *lie on the interval* $[a, b]$*, the distance between them cannot be greater than the width interval. Thus,*

$$|\alpha - c_n| \leq \frac{b_{n-1} - a_{n-1}}{2}, \qquad \text{for all } n.$$

Observe that

$$b_1 - a_1 = \frac{b_0 - a_0}{2},$$

then

$$b_2 - a_2 = \frac{b_1 - a_1}{2} = \frac{b_0 - a_0}{2^2}.$$

Finite mathematical induction is used to conclude that

$$b_{n-1} - a_{n-1} = \frac{b_0 - a_0}{2^{n-1}}.$$

Therefore, the error is bound as follows

$$|\alpha - c_n| \leq \frac{b_{n-1} - a_{n-1}}{2} = \frac{b_0 - a_0}{2^n},$$

and gives the estimate.

Now to establish the bound on the number of bisections n (or iterations), we simply observe that

$$\frac{b-a}{2^n} \leq 10^{-k},$$

together with (2.3) implies that

$$|\alpha - c_n| \leq 10^{-k};$$

that is, we wish to calculate a root to within 10^{-k}*. Since*

$$2^n \geq 10^k(b-a),$$

taking logarithms, we get

$$n \ln 2 \geq \ln\left\{10^k(b-a)\right\},$$

from this we obtain (2.4).

Theorem 2.1 gives us information about bounds for errors in approximation and the number of bisections needed to obtain any given accuracy.

Example 2.2 *Determine the number of iterations needed to achieve an approximation within accuracy* 10^{-2} *to the solution of* $x^3 - 2x - 1 = 0$ *lying on the interval* $[1.5, 2.0]$ *using the bisection method.*

Solution. *Here,* $a = 1.5$, $b = 2.0$, *and* $k = 2$. *Then using inequality (2.4), we get*

$$n \geq \frac{\ln[10^2(2.0 - 1.5)]}{\ln 2} \approx 5.644.$$

Hence, no more than six iterations are required to obtain an approximation accurate to within 10^{-2}. *Table 2.1 shows that the value of the approximation* $c_6 = 1.617188$ *is accurate to within* 10^{-2}.

It is important to keep in mind that the error analysis gives only a bound for the number of iterations necessary, and in many cases this bound is much larger than the actual number required.

Example 2.3 *Use the bisection method to compute the first three approximate values for* $\sqrt[4]{18}$. *Also, compute an error bound and exact error for your approximation.*

Solution. *Consider*

$$x = \sqrt[4]{18} = (18)^{1/4} \qquad or \qquad x^4 - 18 = 0.$$

Choose the interval $[2, 2.5]$ *on which the function* $f(x) = x^4 - 18$ *is continuous and the function* $f(x)$ *satisfies the sign property; that is,*

$$f(2).f(2.5) = (-2)(21.0625) = -42.125 < 0.$$

Hence, root $\alpha = \sqrt[4]{18} = 2.0598 \in [2, 2.5]$ *and we compute its first approximate value using formula (2.2) (when* $n = 1$*) as follows:*

$$c_1 = \frac{2.0 + 2.5}{2} = 2.2500; \quad and \quad f(2.25) = 7.6289.$$

Since the function $f(x)$ *changes sign on* $[2.0, 2.25]$, *to continue, we squeeze from the right and use formula (2.2) again to get the following second approximate value of the root* α *as:*

$$c_2 = \frac{2.0 + 2.25}{2} = 2.1250; \quad and \quad f(2.1250) = 2.3909.$$

Then continuing in a similar way, we find that the third approximate value of the root α *is* $c_3 = 2.0625$ *with* $f(2.0625) = 0.0957$.

Note that the value of the function at each new approximate value is decreasing, which shows that the approximate values are coming closer to the root α. *Now to compute the error bound for the approximation we use formula (2.3) and get*

$$|\alpha - c_3| \leq \frac{2.5 - 2.0}{2^3} = 0.0625,$$

The exact error E *can be calculated as follows:*

$E = 2.0598 - 2.0625 = -0.0027.$

which is the possible maximum error in our approximation.

One drawback of the bisection method is the convergence rate is raster slow. However, the rate of convergence is guaranteed. So for this reason it is often used as a start for the more efficient method used to find roots of the nonlinear equations. The method may give a false root if $f(x)$ is discontinuous on the given interval $[a, b]$.

Procedure 2.1 (Bisection Method)

1. *Establish an interval* $a \leq x \leq b$ *such that* $f(a)$ *and* $f(b)$ *are of opposite sign; that is,* $f(a).f(b) < 0$.

2. *Choose an error tolerance* ($\epsilon > 0$) *value for the function.*

3. *Compute a new approximation for the root:*

$$c_n = \frac{(a_n + b_n)}{2}; \qquad n = 1, 2, 3, \ldots.$$

4. *Check the tolerance. If* $|f(c_n)| \leq \epsilon$, *then use* $c_n (n = 1, 2, 3, \ldots)$ *for the desired root; otherwise continue.*

5. *Check if* $f(a_n).f(c_n) < 0$, *then set* $b_n = c_n$; *otherwise, set* $a_n = c_n$.

6. *Go back to step 3 and repeat the process.*

An iterative process may converge or diverge. If divergence occurs, the procedure should be terminated because there may be no solution. We can restart the procedure by changing the initial approximation if necessary. But in the case of convergence we have to apply some stopping procedures to end the computations, like we did in step *4* of Procedure 2.1. In the following there are some stopping criterion that can be used in step *4* of Procedure 2.1; each of them can be applied to any iterative technique considered in this chapter. By selecting tolerance $\epsilon > 0$ and generating approximate solutions $x_1, x_2, \ldots, x_n$ until one of the following conditions is satisfied:

$$|x_n - x_{n-1}| < \epsilon \quad \text{or} \quad \frac{|x_n - x_{n-1}|}{|x_n|} < \epsilon, \qquad x_n \neq 0.$$

Sometimes difficulties can arise using any of these stopping criteria. For example, there exists the sequence $\{x_n\}_0^\infty$ with the property that the differences $(x_n - x_{n-1})$ converge to zero while the sequence itself diverges. It is also possible for $f(x_n)$ to be close to zero while x_n differs significantly from α. Without additional knowledge about $f(x)$ or α, the above second inequality is the best stopping criteria to apply because it tests relative error. Also, one of the other stopping criteria is to use a fixed number of iterations, and then the final approximation x_n may be considered the value of the required root. This type of stopping criteria is helpful when convergence is very slow. It is important to note that in considering whether an iteration *converges* or not, it may be necessary to ignore the first few iterations since the procedure may appear to diverge initially, even though it ultimately converges.

2.3 False Position Method

Another popular algorithm is the method of false position or the *regula falsi method.* This is also a very old technique and may be thought of as an attempt to improve the convergence characteristic of the bisection method. It provides a test to ensure that the root is bracketed between successive iterations. This method also starts like the bisection method with two initial approximations at which the function has the opposite sign; that is, we begin with limiting values a and b such that $f(x)$ changes sign only once on the interval $a \leq x \leq b$. We seek to decrease that interval. This can be possible by extending a straight line through the two points to replace the function, and the zero of the straight line is used to approximate the zero of the function (see Figure 2.2).

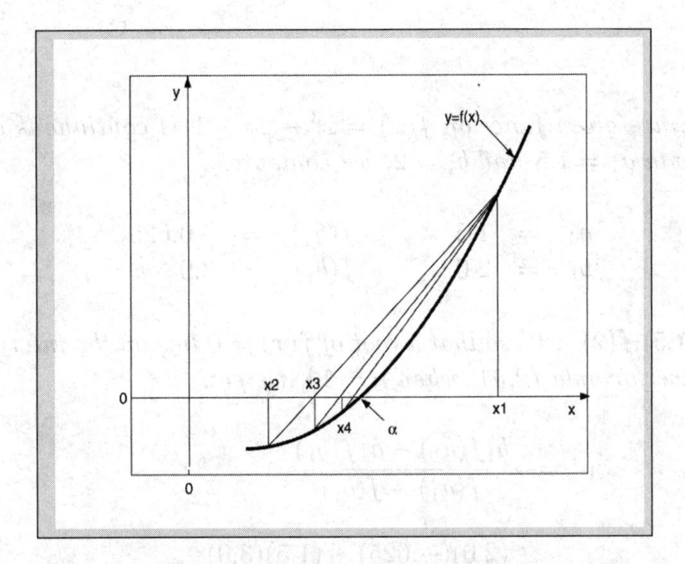

Figure 2.2: False position method.

Given two points $(a, f(a))$ and $(b, f(b))$ such that $f(a).f(b) < 0$, the equation of the line passing through these points is determined as

$$\frac{f(a)}{a-c} = \frac{f(b)}{c-b},$$

then solving for c, we get the zero of the line; that is,

$$c = \frac{bf(a) - af(b)}{f(a) - f(b)},$$

and in general, we have

$$c_n = \frac{b_n f(a_n) - a_n f(b_n)}{f(a_n) - f(b_n)}, \qquad n \geq 1. \tag{2.5}$$

Iterative formula (2.5) is called the method of false position, which requires two initial approximations, $a_1 = a$ and $b_1 = b$. It is also known as the method of linear interpolation. The method may give a false root if $f(x)$ is discontinuous on the given interval $[a, b]$.

Example 2.4 *Use the false position method to find the approximation to the root of* $x^3 = 2x + 1$ *that is located on the interval* $[1.5, 2.0]$ *accurate to within*

10^{-2}.

Solution. *Since given function* $f(x) = x^3 - 2x - 1$ *is continuous on* $[1.5, 2.0]$, *so starting with* $a_1 = 1.5$ *and* $b_1 = 2$, *we compute:*

$$\begin{array}{rclrcl} a_1 & = & 1.5: & f(a_1) & = & -0.625 \\ b_1 & = & 2.0: & f(b_1) & = & 3.0 \end{array}$$

and since $f(1.5).f(2) < 0$, *so that a root of* $f(x) = 0$ *lies on the interval* $[1.5, 2.0]$. *Using iterative formula (2.5) (when* $n = 1$*), we get:*

$$c_1 = \frac{b_1 f(a_1) - a_1 f(b_1)}{f(a_1) - f(b_1)}$$

$$c_1 = \frac{(2.0)(-.625) - (1.5)(3.0)}{-0.625 - 3.0}$$

$$c_1 = 1.586207 \quad and \quad f(c_1) = -0.18434$$

The function changes sign on the interval $[a_1, c_1] = [1.5, 1.586207]$. *To continue, we squeeze from right and set* $a_2 = a_1$ *and* $b_2 = c_1$. *Then iterative formula (2.5) produces the next approximation (when* $n = 2$*):*

$$c_2 = \frac{(2.0)(-.1814342) - (1.586207)(3.0)}{-0.1814342 - 3.0}$$

$$c_2 = 1.609805 \quad and \quad f(c_2) = -0.0478446.$$

Now $f(x)$ *changes sign on the interval* $[a_2, c_2] = [1.5, 1.609805]$, *and the next decision is again to squeeze from the right and set* $a_3 = a_2$ *and* $b_3 = c_2$. *A summary of the calculations is given in Table 2.2.*

The MATLAB single command for the false position method using m-file fn.m, which is defined above, is as follows:

```
>> sol = falseP('fn', 1.5, 2, 1e-2, 10)
s =
    1.617897
```

Table 2.2: Solution of $x^3 = 2x + 1$ by false position method

n	Left Endpoint a_n	Point of Intersection c_n	Right Endpoint b_nk	Function Value $f(c_n)$
01	1.500000	1.586207	2.000000	-0.1814342
02	1.586207	1.609805	2.000000	-0.0478446
03	1.609805	1.615930	2.000000	-0.0122936
04	1.615930	1.617498	2.000000	-0.0031375
05	1.617498	1.617897	2.000000	-0.0007996

Program 2.2

MATLAB m-file for the False Position Method

```
function sol=falseP(fn,a,b,tol,MaxI)
x0 = a; x1 = b; fa = feval(fn, a); fb = feval(fn, b);
if (fa * fb > 0); fprintf('Endpoints have same sign') return end
for i = 1:MaxI; c = (fb * a - fa * b)/(fb - fa);
fc = feval(fn, c);
if (abs(fc) < tol) break end
if (fa * fc < 0) b = c;fb=fc; else a = c; fa = fc; end
end; sol=c;
```

We see that the functional values approach zero as the number of iterations is increased. We see that this method converges *faster than* the method of bisection and got the desired approximation $c_5 = 1.617897$ after five iterations with accuracy, $\epsilon = 10^{-2}$.

Table (2.2) disclosed a serious fault of the false position method; the approach to the root is one-sided. If $f(x)$ has significant curvature between point a and point b, this can be most damaging to the speed with which we approach the root.

There is another way that we can improve the method of false position. Instead of requiring that the function have opposite signs at the endpoint of the interval, we can choose the two values nearest the root. Usually, the nearest values to the root will be the last two values calculated. We call this type of scheme the modified false position method.

Procedure 2.2 (False Position Method)

1. *Establish an interval* $a \leq x \leq b$ *such that* $f(a)$ *and* $f(b)$ *are of opposite signs; that is,* $f(a).f(b) < 0$.

2. *Choose an error tolerance* $\epsilon(> 0)$ *value for the function.*

3. *Compute a new approximation for the root:*

$$c_n = \frac{b_n f(a_n) - a_n f(b_n)}{f(a_n) - f(b_n)}; \qquad n = 1, 2, 3, \ldots.$$

4. *Check the tolerance. If* $|f(c_n)| \leq \epsilon$, *then use* $c_n, (n = 1, 2, 3, \ldots)$ *for the desired root; otherwise, continue.*

5. *Check if* $f(a_n).f(c_n) < 0$, *then set* $b_n = c_n$; *otherwise, set* $a_n = c_n$.

6. *Go back to step 3 and repeat the process.*

2.4 Fixed-point Method

This is another iterative method to solve nonlinear equation (2.1) and needs one initial approximation to start. This is a very general method for finding the root of (2.1) and it provides us with a theoretical framework within which the convergence properties of subsequent methods can be evaluated. The basic idea of this method, which is also called the successive approximation method or function iteration is to rearrange the original equation

$$f(x) = 0, \tag{2.6}$$

into an equivalent expression of the form

$$x = g(x). \tag{2.7}$$

Any solution of (2.7) is called a fixed-point for the iteration function $g(x)$ and hence a root of (2.6).

Definition 2.2 (Fixed-point)

A fixed-point of a function $g(x)$ *is a real number* α *such that* $\alpha = g(\alpha)$.

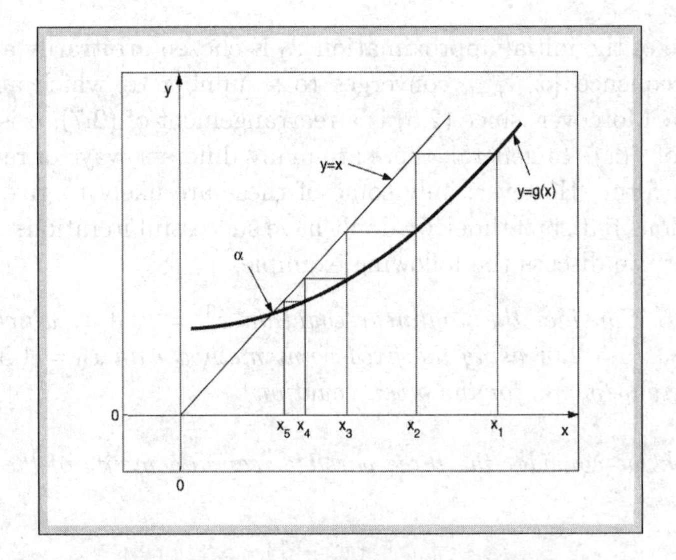

Figure 2.3: Fixed-point method.

The task of solving (2.6) is therefore reduced to that of finding a point satisfying the fixed-point condition (2.7). The fixed-point method essentially solves two functions simultaneously; $y = x$ and $y = g(x)$. The point of intersection of these two functions is the solution to $x = g(x)$, and thus to $f(x) = 0$ (see Figure 2.3).

This method is conceptually very simple. Since $g(x)$ is also nonlinear, the solution must be obtained iteratively. An initial approximation to the solution, say, x_0, must be determined. For choosing the best initial value x_0 for using this iterative method, we have to find an interval $[a, b]$ on which the original function $f(x)$ satisfies the sign property and then use the midpoint $\frac{a+b}{2}$ as the initial approximation x_0. Then this initial value x_0 is substituted in the function $g(x)$ to determine the next approximation, x_1, and so on.

Definition 2.3 (Fixed-point Method)

The iteration defined in the following

$$x_{n+1} = g(x_n); \qquad n = 0, 1, 2, \ldots \tag{2.8}$$

is called the fixed-point method or the fixed-point iteration.

The value of the initial approximation x_0 is chosen arbitrarily and the hope is that the sequence $\{x_n\}_{n=0}^{\infty}$ converges to a number α, which automatically satisfies (2.6). Moreover, since (2.6) is a rearrangement of (2.7), α is guaranteed to be a zero of $f(x)$. In general, there are many different ways of rearranging of (2.7) in (2.6) form. However, only some of these are likely to give rise to successful iterations but sometimes we don't have successful iterations. To describe such behavior, we discuss the following example.

Example 2.5 *Consider the nonlinear equation* $x^3 = 2x + 1$, *which has a root on the interval* $[1.5, 2.0]$ *using the fixed-point method with* $x_0 = 1.5$ *(take three different rearrangements for the given equation).*

Solution. *Let us consider the three possible rearrangements of the given equation as follows:*

$$(i) \quad x_{n+1} = g_1(x_n) = \frac{(x_n^3 - 1)}{2}; \qquad n = 0, 1, 2, \ldots$$

$$(ii) \quad x_{n+1} = g_2(x_n) = \frac{1}{(x_n^2 - 2)}; \qquad n = 0, 1, 2, \ldots$$

$$(iii) \quad x_{n+1} = g_3(x_n) = \sqrt{\frac{(2x_n + 1)}{x_n}}; \qquad n = 0, 1, 2, \ldots$$

then the numerical results for the corresponding iterations, starting with the initial approximation $x_0 = 1.5$ *with accuracy* 5×10^{-2} *are given in Table 2.3.*

Table 2.3: Solution of $x^3 = 2x + 1$ by fixed-point method.

n	$x_{n+1} = g_1(x_n)$ $= (x_n^3 - 1)/2$	$x_{n+1} = g_2(x_n)$ $= 1/(x_n^2 - 2)$	$x_{n+1} = g_3(x_n)$ $= \sqrt{(2x_n + 1)/x_n}$
00	1.500000	1.500000	1.500000
01	1.187500	4.000000	1.632993
02	0.337280	0.071429	1.616284
03	-0.480816	-0.501279	—
04	-0.555579	-0.571847	—
05	-0.585745	-0.597731	—

We note that the first two considered sequences diverge and the last one converges. This example presents the need for mathematical analysis of the method. The following theorem gives sufficient conditions for the convergence of the fixed-point iteration.

Theorem 2.2 (Fixed-point Theorem)

If g is continuously differentiable on the interval $[a, b]$ and $g(x) \in [a, b]$ for all $x \in [a, b]$, then

(a) g has at least one fixed-point on the given interval $[a, b]$.

Moreover, if the derivative $g'(x)$ of the function $g(x)$ exists on an interval $[a, b]$, which contains the starting value x_0, with

$$|g'(x)| \leq k < 1; \qquad \textit{for all} \qquad x \in [a, b] \tag{2.9}$$

then

(b) Sequence (2.8) will converge to the attractive (unique) fixed-point α in $[a, b]$.

(c) Iteration (2.8) will converge to α for any initial approximation.

(d) We have the error estimate

$$|\alpha - x_n| \leq \frac{k^n}{1-k}|x_1 - x_0|, \qquad \textit{for all} \quad n \geq 1. \tag{2.10}$$

(e) The limit holds:

$$\lim_{n \to \infty} \frac{\alpha - x_{n+1}}{\alpha - x_n} = g'(\alpha). \tag{2.11}$$

Proof

(a) If $g(a) = a$ and $g(b) = b$, then the function g has a fixed point at the endpoints. Suppose that it is not happening, that is, $g(a) \neq a$ and $g(b) \neq b$, and define a function $f(x) = g(x) - x$, which is continuous on $[a, b]$. Then

$$f(b) = g(b) - b \leq 0 \qquad \textit{and} \qquad f(a) = g(a) - a \geq 0.$$

Therefore, the Intermediate Value Theorem imposing that the function $f(x)$ has a root α on the interval $[a, b]$ and thus $f(\alpha) = 0$, which implies that $\alpha = g(\alpha)$. Thus, the function $g(x)$ has at least one fixed-point in $[a, b]$. This proves (a).

(b) *Suppose now that (2.9) holds, and α and β are two fixed points of the function g in $[a,b]$. Then we have*

$$\alpha = g(\alpha), \qquad \text{and} \quad \beta = g(\beta).$$

In addition, by the Mean Value Theorem, we have that for any two points α and β in $[a,b]$, there exists a number η such that

$$\begin{aligned} |\alpha - \beta| &= |g(\alpha) - g(\beta)| = |g'(\eta)||\alpha - \beta|, \\ &\leq k|\alpha - \beta|, \end{aligned}$$

where $\eta \in (a,b)$. Thus,

$$(1-k)|\alpha - \beta| \leq 0.$$

Since $k < 1$, we must have $\alpha = \beta$; and thus, the function g has a unique fixed point α on the interval $[a,b]$. This proves (b).

(c) *For convergence, consider the iteration*

$$x_n = g(x_{n-1}), \qquad \text{for all} \qquad n \geq 1, 2 \ldots$$

and the definition of the fixed point; that is,

$$\alpha = g(\alpha).$$

If we subtract the last two equations and take the absolute values, we get

$$|\alpha - x_n| = |g(\alpha) - g(x_{n-1})| \leq k|\alpha - x_{n-1}|.$$

The recursion can be solved readily to get

$$\begin{aligned} |\alpha - x_n| &\leq k|\alpha - x_{n-1}| \\ &\leq k^2|\alpha - x - n - 2| \\ &\vdots \\ &\leq k^n|\alpha - x_0|, \end{aligned} \qquad (2.12)$$

from which it follows that

$$\text{as} \quad n \to \infty, \quad k^n \to 0, \quad (\text{since} \quad k < 1);$$

therefore, $x_n \to \alpha$. Hence, the iteration converges. This proves (c).

(d) *We note that*

$$\begin{aligned} |\alpha - x_0| &= |\alpha - x_1 + x_1 - x_0| \\ &\leq |\alpha - x_1| + |x_1 - x_0| \\ &\leq |g(\alpha) - g(x_0)| + |x_1 - x_0| \\ &\leq k|\alpha - x_0| + |x_1 - x_0|, \end{aligned}$$

or

$$(1-k)|\alpha - x_0| \leq |x_1 - x_0|,$$

from which it follows that

$$|\alpha - x_0| \leq \frac{1}{1-k}|x_1 - x_0|.$$

From (2.12), we can write the above equation as follows

$$|\alpha - x_n| \leq \frac{k^n}{1-k}|x_1 - x_0|,$$

which proves (d).

(**e**) *Finally, by subtracting iterations $x_{n+1} = g(x_n)$ and $\alpha = g(\alpha)$, we have*

$$\alpha - x_{n+1} = g(\alpha) - g(x_n) = g'(\eta(x))(\alpha - x_n),$$

which implies that

$$\frac{\alpha - x_{n+1}}{\alpha - x_n} = g'(\eta(x)),$$

or

$$\lim_{n\to\infty} \frac{\alpha - x_{n+1}}{\alpha - x_n} = \lim_{n\to\infty} g'(\eta(x)) = g'(\alpha),$$

since $\eta(x) \rightarrow \alpha$ is forced by the convergence of x_n to α.

Now we come back to Example 2.5 and discuss why the first two rearrangements we considered do not converge, but on the other hand, the last sequence has a fixed point and converges.

Since, we observe that $f(1.5).f(2) < 0$, then the solution we seek is on the interval $[1.5, 2]$.

(i) For $g_1(x) = \frac{x^3 - 1}{3}$, we have $g_1'(x) = x^2$, which is greater than unity throughout the interval $[1.5, 2]$. So by the Fixed-Point Theorem 2.2 this iteration will fail to converge.

(ii) For $g_2(x) = \frac{1}{x^2 - 2}$, we have $g_2'(x) = \frac{-2x}{(x^2 - 2)^2}$, and $|g_2'(1.5)| > 1$, so by the Fixed-Point Theorem 2.2 this iteration will fail to converge.

(iii) For $g_3(x) = \sqrt{\frac{2x + 1}{x}}$, we have $g_3'(x) = x^{-3/2}/2\sqrt{2x + 1} < 1$, for all x on the given interval $[1.5, 2]$. Also, g_3 is the decreasing function of x, and $g_3(1.5) = 1.63299$ and $g_3(2) = 1.58114$ and both lie on the interval $[1.5, 2]$. Thus, $g_3(x) \in [1.5, 2]$, for all $x \in [1.5, 2]$, so from the Fixed-Point Theorem 2.2 the iteration will converge.

Note 2.1 *From (2.10) note that the rate of convergence of the fixed-point method depends on the factor* $\frac{k^n}{(1-k)}$*; the smaller the value of k, the faster the convergence. The convergence may be very slow if the value of k is very close to 1.*

Note 2.2 *Assume that* $g(x)$ *and* $g'(x)$ *are continuous functions of* x *for some open interval* I*, with the fixed point* α *contained in this interval. Moreover, assume that*

$$|g'(\alpha)| < 1, \qquad \text{for} \quad \alpha \in I,$$

then there exists an interval $[a, b]$*, around the solution* α *for which all the conditions of Theorem 2.2 are satisfied. But if*

$$|g'(\alpha)| > 1, \qquad \text{for} \quad \alpha \in I,$$

then the sequence (2.8) will not converge to the root α*. In this case* α *is called a repulsive fixed point.*

Example 2.6 *Show that the equation* $x = N^{1/3}$ *can be written as* $x = Nx^{-2}$ *and the associated iterative scheme*

$$x_{n+1} = Nx_n^{-2}, \qquad n \geq 0$$

will not be successful in finding the third root of the positive number N*.*

Solution. *Given* $x = N^{1/3}$ *and it can be written as*

$$x^3 - N = 0 \quad or \quad x = \frac{N}{x^2} = Nx^{-2},$$

which gives the iterative scheme

$$x_{n+1} = Nx_n^{-2} = g(x_n). \qquad n \geq 0$$

From this, we have

$$g(x) = Nx^{-2} \qquad and \qquad g'(x) = -2Nx^{-3}.$$

Since $\alpha = x = N^{1/3}$*, therefore,*

$$g'(\alpha) = -2N\alpha^{-3} \qquad and \qquad g'(N^{1/3}) = -2N(N^{1/3})^{-3}$$

or

$$g'(N^{1/3}) = -2NN^{-1} = -2.$$

Thus,

$$|g'(N^{1/3})| = |-2| = 2 > 1,$$

which shows the divergence.

If $|g'(\alpha)| = 1$, then convergence is not guaranteed and if convergence happened, it would be very slow. So to get faster convergence, the value of $|g'(\alpha)|$ should be equal to zero or very close to zero.

Example 2.7 *Which of the following sequences will converge faster to* $\sqrt{5}$*?*

$$(a) \quad x_{n+1} = x_n + 1 - \frac{x_n^2}{5}, \quad (b) \quad x_{n+1} = \frac{1}{3}\left[3x_n + 1 - \frac{x_n^2}{5}\right].$$

Solution. *It can be easily verified using Note 2.2. From the first sequence, we have*

$$g_1(x) = x + 1 - \frac{x^2}{5} \quad and \quad g_1'(x) = 1 - \frac{2x}{5},$$

which implies that

$$|g_1'(\sqrt{5})| = \left|1 - \frac{2\sqrt{5}}{5}\right| = 0.1056 < 1.$$

Similarly, from the second sequence, we have

$$g_2(x) = \frac{1}{3}\left[3x + 1 - \frac{x^2}{5}\right] \quad and \quad g_2'(x) = \frac{1}{3}\left[3 - \frac{2x}{5}\right],$$

which gives

$$|g_2'(\sqrt{5})| = \left|\frac{1}{3}\left[3 - \frac{2\sqrt{5}}{5}\right]\right| = 0.701186 < 1.$$

We note that both sequences are converging to $\sqrt{5}$, but sequence (a) will converge faster than sequence (b) because the value of $|g_1'(\sqrt{5})|$ is closer to zero than the value obtained by $|g_2'(\sqrt{5})|$.

The graphical interpretation of the fixed-point method is illustrated in Figure 2.3.1(a) and in Figure 2.3.1(b). The fixed point α is the abscissa of the intersection of the graph of the function $g(x)$ with the line $y = x$. The ordinate of the function $g(x)$ at x_0 is the value of x_1. To turn this ordinate into an abscissa, reflect it in line $y = x$. We may repeat this process to get x_2, x_3, and so on. It is seen that the iterates (see Figure 2.3.1(a)) move (zigzag) toward the fixed point, while in Figure 2.3.2(b) they are going away: the iterations in Figure 2.3.2(a) converge if you start near enough to the fixed point, whereas the others diverge no matter how close you start. The fixed point in Figure 2.3.2(a) is said to be *attractive*, and the one in Figure 2.3.2(b) is said to be *repulsive.*

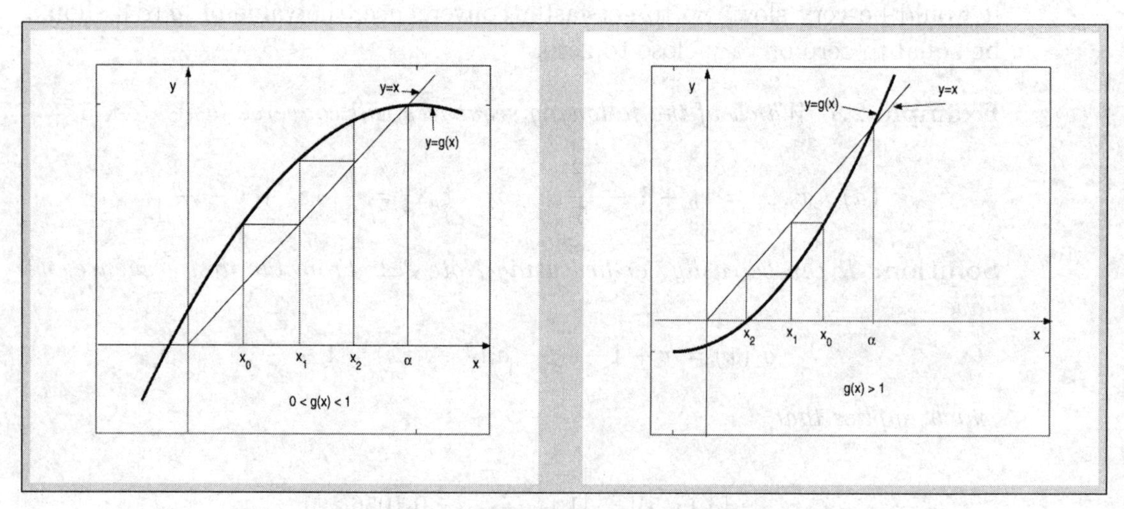

Figure 2.3.1. Convergent function iterations.

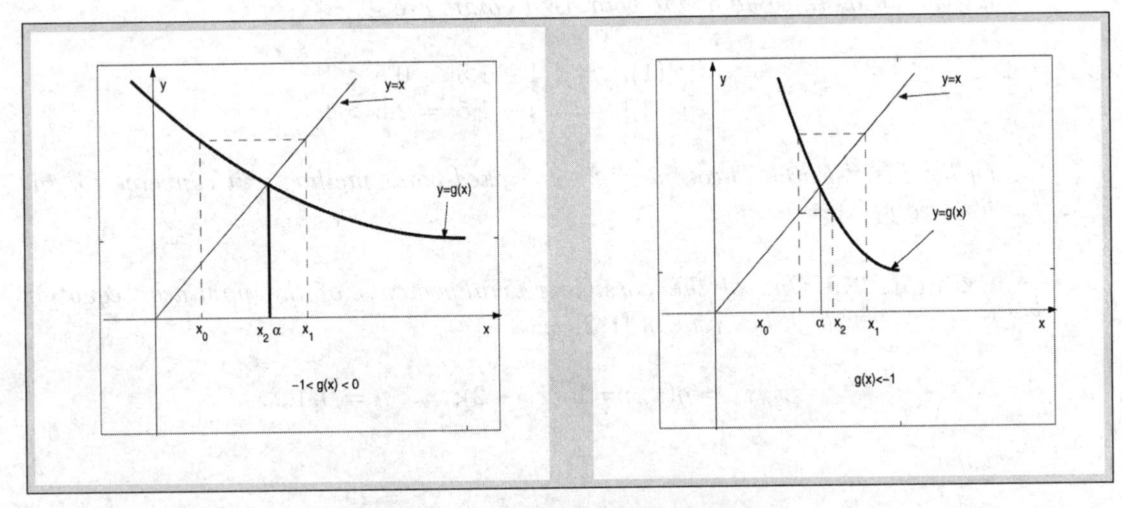

Figure 2.3.2. Divergent function iterations.

Example 2.8 *Let α_1 and α_2 be two fixed points of the function*

$$g(x) = 0.5x^2 - 1.5x + 2.$$

Then:
(a) Find the values of both fixed points.
(b) For which fixed point in part (a), will the fixed-point method will converge?

Solution. *Since $f(x) = g(x) - x = 0$,*

$$0.5x^2 - 2.5x + 2 = 0.$$

Solving this quadratic equation, we get

$$x_1 = 1 \quad and \quad x_2 = 4.$$

Then

$$\begin{aligned} g(1) &= 0.5(1^2) - 1.5(1) + 2 = 1 and \\ g(4) &= 0.5(4^2) - 1.5(4) + 2 = 4 \end{aligned}$$

showing that $\alpha_1 = 1$ and $\alpha_2 = 4$ are the two fixed points of the given function. Since the first derivative of the given function is

$$g'(x) = x - 1.5$$

and its absolute value at the both fixed points are

$$\begin{aligned} |g'(1)| &= |1-1.5| = 0.5 < 1 \\ |g'(4)| &= |4-1.5| = 2.5 > 1 \end{aligned}$$

by the Fixed-Point Theorem (2.2), the fixed-point method will converge for the fixed point $\alpha_1 = 1$ only.

Example 2.9 *One of the possible rearrangements of the nonlinear equation $e^x = x + 2$, which has root in $[1, 2]$ is*

$$x_{n+1} = g(x_n) = \ln(x_n + 2); \qquad n = 0, 1, \ldots.$$

Then

(a) *Show that $g(x)$ has a unique fixed point on the interval $[1, 2]$.*

(b) *Use fixed point iteration formula (2.8) to compute third approximation x_3, starting with $x_0 = 1.5$.*

(c) *Compute an error estimate $|\alpha - x_3|$ for your approximation.*

(d) *Determine the number of iterations needed to achieve an approximation with accuracy 10^{-2} to the solution of $g(x) = \ln(x+2)$ lying on the interval $[1, 2]$ using the fixed-point iteration method.*

Solution. *Since we observe that $f(1).f(2) < 0$, then the solution we seek is on the interval $[1, 2]$.*

(a) *For $g(x) = \ln(x+2)$, we have $g'(x) = 1/(x+2) < 1$, for all x in the given interval $[1, 2]$. Also, g is the increasing function of x, and $g(1) = \ln(3) = 1.0986123$ and $g(2) = \ln(4) = 1.3862944$ and both lie on the interval $[1, 2]$. Thus, $g(x) \in [1, 2]$, for all $x \in [1, 2]$, so from the fixed-point theorem $g(x)$ has a unique fixed point.*

(b) *Since $x_0 = 1.5$ is given, we have*

$$\begin{aligned} x_1 &= g(x_0) = \ln(x_0 + 2) = 1.252763, \\ x_2 &= g(x_1) = \ln(x_1 + 2) = 1.179505, \\ x_3 &= g(x_2) = \ln(x_2 + 2) = 1.156725. \end{aligned}$$

(c) *Since $a = 1$, $b = 2$, the value of k can be found as follows*

$$\begin{aligned} k_1 &= |g'(1)| = |1/3| = 0.333 \\ k_2 &= |g'(2)| = |1/4| = 0.25, \end{aligned}$$

which gives $k = \max\{k_1, k_2\} = 0.333$; therefore, error estimate (2.10) is

$$|\alpha - x_3| \leq \frac{(0.333)^3}{1 - 0.333}|1.252763 - 1.5|$$

or

$$|\alpha - x_3| \leq (0.333)^3(0.37067) = 0.013687.$$

(d) *From error bound formula (2.10), we have*

$$\frac{k^n}{1-k}|x_1 - x_0| \leq 10^{-2}.$$

Using above parts (b) and (c), we have

$$\frac{(0.333)^n}{1 - 0.333}|1.252763 - 1.5| \leq 10^{-2}.$$

Solving this inequality, we obtain

$$n \ln(0.333) \leq \ln(0.02698)$$

and it gives

$$n \geq 3.28539.$$

So we need four approximations to get the desired accuracy for the given problem.

The MATLAB command for the above given rearrangement $x = g(x)$ of $f(x) = x^3 - 2x - 1$ using the initial approximation $x_0 = 1.5$ can be written as follows:

```
function y = fn(x)
y = log(x + 2);
>> x0 = 1.5; tol = 0.01
>> sol = fixpt('fn', x0, tol);
```

Program 2.3

MATLAB m-file for the Fixed-point Method

```
function sol=fixpt(fn,x0,tol)
old= x0+1;
while abs(x0-old) > tol; old=x0;
x0 = feval(fn, old); end; sol=x0;
```

Procedure 2.3 (Fixed-point Method)

1. *Choose an initial approximation* x_0 *such that* $x_0 \in [a, b]$.
2. *Choose a convergence parameter* $\epsilon > 0$.
3. *Compute a new approximation* x_{new} *using iterative formula (2.8).*
4. *Check if* $|x_{new} - x_0| < \epsilon$ *then* x_{new} *is the desire approximate root; otherwise, set* $x_0 = x_{new}$ *and go to step 3.*

2.5 Newton's Method

This is one of the most popular and powerful iterative methods for finding roots of nonlinear equation (2.1). It is also known as the method of tangents because after estimating the actual root, the zero of the tangent to the function at that point is determined. It always converges if the initial approximation is sufficiently close to the exact solution. This method is distinguished from the methods of previous sections by the fact that it requires the evaluation of both the function $f(x)$ and the derivative of the function $f'(x)$, at arbitrary point x. Newton's method consists geometrically of expanding the tangent line at a current point x_i until it crosses zero, then setting the next guess x_{i+1} to the abscissa of that zero crossing (see Figure 2.4). This method is also called the *Newton-Raphson method.*

There are many description of Newton's method. We shall derive the method from the familiar Taylor's series expansion of a function in the neighborhood of a point.

Let $f \in C^2[a, b]$ and let x_n be the *nth* approximation to the root α such that $f'(x_n) \neq 0$ and $|\alpha - x_n|$ is small. Consider the first Taylor polynomial for $f(x)$ expanded about x_n, so we have

$$f(x) = f(x_n) + (x - x_n)f'(x_n) + \frac{(x - x_n)^2}{2} f''(\eta(x)), \tag{2.13}$$

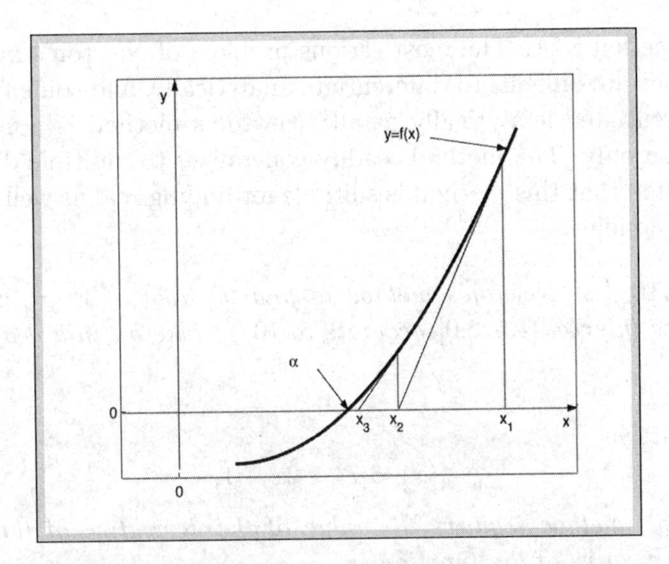

Figure 2.4: Newton's method.

where $\eta(x)$ lies between x and x_n. Since $f(\alpha) = 0$, then (2.13), with $x = \alpha$, gives

$$f(\alpha) = 0 = f(x_n) + (\alpha - x_n)f'(x_n) + \frac{(\alpha - x_n)^2}{2} f''(\eta(\alpha)).$$

Since $|\alpha - x_n|$ is small, then we neglect the term involving $(\alpha - x_n)^2$, and so

$$0 \approx f(x_n) + (\alpha - x_n)f'(x_n).$$

Solving for α, we get

$$\alpha \approx x_n - \frac{f(x_n)}{f'(x_n)}, \tag{2.14}$$

which should be a better approximation to α than is x_n. We call this approximation x_{n+1}, then we get

$$x_{n+1} = x_n - \frac{f(x_n)}{f'(x_n)}, \quad f'(x_n) \neq 0, \qquad \text{for all} \quad n \geq 0. \tag{2.15}$$

Iterative method (2.15) is called Newton's method. Usually, Newton's method converges well and quickly but its convergence cannot, however, be guaranteed and it may sometimes converge to a different root from the one expected. In particular, there may be difficulties if initial approximation is not sufficiently close to the actual root. The most serious problem of Newton's method is that some functions are difficult to differentiate analytically, and some functions cannot be differentiated analytically at all. Newton's method is not restricted to one dimension only. The method readily generalizes to multiple dimensions. It should be noted that this method is suitable for finding real as well as imaginary roots of polynomials.

Example 2.10 *Use Newton's method to find the root of* $x^3 = 2x + 1$ *that is located on the interval* $[1.5, 2.0]$ *accurate to* 10^{-2}; *take an initial approximation* $x_0 = 1.5$.

Solution. *Given*

$$f(x) = x^3 - 2x - 1,$$

since Newton's method requires the value of the derivative of the function be found, the derivative of the function is

$$f'(x) = 3x^2 - 2.$$

Now evaluating $f(x)$ *and* $f'(x)$ *at the give approximation* $x_0 = 1.5$ *gives*

$$x_0 = 1.5, \qquad f(1.5) = -0.625, \textit{and} \qquad f'(1.5) = 4.750.$$

Using Newton's iterative formula (2.15), we get

$$x_1 = x_0 - \frac{f(x_0)}{f'(x_0)} = 1.5 - \frac{(-0.625)}{4.75} = 1.631579.$$

Now evaluating $f(x)$ *and* $f'(x)$ *at the new approximation* x_1 *gives*

$$x_1 = 1.631579, \qquad f(1.631579) = 0.0801869, \textit{and} \qquad f'(1.631579) = 5.9861501,$$

using iterative formula (2.15) again to get the other new approximation. The successive iterates were shown in Table 2.4. Just after the third iterations the root is approximated to be $x_3 = 1.618034$ *and the functional value is reduced to* 6.57×10^{-8}. *Since the exact solution is* 1.6180339, *the actual error is* 1×10^{-7}. *We see that convergence is quite faster compared to the methods considered previously.*

To get the above results using MATLAB commands, firstly the function $x^3 - 2x - 1$ and its derivative $3x^2 - 2$ were saved in m-files called fn.m and dfn.m, respectively, written as follows:

Table 2.4: Solution of $x^3 = 2x + 1$ by Newton's method.

n	x_n	$f(x_n)$	$f'(x_n)$	Error $x - x_n$
00	1.500000	-0.625000	4.750000	0.1180339
01	1.631579	0.0801869	5.9861501	-0.0135451
02	1.618184	0.000878	5.855558	-0.0001501
03	1.618034	0.00000007	5.854102	-0.0000001

```
function y = fn(x)
y = x.^3 - 2*x - 1;
```

and

```
function dy = dfn(x)
dy = 3*x.^2 - 2;
```

.

after which we do the following:

```
>> x0 = 1.5; tol = 0.01;
>> sol = newton('fn','dfn',x0,tol);
```

.

Program 2.4

MATLAB m-file for Newton's Method

```
function sol=newton(fn,dfn,x0,tol)
old = x0+1;
while abs (x0 - old) > tol; old = x0;
x0 = old - feval(fn,old)/feval(dfn,old);
end; sol=x0;
```

Example 2.11 *Successive approximations x_n to the desired root are generated by the scheme*

$$x_{n+1} = \frac{1 + 3x_n^2}{4 + x_n^3}, \qquad n \geq 0.$$

Find $f(x_n)$ and $f'(x_n)$ and then use Newton's method to find the approximation of the root accurate to 10^{-2}, starting with $x_0 = 0.5$.

Solution. *Given*

$$x = \frac{1+3x^2}{4+x^3} = g(x)$$

and

$$x - g(x) = x - \frac{1+3x^2}{4+x^3} = \frac{x^4 - 3x^2 + 4x - 1}{4+x^3}.$$

Since

$$f(x) = x - g(x) = 0,$$

we have

$$f(x_n) = x_n^4 - 3x_n^2 + 4x_n - 1 \quad \text{and} \quad f'(x_n) = 4x_n^3 - 6x_n + 4.$$

Using these function values in Newton's iterative formula (2.15), we have

$$x_{n+1} = x_n - \frac{x_n^4 - 3x_n^2 + 4x_n - 1}{4x_n^3 - 6x_n + 4}.$$

Finding the first approximation of the root using the initial approximation $x_0 =$ 0.5, *we get*

$$x_1 = x_0 - \frac{x_0^4 - 3x_0^2 + 4x_0 - 1}{4x_0^3 - 6x_0 + 4} = 0.5 - \frac{0.3125}{1.5} = 0.2917.$$

Similarly, the other approximations can be obtained as

$$x_2 = x_1 - \frac{f(x_1)}{f'(x_1)} = 0.2917 - \frac{(-0.0813}{2.3491} = 0.3263$$

$$x_3 = x_2 - \frac{f(x_2)}{f'(x_2)} = 0.3263 - \frac{(-0.0029}{2.1812} = 0.3276.$$

Note that $|x_3 - x_2| = 0.0013$.

Newton's method is widely used in computers as a basis for square root and reciprocal evaluation.

Example 2.12 *Develop an iterative procedure for evaluating the root of a number using Newton's method.*

Solution. *We shall compute* $x = N^{1/p}$ *by finding a positive root for the nonlinear equation*

$$x^p - N = 0,$$

where p is any positive integer and $N > 0$ is the number whose root is to be found. Therefore, if $f(x) = 0$, then $x = N^{1/p}$ is the exact root. Let

$$f(x) = x^p - N \quad \textit{and} \quad f'(x) = px^{p-1}.$$

Hence, assuming an initial estimate to the root, say, $x = x_0$ and using iterative formula (2.15), we get

$$x_1 = x_0 - \frac{(x_0^p - N)}{px_0^{p-1}} = (1 - \frac{1}{p})x_0 + \frac{N}{p}x_0^{1-p}.$$

In general, we have

$$x_{n+1} = (1 - \frac{1}{p})x_n + \frac{N}{p}x_n^{1-p}, \tag{2.16}$$

where $p = 2, 3, \ldots,$ and $n = 0, 1, \ldots,$. For example, suppose we want the square root of number 19, then we take $N = 19$, and $p = 2$. Assuming an initial approximation $x_0 = 5$, then using iterative formula (2.16), we get

$$n = 0, \; x_1 = (1 - \frac{1}{2})x_0 + \frac{19}{2}x_0^{1-2} = 4.4$$

$$n = 1, \; x_2 = (1 - \frac{1}{2})x_1 + \frac{19}{2}x_1^{1-2} = 4.3590909.$$

After just two iterations the estimated value compares rather favorably with the exact value of $\sqrt{19} \approx 4.3588989$. We can calculate higher roots of a number using general iterative formula (2.16).

Example 2.13 *Develop an iterative procedure for evaluating the reciprocal of a number using Newton's method.*

Solution. *Consider $x = 1/N$. This problem can be easily solved by noting that we seek to find a root to the nonlinear equation*

$$1/x - N = 0,$$

where $N > 0$ is the number whose reciprocal is to be found. Therefore, if $f(x) = 0$, then $x = 1/N$ is the exact root. Let

$$f(x) = 1/x - N \quad \textit{and} \quad f'(x) = -1/x^2.$$

Hence, assuming an initial estimate to the root, say, $x = x_0$ *and using iterative formula (2.15), we get*

$$x_1 = x_0 - \frac{(1/x_0 - N)}{(-1/x_0^2)} = x_0(2 - Nx_0).$$

In general, we have

$$x_{n+1} = x_n(2 - Nx_n), \qquad n = 0, 1, \ldots, \tag{2.17}$$

For example, suppose we want the reciprocal of number $N = 3$*. Assuming an initial gauss of say* $x_0 = 0.4$*, then using iterative formula (2.17), we get*

$$\begin{array}{rcl} n & = & 0, \quad x_1 = x_0(2 - 3x_0) = 0.4(2 - 3(0.4)) = 0.3200 \\ n & = & 1, \quad x_2 = x_1(2 - 3x_1) = 0.32(2 - 3(0.32)) = 0.3328 \\ n & = & 2, \quad x_3 = x_2(2 - 3x_2) = 0.3328(2 - 3(0.3328)) = 0.3333. \end{array}$$

After just three iterations the estimated value compares rather favorably with the exact value of $1/3 \approx 0.33333$*. We can calculate the other reciprocal of the number in the same way using general iterative formula (2.17).*

Lemma 2.1 *Assume that* $f \in C^2[a, b]$ *and there exists a number* $\alpha \in [a, b]$*, where* $f(\alpha) = 0$*. If* $f'(\alpha) \neq 0$*, then there exists a number* $\delta > 0$ *such that the sequence* $\{x_n\}_{n=0}^{\infty}$ *defined by the iteration*

$$x_{n+1} = g(x_n) = x_n - \frac{f(x_n)}{f'(x_n)}, \quad \text{for} \quad n = 0, 1, \ldots \tag{2.18}$$

will converge to α *for any initial approximation* $x_0 \in [\alpha - \delta, \alpha - \delta]$.

Newton's method uses the iteration function

$$g(x) = x - \frac{f(x)}{f'(x)}, \tag{2.19}$$

and is called *Newton's iteration function.* Since $f(\alpha) = 0$, it is easy to see that $g(\alpha) = \alpha$. Thus, Newton's iteration for finding the root of the equation $f(x) = 0$ is accomplished by finding a fixed point of the equation $g(x) = x$.

Procedure 2.4 (Newton's Method)

1. *Find the initial approximation x_0 for the root by sketching the graph of the function.*

2. *Evaluate the function $f(x)$ and the derivative $f'(x)$ at initial approximation.*
   Check if $f(x_0) = 0$ then x_0 is the desired approximation to a root. But if $f'(x_0) = 0$, then go back to step 1 to choose a new approximation.

3. *Establish the tolerance ($\epsilon > 0$) value for the function.*

4. *Compute a new approximation for the root using iterative formula (2.15).*

5. *Check the tolerance. If $|f(x_n)| \leq \epsilon$, for $n \geq 0$, then end; otherwise, go back to step 4 and repeat the process.*

2.6 Secant Method

Since we know the main obstacle to using Newton's method is that it may be difficult or impossible to differentiate the function $f(x)$, the calculation of $f'(x_n)$ may be avoided by approximating the slope of the tangent at $x = x_n$ by that of the chord joining the two points $(x_{n-1}, f(x_{n-1}))$ and $(x_n, f(x_n))$ (see Figure 2.5).

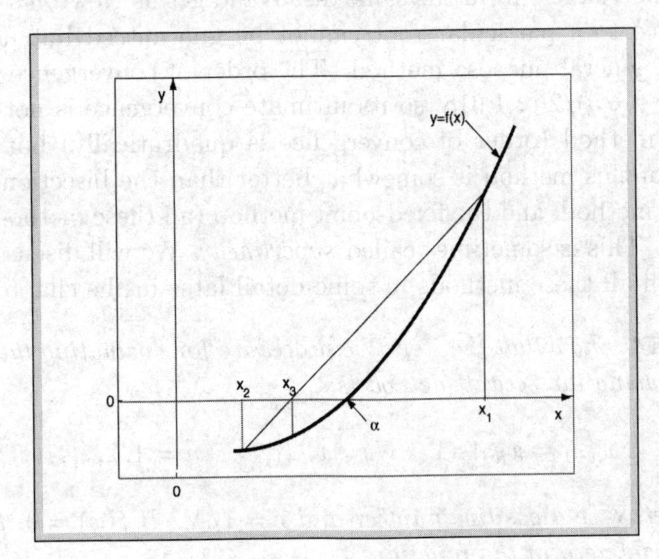

Figure 2.5: Secant method.

The slope of the chord (or secant) is

$$f'(x_n) \approx \frac{f(x_n) - f(x_{n-1})}{x_n - x_{n-1}}. \tag{2.20}$$

Then using this approximation of the derivative of the function in Newton's iterative formula (2.15), we get

$$x_{n+1} = x_n - \frac{(x_n - x_{n-1})f(x_n)}{f(x_n) - f(x_{n-1})}, \qquad n \geq 1$$

which can be also written as

$$x_{n+1} = \frac{x_{n-1}f(x_n) - x_n f(x_{n-1})}{f(x_n) - f(x_{n-1})}, \qquad n \geq 1. \tag{2.21}$$

Note that when $f(x_n) = f(x_{n-1})$, the calculation of x_{n+1} fails. This is because the chord is horizontal. Iterative formula (2.21) is known as the *secant method.* It needs two initial approximations to start. This method is very similar to the false position method described in Section 2.3. However, for the secant method it is not necessary for the interval to contain a root and no account is taken of signs of the numbers $f(x_n)$.

This method suffers from the same disadvantages as Newton's method; that is, convergence to a particular root cannot be guaranteed but nevertheless it is a powerful general purpose method. The order of convergence of the secant method is $(1+\sqrt{5})/2 \approx 1.618$, so its ultimate convergence is not quite as fast as Newton's method (order of convergence is quadratically) but the order of convergence of this method is somewhat better than the bisection method, the false position method, and the fixed-point method (all these methods have linear convergence). This is sometimes called *superlinear.* We will discuss the order of convergence of all these methods in some detail later in the chapter.

Example 2.14 *Show that the iterative procedure for evaluating the reciprocal of a number N using the secant method is:*

$$x_{n+1} = x_n + (1 - Nx_n)x_{n-1}, \qquad n = 1, 2, \ldots.$$

Solution. *Let N be a positive number and $x = 1/N$. If $f(x) = 0$, then $x = \alpha = 1/N$ is the exact zero of the function*

$$f(x) = 1/x - N.$$

Since the secant formula is

$$x_{n+1} = x_n - \frac{(x_n - x_{n-1})f(x_n)}{f(x_n) - f(x_{n-1})}, \qquad n \geq 1.$$

Hence, assuming the initial estimates to the root, say, $x = x_0, x = x_1$ *and using the secant iterative formula, we have*

$$x_2 = x_1 - \frac{(x_1 - x_0)(1/x_1 - N)}{(1/x_1 - N) - (1/x_0 - N)}.$$

It gives

$$x_2 = x_1 + (1 - Nx_1)x_0.$$

In general, we have

$$x_{n+1} = x_n + (1 - Nx_n)x_{n-1}, \qquad n = 1, 2, \ldots.$$

For example, suppose we want the reciprocal of number $N = 5$. *Assuming the initial approximations of say* $x_0 = 0.05$ *and* $x_1 = 0.1$, *then using the above iterative formula, we get the first three approximations as follows:*

$$\begin{aligned} n &= 1, \quad x_2 = x_1 + (1 - 5x_1)x_0 = 0.125 \\ n &= 2, \quad x_3 = x_2 + (1 - 5x_2)x_1 = 0.1625 \\ n &= 3, \quad x_4 = x_3 + (1 - 5x_3)x_2 = 0.1859. \end{aligned}$$

After just three iterations the estimated value compares rather favorably with the exact value of $1/5 = 0.2$.

Example 2.15 *Use the secant method to find the approximate root of the following equation within the accuracy* 10^{-2}

$$x^3 = 2x + 1,$$

take $x_0 = 1.5$ *and* $x_1 = 2.0$ *as starting values.*

Solution. *Since* $f(x) = x^3 - 2x - 1$ *and*

$$\begin{aligned} x_0 &= 1.5, \qquad f(x_0) = -0.625 \quad \text{and} \\ x_1 &= 2.0, \qquad f(x_1) = 3.0, \end{aligned}$$

we see that $f(x_0) \neq f(x_1)$. *Hence, one can use iterative formula (2.21) to get new approximations:*

$$x_2 = \frac{x_0 f(x_1) - x_1 f(x_0)}{f(x_1) - f(x_0)},$$

$$x_2 = \frac{(1.5)(3.0) - (2.0)(-0.625)}{3.0 - (-0.625)}, \text{ and}$$

$$x_2 = 1.586207, \quad \text{and} \quad f(x_2) = -0.18434.$$

In a similar way, we can find the other possible approximations of the root. A summary of the calculations is given in Table 2.5.

Table 2.5: Solution of $x^3 = 2x + 1$ by the secant method.

n	x_{n-1}	x_n	x_{n+1}	$f(x_{n+1})$
01	1.500000	2.000000	1.586207	-0.1814342
02	2.000000	1.586207	1.609805	-0.0478446
03	1.586207	1.609805	1.618257	0.0013040

To use MATLAB commands for the secant method, the function has been used in the m-file as fn.m, then the first few iterations are easily performed by the following sequence of MATLAB commands:

```
>> x0 = 1.5; x1 = 2;
>> x2 = x1 - (x1 - x0)/(fn(x1) - fn(x0)) * fn(x1)
x2 =
    1.586207
>> x0 = x1; x1 = x2;
>> x3 = x2 - (x2 - x1)/(fn(x2) - fn(x1)) * fn(x2)
x3 =
    1.609805
```

The last two commands can be repeated to generate the subsequent iterates shown in Table 2.5.

Program 2.5

MATLAB m-file for the Secant Method

```
function sol=secant(fn,a,b,tol)
x0 = a; x1 = b; fa = feval(fn, x0); fb = feval(fn, x1);
while abs(x1-old)> tol
new = x1 - fb * (x1 - x0)/(fb - fa);
x0 = x1; fa = fb; x1 == new; fb = feval(fn, new);
end; sol=new;
```

We see that the functional values approach zero as the number of iterations is increased. We see that this method converges *faster than* the methods discussed in Sections 2.2 to 2.4, and got the desired approximation $c_3 = 1.618032$ just after four iterations with accuracy $\epsilon = 10^{-2}$.

One can use a MATLAB single command to get the same above results by the secant method as follows:

```
>> a = 1.5; b = 2.0; tol = 1e - 2;
>> sol = secant('fn', a, b, tol);
```

Procedure 2.5 (Secant Method)

1. *Choose the two initial approximations* x_0 *and* x_1.
2. *Check, if* $f(x_0) = f(x_1)$, *go to step 1; otherwise; continue.*
3. *Establish the tolerance* ($\epsilon > 0$) *value for the function.*
4. *Compute the new approximation for the root using iterative formula (2.21).*
5. *Check the tolerance. If* $|x_n - x_{n-1}| \leq \epsilon$, *for* $n \geq 1$, *then end; otherwise, go back to step 4 and repeat the process.*

In the preceding discussion the restriction was made that $f'(\alpha) \neq 0$, where α is the solution to $f(x) = 0$. The rapid convergence of both Newton's method and the secant method depends on this restriction. From the definition of Newton's method, it is clear that difficulties might occur if $f'(x_n)$ goes to zero simultaneously with $f(x_n)$. In particular, Newton's method and the secant method will generally give problems if $f'(\alpha) = 0$ when $f(\alpha) = 0$. In the following section we investigate this situation and uncover an interesting fact, namely, how fast the iteration converges. We will see that both Newton's method and the secant method will continue to converge, but not as rapidly as we expect.

2.7 Multiplicity of a Root

So far we have discussed a function, which has a simple root. Now we will discuss a function which has multiple roots. A root is called a *simple root* if it is distinct, otherwise roots that are of the same order of magnitude are called *multiple*.

Definition 2.4 (Order of a Root)

The equation $f(x) = 0$ has a root α of order m, if there exists a continuous function $h(x)$, and $f(x)$ can be expressed as the product

$$f(x) = (x - \alpha)^m h(x), \quad \textit{where} \quad h(\alpha) \neq 0. \tag{2.22}$$

Thus, $h(x)$ can be used to obtain the remaining roots of $f(x) = 0$. This is called polynomial deflation.

A root of order $m = 1$ is called a *simple root* and if $m > 1$ it is called a *multiple root*. In particular, a root of order $m = 2$ is sometimes called a *double root*, and so on.

The behavior of the graph of $f(x)$ near a root of multiplicity m ($m = 1, 2, 3$) is shown in Figure 2.6.

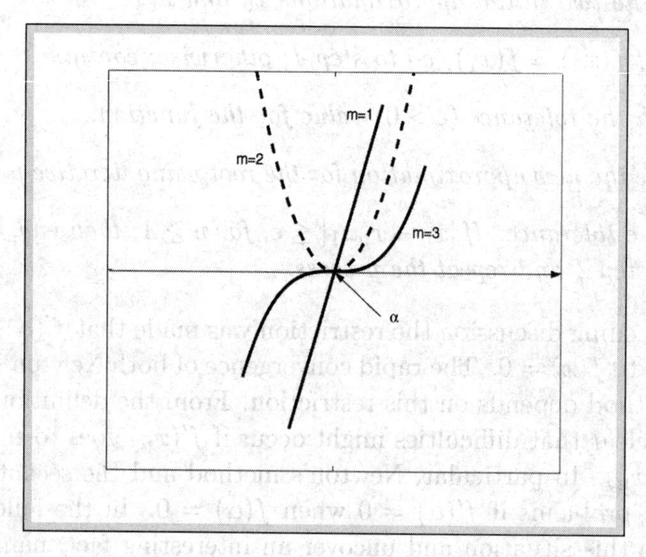

Figure 2.6: Multiple roots of $f(x) = 0$.

It can be seen that when α is a root of odd multiplicity, the graph of $f(x)$ will cross the x-axis at $(\alpha, 0)$; and when α has even multiplicity the graph will be tangent to but will not cross the x-axis at $(\alpha, 0)$. Moreover, the higher the value of m the flatter the graph will be near the point $(\alpha, 0)$.

Sometimes it is more difficult to deal with Definition 2.4 concerning the order of the root. We will use the following Lemma, which will illuminate these concepts.

Lemma 2.2 *Assume that $f(x)$ and its derivatives $f'(x), f''(x), \cdots, f^{(m)}(x)$ are defined and continuous on an interval $x = \alpha$. We say that $f(x) = 0$ has a root α of order m if and only if*

$$f(\alpha) = f'(\alpha) = f''(\alpha) = \cdots = f^{(m-1)}(\alpha) = 0, \qquad f^{(m)}(\alpha) \neq 0. \tag{2.23}$$

For example, consider the function $f(x) = x^3 - x^2 - 21x + 45$, which has two roots; a simple root at $\alpha = -5$ and a double root at $\alpha = 3$. This can be verified by considering the derivatives of the function as follows:

$$f'(x) = 3x^2 - 2x - 21, \qquad f''(x) = 6x - 2.$$

At the value $\alpha = -5$, we have

$$f(5) = 0, \qquad f'(5) = 64 \neq 0$$

so by (2.22), we see that $m = 1$, hence $\alpha = -5$ is a simple root of the function. For the value $\alpha = 3$, we have

$$f(3) = 0, \qquad f'(3) = 0, \qquad f''(3) = 16 \neq 0,$$

so that $m = 2$ by (2.23), hence $\alpha = 3$ is a double root of the function. Also, note that this function $f(x)$ has the factorization and can be written in the form of (2.22) as

$$f(x) = (x - 3)^2(x + 5).$$

Note that for a *simple root* α of the nonlinear equation $f(x) = 0$ means that

$$f(\alpha) = 0 \quad \text{and} \quad f'(\alpha) \neq 0.$$

But for a *multiple root* α of the nonlinear equation, we must have

$$f(\alpha) = 0 \quad \text{and} \quad f'(\alpha) = 0.$$

The order of multiplicity of the multiple root can be easily found by taking the higher derivatives of the function at α unless the higher derivative becomes nonzero at α. Then the order of the nonzero higher derivative will be the order of multiplicity of the multiple root.

Example 2.16 *Find the multiplicity of the root $\alpha = 1$ of the equation $x \ln x = \ln x$.*

Solution. *From the given equation, we have*

$$\begin{aligned} f(x) &= x\ln x - \ln x & \text{and} \quad f(1) = & \ 0, \\ f'(x) &= \ln x + 1 - \frac{1}{x} & \text{and} \quad f'(1) = & \ 0, \\ f''(x) &= \frac{1}{x} + \frac{1}{x^2} & \text{and} \quad f''(1) \neq & \ 0. \end{aligned}$$

Thus, the multiplicity of the root $\alpha = 1$ of the given equation is 2.

Usually we don't know in advance that an equation has multiple roots, although we might suspect it from sketching the graph. Many problems that lead to multiple roots are in fact ill-posed. The methods we discussed so far cannot be guaranteed to converge efficiently for all problems. In particular, when a given function has a multiple root, which we require, the methods we have described will either not converge at all or converge more slowly. For example, Newton's method converges very fast to a simple root but converges more slowly when used for functions involving multiple roots.

Example 2.17 *Consider the following two nonlinear equations*

$$(1) \qquad xe^x = 0, \qquad\qquad (2) \qquad x^2e^x = 0.$$

(a) Find Newton's method for the solutions of the given equations.
(b) Explain why one of the sequences converges much faster than the other to the root $\alpha = 0$.

Solution. *(a) For the first equation, we have*

$$f(x) = xe^x \qquad \text{and} \qquad f'(x) = (1 + x)e^x.$$

Then Newton's method for the solution of the first equation is

$$x_{n+1} = g_1(x_n) = x_n - \frac{f(x_n)}{f'(x_n)} = \frac{x_n^2}{(1 + x_n)}, \qquad n \geq 0.$$

which is the first sequence. Similarly, we can find Newton's method for the solution of the second equation as follows:

$$x_{n+1} = g_2(x_n) = x_n - \frac{x_n^2e^{x_n}}{(2x_n + x_n^2)e^{x_n}} = \frac{x_n + x_n^2}{(2 + x_n)}, \qquad n \geq 0,$$

and it is the second sequence.
(b) From the first sequence, we have

$$g_1(x) = \frac{x^2}{(1+x)},$$

and its derivative can be obtained as

$$g_1'(x) = \frac{x^2+2x}{(1+x)^2},$$

Then

$$|g_1'(\alpha)| = |g_1'(0)| = \left|\frac{0}{1}\right| = 0,$$

which shows that the first sequence converges to zero. Similarly, from the second sequence, we have

$$g_2(x) = \frac{x_+x^2}{(2+x)},$$

and its derivative can be obtained as

$$g_2'(x) = \frac{x^2+4x+2}{(2+x)^2}.$$

Thus,

$$|g_2'(0)| = \left|\frac{2}{4}\right| = \frac{1}{2} < 1,$$

which shows that the second sequence also converges to zero. Since the value of $|g_1'(0)|$ *is smaller than* $|g_2'(0)|$*, the first sequence converges faster than the second one.*

Note that in Example 2.17 the root $\alpha = 0$ is the simple root for the first equation because

$$f(0) = 0 \quad \text{and} \quad f'(0) = 1 \neq 0,$$

and for the second equation it is a multiple root because

$$f(0) = 0 \quad \text{and} \quad f'(0) = 0.$$

Therefore, Newton's method converges very fast for the first equation and converges very slow for the second equation. However, in some cases simple modifications can be made to the methods to maintain the rate of convergence.

Two such modified methods are considered here and are called Newton modified methods. If we wish to determine a root of known multiplicity m for the equation $f(x) = 0$, then the *first Newton's modified method* (also called the *Schroeder's method*) may be used. It has the form

$$x_{n+1} = x_n - m\frac{f(x_n)}{f'(x_n)}, \qquad n = 0, 1, 2, \ldots. \tag{2.24}$$

It is assumed that we have an initial approximation x_0. The similarity to Newton's method is obvious and like Newton's method it converges very fast for the multiple roots. The major disadvantage of this method is that the multiplicity of the root must be known in advance and this is generally not the case in practice. An alternative approach to this problem that does not require any knowledge of the multiplicity of the root is to replace the function $f(x)$ in the equation by $q(x)$, where

$$q(x) = \frac{f(x)}{f'(x)}.$$

One can show that $q(x)$ has only a simple root at $x = \alpha$. Thus, Newton's method applied to find a root of $q(x)$ will avoid any problems of multiple roots. If

$$f(x) = (x - \alpha)^m h(x),$$

then

$$f'(x) = m(x - \alpha)^{m-1} h(x) + (x - \alpha)^m h'(x).$$

Thus,

$$q(x) = \frac{(x - \alpha)h(x)}{[mh(x) + (x - \alpha)h'(x)]}.$$

Obviously, we find that $q(x)$ has the root α to multiplicity one. So with this modification, Newton's method becomes

$$x_{n+1} = x_n - \frac{q(x_n)}{q'(x_n)},$$

which gives

$$x_{n+1} = x_n - \frac{f(x_n)f'(x_n)}{[f'(x_n)]^2 - [f(x_n)][f''(x_n)]}, \qquad n = 0, 1, 2, \ldots. \tag{2.25}$$

Iterative formula (2.25) is known as the *second modified Newton's method.* The disadvantage of this method is that we must calculate a further higher derivative. A similar modification can be made to the secant method.

Example 2.18 *Show that the function* $f(x) = e^x - \frac{x^2}{2} - x - 1$ *has (zero of multiplicity 3) at* $\alpha = 0$ *and then find the approximate solution of the zero of the function with the help of Newton's method and first and second modified Newton's methods by taking an initial approximation* $x_0 = 1.5$ *within an accuracy of* 10^{-4}.

Solution. *Since*

$$\begin{array}{rclrcl} f(x) & = & e^x - \frac{x^2}{2} - x - 1, & f(0) & = & 0 \\ f'(x) & = & e^x - x - 1, & f'(0) & = & 0 \\ f''(x) & = & e^x - 1, & f''(0) & = & 0 \\ f'''(x) & = & e^x, & f'''(0) & = & 1 \neq 0, \end{array}$$

the function has a zero of multiplicity 3. In Table 2.6 we show a comparison of the three methods.

Table 2.6: Comparison results of three methods for Example 2.18.

	Newton's Method	1st. M.N. Method	2nd. M.N. Method
n	x_n	x_n	x_n
00	1.500000	1.500000	1.500000
01	1.067698	0.2030926	-0.297704
02	0.745468	3.482923e-03	-6.757677e-03
03	0.513126	1.010951e-06	-3.798399e-06
..	....		
..	....		
25	7.331582e-05		

To use the MATLAB command for the first modified Newton's method (2.24), we define a function m-file as fn1.m and its derivative m-file as dfn1.m for the equation as follows:

```
function y = fn1(x)
y = exp(x) - x.^2/2 - x - 1;
```

and

```
function dy = dfn1(x)
dy = exp(x) - x - 1;
```

then use the following commands:

```
>> x0 = 1.5; m = 3; tol = 1e - 4;
>> sol = mnewton1('fn1', 'dfn1', x0, m, tol)
sol =
     1.01095e - 06
```

To use the MATLAB command for second modified Newton's method (2.25), the function $e^x - \frac{x^2}{2} - x - 1$ and its first derivative $e^x - x - 1$ has been already used in the m-files fn1.m and dfn1.m, respectively. Now we define a m-file ddfn1.m for the second derivative $e^x - 1$ of the function as follows:

```
function ddy = ddfn1(x)
ddy = exp(x) - 1;
```

and then use the following MATLAB command:

```
>> x0 = 1.5; tol = 1e - 4;
>> sol = mnewton2('fn1', 'dfn1', 'ddfn1', x0, tol)
sol =
     -3.798399e - 06
```

We note that for the multiple root both modified Newton's methods converge very fast as they took four iterations to converge while Newton's method converges very slow and took 25 iterations to converge for the same accuracy.

Program 2.6

MATLAB m-file for first Modified Newton's Method

```
function sol=mnewton1(fn1,dfn1,x0,m,tol)
old = x0+1;
while abs (x0 - old) > tol; old = x0;
fa=feval(fn,old); fb=feval(dfn,old);
x0 = old - (m * fa)/fb;
end; sol=x0;
```

Example 2.19 *Show that $\alpha = 0$ is a zero of multiplicity $m = 2$ of $f(x) = 1 - \cos(x)$. Use the quadratic convergent method to find the first approximation x_1 if $x_0 = 0.1$.*

Solution. *Since* $f(x) = 1 - \cos x$, *first we show that* $\alpha = 0$ *is the zero of the given function as*

$$f(\alpha) = f(0) = 1 - \cos 0 = 1 - 1 = 0.$$

To check whether it is the simple or multiple zero of $f(x)$, *we do the following*

$$f'(x) = \sin x, \qquad f'(\alpha) = f'(0) = \sin 0 = 0,$$

which means that $\alpha = 0$ *is the multiple zero of the given function. To find its order of multiplicity, we do*

$$f''(x) = \cos x, \qquad f''(\alpha) = f''(0) = \cos 0 = 1 \neq 0,$$

hence $\alpha = 0$ *is a zero of multiplicity 2 of the given function. Now we have to find the first approximation to the multiple zero* $\alpha = 0$ *of the given function using the quadratic convergence method. We know that the modified Newton's method is a quadratic convergent method for the multiple zero of the function (or multiple root of the equation* $f(x) = 0$*), which can be written as*

$$x_{n+1} = x_n - m\frac{f(x_n)}{f'(x_n)}, \qquad n = 0, 1, 2, \ldots,$$

where m *is the order of multiplicity of the zero of the function.*
For $n = 0$, $m = 2$, *and an initial approximation* $x_0 = 0.1$, *we have*

$$x_1 = x_0 - 2\frac{f(x_0)}{f'(x_0)} = 0.1 - 2\frac{(1 - \cos 0.1)}{\sin 0.1} = 0.098,$$

which is the required first approximation to $\alpha = 0$.

Note that when the order of multiplicity of a root of the equation $f(x) = 0$ is not known, then second modified Newton's formula (2.25) can be used. The MATLAB m-file can be written as follows:

Program 2.7

MATLAB m-file for the second Modified Newton's Method

```
function sol=mnewton2(fn1,dfn1,ddfn1,x0,tol)
old = x0+1;
while abs (x0 - old) > tol; old = x0;
fa=feval(fn,old); fb=feval(dfn,old); fc=feval(ddfn,old);
x0 = old - (fa * fb)/((fb). ^ 2 - (fa * fc));
end; sol=x0;
```

2.8 Convergence of Iterative Methods

Now we define the order of the convergence of functional iteration schemes discussed in the previous sections. This is a measure of how rapidly a sequence converges.

Definition 2.5 (Order of Convergence)

Suppose that the sequence $\{x_n\}_{n=0}^{\infty}$ *converges to* α. *Let* $e_n = \alpha - x_n$ *and define the error of the nth iterate. If two positive constants* $\beta \neq 0$ *and* $R > 0$ *exist, and*

$$\lim_{n\to\infty} \frac{|\alpha - x_{n+1}|}{|\alpha - x_n|^R} = \lim_{n\to\infty} \frac{|e_{n+1}|}{|e_n|^R} = \beta, \qquad (2.26)$$

then the sequence is said to converge to α *with order of convergence* R. *The number* β *is called the asymptotic error constant. The cases* $R = 1, 2$ *are given special consideration.*
If $R = 1$, *the convergence of the sequence* $\{x_n\}_{n=0}^{\infty}$ *is called linear.*
If $R = 2$, *the convergence of the sequence* $\{x_n\}_{n=0}^{\infty}$ *is called quadratic.*

If R is large, the sequence $\{x_n\}$ converges rapidly to α; that is, (2.26) implies that for large values of n we have the approximation $|e_{n+1}| \approx \beta |e_n|^R$. For example, suppose that $R = 2$ and $|e_n| \approx 10^{-3}$; then we could expect that $|e_{n+1}| \approx \beta \times 10^{-6}$.
Quadratically convergent sequences generally converge much faster than those that converge only linearly, but many techniques that generate convergent sequences do so only linearly. The following two lemmas tell us about the conditions of the linear convergence and the quadratic convergence of the sequences.

Lemma 2.3 (Linear Convergence)

Let g *be continuously differentiable on the interval* $[a, b]$ *and suppose that* $g(x) \in [a, b]$ *for all* $x \in [a, b]$. *Suppose that* $g'(x)$ *is continuous on* (a, b) *with*

$$|g'(x)| \leq k < 1; \qquad \text{for all} \quad x \in (a, b).$$

If $g'(\alpha) \neq 0$, *then for any* $x_0 \in [a, b]$, *the sequence* $x_{n+1} = g(x_n)$, *for* $n \geq 0$, *converges only linearly to the unique fixed point* α *on* $[a, b]$.

Example 2.20 *Consider an iterative scheme*

$$x_{n+1} = 0.4 + x_n - 0.1x_n^2, \quad n \geq 0.$$

Will this scheme converge to the fixed point $\alpha = 2$? If yes, find its rate of convergence.

Solution. *Since*

$$g(x) = 0.4 + x - 0.1x^2.$$

Then

$$g(2) = 0.4 + 2 - 0.1(2)^2 = 2,$$

which shows that the scheme converges to $\alpha = 2$. Also,

$$g'(x) = 1 - 0.2x, \quad gives \quad g'(2) = 1 - 0.4 = 0.6 \neq 0.$$

Therefore, the scheme converges linearly.

Lemma 2.4 (Quadratic Convergence)

Let α be a solution of the equation $x = g(x)$. Suppose that $g'(\alpha) = 0$ and g'' is continuous on an open interval (a, b) containing α. Then there exists a $\delta > 0$ such that, for $x_0 \in [\alpha - \delta, \alpha + \delta]$, the sequence $\{x_n\}_{n=0}^{\infty}$ defined by the iteration $x_{n+1} = g(x_n)$, for $n \geq 0$, converges at least quadratically to α.

Example 2.21 *The iterative scheme*

$$x_{n+1} = 2 - (1 + a)x_n + ax_n^2, \quad n \geq 0$$

converges to $\alpha = 1$ for some values of a. Find the value of a for which the convergence is at least quadratic.

Solution. *Given*

$$g(x) = 2 - (1 + a)x + ax^2.$$

Then

$$g(1) = 2 - (1 + a) + a = 1.$$

Thus, the given iterative scheme converges to 1. Also

$$g'(x) = -(1 + a) + 2ax$$

and so

$$g'(1) = 0 = -(1 + a) + 2a, \quad \text{gives} \quad a = 1.$$

Thus, the convergence of the given iterative scheme is at least quadratic for the value of $a = 1$.

Note 2.3 *The sequence $\{x_n\}_{n=0}^{\infty}$ defined by the iteration*

$$x_{n+1} = g(x_n), \quad \text{for} \quad n \geq 0$$

converges only quadratically to α if

$$g'(\alpha) = 0 \quad \text{but} \quad g''(\alpha) \neq 0.$$

and cubically (order three) to α if

$$g''(\alpha) = 0 \quad \text{but} \quad g'''(\alpha) \neq 0.$$

In a similar manner the higher order of convergence can be achieved.

Example 2.22 *Show that the following iterative scheme*

$$x_{n+1} = x_n - \ln x_n, \quad \text{for} \quad n \geq 0,$$

converges quadratically to $\alpha = 1$.

Solution. *Given*

$$g(x) = x - \ln x, \quad \text{and} \quad g(1) = 1,$$

to find the derivative of $g(x)$, we have

$$g'(x) = 1 - \frac{1}{x}.$$

Taking $x = \alpha = 1$, we have

$$g'(1) = 1 - \frac{1}{1} = 1 - 1 = 0.$$

Also,

$$g''(x) = \frac{1}{x^2},$$

which gives

$$g''(1) = \frac{1}{1^2} = 1 \neq 0.$$

Example 2.23 *(a) What is the order of convergence of the iteration*

$$x_{n+1} = \frac{x_n(x_n^2 + 3b)}{3x_n^2 + b}, \qquad b > 0$$

as it converges to the fixed point $\alpha = \sqrt{b}$*?*
(b) Find the values of a and b such that the iterative scheme

$$x_{n+1} = ax_n^2 + \frac{b}{x_n} - 5, \quad n \geq 0$$

converges quadratically to $\alpha = 1$.

Solution. *(a) Since the given iteration is*

$$x_{n+1} = \frac{x_n(x_n^2 + 3b)}{3x_n^2 + b} = g(x_n),$$

which gives

$$g(x) = \frac{x(x^2 + 3b)}{3x^2 + b},$$

the first derivative of $g(x)$ *can be found as*

$$g'(x) = \frac{3(x^2 - b)^2}{(3x^2 + b)^2}.$$

To find the order of convergence of the iteration, we have to check the derivative $g'(x)$ *at the fixed point* $x = \alpha = \sqrt{b}$*. If it is equal to zero, then order is at least quadratic, otherwise linear. So*

$$g'(\sqrt{b}) = \frac{3((\sqrt{b})^2 - b)^2}{(3(\sqrt{b})^2 + b)^2} = 0.$$

Therefore, the order of convergence for the given iteration is at least quadratic. One can find the second derivative of $g(x)$ *as*

$$g''(x) = \frac{48xb(x^2 - b)}{(3x^2 + b)^3},$$

which gives

$$g''(\sqrt{b}) = 0, \quad but \quad g'''(\sqrt{b}) \neq 0.$$

Hence, the order of convergence for the given iteration is exactly cubic.
(b) Given

$$g(x) = ax^2 + \frac{b}{x} - 5$$

and at the fixed point $\alpha = 1$*, we have*

$$g(1) = 1 = a + b - 5, \quad \textit{which gives} \quad a + b = 6.$$

Also,

$$g'(x) = 2ax - \frac{b}{x^2}$$

and the convergence is quadratic at $\alpha = 1$*, we have*

$$g'(1) = 0 = 2ax - b, \quad \textit{gives} \quad 2a - b = 0.$$

Solving these two equations for unknowns a *and* b*, we obtain*

$$a = 2, \quad \textit{and} \quad b = 4.$$

Note that

$$g''(x) = 2a + \frac{2b}{x^3}, \quad \textit{and} \quad g''(1) = 12 \neq 0.$$

Now we discuss the rate of convergence of all the iterative methods for the nonlinear equations, which we discussed in the previous sections.

Case 2.1 (Bisection Method)

The convergence of the bisection method is very slow. At each step we gain one binary digit in accuracy. Since $10^{-1} \approx 2^{-3.3}$*, we gain on the average one decimal digit per 3.3 steps. Note that the rate of convergence is completely independent of the function* $f(x)$*. This is because we only make use of the sign of the computed function values. To investigate the rate of convergence of the bisection method, we consider the following example.*

Example 2.24 *If* α *is the fixed point of the equation* $x = g(x)$ *in* $[x, b]$*, then show that the rate of convergence of the bisection method is linear.*

Solution. *Since the bisection iteration function is defined on the interval* $[x, b]$*, using the bisection formula (2.2), we have*

$$g(x) = \frac{x + b}{2}.$$

Then

$$g'(x) = \frac{1}{2}$$

so, at $x = \alpha$, we have

$$g'(\alpha) = \frac{1}{2} \neq 0,$$

therefore, by Lemma 2.3, the convergence is linear.

Case 2.2 (False Position Method)

To investigate the rate of convergence of the false position method, it is convenient to rearrange the equation as

$$x_{n+1} = x_n + \frac{f(x_n)(x_0 - x_n)}{f(x_n) - f(x_0)}. \tag{2.27}$$

Let $e_n = x_n - \alpha$ be the error to the nth iterate. Then

$$\begin{aligned} e_{n+1} &= e_n + \frac{f(\alpha + e_n)(x_0 - \alpha - e_n)}{f(\alpha + e_n) - f(x_0)} \\ &= e_n + \frac{e_n f'(\alpha)(x_0 - \alpha)}{-f(x_0)} + O(e^2) \qquad (2.28) \\ &= e_n \left(1 - \frac{f'(\alpha)(x_0 - \alpha)}{f(x_0)}\right) + O(e^2). \end{aligned}$$

By noting that the ratio $\dfrac{(x_0 - \alpha)}{f(x_0)}$ is approximately $1/f'(\alpha)$ as $f(\alpha) = 0$ it is clear that

$$\left|1 - \frac{f'(\alpha)(x_0 - \alpha)}{f(x_0)}\right| < 1, \tag{2.29}$$

and so the convergence of this form is said to be linear or order one.

Case 2.3 (Fixed-point Method)

The convergence rate of the fixed-point iteration can be analyzed as follows. The general procedure is given by

$$x_{n+1} = g(x_n), \qquad n = 0, 1, 2, \ldots. \tag{2.30}$$

Let $x = \alpha$ denote the solution to $f(x) = 0$, so $f(\alpha) = 0$ and $\alpha = g(\alpha)$. Then

$$x_{n+1} - \alpha = e_{n+1} = g(x_n) - g(\alpha), \tag{2.31}$$

where e_{n+1} denotes the error of the (n+1)th iterate. Expressing $g(\alpha)$ in the Taylor series about x_n gives

$$g(\alpha) = g(x_n) + g'(\eta)(\alpha - x_n), \qquad x_n \leq \eta \leq \alpha. \tag{2.32}$$

Solving (2.32) for $g(x_n) - g(\alpha)$ and substituting into (2.31), we get

$$e_{n+1} = g'(\eta)e_n, \tag{2.33}$$

or

$$|e_{n+1}| = |g'(\eta)||e_n|. \tag{2.34}$$

Now suppose that $|g'(x)| \leq k < 1$ for all values of x in an interval. If x_1 is chosen in this interval, x_2 will also be in the interval and the fixed-point iteration method will converge, since

$$\left|\frac{e_{n+1}}{e_n}\right| = |g'(\eta)| < 1. \tag{2.35}$$

Convergence is linear since e_{n+1} is linearly dependent on e_n. If $|g'(\eta)| > 1$, the procedure diverges. If $|g'(\eta)| < 1$, but close to one, convergence is quite slow.

Example 2.25 *(a) Show that $\alpha = 1$ is a unique fixed point of*

$$g(x) = \frac{x^2 - 4x + 7}{4}.$$

(b) Find the rate of convergence of the sequence

$$x_n = \frac{x_{n-1}^2 - 4x_{n-1} + 7}{4}.$$

(c) Find x_3 using $x_0 = 1.5$, and a bound for the error $|\alpha - x_3|$.

Solution. *(a) Firstly, we show that $\alpha = 1$ is a fixed point of $g(x)$ by showing that $g(1) = 1$ and it happened because*

$$g(1) = \frac{1 - 4 + 7}{4} = 1.$$

It is unique also because

$$g'(x) = \frac{2x-4}{4} \quad and \quad |g'(1)| = 0.5 < 1.$$

(b) To find the rate of convergence of the given sequence, we have

$$g(x) = \frac{x^2 - 4x + 7}{4}$$

and its derivative

$$g'(x) = \frac{2x-4}{4}.$$

Taking $x = \alpha = 1$, gives

$$g'(1) = \frac{2-4}{4} = -\frac{1}{2} \neq 0.$$

Hence, the rate of the convergence of the given sequence is linear.
(c) To find the third approximation using the fixed point iteration method

$$x_{n+1} = g(x_n) = \frac{x_n^2 - 4x_n + 7}{4}, \quad n = 0, 1, 2$$

using $x_0 = 1.5$, we have

$$\begin{aligned}
x_1 &= g(x_0) = \frac{x_0^2 - 4x_0 + 7}{4} = \frac{2.25 - 6 + 7}{4} = 0.8125 \\
x_2 &= g(x_1) = \frac{x_1^2 - 4x_1 + 7}{4} = \frac{0.660 - 3.25 + 7}{4} = 1.1025 \\
x_3 &= g(x_2) = \frac{x_2^2 - 4x_2 + 7}{4} = \frac{1.216 - 4.41 + 7}{4} = 0.952.
\end{aligned}$$

Now to find an error bound for our approximation, we have to use

$$|\alpha - x_3| \leq \frac{k^3}{1-k}|x_1 - x_0|,$$

where

$$k = \max_{x=1} |g'(x)| = \max_{x=1} \left| \frac{2x-4}{4} \right| = 0.5.$$

Thus,

$$|\alpha - x_3| \leq \frac{(0.5)^3}{1-0.5} |0.8125 - 1.5| = 0.172,$$

which is the required error bound of our approximation.

Case 2.4 (Newton's Method)

The convergence rate of Newton's method can be analyzed as follows. The general procedure

$$x_{n+1} = x_n - \frac{f(x_n)}{f'(x_n)}, \qquad n = 0, 1, 2, \ldots$$

is of the form $x_{n+1} = g(x_n)$. *Consequently, if the method converges, then the absolute value of the derivative of the function* $g(x)$ *with respect to* x *must be less than one; that is,* $|g'(x)| < 1$. *Since*

$$g(x) = x - \frac{f(x)}{f'(x)},$$

then

$$g'(x) = 1 - \frac{f'(x)f'(x) - f(x)f''(x)}{[f'(x)]^2} = \frac{f(x)f''(x)}{[f'(x)]^2}.$$

Hence, if

$$\left| \frac{f(x)f''(x)}{[f'(x)]^2} \right| < 1, \tag{2.36}$$

on an interval about the root α, *the method will converge for any initial approximation in the interval. The (2.36) represents a sufficient condition for convergence. It is evident that* $f'(x)$ *must not be zero. This is an important factor to consider when choosing the initial* x *value.*

Now we show that Newton's method is quadratically convergent for the simple root. Let $x = \alpha$ *denote the solution to* $f(x) = 0$, *so* $f(\alpha) = 0$ *and* $\alpha = g(\alpha)$. *Since* $x_{n+1} = g(x_n)$, *we can write*

$$x_{n+1} - \alpha = e_{n+1} = g(x_n) - g(\alpha), \tag{2.37}$$

where e_n *denotes the error of the nth iterate. Let us expand* $g(x_n)$ *as a Taylor series in terms of* $(x_n - \alpha)$ *with the second derivative term as the remainder:*

$$g(x_n) = g(\alpha) + g'(\alpha)(x_n - \alpha) + \frac{g''(\eta)}{2}(x_n - \alpha)^2, \quad x_n \leq \eta \leq \alpha.$$

Since

$$g'(\alpha) = \frac{f(\alpha)f''(\alpha)}{[f'(\alpha)]^2} = 0,$$

because $f(\alpha) = 0$, *we have*

$$g(x_n) = g(\alpha) + \frac{g''(\eta)}{2}(x_n - \alpha)^2.$$

Solving the above equation for $(g(x_n) - g(\alpha))$ *and substituting into (2.37), we get*

$$e_{n+1} = g(\alpha) - g(x_n) = -\frac{g''(\eta)}{2}(e_n)^2. \qquad (2.38)$$

This implies that each error is (in the limit) proportional to the square of the previous error; that is, Newton's method is quadratically convergent.

Example 2.26 *Let* $x = \sqrt{N}$. *Find the fixed-point iteration form of Newton's method and show that* e_{n+1} *is proportional to* e_n^2, *where* $e_n = x_n - \alpha$.

Solution. *Since* $x^2 - N = 0)$,

$$f(x) = x^2 - N \qquad \text{and} \qquad f'(x) = 2x.$$

Newton's iteration function is defined as follows:

$$x_{n+1} = g(x_n) = x_n - \frac{f(x_n)}{f'(x_n)}, \quad n \geq 0$$

or

$$x_{n+1} = g(x_n) = \left(\frac{x_n^2 + N}{2x_n}\right).$$

Then

$$e_{n+1} = x_{n+1} - \alpha = \left(\frac{x_n^2 + N}{2x_n}\right) - \alpha = \frac{(x_n - \alpha)^2}{2x_n} = \frac{e_n^2}{2x_n}.$$

Thus,

$$e_{n+1} \propto e_n^2.$$

Since we know that the rate of convergence of Newton's method is linear if the function has a multiple root, in the following example, we discuss the rate of convergence of Newton's method for the multiple roots.

Example 2.27 *If* $x = \alpha$ *is a root of multiplicity* m *of* $f(x) = 0$, *then show that the rate of convergence of Newton's method is linear.*

Solution. *Consider Newton's iteration function, which is defined as follows:*

$$g(x) = x - \frac{f(x)}{f'(x)}.$$

Since the function $f(x)$ *has a multiple root,*

$$f(x) = (x - \alpha)^m h(x),$$

and its derivative is

$$f'(x) = m(x-\alpha)^{m-1}h(x) + (x-\alpha)^m h'(x).$$

Substituting the values of $f(x)$ and $f'(x)$ in the above equation, we get

$$g(x) = x - \frac{(x-\alpha)^m h(x)}{(m(x-\alpha)^{m-1}h(x) + (x-\alpha)^m h'(x))}$$

or

$$g(x) = x - \frac{(x-\alpha)h(x)}{(mh(x) + (x-\alpha)h'(x))}.$$

Then

$$\begin{aligned} g'(x) = 1 \quad - \quad & \{([mh(x) + (x-\alpha)][h(x) + (x-\alpha)h'(x)] - [(x-\alpha)h(x)] \\ & [mh'(x) + h'(x) + (x-\alpha)h''(x)])\}/([mh(x) + (x-\alpha)h'(x)]^2). \end{aligned}$$

At $x = \alpha$, and since $f(\alpha) = 0$, we have

$$g'(\alpha) = 1 - \frac{[mh(\alpha)][h(\alpha)]}{[mh(\alpha)]^2}$$

or

$$g'(\alpha) = 1 - \frac{1}{m} \neq 0, \qquad (m > 1).$$

Therefore, Newton's method converges to a multiple zero from any sufficiently close approximation and the convergence is linear (by Lemma 2.3), with ration $(1 - \frac{1}{m})$. In particular, for a double root, the ration is $\frac{1}{2}$, which is comparable with the convergence of the bisection method.

Case 2.5 (Secant Method)

The convergence rate of the secant method can be analyzed as follows. The general procedure is

$$x_{n+1} = x_n - \frac{f(x_n)(x_n - x_{n-1})}{f(x_n) - f(x_{n-1})}. \tag{2.39}$$

As before, let $x_{n-1} = \alpha - e_{n-1}, x_n = \alpha - e_n$, and $x_{n+1} = \alpha - e_{n+1}$. Then

$$e_{n+1} = e_n - \frac{f(\alpha - e_n)(e_n - e_{n-1})}{f(\alpha - e_n) - f(\alpha - e_{n-1})}$$

since using the Taylor's theorem

$$y_n = f(\alpha - e_n) = f(\alpha) - e_n f'(\alpha) + \frac{e_n^2}{2!} f''(\alpha) - \cdots$$

and

$$y_{n-1} = f(\alpha - e_{n-1}) = f(\alpha) - e_{n-1} f'(\alpha) + \frac{e_{n-1}^2}{2!} f''(\alpha) - \cdots,$$

we have

$$\begin{aligned}
e_{n+1} &= e_n - \frac{(e_n - e_{n-1})[f(\alpha) - e_n f'(\alpha) + 1/2 e_n^2 f''(\alpha) - \cdots]}{[-(e_n - e_{n-1}) f'(\alpha) + 1/2(e_n^2 - e_{n-1}^2) f''(\alpha) - \cdots]} \\
&= e_n - \left[\frac{-e_n f'(\alpha) + 1/2 e_n^2 f''(\alpha) - \cdots}{-f'(\alpha) + 1/2(e_n + e_{n-1}) f''(\alpha) - \cdots}\right] \qquad (because f(\alpha) = 0) \\
&= e_n - \frac{1}{f'(\alpha)}[-e_n f'(\alpha) + 1/2 e_n^2 f''(\alpha) - \cdots] \\
&\times \ [-1 + 1/2(e_n + e_{n-1}) \tfrac{f''(\alpha)}{f'(\alpha)} - \cdots]^{-1} \\
&= e_n - \frac{1}{f'(\alpha)}[e_n f'(\alpha) + 1/2 e_n^2 f''(\alpha) + \cdots] \\
&\times \ [1 - 1/2(e_n + e_{n-1}) \frac{f''(\alpha)}{f'(\alpha)} + \cdots]^{-1} \\
&= e_n - \frac{1}{f'(\alpha)}[e_n f'(\alpha) - 1/2 e_n^2 f''(\alpha) + 1/2 e_n(e_n + e_{n-1}) f''(\alpha) - \cdots] \\
&= e_n - \frac{1}{f'(\alpha)}[-e_n f'(\alpha) + 1/2 e_n e_{n-1} f''(\alpha) - \cdots] \\
&= -\frac{f''(\alpha)}{2f'(\alpha)} e_n e_{n-1} + \cdots
\end{aligned}$$

Hence,

$$e_{n+1} \approx K e_n e_{n-1}, \quad where \quad K = -\frac{f''(\alpha)}{2f'(\alpha)} \tag{2.40}$$

so that the each error is proportional to the product of the previous two errors. By comparison with Newton's method convergence we expect the rate of convergence of the secant method will be inferior to that of Newton's method. If we put

$$e_n = \beta e_{n-1}^R,$$

where R *is the order of convergence and the constant* β *is the asymptotic error constant, then we obtain*

$$e_{n+1} = \beta e_n^R = \beta(\beta e_{n-1}^R)^R$$

and

$$\beta(\beta e_{n-1}^R)^R \approx K\beta e_{n-1}^R e_{n-1}.$$

Thus,

$$e_{n-1}^{R^2} \approx \lambda e_{n-1}^{R+1}$$

for some constant λ*, it follows that*

$$R^2 = R + 1.$$

Solving this quadratic equation, we get

$$R = \frac{1 \pm \sqrt{5}}{2} = 1.61803$$

neglecting the negative value. This formula for R *tells us that*

$$|e_n| \approx \beta |e_{n-1}|^{1.61803}.$$

Thus, the error of the secant method is of order 1.61803*, which is between* 1 *and* 2*. This shows that the order of the secant method is better than the bisection method and fixed-point method but less than Newton's method.*
Remember, however, that the secant method does not require the derivative of $f(x)$ *to be evaluated at each step, so that in many ways the secant method is a very attractive alternative to the standard Newton's method.*

Example 2.28 *If* $x = \alpha$ *is a root of multiplicity* m *of* $f(x) = 0$*, then show that the rate of convergence of the modified Newton's method is at least quadratic.*

Solution. *The first modified Newton's iteration function is defined as follows:*

$$g(x) = x - m\frac{f(x)}{f'(x)}. \tag{2.41}$$

Since the function $f(x)$ *has a multiple root*

$$f(x) = (x-\alpha)^m h(x)$$

and its derivative is

$$f'(x) = m(x-\alpha)^{m-1}h(x) + (x-\alpha)^m h'(x).$$

Substituting the values of $f(x)$ *and* $f'(x)$ *in (2.41), we get*

$$g(x) = x - \frac{m(x-\alpha)^m h(x)}{(m(x-\alpha)^{m-1}h(x) + (x-\alpha)^m h'(x))}$$

or

$$g(x) = x - \frac{m(x-\alpha)h(x)}{(mh(x) + (x-\alpha)h'(x))}.$$

Then

$$g'(x) = 1 - m\{([mh(x) + (x-\alpha)][h(x) + (x-\alpha)h'(x)] - [(x-\alpha)h(x)] [mh'(x) + h'(x) + (x-\alpha)h''(x)])\}/([mh(x) + (x-\alpha)h'(x)]^2).$$

At $x = \alpha$*, and since* $f(\alpha) = 0$*, we have*

$$g'(\alpha) = 1 - \frac{[m^2h^2(\alpha)]}{[mh(\alpha)]^2},$$

which gives

$$g'(\alpha) = 0.$$

Therefore, modified Newton's method converges to a multiple root α and convergence is at least quadratically (by Lemma 2.4). Similarly, if $x = \alpha$ is a root of multiplicity m of $f(x) = 0$, then using Example 2.28 we can easily show that the rate of convergence of Newton's method is linear. As the Newton iteration function is defined by

$$g(x) = x - m\frac{f(x)}{f'(x)},$$

and proceeding in the same way as we did in Example 2.28, we get

$$g'(\alpha) = 1 - \frac{1}{m} \neq 0, \quad \text{because} \quad m > 1.$$

Hence, Newton's method converges to a multiple root α from any sufficiently close approximation and the convergence is linear (by Lemma 2.3) with ration $(1 - \frac{1}{m})$. In particular, for a double root, the ration is $\frac{1}{2}$, which is comparable with the convergence of the bisection method.

2.9 Acceleration of Convergence

In this section we discuss the possibility of improving the rate of convergence for a slow convergence sequence of iterates. The approach we adopt is due to *Aitken's* Δ^2 *method*, or simply, *Aitken's method.*

Suppose $\{x_n\}_{n=0}^{\infty}$ is a linear convergent sequence with limit α and an asymptotic error constant less than 1. Now to construct a sequence $\{x'_n\}$ that converges more rapidly to α than does $\{x_n\}$, let $e_n = \alpha - x_n$ be the error to the *nth* iterate, and α be the converged value, then the linear convergence gives

$$e_{n+1} \approx ke_n.$$

Hence, if three successive iterates are known, then

$$\begin{array}{rcl} \alpha - x_{n+1} & \approx & k(\alpha - x_n) \\ \alpha - x_{n+2} & \approx & k(\alpha - x_{n+1}), \end{array}$$

and k can be eliminated to yield an estimate of α. The resulting formula is

$$\alpha \approx \frac{x_{n+2}x_n - x_{n+1}^2}{x_{n+2} - 2x_{n+1} + x_n},$$

or adding and subtracting the terms x_n^2 and $2x_nx_{n+1}$ in the numerator gives

$$\alpha \approx x_n - \frac{(x_{n+1} - x_n)^2}{x_{n+2} - 2x_{n+1} + x_n}.$$

Aitken's method is based on the assumption that the sequence $\{x'_n\}_{n=0}^{\infty}$, defined by

$$x'_n = x_n - \frac{(x_{n+1} - x_n)^2}{x_{n+2} - 2x_{n+1} + x_n}, \quad n \geq 0, \tag{2.42}$$

converges faster to α compared to the original sequence $\{x_n\}_{n=0}^{\infty}$. By making use of this iterative formula, we can effectively convert any first-order process into a second-order process.

Example 2.29 *Consider the function*

$$f(x) = e^x - x - 2,$$

which has an equivalent direct iteration

$$x_{n+1} = ln(x_n + 2), \qquad n \geq 0.$$

It has been solved and shown to be convergent in Example 2.9. Nevertheless, computing two iterates and using Aitken's method yields the set of results in Table 2.7. We see after just two iterations that we got the better approximations.

Table 2.7: Solution of $x = \ln(x+2)$ by Aitken's method.

n	x_n	x_{n+1}	x_{n+2}	x_{n+3}
00	1.500000	1.252763	1.179505	1.148658
01	1.1252763	1.179505	1.156725	1.146442

To use MATLAB for Aitken's method, starting with the first three iterations (compute by the fixed-point method), we use the following sequence of MATLAB commands:

```
>> x0 = 1.5; x1 = 1.252763; x2 = 1.179505;
>> x3 = x0 - (x1 - x0)^2/(x2 - 2 * x1 + x0)
x3 =
    1.148658
>> x0 = 1.148658; x1 = 1.146976; x2 = 1.146442;
>> x3 = x0 - (x1 - x0)^2/(x2 - 2 * x1 + x0)
x3 =
    1.146193
```

To write formula (2.42) in the difference operator form, we define the forward difference Δx_n of the sequence $\{x_n\}$ as

$$\Delta x_n = x_{n+1} - x_n, \qquad \text{for all} \quad n \geq 0.$$

Similarly, for higher power

$$\Delta^k x_n = \Delta(\Delta^{k-1} x_n), \qquad \text{for all} \quad k \geq 2.$$

From this, we can have

$$\begin{aligned} \Delta^2 x_n &= \Delta(\Delta x_n) = \Delta x_{n+1} - \Delta x_n \\ &= x_{n+2} - 2x_{n+1} + x_n, \end{aligned}$$

which is the denominator of the expression of formula (2.42), is the second difference $\Delta^2 x_n$. So formula (2.42) can be written as

$$x'_n = x_n - \frac{(\Delta x_n)^2}{\Delta^2 x_n}, \qquad n \geq 0. \tag{2.43}$$

This transformation is so simple that it should be regularly used if the convergence is known to be linear; otherwise, sequence $\{x'_n\}$ will usually converge slower than sequence $\{x_n\}$.

Example 2.30 *Apply Aitken's method to the iteration* $x_{n+1} = e^{-x_n}$ *for the solution of the equation* $e^{-x} - x = 0$ *with* $x_0 = 0.5$.

Solution. *The results of the first five simple iteration methods together with their differences and the values of* x'_n *are given in Table 2.8.*

Table 2.8: Solution of $x = e^{-x}$ by Aitken's method.

n	x_n	Δx_n	$\Delta^2 x_{n-1}$	x'_n
00	0.5000000			
		.1065307		
01	0.6065307		-.1678222	
		-.0612915		
02	0.5452392		.0957554	0.5676238
		.0344639		
03	0.5797031		-.0541024	0.5672989
		-.0196385		
04	0.5600646		.0307461	0.5671931
		.0111076		
05	0.5711722		-.0174169	0.5671593
		-.0063093		
06	0.5648629		.0098844	0.5671484

To use MATLAB for Aitken's Δ^2 method (2.42), we define a function m-file as fn.m to the equation as follows:

```
function y = fn(x)
y = exp(-x);
```

and then use the following commands:

```
>> x0 = 0.5; MaxI = 5;
>> sol = AitK('fn', x0, MaxI)
sol =
    0.5671484
```
.

Program 2.8

MATLAB m-file for Aitken Δ^2 Method

```
function sol=AitK(fn,x0,MaxI)
xold = x0; x1 = feval(fn,xold);
for i = 1:MaxI
xnew=feval(fn,x1);
aitken = xnew - (xnew - x1).^2/(xnew - 2*x1 + xold);
xold = x1; x1 = xnew;
end; sol = aitken;
```

It is immediately clear that the right-hand column of values of the accelerated sequence are indeed settling much more quickly to a value close to the true solution, which is 0.567148 to six decimal places.

The approach used in Example 2.30 is a very inefficient use of the acceleration available through this extrapolation process. Once the entry x_2' has been evaluated, the obvious practical approach is to use this value in place of x_3 to generate the next iterates.

2.10 Systems of Nonlinear Equations

A system of nonlinear algebraic equations may arise when one is dealing with problems involving optimization and numerical integration (Gauss quadratures). Generally, the system of equations may not be of the polynomial variety. Therefore, a system of n equations in n unknowns is called nonlinear if one or more of the equations in the systems is/are nonlinear.

The numerical methods we discussed so far have been concerned with finding a root of a nonlinear algebraic equation with one independent variable. We now consider methods for solving systems of nonlinear algebraic equations in which each equation is a function of a specified number of variables.

Consider the system of two nonlinear equations with two variables

$$f_1(x, y) = 0 \tag{2.44}$$

and

$$f_2(x, y) = 0. \tag{2.45}$$

The problem can be stated as follows:

Given continuous functions $f_1(x, y)$ and $f_2(x, y)$, find values $x = \alpha$ and $y = \beta$ such that

$$\begin{aligned} f_1(\alpha, \beta) &= 0 \\ f_2(\alpha, \beta) &= 0. \end{aligned} \tag{2.46}$$

Functions $f_1(x, y)$ and $f_2(x, y)$ may be algebraic equations, transcendental, or any nonlinear relationship between the input x and y and the output $f_1(x, y)$ and $f_2(x, y)$. The solutions to (2.44) and (2.45) are the intersections of $f_1(x, y) = f_2(x, y) = 0$ (see Figure 2.7). This problem is considerably more complicated than the solution of a single nonlinear equation. The one-point iterative method discussed in Section 2.5 for the solution of a single equation may be extended to the system. So to solve the system of nonlinear equations we have many methods but we will use Newton's method.

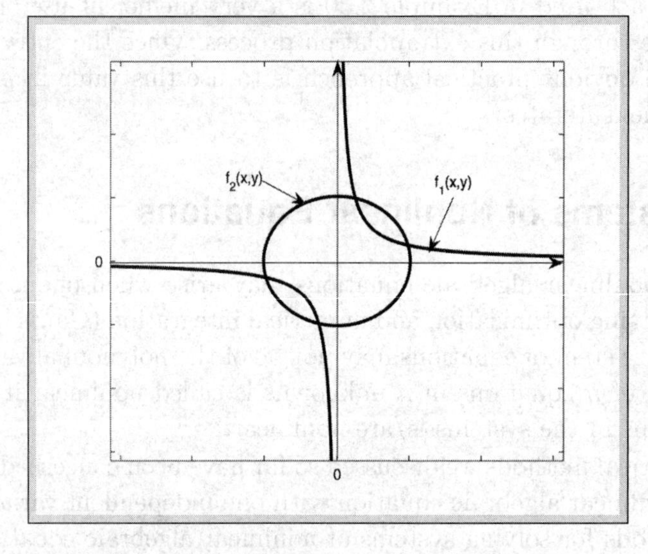

Figure 2.7: Nonlinear equation in two variables.

2.10.1 Newton's Method

Consider the two nonlinear equations specified by equations (2.44) and (2.45). Suppose that (x_n, y_n) is an approximation to a root (α, β), then using the Taylor's theorem for functions of two variables for $f_1(x, y)$ and $f_2(x, y)$ expanding about (x_n, y_n), we have

$$\begin{aligned} f_1(x,y) &= f_1(x_n + (x - x_n), y_n + (y - y_n)) \\ &= f_1(x_n, y_n) + (x - x_n)\frac{\partial f_1(x_n, y_n)}{\partial x} + (y - y_n)\frac{\partial f_1(x_n, y_n)}{\partial y} + \cdots \end{aligned}$$

and

$$\begin{aligned} f_2(x,y) &= f_2(x_n + (x - x_n), y_n + (y - y_n)) \\ &= f_2(x_n, y_n) + (x - x_n)\frac{\partial f_2(x_n, y_n)}{\partial x} + (y - y_n)\frac{\partial f_2(x_n, y_n)}{\partial y} + \cdots. \end{aligned}$$

Since $f_1(\alpha, \beta) = 0$ and $f_2(\alpha, \beta) = 0$, these equations, with $x = \alpha$ and $y = \beta$, give

$$\begin{aligned} 0 &= f_1(x_n, y_n) + (\alpha - x_n)\frac{\partial f_1(x_n, y_n)}{\partial x} + (\beta - y_n)\frac{\partial f_1(x_n, y_n)}{\partial y} + \cdots \\ 0 &= f_2(x_n, y_n) + (\alpha - x_n)\frac{\partial f_2(x_n, y_n)}{\partial x} + (\beta - y_n)\frac{\partial f_2(x_n, y_n)}{\partial y} + \cdots. \end{aligned}$$

Newton's method has a condition that initial approximation (x_n, y_n) should sufficiently close to exact root (α, β), therefore, the higher order terms may be neglected to obtain

$$\begin{aligned} 0 &\approx f_1(x_n, y_n) + (\alpha - x_n)\frac{\partial f_1(x_n, y_n)}{\partial x} + (\beta - y_n)\frac{\partial f_1(x_n, y_n)}{\partial y} \\ 0 &\approx f_2(x_n, y_n) + (\alpha - x_n)\frac{\partial f_2(x_n, y_n)}{\partial x} + (\beta - y_n)\frac{\partial f_2(x_n, y_n)}{\partial y}. \end{aligned} \tag{2.47}$$

We see that this represents a system of two linear algebraic equations for α and β. Of course, since the higher order terms are omitted in the derivation of these equations, their solution (α, β) is no longer an exact root of (2.46) and (2.47). However, it will usually be a better approximation than (x_n, y_n), so replacing (α, β) by (x_{n+1}, y_{n+1}) in (2.46) and (2.47) gives the iterative scheme

$$\begin{aligned} 0 &= f_1(x_n, y_n) + (x_{n+1} - x_n)\frac{\partial f_1(x_n, y_n)}{\partial x} + (y_{n+1} - y_n)\frac{\partial f_1(x_n, y_n)}{\partial y} \\ 0 &= f_2(x_n, y_n) + (x_{n+1} - x_n)\frac{\partial f_2(x_n, y_n)}{\partial x} + (y_{n+1} - y_n)\frac{\partial f_2(x_n, y_n)}{\partial y}. \end{aligned}$$

Then writing in matrix form, we have

$$\begin{pmatrix} \dfrac{\partial f_1}{\partial x} & \dfrac{\partial f_1}{\partial y} \\ \dfrac{\partial f_2}{\partial x} & \dfrac{\partial f_2}{\partial y} \end{pmatrix} \begin{pmatrix} x_{n+1} - x_n \\ y_{n+1} - y_n \end{pmatrix} = - \begin{pmatrix} f_1 \\ f_2 \end{pmatrix}, \tag{2.48}$$

where f_1, f_2 and their partial derivatives f_{1x}, f_{1y} are evaluated at (x_n, y_n). Hence,

$$\begin{pmatrix} x_{n+1} \\ y_{n+1} \end{pmatrix} = \begin{pmatrix} x_n \\ y_n \end{pmatrix} - \begin{pmatrix} \dfrac{\partial f_1}{\partial x} & \dfrac{\partial f_1}{\partial y} \\ \dfrac{\partial f_2}{\partial x} & \dfrac{\partial f_2}{\partial y} \end{pmatrix}^{-1} \begin{pmatrix} f_1 \\ f_2 \end{pmatrix}. \tag{2.49}$$

We call the matrix

$$J = \begin{pmatrix} \dfrac{\partial f_1}{\partial x} & \dfrac{\partial f_1}{\partial y} \\ \dfrac{\partial f_2}{\partial x} & \dfrac{\partial f_2}{\partial y} \end{pmatrix} \tag{2.50}$$

the *Jacobian matrix.*

Note that (2.48) can be written in simplified form as follows

$$\begin{pmatrix} \dfrac{\partial f_1}{\partial x} & \dfrac{\partial f_1}{\partial y} \\ \dfrac{\partial f_2}{\partial x} & \dfrac{\partial f_2}{\partial y} \end{pmatrix} \begin{pmatrix} h \\ k \end{pmatrix} = - \begin{pmatrix} f_1 \\ f_2 \end{pmatrix},$$

where h and k can be evaluated as

$$\begin{aligned} h &= \frac{\left(-f_1\dfrac{\partial f_2}{\partial y} + f_2\dfrac{\partial f_1}{\partial y}\right)}{\left(\dfrac{\partial f_1}{\partial x}\dfrac{\partial f_2}{\partial y} - \dfrac{\partial f_1}{\partial y}\dfrac{\partial f_2}{\partial x}\right)} \\ k &= \frac{\left(f_1\dfrac{\partial f_2}{\partial x} - f_2\dfrac{\partial f_1}{\partial x}\right)}{\left(\dfrac{\partial f_1}{\partial x}\dfrac{\partial f_2}{\partial y} - \dfrac{\partial f_1}{\partial y}\dfrac{\partial f_2}{\partial x}\right)}, \end{aligned} \tag{2.51}$$

where all functions are to be evaluated at (x, y). Newton's method for a pair of equations in two unknowns is therefore

$$\begin{pmatrix} x_{n+1} \\ y_{n+1} \end{pmatrix} = \begin{pmatrix} x_n \\ y_n \end{pmatrix} + \begin{pmatrix} h \\ k \end{pmatrix}, \quad n = 0, 1, 2, \ldots, \tag{2.52}$$

where (h, k) are given by (2.51) evaluated at (x_n, y_n).

At a starting approximation (x_0, y_0), the functions $f_1, f_{1x}, f_{1y}, f_2, f_{2x}$, and f_{2y} are evaluated. The linear equations are then solved for (x_1, y_1) and the whole process is repeated until convergence is obtained. Comparison of (2.15) and (2.49) shows that the above procedure is indeed an extension of Newton's method in one variable, where division by f' generalized to pre-multiplication by J^{-1}.

Example 2.31 *Solve the following system of two equations using Newton's method with accuracy* $\epsilon = 0.5 \times 10^{-4}$.

$$\begin{aligned} x^2 + y^2 &= 4 \\ xy &= 1 \end{aligned}$$

Assume $x_0 = 3.0$ *and* $y_0 = -1.5$ *are the starting values.*

Solution. *Obviously, this system of nonlinear equations has an exact solution of* $x = 1.9318517$ *and* $y = 0.517638$. *Let us see how Newton's method is used to approximate these roots.*

$$\begin{aligned} f_1(x,y) &= x^2 + y^2 - 4, \quad f_{1x} = 2x, \quad f_{1y} = 2y \\ f_2(x,y) &= xy - 1, \quad\quad f_{2x} = y, \quad f_{2y} = x \end{aligned}$$

At the given initial approximations $x_0 = 3.0$ *and* $y_0 = -1.5$, *we get*

$$\begin{aligned} f_1(3.0, -1.5) &= 7.25, \quad \frac{\partial f_1}{\partial x} = f_{1x} = 6, \quad \frac{\partial f_1}{\partial y} = f_{1y} = -3.0 and \\ f_2(3.0, -1.5) &= -5.5, \quad \frac{\partial f_1}{\partial x} = f_{2x} = -1.5, \quad \frac{\partial f_2}{\partial y} = f_{2y} = 3.0. \end{aligned}$$

The Jacobian matrix J *and its inverse* J^{-1} *at the given initial approximation can be calculated as*

$$J = \begin{pmatrix} \dfrac{\partial f_1}{\partial x} & \dfrac{\partial f_1}{\partial y} \\ \dfrac{\partial f_2}{\partial x} & \dfrac{\partial f_2}{\partial y} \end{pmatrix} = \begin{pmatrix} 6.0 & -3.0 \\ -1.5 & 3.0 \end{pmatrix}$$

and one can find its inverse as

$$J^{-1} = \frac{1}{13.5} \begin{pmatrix} 3.0 & 3.0 \\ 1.5 & 6.0 \end{pmatrix}.$$

Substituting all these values in (2.50), we get the first approximation as follows

$$\begin{pmatrix} x_1 \\ y_1 \end{pmatrix} = \begin{pmatrix} 3.0 \\ -1.5 \end{pmatrix} - \frac{1}{13.5}\begin{pmatrix} 3.0 & 3.0 \\ 1.5 & 6.0 \end{pmatrix}\begin{pmatrix} 7.25 \\ -5.5 \end{pmatrix} = \begin{pmatrix} 2.611112 \\ 0.138889 \end{pmatrix}.$$

This and further steps of the method are listed in Table 2.9.

Table 2.9: Solution of a system of two nonlinear equations.

n	x-approx. x_n	y-approx. y_n	1st. func. $f_1(x_n, y_n)$	2nd. func. $f_2(x_n, y_n)$
00	3.000000	-1.50000	7.250000	-5.5000000
01	2.611112	0.138889	2.837240	-0.637344
02	2.053259	0.412649	0.386153	-0.152725
03	1.939689	0.509856	0.0223458	-0.011039
04	1.931894	0.5175954	0.0001194	-0.00000606
05	1.931852	0.5176381	0.00000135	0.0000002

Note that a typical iteration of this method for this pair of equations can be implemented in the MATLAB command window using:

```
>> f1 = x0^2 + y0^2 - 4; f2 = x0 * y0 - 1;
>> f1x = 2 * x0; f1y = 2 * y0; f2x = y0; f2y = x0;
>> D = f1x * f2y - f1y * f2x;
>> h = (f2 * f1y - f1 * f2y)/D; k = (f1 * f2x - f2 * f1x)/D;
>> x0 = x0 + h; y0 = y0 + k;
```

Using the starting value $(3, -1.5)$, we get possible approximations as shown in Table 2.9.

We see that the values of both the functions approach zero as the number of iterations is increased. We got the desired approximations to the roots after five iterations with accuracy $\epsilon = 0.5 \times 10^{-4}$.

Newton's method is fairly easy to implement for the case of two equations in two unknowns. We first need the function m-files for the equations and the partial derivatives. For the equations in Example 2.31, we do the following:

```
function f = fn2(v)
%Here f and v are vector quantities
x = v(1); y = v(2);
f(1) = x.^2 + y.^2 - 4;
f(2) = x * y - 1;
```

```
function J = dfn2(v)
%Jacobian matrix for fn2.m
x = v(1); y = v(2);
J(1,1) = 2*x; J(1,2) = 2*y;
J(2,1) = y; J(2,2) = x;
```

Then the following MATLAB command can be used to generate the solution of Example 2.31:

```
>> s = newton2('fn2','dfn2',[3.0,-1.5],2*1e-5)
s =
    1.931852  0.5176381
```

The m-file Newton2.m will need both the function and its partial derivatives as well as a starting vector and a tolerance. The following code can be used:

Program 2.9

MATLAB m-file for Newton's Method for a Nonlinear System

```
function sol=newton2(fn2,dfn2,x0,tol)
old=x0+1; while max(abs(x0-old))>tol; old=x0;
f = feval(fn2,old); f1 = f(1); f2 = f(2);
J=feval(dfn2,old);
f1x = J(1,1); f1y = J(1,2); f2x = J(2,1); f2y = J(2,2);
D = f1x*f2y - f1y*f2x;
h = (f2*f1y - f1*f2y)/D; k = (f1*f2x - f2*f1x)/D;
x0 = old + [h,k]; end; sol=x0;
```

Similarly, for a large system of equations it is convenient to use vector notation. Consider the system

$$\mathbf{f}(\mathbf{x}) = \mathbf{0},$$

where $\mathbf{f} = (f_1, f_2, \ldots, f_n)^T$ and $\mathbf{x} = (x_1, x_2, \ldots, x_n)^T$. Denoting the *nth* iterate by $\mathbf{x}^{[n]} = (x_1^{[n]}, x_2^{[n]}, x_3^{[n]}, \ldots, x_n^{[n]})^T$, the Newton's method is defined by

$$\mathbf{x}^{[n+1]} = \mathbf{x}^{[n]} - \left[J(\mathbf{x}^{[n]})\right]^{-1} \mathbf{f}(\mathbf{x}^{[n]}), \tag{2.53}$$

where the Jacobian matrix J is defined as

$$J = \begin{pmatrix} \frac{\partial f_1}{\partial x_1} & \frac{\partial f_1}{\partial x_2} & \cdots & \frac{\partial f_1}{\partial x_n} \\ \cdot & & & \cdot \\ \cdot & & & \cdot \\ \cdot & & & \cdot \\ \frac{\partial f_n}{\partial x_1} & \frac{\partial f_n}{\partial x_2} & \cdots & \frac{\partial f_n}{\partial x_n} \end{pmatrix}.$$

Since iterative formula (2.53) involves the inverse of Jacobian J, in practice we do not attempt to find this explicitly. Instead of using the form of (2.53) we use the following form

$$J(\mathbf{x}^{[\mathbf{n}]})\mathbf{Z}^{[\mathbf{n}]} = -\mathbf{f}(\mathbf{x}^{[n]}), \tag{2.54}$$

where $\mathbf{Z}^{[n]} = \mathbf{x}^{[n+1]} - \mathbf{x}^{[n]}$.

This represents a system of linear equations for $\mathbf{Z}^{[n]}$ and can be solved by any method described in Chapter 3. Once $\mathbf{Z}^{[n]}$ has been found, the next iterate is calculated from

$$\mathbf{x}^{[n+1]} = \mathbf{Z}^{[n]} + \mathbf{x}^{[n]}. \tag{2.55}$$

There are two major disadvantages with this method:

1. The method may not converge unless the initial approximation is a good one. Unfortunately, there are no general means by which an initial solution can be obtained. One can assume such values for which $\det(J) \neq 0$. This does not guarantee convergence but it does provide some guidance as to the appropriateness of one's initial approximation.

2. The method requires the user to provide the derivatives of each function with respect to each variable. Therefore, one must evaluate the n functions and the n^2 derivatives at each iteration. So solving systems of nonlinear equations is a difficult task. For systems of nonlinear equations that have analytical partial derivatives, Newton's method can be used; otherwise, multi-dimensional minimization techniques should be used.

Procedure 2.6 (Newton's Method for Two Nonlinear Equations)

1. *Choose the initial guess for the roots of the system, so that the determinant of the Jacobian matrix is not zero.*

2. *Establish a tolerance $\epsilon(> 0)$.*

3. *Evaluate the Jacobian at initial approximations and then find the inverse of the Jacobian.*

4. *Compute a new approximation to the roots using iterative formula (2.55).*

5. *Check the tolerance limit. If* $\|(x_n, y_n) - (x_{n-1}, y_{n-1})\| \leq \epsilon$*, for* $n \geq 0$*, then end; otherwise, go back to step 3 and repeat the process.*

2.11 Roots of Polynomials

The methods of solving nonlinear equations presented in the previous sections apply to any form of nonlinear equation. A very common problem in nonlinear equations is finding the roots of a polynomial. Polynomials are used in nearly all areas of numerical analysis and applied mathematics and sciences. Polynomial equations arise frequently in practice and vast literature is available to find their zeros. The methods discussed so far are used to find a zero of a polynomial, but in this section we will consider a method specifically designed to find all the roots of polynomial equations.

The general form of an *nth*-degree polynomial is

$$p(x) = a_0 x^n + a_1 x^{n-1} + \cdots + a_{n-1} x + a_n, \tag{2.56}$$

where n denotes the degree of polynomial and $a_0, a_1, \ldots, a_{n-1}, a_n$ are real constant coefficients and $a_0 \neq 0$.

The fundamental theorem of algebra states that an *nth*-degree polynomial has exactly n roots, or zeros. The zeros may be real and complex. If the coefficients are all real, the complex zeros always occur in a conjugate pair (that is, if $x_1 = a + ib$ is a root then $x_2 = a - ib$). When the coefficients are complex, the complex roots need not be related. The zeros may be distinct or repeated. The single zero of a linear polynomial can be determined directly. The two zeros of a second-degree polynomial can be determined from the quadratic formula. Exact formulas also exist for the zeros of the third-degree polynomial and fourth-degree polynomial, but they are quite complicated and rarely used. Iterative methods must be used for higher-degree polynomials. Clearly, any of the above methods discussed so far cannot be employed, but these all involve only real arithmetic. Newton's method and the secant method work well for finding the complex zeros of the polynomials, provided that complex arithmetic is used and reasonable complex initial approximations are specified. Several methods that do not require complex arithmetic exist for extracting complex roots of polynomials that have real coefficients. One of them is Muller's method, which can find complex roots even when starting with real approximations. The other one is Bairstow's method, which extracts quadratic factors, which can then be solved by quadratic formula.

2.11.1 Horner's Method

For the efficient evaluation of a polynomial, *Horner's method* is one of the best methods. It is not only useful to evaluate a polynomial at a given point but also can be used for other purposes as well. Horner's method is also known as the *nested multiplication method.*

Suppose we wish to evaluate

$$p(x) = a_0x^n + a_1x^{n-1} + \cdots + a_{n-1}x + a_n.$$

A nested multiplication method states that

$$p(x) = \{\cdots[(a_0x + a_1)x + a_2]x + \cdots + a_n\}x.$$

For example, to evaluate

$$p(x) = x^4 - 7x^3 + 16x^2 - 15x - 4$$

using the nested multiplication method

$$p(x) = \{[(x-7)x + 16]x - 15\}x - 4$$

at $x_0 = 2$, we get

$$\begin{aligned} p(2) &= \{[(2-7)2 + 16]2 - 15\}2 - 4 \\ &= \{[-10 + 16]2 - 15\}2 - 4 \\ &= \{12 - 15\}2 - 4 \\ &= (-3)(2) - 4 \\ &= -6 - 4 = -10. \end{aligned}$$

For finding the above result using MATLAB commands, we first define vector **a**, which consists of coefficients of the given polynomial function, then we do the following:

```
>> a = [1 - 716 - 15 - 4];
>> x0 = 2;
>> sol = HornM(a, x0);
```

Program 2.10

MATLAB m-file for Horner's Method

```
function sol=HornM(a,x0)
n = length(a); p = a(n);
for i = n-1:-1:1;
p = p*x0 + a(i); end; sol = p;
```

Horner's method is also known as *synthetic division.*

Theorem 2.3 (Synthetic Division Theorem)

Let $p(x) = a_0x^n + a_1x^{n-1} + \cdots + a_{n-1}x + a_n$ *and* x_0 *be the given point at which we wish to evaluate a polynomial* $p(x)$. *If* $b_n = a_0$ *and*

$$b_i = a_{n-i} + b_{i+1}x_0, \quad \text{for } i = n-1, n-2, \ldots, 1, 0,$$

then $b_0 = p(x_0)$. *Moreover, if*

$$q(x) = b_nx^{n-1} + b_{n-1}x^{n-2} + \cdots + b_2x + b_1,$$

then

$$p(x) = (x - x_0)q(x) + b_0.$$

Note that the degree of polynomial $q(x)$ is one degree less than the degree of polynomial $p(x)$. To carry out the calculation using synthetic division, the following arrangement can be used:

$$\begin{array}{c|ccccc} x_0 & a_0 & a_1 & a_2 & \cdots & a_n \\ & & x_0b_n & x_0b_{n-1} & \cdots & x_0b_1 \\ \hline & b_n & b_{n-1} & b_{n-2} & \cdots & \boxed{b_0} \end{array}.$$

The boxed number satisfies $p(x_0) = b_0$.

Example 2.32 *Use the synthetic division method to evaluate* $p(2)$, *where polynomial* $p(x)$ *is given as*

$$p(x) = x^4 - 7x^3 + 16x^2 - 15x - 4.$$

Solution. *Consider the following table*

$$\begin{array}{c|ccccc} x_0 = 2 & 1 & -7 & 16 & -15 & -4 \\ & & 2 & -10 & 12 & -6 \\ \hline & 1 & -5 & 6 & -3 & \boxed{-10} \end{array}.$$

Thus, $p(2) = -10$, *and using Theorem 2.3, we can write it as*

$$p(x) = (x-2)(x^3 - 5x^2 + 6x - 3) - 10.$$

Horner's method can be used for *deflation*, the process of deleting a linear factor from a given polynomial. If x_0 is a zero of a polynomial $p(x)$, then $(x - x_0)$ is a linear factor of $p(x)$, and the converse is also true. The remaining zeros of $p(x)$ are the $n-1$ zero of $p(x)/(x - x_0)$. For example, if 4 is one of the zeros of the polynomial $p(x) = x^4 - 7x^3 + 16x^2 - 15x - 4$, then deflate the polynomial as:

$$\begin{array}{r|rrrrr} x_0 = 4 & 1 & -7 & 16 & -15 & -4 \\ & & 4 & -12 & 16 & 4 \\ \hline & 1 & -3 & 4 & 1 & \boxed{0} \end{array}.$$

Thus, $p(4) = 0$, and the given polynomial can be written as

$$p(x) = (x-4)(x^3 - 3x^2 + 4x + 1).$$

The MATLAB function *deconv* can be used to divide one polynomial $r(x) = x - 4$ into the other polynomial $p(x) = x^4 - 7x^3 + 16x^2 - 15x - 4$. The $p(x)$ and $r(x)$ are represented in MATLAB by the vectors (a and b) of their coefficients, beginning with the highest degree term. Any zero coefficients must be included. To divide the polynomials, we use the following MATLAB commands:

```
>> a = [1 -7 16 -15 -4]; b = [1 -4];
>> [c, d] = deconv(a, b)
c =
    1.0000 -3.0000 4.0000 1.0000
d =
    0.0000 0.0000 0.0000 0.0000
```

where the vectors c and d are called the quotient (a third-degree polynomial) and the remainder, respectively. Since the polynomial is a perfect divisor, the remainder polynomial has zero coefficients.

So multiplying the polynomial $r(x) = x - 4$ to the quotient polynomial $q(x) = x^3 - 3x^2 + 4x + 1$, we will get the original polynomial $p(x) = x^4 - 7x^3 + 16x^2 - 15x - 4$. The MATLAB function *conv* can be used to multiply one polynomial to other polynomial as follows:

```
>> sol = conv(c, b)
sol =
    1.0000 -7.0000 16.0000 -15.0000 -4.0000
```

Horner's method is also useful in finding the approximation of the zero of a polynomial $p(x)$ using Newton's method

$$x_{n+1} = x_n - \frac{p(x_n)}{p'(x_n)}, \quad n \geq 0$$

and the value of $p'(x_n)$ can be easily calculated using Theorem 2.3. Since

$$p(x) = (x - x_0)q(x) + b_0$$

then by taking derivative with respect to x, we get

$$p'(x) = q(x) + (x - x_0)q'(x)$$

and putting $x = x_n$ gives

$$p'(x_n) = q(x_n), \quad \text{for} \quad n \geq 0.$$

Example 2.33 *Find the first three approximations to one of the zeros of*

$$p(x) = x^4 - 7x^3 + 16x^2 - 15x - 4$$

using Newton's method, starting with $x_0 = -1$.

Solution. *To find the first approximation by Newton's method*

$$x_1 = x_0 - \frac{p(x_0)}{p'(x_0)}$$

we have to calculate the values of $p(x_0)$ *and* $p'(x_0) = q(x_0)$ *using the same procedure as discussed above. Consider the table*

$$\begin{array}{r|rrrrr} x_0 = -1 & 1 & -7 & 16 & -15 & -4 \\ & & -1 & 8 & -24 & 39 \\ \hline & 1 & -8 & 24 & -39 & \boxed{35} \end{array},$$

which gives $p(-1) = 35$ *and* $q(x) = x^3 - 8x^2 + 24x - 39$. *Similarly, we can find the value of* $q(-1)$ *as follows*

$$\begin{array}{r|rrrr} x_0 = -1 & 1 & -8 & 24 & -39 \\ & & -1 & 9 & -33 \\ \hline & 1 & -9 & 33 & \boxed{-72} \end{array},$$

which gives $q(-1) = -72 = p'(-1)$. *Thus,*

$$x_1 = -1 - \frac{35}{-72} = -0.5138.$$

The MATLAB function polyval can be used to evaluate the polynomial $p(x) = x^4 - 7x^3 + 16x^2 - 15x - 4$ *at the given point* $x_0 = -1$ *as follows:*

```
>> a = [1 -7 16 -15 -4]; x0 = -1;
>> sol = polyval(a, x0)
sol =
    35.0000
```

We can also evaluate the derivative $p'(x) = 4x^3 - 21x^2 + 32x - 15$ *using the MATLAB function polyder as follows:*

```
>> a = [1 -7 16 -15 -4];
>> d = polyder(a)
d =
    4.0000 -21.0000 32.0000 -15.0000
```

and $p'(-1)$ *can be obtained as follows:*

```
>> sol = polyval(d, -1)
sol =
    -72.0000
```

To find the second approximation x_2, *we will repeat the same procedure as follows*

$$\begin{array}{r|rrrrr} x_1 = -0.5138 & 1 & -7 & 16 & -15 & -4 \\ & & -0.5138 & 3.8606 & -10.2044 & 12.9500 \\ \hline & 1 & -7.5138 & 19.8606 & -25.2044 & \boxed{8.9500} \end{array}$$

gives $p(-0.5138) = 8.9500$ *and* $q(x) = x^3 - 7.5138x^2 + 19.8606x - 25.2044$. *Similarly, we can find the value of* $q(-0.5138)$ *as follows*

$$\begin{array}{r|rrrr} x_1 = -0.5138 & 1 & -7.5138 & 19.8606 & -25.2044 \\ & & -0.5138 & 4.1246 & -12.3236 \\ \hline & 1 & -8.0276 & 23.9852 & \boxed{-37.5280} \end{array}$$

gives $q(-0.5138) = -37.5280 = p'(-0.5138)$. *Thus,*

$$x_2 = x_1 - \frac{p(x_1)}{p'(x_1)} = -0.5138 - \frac{8.9500}{-37.5280} = -0.2753.$$

In a similar way, we will get the third approximation x_3 *as follows*

$$x_3 = x_2 - \frac{p(x_2)}{p'(x_2)} = -0.2753 - \frac{1.4939}{-25.4847} = -0.2167,$$

which is close to the exact zero $\alpha = -0.2134$ *of the given polynomial.*

The MATLAB function *roots* can be used for finding all the roots of the given polynomial $p(x)$ as follows:

```
>> sol = roots(a)
sol =
      4.0000
      1.6067 + 1.4506i
      1.6067 - 1.4506i
     -0.2134
```

Another application of Horner's method is in finding Taylor's expansion of a polynomial about the given point x_0. Consider the polynomial

$$p(x) = a_0 x^n + a_1 x^{n-1} + \cdots + a_{n-1} x + a_n,$$

which can be written as

$$p(x) = c_n (x - x_0)^n + c_{n-1} (x - x_0)^{n-1} + \cdots + c_1 (x - x_0) + c_0.$$

Note that $p(x_0) = c_0$. In this case we will repeat Horner's method until all the coefficients c_k are found. Consider the polynomial

$$p(x) = x^4 - 7x^3 + 16x^2 - 15x - 4$$

and $x_0 = 2$, then using the following tables we have

$x_0 = 2$	1	-7	16	-15	-4
		2	-10	12	-6
$x_0 = 2$	1	-5	6	-3	$\boxed{-10}$
		2	-6	0	
$x_0 = 2$	1	-3	0	$\boxed{-3}$	
		2	-1		
$x_0 = 2$	1	-1	$\boxed{-1}$		
		2			
	$\boxed{1}$	$\boxed{1}$			

.

All the boxed numbers are the values of the unknown coefficients. Thus, Taylor's expansion of the polynomial about the point $x_0 = 2$ is

$$p(x) = (x-2)^4 + (x-2)^3 - (x-2)^2 - 3(x-2) - 10.$$

2.11.2 Muller's Method

Most of the methods we have discussed so far to find a root of a nonlinear equation have approximated the function in the neighborhood of the root by a straight line. We noted that the secant method used a linear approximated line through the two points x_0 and x_1. Here, we discuss a method called *Muller's method*, which is a generalization of the secant method and it uses quadratic interpolation among three points.

Muller's method is an iterative method, which needs three initial approximations x_0, x_1, and x_2 and determines the next approximation, x_3, by considering the intersection of the x-axis with parabola through $(x_0, f(x_0)), (x_1, f(x_1))$, and $(x_2, f(x_2))$ (see Figure 2.8).

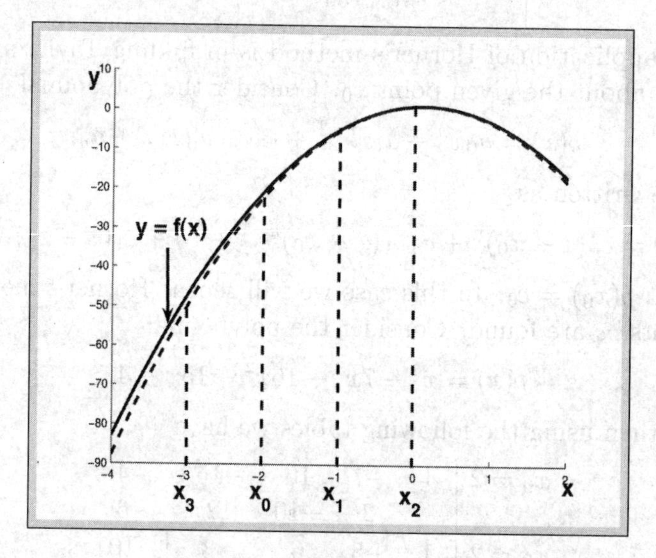

Figure 2.8: Muller's method.

The derivation of Muller's method begins by considering the quadratic polynomial, which can be fitted to $f(x)$

$$y = a(x - x_2)^2 + b(x - x_2) + c. \tag{2.57}$$

The constants a, b, and c can be determined by evaluating (2.57) at $x = x_0, x_1$, and x_2. At $x = x_2$, we have

$$f(x_2) = a(0) + b(0) + c,$$

which gives

$$f(x_2) = c \tag{2.58}$$

and at $x = x_0$, we obtain

$$f(x_0) = a(x_0 - x_2)^2 + b(x_0 - x_2) + c,$$

which gives

$$f(x_0) - f(x_2) = a(x_0 - x_2)^2 + b(x_0 - x_2).$$

Finally, at $x = x_1$, we get

$$f(x_1) = a(x_1 - x_2)^2 + b(x_1 - x_2) + c,$$

which gives

$$f(x_1) - f(x_2) = a(x_1 - x_2)^2 + b(x_1 - x_2).$$

Letting $h_0 = x_0 - x_2$ and $h_1 = x_1 - x_2$, we get

$$f(x_0) - f(x_2) = ah_0^2 + bh_0 \tag{2.59}$$
$$f(x_1) - f(x_2) = ah_1^2 + bh_1. \tag{2.60}$$

Writing system (2.59) in matrix form, we have

$$\begin{bmatrix} h_0^2 & h_0 \\ h_1^2 & h_1 \end{bmatrix} \begin{bmatrix} a \\ b \end{bmatrix} = \begin{bmatrix} f(x_0) - f(x_2) \\ f(x_1) - f(x_2) \end{bmatrix}. \tag{2.61}$$

Solving system (2.61) for a and b, we obtain

$$a = \frac{h_1(f(x_0) - f(x_2)) - h_0(f(x_1) - f(x_2))}{h_2} \tag{2.62}$$

and

$$b = \frac{h_0^2(f(x_1) - f(x_2)) - h_1^2(f(x_0) - f(x_2))}{h_2}, \tag{2.63}$$

where $h_2 = h_0h_1(h_0 - h_1)$.

After finding the values for the unknown coefficients a, b, and c of quadratic polynomial (2.57), we must now find its roots in order to generate a new approximation of the root of $f(x)$. Setting $y = 0$ in (2.57) and solving for $(x - x_2)$ using the quadratic formula, we have

$$x - x_2 = \frac{-b \pm \sqrt{b^2 - 4ac}}{2a}. \tag{2.64}$$

However, because of potential round-off error, rather than use this form, we use the following alternative form (which can be obtained by multiplying on the top and bottom of formula (2.64) by $(-b \mp \sqrt{b^2 - 4ac})$)

$$x_3 = x_2 - \frac{2c}{b \pm \sqrt{b^2 - 4ac}}, \tag{2.65}$$

where the sign in the denominator is chosen to make its absolute value or modulus as large as possible (that is, if $b > 0$, choose plus, if $b < 0$, choose minus; if $b = 0$, choose either). Once the new approximation x_3 is found, the oldest point x_0 is ignored, the new points are taken as x_1, x_2, x_3, and the process is repeated.

Example 2.34 *Use Muller's method with initial approximations $x_0 = 4.4, x_1 = 5.2$ and $x_2 = 4.8$ to determine a root of the following equation*

$$y = x^3 - 5x^2 + 4x.$$

Solution. *Using the given initial approximations, we calculate*

$$h_0 = 4.4 - 4.8 = -0.4, \quad h_1 = 5.2 - 4.8 = 0.4, \quad h_2 = (-0.4)(0.4)(-0.4 - 0.4) = 0.1280$$

and evaluate the functions

$$f(x_0) = f(4.4) = 5.9840, \quad f(x_1) = f(5.2) = 26.2080, and f(x_2) = f(4.8) = 14.5920.$$

Using these calculated values, we can find the values of a, b, and c as follows:

$$\begin{aligned} a &= \frac{(0.4)(5.9840 - 14.5920) - (-0.4)(26.2080 - 14.5920)}{0.1280} = 9.4000, \\ b &= \frac{(-0.4)^2(26.2080 - 14.5920) - (0.4)^2(5.9840 - 14.5920)}{0.1280} = 25.2800, \textit{and} \\ c &= f(4.8) = 14.5920. \end{aligned}$$

The square root of the discriminant can be evaluated as

$$D = \sqrt{(25.2800)^2 - 4(9.4000)(14.5920)} = 9.5089.$$

Since $b > 0$, *a positive sign is employed in the denominator of (2.65), and the new approximation of the zero of* $f(x)$ *can be obtained as*

$$x_3 = 4.8 - \frac{2(14.5920)}{25.2800 + 9.5089} = 3.9611 \quad and \quad y = 14.5920.$$

For the next approximation, we start with the following three approximations

$$x_0 = x_1 = 5.2, \quad x_1 = x_2 = 4.8, \quad and \quad x_2 = x_3 = 3.9611$$

and calculate

$$f(5.2) = 26.2080, \; f(4.8) = 14.5920, \quad and \quad f(3.9611) = -0.4561$$

and

$$h_0 = 1.2389, \; h_1 = 0.8389, \; h_2 = 0.4157.$$

The values of a, b, c, *and* D *are as follows:*

$$a = 8.9611, \quad b = 10.4208, \quad c = -0.4561, \quad D = 11.1778.$$

Thus, the new approximation is $x_3 = 4.0033$ *and* $y = -0.4561$.
After the fifth iteration, the new approximation to the root is $x_3 = 4$ *and* $y = 1.1280e-4$ *and this shows that the method converges rapidly on root 4.*

Note that Muller's method has the advantage of being able to generate approximations to complex zero even if the initial approximations are real. It has been proven that the near simple root of a nonlinear equation, Muller's method converges faster (with order of convergence 1.85) than the secant method but the convergence is not guaranteed.

One can use the MATLAB single command to get the same above results by Muller's method as follows:

```
>> x0 = 4.4; x1 = 5.2; x2 = 4.8; MaxI = 5;
>> sol = Muller('fn', x0, x1, x2, MaxI);
```

Program 2.11

MATLAB m-file for Muller's Method

```
function sol=Muller(fn,x0,x1,x2,MaxI)
Iter=0; for i=1:MaxI
y0=feval(fn,x0); y1=feval(fn,x1); y2=feval(fn,x2);
h0=x0-x2; h1=x1-x2; h2=h0*h1*(h0-h1);
c=y2; d1= y0 - y2; d2= y1 - y2;
a = (h1*d1-h0*d2)/h2; b = (h1.^2*d2-h1.^2*d1)/h2;
D=sqrt(b.^2 - 4*a*c); R1=b+D;R2=b-D;
if abs(R1) > abs(R2), x3 = x2 - 2*c/R1;
else x3 = x2 - 2*c/R2; end;
x0 = x1;x1 = x2;x2 = x3;y0 = y1;y1 = y2;y2 =
feval(fn,x3);
Iter=Iter+1; if Iter > MaxI break end end;
```

Procedure 2.7 (Muller's Method)

1. *Choose the three initial approximations* x_0, x_1, *and* x_2, *and MaxI (maximum number of iterations).*

2. *Compute* $h_0 = x_0 - x_2, h_1 = x_1 - x_2$, *and* $h_2 = h_0 h_1 (h_0 - h_1)$.

3. *Compute* $f(x_0), f(x_1)$, *and* $f(x_2)$.

4. *Compute unknown coefficients* a, b, c *using (2.62), (2.63), and (2.58), respectively.*

5. *Compute discriminant* $D = \sqrt{b^2 - 4ac}$.

6. *Define* $R_1 = b + D$ *and* $R_2 = b - D$ *and check if* $|b + D| \geq |b - D|$, *then set* $x_{new} = x2 - 2 * c/R_1$; *otherwise, set* $x_{new} = x2 - 2 * c/R_2$.

7. *Repeat steps 1-6 until the best approximation is achieved.*

2.11.3 Bairstow's Method

One of the principal difficulties in the solutions of polynomial equations is the occurrence of complex roots or other repetitions of absolute values that may arise from repeated or equal and opposite real roots. One of the better approaches to isolating roots of these types is finding the appropriate quadratic factors of the original polynomial. This is the basic idea of the iteration of Bairstow's, which assumes good initial approximations. Let $p(x)$ be a polynomial function of degree n

$$p(x) = a_0x^n + a_1x^{n-1} + \cdots + a_{n-1}x + a_n, \tag{2.66}$$

which has n zeros. Assume the following quadratic factor

$$x^2 + ux + v, \tag{2.67}$$

where u and v are the constants we wish to determine. Once the quadratic factors are found it is easy to solve the quadratics to find the roots we require. To develop the equations to find u and v we divide (2.66) by (2.67) to get

$$p(x) = (x^2 + ux + v)(b_0x^{n-2} + b_1x^{n-3} + \cdots + b_{n-2}) + R(x), \tag{2.68}$$

where a_0 has been taken as one and R(x) is a remainder having the following form:

$$R(x) = (x + u)b_{n-1} + b_n. \tag{2.69}$$

Equating (2.66) to (2.68) gives

$$\begin{aligned} a_0x^n &+ a_1x^{n-1} + \cdots + a_{n-1}x + a_n = (x^2 + ux + v)(b_0x^{n-2} \\ &+ b_1x^{n-3} + \cdots + b_{n-2}) + (x + u)b_{n-1} + b_n. \end{aligned} \tag{2.70}$$

Obviously, if the quadratic factor is exact, the remainder should be zero; that is, $b_{n-1} = b_n = 0$. Equating the coefficients of both sides of (2.70) we obtain the following system of equations

$$\left.\begin{aligned} b_0 &= a_0 = 1 \\ b_1 &= a_1 - u \\ b_2 &= a_2 - b_1u - v \\ \cdots & \ \cdots \ \cdots \\ b_k &= a_k - b_{k-1}u - b_{k-2}v \\ \cdots & \ \cdots \ \cdots \\ b_n &= a_n - b_{n-1}u - b_{n-2}v \end{aligned}\right\}. \tag{2.71}$$

We know from (2.71) that b_{n-1} and b_n are functions of u and v. Consequently, the problem is reduced to the solution of the system of equations:

$$\left.\begin{aligned} b_{n-1}(u, v) &= 0 \\ b_n(u, v) &= 0 \end{aligned}\right\}. \tag{2.72}$$

Assuming some initial approximation to u and v we require improved values $u+du$ and $v+dv$, which take system (2.72) closer to zero. Using Taylor's series we have

$$\left.\begin{array}{rcl} 0 & = & b_{n-1}(u+du, v+dv) = b_{n-1}(u,v) + \{\partial b_{n-1}/\partial u\}du \\ & + & \{\partial b_{n-1}/\partial v\}dv + \cdots \\ 0 & = & b_n(u+du, v+dv) = b_n(u,v) + \{\partial b_n/\partial u\}du \\ & + & \{\partial b_n/\partial v\}dv + \cdots \end{array}\right\}. \tag{2.73}$$

By ignoring the terms involving the higher powers of du and dv, system (2.73) is linear in du and dv, and can be written as

$$\left.\begin{array}{rcl} 0 & = & b_{n-1}(u,v) + c_{n-2}du + c_{n-3}dv \\ 0 & = & b_n(u,v) + c_{n-1}du + c_{n-2}dv \end{array}\right\}, \tag{2.74}$$

where

$$\begin{array}{rclc} c_{n-2} & = & \partial b_{n-1}/\partial u, & c_{n-3} = \partial b_{n-1}/\partial v \\ c_{n-1} & = & \partial b_n/\partial u, & c_{n-2} = \partial b_n/\partial v. \end{array}$$

The matrix form of system (2.74) can be written as

$$\begin{pmatrix} c_{n-2} & c_{n-3} \\ c_{n-1} & c_{n-2} \end{pmatrix} \begin{pmatrix} du \\ dv \end{pmatrix} = -\begin{pmatrix} b_{n-1} \\ b_n \end{pmatrix}. \tag{2.75}$$

Obviously, the c's values are related to the b's values in the same manner as the b's values are related to the a's values. Consequently, system (2.75) is used to determine changes in the assumed u and v values. Having found du and dv we can use the improved values of u and v to calculate new coefficients $c_{n-1}, c_{n-2}, c_{n-3}, b_{n-1}$, and b_n and the process is repeated until the values of b_{n-1} and b_n become sufficiently small. Note that a solution is possible if the determinant of the coefficient matrix in system (2.75) is not zero; that is, if $c_{n-2}^2 = c_{n-1}c_{n-3}$. To avoid such a problem, one can be very careful in choosing proper initial values for u and v. Now to find the coefficients $c_{n-1}, c_{n-2}, c_{n-3}, b_{n-1}$, and b_n in terms of u and v, we differentiate (2.71) with respect to u and v, and denoting $\partial b_k/\partial u$ by c_k and $\partial b_{k+1}/\partial v$ by d_k, for all k, we have

$$c_k = -b_k - uc_{k-1} - vc_{k-2}, \tag{2.76}$$

where $c_{-2} = c_{-1} = 0$ and

$$d_{k-1} = -b_{k-1} - ud_{k-2} - vd_{k-3}. \tag{2.77}$$

By comparing the above two recurrence formulae, we have

$$c_k = d_k. \tag{2.78}$$

So the recurrence formula for c_k is

$$c_k = -b_k - uc_{k-1} - vc_{k-2}, \qquad k = 0, 1, 2, \ldots, n-1, \tag{2.79}$$

where $c_{-2} = c_{-1} = 0$. Having found a quadratic factor by applying this process iteratively until the required accuracy is obtained we can repeat this process with the residual polynomial determined by the b_k coefficients until all the quadratic factors have been found.

Example 2.35 *Use Bairstow's method to determine all the roots of the polynomial*

$$x^3 - 6x^2 + 11x - 6 = 0$$

by taking initial values $u = -3$ *and* $v = 3$.

Solution. *Firstly, we determine the b's values as follows:*

$$\begin{aligned} b_0 &= a_0 = 1 \\ b_1 &= a_1 - u = -6 - (-3) = -3 \\ b_2 &= a_2 - b_1 u - v = 11 - (-3)(-3) - 3 = -1 \\ b_3 &= a_3 - b_2 u - b_1 v = -6 - (-1)(-3) - (-3)(3) = 0. \end{aligned}$$

Now to compute the c's values with the help of b's values using (2.79), we have

$$\begin{aligned} k &= 0, \quad c_0 = -b_0 - uc_{-1} - vc_{-2} = -1 + (-3)(0) - (3)(0) = -1 \\ k &= 1, \quad c_1 = -b_1 - uc_0 - vc_{-1} = -(-3) - (-3)(-1) - (3)(0) = 0 \\ k &= 2, \quad c_2 = -b_2 - uc_1 - vc_0 = -(-1) - (-3)(0) - (-1)(3) = 4. \end{aligned}$$

To find the changes in the assumed values of u and v, use (2.75), which becomes

$$\begin{pmatrix} c_1 & c_0 \\ c_2 & c_1 \end{pmatrix} \begin{pmatrix} du \\ dv \end{pmatrix} = -\begin{pmatrix} b_2 \\ b_3 \end{pmatrix},$$

which implies that

$$\begin{pmatrix} 0 & -1 \\ 4 & 0 \end{pmatrix} \begin{pmatrix} du \\ dv \end{pmatrix} = -\begin{pmatrix} -1 \\ 0 \end{pmatrix}.$$

By solving this system, we have $du = 0$ *and* $dv = -1$*. Therefore, the improved values are* $u1 = u + du = -3$ *and* $v1 = v + dv = 2$*. Repeating the procedure once again by computing the b's values for* $u = u1 = -3$ *and* $v = v1 = 2$*, we get*

$$\begin{aligned} b_0 &= a_0 = 1 \\ b_1 &= a_1 - u = -6 - (-3) = -3 \\ b_2 &= a_2 - b_1 u - v = 11 - (-3)(-3) - 2 = 0 \\ b_3 &= a_3 - b_3 u - b_1 v = -6 - (0)(-3) - (-3)(2) = 0. \end{aligned}$$

Clearly, both b_2 *and* b_3 *are equal to zero. This means that the assumed values of* $u = u1 = -3$ *and* $v = v1 = 2$ *are exact. Thus, the quadratic factor is* $x^2 - 3x + 2$*. This factor has roots* $x_1 = 1$ *and* $x_2 = 2$*. Now to find the third root, one may then use synthetic division, which gives* $x_3 = 3$*. Alternatively, we can find the third root by noting that* $(-1)(-2)(-x_3) = -6$*, which implies that* $x_3 = 3$*.*

To use MATLAB for finding the approximate values of u and v for the quadratic factor using Bairstow's method, we use the following commands:

```
>> a = [-6, 11, -6, 1];
>> u = -3; v = 3; MaxI = 2;
>> BairSM(a, u, v, MaxI);
```

Program 2.10

MATLAB m-file for Bairstow's Method

```
function sol=BairSM(a,u,v,MaxI)
n=length(a);
for i = 1:MaxI
b(1) = a(1) - u; b(2) = a(2) - b(1) * u - v;
for k=3:n  b(k) = a(k) - b(k - 1) * u - b(k - 2) * v; end
c(1) = b(1) - u; c(2) = b(2) - c(1) * u - v;
for k=3:n-1  c(k) = b(k) - c(k - 1) * u - c(k - 2) * v; end
d1 = c(n - 1); b1 = b(n); d2 = c(n - 1) * b(n - 1);
d3 = c(n - 2) * c(n - 2); d4 = b(n - 1) * c(n - 2);
if (n > 3) d1 = d1 * c(n - 3); b1 = b1 * c(n - 3); end
d = d1 - d3; du = (b1 - d4)/d; dv = (d2 - c(n - 2) * b(n))/d;
u = u + du; v = v + dv; end
```

It is important to note that Bairstow's method can only be applied when every term in a polynomial appears in the expression. If not, then one can transform the polynomial so that Bairstow's method can be applied. For example, if the given polynomial is of the form

$$x^3 + x + 1 = 0$$

then one can assume the following transformation:

$$x = y + 1,$$

then putting the value of x in the given polynomial equation, we get

$$y^3 + 3y^2 + 4y + 2 = 0.$$

After finding the roots y_1, y_2, and y_3, then solve for roots x_1, x_2, and x_3 as follows:

$$x_1 = y_1 + 1, \qquad x_2 = y_2 + 1, \qquad x_3 = y_3 + 1.$$

Also note that, in Bairstow's method, the polynomial is reduced by degree 2 as long as n is greater than 2. The last factor is either the quadratic with a final n value of 2 or in the linear term with a final n value of 1.

Procedure 2.8 (Bairstow's Method)

1. *Choose the initial estimates u_0 and v_0 for u and v, respectively.*
2. *Compute the values of b's using (2.73).*
3. *Establish the tolerance ($\epsilon > 0$). Check if $\|[b_{n-1}\ b_n]^T\| \leq \epsilon$, then u_0 and v_0 are the estimated coefficients; otherwise, continue.*
4. *Compute the values of c's using (2.79).*
5. *Compute the changes du and dv required to improve u_0, and v_0, respectively, by solving (2.75).*
6. *Reset u_0 to $u_0 + du$ and v_0 to $v_0 + dv$, and go back to step 2.*

There is a well-known MATLAB function called *fzero*. This one is used in solving nonlinear equations of a single variable, but for systems of nonlinear equations we cannot use it. This function requires three arguments: the polynomial $p(x)$, initial approximation x_0 of the root of $p(x) = 0$, and the parameter Tol for the accuracy. For example, using the *fzero* function to solve the nonlinear equation

$$x^3 - 2x^2 - 5 = 0$$

with $x_0 = 2.5$ within the accuracy 10^{-4}, we do the following:

```
function y = f1(x)
y = x.^3-2*x.^2-5;
```

and

```
>> x0 = 2.5;
>> tol = 1e-4;
>> sol = fzero('f1', x0, tol)
```

,

which gives

```
sol =
    2.6906
```

.

2.12 Summary

In this chapter we discussed several methods for solving nonlinear equations for one variable and one method for several variables. For the case of one variable we used the bisection method, the false position method, the fixed point method, Newton's method, secant method, and the modified Newton's method. The first four methods we used for finding the approximate of the simple root of the nonlinear equation and the last one we used for the multiple root of nonlinear equations.

The bisection method is a simple, reliable method that gives linear convergence. It reduces the interval of uncertainty by a factor of two with each iteration. The bisection method converges very slowly, but the convergence is guaranteed because the zero lies on the interval being halved. The false position method is similar to the bisection method in that it requires two initial points, which must bracket a root of $f(x)$. It uses a straight line fit through the two endpoints to generate a new point. The false position method converges faster than the bisection method, but there are examples where its performance is inferior to the bisection method. The fixed point method is based on finding a fixed point of $x = g(x)$. It converges only if the derivative of the nonlinear function $g(x)$ is less than unity in magnitude and convergence is linear. Newton's method and the secant method are both effective methods for solving nonlinear equations. Both methods need reasonable initial estimates. Newton's method converges (quadratically) faster than the secant method (superlinear, to 1.62 order). For function whose derivative cannot be evaluated, the secant method is recommended. The secant method is recommended as the best general-purpose method. For the multiple roots of the nonlinear equations, all these methods are very slow so we used a faster convergence (quadratically) modified Newton method. The higher-order methods tend to converge faster, but they can involve more computational effort per iteration.

Aitken's method starting with the third estimate can be applied to any linear iterative method and acceleration (improvement) of convergence can be achieved by it. Solving systems of nonlinear equations is a difficult task. For a system of nonlinear equations that have analytical partial derivatives, Newton's method can be used. Otherwise, multi-dimensional minimization techniques should be used. No single approach has proven to be the most effective.

For the case of polynomials, first we used Horner's method, which is one of the most efficient ways to evaluate polynomials and their derivatives. It is also helpful for finding the initial approximation for a solution by Newton's method. It is also quite stable. We used Muller's method, which is a fast convergence method in finding the approximation of the simple zero of a polynomial equation. It does not require the evaluation of derivatives and needs only one function evaluation per iteration. This method can be used for any root-finding problem, but is most useful for approximating the roots of polynomials. We also discussed Bairstow's method for finding all zeros of polynomials. Bairstow's method of quadratic factors is one of the most efficient for determining real and complex roots of polynomials with real coefficients. By repeating application of the method to the deflated polynomial, one can find all the quadratic factors. Error of deflated polynomials and quadratic factors increase as the method is repeatedly applied. Accuracy of the roots found can be poor, so accuracy must be improved by another method. The iteration may not converge at all for certain problems.

2.13 Exercises

1. Find the root of $f(x) = e^x - 2 - x$ on the interval $[-2.4, -1.6]$ accurate to 10^{-4} using the bisection method.

2. Use the bisection method to find solutions accurate to within 10^{-4} on the interval $[-5, 5]$ of the following functions:

   (a) $f(x) = x^5 - 10x^3 - 4$
   (b) $f(x) = 2x^2 + \ln(x) - 3$
   (c) $f(x) = \ln(x) + 30e^{-x} - 3$

3. The following equations have a root on the interval $[0, 1.6]$. Determine these with an error less than 10^{-4} using the bisection method.

   (*a*) $2x - e^{-x} = 0$; (*b*) $e^{-3x} + 2x - 2 = 0$

4. Estimate the number of iterations needed to achieve an approximation with accuracy 10^{-4} to the solution of $f(x) = x^3 + 4x^2 + 4x - 4$ lying on the interval $[0, 1]$ using the bisection method.

5. Use the bisection method for $f(x) = x^3 - 3x + 1$ on $[1, 3]$ to find:

   (*a*) The first eight approximations to the root of the given equation.
   (*b*) Find an error estimate $|\alpha - x_8|$.

6. Solve Problem 1 by the false position method.

7. Use the false position method to find the root of $f(x) = x^3 + 4x^2 + 4x - 4$ on the interval $[0, 1]$ accurate to 10^{-4}.

8. Use the false position method to find the solution accurate to within 10^{-4} on the interval $[1, 1.5]$ of the equation $2x^3 + 4x^2 - 2x - 5 = 0$.

9. Use the false position method to find the solution accurate to within 10^{-4} on the interval $[3, 4]$ of the equation $e^x - 3x^2 = 0$.

10. The cubic equation $x^3 - 3x - 20 = 0$ can be written as

    (*a*) $x = \dfrac{(x^3 - 20)}{3}$

    (*b*) $x = \dfrac{20}{(x^2 - 3)}$

    (*c*) $x = (3x + 20)^{1/3}$

    Choose the form which satisfies the condition $|g'(x)| < 1$ on $[1, 2]$ and then find the third approximation x_3 when $x_0 = 1.5$.

11. Consider the nonlinear equation $g(x) = \dfrac{1}{2}e^{0.5x}$ defined on the interval $[0, 1]$. Then

    (*a*) Show that there exists a unique fixed point for g on $[0, 1]$.
    (*b*) Use the fixed-point iterative method to compute x_3, set $x_0 = 0$.
    (*c*) Compute an error bound for your approximation in part (*b*).

12. An equation $x^3 - 2 = 0$ can be written in form $x = g(x)$ in two ways:

    (a) $x = g_1(x) = x^3 + x - 2$
    (b) $x = g_2(x) = \dfrac{(2 + 5x - x^3)}{5}$

    Generate the first four approximations from $x_{n+1} = g_i(x_n)$, $i = 1, 2$ using $x_0 = 1.2$. Show which sequence converged to $2^{1/3}$ and why?

13. Find the value of k such that the iterative scheme $x_{n+1} = \dfrac{x_n^2 - 4kx_n + 7}{4}$, $n \geq 0$ converges to 1. Also, find the rate of convergence of the iterative scheme.

14. Write the equation $x^2 - 6x + 5 = 0$ in the form $x = g(x)$, where $x \in [0, 2]$, so that the iteration $x_{n+1} = g(x_n)$ will converge to the root of the given equation for any initial approximation $x_0 \in [0, 2]$.

15. Which of the following iterations

    (a) $x_{n+1} = \dfrac{1}{4}\left(x_n^2 + \dfrac{6}{x_n}\right)$
    (b) $x_{n+1} = \left(4 - \dfrac{6}{x_n^2}\right)$

    is suitable to find a root of the equation $x^3 = 4x^2 - 6$ on the interval $[3, 4]$? Estimate the number of iterations required to achieve 10^{-3} accuracy, starting from $x_0 = 3$.

16. An equation $e^x = 4x^2$ has a root on $[4, 5]$. Show that we cannot find that root using $x = g(x) = \dfrac{1}{2}e^{x/2}$ for the fixed-point iteration method. Can you find another iterative formula, which will locate that root? If yes, then find the third iterations. Also find the error bound.

17. Solve Problem 1 by Newton's method by taking initial approximation $x_0 = -2$ in each part .

18. Let $f(x) = e^x + 3x^2$:

    (a) Find the Newton's formula $g(x_k)$.
    (b) Start with $x_0 = 4$ and compute x_4.
    (c) Start with $x_0 = -0.5$ and compute x_4.

19. Use Newton's formula for the reciprocal of square root of a number 15 and then find the third approximation of the number, with $x_0 = 0.05$.

20. Use Newton's method to find the solution accurate to within 10^{-4} of the equation $\tan(x) - 7x = 0$, with initial approximation $x_0 = 4$.

21. Find Newton's formula for $f(x) = x^3 - 3x + 1$ on $[1, 3]$ to calculate x_3, if $x_0 = 1.5$. Also, find the rate of convergence of the method.

22. Rewrite the nonlinear equation $g(x) = \frac{1}{2}e^{0.5x}$, which defined on the interval $[0, 1]$ in the equivalent form $f(x) = 0$ and then use Newton's method with $x_0 = 0.5$ to find the third approximation x_3.

23. Given the iterative scheme $x_{n+1} = x_n - \frac{f(x_n)}{f'(x_n)}$, $n \geq 0$ with $f(\alpha) = f'(\alpha) = 0$ and $f''(\alpha) \neq 0$, find the rate of convergence for this scheme.

24. Halley's iterative method

$$x_{n+1} = x_n - \frac{f(x)}{f'(x)}\left[1 - \frac{f(x_n)f''(x_n)}{2[f'(x_n)]^2}\right]^{-1}, \quad n \geq 0$$

has an order of convergence 3 for a simple root of $f(x) = 0$. Use it to find an iterative procedure to approximate the square root of a number N. Then use it to approximate $\sqrt{8}$, using $x_0 = 2.5$, and compute the first three iterations.

25. Find the positive root of $f(x) = x^{10} - 1$ by the secant method using the starting values $x_0 = 1.2$ and $x_1 = 1.1$.

26. Find the first three estimates for the equation $x^3 - 2x - 5 = 0$ by the secant method using $x_0 = 2$ and $x_1 = 3$.

27. Solve the equation $e^{-x} - x = 0$ using the secant method, starting with $x_0 = 0$ and $x_1 = 1$, accurate to 10^{-4}.

28. Use the secant method to find a solution accurate to within 10^{-4} for $\ln(x) + x - 5 = 0$ on $[3, 4]$.

29. Find the root of multiplicity of the function $f(x) = (x-1)^2 \ln(x)$ at $\alpha = 1$.

30. Show that if $f(x)$ has a root of multiplicity m at $x = \alpha$, then

$$f^{(n)}(x) = 0, \qquad n = 1, 2, \ldots, m - 1.$$

31. Show that the root of multiplicity of the function $f(x) = x^4 - x^3 - 3x^2 + 5x - 2$ is 3 at $\alpha = 1$. Estimate the number of iterations required to solve the problem with accuracy 10^{-4}, with the starting value $x_0 = 0.5$, using:

    (a) Newton's method.
    (b) first modified Newton's method.
    (c) second modified Newton's method.

32. If $f(x), f'(x)$, and $f''(x)$ are continuous and bounded on a certain interval containing $x = \alpha$ and if both $f(\alpha) = 0$ and $f'(\alpha) = 0$ but $f''(\alpha) \neq 0$, show that

$$x_{n+1} = x_n - 2\frac{f(x_n)}{f'(x_n)}$$

will converge quadratically if x_n is on the interval.

33. Show that iterative scheme $x_{n+1} = 1 + x_n - \dfrac{x_n^2}{2}$, $n \geq 0$ converges to $\sqrt{2}$. Find the rate of convergence of the sequence.

34. Let α be the exact solution of the function $f(x) = 0$ such that $f'(\alpha) \neq 0$, $f''(\alpha) \neq 0$, then find the conditions of the constant K under which the rate of convergence of the sequence $x_{n+1} = x_n^2 - Kf(x_n)$, $n = 0, 1, 2, \ldots$ is quadratic.

35. The following sequence is linearly convergent. Use Aitken's method to generate the first four terms of the sequence $\{x_n'\}$:

$$x_0 = 0.5, \quad x_{n+1} = \cos x_n, \qquad n \geq 0.$$

36. Use Newton's method to approximate the root of the equation $x = e^x$, $x_0 = 0.2$ within accuracy 10^{-4}. Use Aitken's accelerated method to speed up the convergence of the iteration.

37. Use the fixed-point iteration to find the first four approximations to the root of $x_{n+1} = \dfrac{1}{2}\ln(1 + x_n)$, $n \geq 0$, using $x_0 = 0.5$. Use Aitken's accelerated method to speed up the convergence of the iteration method.

38. Apply Aitken's method to the iteration $x_{n+1} = (x_n^{10} + x_n - 1)$ for the solution of $x^{10} - 1 = 0$ with $x_0 \in [1.1, 1.2]$. Compare the performance with the results obtained by the secant method.

39. Solve the following system using Newton's method:

$$\begin{aligned} 4x^3 \quad + y &= 6 \\ x^2 y &= 1 \end{aligned}$$

Start with initial approximation $x_0 = y_0 = 1$. Stop when successive iterates differ by less than 10^{-7}.

40. Solve the following system using Newton's method:

$$\begin{aligned} x \quad + e^y &= 68.1 \\ \sin x \quad - y &= -3.6 \end{aligned}.$$

Start with the initial approximation $x_0 = 2.5$, $y_0 = 4$ and compute the first three approximations.

41. Solve the following system using Newton's method:

$$\begin{aligned} x \quad + y^2 &= -2.26 \\ 2x^3 \quad + y &= -6.45 \end{aligned}.$$

Start with the initial approximation $x_0 = y_0 = -1$ and compute the first three approximations.

42. Solve the following system using Newton's method:

$$\begin{aligned} x^2 \quad + y^2 &= 1 \\ 2x^2 \quad + y &= 2 \end{aligned}.$$

Start with the initial approximation $x_0 = y_0 = 0.5$ and compute the first three approximations.

43. Use Horner's method to evaluate each of the following polynomials at the given points:
(a) $p(x) = 6x^5 + 8x^4 + 18x^3 - 15x^2 + 22x + 28, \quad x_0 = -6.$
(b) $p(x) = 2.2x^5 + 1.8x^4 - 3.3x^3 + 4.8x^2 + 10x + 1, \quad x_0 = 1.9.$
(c) $p(x) = x^7 + 3x^5 + 2x^4 + x^3 + 5x^2 + x - 1, \quad x_0 = 4.$
(d) $p(x) = x^7 + 2x^5 + 12x^4 - 3x^3 + 17x^2 - 23x - 301, \quad x_0 = -12.$

44. Use Horner's method to evaluate each of the following polynomials at the given points:
(a) $p(x) = 6x^5 + 8x^4 + 18x^3 - 15x^2 + 22x + 28, \quad x_0 = -6.$
(b) $p(x) = 2.2x^5 + 1.8x^4 - 3.3x^3 + 4.8x^2 + 10x + 1, \quad x_0 = 1.9.$
(c) $p(x) = x^7 + 3x^5 + 2x^4 + x^3 + 5x^2 + x - 1, \quad x_0 = 4.$
(d) $p(x) = x^7 + 2x^5 + 12x^4 - 3x^3 + 17x^2 - 23x - 301, \quad x_0 = -12.$

45. Use Problem 44 to find the first two approximations of the zero of the polynomial using Newton's method.

46. Find the first three approximations of the zero of each of the following polynomials using Newton's method:
   (a) $p(x) = 1.3x^5 + 4.8x^4 - 13x^3 + 25x^2 + 2x + 2, \quad x_0 = 1.5.$
   (b) $p(x) = x^5 + 8x^4 + 4x^3 + 7x^2 + 3x + 11, \quad x_0 = 2.$
   (c) $p(x) = x^6 + 3x^5 + 11x^4 - 3x^3 + 2x^2 + 3x - 8, \quad x_0 = -3.$
   (d) $p(x) = x^7 + x^6 + 2x^4 - 5x^3 + x^2 - 3x - 4, \quad x_0 = -1.$

47. Use Muller's method to solve each of the following equations. Apply the first three iterations only.
   (a) $3x^4 + 2x^3 + x^2 + 34x + 7 = 0,$ root near $-2.5.$
   (b) $4x^3 + 5x^2 + 2x + 1 = 0,$ root near $-1.$
   (c) $x^2 + e^x + x - 1 = 0,$ root near $0.$
   (d) $\sin x + e^x + x - 2 = 0,$ root near $0.$

48. Use Muller's method to solve each of the following equations. Apply the first three iterations only.
   (a) $2x^3 + 12x^2 + 11x + 5 = 0,$ root near $-5.$
   (b) $13x^3 - 15x^2 + 14x - 21 = 0,$ root near $1.5.$
   (c) $x^4 + x^3 + 7x^2 + 3x - 23 = 0,$ root near $1.5.$
   (d) $\cos x + 3e^x + x - 6 = 0,$ root near $0.$

49. Use Muller's method to solve each of the following equations. Apply the first three iterations only.
   (a) $x^3 + 5x^2 + 6x - 4 = 0,$ root near $0.5.$
   (b) $2x^3 - 3x^2 - 4x + 5 = 0,$ root near $2.$
   (c) $x^4 + 3x^3 + 2x^2 + x - 5 = 0,$ root near $-2.7.$
   (d) $2x^4 - 5x^3 - 4x^2 - 13x + 26 = 0,$ root near $3.5.$

50. Approximate all the roots of the polynomial using Bairstow's method

$$4x^4 + 2x^3 + 3x^2 + x + 1 = 0.$$

Take initial values as $u_0 = v_0 = 1.$

51. Start with $u_0 = 1$ and $v_0 = 1$ and use Bairstow's method to find the next two approximations (u_1, v_1) and (u_2, v_2) of the polynomial function $p(x) = x^3 - x - 1$.

52. Start with the trial factor $x^2 - 4x + 4$ and improve by successive approximations of Bairstow's method to find the quadratic factors of

$$x^4 - 3.1x^3 + 2.1x^2 + 1.1x + 5.2 = 0.$$

Find the four zeros of this polynomial.

Chapter 3

Systems of Linear Equations

3.1 Introduction

When engineering systems are modeled, the mathematical description is frequently developed in terms of sets of algebraic simultaneous equations. Sometimes these equations are nonlinear and sometimes linear. In this chapter we discuss systems of simultaneous linear equations and describe the numerical methods for the approximate solutions of such systems. The solution of a system of simultaneous linear algebraic equations is probably one of the most important topics in engineering computation. Problems involving simultaneous linear equations arise in the areas of elasticity, electric-circuit analysis, heat transfer, vibrations, and so on. Also, the numerical integration of some types of ordinary and partial differential equations may be reduced to the solutions of such systems of equations. It has been estimated, for example, that about 75% of all scientific problems require the solution of a system of linear equations at one stage or another. It is therefore important to be able to solve linear problems efficiently and accurately.

Definition 3.1 (Linear Equation)

A linear equation is an equation in which the highest exponent in a variable term is no more than one. The graph of such equation is a straight line.

A linear equation with two variables x and y is an equation that can be written in the form

$$ax + by = c,$$

where a, b, and c are real numbers. Note that this is the equation of a straight line in the plane. For example, the equations

$$2x + 3y = -4, \qquad \frac{1}{3}x - 2y = 1, \text{and} \qquad x + \pi y = \sqrt{3}$$

are all linear equations in two variables.
A linear equation in n variables $x_1, x_2, \ldots, x_n$ is an equation that can be written as

$$a_1x_1 + a_2x_2 + \cdots + a_nx_n = b,$$

where $a_1, a_2, \ldots, a_n$ are real numbers and called the *coefficients* of unknown variables $x_1, x_2, \ldots, x_n$ and the real number b, the right-hand side of equation, is called the *constant term* of the equation.

Definition 3.2 (System of Linear Equations)

A system of linear equations (or a linear system) is simply a finite set of linear equations.

For example,

$$\begin{array}{rcrcr} 2x & + & y & = & 7 \\ 6x & + & 5y & = & 27 \end{array}$$

is a system of two equations in two variables x and y, and

$$\begin{array}{rcrcrcrcr} 2x & + & y & - & 3z & + & w & = & 2 \\ x & + & 2y & - & 5z & + & w & = & 2 \\ x & + & y & - & 2z & + & 3w & = & 1 \end{array}$$

is a system of three equations in the four variables x, y, z, and w.

In order to write a general system of m linear equations in the n variables $x_1, x_2, \ldots, x_n$, we have

$$\begin{array}{ccccccccc} a_{11}x_1 & + & a_{12}x_2 & + & \cdots & + & a_{1n}x_n & = & b_1 \\ a_{21}x_1 & + & a_{22}x_2 & + & \cdots & + & a_{2n}x_n & = & b_2 \\ \vdots & & \vdots & & \cdots & & \vdots & & \vdots \\ a_{m1}x_1 & + & a_{m2}x_2 & + & \cdots & + & a_{mn}x_n & = & b_m \end{array} \tag{3.1}$$

or, in compact form system (3.1) can be written

$$\sum_{j=1}^{n} a_{ij}x_j = b_i, \qquad i = 1, 2, \ldots, m. \tag{3.2}$$

For such systems we seek all possible ordered sets of numbers $c_1, c_2, \ldots, c_n$, which satisfies all m equations when they are substituted for the variables $x_1, x_2, \ldots, x_n$. Any such set $\{c_1, c_2, \ldots, c_n\}$ is called a *solution* of the system of linear equations (3.1) or (3.2).

There are three possible types of linear systems that arise in engineering problems, and they are described as follows:

1. If there are more equations than unknown variables $(m > n)$, then the system is usually called *overdetermined.* Typically, an overdetermined system has no solution. For example, the following system

$$\begin{array}{ccccc} 3x_1 & & & = & 6 \\ 4x_1 & + & 5x_2 & = & 12 \\ & & 4x_2 & = & 4 \end{array}$$

has no solution.

2. If there are more unknown variables than the number of the equations $(n > m)$, then the system is usually called *underdetermined.* Typically, an underdetermined system has an infinite number of solutions. For example, the following system

$$\begin{array}{ccccccc} 3x_1 & + & 4x_2 & & & = & 70 \\ & & 5x_2 & + & 4x_3 & = & 81 \end{array}$$

has infinitely many solutions.

3. If there are the same number of equations as unknown variables ($m = n$), then the system is usually called a *simultaneous* system. It has a unique solution if the system satisfies certain conditions (which we will discuss below). For example, the system

$$\begin{array}{rcrcrcr} 2x_1 & + & 4x_2 & + & x_3 & = & -11 \\ -x_1 & + & 3x_2 & - & 2x_3 & = & -16 \\ 2x_1 & - & 3x_2 & + & 5x_3 & = & 21 \end{array}$$

has the unique solution $x_1 = 2, x_2 = -4, x_3 = 1$.

Most engineering problems fall into this category. In this chapter we will solve simultaneous linear systems using many numerical methods.

A simultaneous system of linear equations is said to be *linear independent* if no equation in the system can be expressed as a linear combination of the others. Under these circumstances a unique solution exists. For example, the following system of linear equations

$$\begin{array}{rcrcrcr} 2x_1 & + & x_2 & - & x_3 & = & 1 \\ 3x_1 & - & x_2 & - & x_3 & = & -2 \\ x_1 & + & x_2 & & & = & 3 \end{array}$$

is linear independent and therefore has unique solution $x_1 = 1, x_2 = 2, x_3 = 3$.

However, the system

$$\begin{array}{rcrcrcr} 5x_1 & + & x_2 & + & x_3 & = & 4 \\ 3x_1 & - & x_2 & + & x_3 & = & -2 \\ x_1 & + & x_2 & & & = & 3 \end{array}$$

does not have a unique solution since the equations are not linear independent; the first equation is equal to the second equation plus twice the third equation.

Theorem 3.1 (Solution of a Linear System)

Every system of linear equations has either no solution, exactly one solution, or infinitely many solutions.

For example, in the case of a system of two equations with two variables, we can have three possibilities for the solutions of the linear system. Firstly, the two lines (since the graph of the linear equation is a straight line) may be parallel and distinct; in this case there is no solution to the system because the two lines do not intersect each other at any point. For example, consider the following system

$$\begin{array}{rcrcl} x & + & y & = & 1 \\ 2x & + & 2y & = & 3. \end{array}$$

The graphs (see Figure 3.1(a)) of the given two equations show that the lines are parallel so the given system has no solution. It can be proved algebraically, simply, by multiplying the first equation of the system by 2, to get a system of the form

$$\begin{array}{rcl} 2x + 2y & = & 2 \\ 2x + 2y & = & 3, \end{array}$$

which is not possible.

Secondly, the two lines may not be parallel, and they meet at exactly one point, so in this case the system has exactly one solution. For example, consider the following system

$$\begin{array}{rcrcr} x & - & y & = & -1 \\ 3x & - & y & = & 3. \end{array}$$

From the graphs (see Figure 3.1(b)) of these two equations, we can see that the lines intersect at exactly one point, namely, the (2,3), and so the system has exactly one solution, $x = 2, \quad y = 3$. To show algebraically, if we substitute $y = x + 1$ in the second equation, we have $3x - x - 1 = 3$, or $x = 2$ and using this value of x in $y = x + 1$ gives $y = 3$.

Finally, the two lines may actually be the same line, and so in this case, every point on the lines gives a solution to the system, and therefore, there are infinitely many solutions. For example, consider the following system

$$\begin{array}{rcrcl} x & + & y & = & 1 \\ 2x & + & 2y & = & 2. \end{array}$$

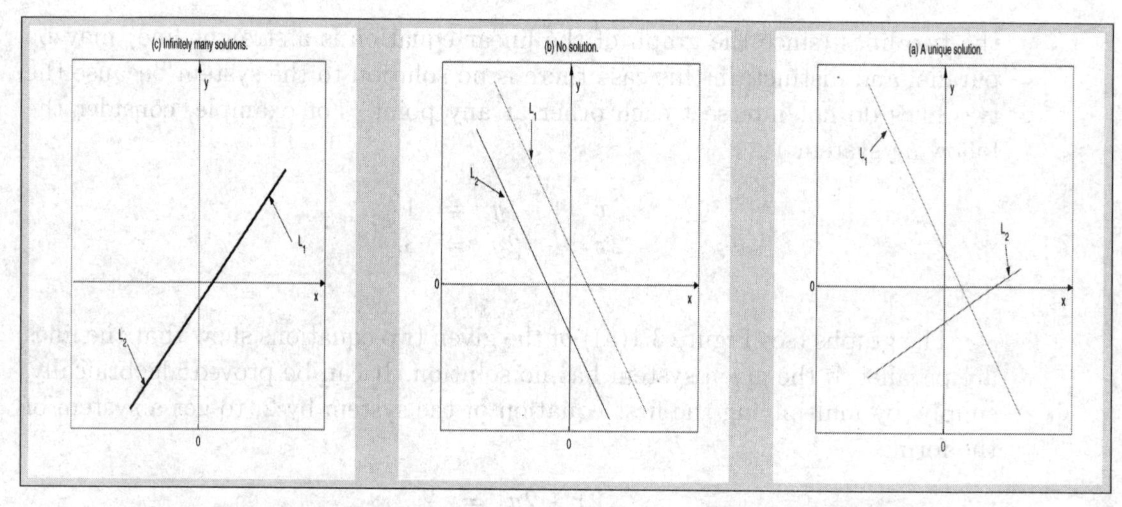

Figure 3.1: Three possible solutions of simultaneous systems.

Here, both equations have the same line for their graph (see Figure 3.1(c)). So this system has infinitely many solutions because any point on this line gives a solution to this system. Any solution of the first equation is also a solution of the second equation. For example, if we set $y = x - 1$, and choose $x = 0, y = 1$, and $x = 1, y = 0$, and so on.

Note that a system of equations with no solution is said to be *inconsistent* and if it has at least one solution, it is said to be *consistent.*

3.1.1 Linear System in Matrix Notation

The general simultaneous system of n linear equations in the n unknown variables $x_1, x_2, \ldots, x_n$ is

$$\begin{array}{ccccccccc} a_{11}x_1 & + & a_{12}x_2 & + & \cdots & + & a_{1n}x_n & = & b_1 \\ a_{21}x_1 & + & a_{22}x_2 & + & \cdots & + & a_{2n}x_n & = & b_2 \\ \vdots & & \vdots & & & & \vdots & & \vdots \\ a_{n1}x_1 & + & a_{n2}x_2 & + & \cdots & + & a_{nn}x_n & = & b_n. \end{array} \tag{3.3}$$

The system of linear equations (3.3) can be written as a single matrix equation

$$\begin{pmatrix} a_{11} & a_{12} & \cdots & a_{1n} \\ a_{21} & a_{22} & \cdots & a_{2n} \\ \vdots & \vdots & \vdots & \vdots \\ a_{n1} & a_{n2} & \cdots & a_{nn} \end{pmatrix} \begin{pmatrix} x_1 \\ x_2 \\ \vdots \\ x_n \end{pmatrix} = \begin{pmatrix} b_1 \\ b_2 \\ \vdots \\ b_n \end{pmatrix}. \tag{3.4}$$

If we compute the product of the two matrices on the left-hand side of (3.4), we have

$$\begin{pmatrix} a_{11}x_1 & + & a_{12}x_2 & + & \cdots & + & a_{1n}x_n \\ a_{21}x_1 & + & a_{22}x_2 & + & \cdots & + & a_{2n}x_n \\ \vdots & & \vdots & & \vdots & & \vdots \\ a_{n1}x_1 & + & a_{n2}x_2 & + & \cdots & + & a_{nn}x_n \end{pmatrix} = \begin{pmatrix} b_1 \\ b_2 \\ \vdots \\ b_n \end{pmatrix}. \tag{3.5}$$

But two matrices are equal if and only if their corresponding elements are equal.

Hence, single matrix equation (3.4) is equivalent to linear system (3.3). If we define

$$A = \begin{pmatrix} a_{11} & a_{12} & \cdots & a_{1n} \\ a_{21} & a_{22} & \cdots & a_{2n} \\ \vdots & \vdots & \vdots & \vdots \\ a_{n1} & a_{n2} & \cdots & a_{nn} \end{pmatrix}, \quad \mathbf{x} = \begin{pmatrix} x_1 \\ x_2 \\ \vdots \\ x_n \end{pmatrix}, \quad \mathbf{b} = \begin{pmatrix} b_1 \\ b_2 \\ \vdots \\ b_n \end{pmatrix}$$

the coefficient matrix, the column matrix of unknowns, and the column matrix of constants, respectively, then system (3.3) can be written very compactly as

$$A\mathbf{x} = \mathbf{b}, \tag{3.6}$$

which is called the *matrix form* of system of linear equations (3.3). The column matrices $\mathbf{x}$ and $\mathbf{b}$ are called *vectors.*

If the constants $\mathbf{b}$ of (3.6) are added to the coefficient matrix A as a column of elements in the position shown below

$$[A|\mathbf{b}] = \begin{pmatrix} a_{11} & a_{12} & \cdots & a_{1n} & \vdots & b_1 \\ a_{21} & a_{22} & \cdots & a_{2n} & \vdots & b_2 \\ \vdots & \vdots & \vdots & \vdots & \vdots & \vdots \\ a_{n1} & a_{n2} & \cdots & a_{nn} & \vdots & b_n \end{pmatrix} \tag{3.7}$$

then matrix $[A|\mathbf{b}]$ is called the *augmented matrix* of system (3.6). In many instances, it is convenient to operate on the augmented matrix instead of manipulating the equations. It is customary to put a bar between the last two columns of the augmented matrix to remind us where the last column came from. However, the bar is not absolutely necessary. The coefficient and augmented matrices of a linear system will play key roles in our methods of solving linear systems.

Using MATLAB commands we can define the augmented matrix as follows:

$$\begin{array}{l} >> A = [1\ 2\ 3; 4\ 5\ 6; 7\ 8\ 9]; \\ >> b = [10; 11; 12]; \\ >> Aug = [A\ b] \\ Aug = \\ \quad \begin{bmatrix} 1 & 2 & 3 & 10 \\ 4 & 5 & 6 & 11 \\ 7 & 8 & 9 & 12 \end{bmatrix} \end{array}.$$

Also,

$$\begin{array}{l} >> Aug = [A\ eye(3)] \\ Aug = \\ \quad \begin{bmatrix} 1 & 2 & 3 & 1 & 0 & 0 \\ 4 & 5 & 6 & 0 & 1 & 0 \\ 7 & 8 & 9 & 0 & 0 & 1 \end{bmatrix} \end{array}.$$

If all of the constant terms on the right-hand sides of the equal signs of linear system (3.6) are zero, then system (3.6) is called a *homogeneous system*, and it can be written as

$$A\mathbf{x} = \mathbf{0}. \tag{3.8}$$

It can be seen by inspection of homogeneous system (3.8) that one of its solution is $\mathbf{x} = \mathbf{0}$, such solution, in which all of the unknowns are zero, is called the *trivial solution*. Of course, it is usually non-trivial solutions that are of interest in physical problems. A non-trivial solution to the homogeneous system can occur with certain conditions on the coefficient matrix A, which we will discuss later.

If the right-hand sides of the equal signs of (3.6) are not zero, then linear system (3.6) is called a *nonhomogeneous system*, and we will find that all the equations must be independent to obtain unique solution.

3.2 Properties of Matrices and Determinants

To discuss the solution of linear systems, it will be necessary to introduce the basic algebraic properties of matrices, which make it possible to describe linear systems in a concise way that make solving a system of n linear equations as easy as possible.

3.2.1 Introduction of Matrices

A matrix can be described as a rectangular array of elements that can be represented as follows:

$$A = \begin{pmatrix} a_{11} & a_{12} & \cdots & a_{1n} \\ a_{21} & a_{22} & \cdots & a_{2n} \\ \vdots & \vdots & \vdots & \vdots \\ a_{m1} & a_{m2} & \cdots & a_{mn} \end{pmatrix}. \tag{3.9}$$

The numbers $a_{11}, a_{12}, \ldots, a_{mn}$ that make up the array are called the *elements* of the matrix. The first subscript for the element denotes the row and the second denotes the column in which the element appears. The elements of a matrix may take many forms. They could be all numbers (real or complex), variables, functions, integrals, derivatives, or even matrices themselves.

The *order* or *size* of a matrix is specified by the number of rows (m) and columns (n); thus matrix A in (3.9) is of order m by n, usually written as $m \times n$.

A vector can be considered a special case of a matrix having only one row or one column. A row vector containing n elements is a $1 \times n$ matrix, called a *row matrix*, and a column vector of n elements is an $n \times 1$ matrix, called a *column matrix*. A matrix of order 1×1 is called a *scalar*.

Definition 3.3 (Matrix Equality)

Two matrices $A = (a_{ij})$ and $B = (b_{ij})$ are equal if they are the same size and the corresponding elements in A and B are equal; that is,

$$A = B \quad \textit{if and only if} \quad a_{ij} = b_{ij}$$

for $i = 1, 2, \ldots, m$ and $j = 1, 2, \ldots, n$. For example, the following matrices

$$A = \begin{pmatrix} 1 & -1 & 2 \\ 1 & 3 & 2 \\ 2 & 4 & 3 \end{pmatrix} \quad \textit{and} \quad B = \begin{pmatrix} 1 & -1 & z \\ 1 & 3 & 2 \\ x & y & w \end{pmatrix}$$

are equal if and only if $x = 2, y = 4, z = 2,$ and $w = 3$.

Definition 3.4 (Addition of Matrices)

Let $A = (a_{ij})$ and $B = (b_{ij})$ be both $m \times n$ matrices. The **sum** *of $A + B$ of two matrices of the same size is a new matrix $C = (c_{ij})$ each of whose elements is the sum of the two corresponding elements in the original matrices; that is,*

$$c_{ij} = a_{ij} + b_{ij}, \quad \text{for } i = 1, 2, \ldots, m, \quad \text{and} \quad j = 1, 2, \ldots, n.$$

For example, let

$$A = \begin{pmatrix} 1 & 2 \\ 3 & 4 \end{pmatrix} \quad \text{and} \quad B = \begin{pmatrix} 4 & 1 \\ 5 & 2 \end{pmatrix}.$$

Then

$$A + B = \begin{pmatrix} 1 & 2 \\ 3 & 4 \end{pmatrix} + \begin{pmatrix} 4 & 1 \\ 5 & 2 \end{pmatrix} = \begin{pmatrix} 5 & 3 \\ 8 & 6 \end{pmatrix} = C.$$

Using the MATLAB command adding two matrices A and B of the same size results in answer C of another matrix of the same size:

```
>> A = [1 2; 3 4];
>> B = [4 1; 5 2];
>> C = A + B;
```

Definition 3.5 (Difference of Matrices)

Let A and B be $m \times n$ matrices. We write $A + (-1)B$ as $A - B$ and call this the difference of two matrices of the same size is a new matrix C each of whose elements is the difference of the two corresponding elements in the original matrices. For example, let

$$A = \begin{pmatrix} 1 & 2 \\ 3 & 4 \end{pmatrix} \quad \text{and} \quad B = \begin{pmatrix} 4 & 1 \\ 5 & 2 \end{pmatrix}.$$

Then

$$A - B = \begin{pmatrix} 1 & 2 \\ 3 & 4 \end{pmatrix} - \begin{pmatrix} 4 & 1 \\ 5 & 2 \end{pmatrix} = \begin{pmatrix} -3 & 1 \\ -2 & 2 \end{pmatrix} = C.$$

Note that $(-1)B = -B$ is obtained by multiplying each entry of matrix B by (-1), and is called the scalar multiple of matrix B by -1. Matrix $-B$ is called the negative of B.

Definition 3.6 (Multiplication of Matrices)

The multiplication of two matrices is defined only when the number of columns in the first matrix is equal to the number of rows in the second. If an $m \times n$ matrix A is multiplied by an $n \times p$ matrix B, then the product matrix C is an $m \times p$ matrix, where each term is defined by

$$c_{ij} = \sum_{k=1}^{n} a_{ik} b_{kj}$$

for each $i = 1, 2, \ldots, m$ and $j = 1, 2, \ldots, p$. For example, let

$$A = \begin{pmatrix} 1 & 2 \\ 3 & 4 \end{pmatrix} \quad \text{and} \quad B = \begin{pmatrix} 4 & 1 \\ 5 & 2 \end{pmatrix}.$$

Then

$$\begin{pmatrix} 1 & 2 \\ 3 & 4 \end{pmatrix} \begin{pmatrix} 4 & 1 \\ 5 & 2 \end{pmatrix} = \begin{pmatrix} 4+10 & 1+4 \\ 12+20 & 3+8 \end{pmatrix} = \begin{pmatrix} 14 & 5 \\ 32 & 11 \end{pmatrix} = C.$$

Note that even when AB is defined, the product BA may not be defined. Moreover, a simple multiplication of two square matrices of the same size will show that even BA is defined and it need not equal to AB; that is, they do not commute. For example, if

$$A = \begin{pmatrix} 1 & 2 \\ -1 & 3 \end{pmatrix} \quad \text{and} \quad B = \begin{pmatrix} 2 & 1 \\ 0 & 1 \end{pmatrix}$$

then

$$AB = \begin{pmatrix} 2 & 3 \\ -2 & 2 \end{pmatrix} \quad \text{while} \quad BA = \begin{pmatrix} 1 & 7 \\ -1 & 3 \end{pmatrix}.$$

Thus, $AB \neq BA$.

Using the MATLAB command for matrix multiplication has the standard meaning as well. Multiplying two matrices A and B of size $m \times p$ and $p \times n$, respectively, results in answer C, another matrix of size $m \times n$ is:

```
>> A = [1 2; -1 3];
>> B = [2 1;0 1];
>> C = A * B;
```

MATLAB also has component-wise operations for multiplication, division, and exponentiation. These three operations are a combination of a period (.) and one of the operators, $*$, / and ^ , which perform operations on a pair of matrices (or vectors) with equal numbers of rows and columns. For example, consider two row vectors:

```
>> u = [1 2 3 4];
>> v = [5 3 0 2];
>> x = u. * v
x =
    5 6 0 8
```

Warning: Divide by zero.

```
>> y = u./v;
y =
    0.2000 0.6667 Inf 2.0000
```

These operations apply to matrices as well as vectors:

```
>> A = [1 2 3; 4 5 6; 7 8 9];
>> B = [9 8 7; 6 5 4; 3 2 1];
>> C = A. * B
C =
    9  16  21
    24 25  24
    21 16  9
```

Note that $A. * B$ is not the same as $A * B$.

The array exponentiation operator, . ^ , raises the individual elements of a matrix to a power:

```
>> A = [1 2 3; 4 5 6; 7 8 9];
>> D = A.^ 2
D =
    1   4   9
    16  25  36
    49  64  81
```

```
>> E = A.^ (1/2)
D =
     1.0000   1.4142   1.7321
     2.0000   2.2361   2.4495
     2.6458   2.8284   3.0000
```
.

The syntax of array operators requires the correct placement of a typographically small symbol, a period, in what might be a complex formula. Although MATLAB will catch syntax errors, it is still possible to make computational mistakes with legal operations. For example, $A.\ \hat{}\ 2$ and $A\ \hat{}\ 2$ are both legal, but not at all equivalent. In linear algebra, the addition and subtraction of matrices and vectors are element-by-element operations. Thus, there are no special array operators for addition and subtraction.

3.2.2 Some Special Matrix Forms

There are many special types of matrices that are encountered frequently in engineering analysis. We discuss some of them here.

Definition 3.7 (Square Matrix)

A matrix A, which has the same number of rows m and columns n, that is, $m = n$, defined as

$$A = (a_{ij}), \qquad \textit{for} \quad i = 1, 2, \ldots, n, \quad \textit{and} \quad j = 1, 2, \ldots, n$$

is called a square matrix. For example, the following matrices

$$A = \begin{pmatrix} 1 & 2 \\ -1 & 3 \end{pmatrix} \quad \textit{and} \quad B = \begin{pmatrix} 2 & 1 & 2 \\ 1 & 2 & 3 \\ 0 & 1 & 5 \end{pmatrix}$$

are square matrices because both matrices have the same numbers of rows and columns.

Definition 3.8 (Null Matrix)

A null matrix is a matrix in which all elements are zero; that is,

$$A = (a_{ij}) = \mathbf{0}, \qquad \textit{for} \quad i = 1, 2, \ldots, n, \quad \textit{and} \quad j = 1, 2, \ldots, n.$$

It is also called a zero matrix. It may be either rectangular or square. For example, the following matrices

$$A = \begin{pmatrix} 0 & 0 & 0 \\ 0 & 0 & 0 \end{pmatrix} \quad \textit{and} \quad B = \begin{pmatrix} 0 & 0 & 0 \\ 0 & 0 & 0 \\ 0 & 0 & 0 \end{pmatrix}$$

are zero matrices.

Definition 3.9 (Identity Matrix)

An identity matrix is a square matrix in which the main diagonal elements are equal to 1, and is defined as follows

$$\mathbf{I} = (a_{ij}) = \begin{cases} a_{ij} = 0, & \textit{if } i \neq j, \\ a_{ij} = 1, & \textit{if } i = j \end{cases}.$$

An example of a 4×4 *identity matrix may be written as*

$$\mathbf{I}_4 = \begin{pmatrix} 1 & 0 & 0 & 0 \\ 0 & 1 & 0 & 0 \\ 0 & 0 & 1 & 0 \\ 0 & 0 & 0 & 1 \end{pmatrix}.$$

The identity matrix (also called a unit matrix) serves somewhat the same purpose in matrix algebra as does the number one (unity) in scalar algebra. It is called the identity matrix because multiplication of a matrix by it will result in the same matrix. For a square matrix A of order n, it can be seen that

$$\mathbf{I}_n A = A\mathbf{I}_n = A.$$

Similarly, for a rectangular matrix B *of order* $m \times n$, *we have*

$$\mathbf{I}_m B = B\mathbf{I}_n = B.$$

The multiplication of an identity matrix by itself results in same identity matrix.

In MATLAB identity matrices are created with the *eye* function, which can take either one or two input arguments:

```
>> I = eye(n)
>> I = eye(m, n)
```

Definition 3.10 (Transpose Matrix)

The transpose of a matrix A is a new matrix formed by interchanging the rows and columns of the original matrix. If the original matrix A is of order $m \times n$, then the transpose matrix, as A^T, will be of order $n \times m$; that is,

$$\text{If} \quad A = (a_{ij}), \quad \text{for} \quad i = 1, 2, \ldots, m \text{ and } j = 1, 2, \ldots, n$$

then

$$A^T = (a_{ji}), \quad \text{for} \quad i = 1, 2, \ldots, n \text{ and } j = 1, 2, \ldots, m.$$

The transpose of a matrix A can be found using the MATLAB command as follows:

```
>> A = [1 2 3; 4 5 6; 7 8 9]
>> B = A'
B =
      1.0000  4.0000  7.0000
      2.0000  5.0000  8.0000
      3.0000  6.0000  9.0000
```

It is to be noted that:

1. $(A^T)^T = A$
2. $(A_1 + A_2)^T = A_1^T + A_2^T$
3. $(A_1 A_2)^T = A_2^T A_1^T$
4. $(\alpha A)^T = \alpha A^T, \quad \alpha$ is a scalar.

Definition 3.11 (Inverse Matrix)

An $n \times n$ matrix A has an inverse or is invertible if there exists an $n \times n$ matrix B such that

$$AB = BA = I_n.$$

Then matrix B is called the inverse of A and is denoted by A^{-1}. For example, let

$$A = \begin{pmatrix} 2 & 3 \\ 2 & 2 \end{pmatrix} \quad \text{and} \quad B = \begin{pmatrix} -1 & \frac{3}{2} \\ 1 & -1 \end{pmatrix}.$$

Then we have

$$AB = BA = \mathbf{I}_2.$$

which means that B is an inverse of A. Note that the invertible matrix is also called the nonsingular matrix.

To find the inverse of square matrix A using MATLAB commands, we do the following:

```
>> A = [2 1 0 0; -1 2 -1 0; 0 -1 2 -1; 0 0 -1 2]
>> Ainv = INVMAT(A)
Ainv =
    0.8000  0.6000  0.4000  0.2000
    0.6000  1.2000  0.8000  0.4000
    0.4000  0.8000  1.2000  0.6000
    0.2000  0.4000  0.6000  0.8000
```

Program 3.1

MATLAB m-file for finding inverse of a matrix

```
function [Ainv]=INVMAT(A)
[n,n]=size(A); I=zeros(n,n);
for i=1:n; I(i,i)=1; end
m(1:n,1:n)=A; m(1:n,n+1:2*n) = I;
for i=1:n; m(i,1:2*n) = m(i,1:2*n)/m(i,i);
for k=1:n; if i~ =k
m(k,1:2*n) = m(k,1:2*n) - m(k,i)*m(i,1:2*n);
end; end; end
invrs = m(1:n,n+1:2*n);
```

The MATLAB built-in function *inv* can be also used to calculate the inverse of a square matrix A if A is invertible:

```
>> I = Ainv * A;
>> format short e
>> disp(I)
I =
    1.0000e+00  -1.1102e-16  0           0
    0           1.0000e+00   0           0
    0           0            1.0000e+00  2.2204e-16
    0           0            0           1.0000e+00
```

The values of $I(2,1)$ and $I(3,4)$ are very small, but nonzero due to round-off errors in the computation of *Ainv* and I. It is often preferable to use rational numbers rather than decimal numbers. Function *frac*(x) returns the rational approximation to x or we can use other MATLAB commands as follows:

```
>> format rat
```

If matrix A is not invertible, then matrix A is called *singular*.

There are some well-known properties of the invertible matrix, which are defined as follows:

Theorem 3.2 *If matrix A is invertible, then*

1. *It has exactly one inverse. If B and C are the inverses of A, then $B = C$.*

2. *Its inverse matrix A^{-1} is also invertible and $(A^{-1})^{-1} = A$.*

3. *Its product with another invertible matrix is invertible, and the inverse of the product is the product of the inverses in reverse order. If A and B are invertible matrices of the same size, then AB is invertible and $(AB)^{-1} = B^{-1}A^{-1}$.*

4. *Its transpose matrix A^T is invertible and $(A^T)^{-1} = (A^{-1})^T$.*

5. *The kA for any non zero k is invertible; that is, $(kA)^{-1} = \frac{1}{k}A^{-1}$.*

6. *The A^k for any k is also invertible; that is, $(A^k)^{-1} = (A^{-1})^k$.*

7. *Its size 1×1 is invertible when it is nonzero. If $A = (a)$, then $A^{-1} = (\frac{1}{a})$.*

8. *The formula for A^{-1} when $n = 2$ is*

$$A^{-1} = \begin{pmatrix} a_{11} & a_{12} \\ a_{21} & a_{22} \end{pmatrix}^{-1} = \frac{1}{a_{11}a_{22} - a_{12}a_{21}} \begin{pmatrix} a_{22} & -a_{12} \\ -a_{21} & a_{11} \end{pmatrix}$$

provided that $a_{11}a_{22} - a_{12}a_{21} \neq 0$.

Definition 3.12 (Diagonal Matrix)

It is a square matrix having all elements equal to zero except those on the main diagonal; that is,

$$A = (a_{ij}) = \begin{cases} a_{ij} = 0, & \text{if } i \neq j \\ a_{ij} \neq 0, & \text{if } i = j \end{cases}.$$

Note that all diagonal matrices are invertible if all diagonal entries are nonzero.

The MATLAB function *diag* is used to either create a diagonal matrix from a vector or to extract the diagonal entries of a matrix. If the input argument of the *diag* function is a vector, MATLAB uses the vector to create a diagonal matrix:

```
>> x = [2, 2, 2];
>> A = diag(x)
A =
    2 0 0
    0 2 0
    0 0 2
```

Matrix A is called the *scalar* matrix because it has all the elements on the main diagonal equal to the same scalars, 2. Multiplication of a square matrix and a scalar matrix is commutative, and the product is also a diagonal matrix.

If the input argument of the *diag* function is a matrix, the result is a vector of the diagonal elements:

```
>> B = [2 -4 1; 6 10 -3; 0 5 8]
>> M = diag(B);
M =
   2
   10
   8
```

Definition 3.13 (Upper-Triangular Matrix)

An upper-triangular matrix is a square matrix, which has zero elements below and to the left of the main diagonal. The diagonal as well as the above diagonal elements can take on any value; that is,

$$\mathbf{U} = (u_{ij}), \quad \textit{where} \quad u_{ij} = 0, \quad \textit{if} \quad i > j.$$

An example of such a matrix is

$$U = \begin{pmatrix} 1 & 2 & 3 \\ 0 & 4 & 5 \\ 0 & 0 & 6 \end{pmatrix}.$$

The upper-triangular matrix is called an upper-unit-triangular matrix if the diagonal elements are equal to one. This type of matrix is used in solving linear algebraic equations by LU decomposition with Crout's method. Also, if the main diagonal elements of the upper-triangular matrix are zero, then the matrix

$$A = \begin{pmatrix} 0 & a_{12} & a_{13} \\ 0 & 0 & a_{23} \\ 0 & 0 & 0 \end{pmatrix}$$

is called strictly upper-triangular matrix. This type of matrix will be used in solving linear systems by iterative methods.

Using the MATLAB command *triu* (A) we can create an upper triangular matrix from a given matrix A as follows:

```
>> A = [1 2 3; 4 5 6; 7 8 9];
>> U = triu(A)
U =
    1  2  3
    0  4  5
    0  0  6
```

Also, we can create a strictly upper-triangular matrix, that is, an upper-triangular matrix with a zero diagonal, from a given matrix A using the MATLAB built-in function *triu*(A,I) as follows:

```
>> A = [1 2 3; 4 5 6; 7 8 9];
>> U = triu(A, I)
U =
    0  2  3
    0  0  5
    0  0  0
```

Definition 3.14 (Lower-triangular Matrix)

A lower-triangular matrix is a square matrix which has zero elements above and to the right of the main diagonal, where the rest of the elements can take on any value; that is,

$$L = (l_{ij}), \quad \text{where} \quad l_{ij} = 0, \quad \text{if} \quad i < j.$$

An example of such a matrix is

$$L = \begin{pmatrix} 2 & 0 & 0 \\ 3 & 1 & 0 \\ 4 & 5 & 3 \end{pmatrix}.$$

A lower-triangular matrix is called a lower-unit-triangular matrix if the diagonal elements are equal to one. This type of matrix is used in solving linear algebraic equations by LU decomposition with Doolittle's method. Also, if the main diagonal elements of the lower-triangular matrix are zero, then the matrix

$$A = \begin{pmatrix} 0 & 0 & 0 \\ a_{21} & 0 & 0 \\ a_{31} & a_{32} & 0 \end{pmatrix}$$

is called a strictly lower-triangular matrix. We will use this type of matrix in solving linear systems by using iterative methods.

In a similar way, we can create a lower-triangular matrix and a strictly lower-triangular matrix from a given matrix A using the MATLAB built-in functions *tril*(A) and *tril*(A,I), respectively.

Note that all triangular matrices (upper or lower) with nonzero diagonal entries are invertible.

Definition 3.15 (Symmetric Matrix)

A symmetric matrix is one in which the elements a_{ij} of a matrix A in the ith row and jth column are equal to the element a_{ji} in the jth row and the ith column, which means that

$$A^T = A, \quad \text{that is,} \quad a_{ij} = a_{ji}, \quad \text{for} \quad i \neq j.$$

One way to generate a symmetric matrix is to multiply a matrix by its transpose, since $A^T A$ is symmetric for any A. To generate a symmetric matrix using MATLAB commands we do the following:

```
>> A = [1 : 4; 5 : 8; 9 : 12]
%A is not symmetric
>> B = A' * A
B =
     107  122  137  152
     122  140  158  176
     137  158  179  200
     152  176  200  224
>> C = A * A'
C =
     30   70   110
     70   174  278
     110  278  446
```

If for a matrix A, the $a_{ij} = -a_{ji}$ *for* $i \neq j$ *and the main diagonal elements are not all zero, then matrix A is called a skew matrix. If all the elements on the main diagonal of a skew matrix are zero, then the matrix is called skew symmetric; that is,*

$$A = -A^T, \quad \text{with} \quad a_{ij} = -a_{ji} \quad \text{for} \quad i \neq j, \quad \text{and} \quad a_{ii} = 0.$$

Any square matrix may be split into the sum of a symmetric and a skew symmetric matrix. Thus,

$$A = \frac{1}{2}(A + A^T) + \frac{1}{2}(A - A^T),$$

where $\frac{1}{2}(A + A^T)$ *is a symmetric matrix and* $\frac{1}{2}(A - A^T)$ *is a skew symmetric matrix.*

The following matrices

$$\begin{pmatrix} 1 & 2 & 3 \\ 2 & 4 & 5 \\ 3 & 5 & 6 \end{pmatrix}, \quad \begin{pmatrix} 1 & 2 & 3 \\ -2 & 4 & -5 \\ -3 & 5 & 6 \end{pmatrix}, \quad \begin{pmatrix} 0 & 2 & 3 \\ -2 & 0 & 5 \\ -3 & 5 & 0 \end{pmatrix}$$

are examples of symmetric, skew, and skew symmetric matrices, respectively.

Definition 3.16 (Partitioned Matrix)

A matrix can be subdivided or partitioned into smaller matrices by inserting horizontal and vertical rules between selected rows and columns. For example,

$$A = \begin{pmatrix} a_{11} & a_{12} & a_{13} & \vdots & a_{14} \\ \cdots & \cdots & \cdots & \vdots & \cdots \\ a_{21} & a_{22} & a_{23} & \vdots & a_{24} \\ a_{31} & a_{32} & a_{33} & \vdots & a_{34} \end{pmatrix} = \begin{pmatrix} A_{11} & A_{12} \\ A_{21} & A_{22} \end{pmatrix}.$$

The other example may be an augmented matrix, which can be partitioned in the form

$$B = [A|\mathbf{b}].$$

Definition 3.17 (Band Matrix)

An $n \times n$ square matrix A is called a band matrix if there exists positive integers p and q, with $1 < p$ and $q < n$, such that

$$a_{ij} = 0 \quad \textit{for} \quad p \leq j - i \quad \textit{or} \quad q \leq i - j.$$

The number p describes the number of diagonals above, and including, the main diagonal on which nonzero entries may lie. The number q describes the number of diagonals below, and including, the main diagonal on which nonzero entries may lie. The number $p+q-1$ is called the bandwidth of a matrix A, which tells us how many of the diagonals can contain nonzero entries. For example, the following matrix

$$A = \begin{pmatrix} 1 & 2 & 3 & \\ 2 & 3 & 4 & 5 \\ 0 & 5 & 6 & 7 \\ 0 & 0 & 7 & 8 \end{pmatrix}$$

is banded with $p = 3$ and $q = 2$, and so the bandwidth is equal to 4. An important property of the band matrix is called the tridiagonal matrix, in this case $p = q = 2$; that is, all nonzero elements lie either on or directly above or below the main diagonal. For such type of matrix, the Gaussian elimination is particularly simple. In general, the nonzero elements of a tridiagonal matrix lie *in three bands: the superdiagonal, diagonal, and subdiagonal. For example, the following matrix*

$$A = \begin{pmatrix} 1 & 2 & & & & & \\ 2 & 3 & 1 & & & & \\ & 3 & 2 & 1 & & & \\ & 2 & 4 & 3 & & & \\ & & 1 & 2 & 3 & & \\ & & & 1 & 6 & 4 & \\ & & & & 3 & 4 & \end{pmatrix}$$

is a tridiagonal matrix.
A matrix which is predominantly zero is called sparse. A band matrix or a tridiagonal matrix is sparse but the nonzero elements of a sparse matrix are not necessarily near the diagonal.

Definition 3.18 (Permutation Matrix)

A permutation matrix P has only 0s and 1s and their is exactly one 1 in each row and column of P. For example, the following matrices are permutation matrices

$$P = \begin{pmatrix} 1 & 0 & 0 \\ 0 & 0 & 1 \\ 0 & 1 & 0 \end{pmatrix} \text{ and } \quad P = \begin{pmatrix} 0 & 1 & 0 & 0 \\ 1 & 0 & 0 & 0 \\ 0 & 0 & 1 & 0 \\ 0 & 0 & 0 & 1 \end{pmatrix}.$$

The product PA has the same rows as A but in a different order (permuted), while AP is just A with the columns permuted.

3.2.3 The Determinant of Matrix

The determinant is a certain kind of function that associates a real number with a square matrix. We will denote the determinant of a square matrix A by $\det(A)$ or $|A|$.

Definition 3.19 (Determinant of Matrix)

Let $A = (a_{ij})$ be an $n \times n$ square matrix and then a determinant of A is given by:

1. $\det(A) = a_{11}$, *if* $n = 1$.

2. $\det(A) = a_{11}a_{22} - a_{12}a_{21}$, *if* $n = 2$.

For example, if

$$A = (-6) \quad \text{and} \quad B = \begin{pmatrix} 5 & 1 \\ 0 & -2 \end{pmatrix}$$

then

$$\det(A) = -6 \quad \text{and} \quad \det(B) = (5)(-2) - (1)(0) = -10.$$

The MATLAB function *det*(A) calculates the determinant of a square matrix as follows:

```
>> A = [1 2; 3 4];
>> B = det(A)
B =
    -2.0000
```

More general, for $n \times n$ square matrix, we define

$$\det(A) = |A| = \sum_{1}^{n} a_{ij} A_{ij}, \quad n > 2, \tag{3.10}$$

where the summation is on i for any fixed value of jth column $(1 \leq j \leq n)$, or on j for any fixed value of ith row $(1 \leq i \leq n)$ and A_{ij} is called the *cofactor* of element a_{ij} and is defined as

$$A_{ij} = (-1)^{i+j} M_{ij}.$$

We define the *minor* M_{ij} of all elements a_{ij} of a matrix A as the determinant of the sub-matrix of order $(n-1)$ obtained from A by deleting the ith row and jth column.

Example 3.1 *Find the minors and cofactors of matrix A and use it to evaluate the determinant of the matrix*

$$A = \begin{pmatrix} 3 & 1 & -4 \\ 2 & 5 & 6 \\ 1 & 4 & 8 \end{pmatrix}.$$

Solution. *The minors are calculated as follows:*

$$M_{11} = \begin{vmatrix} 5 & 6 \\ 4 & 8 \end{vmatrix} = 40 - 24 = 16,$$

$$M_{12} = \begin{vmatrix} 2 & 6 \\ 1 & 8 \end{vmatrix} = 16 - 6 = 10,$$

$$M_{13} = \begin{vmatrix} 2 & 5 \\ 1 & 4 \end{vmatrix} = 8 - 5 = 3.$$

From these we can have the cofactors of the matrix as follows

$$\begin{aligned} A_{11} &= (-1)^{1+1}M_{11} = M_{11} = 16, \\ A_{12} &= (-1)^{1+2}M_{12} = -M_{11} = -10, \\ A_{13} &= (-1)^{1+3}M_{13} = M_{13} = 3. \end{aligned}$$

Now using the cofactor expansion along the first row, we can find the determinant of the matrix as follows

$$\det(A) = a_{11}A_{11} + a_{12}A_{12} + a_{13}A_{13} = 48 - 10 - 12 = 26.$$

Note that the cofactor and minor of an element a_{ij} differs only in sign; that is, $A_{ij} = \pm M_{ij}$. A quick way for determining whether to use the $+$ or $-$ is to use the fact that the sign relating A_{ij} and M_{ij} is in the *ith* row and *jth* column of the *checkerboard* array

$$\begin{pmatrix} + & - & + & - & + & \cdots \\ - & + & - & + & - & \cdots \\ + & - & + & - & + & \cdots \\ - & + & - & + & - & \cdots \\ \vdots & \vdots & \vdots & \vdots & \vdots & \ddots \end{pmatrix}.$$

For example, $A_{11} = M_{11}, A_{21} = -M_{21}, A_{12} = -M_{12}, A_{22} = M_{22}$, and so on.

Definition 3.20 (Cofactor Matrix)

If A is any $n \times n$ matrix and A_{ij} is the cofactor of a_{ij}, then the matrix

$$\begin{pmatrix} A_{11} & A_{21} & \cdots & A_{n1} \\ A_{12} & A_{22} & \cdots & A_{n2} \\ \vdots & \vdots & \cdots & \vdots \\ A_{1n} & A_{2n} & \cdots & A_{nn} \end{pmatrix}$$

is called the matrix of cofactor from A. The transpose of this matrix is called the adjoint of A and is denoted by Adj(A). For example, the cofactors of the matrix

$$A = \begin{pmatrix} 3 & 2 & -1 \\ 1 & 6 & 3 \\ 2 & -4 & 0 \end{pmatrix}$$

can be calculated as follows:

$$\begin{array}{llllll} A_{11} = & 12, & A_{12} = & 6, & A_{13} = & -16, \\ A_{21} = & 4, & A_{22} = & 2, & A_{23} = & 16, \\ A_{31} = & 12, & A_{32} = & -10, & A_{33} = & 16. \end{array}$$

So that the matrix of the cofactors is

$$\begin{pmatrix} 12 & 6 & -16 \\ 4 & 2 & 16 \\ 12 & -10 & 16 \end{pmatrix}$$

and the adjoint of A is

$$Adj(A) = \begin{pmatrix} 12 & 4 & 12 \\ 6 & 2 & -10 \\ -16 & 16 & 16 \end{pmatrix}.$$

The following are special properties, which will be helpful in reducing the amount of work involved in evaluating determinants.

Theorem 3.3

1. *The determinant of a matrix is zero if any row or column is zero or equal to a linear combination of other rows and columns.*

2. *A determinant of a matrix is changed in sign if the two rows or two columns are interchanged.*

3. *The determinant of a matrix is equal to the determinant of its transposed matrix.*

4. *The determinant of a product of matrices is the product of the determinants of all matrices.*

5. *The determinant of upper-triangular or lower-triangular matrices is equal to the product of all their main diagonal elements.*

6. *The determinant of a scalar matrix* (1×1) *is equal to the element itself.*

Theorem 3.4 *If A is an invertible matrix, then*

1. $\det(A) \neq 0.$

2. $\det(A^{-1}) = \dfrac{1}{\det(A)}.$

3. $A^{-1} = \dfrac{Adj(A)}{\det(A)}.$

Using Theorem 3.4 we can find the inverse of the matrix using the adjoint and determinant of matrix A. For example, the determinant of the following matrix

$$A = \begin{pmatrix} 1 & 2 & -1 \\ 2 & -1 & 1 \\ 1 & 2 & 2 \end{pmatrix}$$

is -15, and one can easily compute the adjoint of A as follows

$$Adj(A) = \begin{pmatrix} -4 & -6 & 1 \\ -3 & 3 & -3 \\ 5 & 0 & -5 \end{pmatrix}.$$

Then using Theorem 3.4 we can have the inverse of the matrix as follows

$$A^{-1} = -\frac{1}{15} \begin{pmatrix} -4 & -6 & 1 \\ -3 & 3 & -3 \\ 5 & 0 & -5 \end{pmatrix}.$$

Now we consider the implementation of finding the inverse of a matrix using the adjoint of a matrix and the determinant of a matrix in the MATLAB command window:

```
>> A = [1 -1 1 2; 1 0 1 3; 0 0 2 4; 1 1 -1];
```

To find the cofactors A_{ij} of the elements of given matrix A, we do the following:

$$
\begin{aligned}
&>> A11 = (-1)\hat{}\ (1+1) * det(A([2:4],[2:4]));\\
&>> A12 = (-1)\hat{}\ (1+2) * det(A([2:4],[1,3:4]));\\
&>> A13 = (-1)\hat{}\ (1+3) * det(A([2:4],[1,2:4]));\\
&>> A14 = (-1)\hat{}\ (1+4) * det(A([2:4],[1:3]));\\
&>> A21 = (-1)\hat{}\ (2+1) * det(A([1,3:4],[2:4]));\\
&>> A22 = (-1)\hat{}\ (2+2) * det(A([1,3:4],[1,3:4]));\\
&>> A23 = (-1)\hat{}\ (2+3) * det(A([1,3:4],[1:2,4]));\\
&>> A24 = (-1)\hat{}\ (2+4) * det(A([1,3:4],[1:3]));\\
&>> A31 = (-1)\hat{}\ (3+1) * det(A([1:2,4],[2:4]));\\
&>> A32 = (-1)\hat{}\ (3+2) * det(A([1:2,4],[1,3:4]));\\
&>> A33 = (-1)\hat{}\ (3+3) * det(A([1:2,4],[1:2,4]));\\
&>> A34 = (-1)\hat{}\ (3+4) * det(A([1:2,4],[1:3]));\\
&>> A41 = (-1)\hat{}\ (4+1) * det(A([1:3],[2:4]));\\
&>> A42 = (-1)\hat{}\ (4+2) * det(A([1:3],[1,3:4]));\\
&>> A43 = (-1)\hat{}\ (4+3) * det(A([1:3],[1:2,4]));\\
&>> A44 = (-1)\hat{}\ (4+4) * det(A([1:3],[1:3]));
\end{aligned}
$$

Now form the cofactor matrix B using the A_{ij}s as follows:

$$>> B = [A11\ A12\ A13\ A14; A21\ A22\ A23\ A24; A31\ A32\ A33\ A34; A41\ A42\ A43\ A44];.$$

The adjoint matrix is the transpose of the cofactor matrix:

$$>> adjA = B';.$$

The inverse of A is the adjoint matrix divided by the determinant of A:

$$
\begin{aligned}
&>> invA = (1/det(A)) * adjA;\\
&invA =\\
&\begin{matrix} -1 & 3 & -1.5 & -1 \\ -2 & 3 & -1 & -1 \\ -2 & 4 & -1.5 & -2 \\ 1 & -2 & 1 & 1 \end{matrix}
\end{aligned}
$$

Verify the results by finding A^{-1} directly using the MATLAB command as follows:

$$>> inv(A).$$

Theorem 3.5 *For an $n \times n$ matrix A, the following properties are equivalent:*

1. *The inverse of matrix A exists; that is, A is nonsingular.*

2. *The determinant of matrix A is nonzero.*

3. *The homogeneous system $A\mathbf{x} = \mathbf{0}$ has a trivial solution $\mathbf{x} = \mathbf{0}$.*

4. *The nonhomogeneous system $A\mathbf{x} = \mathbf{b}$ has a unique solution.*

A non-trivial solution to the homogeneous system occurs only when the coefficient matrix A is a *singular*, which means that the determinant of matrix A is zero. The following basic theorems on the solvability of linear systems are proved in linear algebra.

Theorem 3.6 *A homogeneous system of n equations in n unknowns has a solution other than trivial solution if and only if the determinant of coefficients matrix A vanishes; that is, matrix A is singular.*

Theorem 3.7 (Necessary and Sufficient Condition for a Unique Solution)

A nonhomogeneous system of n equations in n unknowns has a unique solution if and only if the determinant of a coefficient matrix A is nonsingular.

If matrix A is nonsingular, then linear system (3.6) always has a unique solution for each $\mathbf{b}$, since the inverse matrix A^{-1} exists, so the solution of linear system (3.6) can formally expressed as

$$\mathbf{x} = A^{-1}\mathbf{b}. \tag{3.11}$$

For example, the solution of the system $A\mathbf{x} = \mathbf{b}$ with

$$A = \begin{pmatrix} 1 & 2 & -1 \\ 2 & -1 & 1 \\ 1 & 2 & 2 \end{pmatrix} \quad \text{and} \quad \mathbf{b} = \begin{pmatrix} 1 \\ 0 \\ 1 \end{pmatrix}$$

is given by

$$\mathbf{x} = A^{-1}\mathbf{b} = -\frac{1}{15}\begin{pmatrix} -4 & -6 & 1 \\ -3 & 3 & -3 \\ 5 & 0 & -5 \end{pmatrix}\begin{pmatrix} 1 \\ 0 \\ 1 \end{pmatrix} = -\frac{1}{15}\begin{pmatrix} -3 \\ -6 \\ 0 \end{pmatrix}$$

or

$$x_1 = \frac{1}{5}, \qquad x_2 = \frac{2}{5}, \qquad x_3 = 0.$$

It is called the *matrix inverse method*. Thus, when the matrix inverse A^{-1} of the coefficient matrix A is computed, the solution vector $\mathbf{x}$ of linear system (3.6) is simply the product of inverse matrix A^{-1} and the right-hand side vector $\mathbf{b}$.

The solution of a linear system of equations defined by the coefficient matrix A and the right hand-side vector $\mathbf{b}$ by the matrix inverse method can be obtained using the following MATLAB commands:

```
>> A = [1 2 4; 1 3 9; 1 4 9];
>> b = [2; 4; 7];
>> x = A \ b
x =
    1.0000
   -0.5000
    0.5000
```

Not all matrices have inverses. Singular matrices don't have inverses and thus the corresponding system of equations does not have a unique solution. The best and simplest method for finding the inverse of a matrix is to perform the Gauss-Jordan method (which we will discuss later in the chapter) on the augmented matrix with an identity matrix of the same size. We will also discuss some other methods to find the inverse of a matrix in this chapter.

3.3 Numerical Methods for Linear Systems

To solve the systems of linear equations using numerical methods, there are two types of methods available. Methods of the first type are called *direct methods* or *elimination methods*. This type of method finds the solution in a finite number of steps. These methods are guaranteed to succeed and are recommended for general purposes. Here, we will consider *Cramer's rule*, the *Gaussian elimination method* and its variants, the *Gauss-Jordan method*, and *LU decomposition* (by Doolittle's, Crout's, and Cholesky methods).

Methods of the second type are called *indirect* or *iterative methods*. Iterative methods start with an arbitrary first approximation to the unknown solution $\mathbf{x}$ of linear system (3.6) and then improve this estimate in an infinite but convergent sequence of steps. These methods are used for solving large systems of equations. The widely used iterative methods, the *Jacobi method*, the *Gauss-Seidel method*, the *successive over-relaxation method (SOR)*, and the *conjugate gradient method* will be presented in this chapter.

3.4 Direct Methods for Linear Systems

This type of method refers to a procedure for computing a solution from a form that is mathematically exact. We shall begin with a simple method called Cramer's rule with determinants. We shall then continue with the Gaussian elimination method and its variants, methods involving triangular matrices, symmetric and tridiagonal matrices.

3.4.1 Cramer's Rule

This is our first direct method for solving linear systems using determinants. This method is one of the least efficient for solving a large number of linear equations. It is, however, very useful for explaining some problems inherent in the solutions of linear equations.

Consider a system of two linear equations

$$\begin{array}{rcrcl} a_{11}x_1 & + & a_{12}x_2 & = & b_1 \\ a_{21}x_1 & + & a_{22}x_2 & = & b_2 \end{array}$$

with the condition that $a_{11}a_{22} - a_{12}a_{21} \neq 0$; that is, the determinant of the given matrix must not be equal to zero or the matrix must be nonsingular. Solving the above system using systematic elimination by multiplying the first equation of the system with a_{22} and the second equation by a_{12}, and subtracting, gives

$$(a_{11}a_{22} - a_{12}a_{21})x_1 = a_{22}b_1 - a_{12}b_2$$

and now solving for x_1, gives

$$x_1 = \frac{a_{22}b_1 - a_{12}b_2}{a_{11}a_{22} - a_{12}a_{21}}$$

and putting the value of x_1 in any equation of the given system, we have x_2, as follows

$$x_2 = \frac{a_{22}b_2 - a_{12}b_1}{a_{11}a_{22} - a_{12}a_{21}}.$$

Then writing in determinant form, we have

$$x_1 = \frac{|A_1|}{|A|}, \quad \text{and} \quad x_2 = \frac{|A_2|}{|A|},$$

where

$$|A_1| = \begin{vmatrix} b_1 & a_{12} \\ b_2 & a_{22} \end{vmatrix}, \quad |A_2| = \begin{vmatrix} a_{11} & b_1 \\ a_{21} & b_2 \end{vmatrix}, \quad \text{and} \quad |A| = \begin{vmatrix} a_{11} & a_{12} \\ a_{21} & a_{22} \end{vmatrix}.$$

In a similar way, one can have Cramer's rule for a set of n linear equations as follows:

$$x_i = \frac{|A_i|}{|A|}, \qquad i = 1, 2, 3, \ldots, n, \tag{3.12}$$

that is, the solution for any one of the unknowns x_i in a set of simultaneous equations is equal to the ratio of two determinants; the determinant in the denominator is the determinant of the coefficient matrix A, while the determinant in the numerator is that same determinant with the *ith* column replaced by the elements from the right-hand sides of the equation.

Example 3.2 *Solve the following linear system using Cramer's rule*

$$\begin{aligned} 3x + 4y &= 4 \\ 2x + 3y &= 5. \end{aligned}$$

Solution. *Writing the given system in matrix form*

$$\begin{pmatrix} 3 & 4 \\ 2 & 3 \end{pmatrix} \begin{pmatrix} x_1 \\ x_2 \end{pmatrix} = \begin{pmatrix} 4 \\ 5 \end{pmatrix}$$

gives

$$A = \begin{pmatrix} 3 & 4 \\ 2 & 3 \end{pmatrix} \quad \textit{and} \quad \mathbf{b} = \begin{pmatrix} 4 \\ 5 \end{pmatrix}.$$

Then

$$|A| = \begin{vmatrix} 3 & 4 \\ 2 & 3 \end{vmatrix} = 9 - 8 = 1 \neq 0,$$

which shows that the given system is nonsingular. Then

$$A_1 = \begin{pmatrix} 4 & 4 \\ 5 & 3 \end{pmatrix} \quad \textit{and} \quad |A_1| = \begin{vmatrix} 4 & 4 \\ 5 & 3 \end{vmatrix} = 12 - 20 = -8$$

also,

$$A_2 = \begin{pmatrix} 3 & 4 \\ 2 & 5 \end{pmatrix} \quad \textit{and} \quad |A_2| = \begin{vmatrix} 3 & 4 \\ 2 & 5 \end{vmatrix} = 15 - 8 = 7.$$

Therefore, by Cramer's rule

$$x_1 = \frac{-8}{1} = -8, \quad \textit{and} \quad x_2 = \frac{7}{1} = 7,$$

which is the required solution of the given system.

Thus, Cramer's rule is useful in hand calculations only if the determinants can be evaluated easily; that is, for $n = 3$. The solution of a system of n linear equations by Cramer's rule will require $N = (n+1)\frac{n^3}{3}$ multiplications. Therefore, this rule is much less efficient for large values of n and is almost never used for computational purposes. When the number of equations is large $(n > 3)$, other methods of solutions are more desirable.

The MATLAB command to find the solution of a linear system by Cramer's rule is as follows:

```
>> A = [2 3 1 -1; 1 2 5 3; -1 0 3 1; 1 -2 10];
>> b = [2; 5; 1; -2];
>> A1 = [b A(:, [2 : 4])];
>> x1 = det(A1)/det(A)
x1 = 0
>> A2 = [A(:, 1) b A(:, [3 : 4])];
>> x2 = det(A2)/det(A)
x2 = 1
>> A3 = [A(:, [1 : 2]) b A(:, 4)];
>> x3 = det(A3)/det(A)
x3 = 0
>> A4 = [A(:, [1 : 3]) b];
>> x4 = det(A4)/det(A)
x4 = 1
```

Procedure 3.1 [Cramer's Rule]

1. *Form the coefficient matrix* A *and column matrix* $\mathbf{b}$.

2. *Compute the determinant of* A. *If* $\det A = 0$, *then the system has no solution; otherwise go to next step 3.*

3. *Compute the determinant of new matrix* A_i *by replacing the ith matrix with the column vector* $\mathbf{b}$.

4. *Repeat step 3 for* $i = 1, 2, \ldots, n$.

5. *Solve for the unknown variables* $\mathbf{x}_i$ *using*

$$\mathbf{x}_i = \frac{\det(A_i)}{\det(A)}, \quad for \quad i = 1, 2, \ldots, n.$$

3.4.2 Gaussian Elimination Method

The Gaussian elimination method is one of the most popular and widely used direct methods for solving linear systems of algebraic equations. No method of solving linear systems requires fewer operations than the Gaussian procedure. The goal of the Gaussian elimination method for solving linear systems is to convert the original system into the equivalent upper-triangular system from which each unknown is determined by backward substitution.

The Gaussian elimination procedure starts with *forward elimination*, in which the first equation in the linear system is used to eliminate the first variable from the rest of the $(n-1)$ equations. Then the new second equation is used to eliminate the second variable from the rest of the $(n-2)$ equations, and so on. If $(n-1)$ such elimination is performed, then the resulting system will be the triangular form. Once this forward elimination is completed, we can determine whether the system is overdetermined or underdetermined or has a unique solution. If it has a unique solution, then *backward substitution* is used to solve the triangular system easily and one can find the unknown variables involved in the system.

Now we shall describe the method in detail for a system of n linear equations. Consider the following system of n linear equations:

$$\begin{array}{ccccccccccc} a_{11}x_1 & + & a_{12}x_2 & + & a_{13}x_3 & + & \cdots & + & a_{1n}x_n & = & b_1 \\ a_{21}x_1 & + & a_{22}x_2 & + & a_{23}x_3 & + & \cdots & + & a_{2n}x_n & = & b_2 \\ a_{31}x_1 & + & a_{32}x_2 & + & a_{33}x_3 & + & \cdots & + & a_{3n}x_n & = & b_3 \\ \vdots & & \vdots & & \vdots & & \vdots & & \vdots & & \vdots \\ a_{n1}x_1 & + & a_{n2}x_2 & & a_{n3}x_3 & + & \cdots & + & a_{nn}x_n & = & b_n. \end{array} \tag{3.13}$$

Forward Elimination

Consider the first equation of given system (3.13)

$$a_{11}x_1 + a_{12}x_2 + a_{13}x_3 + \cdots + a_{1n}x_n = b_1 \tag{3.14}$$

as the first pivotal equation with first pivot element a_{11}. Then the first equation times multiples $m_{i1} = (a_{i1}/a_{11})$ and $i = 2, 3, \ldots, n$, is subtracted from the *ith* equation to eliminate the first the variable x_1, producing an equivalent system

$$\begin{array}{ccccccccccccc} a_{11}x_1 & + & a_{12}x_2 & + & a_{13}x_3 & + & \cdots & + & a_{1n}x_n & = & b_1 \\ & & a_{22}^{(1)}x_2 & + & a_{23}^{(1)}x_3 & + & \cdots & + & a_{2n}^{(1)}x_n & = & b_2^{(1)} \\ & & a_{32}^{(1)}x_2 & + & a_{33}^{(1)}x_3 & + & \cdots & + & a_{3n}^{(1)}x_n & = & b_3^{(1)} \\ & & \vdots & & \vdots & & \vdots & & \vdots & & \vdots \\ & & a_{n2}^{(1)}x_2 & + & a_{n3}^{(1)}x_3 & + & \cdots & + & a_{nn}^{(1)}x_n & = & b_n^{(1)}. \end{array} \tag{3.15}$$

Now consider the second equation of system (3.15), which is

$$a_{22}^{(1)}x_2 + a_{23}^{(1)}x_3 + \cdots + a_{2n}^{(1)}x_n = b_2^{(1)} \tag{3.16}$$

the second pivotal equation with the second pivot element $a_{22}^{(1)}$. Then the second equation times multiples $m_{i2} = (a_{i2}^{(1)}/a_{22}^{(1)})$ and $i = 3, \ldots, n$, is subtracted from the *ith* equation to eliminate the second variable x_2, producing an equivalent system

$$\begin{array}{ccccccccccccc} a_{11}x_1 & + & a_{12}x_2 & + & a_{13}x_3 & + & \cdots & + & a_{1n}x_n & = & b_1 \\ & & a_{22}^{(1)}x_2 & + & a_{23}^{(1)}x_3 & + & \cdots & + & a_{2n}^{(1)}x_n & = & b_2^{(1)} \\ & & & & a_{33}^{(2)}x_3 & + & \cdots & + & a_{3n}^{(2)}x_n & = & b_3^{(2)} \\ & & & & \vdots & & \vdots & & \vdots & & \vdots \\ & & & & a_{n3}^{(2)}x_3 & + & \cdots & + & a_{nn}^{(2)}x_n & = & b_n^{(2)}. \end{array} \tag{3.17}$$

Now consider a third equation of system (3.17), which is

$$a_{33}^{(2)}x_3 + \cdots + a_{3n}^{(2)}x_n = b_3^{(2)} \tag{3.18}$$

the third pivotal equation with the third pivot element $a_{33}^{(2)}$. Then the third equation times multiples $m_{i3} = (a_{i3}^{(2)}/a_{33}^{(2)})$ and $i = 4, \ldots, n$, is subtracted from the *ith* equation to eliminate the third variable x_3. Similarly, after *(n-1)th* steps, we have the *nth* pivotal equation, which has only one unknown variable x_n; that is,

$$\begin{array}{ccccccccccccc} a_{11}x_1 & + & a_{12}x_2 & + & a_{13}x_3 & + & \cdots & + & a_{1n}x_n & = & b_1 \\ & + & a_{22}^{(1)}x_2 & + & a_{23}^{(1)}x_3 & + & \cdots & + & a_{2n}^{(1)}x_n & = & b_2^{(1)} \\ & & & + & a_{33}^{(2)}x_3 & + & \cdots & + & a_{3n}^{(2)}x_n & = & b_3^{(2)} \\ & & & & & & & & \vdots & & \vdots \\ & & & & & & & & a_{nn}^{(n-1)}x_n & = & b_n^{(n-1)} \end{array} \tag{3.19}$$

with the nth pivotal element $a_{nn}^{(n-1)}$. After getting the upper-triangular system, which is equivalent to the original system, forward elimination is completed.

Backward Substitution

After the triangular set of equations has been obtained, the last equation of system (3.19) yields the value of x_n directly. The value is then substituted into the equation next to the last one of system (3.19) to obtain a value of x_{n-1}, which is, in turn, used along with the value of x_n in the second to the last equation to obtain a value of x_{n-2}, and so on. Mathematical formula can be obtained for the backward substitution

$$\left.\begin{aligned} x_n &= \frac{b_n^{(n-1)}}{a_{nn}^{(n-1)}} \\ x_{n-1} &= \frac{1}{a_{n-1n-1}^{(n-2)}}\left(b_{n-1}^{(n-2)} - a_{n-1n}^{(n-2)} x_n\right) \\ &\ \vdots \\ x_1 &= \frac{1}{a_{11}}\left(b_1 - \sum_{j=2}^{n} a_{1j} x_j\right) \end{aligned}\right\}. \tag{3.20}$$

The Gaussian elimination can be carried out by writing only the coefficients and the right-hand side terms in matrix form, which means augmented matrix form. Indeed, this is exactly what a computer program for Gaussian elimination does. Even for hand calculation, the augmented matrix form is more convenient than writing all sets of equations. The augmented matrix is formed as follows

$$\left(\begin{array}{ccccc|c} a_{11} & a_{12} & a_{13} & \cdots & a_{1n} & b_1 \\ a_{21} & a_{22} & a_{23} & \cdots & a_{2n} & b_2 \\ a_{31} & a_{32} & a_{33} & \cdots & a_{3n} & b_3 \\ \vdots & \vdots & \vdots & \vdots & \vdots & \\ a_{n1} & a_{n2} & a_{n3} & \cdots & a_{nn} & b_n \end{array}\right). \tag{3.21}$$

The operations used in the Gaussian elimination method can now be applied to the augmented matrix. Consequently, system (3.19) is now written directly as follows:

$$\left(\begin{array}{cccccc|c} a_{11} & a_{12} & a_{13} & \cdots & a_{1n} & & b_1 \\ & a_{22}^{(1)} & a_{23}^{(1)} & \cdots & a_{2n}^{(1)} & & b_2^{(1)} \\ & & a_{33}^{(2)} & \cdots & a_{3n}^{(2)} & & b_3^{(2)} \\ & & & \vdots & \vdots & & \\ & & & & a_{nn}^{(n-1)} & & b_n^{(n-1)} \end{array}\right) \tag{3.22}$$

from which the unknowns are determined as before using backward substitution. The number of multiplications and divisions for the Gaussian elimination method for one **b** vector is approximately

$$N = \left(\frac{n^3}{3}\right) + n^2 - \left(\frac{n}{3}\right). \tag{3.23}$$

Simple Gaussian Elimination Method

Firstly, we will solve the linear system using the simplest variation of the Gaussian elimination method called the *simple* Gaussian elimination or the Gaussian elimination *without pivoting.* The basis of this variation is that all the possible diagonal elements (called *pivot elements*) should be nonzero. If at any stage it becomes zero, then interchange that row with any below row with the nonzero element at that position. After getting the upper-triangular matrix, we use backward substitution to get the solution of the given linear system.

Example 3.3 *Solve the following linear system using the simple Gaussian elimination method*

$$\begin{array}{rcrcrcr} x_1 & + & 2x_2 & & & = & 3 \\ -x_1 & & & - & 2x_3 & = & -5 \\ -3x_1 & - & 5x_2 & + & x_3 & = & -4. \end{array}$$

Solution. *The process begins with the augmented matrix form*

$$\left(\begin{array}{ccc} 1 & 2 & 0 & \vdots & 3 \\ -1 & 0 & -2 & \vdots & -5 \\ -3 & -5 & 1 & \vdots & -4 \end{array}\right).$$

Since $a_{11} = 1 \neq 0$, we wish to eliminate the elements a_{21} and a_{31} by subtracting from the second and third rows the appropriate multiples of the first row. In this case the multiples are given

$$m_{21} = \frac{-1}{1} = -1, \quad \textit{and} \quad m_{31} = \frac{-3}{1} = -3.$$

Hence,

$$\begin{pmatrix} 1 & 2 & 0 & \vdots & 3 \\ 0 & 2 & -2 & \vdots & -2 \\ 0 & 1 & 1 & \vdots & 5 \end{pmatrix}.$$

$a_{22}^{(1)} = 2 \neq 0$, *therefore, we eliminate the entry in the* $a_{32}^{(1)}$ *position by subtracting the multiple* $m_{32} = \frac{1}{2}$ *of the second row from the third row, to get*

$$\begin{pmatrix} 1 & 2 & 0 & \vdots & 3 \\ 0 & 2 & -2 & \vdots & -2 \\ 0 & 0 & 2 & \vdots & 6 \end{pmatrix}.$$

Obviously, the original set of equations has been transformed to an upper-triangular form. Since all the diagonal elements of the obtaining upper-triangular matrix are nonzero, which means that the coefficient matrix of the given system is nonsingular the given system has a unique solution. Now expressing the set in algebraic form yields

$$\begin{array}{rcrcrcr} x_1 & + & 2x_2 & & & = & 3 \\ & & 2x_2 & - & 2x_3 & = & -2 \\ & & & & 2x_3 & = & 6. \end{array}$$

Using backward substitution to give

$$\begin{array}{rcll} 2x_3 & = & 6 & \textit{gives} \quad x_3 = 3. \\ 2x_2 & = & 2x_3 - 2 = 2(3) - 2 = 4 & \textit{gives} \quad x_2 = 2. \\ x_1 & = & 3 - 2x_2 = 3 - 2(2) & \textit{gives} \quad x_1 = -1. \end{array}$$

The above results can be obtained using MATLAB commands as follows:

```
>> B = [1 2 0 3; -1 0 -2 -5; -3 -5 1 -4];
%B = [A|b] = Augmented matrix
>> x = WP(B);
>> disp(x)
```

Program 3.2

MATLAB m-file for Gaussian Elimination Without Pivoting

```
function x=WP(B)
[n,t]=size(B); U=B;
for k=1:n-1; for i=k:n-1; m=U(i+1,k)/U(k,k);
for j=1:t; U(i+1,j)=U(i+1,j)-m*U(k,j);end;end end
i=n; x(i,1)=U(i,t)/U(i,i);
for i=n-1:-1:1; s=0;
for k=n:-1:i+1; s = s + U(i,k) * x(k,1); end
x(i,1)=(U(i,t)-s)/U(i,i); end; B; U; x; end
```

In the simple description of Gaussian elimination without pivoting just given, we used the *kth* equation to eliminate variable x_k from equations $k+1,\ldots,n$ during the *kth* step of the procedure. This is possible only if at the beginning of the *kth* step, the coefficient $a_{kk}^{(k-1)}$ of x_k in equation k is not zero, since these coefficients are used as denominators both in the multipliers m_{ij} and in the backward substitution equations. But this does not necessarily mean that the linear system is not solvable, only that the procedure of the solution must be altered.

Example 3.4 *Solve the following linear system using the simple Gaussian elimination method*

$$\begin{array}{rcrcrcl} & & 2x_2 & - & x_3 & = & 1 \\ 3x_1 & - & x_2 & + & 2x_3 & = & 4 \\ x_1 & + & 3x_2 & - & 5x_3 & = & 1. \end{array}$$

Solution. *Write the given system in the augmented matrix form*

$$\begin{pmatrix} 0 & 2 & -1 & \vdots & 1 \\ 3 & -1 & 2 & \vdots & 4 \\ 1 & 3 & -5 & \vdots & 1 \end{pmatrix}.$$

To solve this system, the simple Gaussian elimination method will fail immediately because the element in the first row on the leading diagonal, the pivot, is zero. Thus, it is impossible to divide that row by the pivot value. Clearly, this difficulty can be overcome by rearranging the order of the rows; for example, making the first row the third, gives

$$\begin{pmatrix} 1 & 3 & -5 & \vdots & 1 \\ 3 & -1 & 2 & \vdots & 4 \\ 0 & 2 & -1 & \vdots & 1 \end{pmatrix}.$$

Now we use the usual elimination process. The first elimination step is to eliminate the element $a_{21} = 3$ *from the second row by subtracting a multiple* $m_{21} = \frac{3}{1} = 3$ *of row 1 from row 2, which gives*

$$\begin{pmatrix} 1 & 3 & -5 & \vdots & 1 \\ 0 & -10 & 17 & \vdots & 1 \\ 0 & 2 & -1 & \vdots & 1 \end{pmatrix}.$$

We finished with the first elimination step since the element a_{31} *is already eliminated from the third row. The second elimination step is to eliminate the element* $a_{32}^{(1)} = 2$ *from the third row by subtracting a multiple* $m_{32} = \frac{2}{-10}$ *of row 2 from row 3, which gives*

$$\begin{pmatrix} 1 & 3 & -5 & \vdots & 1 \\ 0 & -10 & 17 & \vdots & 1 \\ 0 & 0 & \frac{12}{5} & \vdots & \frac{6}{5} \end{pmatrix}.$$

Obviously, the original set of equations has been transformed to an upper-triangular form. Now expressing the set in algebraic form yields

$$\begin{array}{rcrcrcl} x_1 & + & 3x_2 & - & 5x_3 & = & 1 \\ & - & 10x_2 & + & 17x_3 & = & 1 \\ & & & & \frac{12}{5}x_3 & = & \frac{6}{5} \end{array}.$$

Now use backward substitution to get the solution of the system

$$\frac{12}{5}x_3 = \frac{6}{5} \quad \textit{gives} \quad x_3 = \frac{1}{2}.$$

$$-10x_2 + 17x_3 = 1 \quad \textit{gives} \quad x_2 = \frac{3}{4}.$$

$$x_1 + 3x_2 - 5x_3 = 1 \quad \textit{gives} \quad x_1 = \frac{5}{4}.$$

Example 3.5 *Solve the linear system using the simple Gaussian elimination method*

$$\begin{array}{rcrcrcl} 2x_1 & + & 2x_2 & - & 4x_3 & = & 0 \\ 2x_1 & + & 2x_2 & - & x_3 & = & 1 \\ 3x_1 & + & 2x_2 & - & 3x_3 & = & 3. \end{array}$$

Solution. *Write the given system in the augmented matrix form*

$$\left(\begin{array}{ccccc} 2 & 2 & -4 & \vdots & 0 \\ 2 & 2 & -1 & \vdots & 1 \\ 3 & 2 & -3 & \vdots & 3 \end{array}\right).$$

The first elimination step is to eliminate the elements $a_{21} = 2$ *and* $a_{31} = 3$ *from the second and third rows by subtracting multiples* $m_{21} = \frac{2}{2} = 1$ *and* $m_{31} = \frac{3}{2}$ *of row 1 from row 2 and row 3, respectively, which gives*

$$\left(\begin{array}{ccccc} 2 & 2 & -4 & \vdots & 0 \\ 0 & 0 & 3 & \vdots & 1 \\ 0 & -1 & 3 & \vdots & 3 \end{array}\right).$$

We finished with the first elimination step. To start the second elimination step, since we noted that the element $a_{22}^{(1)} = 0$ *is called the second pivot element, the simple Gaussian elimination cannot continue in its present form. Therefore, we interchange rows 2 and 3 to get*

$$\left(\begin{array}{ccccc} 2 & 2 & -4 & \vdots & 0 \\ 0 & -1 & 3 & \vdots & 3 \\ 0 & 0 & 3 & \vdots & 1 \end{array}\right).$$

We finished with the second elimination step since the element $a_{32}^{(1)}$ *is already eliminated from the third row. Obviously, the original set of equations has been transformed to an upper-triangular form. Now, expressing the set in algebraic form yields*

$$\begin{array}{rcrcrcl} 2x_1 & + & 2x_2 & - & 4x_3 & = & 0 \\ & - & x_2 & + & 3x_3 & = & 3 \\ & & & & 3x_3 & = & 1. \end{array}$$

Now we use backward substitution to get the solution of the system

$$x_3 = \frac{1}{3}, \quad x_2 = -2, \quad x_1 = \frac{8}{3}.$$

Example 3.6 *Using the simple Gaussian elimination method, find all values of bs for which the following linear system is consistent or inconsistent:*

$$\begin{array}{rcrcrcl} 2x_1 & + & x_2 & + & x_3 & = & 2 \\ -2x_1 & + & x_2 & + & 3x_3 & = & 2 \\ 2x_1 & & & - & x_3 & = & b_3. \end{array}$$

Solution. *Write the given system in augmented matrix form*

$$\begin{pmatrix} 2 & 1 & 1 & 2 \\ -2 & 1 & 3 & 2 \\ 2 & 0 & -1 & b_3 \end{pmatrix}$$

in which we wish to eliminate the elements a_{21} and a_{31} by subtracting from the second and third rows the appropriate multiples of the first row. In this case the multiples are given

$$m_{21} = \frac{-2}{2} = -1 \quad and \quad m_{31} = \frac{2}{2} = 1.$$

Hence,

$$\begin{pmatrix} 2 & 1 & 1 & 2 \\ 0 & 2 & 4 & 4 \\ 0 & -1 & -2 & b_3 - 2 \end{pmatrix}.$$

We finished with the first elimination step. The second elimination step is to eliminate the element $a_{32}^{(1)} = -1$ by subtracting multiple $m_{32} = \frac{-1}{2}$ of row 2 from row 3, which gives

$$\begin{pmatrix} 2 & 1 & 1 & 2 \\ 0 & 2 & 4 & 4 \\ 0 & 0 & 0 & b_3 \end{pmatrix}$$

We finished with the second column. The third row of the equivalent upper-triangular system is

$$0x_1 + 0x_2 + 0x_3 = b_3 \tag{3.24}$$

Firstly, if (3.24) has no constraint on unknowns $x_1, x_2,$ and x_3, then the upper-triangular system represents only two non-trivial equations, namely

$$\begin{array}{rcrcrcl} 2x_1 & + & x_2 & + & x_3 & = & 2 \\ & & 2x_2 & + & 4x_3 & = & 4 \end{array}$$

in three unknowns. As a result, one of the unknowns can be chosen arbitrarily, say $x_3 = x_3^*$*, then* x_2^* *and* x_1^* *can be obtained using backward substitution*

$$x_2^* = 2 - 2x_3^*; \quad x_1^* = \frac{1}{2}[2 - x_3^* - (2 - 2x_3^*)] = \frac{1}{2}x_3^*.$$

Hence,

$$\mathbf{x}^* = [\frac{1}{2}x_3^*, 2 - 2x_3^*, x_3^*]^t$$

is an approximation solution of a given system for any value of x_3^**. Hence, the given linear system is consistent. Note that the third equation* $E_3 = \frac{1}{2}[E_1 - E_2]$ *when* $b_3 = 0$ *in the given system. More generally, the system* $A\mathbf{x} = \mathbf{b}$ *will have infinitely many solutions whenever one or more of the given equations* $(E_1, \ldots, E_n)$ *can be expressed as a weighted sum of the others. Such systems are called undetermined systems.*

Secondly, when $b_3 \neq 0$*, in this case (3.24) puts a restriction on unknowns* $x_1, x_2,$ *and* x_3 *that is impossible to satisfy. So the given system cannot have any solutions and therefore, it is inconsistent.*

More generally, the system $A\mathbf{x} = \mathbf{b}$ will have no solutions when one of equations $(E_1, \ldots, E_n)$ puts a restriction on unknowns $x_1, \ldots, x_n$ that is incompatible with the restriction posed by the other equations. Such linear systems are called inconsistent.

Theorem 3.8 *An upper-triangular matrix A is nonsingular if and only if all its diagonal elements are different from zero.*

Example 3.7 *Use the simple Gaussian elimination method to find all the values of* α*, which make the following matrix singular.*

$$A = \begin{pmatrix} 1 & -1 & \alpha \\ 2 & 2 & 1 \\ 0 & \alpha & -1.5 \end{pmatrix}$$

Solution. *Apply the forward elimination step of the simple Gaussian elimination on given matrix A and eliminate the element* a_{21} *by subtracting from the second row the appropriate multiple of the first row. In this case the multiple is given as*

$$\begin{pmatrix} 1 & -1 & \alpha \\ 0 & 4 & 1 - 2\alpha \\ 0 & \alpha & -1.5 \end{pmatrix}.$$

We finished with the first elimination step. The second elimination step is to eliminate the element $a_{32}^{(1)} = \alpha$ *by subtracting the multiple* $m_{32} = \dfrac{\alpha}{4}$ *of row 2 from row 3, which gives*

$$\begin{pmatrix} 1 & -1 & \alpha \\ 0 & 4 & 1-2\alpha \\ 0 & 0 & -1.5 - \dfrac{\alpha(1-2\alpha)}{4} \end{pmatrix}.$$

To show that the given matrix is singular, we have to set the third diagonal element equal to zero (by Theorem 3.8); that is,

$$-1.5 - \frac{\alpha(1-2\alpha)}{4} = 0.$$

After simplifying, we obtain

$$2\alpha^2 - \alpha - 6 = 0.$$

Solving the above quadratic equation, we get

$$\alpha = -\frac{3}{2} \quad \textit{and} \quad \alpha = 2,$$

which are the possible values of α*, which make the given matrix singular.*

Note that the inverse of the nonsingular matrix A can be easily determined using the simple Gaussian elimination method. Here, we have to consider the augmented matrix as a combination of the given matrix A and the identity matrix $\mathbf{I}$ (same size as A). To find the inverse matrix BA^{-1} we must solve the linear system in which the *jth* column of matrix B is the solution of the linear system with the right-hand side of the *jth* column of matrix $\mathbf{I}$.

Example 3.8 *Use the simple Gaussian elimination method to find the inverse of the following matrix*

$$A = \begin{pmatrix} 2 & -3 \\ 1 & 1 \end{pmatrix}.$$

Solution. *Suppose that the inverse* $A^{-1} = B$ *of the given matrix exists and let*

$$AB = \begin{pmatrix} 2 & -3 \\ 1 & 1 \end{pmatrix}\begin{pmatrix} b_{11} & b_{12} \\ b_{21} & b_{22} \end{pmatrix} = \begin{pmatrix} 1 & 0 \\ 0 & 1 \end{pmatrix} = \mathbf{I}.$$

Now to find the elements of matrix B, we apply the simple Gaussian elimination on the augmented matrix

$$[A|\mathbf{I}] = \begin{pmatrix} 2 & -3 & \vdots & 1 & 0 \\ 1 & 1 & \vdots & 0 & 1 \end{pmatrix}.$$

Apply the forward elimination step of the simple Gaussian elimination on given matrix A and eliminate the element $a_{21} = 1$ by subtracting from the second row the appropriate multiple $m_{21} = \frac{1}{2}$ of the first row. It gives

$$\begin{pmatrix} 2 & -3 & \vdots & 1 & 0 \\ 0 & \frac{5}{2} & \vdots & -\frac{1}{2} & 1 \end{pmatrix}.$$

Using backward substitution, we solve the first system

$$\begin{array}{rcrcr} 2b_{11} & - & 3b_{21} & = & 1 \\ & & \frac{5}{2}b_{21} & = & -\frac{1}{2}, \end{array}$$

which gives

$$b_{11} = \frac{1}{5} \quad \textit{and} \quad b_{21} = -\frac{1}{5}.$$

Similarly, the solution of the second linear system

$$\begin{array}{rcrcr} 2b_{12} & - & 3b_{22} & = & 0 \\ & & \frac{5}{2}b_{22} & = & 1 \end{array}$$

is

$$b_{12} = \frac{3}{5} \quad \textit{and} \quad b_{22} = \frac{2}{5}.$$

Hence, the elements of inverse matrix B are

$$B = A^{-1} = \begin{pmatrix} \frac{1}{5} & -\frac{1}{5} \\ \frac{3}{5} & \frac{2}{5} \end{pmatrix},$$

which is the required inverse of given matrix A.

Procedure 3.2 [Gaussian Elimination Method]

1. *Form the augmented matrix* $B = [A|\mathbf{b}]$.

2. *Check first the pivot element* $a_{11} \neq 0$, *then move to the next step; otherwise, interchange rows so that* $a_{11} \neq 0$.

3. *Multiply row one by multiple* $m_{i1} = \dfrac{a_{i1}}{a_{11}}$ *and subtract to the ith row for* $i = 2, 3, \ldots, n$.

4. *Repeat steps 2 and 3 for the remaining pivots elements unless coefficient matrix* A *becomes upper-triangular matrix* U.

5. *Use backward substitution to solve* x_n *from the nth equation* $x_n = \dfrac{b_n^{n-1}}{a_{nn}}$ *and solve the other (n-1) unknowns variables using (3.20).*

We now introduce the most important numerical quantity associated with a matrix.

Definition 3.21 (Rank of a Matrix)

The rank of a matrix A is the number of pivots.

An $m \times n$ matrix will, in general, have a rank r, where r is an integer and $r \leq min\{m, n\}$. If $r = min\{m, n\}$, then the matrix is said to be *full rank.* If $r < min\{m, n\}$, then the matrix is said to be *rank deficient.*

In principle, the rank of a matrix can be determined using the Gaussian elimination process in which the coefficient matrix A is reduced to upper-triangular form U. After reducing the matrix to triangular form, we find that the rank is the number of columns with nonzero values on the diagonal of U. In practice, especially for large matrices, round-off errors during the row operation may cause a loss of accuracy in this method of rank computation.

Theorem 3.9 *For a linear system of n equations in n unknown variables written in the form* $A\mathbf{x} = \mathbf{b}$, *solution* $\mathbf{x}$ *exists and is unique for any* $\mathbf{b}$ *if and only if* $rank(A) = n$.

Conversely, if rank$(A) < n$ for an $n \times n$ matrix A, the system of equations $A\mathbf{x} = \mathbf{b}$ may or may not be consistent. Such a system may not have a solution, or the solution, if it exists, will not be unique.

Example 3.9 *Find the rank of the following matrix*

$$A = \begin{pmatrix} 1 & 2 & \alpha \\ 1 & 1 & 5 \\ 1 & 1 & 6 \end{pmatrix}.$$

Solution. *Apply the forward elimination step of the simple Gaussian elimination on the given matrix A and eliminate elements below the first pivot (first diagonal element) to*

$$\begin{pmatrix} 1 & 2 & 4 \\ 0 & -1 & 1 \\ 0 & -1 & 2 \end{pmatrix}.$$

We finished with the first elimination step. The second pivot is in the $(2, 2)$ *position but after eliminating the element below it we find the triangular form to be*

$$\begin{pmatrix} 1 & 2 & 4 \\ 0 & -1 & 1 \\ 0 & 0 & 3 \end{pmatrix}.$$

Since the number of pivots are three, the rank of the given matrix is 3. *Note that the original matrix is nonsingular as the rank of* 3×3 *matrix is* 3.

The MATLAB command, the built-in *rank* function, can be used to estimate the rank of a matrix:

```
>> A = [1 2 4; 1 1 5; 1 1 6];
>> rank(A)
ans =
   3
```

Note that:

$$\begin{array}{lcl} \text{rank}(AB) & \leq & min(\text{rank}(A), \text{rank}(B)) \\ \text{rank}(A+B) & \leq & \text{rank}(A) + \text{rank}(B) \\ \text{rank}(AA^T) & = & \text{rank}(A) = \text{rank}(A^T A). \end{array}$$

Although the rank of a matrix is very useful to categorize the behavior of matrices and systems of equations, the rank of a matrix is usually not computed. The use of nonzero pivots is sufficient for the theoretical correctness of simple Gaussian elimination, but more care must be taken if one is to obtain reliable results. For example, consider a linear system

$$\begin{array}{rcrcl} 0.000100x_1 & + & x_2 & = & 1 \\ x_1 & + & x_2 & = & 2, \end{array}$$

which has an exact solution $\mathbf{x} = [1.00010, 0.99990]^T$. Now we solve this system by simple Gaussian elimination. The first elimination step is to eliminate the first variable x_1 from the second equation by subtracting the multiple $m_{21} = 10000$ of the first equation from the second equation, which gives

$$\begin{array}{rcrcl} 0.000100x_1 & + & x_2 & = & 1 \\ & - & 10000x_2 & = & -10000. \end{array}$$

Use backward substitution to get the solution $\mathbf{x}^* = [0, 1]^T$. Thus, a computational disaster has occurred. But if we interchange the equations, we obtain

$$\begin{array}{rcrcl} x_1 & + & x_2 & = & 2 \\ 0.000100x_1 & + & x_2 & = & 1. \end{array}$$

Applying Gaussian elimination again, we got the solution $\mathbf{x}^* = [1, 1]^T$. This solution is as good as one would hope.

So, we conclude from this example that it is not enough just to avoid zero pivot, one must also avoid a relatively small one.

Here we need some pivoting strategies, which will help us overcome these difficulties faced during simple Gaussian elimination.

3.4.3 Pivoting Strategies

We know that simple Gaussian elimination is applied to a problem with no pivotal elements zero. However, the method does not work if the first coefficient of the first equation or if a diagonal coefficient becomes zero in the process of the solution because they are used as denominators in a forward elimination.

Pivoting is used to change sequential order of the equations for two purposes. First to prevent diagonal coefficients from becoming zero, and second, to make each diagonal coefficient larger in magnitude than any other coefficient below it; that is, to decrease the round-off errors. The equations are not mathematical affected by changes of sequential order, but changing the order makes the coefficient become nonzero. Even when all diagonal coefficients are nonzero, the change of order increases accuracy of the computations.

There are two standard pivoting strategies, which handle these difficulties easily and are explained below.

Partial Pivoting

Here we develop an implementation of Gaussian elimination, which utilizes the pivoting strategy discussed above. In using Gaussian elimination by partial pivoting (or row pivoting), the basic approach is to use the largest (in absolute value) element on or below the diagonal in the column of current interest as the pivotal element for elimination in the rest of that column.

One immediate effect of this will be to force all the multiples used to be not greater than 1 in absolute value. This will inhibit the growth of error in the rest of the elimination phase and in subsequent backward substitution.

At stage k of forward elimination, it is necessary, therefore, to be able to identify the largest element from $|a_{kk}|, |a_{k+1,k}|, \ldots, |a_{nk}|$, where these a_{ik}s are the elements in the current partially triangularized coefficient matrix. If this maximum occurs in row p, then pth and kth rows of the augmented matrix are interchanged and the elimination proceeds as usual. In solving n linear equations, a total of $N = \frac{n(n+1)}{2}$ coefficients must be examined.

Example 3.10 *Solve the following linear system using Gaussian elimination with partial pivoting*

$$\begin{array}{rcrcrcr} 2x_1 & + & 2x_2 & - & 2x_3 & = & 8 \\ -4x_1 & - & 2x_2 & + & 2x_3 & = & -14 \\ -2x_1 & + & 3x_2 & + & 9x_3 & = & 9. \end{array}$$

Solution. *For the first elimination step, since -4 is the largest absolute coefficient of first variable x_1, the first row and the second row are interchanged, giving us*

$$\begin{array}{rcrcrcr} -4x_1 & - & 2x_2 & + & 2x_3 & = & -14 \\ 2x_1 & + & 2x_2 & - & 2x_3 & = & 8 \\ -2x_1 & + & 3x_2 & + & 9x_3 & = & 9. \end{array}$$

Eliminate the first variable x_1 from the second and third rows by subtracting the multiples $m_{21} = -\frac{2}{4}$ and $m_{31} = \frac{2}{4}$ of row 1 from row 2 and row 3, respectively, giving us

$$\begin{array}{rcrcrcr} -4x_1 & - & 2x_2 & + & 2x_3 & = & -14 \\ & & x_2 & - & x_3 & = & 1 \\ & & 4x_2 & + & 8x_3 & = & 16. \end{array}$$

For the second elimination step, 4 is the largest absolute coefficient of the second variable x_2, so the second and third row are interchanged, giving us

$$\begin{array}{rcrcrcr} -4x_1 & - & 2x_2 & + & 2x_3 & = & -14 \\ & & 4x_2 & + & 8x_3 & = & 16 \\ & & x_2 & - & x_3 & = & 1. \end{array}$$

Eliminate the second variable x_2 *from the third row by subtracting the multiple* $m_{32} = \frac{1}{4}$ *of row 2 from row 3, which gives*

$$\begin{array}{rcrcrcr} -4x_1 & - & 2x_2 & + & 2x_3 & = & -14 \\ & & 4x_2 & + & 8x_3 & = & 16 \\ & & & - & 3x_3 & = & -3. \end{array}$$

Obviously, the original set of equations has been transformed to an equivalent upper-triangular form. Now use backward substitution to get the solution $\mathbf{x}^* = [3, 2, 1]^T$ *of the linear system.*

The following MATLAB commands will give the same results as we obtained in Example 3.10 using the Gaussian elimination method with partial pivoting:

```
>> B = [2 2 -2 8; -4 -2 2 -14; -2 3 9 9];
>> x = PP(B);
>> disp(x)
```

Program 3.3

MATLAB m-file for Gaussian Elimination by Partial Pivoting

```
function x=PP(B)
% B = input('input matrix in form[A : b]');
[n, t] = size(B); U = B;
for M = 1:n-1
mx(M) = abs(U(M, M)); r = M;
for i = M+1:n
if mx(M) < abs(U(i, M))
mx(M)=abs(U(i,M)); r = i; end; end
rw1(1,1:t)=U(r,1:t); rw2(1,1:t)=U(M,1:t);
U(M,1:t)=rw1 ; U(r,1:t)=rw2 ;
for k=M+1:n; m=U(k,M)/U(M,M);
for j=M:t; U(k,j)=U(k,j)-m*U(M,j); end;end
i=n; x(i)=U(i,t)/U(i,i);
for i=n-1:-1:1; s=0;
for k=n:-1:i+1; s = s + U(i, k) * x(k); end
x(i)=(U(i,t)-s)/U(i,i); end; B; U; x; end
```

Procedure 3.3 [Partial Pivoting]

1. *Suppose we are about to work on the ith column of the matrix. Then we search that portion of the ith column below, including the diagonal, and find the element that has the largest absolute value. Let p denote the index of the row that contains this element.*

2. *Interchange row i and p.*

3. *Proceed with elimination (Procedure 3.2).*

Total Pivoting

In the case of total pivoting (or complete pivoting), we search for the largest number (in absolute value) in the entire array instead of just in the first column, and this number is the pivot. This means that we shall probably need to interchange the columns as well as the rows. When solving a system of equations using complete pivoting, each row interchange is equivalent to interchanging two equations while each column interchange is equivalent to interchanging the two unknowns.

At the *kth* step, interchange both the rows and columns of the matrix so that the largest number in the remaining matrix is used as the pivot. That is, after pivoting

$$|a_{kk}| = max|a_{ij}|, \quad \text{for} \quad i = k, k+1, \ldots, n, \quad j = k, k+1, \ldots, n.$$

There are times when the partial pivoting procedure is inadequate. When some rows have coefficients that are very large in comparison to those in other rows, partial pivoting may not give a correct solution. Therefore, when in doubt, use total pivoting. No amount of pivoting will remove inherent ill-conditioning (we discuss later in the chapter) from a set of equations, but it helps to ensure that no further ill-conditioning is introduced in the course of computation.

Example 3.11 *Solve the following linear system using Gaussian elimination with total pivoting*

$$\begin{array}{rcrcrcr} 2x_1 & + & 2x_2 & - & 2x_3 & = & 8 \\ -4x_1 & - & 2x_2 & + & 2x_3 & = & -14 \\ -2x_1 & + & 3x_2 & + & 9x_3 & = & 9. \end{array}$$

Solution. *For the first elimination step, since 9 is the largest absolute coefficient of the variable x_3 in the given system, the first row and the third row are interchanged as well as the first column and third column, giving us*

$$\begin{array}{rcrcrcr} 9x_3 & + & 3x_2 & - & 2x_1 & = & 9 \\ 2x_3 & - & 2x_2 & - & 4x_1 & = & -14 \\ -2x_3 & + & 2x_2 & + & 2x_1 & = & 8. \end{array}$$

Eliminate the third variable x_3 from the second and third rows by subtracting the multiples $m_{21} = \frac{2}{9}$ and $m_{31} = -\frac{2}{9}$ of row 1 from rows 2 and 3, respectively, which gives

$$\begin{array}{rcrcrcr} 9x_3 & + & 3x_2 & - & 2x_1 & = & 9 \\ & - & \frac{8}{3}x_2 & - & \frac{32}{9}x_1 & = & -16 \\ & & \frac{8}{3}x_2 & + & \frac{14}{9}x_1 & = & 10. \end{array}$$

For the second elimination step, $-\frac{32}{9}$ is the largest absolute coefficient of the first variable x_1 in the remaining system of equations, so the second and third rows are interchanged as well as the second and third columns, which gives

$$\begin{array}{rcrcrcr} 9x_3 & - & 2x_1 & + & 3x_2 & = & 9 \\ & - & \frac{32}{9}x_1 & - & \frac{8}{3}x_2 & = & -16 \\ & & \frac{14}{9}x_1 & + & \frac{8}{3}x_2 & = & 10. \end{array}$$

Eliminate the first variable x_1 from the third row by subtracting the multiple $m_{32} = -\frac{7}{16}$ of row 2 from row 3, which gives

$$\begin{array}{rcrcrcr} 9x_3 & - & 2x_1 & + & 3x_2 & = & 9 \\ & - & \frac{32}{9}x_1 & - & \frac{8}{3}x_2 & = & -16 \\ & & & & \frac{3}{2}x_2 & = & 3. \end{array}$$

Obviously, the original set of equations has been transformed to an equivalent upper-triangular form. Use backward substitution to get the solution of the given system as $\mathbf{x}^* = [3, 2, 1]^T$

Program 3.4

MATLAB m-file for Gaussian Elimination by Total Pivoting

```
function x=TP(B)
% B = input('input matrix in form[A : b]');
[n,m]=size(B);U=B; w=zeros(n,n);
for i=1:n; N(i)=i; end
for M = 1:n-1; r=M; c=M;
for i = M:n; for j = M:n
if max(M) < abs(U(i,j)); max(M)=abs(U(i,j));
r = i; c = j; end; end; end
rw1(1,1:m)=U(r,1:m); rw2(1,1:m)=U(M,1:m);
U(M,1:m)=rw1;U(r,1:m)=rw2 ; cl1(1:n,1)= U(1:n,c);
cl2(1 : n, 1) = U(1 : n, M); U(1 : n, M) = cl1(1 : n, 1);
U(1 : n, c) = cl2(1 : n, 1); p = N(M); N(M) = N(c);
N(c) = p; w(M, 1 : n) = N;
for k = M + 1 : n; e = U(k, M)/U(M, M);
for j = M : m; U(k, j) = U(k, j) - e * U(M, j); end; end
i = n; x(i, 1) = U(i, m)/U(i, i);
for i = n - 1 : -1 : 1; s = 0;
for k = n : -1 : i + 1; s = s + U(i, k) * x(k, 1);end
x(i, 1) = (U(i, m) - s)/U(i, i); end
for i=1:n; X(N(i), 1) = x(i, 1); end; B; U; X; end
```

MATLAB commands can be used to get the same results as we obtained in Example 3.11 with the following command:

```
>> B = [2 2 -2 8; -4 -2 2 -14; -2 3 9 9];
>> x = TP(B);
>> disp(x)
```

Total pivoting offers little advantage over partial pivoting and is significantly slower, requiring $N = \dfrac{n(n+1)(2n+1)}{6}$ elements to be examined in total. It is rarely used in practice because interchanging columns changes the order of the xs and, consequently, adds significant and usually unjustified complexity to the computer program. So for getting good results, partial pivoting has proven to be a very reliable procedure.

3.4.4 Gauss-Jordan Method

This method is a modification of the Gaussian elimination method. The Gauss-Jordan method, however, is inefficient for practical calculation, but is often useful for theoretical purposes. The basis of this method is to convert the given matrix into diagonal form. The forward elimination of the Gauss-Jordan method is identical to that of the Gaussian elimination method. However, Gauss-Jordan elimination uses backward elimination rather than backward substitution. In the Gauss-Jordan method the forward elimination and backward elimination need not be separated. This is possible because a pivot element can be used to eliminate the coefficients not only below but also above at the same time. If this approach is taken, the form of the coefficients matrix becomes diagonal when elimination by the last pivot is completed. The Gauss-Jordan method simply yields a transformation of the augmented matrix of the form

$$[A|\mathbf{b}] \rightarrow [\mathbf{I}|\mathbf{c}],$$

where $\mathbf{I}$ is the identity matrix and $\mathbf{c}$ is the column matrix, which represents the possible solution of the given linear system.

Example 3.12 *Solve the following linear system using the Gauss-Jordan method*

$$\begin{array}{rcrcrcr} x_1 & + & 2x_2 & & & = & 3 \\ -x_1 & & & - & 2x_3 & = & -5 \\ -3x_1 & - & 5x_2 & + & x_3 & = & -4. \end{array}$$

Solution. *Write the given system in augmented matrix form*

$$\left(\begin{array}{rrrcr} 1 & 2 & 0 & \vdots & 3 \\ -1 & 0 & -2 & \vdots & -5 \\ -3 & -5 & 1 & \vdots & -4 \end{array}\right).$$

The first elimination step is to eliminate the elements $a_{21} = -1$ and $a_{31} = -3$ by subtracting the multiples $m_{21} = -1$ and $m_{31} = -3$ of row 1 from rows 2 and

3, respectively, which gives

$$\begin{pmatrix} 1 & 2 & 0 & \vdots & 3 \\ 0 & 2 & -2 & \vdots & -2 \\ 0 & 1 & 1 & \vdots & 5 \end{pmatrix}.$$

The second row is now divided by 2 to give

$$\begin{pmatrix} 1 & 2 & 0 & \vdots & 3 \\ 0 & 1 & -1 & \vdots & -1 \\ 0 & 1 & 1 & \vdots & 5 \end{pmatrix}.$$

The second elimination step is to eliminate the elements in positions $a_{12}^{(1)} = 2$ *and* $a_{32} = 1$ *by subtracting the multiples* $m_{12} = 2$ *and* $m_{32} = 1$ *of row 2 from rows 1 and 3, respectively, which gives*

$$\begin{pmatrix} 1 & 0 & 2 & \vdots & 5 \\ 0 & 1 & -1 & \vdots & -1 \\ 0 & 0 & 2 & \vdots & 6 \end{pmatrix}.$$

The third row is now divided by 2 to give

$$\begin{pmatrix} 1 & 0 & 2 & \vdots & 5 \\ 0 & 1 & -1 & \vdots & -1 \\ 0 & 0 & 1 & \vdots & 3 \end{pmatrix}.$$

The third elimination step is to eliminate the elements in positions $a_{23}^{(1)} = -1$ *and* $a_{13} = 2$ *by subtracting the multiples* $m_{23} = -1$ *and* $m_{13} = 2$ *of row 3 from rows 2 and 1, respectively, which gives*

$$\begin{pmatrix} 1 & 0 & 0 & \vdots & -1 \\ 0 & 1 & 0 & \vdots & 2 \\ 0 & 0 & 1 & \vdots & 3 \end{pmatrix}.$$

Obviously, the original set of equations has been transformed to a diagonal form. Now expressing the set in algebraic form yields

$$\begin{aligned} x_1 &= -1 \\ x_2 &= 2 \\ x_3 &= 3, \end{aligned}$$

which is the required solution of the given system.

The above results can be obtained using MATLAB commands. We do the following:

```
>> Ab = [A|b] = [1 2 0 3; -1 0 -2 -5; -3 -5 1 -4];
>> GaussJ(Ab);
```

Program 3.5

MATLAB m-file for the Gauss-Jordan Method

```
function sol=GaussJ(Ab)
[m,n]=size(Ab);
for i=1:m
Ab(i,:) = Ab(i,:)/Ab(i,i);
for j=1:m
if j == i; continue; end
Ab(j,:) = Ab(j,:) - Ab(j,i) * Ab(i,:);
end; end; sol=Ab;
```

Procedure 3.4 [Gauss-Jordan Method]

1. *Form the augmented matrix,* $[A|\mathbf{b}]$.
2. *Reduce the coefficient matrix A to unit upper-triangular form using the Gaussian procedure.*
3. *Use the nth row to reduce the nth column to an equivalent identity matrix column.*
4. *Repeat step 3 for n-1 through 2 to get the the augmented matrix of the form* $[\mathbf{I}|\mathbf{c}]$.
5. *Solve for the unknown* $x_i = c_i$ *(and) for* $i = 1, 2, \ldots, n$.

The number of multiplications and divisions required for the Gauss-Jordan method is approximately

$$N = \left(\frac{n^3}{2}\right) - n^2 - \left(\frac{n}{2}\right),$$

which is approximately 50% larger than for the Gaussian elimination method. Consequently, the Gaussian elimination method is preferred. The Gauss-Jordan method is particularly well suited to compute the inverse of a matrix through the transformation

$$[A|\mathbf{I}] \to [\mathbf{I}|A^{-1}].$$

Note that if the inverse of the matrix can be found, then the solution of the linear system can be computed easily from the product of matrix A^{-1} and column matrix $\mathbf{b}$; that is,

$$\mathbf{x} = A^{-1}\mathbf{b}. \tag{3.25}$$

Example 3.13 *Apply the Gauss-Jordan method to find the inverse of the following matrix*

$$A = \begin{pmatrix} 10 & 1 & -5 \\ -20 & 3 & 20 \\ 5 & 3 & 5 \end{pmatrix}$$

and then solve the system, with $\mathbf{b} = [1, 2, 6]^T$.

Solution. *Consider the following augmented matrix*

$$[A|I] = \begin{pmatrix} 10 & 1 & -5 & \vdots & 1 & 0 & 0 \\ -20 & 3 & 20 & \vdots & 0 & 1 & 0 \\ 5 & 3 & 5 & \vdots & 0 & 0 & 1 \end{pmatrix}.$$

Divide the first row by 10, which gives

$$= \begin{pmatrix} 1 & 0.1 & -0.5 & \vdots & 0.1 & 0 & 0 \\ -20 & 3 & 20 & \vdots & 0 & 1 & 0 \\ 5 & 3 & 5 & \vdots & 0 & 0 & 1 \end{pmatrix}.$$

The first elimination step is to eliminate the elements in positions $a_{21} = -20$ *and* $a_{31} = 5$ *by subtracting the multiples* $m_{21} = -20$ *and* $m_{31} = 5$ *of row 1 from rows 2 and 3, respectively, which gives*

$$= \begin{pmatrix} 1 & 0.1 & -0.5 & \vdots & 0.1 & 0 & 0 \\ 0 & 5 & 10 & \vdots & 2 & 1 & 0 \\ 0 & 2.5 & 7.5 & \vdots & -0.5 & 0 & 1 \end{pmatrix}.$$

Divide the second row by 5, which gives

$$= \begin{pmatrix} 1 & 0.1 & -0.5 & \vdots & 0.1 & 0 & 0 \\ 0 & 1 & 2 & \vdots & 0.4 & 0.2 & 0 \\ 0 & 2.5 & 7.5 & \vdots & -0.5 & 0 & 1 \end{pmatrix}.$$

The second elimination step is to eliminate the elements in positions $a_{12} = 0.1$ *and* $a_{32}^{(1)} = 2.5$ *by subtracting the multiples* $m_{12} = 0.1$ *and* $m_{32} = 2.5$ *of row 2 from rows 1 and 3 respectively, which gives*

$$= \begin{pmatrix} 1 & 0 & -0.7 & \vdots & 0.06 & -0.02 & 0 \\ 0 & 1 & 2 & \vdots & 0.4 & 0.2 & 0 \\ 0 & 0 & 2.5 & \vdots & -1.5 & -0.5 & 1 \end{pmatrix}.$$

Divide the third row by 2.5, which gives

$$= \begin{pmatrix} 1 & 0 & -0.7 & \vdots & 0.06 & -0.02 & 0 \\ 0 & 1 & 2 & \vdots & 0.4 & 0.2 & 0 \\ 0 & 0 & 1 & \vdots & -0.6 & -0.2 & 0.4 \end{pmatrix}.$$

The third elimination step is to eliminate the elements in positions $a_{23}^{(2)} = 2$ *and* $a_{13}^{(2)} = -0.7$ *by subtracting the multiples* $m_{23} = 2$ *and* $m_{13} = -0.7$ *of row 3 from rows 2 and 1 respectively, which gives*

$$= \begin{pmatrix} 1 & 0 & 0 & \vdots & -0.36 & -0.16 & 0.28 \\ 0 & 1 & 0 & \vdots & 1.6 & 0.6 & -0.8 \\ 0 & 0 & 1 & \vdots & -0.6 & -0.2 & 0.4 \end{pmatrix} = [\mathbf{I}|A^{-1}].$$

Obviously, the original augmented matrix $[A|I]$ has been transformed to the augmented matrix of the form $[I|A^{-1}]$. Hence, the solution of the linear system can be obtained by matrix multiplication (3.25) as

$$\begin{pmatrix} x_1 \\ x_2 \\ x_3 \end{pmatrix} = \begin{pmatrix} -0.36 & -0.16 & 0.28 \\ 1.6 & 0.6 & -0.8 \\ -0.6 & -0.2 & 0.4 \end{pmatrix} \begin{pmatrix} 1 \\ 2 \\ 6 \end{pmatrix} = \begin{pmatrix} 1 \\ -2 \\ 1.4 \end{pmatrix}.$$

Hence, $\mathbf{x}^* = [1, -2, 1.4]^T$ *is the solution of the given system.*

The above results can be obtained using MATLAB commands. We do the following:

```
>> Ab = [A|I] = [10 1 -5 1 0 0; -20 3 20 0 1 0; 5 3 5 0 0 1];
>> [I|inv(A)] = GaussJ(Ab);
>> b = [1 2 6]';
>> x = inv(A) * b;
```

3.4.5 LU Decomposition Method

This is another direct method to find the solution of a system of linear equations. LU decomposition (or the factorization method) is a modification of the elimination method. Here, we decompose or factorize the coefficient matrix A into the product of two triangular matrices in the form

$$A = LU, \tag{3.26}$$

where L is a lower-triangular matrix and U is the upper-triangular matrix. Both are the same size as coefficients matrix A. When solving a number of linear equations sets in which the coefficients matrices are all identical but the right-hand side is different, LU decomposition is more efficient than the elimination method. Specifying the diagonal elements of either L and U makes the factoring unique. The procedure based on unity elements on the diagonal of matrix L is called *Doolittle's method* (or Gauss factorization), while the procedure based on unity elements on the diagonal of matrix U is called *Crout's method.* Another method, called the *Cholesky method,* is based on the constraint that the diagonal elements of L are equal to the diagonal elements of U; that is, $l_{ii} = u_{ii}$, for $i = 1, 2, \ldots, n$.

The general forms of L and U are written as

$$L = \begin{pmatrix} l_{11} & 0 & \cdots & 0 \\ l_{21} & l_{22} & \cdots & 0 \\ \vdots & \vdots & \vdots & \vdots \\ l_{n1} & l_{n2} & \cdots & l_{nn} \end{pmatrix}, \quad U = \begin{pmatrix} u_{11} & u_{12} & \cdots & u_{1n} \\ 0 & u_{22} & \cdots & u_{2n} \\ \vdots & \vdots & \vdots & \vdots \\ 0 & 0 & \cdots & u_{nn} \end{pmatrix} \tag{3.27}$$

such that $l_{ij} = 0$ for $i < j$ and $u_{ij} = 0$ for $i > j$.

Consider a linear system

$$A\mathbf{x} = \mathbf{b} \tag{3.28}$$

and let A be factored into the product of L and U, as shown by (3.27). Then linear system (3.28) becomes

$$LU\mathbf{x} = \mathbf{b}$$

or can be written as

$$L\mathbf{y} = \mathbf{b}, \quad \text{where} \quad \mathbf{y} = U\mathbf{x}.$$

The unknown elements of matrix L and matrix U are computed by equating corresponding elements in matrices A and LU in a systematic way. Once matrices L and U have been constructed, the solution of system (3.28) can be computed in the following two steps:

1. Solve the system $L\mathbf{y} = \mathbf{b}$.

Using *forward elimination*, we will find the components of the unknown vector $\mathbf{y}$ using the following steps:

$$\left.\begin{aligned} y_1 &= b_1, \\ y_i &= b_i - \sum_{j=1}^{i-1} l_{ij} y_j, \qquad i = 2, 3, \ldots, n. \end{aligned}\right\} \tag{3.29}$$

2. Solve the system $U\mathbf{x} = \mathbf{y}$.

Using *backward substitution*, we will find the components of unknown vector $\mathbf{x}$ using the following steps:

$$\left.\begin{aligned} x_n &= \frac{y_n}{u_{nn}}, \\ x_i &= \frac{1}{u_{ii}}\left[y_i - \sum_{j=i+1}^{n} u_{ij} x_j \right], \quad i = n-1, n-2, \ldots, 1 \end{aligned}\right\}. \tag{3.30}$$

Thus, the relationship of matrices L and U to the original matrix A is given by the following theorem.

Theorem 3.10 *If Gaussian elimination can be performed on the linear system* $A\mathbf{x} = \mathbf{b}$ *without row interchanges, then matrix A can be factored into the product of a lower-triangular matrix L and an upper-triangular matrix U; that is,*

$$A = LU.$$

Now we discuss all the three possible variations of LU decomposition.

Doolittle's Method

In Doolittle's method (also called Gauss factorization), the upper-triangular matrix U is obtained by forward elimination of the Gaussian elimination method and the lower-triangular matrix L containing the multiples used in the Gaussian elimination process as the elements below the diagonal with unity elements on the main diagonal.

For matrix A in Example 3.3, we can have the decomposition of matrix A in the form

$$\begin{pmatrix} 1 & 2 & 0 \\ -1 & 0 & -2 \\ -3 & -5 & 1 \end{pmatrix} = \begin{pmatrix} 1 & 0 & 0 \\ -1 & 1 & 0 \\ -3 & \frac{1}{2} & 1 \end{pmatrix} \begin{pmatrix} 1 & 2 & 0 \\ 0 & 2 & -2 \\ 0 & 0 & 2 \end{pmatrix},$$

where the unknown elements of matrix L are the used multiples and matrix U is the same as we obtained in the forward elimination process.

Example 3.14 *Construct the LU decomposition of the following matrix A using Doolittle's method. Find the value(s) of α for which the following matrix A is singular*

$$A = \begin{pmatrix} 1 & -1 & \alpha \\ -1 & 2 & -\alpha \\ \alpha & 1 & 1 \end{pmatrix}.$$

Solution. *We know that*

$$A = \begin{pmatrix} 1 & -1 & \alpha \\ -1 & 2 & -\alpha \\ \alpha & 1 & 1 \end{pmatrix} = \begin{pmatrix} 1 & 0 & 0 \\ m_{21} & 1 & 0 \\ m_{31} & m_{32} & 1 \end{pmatrix} \begin{pmatrix} u_{11} & u_{12} & u_{13} \\ 0 & u_{22} & u_{23} \\ 0 & 0 & u_{33} \end{pmatrix} = LU.$$

Now we will use only the forward elimination step of the simple Gaussian elimination method to convert the given matrix A into the upper-triangular matrix U. Since $a_{11} = 1 \neq 0$, we wish to eliminate the elements $a_{21} = -1$ and $a_{31} = \alpha$ by subtracting from the second and third rows the appropriate multiples of the first row. In this case, the multiples are given

$$m_{21} = \frac{-1}{1} = -1, \quad \textit{and} \quad m_{31} = \frac{\alpha}{1} = \alpha.$$

Hence,

$$\begin{pmatrix} 1 & -1 & \alpha \\ 0 & 1 & 0 \\ 0 & 1+\alpha & 1-\alpha^2 \end{pmatrix}.$$

As $a_{22}^{(1)} = 1 \neq 0$, we eliminate entry in the $a_{32}^{(1)} = 1 + \alpha$ position by subtracting the multiple $m_{32} = \dfrac{1+\alpha}{1}$ of the second row from the third row, to get

$$\begin{pmatrix} 1 & -1 & \alpha \\ 0 & 1 & 0 \\ 0 & 0 & 1-\alpha^2 \end{pmatrix}.$$

Obviously, the original set of equations has been transformed to an upper-triangular form. Thus,

$$\begin{pmatrix} 1 & -1 & \alpha \\ -1 & 2 & -\alpha \\ \alpha & 1 & 1 \end{pmatrix} = \begin{pmatrix} 1 & 0 & 0 \\ -1 & 1 & 0 \\ \alpha & 1+\alpha & 1 \end{pmatrix} \begin{pmatrix} 1 & -1 & \alpha \\ 0 & 1 & 0 \\ 0 & 0 & 1-\alpha^2 \end{pmatrix}.$$

which is the required decomposition of matrix A. The given matrix will be singular if the third diagonal element $1 - \alpha^2$ of the upper-triangular U is equal to zero (by Theorem 3.8); that is, $\alpha = \pm 1$.

We can write a MATLAB m-file to factor a nonsingular matrix A into a unit lower triangular matrix L and an upper triangular matrix U using the *lu – gauss* function. The following MATLAB commands can be used to reproduce the solution of the linear system of Example 3.3 as:

```
>> A = [3 -1 1; -1 3 1; 1 1 3];
>> A = lu - gauss(A);
>> L = eye(size(A)) + tril(A, -1);
>> U = triu(A);
>> b = [3 -5 -4]';
>> y = L \ b;
>> x = U \ y;
```

Program 3.6

MATLAB m-file for the LU decomposition Method

```
function A = lu - gauss(A)
% LU factorization using Gauss Elimination (without pivoting)
[n,n] = size(A); for i=1:n-1; pivot = A(i,i);
for k=i+1:n; A(k,i)=A(k,i)/pivot;
for j=i+1:n; A(k,j)= A(k,j) - A(k,i)*A(i,j); end; end; end
```

There is an other way to find the values of the unknown elements of matrices L and U, which we describe in the following example.

Example 3.15 *Construct LU decomposition of the following matrix using Doolittle's method*

$$A = \begin{pmatrix} 2 & 4 & 1 \\ -1 & 3 & -2 \\ 2 & -3 & 5 \end{pmatrix}.$$

Solution. *Since*

$$A = LU = \begin{pmatrix} 1 & 0 & 0 \\ l_{21} & 1 & 0 \\ l_{31} & l_{32} & 1 \end{pmatrix} \begin{pmatrix} u_{11} & u_{12} & u_{13} \\ 0 & u_{22} & u_{23} \\ 0 & 0 & u_{33} \end{pmatrix},$$

performing the multiplication on the right-hand side gives

$$\begin{pmatrix} 2 & 4 & 1 \\ -1 & 3 & -2 \\ 2 & -3 & 5 \end{pmatrix} = \begin{pmatrix} u_{11} & u_{12} & u_{13} \\ l_{21}u_{11} & l_{21}u_{12} + u_{22} & l_{21}u_{13} + u_{23} \\ l_{31}u_{11} & l_{31}u_{12} + l_{32}u_{22} & l_{31}u_{13} + l_{32}u_{23} + u_{33} \end{pmatrix}.$$

Then we equate elements of the first column to obtain

$$\begin{array}{rcll} 2 & = & u_{11}, & u_{11} = 2 \\ -1 & = & l_{21}u_{11}, & l_{21} = -\frac{1}{2}. \\ 2 & = & l_{31}u_{11}, & l_{31} = 1 \end{array}$$

Now we equate elements of the second column to obtain

$$\begin{array}{rcll} 4 & = & u_{12}, & u_{12} = 4 \\ 3 & = & l_{21}u_{12} + u_{22}, & u_{22} = 3 - (-\frac{1}{2})(4) = 5 \\ -3 & = & l_{31}u_{12} + l_{32}u_{22}, & l_{32} = \frac{1}{5}[-3 - (1)(4)] = -\frac{7}{5}. \end{array}$$

Finally, we equate elements of the third column to obtain

$$
\begin{array}{rclrcl}
1 & = & u_{13}, & u_{13} & = & 1 \\
-2 & = & l_{21}u_{13} + u_{23}, & u_{23} & = & -2 - (-\frac{1}{2})(1) = -\frac{3}{2} \\
5 & = & l_{31}u_{13} + l_{32}u_{23} + u_{33}, & u_{33} & = & 5 - (1)(1) - (-\frac{7}{5})(-\frac{3}{2}) = \frac{19}{10}.
\end{array}
$$

Thus, the factorization obtained is

$$
\begin{pmatrix} 2 & 4 & 1 \\ -1 & 3 & -2 \\ 2 & -3 & 5 \end{pmatrix} = \begin{pmatrix} 1 & 0 & 0 \\ -\frac{1}{2} & 1 & 0 \\ 1 & -\frac{7}{5} & 1 \end{pmatrix} \begin{pmatrix} 2 & 4 & 1 \\ 0 & 5 & -\frac{3}{2} \\ 0 & 0 & \frac{19}{10} \end{pmatrix}.
$$

The general formula for getting elements of L and U to correspond to coefficient matrix A for a set of n linear equations can be written as

$$
\left.
\begin{array}{rclrl}
u_{ij} & = & a_{ij} - \displaystyle\sum_{k=1}^{i-1} l_{ik}u_{kj}, & 2 & \leq i \leq j \\
l_{ij} & = & \dfrac{1}{u_{ii}}\left[a_{ij} - \displaystyle\sum_{k=1}^{j-1} l_{ik}u_{kj}\right], & i & > j \geq 2 \\
u_{ij} & = & a_{1j}, & i & = 1 \\
l_{ij} & = & \dfrac{a_{i1}}{u_{11}} = \dfrac{a_{i1}}{a_{11}}, & j & = 1
\end{array}
\right\}. \tag{3.31}
$$

Example 3.16 *Solve the following linear system by LU decomposition using Doolittle's method*

$$
A = \begin{pmatrix} 2 & 4 & 1 \\ -1 & 3 & -2 \\ 2 & -3 & 5 \end{pmatrix} \quad \text{and} \quad \mathbf{b} = \begin{pmatrix} 2 \\ 5 \\ -2 \end{pmatrix}.
$$

Solution. *The factorization of coefficient matrix A has been already constructed* *in Example 3.15 as*

$$\begin{pmatrix} 2 & 4 & 1 \\ -1 & 3 & -2 \\ 2 & -3 & 5 \end{pmatrix} = \begin{pmatrix} 1 & 0 & 0 \\ -\frac{1}{2} & 1 & 0 \\ 1 & -\frac{7}{5} & 1 \end{pmatrix} \begin{pmatrix} 2 & 4 & 1 \\ 0 & 5 & -\frac{3}{2} \\ 0 & 0 & \frac{19}{10} \end{pmatrix}.$$

Then solve the first system $L\mathbf{y} = \mathbf{b}$ *for the unknown vector* $\mathbf{y}$*; that is,*

$$\begin{pmatrix} 1 & 0 & 0 \\ -\frac{1}{2} & 1 & 0 \\ 1 & -\frac{7}{5} & 1 \end{pmatrix} \begin{pmatrix} y_1 \\ y_2 \\ y_3 \end{pmatrix} = \begin{pmatrix} 2 \\ 5 \\ -2 \end{pmatrix}.$$

Performing forward substitution yields

$$\begin{array}{rcrcrcrlrcl} y_1 & & & & & = & 2 & \textit{gives} & y_1 & = & 2, \\ -\frac{1}{2}y_1 & + & y_2 & & & = & 5 & \textit{gives} & y_2 & = & 4, \\ y_1 & - & \frac{7}{5}y_2 & + & y_3 & = & -2 & \textit{gives} & y_3 & = & \frac{8}{5}. \end{array}$$

Then solve the second system $U\mathbf{x} = \mathbf{y}$ *for the unknown vector* $\mathbf{x}$*; that is,*

$$\begin{pmatrix} 2 & 4 & 1 \\ 0 & 5 & -\frac{3}{2} \\ 0 & 0 & \frac{19}{10} \end{pmatrix} \begin{pmatrix} x_1 \\ x_2 \\ x_3 \end{pmatrix} = \begin{pmatrix} 2 \\ 4 \\ \frac{8}{5} \end{pmatrix}.$$

Performing backward substitution yields

$$\begin{array}{rcrcrcrlrcl} 2x_1 & + & 4x_2 & + & x_3 & = & 2 & \textit{gives} & x_1 & = & -1.526, \\ & & 5x_2 & - & \frac{3}{2}x_3 & = & 4 & \textit{gives} & x_2 & = & 1.053, \\ & & & & \frac{19}{10}x_3 & = & \frac{8}{5} & \textit{gives} & x_3 & = & 0.842, \end{array}$$

which gives the approximate solution $\mathbf{x}^* = [-1.526, 1.053, 0.842]^T$.

We can also write an MATLAB m-file called Doolittle.m to get the solution of the linear system by LU decomposition using Doolittle's method. In order to reproduce the above results using MATLAB, we do the following:

$$>> A = [2\ 4\ 1; -1\ 3\ -2; 2\ -3\ 5];$$
$$>> b = [2\ 3\ -2];$$
$$>> sol = Doll(A, b);$$

Program 3.7

MATLAB m-file for using Doolittle's Method

```
function sol = Doll(A,b)
[n,n]=size(A); u=A;l=zeros(n,n);
for i=1:n-1; if abs(u(i,i))> 0
for i1=i+1:n; m(i1,i)=u(i1,i)/u(i,i);
for j=1:n
u(i1,j) = u(i1,j) - m(i1,i) * u(i,j);end;end;end;end
for i=1:n; l(i,1)=A(i,1)/u(1,1); end
for j=2:n; for i=2:n; s=0;
for k=1:j-1; s = s + l(i,k) * u(k,j); end
l(i,j)=(A(i,j)-s)/u(j,j); end; end y(1)=b(1)/l(1,1);
for k=2:n; sum=b(k);
for i=1:k-1; sum = sum - l(k,i) * y(i); end
y(k)=sum/l(k,k); end
x(n)=y(n)/u(n,n);
for k=n-1:-1:1; sum=y(k);
for i=k+1:n; sum = sum - u(k,i) * x(i); end
x(k)=sum/u(k,k); end; l; u; y; x
```

Procedure 3.5 [LU Decomposition by Doolittle's Method]

1. *Take the nonsingular matrix A.*
2. *If possible, decompose matrix $A = LU$ using (3.31).*
3. *Solve linear system $L\mathbf{y} = \mathbf{b}$ using (3.29).*
4. *Solve linear system $U\mathbf{x} = \mathbf{y}$ using (3.30).*

Crout's Method

Crout's method, in which matrix U has unity on the main diagonal, is similar to Doolittle's method in all other aspects. The L and U matrices are obtained by expanding matrix equation $A = LU$ term by term to determine the elements of the L and U matrices.

Example 3.17 *Construct LU decomposition of the following matrix using Crout's method*

$$A = \begin{pmatrix} 4 & 2 & 1 \\ 4 & 3 & 2 \\ 1 & 5 & -2 \end{pmatrix}.$$

Solution. *Since*

$$A = LU = \begin{pmatrix} l_{11} & 0 & 0 \\ l_{21} & l_{22} & 0 \\ l_{31} & l_{32} & l_{33} \end{pmatrix} \begin{pmatrix} 1 & u_{12} & u_{13} \\ 0 & 1 & u_{23} \\ 0 & 0 & 1 \end{pmatrix},$$

performing multiplication on the right-hand side gives

$$\begin{pmatrix} 4 & 2 & 1 \\ 4 & 3 & 2 \\ 1 & 5 & -2 \end{pmatrix} = \begin{pmatrix} l_{11} & l_{11}u_{12} & l_{11}u_{13} \\ l_{21} & l_{21}u_{12} + l_{22} & l_{21}u_{13} + l_{22}u_{23} \\ l_{31} & l_{31}u_{12} + l_{32} & l_{31}u_{13} + l_{32}u_{23} + l_{33} \end{pmatrix}.$$

Then equate elements of the first column to obtain

$$\begin{aligned} 4 &= l_{11}, & l_{11} &= 4 \\ 4 &= l_{21}, & l_{21} &= 4\,. \\ 1 &= l_{31}, & l_{31} &= 1 \end{aligned}$$

Then equate elements of the second column to obtain

$$\begin{aligned} 2 &= l_{11}u_{12}, & u_{12} &= \frac{1}{2} = 0.5, \\ 3 &= l_{21}u_{12} + l_{22}, & l_{22} &= 3 - (4)(0.5) = 1, \\ 5 &= l_{31}u_{12} + l_{32}, & l_{32} &= 5 - (1)(0.5) = \frac{9}{2} = 4.5. \end{aligned}$$

Finally, equate elements of the third column to obtain

$$\begin{aligned} 1 &= l_{11}u_{13}, & u_{13} &= \frac{1}{4} = 0.25, \\ 2 &= l_{21}u_{13} + l_{22}u_{23}, & u_{23} &= 2 - (4)(0.25)(1) = 1, \\ -2 &= l_{31}u_{13} + l_{32}u_{23} + l_{33}, & l_{33} &= -2 - (1)(0.25) - (4.5)(1). \\ & & &= -6.75 \end{aligned}$$

Thus, the factorization obtained

$$\begin{pmatrix} 4 & 2 & 1 \\ 4 & 3 & 2 \\ 1 & 5 & -2 \end{pmatrix} = \begin{pmatrix} 4 & 0 & 0 \\ 4 & 1 & 0 \\ 1 & 4.5 & -6.75 \end{pmatrix} \begin{pmatrix} 1 & 0.5 & 0.25 \\ 0 & 1 & 1 \\ 0 & 0 & 1 \end{pmatrix}.$$

The general formula for getting the elements of L and U to correspond to coefficient matrix A for a set of n linear equations can be written as

$$\left.\begin{aligned} l_{ij} &= a_{ij} - \sum_{k=1}^{j-1} l_{ik}u_{kj}, \quad i \geq j, \quad i = 1, 2, \ldots, n \\ u_{ij} &= \frac{1}{l_{ii}}[a_{ij} - \sum_{k=1}^{i-1} l_{ik}u_{kj}], \quad i < j, \quad j = 2, 3, \ldots, n \\ l_{ij} &= a_{i1}, \qquad j = 1 \\ u_{ij} &= \frac{a_{ij}}{a_{11}}, \qquad i = 1 \end{aligned}\right\}. \tag{3.32}$$

Example 3.18 *Solve the following linear system by LU decomposition using Crout's method:*

$$A = \begin{pmatrix} 4 & 2 & 1 \\ 4 & 3 & 2 \\ 1 & 5 & -2 \end{pmatrix} \quad \textit{and} \quad \mathbf{b} = \begin{pmatrix} 3 \\ 4 \\ 6 \end{pmatrix}.$$

Solution. *The factorization of coefficient matrix A has been already constructed in Example (3.17) as*

$$\begin{pmatrix} 4 & 2 & 1 \\ 4 & 3 & 2 \\ 1 & 5 & -2 \end{pmatrix} = \begin{pmatrix} 4 & 0 & 0 \\ 4 & 1 & 0 \\ 1 & 4.5 & -6.75 \end{pmatrix} \begin{pmatrix} 1 & 0.5 & 0.25 \\ 0 & 1 & 1 \\ 0 & 0 & 1 \end{pmatrix}.$$

Then solve the first system $L\mathbf{y} = \mathbf{b}$ for the unknown vector $\mathbf{y}$*; that is,*

$$\begin{pmatrix} 4 & 0 & 0 \\ 4 & 1 & 0 \\ 1 & 4.5 & -6.75 \end{pmatrix} \begin{pmatrix} y_1 \\ y_2 \\ y_3 \end{pmatrix} = \begin{pmatrix} 3 \\ 4 \\ 6 \end{pmatrix}.$$

Performing forward substitution yields

$$\begin{array}{rcrcrcl} 4y_1 & & & & & = & 3 \quad \textit{gives} \quad y_1 = 0.75, \\ 4y_1 & + & y_2 & & & = & 4 \quad \textit{gives} \quad y_2 = 1, \\ y_1 & + & 4.5y_2 & - & 6.75y_3 & = & 6 \quad \textit{gives} \quad y_3 = -0.111. \end{array}$$

Then solve the second system $U\mathbf{x} = \mathbf{y}$ for the unknown vector $\mathbf{x}$*; that is,*

$$\begin{pmatrix} 1 & 0.5 & 0.25 \\ 0 & 1 & 1 \\ 0 & 0 & 1 \end{pmatrix} \begin{pmatrix} x_1 \\ x_2 \\ x_3 \end{pmatrix} = \begin{pmatrix} 0.75 \\ 1 \\ -0.111 \end{pmatrix}.$$

Performing backward substitution yields

$$\begin{array}{rcrcrcrl} x_1 & + & 0.5x_2 & + & 0.25x_3 & = & 0.75 & \textit{gives} \quad x_1 = 0.222, \\ & & x_2 & + & x_3 & = & 1 & \textit{gives} \quad x_2 = 1.111, \\ & & & & x_3 & = & -0.111 & \textit{gives} \quad x_3 = -0.111, \end{array}$$

which gives the approximate solution $\mathbf{x}^* = [0.222, 1.111, -0.111]^T$.

The above results can be reproduced using MATLAB commands as follows:

```
>> A = [4 2 1; 4 3 2; 1 5 - 2];
>> b = [3 4 6];
>> sol = Crout(A, b);
```

Program 3.8

MATLAB m-file for Crout's Method

```
function sol = Crout(A, b)
[n,n]=size(A); u=zeros(n,n); l=u;
for i=1:n; u(i,i)=1; end
l(1,1)=A(1,1);
for i=2:n
u(1,i)=A(1,i)/l(1,1); l(i,1)=A(i,1); end
for i=2:n; for j=2:n; s=0;
if i <= j; K=i-1;
else; K=j-1; end
for k=1:K; s = s + l(i,k) * u(k,j); end
if j > i; u(i,j)=(A(i,j)-s)/l(i,i); else
l(i,j)=A(i,j)-s; end;end;end
y(1)=b(1)/l(1,1);
for k=2:n; sum=b(k);
for i=1:k-1; sum = sum - l(k,i) * y(i); end
y(k)=sum/l(k,k); end
x(n)=y(n)/u(n,n);
for k=n-1:-1:1; sum=y(k);
for i=k+1:n; sum = sum - u(k,i) * x(i); end
x(k)=sum/u(k,k); end; l; u; y; x;
```

Procedure 3.6 [LU Decomposition by Crout's Method]

1. *Take the nonsingular matrix A.*
2. *If possible, decompose matrix $A = LU$ using (3.32).*
3. *Solve linear system $L\mathbf{y} = \mathbf{b}$ using (3.29).*
4. *Solve linear system $U\mathbf{x} = \mathbf{y}$ using (3.30).*

Note that the factorization method is also used to invert matrices. Their usefulness for this purpose is based on the fact that triangular matrices are easily inverted. Once the factorization has been affected, the inverse of matrix A is found from the formula

$$A^{-1} = (LU)^{-1} = U^{-1}L^{-1}. \tag{3.33}$$

Then

$$UA^{-1} = L^{-1}, \qquad \text{where} \qquad LL^{-1} = I.$$

A practical way of calculating the determinant is to use the forward elimination process of the Gaussian elimination or, alternatively, LU decomposition. If no pivoting is used, calculation of the determinant using LU decomposition is very easy. Since by one of the properties of the determinant

$$\det(A) = \det(LU) = \det(L)\det(U).$$

So when using LU decomposition by Doolittle's method, then

$$\det(A) = \det(U) = \prod_{i=1}^{n} u_{ii} = (u_{11}u_{22}\cdots u_{nn}),$$

where $\det(L) = 1$ because L is lower-triangular matrix and all its diagonal elements are unity and for LU decomposition by Crout's method, then

$$\det(A) = \det(L) = \prod_{i=1}^{n} l_{ii} = (l_{11}l_{22}\cdots l_{nn}),$$

where $\det(U) = 1$ because U is upper-triangular matrix and all its diagonal elements are unity.

For LU decomposition we have not used pivoting for the sake of simplicity. However, pivoting is important for the same reason as in Gaussian elimination. We know that pivoting in Gaussian elimination is equivalent to interchanging the rows of coefficients matrices together with the terms on the right-hand side. This indicates that pivoting may be applied to LU decomposition as long as the interchanging is applied to the left and right terms in the same way. When performing pivoting in LU decomposition, the changes in the order of the rows are recorded. The same reordering is then applied to the right-hand side terms before starting the solution in accordance with the forward elimination and backward substitution steps.

Indirect LU Decomposition

It is to be noted that a nonsingular matrix A sometimes cannot be directly factored as $A = LU$. For example, the matrix in Example 3.5 is nonsingular but it cannot be factored into the product LU. Let us assume it has an LU form

and

$$\begin{pmatrix} 2 & 2 & -4 \\ 2 & 2 & -1 \\ 3 & 2 & -3 \end{pmatrix} = \begin{pmatrix} u_{11} & u_{12} & u_{13} \\ l_{21}u_{11} & l_{21}u_{12} + u_{22} & l_{21}u_{13} + u_{23} \\ l_{31}u_{11} & l_{31}u_{12} + l_{32}u_{22} & l_{31}u_{13} + l_{32}u_{23} + u_{33} \end{pmatrix}.$$

Then equate elements of the first column to obtain

$$\begin{array}{rclcrcl} 2 & = & u_{11} & \text{gives} & u_{11} & = & 2, \\ 2 & = & l_{21}u_{11} & \text{gives} & l_{21} & = & 1, \\ 3 & = & l_{31}u_{11} & \text{gives} & l_{31} & = & \frac{3}{2}. \end{array}$$

Then equate elements of the second column to obtain

$$\begin{array}{rclcrcl} 2 & = & u_{12} & \text{gives} & u_{12} & = & 2, \\ 2 & = & l_{21}u_{12} + u_{22} & \text{gives} & u_{22} & = & 0, \\ 2 & = & l_{31}u_{12} + l_{32}u_{22} & \text{gives} & 0 & = & -1, \end{array}$$

which is not possible because $0 \neq -1$, a contradiction. Hence, matrix A cannot be directly factored into the product of L and U. The indirect factorization LU of A can be obtained using the permutation matrix P and replacing matrix A by PA. For example, using the above matrix A, we have

$$PA = \begin{pmatrix} 1 & 0 & 0 \\ 0 & 0 & 1 \\ 0 & 1 & 0 \end{pmatrix} \begin{pmatrix} 2 & 2 & -4 \\ 2 & 2 & -1 \\ 3 & 2 & -3 \end{pmatrix} = \begin{pmatrix} 2 & 2 & -4 \\ 3 & 2 & -3 \\ 2 & 2 & -1 \end{pmatrix}.$$

From this multiplication we see that rows 2 and 3 of the original matrix A are interchanged, and the resulting matrix PA has a LU factorization and we have

$$LU = \begin{pmatrix} 1 & 0 & 0 \\ 1.5 & 1 & 0 \\ 1 & 0 & 1 \end{pmatrix} \begin{pmatrix} 2 & 2 & -4 \\ 0 & -1 & 3 \\ 0 & 0 & 3 \end{pmatrix} = \begin{pmatrix} 2 & 2 & -4 \\ 3 & 2 & -3 \\ 2 & 2 & -1 \end{pmatrix} = A.$$

The following theorem is an extension of Theorem 3.10, which includes the cases when interchanged rows are required. Thus, LU factorization can be used to find the solution to any linear system $A\mathbf{x} = \mathbf{b}$ with a nonsingular matrix A.

Theorem 3.11 *Let A be a square $n \times n$ matrix. Assume that Gaussian elimination can be performed successfully to solve the linear system $A\mathbf{x} = \mathbf{b}$, but that row interchanges are required. Then there exists a permutation matrix P so that the PA matrix has a LU factorization; that is,*

$$PA = LU. \tag{3.34}$$

When **pivoting** is used in LU decomposition, its effects should be taken into consideration. First, we recognize that LU decomposition with pivoting is equivalent to performing two separate process is:

1. Transforming A to A' by performing all shifting of rows.

2. Then decomposing A' to LU with no pivoting.

The former step can be expressed by

$$A' = PA, \quad \text{equivalently} \quad A = P^{-1}A',$$

where P is called a permutation matrix and represents the pivoting operation. The second process is

$$A' = PA = LU$$

and so

$$A = P^{-1}LU = (P^T L)U$$

since $P^{-1} = P^T$. The determinant of A may now be written as

$$\det(A) = \det(P^{-1})\det(L)\det(U)$$

or

$$\det(A) = \beta \det(L)\det(U),$$

where $\beta = \det(P^{-1})$ equals -1 or $+1$ depending on whether the number of pivoting is odd or even, respectively.

One can use the MATLAB built-in *lu* function to obtain the permutation matrix P so that the PA matrix has a LU decomposition:

```
>> A = [0 1 2; -1 4 2; 2 2 1];
>> [L, U, P] = lu(A);
```

It will give us permutation matrix P, matrix L, and matrix U as follows:

$$P = \begin{pmatrix} 0 & 0 & 1 \\ 0 & 1 & 0 \\ 1 & 0 & 0 \end{pmatrix}, \quad PA = \begin{pmatrix} 1 & 0 & 0 \\ -0.5 & 1 & 0 \\ 0 & 0.2 & 1 \end{pmatrix} \begin{pmatrix} 2 & 2 & 1 \\ 0 & 5 & 2.5 \\ 0 & 0 & 1.5 \end{pmatrix} = LU.$$

So

$$A = P^{-1}LU = (P^T L)U = \begin{pmatrix} 0 & 0.2 & 1 \\ -0.5 & 1 & 0 \\ 1 & 0 & 0 \end{pmatrix} \begin{pmatrix} 2 & 2 & 1 \\ 0 & 5 & 2.5 \\ 0 & 0 & 1.5 \end{pmatrix}.$$

The major advantage of the LU decomposition methods is the efficiency when multiple unknown **b** vectors must be considered. The number of multiplications and divisions required by the complete Gaussian elimination method is $N = (\frac{n^3}{3}) + n^2 - (\frac{n}{3})$. The forward substitution step required to solve system $L\mathbf{y} = \mathbf{b}$ requires $N = n^2 - (\frac{n}{2})$ operations, and a backward substitution step is required to solve system $U\mathbf{x} = \mathbf{y}$, which requires $N = n^2 + (\frac{n}{2})$ operations. Thus, the total number of multiplications and divisions required by LU decomposition, after L and U matrices have been determined, is $N = 2n^2$, which is much less work than required by the Gaussian elimination method, especially for large systems.

In the analysis of many physical system sets, linear equations arise that have coefficient matrices that are both symmetric and positive definite. Now we factorize such matrix A into the product of lower-triangular and upper-triangular matrices, which have these two properties. Before we do the factorization, we define the following matrix.

Definition 3.22 (Positive Definite Matrix)

The function

$$\mathbf{x}^T A\mathbf{x} = \begin{pmatrix} x_1 & x_2 & \cdots & x_n \end{pmatrix} \begin{pmatrix} a_{11} & a_{12} & \cdots & a_{1n} \\ a_{21} & a_{22} & \cdots & a_{2n} \\ a_{31} & a_{32} & \cdots & a_{3n} \\ \vdots & \vdots & \vdots & \vdots \\ a_{n1} & a_{n2} & \cdots & a_{nn} \end{pmatrix} \begin{pmatrix} x_1 \\ x_2 \\ \vdots \\ x_n \end{pmatrix}$$

or

$$\mathbf{x}^T A\mathbf{x} = \sum_{i=1}^{n}\sum_{j=1}^{n} a_{ij}x_i x_j \tag{3.35}$$

can be used to represent any quadratic polynomial in the variables $x_1, x_2, \ldots, x_n$ and is called a quadratic form. A matrix is said to be positive definite if its quadratic form is positive for all real nonzero vectors $\mathbf{x}$*; that is,*

$$\mathbf{x}^T A\mathbf{x} > 0, \qquad \text{for every } n\text{-dimensional column vector} \quad \mathbf{x} \neq 0.$$

Example 3.19 *The matrix*

$$A = \begin{pmatrix} 4 & -1 & 0 \\ -1 & 4 & -1 \\ 0 & -1 & 4 \end{pmatrix}$$

is positive definite, for suppose $\mathbf{x}$ *is any nonzero three-dimensional column vector, then*

$$\mathbf{x}^T A\mathbf{x} = \begin{pmatrix} x_1 & x_2 & x_3 \end{pmatrix} \begin{pmatrix} 4 & -1 & 0 \\ -1 & 4 & -1 \\ 0 & -1 & 4 \end{pmatrix} \begin{pmatrix} x_1 \\ x_2 \\ x_3 \end{pmatrix}$$

or

$$= \begin{pmatrix} x_1 & x_2 & x_3 \end{pmatrix} \begin{pmatrix} 4x_1 - x_2 \\ -x_1 + 4x_2 - x_3 \\ - x_2 + 4x_3 \end{pmatrix}.$$

Thus,

$$\mathbf{x}^T A\mathbf{x} = 4x_1^2 - 2x_1x_2 + 4x_2^2 - 2x_2x_3 + 4x_3^2.$$

After rearranging the terms, we have

$$\mathbf{x}^T A\mathbf{x} = 3x_1^2 + (x_1 - x_2)^2 + 2x_2^2 + (x_2 - x_3)^2 + 3x_3^2.$$

Hence,

$$3x_1^2 + (x_1 - x_2)^2 + 2x_2^2 + (x_2 - x_3)^2 + 3x_3^2 > 0$$

unless $x_1 = x_2 = x_3 = 0.$

Symmetric positive definite matrices occur frequently in equations derived by minimization or energy principles and their properties can often be utilized in numerical processes.

Theorem 3.12 *If A is positive definite matrix, then*

1. *A is nonsingular.*
2. $a_{ii} > 0, \qquad$ *for each* $\qquad i = 1, 2, \ldots, n.$

Theorem 3.13 *Symmetric matrix A is a positive definite matrix if and only if Gaussian elimination without row interchange can be performed on the linear system* $A\mathbf{x} = \mathbf{b}$ *with all pivot elements positive.*

Cholesky Method

The Cholesky method (or square root method) is of the same form as Doolittle's method and Crout's method except it is limited to equations involving symmetrical coefficient matrices. In the case of a symmetric and positive definite matrix A, it is possible to construct an alternative triangular factorization which saves numerous calculations as compared with previous factorization. Here we decompose the matrix A into the product of LL^T; that is,

$$A = LL^T, \tag{3.36}$$

where L is the lower-triangular matrix and L^T is its transpose. The elements of L are computed by equating successive columns in the relation

$$\begin{pmatrix} a_{11} & a_{12} & \cdots & a_{1n} \\ a_{21} & a_{22} & \cdots & a_{2n} \\ \vdots & \vdots & \vdots & \vdots \\ a_{n1} & a_{n2} & \cdots & a_{nn} \end{pmatrix} = \begin{pmatrix} l_{11} & 0 & \cdots & 0 \\ l_{21} & l_{22} & \cdots & 0 \\ \vdots & \vdots & \vdots & \vdots \\ l_{n1} & l_{n2} & \cdots & l_{nn} \end{pmatrix} \begin{pmatrix} l_{11} & l_{21} & \cdots & l_{n1} \\ 0 & l_{22} & \cdots & l_{n2} \\ \vdots & \vdots & \vdots & \vdots \\ 0 & 0 & \cdots & l_{nn} \end{pmatrix}.$$

After constructing matrices L and L^T, the solution of the system $A\mathbf{x} = \mathbf{b}$ can be computed in the following two steps:

1. Solve $\quad L\mathbf{y} = \mathbf{b}, \quad$ for $\quad \mathbf{y}$.
(using forward substitution)

2. Solve $\quad L^T\mathbf{x} = \mathbf{y}, \quad$ for $\quad \mathbf{x}$.
(using backward substitution)

In this procedure it is necessary to take the square root of the elements on the main diagonal of the coefficient matrix. However, for a positive definite matrix the terms on its main diagonal are positive, and so no difficulty will arise when taking the square root of these terms.

If the symmetric coefficient matrix is not positive definite then the terms on the main diagonal can be zero or negative.

Example 3.20 *Construct the LU decomposition of the following matrix using the Cholesky method*

$$A = \begin{pmatrix} 3 & -1 & 1 \\ -1 & 3 & 1 \\ 1 & 1 & 3 \end{pmatrix}.$$

Solution. *Since*

$$A = LL^T = \begin{pmatrix} l_{11} & 0 & 0 \\ l_{21} & l_{22} & 0 \\ l_{31} & l_{32} & l_{33} \end{pmatrix} \begin{pmatrix} l_{11} & l_{21} & l_{31} \\ 0 & l_{22} & l_{32} \\ 0 & 0 & l_{33} \end{pmatrix},$$

performing the multiplication on the right-hand side gives

$$\begin{pmatrix} 3 & -1 & 1 \\ -1 & 3 & 1 \\ 1 & 1 & 3 \end{pmatrix} = \begin{pmatrix} l_{11}^2 & l_{11}l_{21} & l_{11}l_{31} \\ l_{11}l_{21} & l_{21}^2 + l_{22}^2 & l_{21}l_{31} + l_{22}l_{32} \\ l_{11}l_{31} & l_{31}l_{21} + l_{22}l_{32} & l_{31}^2 + l_{32}^2 + l_{33}^2 \end{pmatrix}.$$

Then equate the elements of the first column to obtain

$$\begin{aligned} 3 &= l_{11}^2 \quad \textit{gives} \quad l_{11} = \sqrt{3} = 1.732. \\ -1 &= l_{11}l_{21} \quad \textit{gives} \quad l_{21} = \frac{-1}{\sqrt{3}} = -0.577. \\ 1 &= l_{11}l_{31} \quad \textit{gives} \quad l_{31} = \frac{1}{\sqrt{3}} = 0.577. \end{aligned}$$

Note that l_{11} could be $-\sqrt{3}$ and so matrix L is not (quite) unique.
Now equating elements of second column to obtain

$$\begin{aligned} 3 &= l_{21}^2 + l_{22}^2 \quad \textit{gives} \quad l_{22} = 1.632. \\ 1 &= l_{31}l_{21} + l_{32}l_{22} \quad \textit{gives} \quad l_{32} = (1.224)[1 + (0.577)^2] = 0.817. \end{aligned}$$

Finally, equating elements of the third column to obtain

$$3 = l_{31}^2 + l_{32}^2 + l_{33}^2 \quad \textit{gives} \quad l_{33} = 1.414.$$

Thus, the factorization obtained

$$\begin{pmatrix} 3 & -1 & 1 \\ -1 & 3 & 1 \\ 1 & 1 & 3 \end{pmatrix} = \begin{pmatrix} 1.732 & 0 & 0 \\ -0.577 & 1.632 & 0 \\ 0.577 & 0.817 & 1.414 \end{pmatrix} \begin{pmatrix} 1.732 & -0.577 & 0.577 \\ 0 & 1.632 & 0.817 \\ 0 & 0 & 1.414 \end{pmatrix}.$$

For the general $n \times n$ matrix, the elements of lower-triangular matrix L are constructed from

$$
\left.
\begin{array}{rcl}
l_{11} & = & \sqrt{a_{11}} \\
l_{j1} & = & \dfrac{a_{j1}}{l_{11}}, \qquad j > 1 \\
l_{ii} & = & \sqrt{\left(a_{ii} - \displaystyle\sum_{k=1}^{i-1} l_{ik}^2\right)}, \qquad 1 < i < n \\
l_{ji} & = & \dfrac{1}{l_{ii}}\left[a_{ji} - \displaystyle\sum_{k=1}^{i-1} l_{jk}l_{ik}\right], \qquad j > i > 1 \\
l_{nn} & = & \sqrt{\left(a_{nn} - \displaystyle\sum_{k=1}^{n-1} l_{nk}^2\right)}
\end{array}
\right\}. \tag{3.37}
$$

The method fails if $l_{jj} = 0$ and the expression inside the square root is negative, in which case all of the elements in column j are purely imaginary. There is, however, a special class of matrices for which these problems don't occur.

The Cholesky method provides a convenient method for investigating the positive definiteness of symmetric matrices. The formal definition $\mathbf{x}^T A\mathbf{x} > 0$, for all $\mathbf{x} \neq 0$, is not easy to verify in practice. However, it is relatively straightforward to attempt the construct of a Cholesky decomposition of a symmetric matrix.

Theorem 3.14 *Matrix A is positive definite if and only if A can be factored in the form $A = LL^T$, where L is lower-triangular matrix with nonzero diagonal entries.*

Example 3.21 *Solve the following linear system by LU decomposition using the Cholesky method*

$$
A = \begin{pmatrix} 3 & -1 & 1 \\ -1 & 3 & 1 \\ 1 & 1 & 3 \end{pmatrix} \quad \textit{and} \quad \mathbf{b} = \begin{pmatrix} 4 \\ -1 \\ 2 \end{pmatrix}.
$$

Solution. *Since the factorization of coefficient matrix A has been already con-*

structed in Example (3.20) as

$$\begin{pmatrix} 3 & -1 & 1 \\ -1 & 3 & 1 \\ 1 & 1 & 3 \end{pmatrix} = \begin{pmatrix} 1.732 & 0 & 0 \\ -0.577 & 1.632 & 0 \\ 0.577 & 0.817 & 1.414 \end{pmatrix} \begin{pmatrix} 1.732 & -0.577 & 0.577 \\ 0 & 1.632 & 0.817 \\ 0 & 0 & 1.414 \end{pmatrix},$$

then solve the first system $L\mathbf{y} = \mathbf{b}$ *for unknown vector* $\mathbf{y}$*; that is,*

$$\begin{pmatrix} 1.732 & 0 & 0 \\ -0.577 & 1.632 & 0 \\ 0.577 & 0.817 & 1.414 \end{pmatrix} \begin{pmatrix} y_1 \\ y_2 \\ y_3 \end{pmatrix} = \begin{pmatrix} 4 \\ -1 \\ 2 \end{pmatrix}.$$

Performing forward substitution yields

$$\begin{array}{rcrcrcrl} 1.732y_1 & & & & & = & 4, & y_1 = 2.310 \\ -0.577y_1 & + & 1.632y_2 & & & = & -1, & y_2 = 0.204 \\ 0.577y_1 & + & 0.817y_2 & + & 1.414y_3 & = & 2, & y_3 = 0.354 \end{array}.$$

Then solve the second system $L^T\mathbf{x} = \mathbf{y}$ *for the unknown vector* $\mathbf{x}$*; that is,*

$$\begin{pmatrix} 1.732 & -0.577 & 0.577 \\ 0 & 1.632 & 0.817 \\ 0 & 0 & 1.414 \end{pmatrix} \begin{pmatrix} x_1 \\ x_2 \\ x_3 \end{pmatrix} = \begin{pmatrix} 2.310 \\ 0.204 \\ 0.354 \end{pmatrix}.$$

Performing backward substitution yields

$$\begin{array}{rcrcrcrl} 1.732x_1 & - & 0.577x_2 & + & 0.577x_3 & = & 2.310 & \textit{gives} \ x_1 = 1.250, \\ & & 1.632x_2 & + & 0.817x_3 & = & 0.204 & \textit{gives} \ x_2 = 0, \\ & & & & 1.414x_3 & = & 0.354 & \textit{gives} \ x_3 = 0.250, \end{array}$$

which gives the approximate solution $\mathbf{x}^* = [1.250, 0, 0.250]^T$.

Now use the following MATLAB commands to obtain the above results:

```
>> A = [3  -1 1; -1 3 1; 1 1 3];
>> b = [4  -1 2];
>> sol = Cholesky(A, b);
```

Program 3.9

MATLAB m-file for the Cholesky Method

```
function sol = Cholesky(A, b)
[n,n]=size(A); l=zeros(n,n); u=l;
l(1,1) = (A(1,1))^ 0.5; u(1,1) = l(1,1);
for i=2:n; u(1,i)=A(1,i)/l(1,1);
l(i,1)=A(i,1)/u(1,1); end
for i=2:n; for j=2:n; s=0;
if i <= j; K=i-1; else; K=j-1; end
for k=1:K; s = s + l(i,k) * u(k,j); end
if j > i; u(i,j)=(A(i,j)-s)/l(i,i);
elseif i == j
l(i,j) = (A(i,j) - s) \ 0.5; u(i,j)=l(i,j);
else; l(i,j)=(A(i,j)-s)/u(j,j); end; end; end
y(1)=b(1)/l(1,1);
for k=2:n; sum=b(k);
for i=1:k-1; sum = sum - l(k,i) * y(i); end
y(k)=sum/l(k,k); end
x(n)=y(n)/u(n,n);
for k=n-1:-1:1; sum=y(k);
for i=k+1:n; sum = sum - u(k,i) * x(i); end
x(k)=sum/u(k,k); end; l; u; y; x;
```

Procedure 3.7 [LU Decomposition by Cholesky Method]

1. *Take the positive definite matrix A.*
2. *If possible, decompose matrix $A = LL^T$ using (3.37).*
3. *Solve linear system $\mathbf{Ly} = \mathbf{b}$ using (3.29).*
4. *Solve linear system $L^T\mathbf{x} = \mathbf{y}$ using (3.30).*

Since we know that not every matrix has a direct LU decomposition, we define the following matrix, which gives the sufficient condition for the LU decomposition of the matrix. It also helps us for the convergence of the iterative methods for solving linear systems.

Definition 3.23 (Strictly Diagonally Dominant Matrix)

A square matrix is said to be strictly diagonally dominant (SDD) if the absolute value of each element on the main diagonal is greater than the sum of the absolute values of all the other elements in that row. Thus, a strictly diagonally dominant matrix is defined as

$$|a_{ii}| > \sum_{\substack{j=1 \\ j\neq i}}^{n} |a_{ij}|, \qquad \textit{for} \quad i = 1, 2, \ldots, n. \tag{3.38}$$

Example 3.22 *The matrix*

$$A = \begin{pmatrix} 5 & 2 & 1 \\ -1 & 6 & -2 \\ 2 & -3 & 8 \end{pmatrix}$$

is strictly diagonally dominant since

$$\begin{array}{lcll} |5| & > & |2| + |1|, & \textit{that is} \quad 5 \ > \ 3 \\ |6| & > & |-1| + |-2|, & \textit{that is} \quad 6 \ > \ 3 \\ |8| & > & |2| + |-3|, & \textit{that is} \quad 8 \ > \ 5 \end{array}$$

but the following matrix

$$B = \begin{pmatrix} 5 & -4 & 1 \\ -3 & 7 & -2 \\ 4 & -3 & 10 \end{pmatrix}$$

is not strictly diagonally dominant since

$$|5| > |-4| + |1|, \quad \textit{that is} \quad 5 > 5,$$

which is not true.

Theorem 3.15 *If matrix A is strictly diagonally dominant, then A is nonsingular. Moreover, Gaussian elimination without row interchange can be performed on the linear system* $A\mathbf{x} = \mathbf{b}$.

3.4.6 Tridiagonal Systems of linear equations

The application of numerical methods to the solution of certain engineering problems may in some cases result in a set of tridiagonal linear algebraic equations.

Heat conduction and fluid flow problems are some of the many applications which generate such system.

A tridiagonal system has a coefficients matrix T of which all elements except those on the main diagonal and the two diagonals just above and below the main diagonal (usually called superdiagonal and subdiagonal, respectively) and is defined as

$$T = \begin{pmatrix} \alpha_1 & c_1 & 0 & \cdots & 0 \\ \beta_2 & \alpha_2 & c_2 & \ddots & 0 \\ 0 & \beta_3 & a_3 & \ddots & 0 \\ \vdots & \ddots & \ddots & \ddots & c_{n-1} \\ 0 & 0 & 0 & \beta_n & \alpha_n \end{pmatrix}. \tag{3.39}$$

This type of matrix can be stored more economically (that is the case for a fully populated matrix). Obviously, one may use any of the methods discussed in the previous sections for solving tridiagonal system

$$T\mathbf{x} = \mathbf{b} \tag{3.40}$$

but the linear system involving nonsingular matrices of form T given in (3.40) are also most easily solved by the LU decomposition method just described for the general linear system. The tridiagonal matrix T can be factored into a lower bidiagonal factor L and an upper bidiagonal factor U having the following forms

$$L = \begin{pmatrix} 1 & 0 & 0 & \cdots & 0 \\ l_2 & 1 & 0 & \ddots & 0 \\ 0 & l_3 & 1 & \ddots & 0 \\ \ddots & \ddots & \ddots & \ddots & 0 \\ 0 & 0 & 0 & l_n & 1 \end{pmatrix}, \quad U = \begin{pmatrix} u_1 & c_1 & 0 & \cdots & 0 \\ 0 & u_2 & c_2 & \ddots & 0 \\ 0 & 0 & u_3 & \ddots & 0 \\ \ddots & \ddots & \ddots & \ddots & c_{n-1} \\ 0 & 0 & 0 & 0 & u_n \end{pmatrix}. \tag{3.41}$$

The unknown elements l_i and u_i of matrices L and U, respectively, can be computed as a special case of Doolittle's method using the LU decomposition method,

$$\left.\begin{aligned} u_1 &= \alpha_1 \\ l_i &= \frac{\beta_i}{u_{i-1}}, \qquad i = 2, 3, \ldots, n \\ u_i &= \alpha_i - l_i c_{i-1}, \qquad i = 2, 3, \ldots, n \end{aligned}\right\}. \tag{3.42}$$

After finding the values for l_i and u_i, then they are used along with the elements c_i, to solve tridiagonal system (3.40) by solving the first bidiagonal system

$$L\mathbf{y} = \mathbf{b} \tag{3.43}$$

for $\mathbf{y}$ using forward substitution

$$\left.\begin{array}{rcl} y_1 &=& b_1 \\ y_i &=& b_i - l_i y_{i-1}, \qquad i = 2, 3, \ldots, n \end{array}\right\} \tag{3.44}$$

followed by solving the second bidiagonal system

$$U\mathbf{x} = \mathbf{y} \tag{3.45}$$

for $\mathbf{x}$ using backward substitution

$$\left.\begin{array}{rcl} x_n &=& y_n/u_n \\ x_i &=& y_i - c_i x_{i+1}, \qquad i = n-1, \ldots, 1 \end{array}\right\}. \tag{3.46}$$

The entire process for solving the original system (3.40) requires $3n$ additions, $3n$ multiplications, and $2n$ divisions. Thus, the total number of multiplications and divisions is approximately $5n$.

Most large tridiagonal systems are strictly diagonally dominant (defined below), so pivoting is not necessary. When solving systems of equations with a tridiagonal coefficients matrix T, iterative methods can sometimes be used to advantage. These methods are introduced later in the chapter.

Example 3.23 *Solve the following tridiagonal system of equations using the LU decomposition method*

$$\begin{array}{rcrcrcrcl} x_1 &+& x_2 & & & & &=& 1 \\ x_1 &+& 2x_2 &+& x_3 & & &=& 0 \\ & & x_2 &+& 3x_3 &+& x_4 &=& 1 \\ & & & & x_3 &+& 4x_4 &=& 1. \end{array}$$

Solution. *Construct the factorization of tridiagonal matrix T as follows*

$$\begin{pmatrix} 1 & 1 & 0 & 0 \\ 1 & 2 & 1 & 0 \\ 0 & 1 & 3 & 1 \\ 0 & 0 & 1 & 4 \end{pmatrix} = \begin{pmatrix} 1 & 0 & 0 & 0 \\ l_2 & 1 & 0 & 0 \\ 0 & l_3 & 1 & 0 \\ 0 & 0 & l_4 & 1 \end{pmatrix} \begin{pmatrix} u_1 & 1 & 0 & 0 \\ 0 & u_2 & 1 & 0 \\ 0 & 0 & u_3 & 1 \\ 0 & 0 & 0 & u_4 \end{pmatrix}.$$

Then the elements of the L and U matrices can be computed using (3.41) as follows

$$\begin{aligned} u_1 &= \alpha_1 = 1, \\ l_2 &= \frac{\beta_2}{u_1} = \frac{1}{1} = 1, \\ u_2 &= \alpha_2 - l_2 c_1 = 2 - (1)1 = 1, \\ l_3 &= \frac{b_3}{u_2} = \frac{1}{1} = 1, \\ u_3 &= \alpha_3 - l_3 c_2 = 3 - (1)1 = 2, \\ l_4 &= \frac{b_4}{u_3} = \frac{1}{2}, \\ u_4 &= \alpha_4 - l_4 c_3 = 4 - (\frac{1}{2})1 = \frac{7}{2}. \end{aligned}$$

After finding the elements of the bidiagonal matrices L and U, we solve the first system $L\mathbf{y} = \mathbf{b}$ *as follows*

$$\begin{pmatrix} 1 & 0 & 0 & 0 \\ 1 & 1 & 0 & 0 \\ 0 & 1 & 1 & 0 \\ 0 & 0 & \frac{1}{2} & 1 \end{pmatrix} \begin{pmatrix} y_1 \\ y_2 \\ y_3 \\ y_4 \end{pmatrix} = \begin{pmatrix} 1 \\ 0 \\ 1 \\ 1 \end{pmatrix}.$$

Using forward substitution, we get

$$[y_1, y_2, y_3, y_4]^T = [1, -1, 2, 0]^T.$$

Now we solve the second system $U\mathbf{x} = \mathbf{y}$ *as follows*

$$\begin{pmatrix} 1 & 1 & 0 & 0 \\ 0 & 1 & 1 & 0 \\ 0 & 0 & 2 & 1 \\ 0 & 0 & 0 & \frac{7}{2} \end{pmatrix} \begin{pmatrix} x_1 \\ x_2 \\ x_3 \\ x_4 \end{pmatrix} = \begin{pmatrix} 1 \\ -1 \\ 2 \\ 0 \end{pmatrix}.$$

Using backward substitution, we get the required solution

$$\mathbf{x}^* = [x_1, x_2, x_3, x_4]^T = [3, -2, 1, 0]^T.$$

The above results can be obtained using MATLAB commands. We do the following:

```
>> Tb = [T|b] = [1 1 0 0 1;1 2 1 0 0;0 1 3 1 1;0 0 1 4 1];
>> TridLU(Tb);
```

Program 3.10

MATLAB m-file for the LU Decomposition for Tridiagonal System

```
function sol=TridLU(Tb)
[m,n]=size(Tb); L=eye(m); U=zeros(m);
U(1,1)=Tb(1,1);
for i=2:m
U(i-1,i) = Tb(i-1,i);
L(i,i-1) = Tb(i,i-1)/U(i-1,i-1);
U(i,i) - L(i,i-1) * Tb(i-1,i); end
disp('The lower-triangular matrix') L;
disp('The upper-triangular matrix') U;
y = inv(L) * Tb(:,n); x = inv(U) * y;
```

Procedure 3.8 [LU Decomposition by Tridiagonal Method]

1. *Take the tridiagonal matrix T.*
2. *Decompose matrix $T = LU$ using (3.42).*
3. *Solve linear system $L\mathbf{y} = \mathbf{b}$ using (3.44).*
4. *Solve linear system $U\mathbf{x} = \mathbf{y}$ using (3.46).*

3.5 Norms of Vectors and Matrices

Before discussing the iterative methods for solving linear systems, we discuss a method for quantitatively measuring the distance between vectors in $\mathbb{R}^n$, the set of all column vectors with real components, to determine whether the sequence of vectors that results from using an iterative method converges to a solution of the system. To define a distance in $\mathbb{R}^n$, we use the notation of the *norm* of a vector.

3.5.1 Vector Norms

It is sometimes useful to have a scalar measure of the magnitude of a vector. Such a measure is called a *vector norm* and for a vector $\mathbf{x}$ is written as $\|\mathbf{x}\|$.

A vector norm on $\mathbb{R}^n$ is a function, from $\mathbb{R}^n$ to $\mathbb{R}$, satisfying:

1. $\|\mathbf{x}\| > 0 \quad \text{for all} \quad \mathbf{x} \in \mathbb{R}^n.$
2. $\|\mathbf{x}\| = 0 \quad \text{if and only if} \quad \mathbf{x} = \mathbf{0}.$
3. $\|\alpha\mathbf{x}\| = |\alpha|\|\mathbf{x}\|, \quad \text{for all} \quad \alpha \in \mathbb{R}, \quad \mathbf{x} \in \mathbb{R}^n.$
4. $\|\mathbf{x}+\mathbf{y}\| \leq \|\mathbf{x}\| + \|\mathbf{y}\|, \quad \text{for all} \quad \mathbf{x}, \mathbf{y} \in \mathbb{R}^n.$

There are three norms in $\mathbb{R}^n$ that are most commonly used in applications called l_1-norm, l_2-norm, and l_∞-norm, and are defined for the given vectors $\mathbf{x} = [x_1, x_2, \ldots, x_n]^T$ as

$$
\begin{aligned}
\|\mathbf{x}\|_1 &= \sum_{i=1}^{n} |x_i|, \\
\|\mathbf{x}\|_2 &= \left(\sum_{i=1}^{n} x_i^2 \right)^{1/2}, \\
\|\mathbf{x}\|_\infty &= \max_{1 \leq i \leq n} |x_i|.
\end{aligned}
$$

The l_1-norm is called the *absolute norm* and the l_2-norm is frequently called the *Euclidean norm* as it is just the formula for distance in ordinary three-dimensional Euclidean space extended to dimension n. Finally, the l_∞-norm is called the *maximum norm* or occasionally the *uniform norm*. All these three norms are also called *natural norms*.

Example 3.24 *Compute l_p-norms ($p = 1, 2, \infty$) of the vector $\mathbf{x} = [-5, 3, -2]^T$ in $\mathbb{R}^3$.*

Solution. *These norms are:*

$$
\begin{aligned}
\|\mathbf{x}\|_1 &= |x_1| + |x_2| + |x_3| = |-5| + |3| + |-2| = 10, \\
\|\mathbf{x}\|_2 &= (x_1^2 + x_2^2 + x_3^2)^{1/2} = \left[(-5)^2 + (3)^2 + (-2)^2 \right]^{1/2} \approx 6.16, \\
\|\mathbf{x}\|_\infty &= \max\{|x_1|, |x_2|, |x_3|\} = \max\{|-5|, |3|, |-2|\} = 5.
\end{aligned}
$$

In MATLAB, the built-in *norm* function computes l_p-norms of vectors. If only one argument is passed to norm, the l_2-norm is returned and for two arguments, the second one is used to specify the value of p:

```
>> x = [-5 3 -2];
>> v = norm(x)
v = 6.16
>> x = [-5 3 -2];
>> v = norm(x, 2)
v = 6.16
>> x = [-5 3 -2];
>> v = norm(x, 1)
v = 10
>> x = [-5 3 -2];
>> v = norm(x, inf)
v = 5
```

The internal MATLAB constant *inf* is used to select the l_∞-norm.

3.5.2 Matrix Norms

A matrix norm is a measure of how well one matrix approximates another, or, more accurately, of how well their difference approximates the zero matrix. An iterative procedure for inverting a matrix produces a sequence of approximate inverses. Since in practices such a process must be terminated, it is desirable to have some measure of the error of approximate inverse.

So a matrix norm on the set of all $n \times n$ matrices is a real-valued function, $\|.\|$, defined on this set, satisfying for all $n \times n$ matrices A and B and all real number α as follows:

1. $\|A\| > 0, \quad A \neq \mathbf{0}.$
2. $\|A\| = 0, \quad A = \mathbf{0}.$
3. $\|I\| = 1, \quad I$ *is the identity matrix.*
4. $\|\alpha A\| = |\alpha| \|A\|, \quad$ for scalar $\alpha \in \mathbb{R}$.
5. $\|A + B\| \leq \|A\| + \|B\|.$
6. $\|AB\| \leq \|A\| \|B\|.$

7. $\|A - B\| \geq \Big| \|A\| - \|B\| \Big|$.

Several norms for matrices have been defined. We shall use the following three natural norms l_1, l_2, and l_∞ for a square matrix of order n:

$$\|A\|_1 = \max_j \left(\sum_{i=1}^{n} |a_{ij}| \right) = \text{maximum column-sum},$$

$$\|A\|_2 = \max_{\|x\|_2=1} \|A\mathbf{x}\|_2 = \text{spectral norm},$$

$$\text{and} \|A\|_\infty = \max_i \left(\sum_{j=1}^{n} |a_{ij}| \right) = \text{row-sum norm.}$$

The l_1-norm and l_∞-norm are widely used because they are easy to calculate. The matrix norm $\|A\|_2$ that corresponds to the l_2-norm is related the eigenvalues of the matrix. It sometimes has special utility because no other norm is smaller than this norm. It, therefore, provides the best measure of the size of a matrix, but is also the most difficult to compute. We will discuss this natural norm later in the chapter.

For an $m \times n$ matrix, we can paraphrase the *Frobenius* (or *Euclidean*) norm (which is not a natural norm) and defined as

$$\|A\|_F = \left(\sum_{i=1}^{m} \sum_{j=1}^{n} |a_{ij}|^2 \right)^{1/2}.$$

It can be shown that

$$\|A\|_F = \sqrt{tr(A^T A)},$$

where $tr(A^T A)$ is the *trace* of a matrix $A^T A$; that is, the sum of the diagonal entries of $A^T A$. The Frobenius norm of a matrix is a good measure of the magnitude of a matrix. It is to be noted that $\|A\|_F \neq \|A\|_2$. For a diagonal matrix, all norms have the same values.

Example 3.25 *Compute l_p-norms $(p = 1, \infty, F)$ of the following matrix*

$$A = \begin{pmatrix} 4 & 2 & -1 \\ 3 & 5 & -2 \\ 1 & -2 & 7 \end{pmatrix}.$$

Solution. *These norms are:*

$$\sum_{i=1}^{3}|a_{i1}| = |4|+|3|+|1| = 8,$$
$$\sum_{i=1}^{3}|a_{i2}| = |2|+|5|+|-2| = 9,$$

and

$$\sum_{i=1}^{3}|a_{i3}| = |-1|+|-2|+|7| = 10,$$

so

$$\|A\|_1 = max\{8, 9, 10\} = 10.$$

Also,

$$\sum_{j=1}^{3}|a_{1j}| = |4|+|2|+|-1| = 7,$$
$$\sum_{j=1}^{3}|a_{2j}| = |3|+|5|+|-2| = 10,$$

and

$$\sum_{j=1}^{3}|a_{3j}| = |1|+|-2|+|7| = 10,$$

so

$$\|A\|_\infty = max\{7, 10, 10\} = 10.$$

Finally, we have

$$\|A\|_F = (16+4+1+9+25+4+1+4+49)^{1/2} \approx 10.6301.$$

Like l_p-norms of vectors, in MATLAB the built-in *norm* function can be used to compute l_p-norms of matrices. The l_1-norm of a matrix can be computed as follows:

```
>> A = [4 2 -1; 3 5 -2; 1 -2 -7];
>> B = norm(A, 1)
B =
   10
```

The l_∞-norm of a matrix A is:

$$\begin{array}{l} >> A = [4\ 2\ \ -1; 3\ 5\ \ -2; 1\ \ -2\ \ -7]; \\ >> B = norm(A, inf) \\ B = \\ \quad 10 \end{array}$$

Finally, the Frobenius norm of a matrix A is:

$$\begin{array}{l} >> A = [4\ 2\ \ -1; 3\ 5\ \ -2; 1\ \ -2\ \ -7]; \\ >> B = norm(A, 'fro') \\ B = \\ \quad 10.6301 \end{array}$$

3.6 Iterative Methods for Solving Linear Systems

The methods discussed so far for the solution of simultaneous systems of linear equations have been direct, which required a finite number of arithmetic operations. The elimination methods for solving such systems usually yield sufficiently accurate solutions for approximately 20 to 25 simultaneous equations, where most of the unknowns are present in all of the equations. When the coefficients matrix is sparse (has many zeros), a considerably large number of equations can be handled by the elimination methods. But these methods are generally impractical when many hundreds or thousands of equations must be solved simultaneously.

There are, however, several methods which can be used to solve large numbers of simultaneous equations. These methods are called *iterative methods* by which an approximation to the solution of a system of linear equations may be obtained. The iterative methods are used most often for large sparse systems of linear equations and efficient in terms of computer storage and time requirements. Systems of this type arise frequently in the numerical solution of boundary value problems and partial differential equations. Unlike the direct methods, iterative methods may not always yield a solution, even if the coefficients matrix is nonsingular.

The iterative methods to solve the system of linear equations

$$A\mathbf{x} = \mathbf{b} \tag{3.47}$$

start with an initial approximation $\mathbf{x}^{(0)} \in \mathbb{R}$ to the solution $\mathbf{x}$ of linear system (3.47), and generate a sequence of vectors $\{\mathbf{x}^{(k)}\}_{k=0}^{\infty}$ that converge to $\mathbf{x}$. Most of these iterative methods involve a process that converts system (3.47) into an equivalent system of the form

$$\mathbf{x} = T\mathbf{x} + \mathbf{c} \tag{3.48}$$

for some square matrix T and vector $\mathbf{c}$. After the initial vector $\mathbf{x}^{(0)}$ is selected, the sequence of approximate solutions vector is generated by computing

$$\mathbf{x}^{(k+1)} = T\mathbf{x}^{(k)} + \mathbf{c}, \qquad \text{for} \quad k = 0, 1, 2, \ldots. \tag{3.49}$$

The sequence is terminated when the error is sufficiently small; that is,

$$\|\mathbf{x}^{(k+1)} - \mathbf{x}^{(k)}\| < \epsilon, \quad \text{for small positive} \quad \epsilon. \tag{3.50}$$

Among them, the most useful methods are the *Jacobi* method, the *Gauss-Seidel* method, and the *Successive over-relaxation* (SOR) method.

Before discussing all these methods, it is convenient to introduce notations for some matrices. Matrix A is written as

$$A = L + D + U, \tag{3.51}$$

where L is strictly lower-triangular, U is strictly upper-triangular, and D is the diagonal parts of coefficients matrix A; that is,

$$L = \begin{pmatrix} 0 & 0 & 0 & \cdots & 0 \\ a_{21} & 0 & 0 & \cdots & 0 \\ a_{31} & a_{32} & 0 & \cdots & 0 \\ \vdots & \vdots & \vdots & \vdots & \vdots \\ a_{n1} & a_{n2} & a_{n3} & \cdots & 0 \end{pmatrix}, \qquad U = \begin{pmatrix} 0 & a_{12} & a_{13} & \cdots & a_{1n} \\ 0 & 0 & a_{23} & \cdots & a_{2n} \\ 0 & 0 & 0 & \cdots & a_{3n} \\ \vdots & \vdots & \vdots & \vdots & \vdots \\ 0 & 0 & 0 & \cdots & 0 \end{pmatrix},$$

and

$$D = \begin{pmatrix} a_{11} & 0 & 0 & \cdots & 0 \\ 0 & a_{22} & 0 & \cdots & 0 \\ 0 & 0 & a_{33} & \cdots & 0 \\ \vdots & \vdots & \vdots & \vdots & \vdots \\ 0 & 0 & 0 & \cdots & a_{nn} \end{pmatrix}.$$

Then linear system (3.47) can be written as

$$(L + D + U)\mathbf{x} = \mathbf{b}. \tag{3.52}$$

Now we discuss the first iterative method to solve linear system (3.52).

3.6.1 Jacobi Iterative Method

This is one of the easiest iterative methods to find the approximate solution of the system of linear equations (3.47). To explain its procedure, consider a system of three linear equations as follows:

$$\begin{array}{rcrcrcl} a_{11}x_1 & + & a_{12}x_2 & + & a_{13}x_3 & = & b_1, \\ a_{21}x_1 & + & a_{22}x_2 & + & a_{23}x_3 & = & b_2, \\ \text{and } a_{31}x_1 & + & a_{32}x_2 & + & a_{33}x_3 & = & b_3. \end{array}$$

The solution process starts by solving for the first the variable x_1 from the first equation, the second variable x_2 from second equation, and the third variable x_3 from the third equation, which gives

$$\begin{array}{rcrcrcl} a_{11}x_1 & = & b_1 & - & a_{12}x_2 & - & a_{13}x_3, \\ a_{22}x_2 & = & b_2 & - & a_{21}x_1 & - & a_{23}x_3, \\ \text{and } a_{33}x_3 & = & b_3 & - & a_{31}x_1 & - & a_{32}x_2 \end{array}$$

or in matrix form

$$D\mathbf{x} = \mathbf{b} - (L + U)\mathbf{x}.$$

Divide both sides of the above three equations by their diagonal elements, a_{11}, a_{22}, and a_{33}, respectively, to have

$$\begin{array}{rcl} x_1 & = & \frac{1}{a_{11}}\Big[b_1 - a_{12}x_2 - a_{13}x_3\Big], \\ x_2 & = & \frac{1}{a_{22}}\Big[b_2 - a_{21}x_1 - a_{23}x_3\Big], \\ \text{and } x_3 & = & \frac{1}{a_{33}}\Big[b_3 - a_{31}x_1 - a_{32}x_2\Big], \end{array}$$

which can be written in matrix form

$$\mathbf{x} = D^{-1}[\mathbf{b} - (L + U)\mathbf{x}].$$

Let $\mathbf{x}^{(0)} = \left[x_1^{(0)}, x_2^{(0)}, x_3^{(0)}\right]^T$ be an initial solution of the exact solution $\mathbf{x}$ of linear system (3.47). Then define an iterative sequence

$$\begin{aligned} x_1^{(k+1)} &= \frac{1}{a_{11}}\left[b_1 - a_{12}x_2^{(k)} - a_{13}x_3^{(k)}\right], \\ x_2^{(k+1)} &= \frac{1}{a_{22}}\left[b_2 - a_{21}x_1^{(k)} - a_{23}x_3^{(k)}\right], \\ \text{and } x_3^{(k+1)} &= \frac{1}{a_{33}}\left[b_3 - a_{31}x_1^{(k)} - a_{32}x_2^{(k)}\right] \end{aligned} \tag{3.53}$$

or in matrix form

$$\mathbf{x}^{(k+1)} = D^{-1}[\mathbf{b} - (L+U)\mathbf{x}^{(k)}], \qquad k = 0, 1, 2, \ldots, \tag{3.54}$$

where k is the number of iterative steps. Then form (3.53) is called the Jacobi formula for the system of three equations and (3.54) is called its matrix form. For a general system of n linear equations, the Jacobi method is defined by

$$\begin{aligned} x_i^{(k+1)} &= \frac{1}{a_{ii}}\left[b_i - \sum_{j=1}^{i-1} a_{ij}x_j^{(k)} - \sum_{j=i+1}^{n} a_{ij}x_j^{(k)}\right] \\ i &= 1, 2, \ldots, n, \quad k = 0, 1, 2, \ldots \end{aligned} \tag{3.55}$$

provided that the diagonal elements $a_{ii} \neq 0$ for each $i = 1, 2, \ldots, n$. If the diagonal elements equal zero, then reordering of the equations can be performed so that no element in the diagonal position equals zero. The matrix form of Jacobi iterative method (3.55) can be written as

$$\mathbf{x}^{(k+1)} = \mathbf{c} + T_J\mathbf{x}^{(k)}, \qquad k = 0, 1, 2, \ldots \tag{3.56}$$

or

$$\begin{pmatrix} x_1 \\ x_2 \\ \vdots \\ x_n \end{pmatrix}_{(k+1)} = \begin{pmatrix} c_1 \\ c_2 \\ \vdots \\ c_n \end{pmatrix} + \begin{pmatrix} 0 & -t_{12} & \cdots & -t_{1n} \\ -t_{21} & 0 & \cdots & -t_{2n} \\ \vdots & \vdots & \vdots & \vdots \\ -t_{n1} & -t_{n2} & \cdots & 0 \end{pmatrix} \begin{pmatrix} x_1 \\ x_2 \\ \vdots \\ x_n \end{pmatrix}_{(k)}, \tag{3.57}$$

where the **Jacobi iteration matrix** T_J and vector $\mathbf{c}$ are defined as follows

$$T_J = -D^{-1}(L+U), \qquad \text{and} \qquad \mathbf{c} = D^{-1}\mathbf{b} \tag{3.58}$$

and their elements are defined by

$$\begin{aligned} t_{ij} &= \frac{a_{ij}}{a_{ii}}, & i,j = 1,2,\ldots,n, \quad i \neq j \\ t_{ij} &= 0, & i = j \\ c_i &= \frac{b_i}{a_{ii}}, & i = 1,2,\ldots,n. \end{aligned}$$

The Jacobi iterative method is sometimes called the method of *simultaneous iteration* because all values of x_i are iterated simultaneously, that is, all values of $x_i^{(k+1)}$ depend only on the values of $x_i^{(k)}$.

Note that the diagonal elements of the Jacobi iteration matrix T_J are always zero. As usual with iterative methods, an initial approximation $x_i^{(0)}$ must be supplied. If we don't have knowledge of the exact solution, it is conventional to start with $x_i^{(0)} = \mathbf{0}$ for all i. The iterations defined by (3.55) are stopped when

$$\|\mathbf{x}^{(k+1)} - \mathbf{x}^{(k)}\| < \epsilon \tag{3.59}$$

or by using other possible stopping criteria

$$\frac{\|\mathbf{x}^{(k+1)} - \mathbf{x}^{(k)}\|}{\|\mathbf{x}^{(k+1)}\|} < \epsilon, \tag{3.60}$$

where ϵ is a preassigned small positive number. For this purpose, any convenient norm can be used, the most usual being the l_∞-norm.

Example 3.26 *Solve the following system of equations using the Jacobi iterative method, using $\epsilon = 0.5 \times 10^{-6}$ in the l_∞-norm:*

$$\begin{array}{rcrcrcr} 5x_1 & - & x_2 & + & x_3 & = & 10 \\ 2x_1 & + & 8x_2 & - & x_3 & = & 11 \\ -x_1 & + & x_2 & + & 4x_3 & = & 3. \end{array}$$

Start with the initial solution $\mathbf{x}^{(0)} = [0,0,0]^T$.

Solution. *The Jacobi method for the given system is*

$$x_1^{(k+1)} = \frac{1}{5}\Big[10 + x_2^{(k)} - x_3^{(k)}\Big],$$

$$x_2^{(k+1)} = \frac{1}{8}\Big[11 - 2x_1^{(k)} + x_3^{(k)}\Big],$$

$$\text{and } x_3^{(k+1)} = \frac{1}{4}\Big[3 + x_1^{(k)} - x_2^{(k)}\Big]$$

and starting with initial approximation $x_1^{(0)} = 0, x_2^{(0)} = 0, x_3^{(0)} = 0$, *then for* $k = 0$, *we obtain*

$$x_1^{(1)} = \frac{1}{5}\Big[10 + x_2^{(0)} - x_3^{(0)}\Big] = \frac{1}{5}\Big[10 + 0 - 0\Big] = 2,$$

$$x_2^{(1)} = \frac{1}{8}\Big[11 - 2x_1^{(0)} + x_3^{(0)}\Big] = \frac{1}{8}\Big[11 - 0 + 0\Big] = 1.375,$$

$$\text{and } x_3^{(1)} = \frac{1}{4}\Big[3 + x_1^{(0)} - x_2^{(0)}\Big] = \frac{1}{4}\Big[3 + 0 - 0\Big] = 0.75.$$

The first and subsequent iterations are listed in Table 3.1.

Table 3.1: Solution of Example 3.26

k	$x_1^{(k)}$	$x_2^{(k)}$	$x_3^{(k)}$
0	0.000000	0.000000	0.000000
1	2.000000	1.375000	0.750000
2	2.125000	0.968750	0.906250
3	2.012500	0.957031	1.039063
4	1.983594	1.001758	1.013867
5	1.997578	1.005835	0.995459
6	2.002075	1.000038	0.997936
⋮	⋮	⋮	⋮
15	2.000000	0.999999	1.000000
16	2.000000	1.000000	1.000000

Note that the Jacobi method converges and after 16 iterations we obtained what is obviously the exact solution. Ideally, the iteration should stop automatically when we obtain the required accuracy using one of the stopping criteria obtained by (3.59) or (3.60).

To get the above results using MATLAB, we do the following:

```
>> Ab = [A|b] = [5 -1 1 10; 2 8 -1 11; -1 1 4 3];
>> x = [0 0 0];
>> acc = 0.5e-6;
>> JacobiM(Ab, x, acc);
```

Example 3.27 *Solve the following system of equations using the Jacobi iterative method:*

$$\begin{array}{rcrcrcl} 2x_1 & + & 8x_2 & - & x_3 & = & 11 \\ 5x_1 & - & x_2 & + & x_3 & = & 10 \\ -x_1 & + & x_2 & + & 4x_3 & = & 3. \end{array}$$

Start with the initial solution $\mathbf{x}^{(0)} = [0, 0, 0]^T$.

Solution. *Results for this linear system are listed in Table 3.2.*

Table 3.2: Solution of Example 3.27

k	$x_1^{(k)}$	$x_2^{(k)}$	$x_3^{(k)}$
0	0.000000	0.000000	0.000000
1	5.500000	-10.0000	0.750000
2	45.87500	18.25000	4.625000
3	-65.1875	224.0000	7.656250
4	-886.672	-328.281	-71.5469
5	1282.852	-4514.91	-138.848

Note that in this case the Jacobi method diverges rapidly. Although the given linear system is the same as the linear system of Example 3.26 except the first and second equations are interchanged, from this example we concluded that the Jacobi iterative method is not always convergent.

Program 3.11

MATLAB m-file for the Jacobi Iterative Method

```
function x=JacobiM(Ab,x,acc)
[n,t]=size(Ab); b=Ab(1:n,t); R=1; k=1;
d(1,1:n+1)=[0 x]; while R > acc
for i=1:n
sum=0;
for j=1:n; if j ~ =i
sum = sum + Ab(i,j) * d(k,j + 1); end;
x(1,i) = (1/Ab(i,i)) * (b(i,1) - sum); end;end
k=k+1; d(k,1:n+1)=[k-1 x];
R=max(abs((d(k,2:n+1)-d(k-1,2:n+1))));
if k > 10 & R > 100
('Jacobi Method is diverges')
break; end; end; x=d;
```

Procedure 3.9 [Jacobi Method]

1. *Check that coefficient matrix A is strictly diagonally dominant (for guaranteed convergence).*
2. *Initialize the first approximation* $\mathbf{x}^{(0)}$ *and pre-assigned accuracy* ϵ.
3. *Compute the constant* $\mathbf{c} = D^{-1}\mathbf{b} = \dfrac{b_i}{a_{ii}}$, *for* $i = 1, 2, \ldots, n$.
4. *Compute the Jacobi iteration matrix* $T_J = -D^{-1}(L + U)$.
5. *Solve for the approximate solutions* $\mathbf{x}_i^{(k+1)} = T_J\mathbf{x}_i^{(k)} + \mathbf{c}$ *and* $i = 1, 2, \ldots, n$ $k = 0, 1, \ldots$
6. *Repeat step 5 until* $\|\mathbf{x}_i^{(k+1)} - \mathbf{x}_i^{(k)}\| < \epsilon$.

3.6.2 Gauss-Seidel Iterative Method

This is one of the most popular and widely used iterative methods to find the approximate solution of the system of linear equations. This iterative method is a modification of the Jacobi iterative method and gives us good accuracy using the most recently calculated values.

From Jacobi iterative formula (3.55), it is seen that the new estimates for solution $\mathbf{x}$ are computed from the old estimates and only when all the new estimates have been determined are they used in the right-hand side of the equation to perform the next iteration. But the Gauss-Seidel method makes use of the new estimates in the right-hand side of the equation as soon as they become available. For example, the Gauss-Seidel formula for the system of three equations can be defined as an iterative sequence

$$\begin{aligned} x_1^{(k+1)} &= \frac{1}{a_{11}}\Big[b_1 - a_{12}x_2^{(k)} - a_{13}x_3^{(k)}\Big], \\ x_2^{(k+1)} &= \frac{1}{a_{22}}\Big[b_2 - a_{21}x_1^{(k+1)} - a_{23}x_3^{(k)}\Big], \\ \text{and } x_3^{(k+1)} &= \frac{1}{a_{33}}\Big[b_3 - a_{31}x_1^{(k+1)} - a_{32}x_2^{(k+1)}\Big]. \end{aligned} \tag{3.61}$$

For a general system of n linear equations, the Gauss-Seidel iterative method is defined as

$$\begin{aligned} x_i^{(k+1)} &= \frac{1}{a_{ii}}\left[b_i - \sum_{j=1}^{i-1} a_{ij}x_j^{(k+1)} - \sum_{j=i+1}^{n} a_{ij}x_j^{(k)}\right] \\ i &= 1, 2, \ldots, n, \quad k = 0, 1, 2, \ldots \end{aligned} \tag{3.62}$$

and in matrix form, the Gauss-Seidel iterative method can be represented by

$$\mathbf{x}^{(k+1)} = (D + L)^{-1}[\mathbf{b} - U\mathbf{x}^{(k)}], \qquad \text{for each} \quad k = 0, 1, 2, \ldots. \tag{3.63}$$

For the lower-triangular matrix $(D + L)$ to be nonsingular, it is necessary and sufficient that the diagonal elements $a_{ii} \neq 0$, for each $i = 1, 2, \ldots, n$. By comparing (3.49) and (3.63), we obtain

$$T_G = -(D + L)^{-1}U \qquad \text{and} \qquad \mathbf{c} = (D + L)^{-1}\mathbf{b}, \tag{3.64}$$

which are called the *Gauss-Seidel iteration matrix* and the vector, respectively.

The Gauss-Seidel iterative method is sometimes called the method of *successive iteration* because the most recent values of all $\mathbf{x}_i$ are used in the calculation.

Example 3.28 *Solve the following system of equations using the Gauss-Seidel iterative method, with $\epsilon = 0.5 \times 10^{-6}$ in l_∞-norm.*

$$\begin{array}{rcrcrcr} 5x_1 & - & x_2 & + & x_3 & = & 10 \\ 2x_1 & + & 8x_2 & - & x_3 & = & 11 \\ -x_1 & + & x_2 & + & 4x_3 & = & 3. \end{array}$$

Start with the initial solution $\mathbf{x}^{(0)} = [0, 0, 0]^T$.

Solution. *The Gauss-Seidel iteration for the given system is*

$$\begin{aligned} x_1^{(k+1)} &= \frac{1}{5}\Big[10 + x_2^{(k)} - x_3^{(k)}\Big] \\ x_2^{(k+1)} &= \frac{1}{8}\Big[11 - 2x_1^{(k+1)} + x_3^{(k)}\Big] \\ x_3^{(k+1)} &= \frac{1}{4}\Big[3 + x_1^{(k+1)} - x_2^{(k+1)}\Big] \end{aligned}$$

and starting with the initial approximation $x_1^{(0)} = 0, x_2^{(0)} = 0, x_3^{(0)} = 0$, *then for* $k = 0$, *we obtain*

$$\begin{aligned} x_1^{(1)} &= \frac{1}{5}\Big[10 + x_2^{(0)} - x_3^{(0)}\Big] = \frac{1}{5}\Big[10 + 0 - 0\Big] = 2 \\ x_2^{(1)} &= \frac{1}{8}\Big[11 - 2x_1^{(1)} + x_3^{(0)}\Big] = \frac{1}{8}\Big[11 - 4 + 0\Big] = 0.875 \\ x_3^{(1)} &= \frac{1}{4}\Big[3 + x_1^{(1)} - x_2^{(1)}\Big] = \frac{1}{4}\Big[3 + 2 - 0.875\Big] = 1.03125. \end{aligned}$$

The first and subsequent iterations are listed in Table 3.3.

The above results can be obtained using MATLAB as follows:

```
>> Ab = [A|b] = [5 -1 1 10; 2 8 -1 11; -1 1 4 3];
>> x = [0 0 0];
>> acc = 0.5e-6;
>> GaussSM(Ab, x, acc);
```

Note that the Gauss-Seidel method converged and required ten iterations to obtain the correct solution for the given system, which is six iterations less than required by the Jacobi method for Example 3.26.

Table 3.3: Solution of Example 3.28

k	$x_1^{(k)}$	$x_2^{(k)}$	$x_3^{(k)}$
0	0.000000	0.000000	0.000000
1	2.000000	0.875000	1.031250
2	1.968750	1.011719	0.989258
3	2.004492	0.997534	1.001740
4	1.999159	1.000428	0.999683
5	2.000149	0.999923	1.000057
6	1.999973	1.000014	0.999990
⋮	⋮	⋮	⋮
9	2.000000	0.999999	1.000000
10	2.000000	1.000000	1.000000

Program 3.12

MATLAB m-file for the Gauss-Seidel Iterative Method

```
function x=GaussSM(Ab,x,acc)
[n,t]=size(Ab); b=Ab(1:n,t);R=1; k=1;
d(1,1:n+1)=[0 x]; k=k+1; while R > acc
for i=1:n; sum=0; for j=1:n
if j <= i - 1; sum = sum + Ab(i,j) * d(k,j + 1);
elseif j >= i + 1
sum = sum + Ab(i,j) * d(k - 1,j + 1); end; end
x(1,i) = (1/Ab(i,i)) * (b(i,1) - sum);
d(k,1)=k-1; d(k,i+1)=x(1,i); end
R=max(abs((d(k,2:n+1)-d(k-1,2:n+1))));
k=k+1; if R > 100 & k > 10; ('Gauss-Seidel method is Diverges')
break;end;end;x=d;
```

Example 3.29 *Solve the following system of equations using the Gauss-Seidel iterative method.*

$$\begin{array}{rcrcrcr} 2x_1 & + & 8x_2 & - & x_3 & = & 11 \\ 5x_1 & - & x_2 & + & x_3 & = & 10 \\ -x_1 & + & x_2 & + & 4x_3 & = & 3 \end{array}$$

Start with the initial solution $\mathbf{x}^{(0)} = [0, 0, 0]^T$.

Solution. *Results for this linear system are listed in Table 3.4. Note that in this case the Gauss-Seidel method diverges rapidly. Although the given linear system is the same as the linear system of Example 3.28 except the first and second equations are interchanged, from this example we concluded that the Gauss-Seidel iterative method is not always convergent.*

Table 3.4: Solution of Example 3.29

k	$x_1^{(k)}$	$x_2^{(k)}$	$x_3^{(k)}$
0	0.000000	0.000000	0.000000
1	5.500000	17.5000	-2.25000
2	-65.6250	-340.375	69.43750
3	1401.719	7068.031	-1415.83
4	-28974.5	-146298.5	29331.75

Procedure 3.10 [Gauss-Seidel Method]

1. *Check that coefficient matrix A is strictly diagonally dominant (for guaranteed convergence).*

2. *Initialize the first approximation $\mathbf{x}^{(0)} \in \mathbb{R}$ and pre-assigned accuracy ϵ.*

3. *Compute the constant $\mathbf{c} = (D+L)^{-1}\mathbf{b}$.*

4. *Compute the Gauss-Seidel iteration matrix $T_G = -(D+L)^{-1}U$.*

5. *Solve for the approximate solutions $x_i^{(k+1)} = T_G x_i^{(k)} + \mathbf{c}$ and $i = 1, 2, \ldots, n$*
   $k = 0, 1, \ldots$

6. *Repeat step 5 until $\|\mathbf{x}_i^{(k+1)} - \mathbf{x}_i^{(k)}\| < \epsilon$.*

From Examples 3.26 and (3.28) we noted that the solution by the Gauss-Seidel method converges more quickly than the Jacobi method. In general, we may state that **if both the Jacobi method and the Gauss-Seidel method converge, then the Gauss-Seidel method converges more quickly**. This is generally the case but not always true. In fact, there are some linear systems for which the Jacobi method converges but the Gauss-Seidel method does not, and others for which the Gauss-Seidel method converges but the Jacobi method does not. For example, from the following linear system

$$\begin{array}{rcrcrcr} x_1 & + & 2x_2 & - & 2x_3 & = & 3 \\ x_1 & + & x_2 & + & x_3 & = & 0 \\ 2x_1 & + & 2x_2 & + & x_3 & = & 1 \end{array}$$

starting with initial approximation $\mathbf{x}^{(0)} = [0,0,0]^T$, we see that the Jacobi method converges to the exact solution $\mathbf{x} = [1,0,-1]^T$ within the accuracy 0.5×10^{-6} just after four iterations while the Gauss-Seidel method diverges. But from the other following linear system

$$\begin{array}{rcrcrcr} 2x_1 & + & x_2 & + & x_3 & = & 4 \\ x_1 & + & 2x_2 & + & x_3 & = & 4 \\ x_1 & + & x_2 & + & 2x_3 & = & 4 \end{array}$$

starting with initial approximation $\mathbf{x}^{(0)} = [0,0,0]^T$, we see that the Jacobi method diverges and the Gauss-Seidel method converges but very slow, to the exact solution $\mathbf{x} = [1,1,1]^T$, within the accuracy 0.5×10^{-6} in 17 iterations.

3.6.3 Convergence Criteria

Note that the Jacobi method and the Gauss-Seidel method do not always converge to the solution of the given system of linear equations. Here we need some conditions that make both methods converge. The sufficient condition for the convergence of both methods is defined in the following theorem.

Theorem 3.16 (Sufficient Condition for Convergence)

If matrix A is strictly diagonally dominant (SDD), then for any choice of initial approximation $\mathbf{x}^{(0)} \in \mathbb{R}$ *both the Jacobi method and the Gauss-Seidel method give the sequence* $\{x^{(k)}\}_{k=0}^{\infty}$ *of approximations that converges to the solution of the linear system.*

There is another sufficient condition for the convergence of both iterative methods, which is defined in the following theorem.

Theorem 3.17 (Sufficient Condition for Convergence)

For any initial approximation $\mathbf{x}^{(0)} \in \mathbb{R}$, *the sequence* $\{x^{(k)}\}_{k=0}^{\infty}$ *of approximations defined by*

$$\mathbf{x}^{(k+1)} = T\mathbf{x}^{(k)} + \mathbf{c}, \quad \text{for each} \quad k \geq 0, \quad \text{and} \quad \mathbf{c} \neq 0 \tag{3.65}$$

converges to the unique solution of $\mathbf{x} = T\mathbf{x} + \mathbf{c}$ *if* $\|T\| < 1$ *for any natural matrix norm, and the following error bounds hold:*

$$\|\mathbf{x} - \mathbf{x}^{(k)}\| \leq \|T\|^k \|\mathbf{x}^{(0)} - \mathbf{x}\|, \tag{3.66}$$

$$\text{and } \|\mathbf{x} - \mathbf{x}^{(k)}\| \leq \frac{\|T\|^k}{1 - \|T\|} \|\mathbf{x}^{(1)} - \mathbf{x}^{(0)}\|. \tag{3.67}$$

Note that the condition $\|T\| < 1$ is equivalent to the condition that matrix A is to be strictly diagonally dominant.

For the Jacobi method for a general matrix A, the norm of the Jacobi iteration matrix is defined as

$$\|T_J\| = \max_{1 \leq i \leq n} \sum_{\substack{j=1 \\ j \neq i}}^{n} \left| \frac{a_{ij}}{a_{ii}} \right|.$$

Thus, $\|T_J\| < 1$ is equivalent to requiring

$$\sum_{\substack{j=1 \\ j \neq i}}^{n} |a_{ij}| < |a_{ii}|;$$

that is, matrix A is strictly diagonally dominant.

Example 3.30 *Consider the following linear system of equations*

$$\begin{array}{rcrcrcr} 4x_1 & - & x_2 & + & x_3 & = & 12 \\ -x_1 & + & 3x_2 & + & x_3 & = & 1 \\ x_1 & + & x_2 & + & 5x_3 & = & -14. \end{array}$$

(a) Show that both iterative methods (Jacobi and Gauss-Seidel) will converge using $\|T\|_\infty < 1$.

(b) Find the second approximation $\mathbf{x}^{(2)}$ *when the initial solution is* $\mathbf{x}^{(0)} = [4, 3, -3]^T$.

(c) Compute the error bounds for your approximations.

(d) How many iterations are needed to get an accuracy within 10^{-4}?

Solution. *From (3.51), we have*

$$A = \begin{pmatrix} 4 & -1 & 1 \\ -1 & 3 & 1 \\ 1 & 1 & 5 \end{pmatrix} = \begin{pmatrix} 0 & 0 & 0 \\ -1 & 0 & 0 \\ 1 & 1 & 0 \end{pmatrix} + \begin{pmatrix} 0 & -1 & 1 \\ 0 & 0 & 1 \\ 0 & 0 & 0 \end{pmatrix} + \begin{pmatrix} 4 & 0 & 0 \\ 0 & 3 & 0 \\ 0 & 0 & 5 \end{pmatrix} = L+U+D.$$

Jacobi Method

(a) Since the Jacobi iteration matrix is defined as

$$T_J = -D^{-1}(L+U)$$

and using the given information, we have

$$T_J = -\begin{pmatrix} \frac{1}{4} & 0 & 0 \\ 0 & \frac{1}{3} & 0 \\ 0 & 0 & \frac{1}{5} \end{pmatrix} \begin{pmatrix} 0 & -1 & 1 \\ -1 & 0 & 1 \\ 1 & 1 & 0 \end{pmatrix} = \begin{pmatrix} 0 & \frac{1}{4} & -\frac{1}{4} \\ \frac{1}{3} & 0 & -\frac{1}{3} \\ -\frac{1}{5} & -\frac{1}{5} & 0 \end{pmatrix}$$

then the l_∞ norm of the matrix T_J is

$$\|T_J\|_\infty = \max\left\{\frac{2}{4}, \frac{2}{3}, \frac{2}{5}\right\} = \frac{2}{3} < 1.$$

Thus, the Jacobi method will converge for the given linear system.

(b) The Jacobi method for the given system is

$$\begin{aligned} x_1^{(k+1)} &= \frac{1}{4}\Big[12 + x_2^{(k)} - x_3^{(k)}\Big] \\ x_2^{(k+1)} &= \frac{1}{3}\Big[1 + x_1^{(k)} - x_3^{(k)}\Big] \\ x_3^{(k+1)} &= \frac{1}{5}\Big[-12 - x_1^{(k)} - x_2^{(k)}\Big]. \end{aligned}$$

Starting with the initial approximation $x_1^{(0)} = 4, x_2^{(0)} = 3, x_3^{(0)} = -3$, *and for* $k = 0, 1$, *we obtain the first and the second approximations as*

$$\mathbf{x}^{(1)} = [4.5, 2.6667, -4.2]^T \quad \textit{and} \quad \mathbf{x}^{(2)} = [4.7167, 3.2333, -4.2333]^T.$$

(c) Using error bound formula (3.66), we obtain

$$\|\mathbf{x} - \mathbf{x}^{(2)}\| \leq \frac{(2/3)^2}{1 - 2/3} \left\| \begin{pmatrix} 4.5 \\ 2.6667 \\ -4.2 \end{pmatrix} - \begin{pmatrix} 4 \\ 3 \\ -3 \end{pmatrix} \right\|$$

or

$$\|\mathbf{x} - \mathbf{x}^{(2)}\| \leq \frac{4}{3}(1.2) = 1.6.$$

(d) To find the number of iterations, we use formula (3.66) as

$$\|\mathbf{x} - \mathbf{x}^{(k)}\| \leq \frac{\|T_J\|^k}{1 - \|T_J\|} \|\mathbf{x}^{(1)} - \mathbf{x}^{(0)}\| \leq 10^{-4}.$$

It gives

$$\frac{(2/3)^k}{1/3}(1.2) \leq 10^{-4}$$

or

$$(2/3)^k \leq \frac{10^{-4}}{3.6}.$$

Taking ln *on both sides, we obtain*

$$k \ln(2/3) \leq \ln\left(\frac{10^{-4}}{3.6}\right)$$

and it gives

$$k \geq 25.8789 \quad \textit{or} \quad k = 26,$$

which is the required number of iterations.

Gauss-Seidel Method

(a) Since the Gauss-Seidel iteration matrix is defined as

$$T_G = -(D + L)^{-1} U$$

and using the given information, we have

$$T_G = -\begin{pmatrix} \frac{1}{4} & 0 & 0 \\ \frac{1}{12} & \frac{1}{3} & 0 \\ -\frac{4}{60} & -\frac{1}{15} & \frac{1}{5} \end{pmatrix} \begin{pmatrix} 0 & -1 & 1 \\ 0 & 0 & 1 \\ 0 & 0 & 0 \end{pmatrix} = \begin{pmatrix} 0 & \frac{1}{4} & -\frac{1}{4} \\ 0 & \frac{1}{12} & -\frac{5}{12} \\ 0 & -\frac{4}{60} & \frac{8}{60} \end{pmatrix},$$

then the l_∞ *norm of the matrix* T_G *is*

$$\|T_G\|_\infty = \max\left\{\frac{2}{4}, \frac{6}{12}, \frac{12}{60}\right\} = \frac{1}{2} < 1.$$

Thus, the Gauss-Seidel method will converge for the given linear system.

(b) The Gauss-Seidel method for the given system is

$$\begin{aligned} x_1^{(k+1)} &= \frac{1}{4}\Big[12 + x_2^{(k)} - x_3^{(k)}\Big] \\ x_2^{(k+1)} &= \frac{1}{3}\Big[1 + x_1^{(k+1)} - x_3^{(k)}\Big] \\ x_3^{(k+1)} &= \frac{1}{5}\Big[-12 - x_1^{(k+1)} - x_2^{(k+1)}\Big]. \end{aligned}$$

Starting with the initial approximation $x_1^{(0)} = 4, x_2^{(0)} = 3, x_3^{(0)} = -3$, *and for* $k = 0, 1$, *we obtain the first and the second approximations as*

$$\mathbf{x}^{(1)} = [4.5, 2.8333, -4.2667]^T \quad and \quad \mathbf{x}^{(2)} = [4.775, 3.3472, -4.4244]^T.$$

(c) Using error bound formula (3.66), we obtain

$$\|\mathbf{x} - \mathbf{x}^{(2)}\| \le \frac{(1/2)^2}{1 - 1/2}\left\|\begin{pmatrix} 4.5 \\ 2.8333 \\ -4.2667 \end{pmatrix} - \begin{pmatrix} 4 \\ 3 \\ -3 \end{pmatrix}\right\|$$

or

$$\|\mathbf{x} - \mathbf{x}^{(2)}\| \le \frac{1}{2}(1.2667) = 0.6334.$$

(d) To find the number of iterations, we use formula (3.66) as

$$\|\mathbf{x} - \mathbf{x}^{(k)}\| \le \frac{\|T_J\|^k}{1 - \|T_J\|}\|\mathbf{x}^{(1)} - \mathbf{x}^{(0)}\| \le 10^{-4}.$$

It gives

$$\frac{(1/2)^k}{1/2}(1.2667) \le 10^{-4}$$

or

$$(1/2)^k \le \frac{10^{-4}}{2.5334}.$$

Taking ln *on both sides, we obtain*

$$k\ln(1/2) \le \ln\left(\frac{10^{-4}}{2.5334}\right)$$

and it gives

$$k \ge 14.6084 \quad or \quad k = 15,$$

which is the required number of iterations.

Theorem 3.18 *If A is a symmetric positive definite matrix with positive diagonal entries, then the Gauss-Seidel method converges to a unique solution of the linear system $A\mathbf{x} = \mathbf{b}$.*

3.7 Eigenvalues and Eigenvectors

Here, we will discuss briefly eigenvalues and eigenvectors of an $n \times n$ matrix. We also show how they can be used to describe solutions of linear systems.

Definition 3.24 *An $n \times n$ matrix A is said to have an eigenvalue λ of A if there exists a nonzero vector, called an eigenvector $\mathbf{x}$, such that*

$$A\mathbf{x} = \lambda\mathbf{x}. \tag{3.68}$$

Then relation (3.68) represents the eigenvalue problem and we will refer to $(\lambda, \mathbf{x})$ as an eigenpair.

The equivalent form of (3.68) is

$$(A - \lambda\mathbf{I})\mathbf{x} = \mathbf{0}, \tag{3.69}$$

where $\mathbf{I}$ is an $n \times n$ identity matrix. The system of equation (3.69) has nontrivial solutions $\mathbf{x}$ if, and only if, $A - \lambda\mathbf{I}$ is singular or, equivalently,

$$\det(A - \lambda\mathbf{I}) = |A - \lambda\mathbf{I}| = 0. \tag{3.70}$$

Relation (3.70) represents a polynomial equation in λ of degree n, which in principle, could be used to obtain the eigenvalues of matrix A. This equation is called the *characteristic equation* of A. There are n roots of (3.70), which we will denote by $\lambda_1, \lambda_2, \ldots, \lambda_n$. For a given eigenvalue λ_i, the corresponding eigenvector $\mathbf{x}_i$ is not uniquely determined. If $\mathbf{x}$ is an eigenvector then so is $\alpha\mathbf{x}$, where α is any nonzero scalar.

Example 3.31 *Find the eigenvalues and eigenvectors of the following matrix*

$$A = \begin{pmatrix} -6 & 0 & 0 \\ 11 & -3 & 0 \\ -3 & 6 & 7 \end{pmatrix}.$$

Solution. *To find the eigenvalues of the given matrix A using (3.70), we have*

$$\begin{vmatrix} -6-\lambda & 0 & 0 \\ 11 & -3-\lambda & 0 \\ -3 & 6 & 7-\lambda \end{vmatrix} = 0,$$

which gives a characteristic equation of the form

$$\lambda^3 + 2\lambda^2 - 45\lambda - 126 = 0.$$

It factorizes to

$$(-6-\lambda)(-3-\lambda)(7-\lambda) = 0$$

and gives us the eigenvalues $\lambda = -6, \lambda = -3$, *and* $\lambda = 7$ *of the given matrix A. One can note that the sum of these eigenvalues is -2, and this agrees with the trace of A. After finding the eigenvalues of the matrix we turn to the problem of finding eigenvectors. The eigenvectors of A corresponding to eigenvalues* λ *are the nonzero vectors* $\mathbf{x}$ *that satisfy (3.69). Equivalently, the eigenvectors corresponding to* λ *are the nonzero vectors in the solution space of (3.69). We call this solution space the eigenspace of A corresponding to* λ.

To find the eigenvectors of the above given matrix A corresponding to each of these eigenvalues, we substitute each of these three eigenvalues in (3.69). When $\lambda = -6$, *we have*

$$\begin{pmatrix} 0 & 0 & 0 \\ 11 & 3 & 0 \\ -3 & 6 & 13 \end{pmatrix} \begin{pmatrix} x_1 \\ x_2 \\ x_3 \end{pmatrix} = \begin{pmatrix} 0 \\ 0 \\ 0 \end{pmatrix},$$

which implies that

$$\begin{array}{rcrcrl} 0x_1 & + & 0x_2 & + & 0x_3 & = 0 \\ 11x_1 & + & 3x_2 & + & 0x_3 & = 0 \\ -3x_1 & + & 6x_2 & + & 13x_3 & = 0. \end{array}$$

Solving this system, we get, $x_1 = 3, x_2 = -11$, *and* $x_3 = 75$. *Hence, the eigenvector* $\mathbf{x}^{(1)}$ *corresponding to the first eigenvalue,* $\lambda_1 = -6$, *is*

$$\mathbf{x}^{(1)} = \alpha[3, -11, 75]^T, \quad \textit{where} \quad \alpha \in \mathbb{R}, \quad \alpha \neq 0.$$

When $\lambda = -3$, *we have*

$$\begin{pmatrix} -3 & 0 & 0 \\ 11 & 0 & 0 \\ -3 & 6 & 10 \end{pmatrix} \begin{pmatrix} x_1 \\ x_2 \\ x_3 \end{pmatrix} = \begin{pmatrix} 0 \\ 0 \\ 0 \end{pmatrix},$$

which implies that

$$\begin{array}{rcrcrl} 3x_1 & + & 0x_2 & + & 0x_3 & = 0 \\ 11x_1 & + & 0x_2 & + & 0x_3 & = 0 \\ -3x_1 & + & 6x_2 & + & 10x_3 & = 0, \end{array}$$

which gives the solution, $x_1 = 0, x_2 = 5$, *and* $x_3 = -3$. *Hence, the eigenvector* $\mathbf{x}^{(2)}$ *corresponding to the second eigenvalue,* $\lambda_2 = -3$ *is*

$$\mathbf{x}^{(2)} = \alpha[0, 5, -3]^T, \quad \textit{where} \quad \alpha \in \mathbb{R}, \quad \alpha \neq 0.$$

Finally, when $\lambda = 7$, *we have*

$$\begin{pmatrix} -13 & 0 & 0 \\ 11 & -10 & 0 \\ -3 & 6 & 0 \end{pmatrix} \begin{pmatrix} x_1 \\ x_2 \\ x_3 \end{pmatrix} = \begin{pmatrix} 0 \\ 0 \\ 0 \end{pmatrix},$$

which implies that

$$\begin{array}{rcrcrl} -13x_1 & + & 0x_2 & + & 0x_3 & = 0 \\ 11x_1 & - & 10x_2 & + & 0x_3 & = 0 \\ -3x_1 & + & 6x_2 & + & 0x_3 & = 0, \end{array}$$

which gives, $x_1 = x_2 = 0$, *and* $x_3 = 1$. *Hence, the eigenvector* $\mathbf{x}^{(3)}$ *corresponding to the third eigenvalue,* $\lambda_3 = 7$, *is*

$$\mathbf{x}^{(3)} = \alpha[0, 0, 1]^T, \quad \textit{where} \quad \alpha \in \mathbb{R}, \quad \alpha \neq 0.$$

The MATLAB command *eig* is a basic eigenvalue and eigenvector routine. The command

$$>> D = eig(A);$$

returns a vector containing all the eigenvalues of matrix A. If the eigenvectors are also wanted, the syntax

$$>> [X, D] = eig(A);$$

will return a matrix X whose columns are eigenvectors of A corresponding to the eigenvalues in diagonal matrix D. To get the results of Example 3.31, we use the MATLAB command window as follows:

```
>> A = [-6 0 0; 11 -3 0; -3 6 7];
>> P = poly(A);
>> [X, D] = eig(A);
>> eigenvalues = diag(D);
```

Definition 3.25 (Spectral Radius of a Matrix)

Let A be an $n \times n$ matrix. Then the spectral radius $\rho(A)$ of matrix A is defined as

$$\rho(A) = \max_{1 \leq i \leq n} |\lambda_i|,$$

where λ_i are the eigenvalues of a matrix A.

For example, the following matrix

$$A = \begin{pmatrix} 4 & 1 & -3 \\ 0 & 0 & 2 \\ 0 & 0 & -3 \end{pmatrix}$$

has the characteristic equation of the form

$$-\lambda^3 + \lambda^2 + 12\lambda = 0,$$

which gives the eigenvalues $\lambda = 4, 0, -3$ of A. Hence, the spectral radius of A is

$$\rho(A) = \max\{|4|, |0|, |-3|\} = 4.$$

The spectral radius of matrix A may be found from the MATLAB command:

```
>> A = [4 1 − 3; 0 0 2; 0 0 − 3];
>> B = max(eig(A))
B =
   4
```

The necessary and sufficient condition for the convergence of the Jacobi iterative method and the Gauss-Seidel iterative method is defined in the following theorem.

Theorem 3.19 (Necessary and Sufficient Condition for Convergence)

For any initial approximation $\mathbf{x}^{(0)} \in \mathbb{R}$, *the sequence* $\{x^{(k)}\}_{k=0}^{\infty}$ *of approximations defined by*

$$\mathbf{x}^{(k+1)} = T\mathbf{x}^{(k)} + \mathbf{c}, \quad \text{for each} \quad k \geq 0, \quad \text{and} \quad \mathbf{c} \neq 0 \tag{3.71}$$

converges to the unique solution of $\mathbf{x} = T\mathbf{x} + \mathbf{c}$ *if and only if* $\rho(T) < 1$.
Note that the condition $\rho(T) < 1$ *is satisfied when* $\|T\| < 1$ *because* $\rho(T) \leq \|T\|$ *for any natural norm.*

No general results exist to help us choose a method among the Jacobi method or the Gauss-Seidel method to solve an arbitrary linear system. However, the following theorem is suitable for the special case.

Theorem 3.20 *If* $a_{ii} \leq 0$ *for each* $i \neq j$ *and* $a_{ii} > 0$ *for each* $i = 1, 2, \ldots, n$, *then one and only one of the following statements holds:*

1. $0 \leq \rho(T_G) < \rho(T_J) < 1.$
2. $1 < \rho(T_J) < \rho(T_G).$
3. $\rho(T_J) = \rho(T_G) = 0.$
4. $\rho(T_J) = \rho(T_G) = 1.$

Since the rate of convergence of an iterative method depends on the spectral radius of the matrix associated with the method, one may choose a method to accelerate convergence or choose a method whose associated matrix T has minimal spectral radius.

Definition 3.26 (Convergent Matrix)

An $n \times n$ *matrix is called a convergent matrix if*

$$\lim_{k\to\infty} (A^k)_{ij} = 0, \quad \text{for each} \quad i, j = 1, 2, \ldots, n.$$

Example 3.32 *Show that the following matrix*

$$A = \begin{pmatrix} \frac{1}{3} & 0 \\ \frac{1}{9} & \frac{1}{3} \end{pmatrix}$$

is convergent.

Solution. *By computing the powers of the given matrix, we obtain*

$$A^2 = \begin{pmatrix} \frac{1}{9} & 0 \\ \frac{2}{27} & \frac{1}{9} \end{pmatrix}, \quad A^3 = \begin{pmatrix} \frac{1}{27} & 0 \\ \frac{3}{81} & \frac{1}{27} \end{pmatrix}, \quad A^4 = \begin{pmatrix} \frac{1}{81} & 0 \\ \frac{4}{243} & \frac{1}{81} \end{pmatrix}.$$

Then in general, we have

$$A^k = \begin{pmatrix} \left(\frac{1}{3}\right)^k & 0 \\ \frac{k}{3^{k+1}} & \left(\frac{1}{3}\right)^k \end{pmatrix}$$

and it gives

$$\lim_{k\to\infty} \left(\frac{1}{3}\right)^k = 0 \quad and \quad \lim_{k\to\infty} \left(\frac{k}{3^{k+1}}\right) = 0.$$

Hence, given matrix A is convergent.

Since the above matrix has the eigenvalue $\frac{1}{3}$ of order two, its spectral radius is $\frac{1}{3}$. This shows the important relation existing between the spectral radius of a matrix and the convergent of a matrix.

Theorem 3.21 *The following statements are equivalent:*

1. *A is a convergent matrix.*
2. $\lim_{n\to\infty} \|A^n\| = 0,$ *for all natural norms.*
3. $\rho(A) < 1.$

4. $\lim_{n\to\infty} A^n\mathbf{x} = \mathbf{0}$, *for every* $\mathbf{x}$.

We will discuss some very important results concerning eigenvalue problems. The proofs of all the results are beyond the text and will be omitted, however, they are very easily understood and can be used.

Theorem 3.22 *If A is an $n \times n$ matrix then*

(a) $[\rho(A^TA)]^{1/2} = \|A\|_2$.

(b) $\rho(A) \leq \|A\|$, *for any natural norm* $\|.\|$.

Example 3.33 *Consider the following matrix*

$$A = \begin{pmatrix} -2 & 1 & 2 \\ 1 & 0 & 0 \\ 0 & 1 & 0 \end{pmatrix},$$

which gives a characteristic equation of the form

$$det(A - \lambda\mathbf{I}) = -\lambda^3 - 2\lambda^2 + \lambda + 2 = 0.$$

Solving this cubic equation, the eigenvalues of A are -2, -1 and 1. Thus, the spectral radius of A is

$$\rho(A) = max\{|-2|, |-1|, |1|\} = 2.$$

Also,

$$A^TA = \begin{pmatrix} -2 & 1 & 0 \\ 1 & 0 & 1 \\ 2 & 0 & 0 \end{pmatrix} \begin{pmatrix} -2 & 1 & 2 \\ 1 & 0 & 0 \\ 0 & 1 & 0 \end{pmatrix} = \begin{pmatrix} 5 & -2 & -4 \\ -2 & 2 & 2 \\ -4 & 2 & 4 \end{pmatrix}$$

and a characteristic equation of A^TA is

$$-\lambda^3 + 11\lambda^2 - 14\lambda + 4 = 0,$$

which gives the eigenvalues 0.4174, 1 and 9.5826. Therefore, the spectral radius of A^TA is 9.5826. Hence,

$$\|A\|_2 = \sqrt{\rho(A^TA)} = \sqrt{9.5826} \approx 3.0956.$$

From this we conclude that

$$\rho(A) = 2 < 3.0956 \approx \|A\|_2.$$

One can also show that

$$\begin{aligned} \rho(A) &= 2 < 5 = \|A\|_\infty \\ \rho(A) &= 2 < 3 = \|A\|_1, \end{aligned}$$

which satisfies Theorem 3.22.

The spectral norm of matrix A may be found from MATLAB as follows:

```
>> A = [-2 1 2; 1 0 0; 0 1 0];
>> B = sqrt(max(eig(A' * A)))
B =
    3.0956
```

Theorem 3.23 *If A is a symmetric matrix then*

$$\|A\|_2 = \sqrt{\rho(A^T A)} = \rho(A).$$

Example 3.34 *Consider a symmetric matrix*

$$A = \begin{pmatrix} 3 & 0 & 1 \\ 0 & -3 & 0 \\ 1 & 0 & 3 \end{pmatrix},$$

which has a characteristic equation of the form

$$-\lambda^3 + 4\lambda^2 + 9\lambda - 36 = 0.$$

Solving this cubic equation, we have the eigenvalues 4, -3, and 3 of the given matrix A. Therefore, the spectral radius of A is 4. Since A is symmetric,

$$A^T A = A^2 = \begin{pmatrix} 10 & 0 & 6 \\ 0 & 9 & 0 \\ 6 & 0 & 10 \end{pmatrix}.$$

Since we know that the eigenvalues of A^2 are the eigenvalues of A raised to the power 2, the eigenvalues of $A^T A$ are, 16, 9, and 9, and its spectral radius is $\rho(A^T A) = \rho(A^2) = [\rho(A)]^2 = 16$. Hence,

$$\|A\|_2 = \sqrt{\rho(A^T A)} = \sqrt{16} = 4 = \rho(A),$$

which satisfies Theorem 3.23.

Theorem 3.24 *If A is nonsingular matrix, then for any eigenvalue of A*

$$\frac{1}{\|A^{-1}\|_2} \leq |\lambda| \leq \|A\|_2.$$

Note that this result is also true for any natural norm.

Example 3.35 *Consider the following matrix*

$$A = \begin{pmatrix} 2 & 1 \\ 3 & 2 \end{pmatrix}$$

and its inverse matrix is

$$A^{-1} = \begin{pmatrix} 2 & -1 \\ -3 & 2 \end{pmatrix}.$$

Firstly, we find the eigenvalues of the matrix

$$A^T A = \begin{pmatrix} 13 & 8 \\ 8 & 5 \end{pmatrix},$$

which can be obtained by solving a characteristic equation

$$\det(A^T A - \lambda I) = \begin{vmatrix} 13-\lambda & 8 \\ 8 & 5-\lambda \end{vmatrix} = \lambda^2 - 18\lambda + 1 = 0,$$

which gives the eigenvalues 17.96 and 0.04. The spectral radius of $A^T A$ is 17.96. Hence,

$$\|A\|_2 = \sqrt{\rho(A^T A)} = \sqrt{17.96} \approx 4.24$$

Since a characteristic equation of $(A^{-1})^T(A^{-1})$ is

$$\det[(A^{-1})^T(A^{-1}) - \lambda I] = \begin{vmatrix} 13-\lambda & 4 \\ 4 & 5-\lambda \end{vmatrix} = \lambda^2 - 18\lambda + 49 = 0,$$

which gives the eigenvalues 14.64 *and* 3.36 *of $(A^{-1})^T(A^{-1})$, its spectral radius is* 14.64. *Hence,*

$$\|A^{-1}\|_2 = \sqrt{\rho((A^{-1})^T(A^{-1}))} = \sqrt{14.64} \approx 3.83.$$

Note that the eigenvalues of A are 3.73 *and* 0.27, *therefore, its spectral radius is* 3.73. *Hence,*

$$\frac{1}{3.83} < |3.73| < 4.24,$$

which satisfies Theorem 3.24.

3.7.1 Successive Over-relaxation Method

We have seen that the Gauss-Seidel method uses updated information immediately and converges more quickly than the Jacobi method. In some large systems of equations the Gauss-Seidel method converges at a very slow rate. Many techniques have been developed in order to improve the convergence of the Gauss-Seidel method. Perhaps one of the simplest and widely used method is *successive over-relaxation (SOR)*. A useful modification to the Gauss-Seidel method is defined by the iterative scheme

$$\begin{aligned} x_i^{(k+1)} &= (1-\omega)x_i^{(k)} + \frac{\omega}{a_{ii}}\left[b_i - \sum_{j=1}^{i-1} a_{ij}x_j^{(k+1)} - \sum_{j=i+1}^{n} a_{ij}x_j^{(k)}\right] \\ i &= 1,2,\dots,n, \quad k = 1,2,\dots \end{aligned} \tag{3.72}$$

or, it can be written as

$$\begin{aligned} x_i^{(k+1)} &= x_i^{(k)} + \frac{\omega}{a_{ii}}\left[b_i - \sum_{j=1}^{i-1} a_{ij}x_j^{(k+1)} - \sum_{j=i}^{n} a_{ij}x_j^{(k)}\right]. \\ i &= 1,2,\dots,n, \quad k = 1,2,\dots \end{aligned} \tag{3.73}$$

The matrix form of the SOR method can be represented by

$$\mathbf{x}^{(k+1)} = (D+\omega L)^{-1}[(1-\omega)D + \omega U]\mathbf{x}^{(k)} + \omega(D-\omega L)^{-1}\mathbf{b}, \tag{3.74}$$

which is equivalent to

$$\mathbf{x}^{(k+1)} = T_\omega \mathbf{x}^{(k)} + \mathbf{c}, \tag{3.75}$$

where

$$T_\omega = (D+\omega L)^{-1}[(1-\omega)D - \omega U] \quad \text{and} \quad \mathbf{c} = \omega(D-\omega L)^{-1}\mathbf{b} \tag{3.76}$$

are called the *SOR iteration matrix* and the vector, respectively.

The quantity ω is called the relaxation factor. It can be formally proved that convergence can be obtained for values of ω in the range $0 < \omega < 2$. For $\omega = 1$, the SOR method (3.72) is simply the Gauss-Seidel method. The methods involving (3.72) are called *relaxation methods.* For choices of $0 < \omega < 1$, the procedures are called *under-relaxation methods* and can be used to obtain convergence of some systems that are not convergent by the Gauss-Seidel method. For choices $1 < \omega < 2$, the procedures are called *over-relaxation methods*, which can used to accelerate the convergence for systems that are convergent by the Gauss-Seidel method. The SOR methods are particularly useful for solving linear systems that occur in the numerical solutions of certain partial differential equations.

Example 3.36 *Solve the following system of linear equations using the SOR method, with $\epsilon = 0.5 \times 10^{-6}$ in l_∞-norm:*

$$\begin{array}{rcrcrcl} 2x_1 & + & 8x_2 & - & x_3 & = & 11 \\ 5x_1 & - & x_2 & + & x_3 & = & 10 \\ -x_1 & + & x_2 & + & 4x_3 & = & 3. \end{array}$$

Start with the initial approximation $\mathbf{x}^{(0)} = [0, 0, 0]^T$, and take $\omega = 0.35$.
Solution. *For the given system, the SOR method with $\omega = 0.35$ is*

$$\begin{aligned} x_1^{(k+1)} &= (1-\omega)x_1^{(k)} + \frac{\omega}{2}\Big[11 - 8x_2^{(k)} + x_3^{(k)}\Big] \\ x_2^{(k+1)} &= (1-\omega)x_2^{(k)} + \frac{\omega}{-1}\Big[10 - 5x_1^{(k+1)} + x_3^{(k)}\Big] \\ x_3^{(k+1)} &= (1-\omega)x_3^{(k)} + \frac{\omega}{4}\Big[3 + x_1^{(k+1)} - x_2^{(k+1)}\Big]. \end{aligned}$$

Starting with the initial approximation $\mathbf{x}^{(0)} = [0, 0, 0]^T$, and for $k = 0$, we obtain

$$\begin{aligned} x_1^{(1)} &= (1-0.35)x_1^{(0)} + \frac{0.35}{2}\Big[11 - 8x_2^{(0)} + x_3^{(0)}\Big] = 1.925000 \\ x_2^{(1)} &= (1-0.35)x_2^{(0)} + \frac{0.35}{-1}\Big[10 - 5x_1^{(1)} + x_3^{(0)}\Big] = -0.131250 \\ x_3^{(1)} &= (1-0.35)x_3^{(0)} + \frac{0.35}{4}\Big[3 + x_1^{(1)} - x_2^{(1)}\Big] = 0.442422. \end{aligned}$$

The first and subsequent iterations are listed in Table 3.5. We know that for the given system the Gauss-Seidel method diverged (see Example 3.29) but the SOR method converged and it converged very slow.

Example 3.37 *Solve the following system of linear equations using the SOR method, with $\epsilon = 0.5 \times 10^{-6}$ in l_∞-norm:*

$$\begin{array}{rcrcrcrcl} 2x_1 & + & x_2 & & & & & = & 4 \\ x_1 & + & 2x_2 & + & x_3 & & & = & 8 \\ & & x_2 & + & 2x_3 & + & x_4 & = & 12 \\ & & & & x_3 & + & 2x_4 & = & 11. \end{array}$$

Table 3.5: SOR method for Example 3.36

k	$x_1^{(k)}$	$x_2^{(k)}$	$x_3^{(k)}$
0.000000	0	0	0
1.000000	1.925000	-0.131250	0.442422
2.000000	3.437424	2.585027	0.624659
3.000000	0.649603	-0.464296	0.765995
4.000000	3.131306	1.946091	0.864103
5.000000	1.387040	0.494714	0.902245
6.000000	2.291869	1.148120	0.949037
⋮	⋮	⋮	⋮
39.000000	2.000000	0.999999	0.999999
40.000000	2.000000	1.000000	1.000000

Start with the initial approximation $\mathbf{x}^{(0)} = [0,0,0,0]^T$, *and take* $\omega = 1.27$.

Solution. *For the given system, the SOR method with* $\omega = 1.27$ *is*

$$
\begin{aligned}
x_1^{(k+1)} &= (1-\omega)x_1^{(k)} + \frac{\omega}{2}\Big[4 - x_2^{(k)}\Big] \\
x_2^{(k+1)} &= (1-\omega)x_2^{(k)} + \frac{\omega}{2}\Big[8 - x_1^{(k+1)} - x_3^{(k)}\Big] \\
x_3^{(k+1)} &= (1-\omega)x_3^{(k)} + \frac{\omega}{2}\Big[12 - x_2^{(k+1)} - x_4^{(k)}\Big] \\
x_4^{(k+1)} &= (1-\omega)x_4^{(k)} + \frac{\omega}{2}\Big[11 - x_3^{(k+1)}\Big].
\end{aligned}
$$

Starting with the initial approximation $\mathbf{x}^{(0)} = [0,0,0,0]^T$, *and for* $k = 0$, *we obtain*

$$
\begin{aligned}
x_1^{(1)} &= (1-1.27)x_1^{(0)} + \frac{1.27}{2}[4 - x_2^{(0)}] = 2.54 \\
x_2^{(1)} &= (1-1.27)x_2^{(0)} + \frac{1.27}{2}[8 - x_1^{(1)} - x_3^{(0)}] = 3.4671 \\
x_3^{(1)} &= (1-1.27)x_3^{(0)} + \frac{1.27}{2}[12 - x_2^{(1)} - x_4^{(0)}] = 5.418392
\end{aligned}
$$

$$x_4^{(1)} = (1-1.27)x_4^{(0)} + \frac{1.27}{2}[11 - x_3^{(1)}] = 3.544321.$$

The first and subsequent iterations are listed in Table 3.6.

Table 3.6: Solution of Example 3.37 by the SOR method.

k	$x_1^{(k)}$	$x_2^{(k)}$	$x_3^{(k)}$	$x_4^{(k)}$
0	0.000000	0.000000	0.000000	0.000000
1	2.540000	3.467100	5.418392	3.544321
2	-0.34741	0.923809	3.319772	3.919978
3	2.047182	1.422556	3.331152	3.811324
4	1.083938	1.892328	3.098770	3.988224
5	1.045709	1.937328	3.020607	3.990094
6	1.027456	1.986402	3.009361	3.996730
⋮	⋮	⋮	⋮	⋮
15	0.999999	2.000000	3.000000	4.000000
16	1.000000	2.000000	3.000000	4.000000

To get the above results using MATLAB, we do the following:

```
>> Ab = [A|b] = [2 1 0 0 4; 1 2 1 0 8; 0 1 2 1 12; 0 0 1 2 11];
>> w = 1.27; acc = 0.5e - 6;
>> SORM(Ab, w, acc);
```

Note that the SOR method converges and required 16 iterations to obtain what is obviously the correct solution for the given system. If we solve Example 3.37 using the Gauss-Seidel method, we find that this method also converges but very slow because it needed 36 iterations to obtain the correct solution, shown by Table 3.7, which is 20 iterations more than required by the SOR method. Also, if we solve the same example using the Jacobi method we find that it needed 73 iterations to get the correct solution.

Comparing the SOR method with the Gauss-Seidel method a large reduction in the number of iterations can be achieved, given an efficient choice of ω.

In practice ω should be chosen in the range $1 < \omega < 2$ but the precise choice of ω is a major problem. Finding the optimum value for ω depends on the particular problem (size of the system of equations and the nature of the equations) and often requires careful calculation. A detailed study for the optimization of

Table 3.7: Solution of Example 3.37 by the Gauss-Seidle Method.

k	$x_1^{(k)}$	$x_2^{(k)}$	$x_3^{(k)}$	$x_4^{(k)}$
0	0.000000	0.000000	0.000000	0.000000
1	2.000000	3.000000	4.500000	3.250000
2	0.500000	1.500000	3.625000	3.687500
3	1.250000	1.562500	3.375000	3.812500
4	1.218750	1.703125	3.242188	3.878906
5	1.148438	1.804688	3.158203	3.920898
6	1.097656	1.872070	3.103516	3.948242
⋮	⋮	⋮	⋮	⋮
35	1.000000	1.999999	3.000000	4.000000
36	1.000000	2.000000	3.000000	4.000000

ω can be found in Isaacson and Keller (1966). The following theorems can be used in certain situations for the convergence of the SOR method.

Program 3.13

MATLAB m-file for the SOR Iterative Method

```
function sol=SORM(Ab,w,acc)
[n,t]=size(Ab); b=Ab(1:n,t); R=1; k=1;
x=zeros(1,n);d(1,1:n+1)=[0  x];
k=k+1; while R > acc
for i=1:n
sum=0;
for j=1:n
if j <= i - 1; sum = sum + Ab(i,j) * d(k,j + 1);
elseif j >= i + 1; sum = sum + Ab(i,j) * d(k - 1,j + 1);
end;end
x(1,i) = (1-w)*d(k-1,i+1)+(w/Ab(i,i))*(b(i,1)-sum);
d(k,1) = k - 1; d(k,i + 1) = x(1,i); end
R = max(abs((d(k,2 : n + 1) - d(k - 1,2 : n + 1))));
if R > 100 & k > 10; break; end
k=k+1; end; x=d;
```

Theorem 3.25 *If all the diagonal elements of matrix A are nonzero, that is, $a_{ii} \neq 0$, for each $i = 1, 2, \ldots, n$, then*

$$\rho(T_\omega) = |\omega - 1|.$$

This implies that the SOR method can converge only if $0 < \omega < 2$.

Theorem 3.26 *If A is a positive definite matrix and $0 < \omega < 2$, then the SOR method converges for any choice of the initial approximation vector $\mathbf{x}^{(0)} \in \mathbb{R}$.*

Theorem 3.27 *If A is a positive definite and tridiagonal matrix, then*

$$\rho(T_G) = [\rho(T_J)]^2 < 1$$

and the optimal choices for relaxation factor ω for the SOR method is

$$\omega = \frac{2}{1 + \sqrt{1 - [\rho(T_J)]^2}}, \tag{3.77}$$

where T_G and T_J are the Gauss-Seidel iteration and the Jacobi iteration matrices, respectively. With this choice of relaxation factor ω, we can have the spectral radius of the SOR iteration matrix T_ω as

$$\rho(T_\omega) = \omega - 1.$$

Note that the optimal value of ω can be found also using (3.77) if the eigenvalues of the Jacobi iteration matrix T_J are real and $0 < \rho(T_J) < 1$.

Example 3.38 *Find the optimal choice for the relaxation factor ω for using it in the SOR method for solving the linear system $A\mathbf{x} = \mathbf{b}$, where the coefficient matrix A is given as follows*

$$A = \begin{pmatrix} 2 & -1 & 0 \\ -1 & 2 & -1 \\ 0 & -1 & 2 \end{pmatrix}.$$

Solution. *Since the given matrix A is positive definite and tridiagonal, we can use Theorem 3.27 to find the optimal choice for ω. Using matrix A, we can find the Jacobi iteration matrix T_J as follows*

$$T_J = \begin{pmatrix} 0 & \frac{1}{2} & 0 \\ \frac{1}{2} & 0 & \frac{1}{2} \\ 0 & \frac{1}{2} & 0 \end{pmatrix}.$$

Now to find the spectral radius of the Jacobi iteration matrix T_J, we use the following characteristic equation

$$\det(T_J - \lambda \mathbf{I}) = |T_J - \lambda \mathbf{i}| = -\lambda^3 + \frac{\lambda}{2},$$

which gives the eigenvalues of matrix T_J, as $\lambda = 0, \pm\frac{1}{\sqrt{2}}$. Thus,

$$\rho(T_J) = \frac{1}{\sqrt{2}} = 0.707107$$

and the optimal value of ω is

$$\omega = \frac{2}{1 + \sqrt{1 - (0.707107)^2}} = 1.171573.$$

Procedure 3.11 [SOR Method]

1. *Find or take ω in the interval $(0, 2)$ (for guaranteed convergence).*
2. *Initialize the first approximation $\mathbf{x}^{(0)}$ and preassigned accuracy ϵ.*
3. *Compute the constant $\mathbf{c} = \omega(D - \omega L)^{-1}\mathbf{b}$.*
4. *Compute the SOR iteration matrix $T_\omega = (D + \omega L)^{-1}[(1 - \omega)D - \omega U]$.*
5. *Solve for the approximate solutions $x_i^{(k+1)} = T_\omega \mathbf{x}_i^{(k)} + \mathbf{c}, \quad i = 1, 2, \ldots, n$ $k = 0, 1, \ldots$.*
6. *Repeat step 5 until $\|\mathbf{x}_i^{(k+1)} - \mathbf{x}_i^{(k)}\| < \epsilon$.*

3.7.2 Conjugate Gradient Method

So far, we have discussed two broad classes of methods for solving linear systems. The first class, known as *direct methods* (Chapter 2) is based on some version of Gaussian elimination or LU decomposition. Direct methods eventually obtain the exact solution, but must be carried through to completion before any useful information is obtained. The second class contains the *iterative methods* discussed in the present chapter that lead to closer and closer approximations to the solution, but almost never reach the exact value.

Now we discuss a method called the *conjugate gradient method*, which was developed in 1952. It was originally developed as a direct method designed to solve an $n \times n$ positive definite linear system. As a direct method it is generally inferior to Gaussian elimination with pivoting since both methods require n major steps to determine a solution, and the steps of the conjugate gradient method are more computationally expansive than those in the Gaussian elimination. However, the conjugate gradient method is very useful when employed as an iterative approximation method for solving large sparse systems. Actually, this method is rarely used as a primary method for solving linear systems; rather, its more common applications arise in solving differential equations and when other iterative methods converge very slowly. We assume the coefficient matrix A of the linear system $A\mathbf{x} = \mathbf{b}$ is positive definite and orthogonality with respect to the inner product notation

$$< \mathbf{x}, \mathbf{y} >= \mathbf{x}^T A\mathbf{y},$$

where $\mathbf{x}$ and $\mathbf{y}$ are n-dimensional vectors. Also, we have for each $\mathbf{x}$ and $\mathbf{y}$,

$$< \mathbf{x}, A\mathbf{y} >=< A\mathbf{x}, \mathbf{y} > .$$

The conjugate gradient method is a variational approach in which we seek the vector $\mathbf{x}^*$, a solution to the linear system $A\mathbf{x} = \mathbf{b}$ if and only if $\mathbf{x}^*$ minimizes

$$E(\mathbf{x}) =< \mathbf{x}, A\mathbf{x} > -2 < \mathbf{x}, \mathbf{b} > . \tag{3.78}$$

In addition, for any $\mathbf{x}$ and $\mathbf{v} \neq \mathbf{0}$ the function $E(\mathbf{x} + t\mathbf{v})$ has its minimum when

$$t = \frac{< \mathbf{v}, \mathbf{b} - A\mathbf{x} >}{< \mathbf{v}, A\mathbf{v} >}.$$

The process is started by specifying an initial estimate $\mathbf{x}^{(0)}$ at iteration zero, and by computing the initial residual vector from

$$\mathbf{r}^{(0)} = \mathbf{b} - A\mathbf{x}^{(0)}.$$

We then obtain improved estimates $\mathbf{x}^{(k)}$ from the iterative process

$$\mathbf{x}^{(k)} = \mathbf{x}^{(k-1)} + t_k \mathbf{v}^{(k)}, \tag{3.79}$$

where $\mathbf{v}^{(k)}$ is a search direction expressed as a vector and the value of

$$t_k = \frac{< \mathbf{v}^{(k)}, \mathbf{b} - A\mathbf{x}^{(k-1)} >}{< \mathbf{v}^{(k)}, A\mathbf{v}^{(k)} >}$$

is chosen to minimize the value of $E(\mathbf{x}^{(k)})$.

In a related method called the method of *steepest descent*, $\mathbf{v}^{(k)}$ is chosen as the residual vector

$$\mathbf{v}^{(k)} = \mathbf{r}^{(k-1)} = \mathbf{b} - A\mathbf{x}^{(k-1)}.$$

This method has merit for nonlinear systems and optimization problems; it is not used for linear systems because of slow convergence. An alternative approach uses a set of nonzero direction vectors $\{\mathbf{v}^{(1)}, \ldots, \mathbf{v}^{(n)}\}$ that satisfy

$$< \mathbf{v}^{(i)}, A\mathbf{v}^{(j)} > = 0, \qquad \text{if} \quad i \neq j.$$

This is called an A-*orthogonality condition*, and the set of vectors $\{\mathbf{v}^{(1)}, \ldots, \mathbf{v}^{(n)}\}$ is said to be A-*orthogonal.*

In the conjugate gradient method, we use $\mathbf{v}^{(1)}$ equal to $\mathbf{r}^{(0)}$ *only* at the beginning of the process. For all later iterations, we choose

$$\mathbf{v}^{(k+1)} = \mathbf{r}^{(k)} + \frac{\|\mathbf{r}^{(k)}\|^2}{\|\mathbf{r}^{(k-1)}\|^2}\mathbf{v}^{(k)}$$

to be conjugated to all previous direction vectors.

Note that the initial approximation $\mathbf{x}^{(0)}$ can be chosen by the user, with $\mathbf{x}^{(0)} = \mathbf{0}$ as the default. The number of iterations $m \leq n$ can be chosen by the user in advance; alternatively, one can impose a stopping criterion based on the size of the residual vector, $\|\mathbf{r}^{(k)}\|$, or, alternatively, the distance between successive iterates, $\|\mathbf{x}^{(k+1)} - \mathbf{x}^{(k)}\|$. If the process is carried on to the bitter end; that is, $m = n$, then, in the absence of round-off errors, the results will be the exact solution to the linear system. More iterations than n may be required in practical applications because of the introduction of round-off errors.

Example 3.39 *The following linear system*

$$\begin{array}{rcrcrcl} 2x_1 & - & x_2 & & & = & 1 \\ -x_1 & + & 2x_2 & - & x_3 & = & 0 \\ & - & x_2 & + & x_3 & = & 1 \end{array}$$

has the exact solution $\mathbf{x} = [2, 3, 4]^T$. *Solve the system by the conjugate gradient method.*

Solution. *Start with the initial approximation* $\mathbf{x}^{(0)} = [0, 0, 0]^T$, *and find the residual vector as*

$$\mathbf{r}^{(0)} = \mathbf{b} - A\mathbf{x}^{(0)} = \mathbf{b} = [1, 0, 1]^T.$$

The first conjugate direction is $\mathbf{v}^{(1)} = \mathbf{r}^{(0)} = [1, 0, 1]^T$. *Since* $\|\mathbf{r}^{(0)}\| = \sqrt{2}$ *and*

$$< \mathbf{v}^{(1)}, \mathbf{v}^{(1)} >= [\mathbf{v}^{(1)}]^T A \mathbf{v}^{(1)} = 3$$

we use (3.79) to obtain the updated approximation to the solution

$$\mathbf{x}^{(1)} = \mathbf{x}^{(0)} + \frac{\|\mathbf{r}^{(0)}\|^2}{< \mathbf{v}^{(1)}, \mathbf{v}^{(1)} >} \mathbf{v}^{(1)} = \frac{2}{3} \begin{pmatrix} 1 \\ 0 \\ 1 \end{pmatrix} = \begin{pmatrix} \frac{2}{3} \\ 0 \\ \frac{2}{3} \end{pmatrix}.$$

Now we compute the next residual vector as

$$\mathbf{r}^{(1)} = \mathbf{b} - A\mathbf{x}^{(1)} = \mathbf{b} = [-1/3, 4/3, 4/3]^T$$

and the conjugate direction as

$$\mathbf{v}^{(2)} = \mathbf{r}^{(1)} + \frac{\|\mathbf{r}^{(1)}\|^2}{\|\mathbf{r}^{(0)}\|^2} \mathbf{v}^{(1)} = \begin{pmatrix} -\frac{1}{3} \\ \frac{4}{3} \\ \frac{1}{3} \end{pmatrix} + \frac{2}{2} \begin{pmatrix} 1 \\ 0 \\ 1 \end{pmatrix} = \begin{pmatrix} \frac{2}{3} \\ \frac{4}{3} \\ \frac{4}{3} \end{pmatrix},$$

which, as designed, satisfies the conjugacy condition $< \mathbf{v}^{(1)}, \mathbf{v}^{(2)} >= [\mathbf{v}^{(1)}]^T A \mathbf{v}^{(2)} = 0$. *Now the new approximation is given*

$$\mathbf{x}^{(2)} = \mathbf{x}^{(1)} + \frac{\|\mathbf{r}^{(1)}\|^2}{< \mathbf{v}^{(2)}, \mathbf{v}^{(2)} >} \mathbf{v}^{(2)} = \begin{pmatrix} \frac{2}{3} \\ 0 \\ \frac{2}{3} \end{pmatrix} + \frac{2}{8/9} \begin{pmatrix} \frac{2}{3} \\ \frac{4}{3} \\ \frac{4}{3} \end{pmatrix} = \begin{pmatrix} \frac{13}{6} \\ 3 \\ \frac{11}{3} \end{pmatrix}.$$

Since we are dealing with a 3×3 *system, we will recover the exact solution by one more iteration of the method. The new residual vector is*

$$\mathbf{r}^{(2)} = \mathbf{b} - A\mathbf{x}^{(2)} = \mathbf{b} = [-1/3, -1/6, 1/3]^T$$

and the final conjugate direction is

$$\mathbf{v}^{(3)} = \mathbf{r}^{(2)} + \frac{\|\mathbf{r}^{(2)}\|^2}{\|\mathbf{r}^{(1)}\|^2}\mathbf{v}^{(2)} = \begin{pmatrix} -\frac{1}{3} \\ -\frac{1}{6} \\ \frac{1}{3} \end{pmatrix} + \frac{1/4}{2}\begin{pmatrix} \frac{2}{3} \\ \frac{4}{3} \\ \frac{4}{3} \end{pmatrix} = \begin{pmatrix} -\frac{1}{4} \\ 0 \\ \frac{1}{2} \end{pmatrix},$$

which, as one can check, is conjugated to both $\mathbf{v}^{(1)}$ *and* $\mathbf{v}^{(2)}$*. Thus, the solution obtained is*

$$\mathbf{x}^{(3)} = \mathbf{x}^{(2)} + \frac{\|\mathbf{r}^{(2)}\|^2}{< \mathbf{v}^{(3)}, \mathbf{v}^{(3)} >}\mathbf{v}^{(3)} = \begin{pmatrix} \frac{13}{6} \\ 3 \\ \frac{11}{3} \end{pmatrix} + \frac{1/4}{3/8}\begin{pmatrix} -\frac{1}{4} \\ 0 \\ \frac{1}{2} \end{pmatrix} = \begin{pmatrix} 2 \\ 3 \\ 4 \end{pmatrix}.$$

Since we applied the method $n = 3$ *times, this is the actual solution.*

Note that in larger examples, one would not carry through the method to the bitter end since an approximation to the solution is typically obtained with only a few iterations. The result can be a substantial savings in computational time and effort required to produce an approximation to the solution.

To get above results using MATLAB, we do the following:

```
>> A = [2 -1 0;-1 2 -1;0 -1 1];
>> b = [1 0 1]';
>> x0 = [0 0 0]';
>> acc = 0.5e-6;
>> maxI = 3;
>> CONJG(A, b, x0, acc, maxI);
```

Program 3.14

MATLAB m-file for the Conjugate Gradient Method

```
function x=CONJG(A,b,x0,acc,maxI)
x = x0; r = b - A * x0; v = r;
alpha=r' * r; iter=0;flag=0;
normb=norm(b); if normb < eps; normb=1; end
while (norm(r)/normb > acc)
u = A * v; t = alpha/(u' * v); x = x + t * v;
r = r - t * u; beta = r' * r;
v = r + beta/alpha * v; alpha = beta;
iter = iter+1; if (iter == maxI); flag= 1;
break; end; end
```

Procedure 3.12 [Conjugate Gradient Method]

1. *Initialize the first approximation* $\mathbf{x}^{(0)} = 0$.
2. *Compute* $\mathbf{r}^{(0)}$ *and set* $\mathbf{v}^{(1)}$ *equal to* $\mathbf{r}^{(0)}$.
3. *For iterations* k *equal to* $(1, 2, \ldots)$ *until convergence (a) Compute* $\mathbf{v}^{(k+1)} = \mathbf{r}^{(k)} + \dfrac{\|\mathbf{r}^{(k)}\|^2}{\|\mathbf{r}^{(k-1)}\|^2}\mathbf{v}^{(k)}$ *(b) Compute* $\mathbf{x}^{(k+1)} = \mathbf{x}^{(k)} + \dfrac{\|\mathbf{r}^{(k)}\|^2}{[\mathbf{v}^{(k+1)}]^T A\mathbf{v}^{(k+1)}}\mathbf{v}^{(k+1)}$.

3.8 Conditioning of Linear Systems

In solving the linear system numerically we have to consider the problem's conditioning, algorithm stability, and cost. In using direct methods for solutions of linear systems we discussed efficient elimination schemes, and these schemes are stable when pivoting is employed. But there are some ill-conditioned systems, which are tough to solve by any method. These types of linear systems are identified here.

Here, we will present a parameter, the *condition number*, which quantitatively measures the conditioning of a linear system. The condition number is greater and equal to one and as a linear system becomes more ill-conditioned, the condition number increases. After factoring a matrix, the condition number can be estimated in roughly the same time as it takes to solve a few factored systems $(LU)\mathbf{x} = \mathbf{b}$. Hence, after factoring a matrix, the extra computer time needed to estimate the condition number is usually insignificant.

3.8.1 Errors in Solving Linear Systems

Any computed solution of a linear system must, because of round-off and other errors, be considered an approximate solution. Here, we shall consider the most natural method for determining the accuracy of a solution of the linear system. One obvious way of estimating the accuracy of the computed solution $\mathbf{x}^*$ is to compute $A\mathbf{x}^*$ and to see how close $A\mathbf{x}^*$ comes to $\mathbf{b}$. Thus, if $\mathbf{x}^*$ is an approximate solution of the given system $A\mathbf{x} = \mathbf{b}$, we compute a vector

$$\mathbf{r} = \mathbf{b} - A\mathbf{x}^*, \tag{3.80}$$

which is called the *residual vector* and can be easily calculated. The quantity

$$\frac{\|\mathbf{r}\|}{\|\mathbf{b}\|} = \frac{\|\mathbf{b} - A\mathbf{x}^*\|}{\|\mathbf{b}\|}$$

is called the *relative residual.*

Program 3.15

MATLAB m-file for finding Residual Vector

```
function r=RESID(A,b,x0)
[n,n]=size(A);
for i=1:n; R(i) = b(i);
for j=1:n
R(i)=R(i)-A(i,j)*x0(j);end
RES(i)=R(i); end
r=RES'
```

The smallness of the residual then provides a measure of the goodness of the approximate solution $\mathbf{x}^*$. If every component of vector $\mathbf{r}$ vanishes, then $\mathbf{x}^*$ is the exact solution. If $\mathbf{x}^*$ is a good approximation then we would expect each component of $\mathbf{r}$ to be small, at least in a relative sense. For example, the following linear system

$$\begin{array}{rcrcl} x_1 & + & 2x_2 & = & 3 \\ 1.0001x_1 & + & 2x_2 & = & 3.0001 \end{array}$$

has the approximate solution $\mathbf{x}^* = [3, 0]^T$. To see how good this solution is, we compute the residual, $\mathbf{r} = [0, -0.0002]^T$.

To get above results using MATLAB, we do the following:

```
>> A = [1 2; 1.0001 2];
>> b = [3 3.0001];
>> x0 = [3 0];
>> RESID(A, b, x0);
```

We can conclude from the residual that the approximate solution is correct to at most three decimal places. Also, the following linear system

$$
\begin{array}{rcrcrcrcl}
1.0000x_1 & + & 0.9600x_2 & + & 0.8400x_3 & + & 0.6400x_4 & = & 3.4400 \\
0.9600x_1 & + & 0.9214x_2 & + & 0.4406x_3 & + & 0.2222x_4 & = & 2.5442 \\
0.8400x_1 & + & 0.4406x_2 & + & 1.0000x_3 & + & 0.3444x_4 & = & 2.6250 \\
0.6400x_1 & + & 0.2222x_2 & + & 0.3444x_3 & + & 1.0000x_4 & = & 2.2066
\end{array}
$$

has the exact solution $\mathbf{x} = [1, 1, 1, 1]^T$ and the approximate solution due to the Gaussian elimination without pivoting is

$$\mathbf{x}^* = [1.0000322, 0.99996948, 0.99998748, 1.0000113]^T$$

and the residual is

$$\mathbf{r} = [0.6 \times 10^{-7}, 0.6 \times 10^{-7}, -0.53 \times 10^{-5}, -0.21 \times 10^{-4}]^T.$$

The approximate solution due to Gaussian elimination with partial pivoting is

$$\mathbf{x}^* = [0.9999997, 0.99999997, 0.99999996, 1.0000000]^T$$

and the residual is

$$\mathbf{r} = [0.3 \times 10^{-7}, 0.3 \times 10^{-7}, 0.6 \times 10^{-7}, 0.1 \times 10^{-8}]^T.$$

We found that all the elements of the residual for the second case (with pivoting) are less than 0.6×10^{-7}, whereas for the first case (without pivoting) they are as large as 0.2×10^{-4}. Even without knowing the exact solution, it is clear that the solution obtained in case 2 is much better than that of case 1. The residual provides a reasonable measure of the accuracy of a solution in those cases where the error is primarily due to the accumulation of round-off errors.

Intuitively it would seem reasonable to assume that when $\|\mathbf{r}\|$ is small for a given vector norm, then the error $\|\mathbf{x} - \mathbf{x}^*\|$ would be small as well. In fact, this is true for some systems. However, there are systems of equations, which do not satisfy this property. Such systems are said to be *ill-conditioned.*

These are systems in which small changes in the coefficients of the system lead to large changes in the solution. For example, consider the linear system

$$\begin{array}{rcrcl} x_1 & + & x_2 & = & 2 \\ x_1 & + & 1.01x_2 & = & 2.01. \end{array}$$

The exact solution is easily verified to be $x_1 = x_2 = 1$. On the other hand, the system

$$\begin{array}{rcrcl} x_1 & + & x_2 & = & 2 \\ 1.001x_1 & + & x_2 & = & 2.01 \end{array}$$

has the solution $x_1 = 10, x_2 = -8$. Thus, a change of one percent in the coefficients has changed the solution by a factor of 10. If in the above given system, we substitute $x_1 = 10, x_2 = 8$, we find that the residual is $r_1 = 0, r_2 = 0.09$, so that this solution looks reasonable although it is grossly in error. In practical problems we can expect the coefficients in the system to be subject to small errors, either because of round-off or because of physical measurement. If the system is ill-conditioned, the resulting solution may be grossly in error. Errors of this type, unlike those caused by round-off error accumulation, cannot be avoided by careful programming.

We have seen that for ill-conditioned systems the residual is not necessarily a good measure of the accuracy of a solution. How then can we tell when a system is ill-conditioned? In the following we discuss some possible indicators of an ill-conditioned system.

Definition 3.27 (Condition Number of a Matrix)

The number $\|A\|\|A^{-1}\|$ is called the condition number of a nonsingular matrix A and is denoted by $K(A)$; that is,

$$cond(A) = K(A) = \|A\|\|A^{-1}\|. \tag{3.81}$$

Note that the condition number $K(A)$ for A depends on the matrix norm used and can, for some matrices, vary considerably as the matrix norm is changed. Since

$$1 = \|I\| = \|AA^{-1}\| \leq \|A\|\|A^{-1}\| = K(A)$$

therefore, the condition number is always in the range $1 \leq K(A) \leq \infty$ regardless of any natural norm. The lower limit is attained for identity matrices and $K(A) = \infty$ if A is singular. So the matrix A is *well-behaved* (or well-conditioned) if $K(A)$ is close to 1 and is increasingly *ill-conditioned* when $K(A)$ is significantly greater than 1; that is, $K(A) \to \infty$.

The condition numbers provide bounds for the sensitivity of the solution of a set of equations to changes in the coefficient matrix. Unfortunately, the evaluation of any of the condition numbers of a matrix A is not a trivial task since it is necessary first to obtain its inverse.

So if the condition number of a matrix is a very large number then this is one of the indicators of an ill-conditioned system. An other indicator of ill-conditioning is when the pivots during the process of elimination suffer a loss of one or more significant figures. Small changes in the right-hand side terms of the system lead to large changes in the solution, another indicator of an ill-conditioned system. Also, when the elements of the inverse of the coefficient matrix are large compared to the elements of the coefficients matrix, this also indicates an ill-conditioned system.

Example 3.40 *Compute the condition number of the following matrix using the l_∞-norm*

$$A = \begin{pmatrix} 2 & -1 & 0 \\ 2 & -4 & -1 \\ -1 & 0 & 2 \end{pmatrix}.$$

Solution. *Since the condition number of a matrix is defined as*

$$K(A) = \|A\|_\infty \|A^{-1}\|_\infty.$$

First, we calculate the inverse of the given matrix, which is

$$A^{-1} = \begin{pmatrix} \frac{8}{13} & -\frac{2}{13} & -\frac{1}{13} \\ \frac{3}{13} & -\frac{4}{13} & -\frac{2}{13} \\ \frac{4}{13} & -\frac{1}{13} & \frac{6}{13} \end{pmatrix}.$$

Now we calculate the l_∞-norm of both matrices A and A^{-1}. Since the l_∞-norm of a matrix is the maximum of the absolute row sums, we have

$$\|A\|_\infty = max\{|2| + |-1| + |0|, |2| + |-4| + |-1|, |-1| + |0| + |2|\} = 7$$

and

$$\|A^{-1}\|_\infty = max\left\{\left|\frac{8}{13}\right| + \left|\frac{-2}{13}\right| + \left|\frac{-1}{13}\right|, \left|\frac{3}{13}\right| + \left|\frac{-4}{13}\right| + \left|\frac{-2}{13}\right|, \left|\frac{4}{13}\right| + \left|\frac{-1}{13}\right| + \left|\frac{6}{13}\right|\right\},$$

which gives

$$\|A^{-1}\|_\infty = \frac{11}{13}.$$

Therefore,

$$K(A) = \|A\|_\infty \|A^{-1}\|_\infty = (7)\left(\frac{11}{13}\right) \approx 5.9231$$

Depending on the application, we might consider this number to be reasonably small and conclude that the given matrix A is reasonably well-conditioned.

To get above results using MATLAB, we do the following:

```
>> A = [2 -1 0; 2 -4 -1; -1 0 2];
>> Ainv = inv(A)
>> K(A) = norm(A, inf) * norm(Ainv, inf);
K(A) =
    5.9231
```

Some matrices are notoriously ill-conditioned. For example, consider the 4×4 Hilbert matrix

$$H = \begin{pmatrix} 1 & \frac{1}{2} & \frac{1}{3} & \frac{1}{4} \\ \frac{1}{2} & \frac{1}{3} & \frac{1}{4} & \frac{1}{5} \\ \frac{1}{3} & \frac{1}{4} & \frac{1}{5} & \frac{1}{6} \\ \frac{1}{4} & \frac{1}{5} & \frac{1}{6} & \frac{1}{7} \end{pmatrix}$$

whose entries are defined by

$$a_{ij} = \frac{1}{(i+j-1)}, \quad \text{for} \quad i, j = 1, 2, \ldots, n.$$

The inverse of matrix H can be obtained as

$$H^{-1} = \begin{pmatrix} 16 & -120 & 240 & -140 \\ -120 & 1200 & -2700 & 1680 \\ 240 & -2700 & 6480 & -4200 \\ -140 & 1680 & -4200 & 2800 \end{pmatrix}.$$

Then the condition number of the Hilbert matrix is

$$K(H) = \|H\|_\infty \|H^{-1}\|_\infty = (2.0833)(13620) \approx 28375,$$

which is quite large. Note that the condition number of Hilbert matrices increases rapidly as the size of the matrices increase. Therefore, large Hilbert matrices are considered to be extremely ill-conditioned.

We might think that if the determinant of a matrix is close to zero, then the matrix is ill-conditioned. However, this is false. Consider the following matrix

$$A = \begin{pmatrix} 10^{-7} & 0 \\ 0 & 10^{-7} \end{pmatrix}$$

for which $\det A = 10^{-14} \approx 0$. One can easily find the condition number of the given matrix as

$$K(A) = \|A\|_\infty \|A^{-1}\|_\infty = (10^{-7})(10^7) = 1.$$

Matrix A is therefore perfectly conditioned. Thus, a small determinant is necessary but not sufficient for a matrix to be ill-conditioned.

The condition number of a matrix $K(A)$ using l_2-norm can be computed by the built-in function *cond* command in MATLAB as follows:

```
>> A = [1 -1 2; 3 1 -1; 2 0 1];
>> K(A) = cond(A);
K(A) =
    19.7982
```

Theorem 3.28 (Error in Linear Systems)

Suppose that $\mathbf{x}^*$ *is an approximation to the solution* $\mathbf{x}$ *of the linear system* $A\mathbf{x} = \mathbf{b}$ *and* A *is a nonsingular matrix and* $\mathbf{r}$ *is the residual vector for* $\mathbf{x}^*$. *Then for any natural norm, the error is*

$$\|\mathbf{x} - \mathbf{x}^*\| \leq \|\mathbf{r}\| \|A^{-1}\| \tag{3.82}$$

and the relative error is

$$\frac{\|\mathbf{x} - \mathbf{x}^*\|}{\|\mathbf{x}\|} \leq K(A) \frac{\|\mathbf{r}\|}{\|\mathbf{b}\|}, \quad \textit{provided} \quad \mathbf{x} \neq 0, \ \mathbf{b} \neq 0. \tag{3.83}$$

Proof. Since $\mathbf{r} = \mathbf{b} - A\mathbf{x}^*$ and A is nonsingular, then

$$A\mathbf{x} - A\mathbf{x}^* = \mathbf{b} - (\mathbf{b} - \mathbf{r}) = \mathbf{r},$$

which implies that

$$A(\mathbf{x} - \mathbf{x}^*) = \mathbf{r} \tag{3.84}$$

or

$$\mathbf{x} - \mathbf{x}^* = A^{-1}\mathbf{r}.$$

Taking norm on both side, gives

$$\|\mathbf{x} - \mathbf{x}^*\| = \|A^{-1}\mathbf{r}\| \leq \|A^{-1}\|\|\mathbf{r}\|.$$

Moreover, since $\mathbf{b} = A\mathbf{x}$, then

$$\|\mathbf{b}\| \leq \|A\|\|\mathbf{x}\|, \qquad \text{or} \quad \|\mathbf{x}\| \geq \frac{\|\mathbf{b}\|}{\|A\|}.$$

Hence,

$$\frac{\|\mathbf{x} - \mathbf{x}^*\|}{\|\mathbf{x}\|} \leq \frac{\|A^{-1}\|\|\mathbf{r}\|}{\|\mathbf{b}\|/\|A\|} \leq K(A)\frac{\|\mathbf{r}\|}{\|\mathbf{b}\|}.$$

Inequalities (3.82) and (3.83) imply that the quantities $\|A^{-1}\|$ and $K(A)$ can be used to give an indication of the connection between the residual vector and the accuracy of the approximation. If the quantity $K(A) \approx 1$, the relative error will be fairly close to the relative residual. But if $K(A) >> 1$, then the relative error could be many times larger than the relative residual.

Example 3.41 *Consider the following linear system*

$$\begin{array}{rcrcrcr} x_1 & + & x_2 & - & x_3 & = & 1 \\ x_1 & + & 2x_2 & - & 2x_3 & = & 0 \\ -2x_1 & + & x_2 & + & x_3 & = & -1. \end{array}$$

(a) Discuss the ill-conditioning of the given linear system.
(b) If $\mathbf{x}^* = [2.01, 1.01, 1.98]^T$ *is an approximate solution of the given system, then find the residual vector* $\mathbf{r}$ *and its norm* $\|\mathbf{r}\|_\infty$.
(c) Estimate the relative error using (3.83).
(d) Use the simple Gaussian elimination method to find the approximate error using (3.84).

Solution. *(a) Given the matrix*

$$A = \begin{pmatrix} 1 & 1 & -1 \\ 1 & 2 & -2 \\ -2 & 1 & 1 \end{pmatrix}$$

and whose inverse can be computed as

$$A^{-1} = \begin{pmatrix} 2 & -1 & 0 \\ 1.5 & -0.5 & 0.5 \\ 2.5 & -1.5 & 0.5 \end{pmatrix},$$

then the l_∞-norm of both matrices are

$$\|A\|_\infty = 5 \quad and \quad \|A^{-1}\|_\infty = 4.5.$$

Using the values of both matrix norms, we can find the value of the condition number of A as follows:

$$K(A) = \|A\|_\infty \|A^{-1}\|_\infty = 22.5 >> 1,$$

which shows that the matrix is ill-conditioned. Thus, the given system is ill-conditioned. Using MATLAB, we do the following:

```
>> A = [1 1 - 1; 1 2 - 2; -2 1 1];
>> K(A) = norm(A, inf) * norm(inv(A), inf);
```

(b) The residual vector can be calculated as

$$\begin{aligned} \mathbf{r} &= \mathbf{b} - A\mathbf{x}^* \\ &= \begin{pmatrix} 1 \\ 0 \\ -1 \end{pmatrix} - \begin{pmatrix} 1 & 1 & -1 \\ 1 & 2 & -2 \\ -2 & 1 & 1 \end{pmatrix} \begin{pmatrix} 2.01 \\ 1.01 \\ 1.98 \end{pmatrix}. \end{aligned}$$

After simplifying, we get

$$\mathbf{r} = \begin{pmatrix} -0.04 \\ -0.07 \\ 0.03 \end{pmatrix}$$

and it gives

$$\|\mathbf{r}\|_\infty = 0.07.$$

```
>> A = [1 1 - 1; 1 2 - 2; -2 1 1];
>> b = [1 0 - 1]';
>> x0 = [2.01 1.01 1.98]';
>> r = RESID(A, b, x0);
>> rnorm = norm(r, inf);
```

(c) From (3.83), we have

$$\frac{\|\mathbf{x} - \mathbf{x}^*\|}{\|\mathbf{x}\|} \leq K(A)\frac{\|\mathbf{r}\|}{\|\mathbf{b}\|}.$$

Using the above parts (a) and (b) and the value $\|\mathbf{b}\|_\infty = 1$, *we obtain*

$$\frac{\|\mathbf{x} - \mathbf{x}^*\|}{\|\mathbf{x}\|} \leq (22.5)\frac{(0.07)}{1} = 1.575$$

```
>> RelErr = (K(A) * rnorm)/norm(b, inf);
```

(d) Solve the linear system $A\mathbf{e} = \mathbf{r}$, *where*

$$A = \begin{pmatrix} 1 & 1 & -1 \\ 1 & 2 & -2 \\ -2 & 1 & 1 \end{pmatrix} \quad \text{and} \quad \mathbf{r} = \begin{pmatrix} -0.04 \\ -0.07 \\ 0.03 \end{pmatrix}$$

and $\mathbf{e} = \mathbf{x} - \mathbf{x}^*$. *Write the above system in the augmented matrix form*

$$\left(\begin{array}{rrrcr} 1 & 1 & -1 & \vdots & -0.04 \\ 1 & 2 & -2 & \vdots & -0.07 \\ -2 & 1 & 1 & \vdots & 0.03 \end{array}\right).$$

After applying the forward elimination step of the simple Gauss elimination method, we obtain

$$\left(\begin{array}{rrrcr} 1 & 1 & -1 & \vdots & -0.04 \\ 0 & 1 & -1 & \vdots & -0.03 \\ 0 & 0 & 2 & \vdots & 0.04 \end{array}\right).$$

Using the backward substitution, we obtain the solution

$$\mathbf{e}^* = [-0.01, -0.01, 0.02]^T,$$

which is the required approximation of the exact error.

```
>> A = [1 1 -1 -0.04; 1 2 -2 -0.07; -2 1 1 0.03];
>> WP(A);
```

Conditioning

Let us consider the conditioning of the linear system

$$A\mathbf{x} = \mathbf{b}. \tag{3.85}$$

Case 3.1 *Suppose that the right-hand side term* $\mathbf{b}$ *is replaced by* $\mathbf{b} + \delta\mathbf{b}$*, where* $\delta\mathbf{b}$ *is an error in* $\mathbf{b}$*. If* $\mathbf{x} + \delta\mathbf{x}$ *is the solution corresponding to the right-hand side* $\mathbf{b} + \delta\mathbf{b}$*, then we have*

$$A(\mathbf{x} + \delta\mathbf{x}) = (\mathbf{b} + \delta\mathbf{b}), \tag{3.86}$$

which implies that

$$\begin{aligned} A\mathbf{x} + A\delta\mathbf{x} &= \mathbf{b} + \delta\mathbf{b}, \\ A\delta\mathbf{x} &= \delta\mathbf{b}. \end{aligned}$$

Multiplying by A^{-1}*, we get*

$$\delta\mathbf{x} = A^{-1}\delta\mathbf{b}.$$

Taking the norm, gives

$$\|\delta\mathbf{x}\| = \|A^{-1}\delta\mathbf{b}\| \leq \|A^{-1}\|\|\delta\mathbf{b}\|. \tag{3.87}$$

Thus, the change $\|\delta\mathbf{x}\|$ *in the solution is bounded by* $\|A^{-1}\|$ *times the change* $\|\delta\mathbf{b}\|$ *in the right-hand side.*
The conditioning of the linear system is connected with the ratio between the relative error $\frac{\|\delta\mathbf{x}\|}{\|\mathbf{x}\|}$ *and the relative change* $\frac{\|\delta\mathbf{b}\|}{\|\mathbf{b}\|}$ *in the right-hand side, which gives*

$$\frac{\|\delta\mathbf{x}\|/\|\mathbf{x}\|}{\|\delta\mathbf{b}\|/\|\mathbf{b}\|} = \frac{\|A^{-1}\delta\mathbf{b}\|/\|\mathbf{x}\|}{\|\delta\mathbf{b}\|/\|A\mathbf{x}\|} = \frac{\|A\mathbf{x}\|\|A^{-1}\delta\mathbf{b}\|}{\|\mathbf{x}\|\|\delta\mathbf{b}\|} \leq \|A\|\|A^{-1}\|,$$

which implies that

$$\frac{\|\delta\mathbf{x}\|}{\|\mathbf{x}\|} \leq K(A)\frac{\|\delta\mathbf{b}\|}{\|\mathbf{b}\|}. \tag{3.88}$$

Thus, the relative change in the solution is bounded by the condition number of the matrix times the relative change in the right-hand side. When the product in the right-hand side is small, the relative change in the solution is small.

Case 3.2 *Suppose that matrix A is replaced by $A + \delta A$, where δA is an error in A while the right-hand side term $\mathbf{b}$ is similar. If $\mathbf{x} + \delta\mathbf{x}$ is the solution corresponding to the matrix $A + \delta A$, then we have*

$$(A + \delta A)(\mathbf{x} + \delta\mathbf{x}) = \mathbf{b}, \tag{3.89}$$

which implies that

$$A\mathbf{x} + A\delta\mathbf{x} + \delta A(\mathbf{x} + \delta\mathbf{x}) = \mathbf{b}$$

or

$$A\delta\mathbf{x} = -\delta A(\mathbf{x} + \delta\mathbf{x}).$$

Multiplying by A^{-1}, we get

$$\delta\mathbf{x} = -A^{-1}\delta A(\mathbf{x} + \delta\mathbf{x}).$$

Taking the norm gives

$$\|\delta\mathbf{x}\| = \| - A^{-1}\delta A(\mathbf{x} + \delta\mathbf{x})\| \leq \|A^{-1}\|\|\delta A\|(\|\mathbf{x}\| + \|\delta\mathbf{x}\|)$$

or

$$\|\delta\mathbf{x}\|(1 - \|A^{-1}\|\|\delta A\|) \leq \|A^{-1}\|\|\delta A\|\|\mathbf{x}\|,$$

which can be written as

$$\frac{\|\delta\mathbf{x}\|}{\|\mathbf{x}\|} \leq \frac{\|A^{-1}\|\|\delta A\|}{(1 - \|A^{-1}\|\|\delta A\|)} = \frac{K(A)\|\delta A\|/\|A\|}{(1 - \|A^{-1}\|\|\delta A\|)}. \tag{3.90}$$

If the product $\|A^{-1}\|\|\delta A\|$ is much smaller than 1, the denominator in (3.90) is near 1. Consequently, when $\|A^{-1}\|\|\delta A\|$ is much smaller than 1, then (3.90) implies that the relative change in the solution is bounded by the condition number of a matrix times the relative change in the coefficient matrix.

Case 3.3 *Suppose that there is a change in the coefficient matrix A and right-hand side term $\mathbf{b}$ together, and if $\mathbf{x} + \delta\mathbf{x}$ is the solution corresponding to coefficient matrix $A + \delta A$ and the right-hand side $\mathbf{b} + \delta\mathbf{b}$, then we have*

$$(A + \delta A)(\mathbf{x} + \delta\mathbf{x}) = (\mathbf{b} + \delta\mathbf{b}), \tag{3.91}$$

which implies that

$$A\mathbf{x} + A\delta\mathbf{x} + \mathbf{x}\delta A + \delta A\delta\mathbf{x} = \mathbf{b} + \delta\mathbf{b}$$

or

$$A\delta\mathbf{x} + \delta\mathbf{x}\delta A = (\delta\mathbf{b} - \mathbf{x}\delta A).$$

Multiplying by A^{-1}, *we get*

$$\delta\mathbf{x}(I + A^{-1}\delta A) = A^{-1}(\delta\mathbf{b} - \mathbf{x}\delta A)$$

or

$$\delta\mathbf{x} = (I + A^{-1}\delta A)^{-1}A^{-1}(\delta\mathbf{b} - \mathbf{x}\delta A). \quad (3.92)$$

Since we know that if A *is nonsingular and* δA *is an error in* A, *we obtain*

$$\|A^{-1}\delta A\| \leq \|A^{-1}\|\|\delta A\| < 1, \quad (3.93)$$

then it follows that (see Fröberg 1969) the matrix $(I + A^{-1}\delta A)$ *is nonsingular and*

$$\|(I + A^{-1}\delta A)^{-1}\| \leq \frac{1}{1 - \|A^{-1}\delta A\|} \leq \frac{1}{1 - \|A^{-1}\|\|\delta A\|}. \quad (3.94)$$

Take the norm of (3.92) and use (3.94), which gives

$$\|\delta\mathbf{x}\| \leq \frac{\|A^{-1}\|}{1 - \|A^{-1}\|\|\delta A\|}[\|\delta\mathbf{b}\| + \|\mathbf{x}\|\|\delta A\|]$$

or

$$\frac{\|\delta\mathbf{x}\|}{\|\mathbf{x}\|} \leq \frac{\|A^{-1}\|}{1 - \|A^{-1}\|\|\delta A\|}\left[\frac{\|\delta\mathbf{b}\|}{\|\mathbf{x}\|} + \|\delta A\|\right]. \quad (3.95)$$

Since we know that

$$\|\mathbf{x}\| \geq \frac{\|\mathbf{b}\|}{\|A\|} \quad (3.96)$$

using (3.96) in (3.95) gives

$$\frac{\|\delta\mathbf{x}\|}{\|\mathbf{x}\|} \leq \frac{K(A)}{1 - K(A)\frac{\|\delta A\|}{\|A\|}}\left[\frac{\|\delta A\|}{\|A\|} + \frac{\|\delta\mathbf{b}\|}{\|\mathbf{b}\|}\right]. \quad (3.97)$$

Estimate (3.97) shows that small relative changes in A *and* $\mathbf{b}$ *cause small relative changes in solution* $\mathbf{x}$ *of linear system (3.85) if the inequality*

$$\frac{K(A)}{1 - K(A)\frac{\|\delta A\|}{\|A\|}} \quad (3.98)$$

is not too large.

3.8.2 Iterative Refinement

In those cases when the left-hand side coefficients a_{ij} of the system are exact but the system is ill-conditioned, an approximate solution can be improved by an iterative technique called the method of *residual correction.* The procedure of the method is defined below.

Let $\mathbf{x}^{(1)}$ be an approximate solution to the system

$$A\mathbf{x} = \mathbf{b} \tag{3.99}$$

and let $\mathbf{y}$ be a correction to $\mathbf{x}^{(1)}$ so that the exact solution $\mathbf{x}$ satisfy

$$\mathbf{x} = \mathbf{x}^{(1)} + \mathbf{y}.$$

Then by substituting into (3.99), we find that $\mathbf{y}$ must satisfy

$$A\mathbf{y} = \mathbf{b} - A\mathbf{x}^{(1)} == \mathbf{r}, \tag{3.100}$$

where $\mathbf{r}$ is the residual. System (3.100) can now be solved to give correction $\mathbf{y}$ the approximation $\mathbf{x}^{(1)}$. Thus, the new approximation

$$\mathbf{x}^{(2)} = \mathbf{x}^{(1)} + \mathbf{y}$$

will be closer to the solution compared to $\mathbf{x}^{(1)}$. If necessary, we compute a new residual

$$\mathbf{r} = \mathbf{b} - A\mathbf{x}^{(2)}$$

and solve again system (3.100) to get new corrections. Normally, two or three iterations are enough for getting an exact solution. This iterative method can be used to obtain an improved solution whenever an approximate solution has been obtained by any means.

Example 3.42 *The following linear system*

$$\begin{array}{rcrcl} 1.01x_1 & + & 0.99x_2 & = & 2 \\ 0.99x_1 & + & 1.01x_2 & = & 2 \end{array}$$

has the exact solution $\mathbf{x} = [1, 1]^T$. *The approximate solution due to the Gaussian elimination method is* $\mathbf{x}^{(1)} = [1.01, 1.01]^T$, *and the residual is* $\mathbf{r}^{(1)} = [-0.02, -0.02]^T$. *Then the solution to the system*

$$A\mathbf{y} = \mathbf{r}^{(1)}$$

using the simple Gaussian elimination method is $\mathbf{y}^{(1)} = [-0.01, -0.01]^T$. *So the new approximation is*

$$\mathbf{x}^{(2)} = \mathbf{x}^{(1)} + \mathbf{y}^{(1)} = [1, 1]^T,$$

which is equal to the exact solution after just one iteration.

To use MATLAB for the above iterative method, the two m-files as RESID.m and WP.m have been used, then the first iteration is easily performed by the following sequence of MATLAB commands:

```
>> A = [1.01 0.99; 0.99 1.01];
>> b = [2 2]';
>> x0 = [1.01 1.01]';
>> r = RESID(A, b, x0);
>> B = [A r];
>> y = WP(B);
>> x0 = x0 + y;
```

If needed, the last four commands can be repeated to generate the subsequent iterates.

Procedure 3.13 [Iterative Refinement Method]

1. *Find or give initial approximation* $\mathbf{x}^{(1)} \in \mathbb{R}$.
2. *Compute the residual vector* $\mathbf{r}^{(1)} = \mathbf{b} - A\mathbf{x}^{(1)}$.
3. *Solve linear system* $A\mathbf{y} = \mathbf{r}^{(1)}$ *for unknown* $\mathbf{y}$.
4. *Set* $\mathbf{x}^{(k+1)} = \mathbf{x}^{(k)} + \mathbf{y}, \qquad \text{for} \quad k = 1, \ldots.$
5. *Repeat steps 2 to 4, unless the best approximation is achieved.*

3.9 Summary

The basic methods for solving systems of linear algebraic equations were discussed in this chapter. Since these methods use matrices and determinants, the basic properties of matrices and determinants were presented.

Several direct solution methods were also discussed. Among them were Cramer's rule, Gaussian elimination and its variants, the Gauss-Jordan method, and the LU decomposition method. Cramer's rule is impractical for solving systems with more than three or four equations. Gaussian elimination is the best choice for solving linear systems. For systems of equations having a constant coefficients matrix but many right-hand side vectors, LU decomposition is the method of choice. The LU decomposition method has been used for the solution of the tridiagonal system. Direct methods are generally used when the number of equations is small, or most of the coefficients of the equations are nonzero, or the system of equations is not diagonally dominant, or the system of equations is ill-conditioned. Several iterative methods were discussed. Among them were the Jacobi method, the Gauss-Seidel method, and the SOR method. All methods converge if the coefficient matrix is strictly diagonally dominant. The SOR is the best method of choice. Although the determination of the optimum value of the relaxation factor ω is difficult, it is generally worthwhile if the system of equations is to be solved many times for right-hand side vectors. The need for estimating parameters is removed in the conjugate gradient method, which, although more complicated to code, can rival the SOR method in efficiency when dealing with large, sparse systems. Iterative methods are generally used when the number of equations is large, the coefficient matrix is sparse, and the coefficient matrix is strictly diagonally dominant. We also discussed conditioning of the linear systems using a parameter, the condition number. We discussed many ill-conditioned systems. The coefficient matrix A of an ill-conditioned system $A\mathbf{x} = \mathbf{b}$ has a large condition number. The numerical solution to a linear system is less reliable when A has a large condition number. If the condition number is small, any stable method will return a solution with small residuals while if the condition number is large, then the return solution may have large errors even though the residuals are small. The best way to deal with ill-conditioning is to avoid it by reformulating the problem. In the end of the chapter we discussed the residual corrector method, which improved the approximate solution.

3.10 Exercises

1. Determine matrix C given by the following expression

$$C = 2A - 3B,$$

if matrices A and B are

$$A = \begin{pmatrix} 2 & -1 & 1 \\ -1 & 2 & 3 \\ 2 & 1 & 2 \end{pmatrix}, \qquad B = \begin{pmatrix} 1 & 1 & 1 \\ 0 & 1 & 3 \\ 2 & 1 & 4 \end{pmatrix}.$$

2. Find the product AB and BA for the matrices of Problem 1.

3. Show that the product AB of the following rectangular matrices is a singular matrix:

$$A = \begin{pmatrix} 6 & -3 \\ 1 & 4 \\ -2 & 1 \end{pmatrix} \quad \text{and} \quad B = \begin{pmatrix} 2 & -1 & -2 \\ 3 & -4 & -1 \end{pmatrix}.$$

4. Let

$$A = \begin{pmatrix} 1 & 2 & 3 \\ 0 & -1 & 2 \\ 2 & 0 & 2 \end{pmatrix}, \quad B = \begin{pmatrix} 1 & 1 & 2 \\ -1 & 1 & -1 \\ 1 & 0 & 2 \end{pmatrix}, \text{ and } \quad C = \begin{pmatrix} 1 & 0 & 1 \\ 0 & 1 & 2 \\ 2 & 0 & 1 \end{pmatrix}.$$

(a) Compute AB and BA and show that $AB \neq BA$.
(b) Find $(A + B) + C$ and $A + (B + C)$.
(c) Show that $(AB)^T = B^T A^T$.

5. Let

$$A = \begin{pmatrix} 1 & 1 \\ 0 & 1 \end{pmatrix}, \quad B = \begin{pmatrix} 1 & 0 \\ 1 & 1 \end{pmatrix},$$

then show that $(AB)^{-1} = B^{-1}A^{-1}$.

6. Find a value of x and a value of y so that $AB^T = 0$, where $A = [1\ x\ 1]$ and $B = [-2\ 2\ y]$.

7. Evaluate the determinant of each matrix

$$A = \begin{pmatrix} 3 & 1 & -1 \\ 2 & 0 & 4 \\ 1 & -5 & 1 \end{pmatrix}, \quad B = \begin{pmatrix} 4 & 1 & 6 \\ -3 & 6 & 4 \\ 5 & 0 & 9 \end{pmatrix}, \quad C = \begin{pmatrix} 17 & 46 & 7 \\ 20 & 49 & 8 \\ 23 & 52 & 9 \end{pmatrix}.$$

8. Find all zeros (values of x such that $f(x) = 0$) of the polynomial $f(x) = \det(A)$, where

$$A = \begin{pmatrix} x-1 & 3 & 2 \\ 3 & x & 1 \\ 2 & 1 & x-2 \end{pmatrix}.$$

9. Compute the adjoint of each matrix A, and find the inverse of it if it exists

$$(a) \begin{pmatrix} 1 & 2 \\ -3 & 4 \end{pmatrix}, \quad (b) \begin{pmatrix} 1 & 2 & -1 \\ 2 & 1 & 4 \\ 1 & 5 & -7 \end{pmatrix}, \quad (c) \begin{pmatrix} 1 & 1 & 0 \\ 1 & 0 & 1 \\ 0 & 1 & 1 \end{pmatrix}.$$

10. Let

$$A = \begin{pmatrix} 2 & 1 & 3 \\ -1 & 2 & 0 \\ 3 & -2 & 1 \end{pmatrix}.$$

Show that $A(Adj\ A) = (Adj\ A)A = \det(A)\mathbf{I}_3$.

11. Use the matrices of Problem 9 and solve the following systems using the matrix inverse method

$$(a) \quad A\mathbf{x} = [1,1]^T, \quad (b) \quad A\mathbf{x} = [2,1,3]^T, \text{and} \quad (c) \quad A\mathbf{x} = [1,0,1]^T.$$

12. Solve the following systems using the matrix inverse method

(a)

$$\begin{array}{rcrcrcr} x_1 & + & 3x_2 & - & x_3 & = & 4 \\ 5x_1 & - & 2x_2 & - & x_3 & = & -2 \\ 2x_1 & + & 2x_2 & + & x_3 & = & 9, \end{array}$$

(b)

$$\begin{array}{rcrcrcr} x_1 & + & x_2 & + & 3x_3 & = & 2 \\ 5x_1 & + & 3x_2 & + & x_3 & = & 3 \\ 2x_1 & + & 3x_2 & + & x_3 & = & -1, \end{array}$$

(c)

$$\begin{array}{rcrcrcr} 4x_1 & + & x_2 & - & 3x_3 & = & 9 \\ 3x_1 & + & 2x_2 & - & 6x_3 & = & -2 \\ x_1 & - & 5x_2 & + & 3x_3 & = & 1. \end{array}$$

13. Solve Problem 12 using Cramer's rule.

14. Use the simple Gaussian elimination method to show that the following system does not have a solution

$$\begin{array}{rcrcrcr} 3x_1 & + & x_2 & & & = & 1.5 \\ 2x_1 & - & x_2 & - & x_3 & = & 2 \\ 4x_1 & + & 3x_2 & + & x_3 & = & 0. \end{array}$$

15. Solve the following systems using the simple Gaussian elimination method:

(a)

$$\begin{array}{rcrcrcr} x_1 & - & x_2 & & & = & 0 \\ -x_1 & + & 2x_2 & - & x_3 & = & 1 \\ & - & x_2 & + & 4x_3 & = & 0, \end{array}$$

(b)

$$\begin{aligned} x_1 + x_2 + x_3 &= 1 \\ 2x_1 + 3x_2 + 4x_3 &= 3 \\ 4x_1 + 9x_2 + 16x_3 &= 11, \end{aligned}$$

(c)

$$\begin{aligned} 3x_1 + 2x_2 - x_3 &= 1 \\ x_1 - 3x_2 + 2x_3 &= 2 \\ 2x_1 - x_2 + x_3 &= 3, \end{aligned}$$

and

(d)

$$\begin{aligned} 2x_1 + x_2 + x_3 - x_4 &= -3 \\ x_1 + 9x_2 + 8x_3 + 4x_4 &= 15 \\ -x_1 + 3x_2 + 5x_3 + 2x_4 &= 10 \\ x_2 + x_4 &= 2. \end{aligned}$$

16. Find a value of k so that the following linear system has a non-trivial solution, and solve it in this case:

$$\begin{aligned} 2x_1 + 2x_2 + 3x_3 &= 0 \\ 3x_1 + kx_2 + 5x_3 &= 0 \\ x_1 + 7x_2 + 3x_3 &= 0. \end{aligned}$$

17. Determine the rank of each matrix

$$A = \begin{pmatrix} 3 & 1 & -1 \\ 2 & 0 & 4 \\ 1 & -5 & 1 \end{pmatrix}, \quad B = \begin{pmatrix} 4 & 1 & 6 \\ -3 & 6 & 4 \\ 5 & 0 & 9 \end{pmatrix}, \quad C = \begin{pmatrix} 17 & 46 & 7 \\ 20 & 49 & 8 \\ 23 & 52 & 9 \end{pmatrix}.$$

18. Let A be an $m \times n$ matrix and B be an $n \times p$ matrix. Show that the rank of AB is less than or equal to the rank of A.

19. Determine the rank of each matrix

$$A = \begin{pmatrix} 2 & -1 & 0 \\ 2 & -1 & 1 \\ 1 & 1 & -1 \end{pmatrix}, \quad B = \begin{pmatrix} 0.1 & 0.2 & 0.3 \\ 0.4 & 0.5 & 0.6 \\ 0.7 & 0.8 & 0.901 \end{pmatrix}, \quad C = \begin{pmatrix} 1 & 2 & 3 & 4 \\ 2 & 4 & 6 & 8 \\ 3 & 5 & 7 & 9 \\ 4 & 6 & 8 & 10 \end{pmatrix}.$$

20. Solve Problem 15 using Gaussian elimination with partial pivoting and complete pivoting.

21. Solve the following linear systems using Gaussian elimination with partial pivoting and without pivoting:

(a)

$$\begin{aligned} 1.001x_1 + 1.5x_2 &= 0 \\ 2x_1 + 3x_2 &= 1, \end{aligned}$$

(b)

$$\begin{aligned} x_1 + 1.001x_2 &= 2.001 \\ x_1 + x_2 &= 2, \end{aligned}$$

(c)

$$\begin{aligned} 6.122x_1 + 1500.5x_2 &= 1506.622 \\ 2000x_1 + 3x_2 &= 2003. \end{aligned}$$

22. The elements of matrix A, the Hilbert matrix, are defined by

$$a_{ij} = 1/(i+j-1), \quad \text{for} \quad i, j = 1, 2, \ldots, n.$$

Find the solution of the system $A\mathbf{x} = \mathbf{b}$ for $n = 3$ and $\mathbf{b} = [1, 2, 3]^T$, using Gaussian elimination by simple pivoting, partial pivoting, and complete pivoting.

23. Solve the following systems using the Gauss-Jordan method:

(a)

$$\begin{aligned} x_1 + 4x_2 + x_3 &= 1 \\ 2x_1 + 4x_2 + x_3 &= 9 \\ 3x_1 + 5x_2 - 2x_3 &= 11, \end{aligned}$$

(b)

$$\begin{aligned} x_1 + x_2 + x_3 &= 1 \\ 2x_1 - x_2 + 3x_3 &= 4 \\ 3x_1 + 2x_2 - 2x_3 &= -2, \end{aligned}$$

(c)

$$\begin{aligned} 2x_1 + 3x_2 + 6x_3 + x_4 &= 2 \\ x_1 + x_2 - 2x_3 + 4x_4 &= 1 \\ 3x_1 + 5x_2 - 2x_3 + 2x_4 &= 11 \\ 2x_1 + 2x_2 + 2x_3 - 3x_4 &= 2. \end{aligned}$$

24. The following set of linear equations have a common coefficients matrix but different right-side terms:

(a)

$$\begin{aligned} 2x_1 + 3x_2 + 5x_3 &= 0 \\ 3x_1 + x_2 - 2x_3 &= -2 \\ x_1 + 3x_2 + 4x_3 &= -3, \end{aligned}$$

(b)

$$\begin{aligned} 2x_1 + 3x_2 + 5x_3 &= 1 \\ 3x_1 + x_2 - 2x_3 &= 2 \\ x_1 + 3x_2 + 4x_3 &= 4, \end{aligned}$$

(c)

$$\begin{aligned} 2x_1 + 3x_2 + 5x_3 &= -5 \\ 3x_1 + x_2 - 2x_3 &= 6 \\ x_1 + 3x_2 + 4x_3 &= -1. \end{aligned}$$

The coefficients and the three sets of right-side terms may be combined into an augmented matrix form

$$\left(\begin{array}{ccccccc} 2 & 3 & 5 & \vdots & 0 & 1 & -5 \\ 3 & 1 & -2 & \vdots & -2 & 2 & 6 \\ 1 & 3 & 4 & \vdots & -3 & 4 & -1 \end{array}\right).$$

If we apply the Gauss-Jordan method to this augmented matrix form and reduce the first three columns to the unity matrix form, the solution for the three problems are automatically obtained in the fourth, fifth, and sixth columns when elimination is completed. Calculate the solution in this way.

25. Calculate the inverse of each matrix using the Gauss-Jordan method:

$$(a)\ \begin{pmatrix} 3 & -9 & 5 \\ 0 & 5 & 1 \\ -1 & 6 & 3 \end{pmatrix}, \quad (b)\ \begin{pmatrix} 1 & 4 & 5 \\ 2 & 1 & 2 \\ 8 & 1 & 1 \end{pmatrix}, \quad (c)\ \begin{pmatrix} 5 & -2 & 0 & 0 \\ -2 & 5 & -2 & 0 \\ 0 & -2 & 5 & -2 \\ 0 & 0 & -2 & 5 \end{pmatrix}.$$

26. Find the inverse of the Hilbert matrix of size 4×4 using the Gauss-Jordan method. Then solve the linear system $A\mathbf{x} = [1, 2, 3, 4]^T$.

27. Find the LU decomposition of each matrix A using Doolittle's method, and then solve the systems:

(a)

$$A = \begin{pmatrix} 2 & -1 & 1 \\ -3 & 4 & -1 \\ 1 & -1 & 1 \end{pmatrix}, \qquad \mathbf{b} = \begin{pmatrix} 4 \\ 5 \\ 6 \end{pmatrix}.$$

(b)

$$A = \begin{pmatrix} 7 & 6 & 5 \\ 5 & 4 & 3 \\ 3 & 7 & 6 \end{pmatrix}, \qquad \mathbf{b} = \begin{pmatrix} 2 \\ 1 \\ 2 \end{pmatrix}.$$

(c)

$$A = \begin{pmatrix} 2 & 2 & 2 \\ 1 & 2 & 1 \\ 3 & 3 & 4 \end{pmatrix}, \qquad \mathbf{b} = \begin{pmatrix} 0 \\ -4 \\ 1 \end{pmatrix}.$$

(d)

$$A = \begin{pmatrix} 2 & 4 & -6 \\ 1 & 5 & 3 \\ 1 & 3 & 2 \end{pmatrix}, \qquad \mathbf{b} = \begin{pmatrix} -4 \\ 10 \\ 5 \end{pmatrix}.$$

(e)

$$A = \begin{pmatrix} 1 & -1 & 0 \\ 2 & -1 & 1 \\ 2 & -2 & -1 \end{pmatrix}, \qquad \mathbf{b} = \begin{pmatrix} 2 \\ 4 \\ 3 \end{pmatrix}.$$

(f)

$$A = \begin{pmatrix} 1 & 5 & 3 \\ 2 & 4 & 6 \\ 1 & 3 & 2 \end{pmatrix}, \qquad \mathbf{b} = \begin{pmatrix} 4 \\ 11 \\ 5 \end{pmatrix}.$$

28. Solve Problem 27 by LU decomposition using Crout's method.

29. Solve the following system by LU decomposition using the Cholesky method.

(a)

$$A = \begin{pmatrix} 2 & -1 & 1 \\ -1 & 2 & -1 \\ 1 & -1 & 2 \end{pmatrix}, \qquad \mathbf{b} = \begin{pmatrix} 1 \\ 2 \\ 3 \end{pmatrix}.$$

(b)

$$A = \begin{pmatrix} 10 & 2 & 1 \\ 2 & 10 & 3 \\ 1 & 3 & 10 \end{pmatrix}, \qquad \mathbf{b} = \begin{pmatrix} 7 \\ -4 \\ 3 \end{pmatrix}.$$

(*c*)

$$A = \begin{pmatrix} 5 & 2 & 3 \\ 2 & 4 & 1 \\ 3 & 1 & 6 \end{pmatrix}, \qquad \mathbf{b} = \begin{pmatrix} 1 \\ 1 \\ 1 \end{pmatrix}.$$

(*d*)

$$A = \begin{pmatrix} 3 & 4 & -6 & 0 \\ 4 & 5 & 3 & 1 \\ -6 & 3 & 3 & 1 \\ 0 & 1 & 1 & 3 \end{pmatrix}, \qquad \mathbf{b} = \begin{pmatrix} 4 \\ 5 \\ 2 \\ 3 \end{pmatrix}.$$

30. Solve the following tridiagonal systems using LU decomposition.

(*a*)

$$A = \begin{pmatrix} 3 & -1 & 0 \\ -1 & 3 & -1 \\ 0 & -1 & 3 \end{pmatrix}, \qquad \mathbf{b} = \begin{pmatrix} 1 \\ 2 \\ 3 \end{pmatrix}.$$

(*b*)

$$A = \begin{pmatrix} 2 & 3 & 0 & 0 \\ 3 & 2 & 3 & 0 \\ 0 & 3 & 2 & 3 \\ 0 & 0 & 3 & 2 \end{pmatrix}, \qquad \mathbf{b} = \begin{pmatrix} 6 \\ 7 \\ 5 \\ 3 \end{pmatrix}.$$

(*c*)

$$A = \begin{pmatrix} 4 & -1 & 0 & 0 \\ -1 & 4 & -1 & 0 \\ 0 & -1 & 4 & -1 \\ 0 & 0 & -1 & 4 \end{pmatrix}, \qquad \mathbf{b} = \begin{pmatrix} 3 \\ -1 \\ 2 \\ 5 \end{pmatrix}.$$

31. Find $\|\mathbf{x}\|_1$, $\|\mathbf{x}\|_2$ and $\|\mathbf{x}\|_\infty$ for the following vectors:

(*a*) $[2, -1, -6, 3]^T$, (*b*) $[\sin k, \cos k, 3^k]^T$ for a fixed integer k, (*c*) $[3, -4, 0, 3/2]^T$.

32. Find $\|.\|_1$, $\|.\|_\infty$ and $\|.\|_e$ for the following matrices:

$$A = \begin{pmatrix} 3 & 1 & -1 \\ 2 & 0 & 4 \\ 1 & -5 & 1 \end{pmatrix}, \quad B = \begin{pmatrix} 4 & 1 & 6 \\ -3 & 6 & 4 \\ 5 & 0 & 9 \end{pmatrix}, \quad C = \begin{pmatrix} 17 & 46 & 7 \\ 20 & 49 & 8 \\ 23 & 52 & 9 \end{pmatrix}.$$

33. Consider the following matrices:

$$A = \begin{pmatrix} -11 & 7 & -8 \\ 5 & 9 & 6 \\ 6 & 3 & 7 \end{pmatrix}, \quad B = \begin{pmatrix} 6 & 2 & 7 \\ -12 & 10 & 8 \\ 3 & -15 & 14 \end{pmatrix}, \quad C = \begin{pmatrix} 5 & -6 & 4 \\ -7 & 8 & 5 \\ 3 & -9 & 12 \end{pmatrix}.$$

Find $\|.\|_1$ and $\|.\|_\infty$ for (*a*) A^3, (*b*) $A^2 + B^2$, (*c*) BC, and (*d*) $C^2 + A^2$.

34. The $n \times n$ Hilbert matrix $H^{(n)}$ is defined by

$$H_{ij}^{(n)} = \frac{1}{i + j - 1}, \qquad 1 \le i,\ j \le n.$$

Find the l_∞-norm of the 10×10 Hilbert matrix.

35. Solve the following linear systems using the Jacobi method. Start with the initial approximation $\mathbf{x}^{(0)} = \mathbf{0}$ and iterate until $\|\mathbf{x}^{(k+1)} - x^{(k)}\|_\infty \le 10^{-5}$ for each system.

(*a*)

$$\begin{aligned} 4x_1 - x_2 + x_3 &= 7 \\ 4x_1 - 8x_2 + x_3 &= -21 \\ -2x_1 + x_2 + 5x_3 &= 15. \end{aligned}$$

(*b*)

$$\begin{aligned} 3x_1 + x_2 + x_3 &= 5 \\ 2x_1 + 6x_2 + x_3 &= 9 \\ x_1 + x_2 + 4x_3 &= 6. \end{aligned}$$

(*c*)

$$\begin{aligned} 4x_1 + 2x_2 + x_3 &= 1 \\ x_1 + 7x_2 + x_3 &= 4 \\ x_1 + x_2 + 20x_3 &= 7. \end{aligned}$$

(*d*)

$$\begin{aligned} 5x_1 + 2x_2 - x_3 &= 6 \\ x_1 + 6x_2 - 3x_3 &= 4 \\ 2x_1 + x_2 + 4x_3 &= 7. \end{aligned}$$

(*e*)

$$\begin{aligned} 6x_1 - x_2 + 3x_3 &= -2 \\ 3x_2 + x_3 &= 1 \\ -2x_1 + x_2 + 5x_3 &= 5. \end{aligned}$$

(f)

$$\begin{aligned} 4x_1 + x_2 \quad &= -1 \\ 2x_1 + 5x_2 + x_3 &= 0 \\ -x_1 + 2x_2 + 4x_3 &= 3. \end{aligned}$$

(g)

$$\begin{aligned} 5x_1 - x_2 + x_3 &= 1 \\ 3x_2 - x_3 &= -1 \\ x_1 + 2x_2 + 4x_3 &= 2. \end{aligned}$$

(h)

$$\begin{aligned} 9x_1 + x_2 + x_3 &= 10 \\ 2x_1 + 10x_2 + 3x_3 &= 19 \\ 3x_1 + 4x_2 + 11x_3 &= 0. \end{aligned}$$

36. Consider the following system of equations:

$$\begin{aligned} 4x_1 + 2x_2 + x_3 &= 1 \\ x_1 + 7x_2 + x_3 &= 4 \\ x_1 + x_2 + 20x_3 &= 7. \end{aligned}$$

(a) Show that the Jacobi method converges using $\|T_J\|_\infty < 1$.

(b) Compute the second approximation $\mathbf{x}^{(2)}$, starting with $\mathbf{x}^{(0)} = [0, 0, 0]^T$.

(c) Compute an error estimate $\|\mathbf{x} - \mathbf{x}^{(2)}\|_\infty$ for your approximation.

37. Solve Problem 35 using the Gauss-Seidel method.

38. Consider the following system of equations:

$$\begin{aligned} 4x_1 + 2x_2 + x_3 &= 11 \\ -x_1 + 2x_2 \quad &= 3 \\ 2x_1 + x_2 + 4x_3 &= 16. \end{aligned}$$

(a) Show that the Gauss-Seidel method converges using $\|T_G\|_\infty < 1$.

(b) Compute the second approximation $\mathbf{x}^{(2)}$, starting with $\mathbf{x}^{(0)} = [1, 1, 1]^T$.

(c) Compute an error estimate $\|\mathbf{x} - \mathbf{x}^{(2)}\|_\infty$ for your approximation.

39. Which of the following matrices is convergent?

$$(a)\ \begin{pmatrix} 2 & 2 & 3 \\ 1 & 2 & 1 \\ 2 & -2 & 1 \end{pmatrix}, \quad (b)\ \begin{pmatrix} 1 & 0 & 0 \\ -1 & 3 & 0 \\ 3 & 2 & -2 \end{pmatrix}.$$

40. Find the eigenvalues and their associated eigenvectors of the matrix

$$A = \begin{pmatrix} 2 & -2 & 3 \\ 0 & 3 & -2 \\ 0 & -1 & 2 \end{pmatrix}.$$

Also, show that $\|A\|_2 > \rho(A)$.

41. Solve Problem 35 using the SOR method by taking $\omega = 1.02$ for each system.

42. Use the parameter $\omega = 1.543$ to solve the linear system by the SOR method within accuracy 10^{-6} in the l_∞-norm, starting $\mathbf{x}^{(0)} = \mathbf{0}$

$$\begin{array}{rcrcrcrcr} x_1 & - & 2x_2 & & & & & = & -3 \\ -2x_1 & + & 5x_2 & - & x_3 & & & = & 5 \\ & - & x_2 & + & 2x_3 & - & 0.5x_4 & = & 2 \\ & & & - & 0.5x_3 & + & 1.25x_4 & = & 3.5. \end{array}$$

43. Find the optimal choice for ω and use it to solve the linear system by the SOR method within accuracy 10^{-4} in the l_∞-norm, starting $\mathbf{x}^{(0)} = \mathbf{0}$. Also, find how many iterations are needed using the Jacobi method and the Gauss-Seidel method:

$$\begin{array}{rcrcrcr} 4x_1 & + & 2x_2 & & & = & 5 \\ 2x_1 & + & 4x_2 & - & 2x_3 & = & 3 \\ & - & x_2 & + & 4x_3 & = & 6 \end{array}$$

44. Consider the following system:

$$\begin{array}{rcrcrcrcr} 4x_1 & - & 2x_2 & - & x_3 & & & = & 1 \\ -x_1 & + & 4x_2 & & & - & x_4 & = & 2 \\ -x_1 & & & + & 4x_3 & - & x_4 & = & 0 \\ & - & x_2 & - & x_3 & + & 4x_4 & = & 1. \end{array}$$

Using $\mathbf{x}^{(0)} = \mathbf{0}$, how many iterations are required to approximate the solution to within five decimal places using: *(a)* Jacobi method, *(b)* Gauss-Seidel method, and *(c)* SOR method (take $\omega = 1.1$).

45. Find the spectral radius of the Jacobi, the Gauss-Seidel, and the SOR ($\omega = 1.25962$) iteration matrices when

$$A = \begin{pmatrix} 2 & 1 & 0 & 0 \\ 1 & 2 & 1 & 0 \\ 0 & 1 & 2 & 1 \\ 0 & 0 & 1 & 2 \end{pmatrix}.$$

46. Perform only two steps of the conjugate gradient method for the following linear systems, taking $\mathbf{x}^{(0)} = \mathbf{0}$

(a)

$$\begin{array}{rcrcrcl} 3x_1 & - & x_2 & + & x_3 & = & 7 \\ -1x_1 & + & 3x_2 & + & 2x_3 & = & 1 \\ x_1 & + & 2x_2 & + & 5x_3 & = & 5. \end{array}$$

(b)

$$\begin{array}{rcrcrcl} 3x_1 & - & 2x_2 & + & x_3 & = & 5 \\ -2x_1 & + & 6x_2 & - & x_3 & = & 9 \\ x_1 & - & x_2 & + & 4x_3 & = & 6. \end{array}$$

(c)

$$\begin{array}{rcrcrcl} 4x_1 & - & 2x_2 & + & x_3 & = & 1 \\ -2x_1 & + & 7x_2 & + & x_3 & = & 4 \\ x_1 & + & x_2 & + & 20x_3 & = & 1. \end{array}$$

(d)

$$\begin{array}{rcrcrcl} 5x_1 & - & 3x_2 & - & x_3 & = & 6 \\ -3x_1 & + & 6x_2 & - & 3x_3 & = & 4 \\ -x_1 & - & 3x_2 & + & 4x_3 & = & 7. \end{array}$$

47. Perform only two steps of the conjugate gradient method for the following linear systems, taking $\mathbf{x}^{(0)} = \mathbf{0}$

(a)

$$\begin{array}{rcrcrcl} 6x_1 & + & 2x_2 & + & x_3 & = & 1 \\ 2x_1 & + & 3x_2 & - & x_3 & = & 0 \\ x_1 & - & x_2 & + & 2x_3 & = & -2. \end{array}$$

(b)

$$\begin{array}{rcrcrcl} 5x_1 & - & 2x_2 & + & x_3 & = & 3 \\ -2x_1 & + & 4x_2 & - & x_3 & = & 2 \\ x_1 & - & x_2 & + & 3x_3 & = & 1. \end{array}$$

(c)

$$\begin{array}{rcrcrcrcl} 6x_1 & - & x_2 & - & x_3 & + & 5x_4 & = & 1 \\ -x_1 & + & 7x_2 & + & x_3 & - & x_4 & = & 2 \\ -x_1 & + & x_2 & + & 3x_3 & - & 3x_4 & = & 0 \\ 5x_1 & - & x_2 & - & 3x_3 & + & 6x_4 & = & -1. \end{array}$$

(d)

$$\begin{array}{rcrcrcrcl} 3x_1 & - & 2x_2 & - & x_3 & + & 3x_4 & = & 1 \\ -2x_1 & + & 7x_2 & + & x_3 & - & x_4 & = & 0 \\ -x_1 & + & x_2 & + & 3x_3 & - & 3x_4 & = & 0 \\ 3x_1 & - & x_2 & - & 3x_3 & + & 6x_4 & = & 0. \end{array}$$

48. Compute the condition numbers of each of the matrices in Problem 47 relative to $\|.\|_\infty$.

49. Compute the condition numbers of the following matrices relative to $\|.\|_\infty$:

$$(a) \begin{pmatrix} \frac{1}{3} & \frac{1}{2} & \frac{1}{5} \\ \frac{1}{2} & \frac{1}{5} & \frac{1}{3} \\ \frac{1}{5} & \frac{1}{3} & \frac{1}{2} \end{pmatrix}, \quad (b) \begin{pmatrix} 0.03 & 0.01 & -0.02 \\ 0.15 & 0.51 & -0.11 \\ 1.11 & 2.22 & 3.33 \end{pmatrix}, \quad (c) \begin{pmatrix} 1.11 & 1.98 & 2.01 \\ 1.01 & 1.05 & 2.05 \\ 0.85 & 0.45 & 1.25 \end{pmatrix}.$$

50. The following linear systems have $\mathbf{x}$ as the exact solution and $\mathbf{x}^*$ as an approximate solution. Compute $\|\mathbf{x} - \mathbf{x}^*\|_\infty$ and $K(A)\dfrac{\|\mathbf{r}\|_\infty}{\|\mathbf{b}\|_\infty}$, where $\mathbf{r} = \mathbf{b} - A\mathbf{x}^*$ is the residual vector.

(a)

$$\begin{array}{rcrcl} 0.89x_1 & + & 0.53x_2 & = & 0.36 \\ 0.47x_1 & + & 0.28x_2 & = & 0.19. \end{array}$$

$$\begin{array}{rcl} \mathbf{x} & = & [1, -1]^T \\ \mathbf{x}^* & = & [0.702, -0.500]^T \end{array}$$

(b)

$$\begin{array}{rcrcl} 0.986x_1 & + & 0.579x_2 & = & 0.235 \\ 0.409x_1 & + & 0.237x_2 & = & 0.107. \end{array}$$

$$\begin{array}{rcl} \mathbf{x} & = & [2, -3]^T \\ \mathbf{x}^* & = & [2.110, -3.170]^T \end{array}$$

(c)

$$\begin{array}{rcrcl} 1.003x_1 & + & 58.090x_2 & = & 68.12 \\ 5.550x_1 & + & 321.8x_2 & = & 377.3. \end{array}$$

$$\begin{array}{rcl} \mathbf{x} & = & [10, 1]^T \\ \mathbf{x}^* & = & [-10, 1]^T \end{array}$$

51. Discuss the ill-conditioning (stability) of the linear system

$$\begin{array}{rcrcl} 1.01x_1 & + & 0.99x_2 & = & 2 \\ 0.99x_1 & + & 1.01x_2 & = & 2. \end{array}$$

If $\mathbf{x}^* = [2, 0]^T$ is an approximate solution of the system, then find the residual vector $\mathbf{r}$ and estimate the relative error.

52. Show that if B is singular, then

$$\frac{1}{K(A)} \leq \frac{\|A - B\|}{\|A\|}.$$

53. Consider the following matrices

$$A = \begin{pmatrix} 0.06 & 0.01 & 0.02 \\ 0.13 & 0.05 & 0.11 \\ 1.01 & 2.02 & 3.03 \end{pmatrix} \text{ and } \quad B = \begin{pmatrix} -0.04 & 0.01 & -0.05 \\ 0.11 & 0.02 & -0.03 \\ 0.89 & 1.94 & 2.99 \end{pmatrix}.$$

Using Problem 52, compute the approximation of the condition number of matrix A relative to $\|.\|_\infty$.

54. Let A and B be nonsingular $n \times n$ matrices. Show that

$$K(AB) \leq K(A) + K(B).$$

55. The exact solution of the following linear system

$$\begin{array}{rcrcl} x_1 & + & x_2 & = & 1 \\ x_1 & + & 1.01x_2 & = & 2 \end{array}$$

is $\mathbf{x} = [-99, 100]^T$. Change the coefficient matrix slightly to

$$\delta A = \begin{pmatrix} 1 & 1 \\ 1 & 0.99 \end{pmatrix}$$

and consider the linear system

$$\begin{array}{rcrcl} x_1 & + & x_2 & = & 1 \\ x_1 & + & 0.99x_2 & = & 2. \end{array}$$

Compute the change solution $\delta\mathbf{x}$ of the system. Is matrix A ill-conditioned?

56. Using Problem 15, compute the relative error and the relative residual.

57. The exact solution of the following linear system

$$\begin{array}{rcrcl} x_1 & + & 3x_2 & = & 4 \\ 1.0001x_1 & + & 3x_2 & = & 4.0001 \end{array}$$

is $\mathbf{x} = [1, 1]^T$. Change the right-hand vector $\mathbf{b}$ slightly to $\delta\mathbf{b} = [4.0001, 4.0003]^T$ and consider the linear system

$$\begin{array}{rcrcl} x_1 & + & 3x_2 & = & 4.0001 \\ 1.0001x_1 & + & 3x_2 & = & 4.0003. \end{array}$$

Find the change solution $\delta\mathbf{x}$ of this system. Is matrix A ill-conditioned?

58. If $\|A\| < 1$, then show that matrix $(\mathbf{I} - A)$ is nonsingular and

$$\|(\mathbf{I} - A)^{-1}\| \leq \frac{1}{1 - \|A\|}.$$

59. The exact solution of the following linear system

$$\begin{array}{rcrcl} x_1 & + & x_2 & = & 3 \\ x_1 & + & 1.0005x_2 & = & 3.0010 \end{array}$$

is $\mathbf{x} = [1, 2]^T$. Change the coefficient matrix and the right-hand vector $\mathbf{b}$ slightly to

$$\delta A = \begin{pmatrix} 1 & 1 \\ 1 & 1.001 \end{pmatrix} \quad \text{and} \quad \delta\mathbf{b} = \begin{pmatrix} 2.99 \\ 3.01 \end{pmatrix}$$

and consider the linear system

$$\begin{array}{rcrcl} x_1 & + & x_2 & = & 2.99 \\ x_1 & + & 1.001x_2 & = & 3.01. \end{array}$$

Find the change solution $\delta\mathbf{x}$ of this system. Is matrix A ill-conditioned?

60. Find the condition number of the following matrix

$$A_n = \begin{pmatrix} 1 & 1 \\ 1 & 1 - \dfrac{1}{n} \end{pmatrix}.$$

Solve the linear system $A_4\mathbf{x} = [2, 2]^T$ and compute the relative residual.

61. The following linear system has the exact solution $\mathbf{x} = [10, 1]^T$. Find the approximate solution of the system

$$\begin{array}{rcrcl} 0.03x_1 & + & 58.9x_2 & = & 59.2 \\ 5.31x_1 & - & 6.10x_2 & = & 47.0 \end{array}$$

using simple Gaussian elimination and then use the residual correction method (one iteration only) to improve the approximate solution.

62. The following linear system has the exact solution $\mathbf{x} = [1, 1]^T$. Find the approximate solution of the system

$$\begin{array}{rcrcl} x_1 & + & 2x_2 & = & 3 \\ x_1 & + & 2.01x_2 & = & 3.01 \end{array}$$

using simple Gaussian elimination and then use the residual correction method (one iteration only) to improve the approximate solution.

Chapter 4

Approximating Functions

4.1 Introduction

In this chapter we describe numerical methods for the approximation of functions other than elementary functions. The main purpose of these techniques is to replace a complicated function by one that is simpler and more manageable. We sometimes know the value of a function $f(x)$ at a set of points (say, $x_0 < x_1 < x_2 \cdots < x_n$) but we do not have an analytic expression for $f(x)$ that lets us calculate its value at an arbitrary point. We concentrate on techniques that may be adapted if, for example, we have a table of values of functions that may have been obtained from some physical measurement or some experiments or long numerical calculations that cannot be cast into a simple functional form. The task now is to estimate $f(x)$ for an arbitrary point x by, in some sense, drawing a smooth curve through (and perhaps beyond) the data point x_i. If the desired x is in between the largest and smallest of the data point, then the problem is called *interpolation*; and if x is outside that range, it is called *extrapolation*. In this chapter we shall restrict our attention to interpolation. It is a rational process generally used in estimating a missing functional value by taking a weighted average of known functional values at neighboring data points.

An interpolation scheme must model the function in between or beyond the known data point by some plausible functional form. The form should be sufficiently general as to be able to approximate large classes of functions that might arise in practice. The functional forms are polynomials, trigonometric functions, rational functions, and exponential functions. However, we shall restrict our attention to *polynomials*. Polynomial functions are widely used in practice, since they are easy to determine, evaluate, differentiate, and integrable. *Polynomial interpolation* provides some mathematical tools that can be used in developing methods for approximation theory, numerical differentiation, numerical integration, and numerical solutions of ordinary differential equations and partial differential equations. A set of data points we consider here may be *equally spaced or unequally spaced* in the independent variable x. Several procedures can be used to fit approximation polynomials for both cases. For example, the Lagrange interpolatory polynomial, the Newton divided-difference interpolatory polynomial, and Aitken's interpolatory polynomial, (for unequally spaced and procedures based on differences) can be used for equally spaced, including the Newton forward and backward-difference polynomials, the Gauss forward and backward-difference polynomials, the Bessel difference polynomial and the Stirling difference polynomial. These methods are quite easy to apply. The general form of an *nth*-degree polynomial is

$$p_n(x) = a_0 + a_1 x + a_2 x^2 + \cdots + a_n x^n, \tag{4.1}$$

where n denotes the degree of the polynomial and $a_0, a_1, \ldots, a_n$ are constant coefficients. Since there are $(n+1)$ coefficients, $(n+1)$ data points are required to obtain unique values for the coefficients. The important property of polynomials which makes them suitable for approximating functions is the following theorem called the *Weierstrass Approximation Theorem.*

Theorem 4.1 (Weierstrass Approximation Theorem)

If $f(x)$ is a continuous function on the closed interval $[a, b]$ then for every $\epsilon > 0$ there exists a polynomial $p_n(x)$, where the value of n depends on the value of ϵ, such that for all x in $[a, b]$,

$$|f(x) - p_n(x)| < \epsilon. \tag{4.2}$$

Consequently, any continuous function can be approximated to any accuracy by a polynomial of high enough degree.

Taylor's Polynomial

One important polynomial is the Taylor polynomial, which is a truncated Taylor series, with an explicit remainder, or error term. The Taylor polynomial is defined as:

$$p_n(x) = f(x_0) + \frac{(x - x_0)}{1!} f'(x_0) + \frac{(x - x_0)^2}{2!} f''(x_0) + \cdots + \frac{(x - x_0)^n}{n!} f^{(n)}(x_0) \quad (4.3)$$

with a remainder term

$$R_{n+1}(x) = \frac{(x - x_0)^{n+1}}{(n+1)!} f^{(n+1)}(\eta(x)), \quad x_0 < \eta(x) < x. \quad (4.4)$$

The Taylor polynomial cannot be used as an approximating function for discrete data because the derivatives required in the coefficients cannot be determined. But it does have great significance for polynomial approximation because it has an explicit error term.

Example 4.1
(a) Find the Taylor polynomial of degree three for $f(x) = e^x \sin x$ expanded about $x_0 = 0$.
(b) Use the polynomial in part (a) to approximate the function $f(0.2)$. Also, compute a bound for the error involved.

Solution.

(a) The third degree Taylor polynomial expanded about x_0 is

$$p_3(x) = f(x_0) + \frac{(x - x_0)}{1!} f'(x_0) + \frac{(x - x_0)^2}{2!} f''(x_0) + \frac{(x - x_0)^3}{3!} f'''(x_0) \quad (4.5)$$

with a remainder term

$$R_4(x) = \frac{f^{(4)}(\eta(x))}{4!}(x - x_0)^4, \quad x_0 \le \eta(x) \le x. \quad (4.6)$$

Since $f(x) = e^x \sin x$, then $f(x_0) = f(0) = 0$. Now calculating the derivatives required for the desired polynomial $p_3(x)$, we get

$$\begin{aligned} f'(x) &= e^x(\sin x + \cos x), & f'(x_0) = f'(0) = 1 \\ f''(x) &= 2e^x \cos x, & f''(x_0) = f''(0) = 2 \\ f'''(x) &= 2e^x(\cos x - \sin x), & f'''(x_0) = f'''(0) = 2, \end{aligned}$$

putting all these values in (4.5) gives

$$p_3(x) = x + x^2 + \frac{x^3}{3}. \tag{4.7}$$

(b) Now taking $x = 0.2$, *in (4.7), we get the estimate value* $p_3(0.2) = 0.2427$ *of* $f(x)$ *at* $x = 0.2$.

Now to find an error bound for the approximation, we will use formula (4.6). For this we have to calculate the value of the fourth derivative of the function as follows

$$f^{(4)}(x) = -4e^x \sin x; \qquad f^{(4)}(\eta(x)) = -4e^{\eta(x)} sin(\eta(x)),$$

where unknown point $\eta(x)$ *lies on the interval* $(0.0, 0.2)$. *The value of* $f^{(4)}(\eta(x))$ *can not be computed exactly because* $\eta(x)$ *is not known. But we can bound the error by computing the largest possible value for* $|f^{(4)}|$. *So bound* $|f^{(4)}|$ *on [0.0,0.2] can be obtained*

$$M = \max_{0.0 \leq x \leq 0.2} |4e^x \sin x| = 0.9706$$

at $x = 0.2$ *and* $|f^{(4)}(\eta(x))| \leq M$. *Hence,*

$$|R_4(x)| \leq \frac{M}{24}|x - x_0|^4$$

and it gives

$$|R_4(0.2)| \leq \frac{0.9706}{24}(0.2)^4$$

$$\leq 6.4707 \times 10^{-5}.$$

Since the true value of the function $f(x)$ *at* $x = 0.2$ *is* 0.2427, *from this we note that the actual error is correct to four decimal places.*

Note that the difficulty associated with approximation by the Taylor polynomial will start if we look for the estimate value of the function at a point that is not very close to x_0. For example, one can find the estimate values of the above function $f(x) = e^x \sin x$ at $x = 1$ and $x = 3$, which are $p_3(1) = 2.3333$ and $p_3(3) = 21$, respectively. This shows that the error will increase as we go away from x_0. The main reason is that Taylor polynomials have the property that all the information used in the approximation is concerned at one point; that is, x_0. For ordinary computational purposes, it is more efficient to use polynomials that include information at various points; these type of polynomials we will consider in the coming sections. The primary use of Taylor polynomials in numerical analysis is not for approximation purposes but for the derivation of numerical techniques.

4.2 Polynomial Interpolation for Uneven Intervals

Suppose we have given a set of $(n+1)$ data points relating dependent variables $f(x)$ to an independent variable x as follows

x	x_0	x_1	$\cdots$	x_n
$f(x)$	$f(x_0)$	$f(x_1)$	$\cdots$	$f(x_n)$

Generally, the data points $x_0, x_1, \ldots, x_n$ are arbitrary and assume the interval between two adjacent points is not the same and assume that the data points are organized in such a way that $x_0 < x_1 < x_2 < \cdots < x_{n-1} < x_n$.

When the data points in a given functional relationship are not equally spaced, the interpolation problem becomes more difficult to solve. The basis for this assertion lies in the fact that the interpolating polynomial coefficient will depend on the functional values as well as on the data points given in the table.

4.2.1 Lagrange Interpolating Polynomials

Lagrange interpolating polynomials are a popular and well-known interpolation method used to approximate the functions at an arbitrary point x. The Lagrange interpolation method provides a direct approach for determining interpolated values regardless of data points spacing; that is, it can be fitted to unequally spaced or equally spaced data. To discuss the Lagrange interpolation method, we start with a simplest form of interpolation; that is, *linear interpolation.* The interpolated value is obtained from the equation of a straight line that passes through two tabulated values, one on each side of the required value. This straight line is a first-degree polynomial. The problem of determining a polynomial of degree one that passes through the distinct points (x_0, y_0) and (x_1, y_1) is the same as approximating the function $f(x)$ for which $f(x_0) = y_0$ and $f(x_1) = y_1$ by means of first degree polynomial interpolation. Let us consider the construction of a linear polynomial $p_1(x)$ passing through two data points $(x_0, f(x_0))$ and $(x_1, f(x_1))$ (see Figure 4.1). Let us consider a linear polynomial of the form

$$p_1(x) = a_0 + a_1 x. \tag{4.8}$$

Since a polynomial of degree one has two coefficients, one might expect to be able to choose two conditions, which satisfy

$$p_1(x_k) = f(x_k); \qquad k = 0, 1.$$

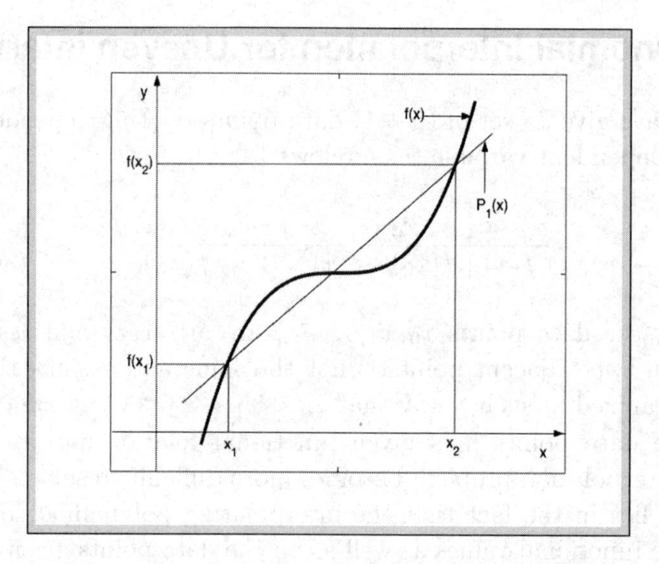

Figure 4.1: Linear Lagrange interpolation.

When $p_1(x)$ passes through point $(x_0, f(x_0))$, we have

$$p_1(x_0) = a_0 + a_1 x_0 = y_0 = f(x_0)$$

and if it passes through point $(x_1, f(x_1))$, we have

$$p_1(x_1) = a_0 + a_1 x_1 = y_1 = f(x_1).$$

Solving the last two equations gives a unique solution

$$a_0 = \frac{x_0 y_1 - x_1 y_0}{x_0 - x_1}; \quad a_1 = \frac{y_1 - y_0}{x_1 - x_0}. \tag{4.9}$$

Putting these values in (4.8), we have

$$p_1(x) = \left(\frac{x - x_1}{x_0 - x_1}\right) y_0 + \left(\frac{x - x_0}{x_1 - x_0}\right) y_1,$$

which can also be written as

$$p_1(x) = L_0(x) f(x_0) + L_1(x) f(x_1), \tag{4.10}$$

where

$$L_0(x) = \frac{x - x_1}{x_0 - x_1} \quad \text{and} \quad L_1(x) = \frac{x - x_0}{x_1 - x_0}. \tag{4.11}$$

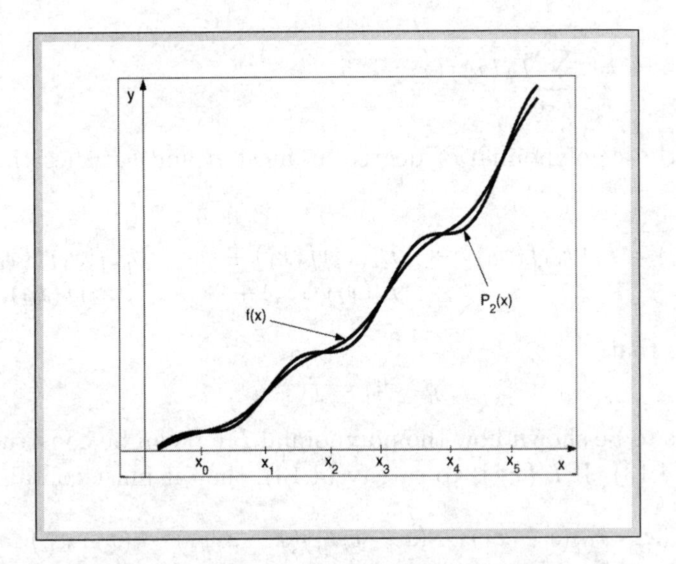

Figure 4.2: General Lagrange interpolation.

Note that when $x = x_0$, $L_0(x_0) = 1$ and $L_1(x_0) = 0$. Similarly, when $x = x_1$, then $L_0(x_1) = 0$ and $L_1(x_1) = 1$. Polynomial (4.10) is known as a *linear Lagrange interpolating polynomial* and (4.11) is called a *Lagrange coefficient polynomial.* To generalize the concept of linear interpolation, consider the construction of a polynomial $p_n(x)$ of degree at most n that passes through $(n+1)$ distinct points $(x_0, f(x_0)), \ldots, (x_n, f(x_n))$ (see Figure 4.2) and satisfies the interpolation conditions

$$p_n(x_k) = f(x_k); \qquad k = 0, 1, 2, \ldots, n. \tag{4.12}$$

Assume that there exists polynomial $L_k(x)$ $(k = 0, 1, 2, \ldots, n)$ of degree n having the property

$$L_k(x_j) = \begin{cases} 0 & \text{for} \quad k \neq j \\ 1 & \text{for} \quad k = j \end{cases} \tag{4.13}$$

and

$$\sum_{k=0}^{n} L_k(x) = 1. \tag{4.14}$$

The polynomial $p_n(x)$ is given by

$$\begin{aligned} p_n(x) &= L_0(x)f(x_0) + L_1(x)f(x_1) + \cdots + L_{i-1}(x)f(x_{i-1}) \\ &+ L_i(x)f(x_i) + \cdots + L_n(x)f(x_n) \end{aligned}$$

$$= \sum_{k=0}^{n} L_k(x) f(x_k). \qquad (4.15)$$

It is clearly a polynomial of degree at most n and satisfies the conditions (4.12) since

$$\begin{aligned} p_n(x_i) = L_0(x_i)f(x_0) &+ L_1(x_i)f(x_1) + \cdots + L_{i-1}(x_i)f(x_{i-1}) \\ &+ L_i(x_i)f(x_i) + \cdots + L_n(x_i)f(x_n), \end{aligned}$$

which implies that

$$p_n(x_i) = f(x_i).$$

It remains to be shown how the polynomial $L_i(x)$ can be constructed so that they satisfy (4.13). If $L_i(x)$ is to satisfy (4.13), then it must contain a factor

$$(x - x_0)(x - x_1) \cdots (x - x_{i-1})(x - x_{i+1}) \cdots (x - x_n). \qquad (4.16)$$

Since this expression has exactly n terms and $L_i(x)$ is a polynomial of degree n, we can deduce that

$$L_i(x) = A_i(x - x_0)(x - x_1) \cdots (x - x_{i-1})(x - x_{i+1}) \cdots (x - x_n) \qquad (4.17)$$

for some multiplicative constant A_i. Let $x = x_i$, then the value of A_i is chosen so that

$$A_i = 1/(x_i - x_0)(x_i - x_1) \cdots (x_i - x_{i-1})(x_i - x_{i+1}) \cdots (x_i - x_n), \qquad (4.18)$$

where none of the terms in the denominator can be zero from the assumption of distinct points. Hence,

$$L_i(x) = \prod_{k=0}^{n} \left(\frac{x - x_k}{x_i - x_k} \right), \quad i \neq k. \qquad (4.19)$$

The interpolating polynomial can now be readily evaluated by substituting (4.19) into (4.15) to give

$$f(x) \approx p_n(x) = \sum_{i=0}^{n} \prod_{k=0}^{n} \left(\frac{x - x_k}{x_i - x_k} \right) f(x_i), \qquad i \neq k. \qquad (4.20)$$

This formula is called the Lagrange interpolation formula of degree n and the terms in (4.19) are called Lagrange coefficient polynomials. To show the *uniqueness* of the interpolating polynomial $p_n(x)$, we suppose that in addition to the polynomial $p_n(x)$ the interpolation problem has another solution $q_n(x)$ of degree $\leq n$ whose graph passes through (x_i, y_i), $i = 0, 1, \ldots, n$. Then define

$$r_n(x) = p_n(x) - q_n(x)$$

of the degree not greater than n. Since

$$r_n(x_i) = p_n(x_i) - q_n(x_i) = f(x_i) - f(x_i) = 0$$

the polynomial $r_n(x)$ vanishes at $n+1$ point. Use the following well-known result from the theory of equations:
"If a polynomial of degree n vanishes at n+1 distinct points, then the polynomial is identically zero."

Hence, $r_n(x)$ vanishes identically, or equivalently, $p_n(x) = q_n(x)$.

Example 4.2 *Let $p_2(x)$ be the quadratic Lagrange interpolating polynomial for the data: $(0,0), (1,y)$, and $(2,3)$. Find the value of y if the coefficient of x^2 is $p_2(x)$ is $\frac{1}{2}$.*

Solution. *Consider the quadratic Lagrange interpolating polynomial as follows:*

$$p_2(x) = L_0(x)f(x_0) + L_1(x)f(x_1) + L_2(x)f(x_2).$$

Using the given data points, we get

$$p_2(x) = L_0(x)(0) + L_1(x)(y) + L_2(x)(3),$$

where the Lagrange coefficients can be calculated as follows:

$$\begin{aligned}
L_0(x) &= \frac{(x-x_1)(x-x_2)}{(x_0-x_1)(x_0-x_2)} = \frac{(x-1)(x-2)}{(0-1)(0-2)} \\
&= \frac{1}{2}(x^2 - 3x + 2) \\
L_1(x) &= \frac{(x-x_0)(x-x_2)}{(x_1-x_0)(x_1-x_2)} = \frac{(x-0)(x-2)}{(1-0)(1-2)} \\
&= -(x^2 - 2x) \\
L_2(x) &= \frac{(x-x_0)(x-x_1)}{(x_2-x_0)(x_2-x_1)} = \frac{(x-0)(x-1)}{(2-0)(2-1)}. \\
&= \frac{1}{2}(x^2 - x)
\end{aligned}$$

Using the values of the Lagrange coefficients and after simplifying, we have

$$p_2(x) = (-y + \frac{3}{2})x^2 + (2y - \frac{3}{2})x.$$

Since the given condition is that the coefficient of x^2 *is* $\frac{1}{2}$,

$$-y + \frac{3}{2} = \frac{1}{2},$$

which gives $y = 1$.

Example 4.3 *Use the quadratic Lagrange interpolation formula to find the numbers* $A, B,$ *and* C *such that* $p_2(1.4) = Af(0) + Bf(1) + Cf(2)$.

Solution. *Consider the quadratic Lagrange interpolating polynomial as follows:*

$$p_2(x) = Af(x_0) + Bf(x_1) + Cf(x_2),$$

where the values of $A, B,$ *and* C *can be defined as follows:*

$$\begin{aligned} A &= \frac{(x - x_1)(x - x_2)}{(x_0 - x_1)(x_0 - x_2)}, \\ B &= \frac{(x - x_0)(x - x_2)}{(x_1 - x_0)(x_1 - x_2)}, \\ C &= \frac{(x - x_0)(x - x_1)}{(x_2 - x_0)(x_2 - x_1)}. \end{aligned}$$

Using the given values as $x_0 = 0, x_1 = 1,$ *and* $x_2 = 2$ *and the interpolating point* $x = 1.4$, *we obtain*

$$\begin{aligned} A &= \frac{(1.4 - 1)(1.4 - 2)}{(0 - 1)(0 - 2)} = -0.12, \\ B &= \frac{(1.4 - 0)(1.4 - 2)}{(1 - 0)(1 - 2)} = 0.84, \\ C &= \frac{(1.4 - 0)(1.4 - 1)}{(2 - 0)(2 - 1)} = 0.28. \end{aligned}$$

Example 4.4 *Consider the following table:*

x	0	3	7
$f(x)$	2	4	19

(a) Construct the interpolating polynomial to approximate the function using the suitable Lagrange interpolation formula.
(b) Use the polynomial in part (a) to interpolate $f(x)$ at $x = 4$.

Solution. *(a) Obviously, a quadratic polynomial can be determined so that it passes through the three points. Consider the quadratic Lagrange interpolating polynomial as follows:*

$$p_2(x) = L_0(x)f(x_0) + L_1(x)f(x_1) + L_2(x)f(x_2) \tag{4.21}$$

or

$$p_2(x) = 2L_0(x) + 4L_1(x) + 19L_2(x). \tag{4.22}$$

The Lagrange coefficients can be calculated as follows:

$$\begin{aligned} L_0(x) &= \frac{(x-x_1)(x-x_2)}{(x_0-x_1)(x_0-x_2)} = \frac{1}{21}(x^2 - 10x + 21), \\ L_1(x) &= \frac{(x-x_0)(x-x_2)}{(x_1-x_0)(x_1-x_2)} = -\frac{1}{12}(x^2 - 7x), \\ L_2(x) &= \frac{(x-x_0)(x-x_1)}{(x_2-x_0)(x_2-x_1)} = \frac{1}{28}(x^2 - 3x). \end{aligned}$$

Putting these values of the Lagrange coefficients in (4.22), we have

$$p_2(x) = \frac{1}{84}(37x^2 - 55x + 168),$$

which is the required quadratic interpolating polynomial.
(b) Now taking $x = 4$ in the above polynomial, we have

$$p_2(4) = \frac{1}{84}\Big[37(4)^2 - 55(4)x + 168\Big] = 6.4286,$$

which is the required estimate value of $f(4)$.

Note that the sum of the Lagrange coefficients is equal to 1 as it should be

$$L_0(4) + L_1(4) + L_2(4) = -\frac{1}{7} + 1 + \frac{1}{7} = 1.$$

Using MATLAB the above results can be reproduced as follows:

```
>> x = [0 3 7];
>> y = [2 4 19];
>> x0 = 4;
>> sol = lint(x, y, x0);
```

Program 4.1

MATLAB m-file for the Lagrange Interpolation Method

```
function fi=lint(x,y,x0)
dxi=x0-x; m=length(x); L=zeros(size(y));
L(1) = prod(dxi(2 : m))/prod(x(1) - x(2 : m));
L(m) = prod(dxi(1 : m - 1))/prod(x(m) - x(1 : m - 1));
for j=2:m-1
num = prod(dxi(1 : j - 1)) * prod(dxi(j + 1 : m));
dem = prod(x(j) - x(1 : j - 1)) * prod(x(j) - x(j + 1 : m));
L(j)=num/dem; end; fi = sum(y. * L);
```

Error Formula

As with any numerical technique, it is important to obtain bounds for the errors involved. Now we discuss the error term when the Lagrange polynomial is used to approximate continuous function $f(x)$. It is similar to the error term for the Taylor polynomial, except that the factor $(x - x_0)^{n+1}$ is replaced with the product $(x - x_0)(x - x_1)\cdots(x - x_n)$. This is expected because interpolation is exact at each of the $(n+1)$ data point x_k, where we have

$$f(x_k) - p_n(x_k) = y_k - y_k = 0, \quad \text{for} \quad k = 1, 2, \ldots, n. \tag{4.23}$$

Theorem 4.2 (Error Formula of Lagrange Polynomial)

If $f(x)$ has $(n+1)$ derivatives on interval I and if it is approximated by a polynomial $p_n(x)$ passing through $(n+1)$ data points on I, then the error E_n is given by

$$E_n = f(x) - p_n(x) = \frac{f^{(n+1)}(\eta(x))}{(n+1)!}(x - x_0)(x - x_1)\cdots(x - x_n), \quad \eta(x) \in I, \tag{4.24}$$

where $p_n(x)$ is Lagrange interpolating polynomial (4.15) and an unknown point $\eta(x) \in (x_0, x_n)$.

Error formula (4.24) is an important theoretical result because Lagrange polynomials are used extensively for deriving numerical differentiation and integration methods. Error bounds for these techniques are obtained from the Lagrange error formula.

Example 4.5 *Find a bound for the error in the linear interpolation.*

Solution. *Consider two points x_0 and x_1. The linear polynomial $p_1(x)$ interpolating $f(x)$ at these points is*

$$p_1(x) = \frac{(x - x_1)}{(x_0 - x_1)} f(x_0) + \frac{(x - x_0)}{(x_1 - x_0)} f(x_1).$$

Using the given data point, error formula (4.24) becomes

$$f(x) - p_1(x) = \frac{(x - x_0)(x - x_1)}{2!} f''(\eta(x)),$$

where $\eta(x)$ is an unknown point between x_0 and x_1. Hence,

$$|f(x) - p_1(x)| = \left| \frac{(x - x_0)(x - x_1)}{2!} \right| |f''(\eta(x))|.$$

The value of $f''(\eta(x))$ cannot be computed exactly because $\eta(x)$ is not known. But we can bound the error by computing the largest possible value for $|f''(\eta(x))|$. So bound $|f''(x)|$ on $[x_0, x_1]$ can be obtained

$$M = \max_{x_0 \leq x \leq x_1} |f''(x)|$$

and so for $|f''(\eta(x))| \leq M$, we have

$$|f(x) - p_1(x)| \leq \frac{M}{2} |(x - x_0)(x - x_1)|.$$

Since the function $|(x - x_0)(x - x_1)|$ attains its maximum in $[x_0, x_1]$ and occurs at the point $x = \dfrac{(x_0 + x_1)}{2}$, the maximum is $\dfrac{(x_1 - x_0)^2}{4}$.
This follows easily by noting that the function $(x - x_0)(x - x_1)$ is a quadratic

and has two roots, x_0 and x_1; hence, its maximum value occurs midway between these roots. Thus, for any $x \in [x_0, x_1]$, we have

$$|f(x) - p_1(x)| \leq \frac{(x_1 - x_0)^2}{8} M,$$

or

$$|f(x) - p_1(x)| \leq \frac{h^2}{8} M,$$

where $h = x_1 - x_0$.

Example 4.6 *Consider the following table having the data for $f(x) = e^{3x} \cos 2x$:*

x	0.1	0.2	0.4	0.5
$f(x)$	1.32295	1.67828	2.31315	2.42147

Find the approximation of $f(0.3)$ using the suitable Lagrange interpolation formula; also estimate an error bound for the approximation.

Solution. *The suitable Lagrange formula to find the interpolating polynomial to approximate the function is*

$$p_3(x) = L_0(x)f(x_0) + L_1(x)f(x_1) + L_2(x)f(x_2) + L_3(x)f(x_3),$$

which implies that

$$\begin{aligned} p_3(x) &= \frac{(x - x_1)(x - x_2)(x - x_3)}{(x_0 - x_1)(x_0 - x_2)(x - x_3)} f(x_0) \\ &+ \frac{(x - x_0)(x - x_2)(x - x_3)}{(x_1 - x_0)(x_1 - x_2)(x_1 - x_3)} f(x_1) \\ &+ \frac{(x - x_0)(x - x_1)(x - x_3)}{(x_2 - x_0)(x_2 - x_1)(x_2 - x_3)} f(x_2) \\ &+ \frac{(x - x_0)(x - x_1)(x - x_2)}{(x_3 - x_0)(x_3 - x_1)(x_3 - x_2)} f(x_3) \end{aligned}$$

or

$$\begin{aligned} p_3(x) &= (-110.2458)[(x - 0.2)(x - 0.4)(x - 0.5)] \\ &+ (279.7133)[(x - 0.1)(x - 0.4)(x - 0.5)] \\ &+ (-385.5250)[(x - 0.1)(x - 0.2)(x - 0.5)] \\ &+ (201.7892)[(x - 0.1)(x - 0.2)(x - 0.4)]. \end{aligned}$$

Thus,

$$p_3(x) = -14.269x^3 + 8.726x^2 + 1.935x + 1.057. \qquad (4.25)$$

Taking $x = 0.3$ in polynomial (4.25), we have

$$f(0.3) \approx p_3(0.3) = 2.03754.$$

The exact value of $f(0.3) = 2.029998$, so, the actual error is 0.00754. Note that the sum of the values of the Lagrange coefficients is equal to one; that is,

$$\sum_{i=0}^{3} L_i(0.3) = -0.16667 + 0.66667 + 0.66667 - 0.16667 = 1.$$

Now to compute an error bound of the approximation, we use the following formula

$$|f(x) - p_3(x)| = \frac{|f^{(4)}(\eta(x))|}{4!}|(x - x_0)(x - x_1)(x - x_2)(x - x_3)|. \qquad (4.26)$$

Taking the fourth derivative of the given function, we have

$$\begin{aligned} f'(x) &= e^{3x}(3\cos 2x - 2\sin 2x) \\ f''(x) &= e^{3x}(5\cos 2x - 12\sin 2x) \\ f'''(x) &= e^{3x}(-9\cos 2x - 46\sin 2x) \\ f^{(4)}(x) &= -e^{3x}(119\cos 2x + 120\sin 2x). \end{aligned}$$

Thus,

$$|f^{(4)}(\eta(x))| = |-e^{3\eta(x)}(119\cos 2(\eta(x)) + 120\sin 2(\eta(x)))|, \quad \text{for} \quad \eta(x) \in (0.1, 0.5)$$

and it gives

$$\begin{aligned} |f^{(4)}(0.1)| &= 189.61229 \\ |f^{(4)}(0.5)| &= 740.69991. \end{aligned}$$

The value of $f^{(4)}(\eta(x))$ cannot be computed exactly because $\eta(x)$ is not known. But we can bound the error by computing the largest possible value for $|f^{(4)}(\eta(x))|$. So bound $|f^{(4)}(x)|$ on $[0.1, 0.5]$ can be obtained

$$M = \max_{0.1 \le x \le 0.5} |f^{(4)}(x)| = 740.69991$$

and so for $|f^{(4)}(\eta(x))| \le M$, we have (4.26) as follows

$$|f(x) - p_3(x)| \le (740.69991)(0.0004)/24 = 0.01235,$$

which is the required error bound for the approximation.

Example 4.7 *Determine the spacing h in a table of equally spaced value of function $f(x) = e^x$ between the smallest point $a = 1$ and the largest point $b = 2$, so that interpolation with a second-degree polynomial in this table will yield a desired accuracy.*

Solution. *Suppose that the given table contains the function values $f(x_i)$ for the points $x_i = 1 + ih$, $i = 0, 1, \ldots, n$, where $n = \frac{(2-1)}{h}$. If $x \in [x_{i-1}, x_{i+1}]$, then we approximate the function $f(x)$ by degree two polynomial $p_2(x)$, which interpolates $f(x)$ at x_{i-1}, x_i, x_{i+1}. Then error formula (4.24) for these data points becomes*

$$|f(x) - p_2(x)| = \left|\frac{(x - x_{i-1})(x - x_i)(x - x_{i+1})}{3!}\right|\left|f'''(\eta(x))\right|,$$

where $\eta(x) \in (x_{i-1}, x_{i+1})$. Since point $\eta(x)$ is unknown, we cannot estimate $f'''(\eta(x))$; therefore, let

$$|f'''(\eta(x))| \leq M = \max_{1 \leq x \leq 2} |f'''(x)|.$$

Then

$$|f(x) - p_2(x)| \leq \frac{M}{6}|(x - x_{i-1})(x - x_i)(x - x_{i+1})|.$$

Since $f(x) = e^x$, and $f'''(x) = e^x$, therefore,

$$|f'''(\eta(x))| \leq M = e^2 = 7.3891.$$

Now to find the maximum value of $|(x - x_{i-1})(x - x_i)(x - x_{i+1})|$, we have

$$\begin{aligned} \max_{x \in [x_{i-1}, x_{i+1}]} |(x - x_{i-1})(x - x_i)(x - x_{i+1}| &= \max_{t \in [-h, h]} |(t - h)t(t + h)| \\ &= \max_{t \in [-h, h]} |t(t^2 - h^2)| \end{aligned}$$

using the linear change of variables $t = x - x_i$. As we see, the function $H(t) = t^3 - th^2$ vanishes at $t = -h$ and $t = h$, so the maximum value of $|H(t)|$ on $[-h, h]$ must occur at one of the extremes of $H(t)$, which can be found by solving the equation

$$H'(t) = 3t^2 - h^2 = 0, \quad \textit{gives}, \quad t = \pm h/\sqrt{3}.$$

Hence,

$$\max_{x \in [x_{i-1}, x_{i+1}]} |(x - x_{i-1})(x - x_i)(x - x_{i+1})| = \frac{2h^3}{3\sqrt{3}}.$$

Thus, for any $x \in [1, 2]$, *we have*

$$|f(x) - p_2(x)| \leq \frac{(2h^3/3\sqrt{3})e^2}{6} = \frac{h^3 e^2}{9\sqrt{3}}$$

if $p_2(x)$ *is chosen as the polynomial of degree two, which interpolates* $f(x) = e^x$ *at the three tabular points nearest* x. *If we wish to obtain six decimal place accuracy this way, we would have to choose h so that*

$$\frac{h^3 e^2}{9\sqrt{3}} < 5 \times 10^{-7},$$

which implies that

$$h^3 < 10.5483 \times 10^{-7},$$

which gives $h = 0.01$.

While the Lagrange interpolation formula is at the heart of polynomial interpolation, it is not, by any stretch of the imagination, the most practical way to use it. Just consider for a moment if we had to add an addition data point in Example 4.4. In order to find cubic polynomial $p_3(x)$, we have to repeat the whole process again because we cannot use the solution of the quadratic polynomial $p_2(x)$ in the construction of the cubic polynomial $p_3(x)$. Therefore, one can note that the Lagrange method is not particularly efficient for large values of n, the degree of the polynomial. When n is large and the data for x is ordered, some improvement in efficiency can be obtained by considering only the data pairs in the vicinity of the x values for which $f(x)$ is sought.

One will be quickly convinced that there must be better techniques available. In the following section we discuss some of the more practical approaches to polynomial interpolation. In using the following scheme, the construction of the difference table plays an important role. In using the Lagrange interpolation scheme, there is no need to construct a difference table.

4.2.2 Newton's General Interpolating Formula

Since we noted in the previous section that for a small number of data points one can easily use the Lagrange formula of the interpolating polynomial. However, for a large number of data points there will be many multiplication and more significantly, whenever a new data point is added to an existing set, the interpolating polynomial has to be completely recalculated. Here, we describe an efficient way of organizing the calculations so as to overcome these disadvantages.

Let us consider the *nth*-degree polynomial $p_n(x)$ that agrees with the function $f(x)$ at the distinct numbers $x_0, x_1, \ldots, x_n$. The divided differences of $f(x)$ with respect to $x_0, x_1, \ldots, x_n$ are derived to express $p_n(x)$ in the form

$$\begin{aligned} p_n(x) = a_0 &+ a_1(x - x_0) + a_2(x - x_0)(x - x_1) + \cdots \\ &+ a_n(x - x_0)(x - x_1)\cdots(x - x_{n-1}) \end{aligned} \tag{4.27}$$

for appropriate constants $a_0, a_1, \ldots, a_n$.
Now to determine the constants, firstly, by evaluating $p_n(x)$ at x_0, we have

$$p_n(x_0) = a_0 = f(x_0). \tag{4.28}$$

Similarly, when $p_n(x)$ is evaluated at x_1, then

$$p_n(x_1) = a_0 + a_1(x_1 - x_0) = f(x_1),$$

which implies that

$$a_1 = \frac{f(x_1) - f(x_0)}{x_1 - x_0}. \tag{4.29}$$

Divided Differences

Now we express the interpolating polynomial in terms of divided difference.

Firstly, we define the *Zeroth divided difference* at the point x_i by

$$f[x_i] = f(x_i), \tag{4.30}$$

which is simply the value of the function $f(x)$ at x_i.

The *first-order* or *first divided difference* at the points x_i and x_{i+1} can be defined by

$$f[x_i, x_{i+1}] = \frac{f[x_{i+1}] - f[x_i]}{x_{i+1} - x_i} = \frac{f(x_{i+1}) - f(x_i)}{x_{i+1} - x_i}. \tag{4.31}$$

In general, the *nth divided difference* $f[x_i, x_{i+1}, \ldots, x_{i+n}]$ is defined by

$$f[x_i, x_{i+1}, \ldots, x_{i+n}] = \frac{f[x_{i+1}, x_{i+2}, \ldots, x_{i+n}] - f[x_i, x_{i+1}, \ldots, x_{i+n-1}]}{x_{i+n} - x_i}. \tag{4.32}$$

Using this definition, (4.28) and (4.29) can be written as

$$a_0 = f[x_0]; \qquad a_1 = f[x_0, x_1],$$

respectively. Similarly, one can have the values of other constants involved in (4.27) such as

$$\begin{aligned} a_2 &= f[x_0, x_1, x_2] \\ a_3 &= f[x_0, x_1, x_2, x_3] \\ \cdots &= \cdots \\ \cdots &= \cdots \\ a_n &= f[x_0, x_1, \ldots, x_n]. \end{aligned}$$

Putting the values of these constants in (4.27), we get

$$\begin{aligned} p_n(x) &= f[x_0] + f[x_0, x_1](x - x_0) + f[x_0, x_1, x_2](x - x_0)(x - x_1) \\ &+ \cdots + f[x_0, x_1, \ldots, x_n](x - x_0)(x - x_1) \cdots (x - x_{n-1}), \end{aligned} \tag{4.33}$$

which can also be written as

$$p_n(x) = f[x_0] + \sum_{k=1}^{n} f[x_0, x_1, \ldots, x_k](x - x_0)(x - x_1) \cdots (x - x_{k-1}). \tag{4.34}$$

This type of polynomial is known as *Newton's interpolatory divided difference polynomial.* Table 4.1 shows the divided difference for a function $f(x)$. One can

Table 4.1: Divided difference table for a function $y = f(x)$.

k	x_k	Zero Divided Difference	First Divided Difference	Second Divided Difference	Third Divided Difference
0	x_0	$f[x_0]$			
1	x_1	$f[x_1]$	$f[x_0, x_1]$		
2	x_2	$f[x_2]$	$f[x_1, x_2]$	$f[x_0, x_1, x_2]$	
3	x_3	$f[x_3]$	$f[x_2, x_3]$	$f[x_1, x_2, x_3]$	$f[x_0, x_1, x_2, x_3]$

easily show that (4.34) is simply a rearrangement of the Lagrange form defined by (4.15). For example, the Newton divided difference interpolation polynomial of degree one is

$$p_1(x) = f[x_0] + f[x_0, x_1](x - x_0),$$

which implies that

$$\begin{aligned} p_1(x) &= f(x_0) + \left(\frac{f(x_1) - f(x_0)}{x_1 - x_0}\right)(x - x_0) \\ &= \frac{(x_1 - x_0)f(x_0) + (x - x_0)f(x_1) - f(x_0)(x - x_0)}{x_1 - x_0} \\ &= \left(\frac{x - x_1}{x_0 - x_1}\right) f(x_0) + \left(\frac{x - x_0}{x_1 - x_0}\right) f(x_1), \end{aligned}$$

which is the Lagrange interpolating polynomial of degree one. Similarly, one can show the equivalent for the *nth*-degree polynomial.

Example 4.8 *Find the Lagrange and the Newton forms of the interpolating polynomial for the following data*

x	0	1	3
$f(x)$	1	2	3

Write both polynomials in the form $a + bx + cx^2$ to verify that they are identical as functions.

Solution. *With $x_0 = 0, x_1 = 1$ and $x_2 = 3$, we obtain the quadratic Lagrange interpolating polynomial*

$$\begin{aligned} p_2(x) &= \frac{(x - x_1)(x - x_2)}{(x_0 - x_1)(x_0 - x_2)} f(x_0) + \frac{(x - x_0)(x - x_2)}{(x_1 - x_0)(x_1 - x_2)} f(x_1) + \frac{(x - x_0)(x - x_1)}{(x_2 - x_0)(x_2 - x_1)} f(x_2) \\ &= \frac{(x - 1)(x - 3)}{(0 - 1)(0 - 3)}(1) + \frac{(x - 0)(x - 3)}{(1 - 0)(1 - 3)}(2) + \frac{(x - 0)(x - 1)}{(3 - 0)(3 - 1)}(3). \end{aligned}$$

After simplifying, we get

$$p_2(x) = \frac{(6 + 7x - x^2)}{6},$$

which is the quadratic Lagrange interpolating polynomial.
Now we construct the divided differences table for the given data points. The result of the divided difference is listed in Table 4.2.

Since Newton's interpolating polynomial of degree two is defined as

$$p_2(x) = f[x_0] + f[x_0, x_1](x - x_0) + f[x_0, x_1, x_2](x - x_0)(x - x_1),$$

using Table 4.2 we have

$$p_2(x) = 1 + (1)(x - 0) + \left(-\frac{1}{6}\right)(x - 0)(x - 1),$$

which gives

$$p_2(x) = \frac{(6 + 7x - x^2)}{6},$$

which shows that both polynomials are identical as functions.

Table 4.2: Divided differences table for Example 4.8.

k	x_k	Zeroth Divided Difference	First Divided Difference	Second Divided Difference
0	0	**1**		
1	1	2	**1**	
2	3	3	$\frac{1}{2}$	$-\frac{\mathbf{1}}{\mathbf{6}}$

Example 4.9 *Show that Newton's interpolating polynomial $p_2(x)$ of degree two satisfies the interpolation conditions*

$$p_2(x_i) = f(x_i), \qquad i = 0, 1, 2.$$

Solution. *Since Newton's interpolating polynomial of degree two is*

$$p_2(x) = f[x_0] + f[x_0, x_1](x - x_0) + f[x_0, x_1, x_2](x - x_0)(x - x_1),$$

first take $x = x_0$ and we have

$$p_2(x_0) = f[x_0] + 0 + 0 = f(x_0).$$

Now take $x = x_1$ and we have

$$p_2(x_1) = f[x_0] + f[x_0, x_1](x_1 - x_0) + 0 = f(x_0) + \frac{f(x_1) - f(x_0)}{x_1 - x_0}(x_1 - x_0),$$

which gives

$$p_2(x_1) = f(x_0) + f(x_1) - f(x_0) = f(x_1).$$

Finally, take $x = x_2$ and we have

$$p_2(x_2) = f[x_0] + f[x_0, x_1](x_2 - x_0) + f[x_0, x_1, x_2](x_2 - x_0)(x_2 - x_1),$$

which can be written as

$$p_2(x_2) = f[x_0] + f[x_0, x_1](x_2 - x_0) + \frac{f[x_1, x_2] - f[x_0, x_1]}{x_2 - x_0}(x_2 - x_0)(x_2 - x_1),$$

which gives

$$p_2(x_2) = f[x_0] + f[x_0, x_1](x_2 - x_1 + x_1 - x_0) + f[x_1, x_2](x_2 - x_1) - f[x_0, x_1](x_2 - x_1)$$

or

$$p_2(x_2) = f[x_0] + f[x_0, x_1](x_1 - x_0) + f[x_1, x_2](x_2 - x_1).$$

From (4.31), we have

$$p_2(x_2) = f[x_0] + \frac{f(x_1) - f(x_0)}{x_1 - x_0}(x_1 - x_0) + \frac{f(x_2) - f(x_1)}{x_2 - x_1}(x_2 - x_1),$$

which gives

$$p_2(x_2) = f(x_0) + f(x_1) - f(x_0) + f(x_2) - f(x_1) = f(x_2).$$

Example 4.10 *Construct the divided differences table for the function $f(x) = x^3 + 7x^2 + 1$ for the values $x = 1, 2, 3, 4, 5$.*

Solution. *The results are listed in Table 4.3.*

From the results in Table 4.3, one can note that the nth divided difference for the nth polynomial equation is always constant and the (n+1)th divided difference is always zero for the nth polynomial equation.

Using the following MATLAB commands one can construct Table 4.3 as follows:

```
>> x = [1 2 3 4 5];
>> y = x.^3+7*x.^2+1;
>> D = divdiff(x,y);
```

Table 4.3: Divided differences table for $f(x) = x^3 + 7x^2 + 1$.

k	x_k	Zeroth Divided Difference	First Divided Difference	Second Divided Difference	Third Divided Difference	Fourth Divided Difference
0	1	**9**				
1	2	37	**28**			
2	3	91	54	**13**		
3	4	177	86	16	**1**	
4	5	301	124	19	1	**0**

The main advantage of the Newton divided difference form over the Lagrange form is that polynomial $p_n(x)$ can be calculated from polynomial $p_{n-1}(x)$ by adding just one extra term, since it follows from (4.34) that

$$p_n(x) = p_{n-1}(x) + f[x_0, x_1, \ldots, x_n](x - x_0)(x - x_1)\cdots(x - x_{n-1}). \tag{4.35}$$

Program 4.2

MATLAB m-file for the Divided Differences

```
function D=divdiff(x,y)
% Construct divided difference table
m = length(x); D = zeros(m,m); D(:,1) = y(:);
for j=2:m; for i=j:m
D(i,j) = (D(i,j-1) - D(i-1,j-1))/(x(i) - x(i-j+1));
end; end
```

Example 4.11

(a) Construct the divided difference table for function $f(x) = \ln(x + 2)$ *in the interval* $0 \leq x \leq 3$ *for stepsize* $h = 1$.

(b) Use the Newton divided difference interpolation formula to construct the interpolating polynomials of degree two and degree three to approximate $\ln(3.5)$.

(c) Compute error bounds for the approximations in part (b).

Solution. *(a) The results of the divided differences are listed in Table 4.4.*

(b) Firstly, we construct the second degree polynomial $p_2(x)$ *using the quadratic Newton interpolation formula as follows*

$$p_2(x) = f[x_0] + f[x_0, x_1](x - x_0) + f[x_0, x_1, x_2](x - x_0)(x - x_1).$$

Then with the help of divided differences Table 4.4, we get

$$p_2(x) = 0.6932 + 0.4055(x-0) - 0.0589(x-0)(x-1),$$

which implies that

$$p_2(x) = -0.0568x^2 + 0.4644x + 0.6932.$$

Then at $x = 1.5$, we have

$$p_2(1.5) = 1.2620$$

with possible actual error

$$f(1.5) - p_2(1.5) = 1.2528 - 1.2620 = -0.0072.$$

Now to construct the cubic interpolatory polynomial $p_3(x)$ that fits at all four points, we only have to add one more term to the polynomial $p_2(x)$:

$$p_3(x) = p_2(x) + f[x_0, x_1, x_2, x_3](x-x_0)(x-x_1)(x-x_2)$$

or

$$\begin{aligned} p_3(x) &= p_2(x) + 0.0089(x-0)(x-1)(x-2) \\ p_3(x) &= p_2(x) + 0.0089(x^3 - 3x^2 + 2x). \end{aligned}$$

Then at $x = 1.5$, we get

$$\begin{aligned} p_3(1.5) &= p_2(1.5) + 0.0089((1.5)^3 - 3(1.5)^2 + 2(1.5)) \\ p_3(1.5) &= 1.2620 - 0.0033 = 1.2587 \end{aligned}$$

with possible actual error

$$f(1.5) - p_3(1.5) = 1.2528 - 1.2587 = -0.0059.$$

We note that the estimated value of $f(1.5)$ by cubic interpolating polynomial is closer to the exact solution than the quadratic polynomial.

(c) Now to compute the error bounds for the approximations in part (b), we use error formula (4.24). For polynomial $p_2(x)$, we have

$$|f(x) - p_2(x)| = \frac{|f'''(\eta(x))|}{3!}|(x-x_0)(x-x_1)(x-x_2)|,$$

since the third derivative of the given function is

$$f'''(x) = \frac{2}{(x+2)^3}$$

Table 4.4: Divided differences table for Example 4.11.

k	x_k	Zeroth Divided Difference	First Divided Difference	Second Divided Difference	Third Divided Difference
0	0	**0.6932**			
1	1	1.0986	**0.4055**		
2	2	1.3863	0.2877	**- 0.0589**	
3	3	1.6094	0.2232	- 0.0323	**0.0089**

and

$$|f'''(\eta(x))| = \left|\frac{2}{(\eta(x)+2)^3}\right|, \qquad \textit{for} \quad \eta(x) \in (0,2).$$

Then

$$M = \max_{0 \le x \le 2} \left|\frac{2}{(x+2)^3}\right| = 0.25$$

and

$$|f(1.5) - p_2(1.5)| \le (0.375)(0.25)/6 = 0.0156,$$

which is the required error bound for approximation $p_2(1.5)$.
Since the error bound for cubic polynomial $p_3(x)$ *is*

$$|f(x) - p_3(x)| = \frac{|f^{(4)}(\eta(x))|}{4!}|(x-x_0)(x-x_1)(x-x_2)(x-x_3)|,$$

taking the fourth derivative of the given function, we have

$$f^{(4)}(x) = \frac{-6}{(x+2)^4}$$

and

$$|f^{(4)}(\eta(x))| = \left|\frac{-6}{(\eta(x)+2)^4}\right|, \qquad \textit{for} \quad \eta(x) \in (0,3),$$

since

$$\begin{aligned} |f^{(4)}(0)| &= 0.375 \\ |f^{(4)}(3)| &= 0.0096 \end{aligned}$$

so $|f^{(4)}(\eta(x))| \le \max_{0 \le x \le 3} \left|\frac{-6}{(x+2)^4}\right| = 0.375$ *and*

$$|f(1.5) - p_3(1.5)| \le (0.5625)(0.375)/24 = 0.0088,$$

which is the required error bound for the approximation $p_3(1.5)$.

Note that in Example 4.11, we used the value of quadratic polynomial $p_2(1.5)$ in calculating cubic polynomial $p_3(1.5)$. It was possible because the initial value for both polynomials was the same as $x_0 = 0$. But the situation will be quite different if the initial point for both polynomials are different. For example, if we have to find the approximate value of $\ln(4.5)$, then the suitable data points for the quadratic polynomial will be $x_0 = 1, x_1 = 2$, and $x_2 = 3$ and for the cubic polynomial $x_0 = 0, x_1 = 1, x_2 = 2$, and $x_3 = 3$. So for getting the best approximation of $\ln(4.5)$ by cubic polynomial $p_3(2.5)$, we cannot use the value of quadratic polynomial $p_2(2.5)$ in cubic polynomial $p_3(2.5)$. The best way is to use the following cubic polynomial form

$$\begin{aligned} p_3(2.5) &= f[0] + f[0,1](2.5-0) + f[0,1,2](2.5-0)(2.5-1) \\ &+ f[0,1,2,3](2.5-0)(2.5-1)(2.5-2), \end{aligned}$$

which gives

$$p_3(2.5) = 0.6932 + 1.0137 - 0.2208 + 0.0166 = 1.5027.$$

Using the following MATLAB commands one can reproduce the results of Example 4.11 as follows:

```
>> x = [0 1 2 3];
>> y = log(x + 2);
>> x0 = 1.5;
>> Y = Ndivf(x, y, x0);
```

Program 4.3

MATLAB m-file for Linear Newton's Interpolation Method

```
function Y=Ndivf(x,y,x0)
m = length(x); D = zeros(m, m); D(:, 1) = y(:);
for j=2:m; for i=j:m;
D(i, j) = (D(i, j - 1) - D(i - 1, j - 1))/(x(i) - x(i - j + 1)); end; end;
Y = D(m, m) * ones(size(x0));
for i = m - 1 : -1 : 1;
Y = D(i, i) + (x0 - x(i)) * Y; end
```

Example 4.12 *Consider the points $x_0 = 0, x_1 = 0.4$, and $x_2 = 0.7$ and for a function $f(x)$, the divided differences are $f[x_2] = 6, f[x_1, x_2] = 10$, and $f[x_0, x_1, x_2] = 50/7$. Use this information and construct the complete divided* *differences table for the given data points.*

Solution. *Since we know the third divided difference is defined as*

$$f[x_0, x_1, x_2] = \frac{f[x_1, x_2] - f[x_0, x_1]}{x_2 - x_0},$$

using the given values, we have

$$\frac{50}{7} = \frac{10 - f[x_0, x_1]}{0.7 - 0}.$$

Solving for $f[x_0, x_1]$, *we have,* $f[x_0, x_1] = 5$. *Now we need to find the values of the zeroth divided differences* $f[x_0]$ *and* $f[x_1]$, *which can be obtained using the first-order divided differences* $f[x_0, x_1]$ *and* $f[x_1, x_2]$. *Firstly, we find the value of* $f[x_1]$ *as follows*

$$\begin{aligned} f[x_1, x_2] &= \frac{f[x_2] - f[x_1]}{x_2 - x_1} \\ 10 &= \frac{6 - f[x_1]}{0.7 - 0.4} \\ f[x_1] &= 6 - 10(0.3) = 3. \end{aligned}$$

The other zeroth divided difference $f[x_0]$ *can be computed as follows*

$$\begin{aligned} f[x_0, x_1] &= \frac{f[x_1] - f[x_0]}{x_1 - x_0} \\ 5 &= \frac{3 - f[x_0]}{0.4 - 0} \\ f[x_0] &= 3 - 5(0.4) = 1, \end{aligned}$$

which completes the divided differences table as shown in Table 4.5.

Example 4.13 *If* $f(x) = p(x)q(x)$, *then show that*

$$f[x_0, x_1] = p(x_1)q[x_0, x_1] + q(x_0)p[x_0, x_1].$$

Also, find the values of $p[0, 1]$ *and* $q[0, 1]$ *when* $f[0, 1] = 4, f(1) = 5$, *and* $p(1) = q(0) = 2$.

Solution. *The first-order divided difference can be written as*

$$f[x_0, x_1] = \frac{f(x_1) - f(x_0)}{x_1 - x_0}.$$

Using $f(x_1) = p(x_1)q(x_1)$ *and* $f(x_0) = p(x_0)q(x_0)$ *in the above formula, we have*

$$f[x_0, x_1] = \frac{p(x_1)q(x_1) - p(x_0)q(x_0)}{x_1 - x_0}.$$

Adding and subtracting the term $p(x_1)q(x_0)$, *we obtain*

$$f[x_0, x_1] = \frac{p(x_1)q(x_1) - p(x_1)q(x_0) + p(x_1)q(x_0) - p(x_0)q(x_0)}{x_1 - x_0},$$

which can be written as

$$f[x_0, x_1] = p(x_1)\frac{q(x_1) - q(x_0)}{x_1 - x_0} + q(x_0)\frac{p(x_1) - p(x_0)}{x_1 - x_0}.$$

Thus,

$$f[x_0, x_1] = p(x_1)q[x_0, x_1] + q(x_0)p[x_0, x_1].$$

Given $x_0 = 0, x_1 = 1, f(1) = 5,$ *and* $f[0, 1] = 4$, *we obtain*

$$f[0, 1] = \frac{f(1) - f(0)}{1 - 0} = f(1) - f(0)$$

or

$$f[0, 1] = 4 = 5 - f(0), \quad \textit{gives} \quad f(0) = 1.$$

Also

$$f(1) = 5 = p(1)q(1) = 2q(1), \quad \textit{gives} \quad q(1) = 5/2$$

and

$$f(0) = 1 = p(0)q(0) = 2p(0), \quad \textit{gives} \quad p(0) = 1/2.$$

Hence,

$$p[0, 1] = \frac{p(1) - p(0)}{1 - 0} = p(1) - p(0) = 2 - 1/2 = 3/2$$

and

$$q[0, 1] = \frac{q(1) - q(0)}{1 - 0} = q(1) - q(0) = 5/2 - 2 = 1/2.$$

Table 4.5: Divided differences table for Example 4.12.

k	x_k	Zeroth Divided Difference	First Divided Difference	Second Divided Difference
0	0	**1**		
1	0.4	**3**	**5**	
2	0.7	6	10	$\frac{50}{7}$

In the case of the Lagrange interpolating polynomial we derived an expression for the truncation error in the form given by (4.24), namely, that

$$R_{n+1}(x) = \frac{f^{(n+1)}(\eta(x))}{(n+1)!} L_n(x),$$

where $L_n(x) = (x - x_0)(x - x_1) \cdots (x - x_n)$.

For Newton's divided difference formula, we obtain, following the same reasoning as above,

$$\begin{aligned} f(x) &= f[x_0] + f[x_0, x_1](x - x_0) + f[x_0, x_1, x_2](x - x_0)(x - x_1) + \cdots \\ &+ f[x_0, x_1, \ldots, x_n](x - x_0)(x - x_1) \cdots (x - x_{n-1}) \\ &+ f[x_0, x_1, \ldots, x_n, x](x - x_0)(x - x_1) \cdots (x - x_{n-1})(x - x_n), \end{aligned}$$

which can also be written as

$$f(x) = p_n(x) + f[x_0, x_1, \ldots, x_n, x](x - x_0)(x - x_1) \cdots (x - x_n) \tag{4.36}$$

or

$$f(x) - p_n(x) = L_n(x) f[x_0, x_1, \ldots, x_n, x]. \tag{4.37}$$

Since the interpolation polynomial agreeing with $f(x)$ at $x_0, x_1, \ldots, x_n$ is unique, it follows that these two error expressions must be equal.

Theorem 4.3 *Let $p_n(x)$ be the polynomial of degree at most n that interpolates a function $f(x)$ at a set of $n + 1$ distinct points $x_0, x_1, \ldots, x_n$. If x is a point different from the points $x_0, x_1, \ldots, x_n$, then*

$$f(x) - p_n(x) = f[x_0, x_1, \ldots, x_n, x] \prod_{j=0}^{n} (x - x_j). \tag{4.38}$$

One can easily show the relationship between the divided differences and the derivative. From (4.32), we have

$$f[x_0, x_1] = \frac{f(x_1) - f(x_0)}{x_1 - x_0}.$$

Now applying the *Mean Value Theorem* to the above equation implies that when the derivative f' exists, then we have

$$f[x_0, x_1] = f'(\eta(x))$$

and unknown point $\eta(x)$ lies between x_0 and x_1. The following theorem generalizes this result.

Theorem 4.4 (Divided Differences and Derivatives)

Suppose that $f \in C^n[a, b]$ and $x_0, x_1, \ldots, x_n$ are distinct numbers in $[a, b]$. Then for some point $\eta(x)$ on the interval (a, b) spanned by $x_0, \ldots, x_n$ exists with

$$f[x_0, x_1, \ldots, x_n] = \frac{f^{(n)}(\eta(x))}{n!}. \tag{4.39}$$

Example 4.14 *Let $f(x) = x \ln x$, and the points $x_0 = 1.1, x_1 = 1.2$, and $x_2 = 1.3$. Find the best approximate value for unknown point $\eta(x)$ using relation (4.39).*

Solution. *Given $f(x) = x \ln x$, then*

$$\begin{aligned} f(1.1) &= 1.1 \ln(1.1) = 0.1048 \\ f(1.2) &= 1.2 \ln(1.2) = 0.2188 \\ f(1.3) &= 1.3 \ln(1.3) = 0.3411. \end{aligned}$$

Since relation (4.39) for the given data points is

$$f[x_0, x_1, x_2] = \frac{f''(\eta(x))}{2!}, \tag{4.40}$$

to compute the value of the left-hand side of relation (4.40), we have to find the values of the first-order divided differences

$$f[x_0, x_1] = \frac{f(x_1) - f(x_0)}{x_1 - x_0} = \frac{0.2188 - 0.1048}{1.2 - 1.1} = 1.1400$$

and

$$f[x_1, x_2] = \frac{f(x_2) - f(x_1)}{x_2 - x_1} = \frac{0.3411 - 0.2188}{1.3 - 1.2} = 1.2230.$$

Using these values, we can compute the second-order divided difference as

$$f[x_0, x_1, x_2] = \frac{f[x_1, x_2] - f[x_0, x_1]}{x_2 - x_0} = \frac{1.2230 - 1.1400}{1.3 - 1.1} = 0.4150.$$

Now we calculate the right-hand side of relation (4.40) for the given points, which gives us

$$\frac{f''(x_0)}{2} = \frac{1}{2x_0} = 0.4546$$

and

$$\frac{f''(x_1)}{2} = \frac{1}{2x_1} = 0.4167$$

and

$$\frac{f''(x_2)}{2} = \frac{1}{2x_2} = 0.3846.$$

We note that the left-hand side of (4.40) is nearly equal to the right-hand side when $x_1 = 1.2$. Hence, the best approximate value of $\eta(x)$ is $x_1 = 1.2$.

Properties of Divided Differences

Now we discuss some of the nice properties of the divided differences as follows:

1. If $p_n(x)$ is a polynomial of degree n, then the divided differences of order n is always constant and $(n+1), (n+2), \ldots$ are identically zero.

2. The divided difference is a symmetric function of its arguments. Thus, if $(t_0, t_1, \ldots, t_n)$ is a permutation of $(x_0, x_1, \ldots, x_n)$, then

$$f[t_0, t_1, \ldots, t_n] = f[x_0, x_1, \ldots, x_n].$$

   This can be verified easily, since the divided differences on both sides of the above equation are the coefficient of x^n in the polynomial of degree at most n that interpolates $f(x)$ at the $n+1$ distinct points $t_0, t_1, \ldots, t_n$ and $x_0, x_1, \ldots, x_n$. These two polynomials are, of course, the same.

3. The interpolating polynomial of degree n can be obtained by adding a single term to the polynomial of degree $(n-1)$ expressed in the Newton form

$$p_n(x) = p_{n-1}(x) + f[x_0, \ldots, x_n] \prod_{j=0}^{n-1} (x - x_j).$$

4. The divided difference $f[x_0, \ldots, x_{n-1}]$ is the coefficient of x^{n-1} in the polynomial that interpolates $(x_0, f_0), (x_1, f_1), \ldots, (x_{n-1}, f_{n-1})$.

5. A sequence of divided differences may be constructed recursively from the formula
$$f[x_0, \ldots, x_n] = \frac{f[x_1, \ldots, x_n] - f[x_0, \ldots, x_{n-1}]}{x_n - x_0}$$
and the zeroth-order divided difference is defined by
$$f[x_i] = f(x_i), \quad i = 0, 1, \ldots, n.$$

6. Another useful property of the divided difference can be obtained using the definitions of divided differences (4.32) and (4.33), which can be extended to the case where some or all of the points x_i are coincident, provided that $f(x)$ is sufficiently differentiable. For example, define
$$f[x_0, x_0] = \lim_{x_1 \to x_0} f[x_0, x_1] = \lim_{x_1 \to x_0} \frac{f(x_1) - f(x_0)}{x_1 - x_0} = f'(x_0). \qquad (4.41)$$
For an arbitrary $n \geq 1$, let all the points in Theorem 4.4 approach x_0. This leads to the definition
$$f[x_0, x_0, \ldots, x_0] = \frac{f^{(n)}(x_0)}{n!},$$
where the left-hand side denotes the *nth* divided difference, all of whose points are x_0.

Example 4.15 *Let $f(x) = e^{-x}$ and let $x_0 = 0, x_1 = 1$. Using (4.39) and the above divide difference property 6, calculate $f[x_0, x_1, x_0]$, $f[x_0, x_0, x_1, x_1]$, and $f[x_0, x_1, x_1, x_1]$.*

Solution. *Using (4.39), we have*
$$f[x_0, x_0] = \frac{1}{1!} f'(x_0) = f'(x_0).$$
Therefore,
$$f[x_0, x_1, x_0] = f[x_0, x_0, x_1] = \frac{f[x_0, x_1] - f[x_0, x_0]}{x_1 - x_0}$$
or
$$f[x_0, x_1, x_0] = \frac{f[x_0, x_1] - f'(x_0)}{x_1 - x_0}.$$

Using the definition of the first-order divided difference of $f(x)$ *at points* x_0 *and* x_1*, we have*

$$f[x_0, x_1] = \frac{f[x_1] - f[x_0]}{x_1 - x_0}$$

and it gives

$$f[0, 1] = \frac{0.368 - 1}{1 - 0} = -0.632.$$

Also,

$$f'(x_0) = -e^{-x_0} \quad and \quad f'(0) = -1.$$

Using these values, we obtain the value of the second divided difference as

$$f[0, 1, 0] = \frac{-0.632 + 1}{1 - 0} = 0.368.$$

Find the value of the third divided difference, which is defined as

$$f[x_0, x_0, x_1, x_1] = \frac{f[x_0, x_1, x_1] - f[x_0, x_0, x_1]}{x_1 - x_0},$$

and after simplifying, we have

$$f[x_0, x_0, x_1, x_1] = \frac{f'(x_1) - 2f[x_0, x_1] + f'(x_0)}{(x_1 - x_0)^2}.$$

Thus,

$$f[0, 0, 1, 1] = \frac{-0.368 - 2(-0.632) - 1}{(1 - 0)^2} = -0.014.$$

Finally, the other third divided difference is defined as

$$f[x_0, x_1, x_1, x_1] = \frac{f[x_1, x_1, x_1] - f[x_0, x_1, x_1]}{x_1 - x_0}$$

or

$$f[x_0, x_1, x_1, x_1] = \frac{f''(x_1)/2! - (f'(x_1) - f[x_0, x_1])/(x_1 - x_0)}{x_1 - x_0}.$$

After simplifying, we have

$$f[x_0, x_0, x_1, x_1] = \frac{(x_1 - x_0)f''(x_1) - 2f'(x_1) + 2f[x_0, x_1]}{2(x_1 - x_0)^2}.$$

By using the value of $f''(1) = e^{-1} = 0.368$*, we have*

$$f[0, 1, 1, 1] = \frac{(1 - 0)(0.368) - 2(-1) + 2(-0.632)}{2(1 - 0)^2} = 1.104.$$

There are many schemes for the efficient implementation of divided difference interpolation, such as that due to *Aitken's Method*, which is designed for the easy evaluation of the polynomial, taking the points closest to the one of interest first and computing only those divided differences which are actually necessary for the computation. The implementation is iterative in nature; additional data points are included one at a time until successive estimates $p_k(x)$ and $p_{k+1}(x)$ of $f(x)$ agree to some specified accuracy or until all data has been used.

4.2.3 Aitken's Method

This is an iterative interpolation process, which is based on the repeated application of a simple interpolation method. This elegant method may be used to interpolate between both equal and unequal spaced data points. The basis of this method is equivalent to generating a sequence of Lagrange polynomials but it is a very efficient formulation. The method is used to compute an interpolated value using successive higher degree polynomials until further increases in the degree of the polynomials give a negligible improvement on the interpolated value.

Suppose we want to fit a polynomial function, for the purpose of interpolation, to the following data points

x	x_0	x_1	$\cdots$	x_n
$f(x)$	$f(x_0)$	$f(x_1)$	$\cdots$	$f(x_n)$

In order to estimate the value of the function $f(x)$ corresponding to any given value of x, we consider the following expression

$$\begin{aligned} P_{01}(x) &= \frac{1}{x_1 - x_0}\begin{vmatrix} x - x_0 & f(x_0) \\ x - x_1 & f(x_1) \end{vmatrix} \\ &= \frac{1}{x_1 - x_0}\Big[(x - x_0)f(x_1) - (x - x_1)f(x_0)\Big] \end{aligned}$$

and

$$\begin{aligned} P_{02}(x) &= \frac{1}{x_2 - x_0}\begin{vmatrix} x - x_0 & f(x_0) \\ x - x_2 & f(x_2) \end{vmatrix} \\ &= \frac{1}{x_2 - x_0}\Big[(x - x_0)f(x_2) - (x - x_2)f(x_0)\Big]. \end{aligned}$$

In general,

$$P_{0m}(x) = \frac{1}{x_m - x_0}\begin{vmatrix} x - x_0 & f(x_0) \\ x - x_m & f(x_m) \end{vmatrix}$$
$$= \frac{1}{x_m - x_0}\Big[(x - x_0)f(x_m) - (x - x_m)f(x_0)\Big]. \tag{4.42}$$

It represents a first degree polynomial and is equivalent to a linear interpolation using the data points $(x_0, f(x_0))$ and $(x_m, f(x_m))$. One can easily verify that

$$P_{0m}(x_0) = f(x_0) \qquad \text{and} \quad P_{0m}(x_m) = f(x_m). \tag{4.43}$$

Similarly, the second degree polynomials are generated as follows:

$$P_{012}(x) = \frac{1}{x_2 - x_1}\begin{vmatrix} x - x_1 & P_{01}(x) \\ x - x_2 & P_{02}(x) \end{vmatrix}$$
$$= \frac{1}{x_2 - x_1}\Big[(x - x_1)P_{02}(x) - (x - x_2)P_{01}(x)\Big]$$

and

$$P_{01m}(x) = \frac{1}{x_m - x_1}\begin{vmatrix} x - x_1 & P_{01}(x) \\ x - x_m & P_{0m}(x) \end{vmatrix}$$
$$= \frac{1}{x_m - x_1}\Big[(x - x_1)P_{0m}(x) - (x - x_m)P_{01}(x)\Big], \tag{4.44}$$

where m can now take any value from 2 to n, and P_{01m} denotes a polynomial of degree two that passes through the three points $(x_0, f(x_0))$, $(x_1, f(x_1))$, and $(x_m, f(x_m))$. By repeated use of this procedure, higher degree polynomials can be generated. In general, one can define as follows

$$P_{012\cdots n}(x) = \frac{1}{x_n - x_{n-1}}\begin{vmatrix} P_{01\cdots(n-1)}(x) & f(x_{n-1}) \\ P_{01\cdots n}(x) & f(x_n) \end{vmatrix}$$
$$= \frac{1}{x_n - x_{n-1}}\Big[P_{01\cdots(n-1)}(x)f(x_n) - P_{01\cdots n}(x)f(x_{n-1})\Big]. \tag{4.45}$$

This is polynomial of degree n and it fits all the data. Table 4.6 shows the construction of $P_{012\cdots n}(x)$. When using Aitken's method in practice only the values

Table 4.6: Aitken's scheme to approximate a function.

k	x_k	$x - x_k$	$f(x_k)$	First Order	Second Order	Third Order		Nth Order
0	x_0	$x - x_0$	$f(x_0)$					
1	x_1	$x - x_1$	$f(x_1)$	$P_{01}(x)$				
2	x_2	$x - x_2$	$f(x_2)$	$P_{02}(x)$	$P_{012}(x)$			
3	x_3	$x - x_3$	$f(x_3)$	$P_{03}(x)$	$P_{013}(x)$			
$\cdots$	$\cdots$	$\cdots$	$\cdots$	$\cdots$	$\cdots$	$\cdots$	$\cdots$	$\cdots$
n	x_n	$x - x_n$	$f(x_n)$	$P_{0n}(x)$	$P_{01n}(x)$	$P_{012n}(x)$	$\cdots$	$P_{012\cdots n}(x)$

of the polynomials for specified values of x are computed and coefficients of the polynomials are not determined explicitly. Furthermore if, for a specified x, the stage is reached when the difference in value between successive degree polynomials is negligible then the procedure can be terminated. It is an advantage of the method when compared with the Lagrange interpolation formula.

Example 4.16 *Apply Aitken's method to the approximate evaluation of* $\ln x$ *at* $x = 4.5$ *from the following data points:*

x	2	3	4	5
$f(x)$	0.6932	1.0986	1.3863	1.6094

Solution. *To find the estimate value of* $\ln(4.5)$, *using the given data points, we have to compute all the unknowns required in the given problem as follows*

$$P_{01}(x) = \frac{1}{x_1 - x_0}\begin{vmatrix} x - x_0 & f(x_0) \\ x - x_1 & f(x_1) \end{vmatrix}$$

$$P_{01}(4.5) = \frac{1}{3-2}\begin{vmatrix} 4.5 - 2 & 0.6932 \\ 4.5 - 3 & 1.0986 \end{vmatrix}$$

$$= (2.5)(1.0986) - (1.5)(0.6932) = 1.7067$$

and

$$P_{02}(x) = \frac{1}{x_2 - x_0}\begin{vmatrix} x - x_0 & f(x_0) \\ x - x_2 & f(x_2) \end{vmatrix}$$

$$P_{02}(4.5) = \frac{1}{4-2}\begin{vmatrix} 4.5 - 2 & 0.6932 \\ 4.5 - 4 & 1.3863 \end{vmatrix}$$

$$= \frac{1}{2}\Big[(2.5)(1.3863) - (0.5)(0.6932)\Big] = 1.5596$$

and

$$P_{03}(x) = \frac{1}{x_3 - x_0}\begin{vmatrix} x - x_0 & f(x_0) \\ x - x_3 & f(x_3) \end{vmatrix}$$

$$P_{03}(4.5) = \frac{1}{5-2}\begin{vmatrix} 4.5 - 2 & 0.6932 \\ 4.5 - 5 & 1.6094 \end{vmatrix}$$

$$= \frac{1}{3}\Big[(2.5)(1.6094) - (-0.5)(0.6932)\Big] = 1.4567.$$

Similarly, the values of the second degree polynomials can be generated as follows:

$$P_{012}(x) = \frac{1}{x_2 - x_1}\begin{vmatrix} x - x_1 & P_{01}(x) \\ x - x_2 & P_{02}(x) \end{vmatrix}$$

$$P_{012}(4.5) = \frac{1}{4-3}\begin{vmatrix} 4.5 - 3 & P_{01}(4.5) \\ 4.5 - 4 & P_{02}(4.5) \end{vmatrix}$$

$$= (1.5)(1.5596) - (0.5)(1.7067) = 1.4860$$

and

$$P_{013}(x) = \frac{1}{x_3 - x_1}\begin{vmatrix} x - x_1 & P_{01}(x) \\ x - x_3 & P_{03}(x) \end{vmatrix}$$

$$P_{013}(4.5) = \frac{1}{5-3}\begin{vmatrix} 4.5 - 3 & P_{01}(4.5) \\ 4.5 - 5 & P_{03}(4.5) \end{vmatrix}$$

$$= \frac{1}{2}\Big[(1.5)(1.4567) - (-0.5)(1.7067)\Big] = 1.5193.$$

Finally, the values of the third degree polynomials can be generated as follows:

$$P_{0123}(x) = \frac{1}{x_3 - x_2} \begin{vmatrix} x - x_2 & P_{012}(x) \\ x - x_3 & P_{013}(x) \end{vmatrix}$$

$$P_{0123}(4.5) = \frac{1}{5-4} \begin{vmatrix} 4.5 - 4 & P_{012}(4.5) \\ 4.5 - 4 & P_{013}(4.5) \end{vmatrix}$$

$$= (0.5)(1.5193) - (-0.5)(1.4860) = 1.5027.$$

The results obtained are listed in Table 4.7. Note that the approximate value of $\ln(4.5)$ *is* $P_{0123}(4.5) = 1.5027$ *and its exact value is* 1.5048.

Table 4.7: Approximate solution for Example 4.16.

k	x_k	$f(x_k)$	$x - x_k$	First Order	Second Order	Third Order
0	2.0	0.6932	2.5			
1	3.0	1.0986	1.5	1.7067		
2	4.0	1.3863	0.5	1.5596	1.4860	
3	5.0	1.6094	-0.5	1.4567	1.5193	1.5027

4.3 Polynomial Interpolation for Even Intervals

In deriving the Lagrange, the Newton divided difference, and Aitken's form of the interpolating polynomials, the only restriction on the points $x_i (i = 0, 1, \ldots, n)$ was that they should be unequally spaced. Now we discuss polynomial interpolation to the situation in which the data points are equally spaced. This is precisely the situation of looking up values of a function in a table and estimating the value at some other non-tabular form. The study of interpolation to equally spaced data used to be an extremely important part of numerical analysis before computers and calculators became available, and virtually all function evaluation were performed using tables. These tables usually give data and equally spaced values of x and special techniques were developed for polynomial interpolation to such data. Although less use is made of tables nowadays, methods based on equally spaced data can be used to derive formulas for numerical differentiation, integration, and for the solutions of ordinary differential equations. Consider the following table:

x	x_0	x_1	$\cdots$	x_n
$f(x)$	$f(x_0)$	$f(x_1)$	$\cdots$	$f(x_n)$

The data points are equally spaced such that

$$h = x_{k+1} - x_k, \qquad k = 0, 1, 2, \ldots, n-1 \tag{4.46}$$

and are arranged in increasing order; that is,

$$x_0 < x_1 < x_2 < \cdots < x_n.$$

The most basic formula for interpolating the polynomial with equally spaced data points is *Newton's forward-difference interpolation formula* (also called the *Newton-Gregory formula*), which can be used for interpolation near the beginning of the table of values. For this we use forward-differences in the definition of the polynomial. These differences have the effect of introducing the points in the order $x_0, x_1, x_2, \ldots, x_n$. Similarly, for interpolation near the end of the table, we use *backward-differences* and for interpolation in the middle, we use *central-differences.*

4.3.1 forward-differences

forward-differences are defined as follows:

$$\begin{aligned} \Delta f(x_0) &= \Delta f_0 = f_1 - f_0 \\ \Delta f(x_1) &= \Delta f_1 = f_2 - f_1 \end{aligned}$$

or, in general

$$\Delta f(x_k) = \Delta f_k = f_{k+1} - f_k, \qquad k = 0, 1, 2, \ldots, n-1. \tag{4.47}$$

These are first forward-differences and Δ is the forward-difference operator. Similarly, the higher-order forward differences are defined recursively by

$$\begin{aligned} \Delta^2 f_0 &= \Delta(\Delta f_0) = \Delta(f_1 - f_0) \\ &= \Delta f_1 - \Delta f_0 = f_2 - 2f_1 + f_0 \end{aligned}$$

or, in general

$$\Delta^k f_k = \Delta(\Delta^{k-1} f_0) = \Delta^{k-1} f_{k+1} - \Delta^{k-1} f_k. \tag{4.48}$$

The forward-difference table is shown in Table 4.8. There are many finite

Table 4.8: forward-difference table.

k	x_k	f_k	Δf_k	$\Delta^2 f_k$	$\Delta^3 f_k$	$\Delta^4 f_k$
0	x_0	f_0				
			Δf_0			
1	x_1	f_1		$\Delta^2 f_0$		
			Δf_1		$\Delta^3 f_0$	
2	x_2	f_2		$\Delta^2 f_1$		$\Delta^4 f_0$
			Δf_2		$\Delta^3 f_1$	
3	x_3	f_3		$\Delta^2 f_2$		
			Δf_3			
4	x_4	f_4				

difference interpolation formulas. They are almost all derived from the Newton divided difference formula

$$\begin{aligned} p_n(x) &= f[x_0] + f[x_0, x_1](x - x_0) + f[x_0, x_1, x_2](x - x_0)(x - x_1) \\ &+ \cdots + f[x_0, x_1, \ldots, x_n](x - x_0)(x - x_1) \cdots (x - x_{n-1}). \end{aligned} \tag{4.49}$$

Introducing the notation $x = x_0 + Sh$ or $S = \dfrac{x - x_0}{h}$, the differences, in general, $x - x_k = (S - k)h$, where $h = x_{k+1} - x_k$, so (4.49) becomes

$$\begin{aligned} p_n(x) &= f[x_0] + Shf[x_0, x_1] + S(S-1)h^2 f[x_0, x_1, x_2] + \cdots \\ &+ S(S-1)(S-2)\cdots(S-n+1)h^n f[x_0, x_1, \ldots, x_n] \end{aligned} \tag{4.50}$$

or

$$\begin{aligned} p_n(x) &= \sum_{k=0}^{n} S(S-1)(S-2)\cdots(S-k+1)h^k f[x_0, x_1, \ldots, x_k] \\ &= \sum_{k=0}^{n} \binom{S}{k} k! h^k f[x_0, x_1, \ldots, x_k]. \end{aligned} \tag{4.51}$$

where the generalized binomial coefficients are defined by

$$\binom{S}{k} = \frac{S(S-1)(S-2)\cdots(S-k+1)}{k!}. \tag{4.52}$$

Then formula (4.51) is called the *Newton forward divided difference formula.* Since we know that the first divided difference $f[x_0, x_1]$ is defined as

$$f[x_0, x_1] = \frac{f(x_1) - f(x_0)}{x_1 - x_0} \tag{4.53}$$

or

$$f(x_1) - f(x_0) = hf[x_0, x_1], \qquad \text{with} \quad h = x_1 - x_0, \tag{4.54}$$

then from the definition of the forward-difference, we find that

$$\Delta f(x_0) = f(x_1) - f(x_0) = hf[x_0, x_1] \tag{4.55}$$

and

$$\Delta^2 f(x_0) = hf[x_1, x_2] - hf[x_0, x_1] = 2h^2 f[x_0, x_1, x_2]$$

or, in general

$$\Delta^k f(x_i) = k!h^k f[x_i, x_{i+1}, \dots, x_{i+k}]. \tag{4.56}$$

Using forward-differences, formula (4.50) can be written as

$$\begin{aligned} p_n(x) &= f(x_0) + \frac{(x - x_0)}{h}\Delta f(x_0) + \frac{(x - x_0)(x - x_1)}{2h^2}\Delta^2 f(x_0) \\ &+ \dots + \frac{(x - x_0)(x - x_1)\cdots(x - x_{n-1})}{n!h^n}\Delta^n f(x_0). \end{aligned} \tag{4.57}$$

Introducing the notation

$$x = x_0 + Sh, \quad \text{or} \quad x - x_k = (S - k)h$$

(4.57) becomes

$$\begin{aligned} p_n(x) = f(x_0) &+ S\Delta f(x_0) + \frac{S(S-1)}{2}\Delta^2 f(x_0) + \cdots \\ &+ \frac{S(S-1)(S-2)\cdots(S-n+1)}{n!}\Delta^n f(x_0). \end{aligned} \tag{4.58}$$

Using the binomial coefficient notation, we get

$$p_n(x) = \sum_{k=0}^{n} \binom{S}{k} \Delta^k f(x_0). \tag{4.59}$$

Formula (4.58) is called the *Newton forward difference formula.*
The error term for the Newton forward-difference polynomial can be obtained from the general error term (4.24), which is

$$f(x) - p_n(x) = \frac{f^{(n+1)}(\eta(x))}{(n+1)!}(x - x_0)(x - x_1)\cdots(x - x_n), \tag{4.60}$$

where $\eta(x)$ lies between x and x_n. Since we know that

$$x - x_0 = Sh \tag{4.61}$$

then

$$\left.\begin{array}{lllll} (x-x_0) & = & (x_0+Sh)-x_0 & = & Sh \\ (x-x_1) & = & (x_0+Sh)-x_1 & = & (S-1)h \\ \cdots & & \cdots & & \cdots \\ \cdots & & \cdots & & \cdots \\ (x-x_n) & = & (x_0+Sh)-x_n & = & (S-n)h \end{array}\right\}. \tag{4.62}$$

Substituting (4.62) into (4.60) gives

$$f(x) - p_n(x) = \binom{S}{n+1} h^{n+1} f^{(n+1)}(\eta(x)). \tag{4.63}$$

From (4.58), the term after *nth* term is

$$\binom{S}{n+1} \Delta^{n+1} f(x_0). \tag{4.64}$$

Error term (4.63) can be obtained from (4.64) by replacement

$$\Delta^{n+1} f(x_0) \longrightarrow h^{n+1} f^{(n+1)}(\eta(x)). \tag{4.65}$$

This procedure can be used to obtain the error term for all points based on a set of equally spaced data points.

Example 4.17 *Let* $f(x) = \dfrac{2}{x+1}$.

(a) Construct the forward-difference table based on the data points $x_0 = 1$, $x_1 = 2, x_2 = 3$, $x_3 = 4$, *and* $x_4 = 5$.

(b) Use the Newton forward-difference formula to find the polynomials $p_3(x)$ *and* $p_4(x)$ *and evaluate the polynomials at* $x = 1.1$.

Solution. *(a) Table 4.9 showed the forward divided differences for* $f(x) = \dfrac{2}{x+1}$.

(b) Since the cubic polynomial $p_3(1.1)$ *interpolates the first four data points, from (4.58) we have*

$$\begin{aligned} p_3(x) &= f(x_0) + S\Delta f(x_0) + \frac{S(S-1)}{2}\Delta^2 f(x_0) \\ &+ \frac{S(S-1)(S-2)}{6}\Delta^3 f(x_0). \end{aligned} \tag{4.66}$$

Table 4.9: forward-difference table for Example 4.17.

k	x_k	f_k	Δf_k	$\Delta^2 f_k$	$\Delta^3 f_k$	$\Delta^4 f_k$
0	1	1.000				
			−0.333			
1	2	0.667		0.166		
			- 0.167		−0.099	
2	3	0.500		- 0.067		0.065
			- 0.1		−0.034	
3	4	0.400		0.033		
			−0.067			
4	5	0.333				

Since it is given, $x_0 = 1, h = 1, x = 1.1,$ $S = \dfrac{x - x_0}{h} = \dfrac{1.1 - 1}{1} = 0.1.$
Now substitution of the tabulated forward-differences (overline) yields the approximation

$$\begin{aligned} p_3(1.1) &= 1 + (0.1)(-0.333) + [(0.1)(-0.9)(0.166)]/2 \\ &+ [(0.1)(-0.9)(-1.9)(-0.099)]/6 \\ &= 0.9564, \end{aligned}$$

which is the estimate value of the function $f(1.1)$*. The exact value of the function* $f(1.1)$ *is 0.9524, and this estimate has an error, 0.0040. Now to find the* $p_4(x)$ *polynomial, one can write*

$$p_4(x) = p_3(x) + \frac{S(S-1)(S-2)(S-3)}{24}\Delta^4 f(x_0) \tag{4.67}$$

and, so

$$\begin{aligned} p_4(1.1) &= p_3(1.1) + [(0.1)(-0.9)(-1.9)(-2.9)(0.065)]/24 \\ &= 0.9564 - 0.0015 \\ &= 0.9549, \end{aligned}$$

which is the estimate value of the function $f(1.1)$ and this estimate has an error, 0.0025. The advantage of higher-order interpolation is obvious.

Newton forward-difference polynomial (4.58) can be applied at the top of a set of tabular data. However, at the bottom of a set of tabular data, the required forward-differences do not exist and the Newton forward-difference polynomial cannot be used. In this case it is convenient to use an alternative notation for the differences called *backward-differences.* In the following we discuss this alternative notation in some detail.

4.3.2 backward-differences

backward-differences are defined as follows:

$$\begin{aligned} \nabla f(x_1) &= \nabla f_1 = f_1 - f_0 \\ \nabla f(x_2) &= \nabla f_2 = f_2 - f_1 \end{aligned}$$

or, in general

$$\nabla f(x_k) = \nabla f_k = f_k - f_{k-1}, \qquad k = 0, 1, 2, \ldots, n. \tag{4.68}$$

These are first backward-differences and ∇ is the backward-difference operator. Similarly, the *nth* backward difference is given as

$$\nabla^n f_k = \nabla^{n-1}(\nabla f_k) = \nabla^{n-1}(f_k - f_{k-1}), \quad \text{for } n \geq 2. \tag{4.69}$$

The backward-difference table is shown in Table 4.10. If the interpolating points

Table 4.10: backward-difference table.

k	x_k	f_k	∇f_k	$\nabla^2 f_k$	$\nabla^3 f_k$	$\nabla^4 f_k$
0	x_0	f_0				
			∇f_1			
1	x_1	f_1		$\nabla^2 f_2$		
			∇f_2		$\nabla^3 f_3$	
2	x_2	f_2		$\nabla^2 f_3$		$\nabla^4 f_4$
			∇f_3		$\nabla^3 f_4$	
3	x_3	f_3		$\nabla^2 f_4$		
			∇f_4			
4	x_4	f_4				

are reordered as $x_n, x_{n-1}, \ldots, x_1, x_0$, then we get a formula similar to formula (4.49), which is

$$\begin{aligned} p_n(x) = f[x_n] &+ f[x_n, x_{n-1}](x - x_n) + f[x_n, x_{n-1}, x_{n-2}](x - x_n)(x - x_{n-1}) \\ &+ \cdots + f[x_n, x_{n-1}, \ldots, x_0](x - x_n) \cdots (x - x_1). \end{aligned} \tag{4.70}$$

This formula is based on the point x_n and then by setting $x = x_n + hS$ and taking

$$\left.\begin{array}{rcl}(x - x_{n-1}) & = & (S+1)h \\ (x - x_{n-2}) & = & (S+2)h \\ \cdots & & \cdots \\ \cdots & & \cdots \\ (x - x_1) & = & (S+n-1)h\end{array}\right\} \quad (4.71)$$

formula (4.70) becomes

$$\begin{aligned} p_n(x) &= f[x_n] + Shf[x_n, x_{n-1}] + S(S+1)h^2 f[x_n, x_{n-1}, x_{n-2}] + \cdots \\ &+ S(S+1)\cdots(S+n-1)h^n f[x_n, x_{n-1}, \cdots, x_0]. \end{aligned} \quad (4.72)$$

Formula (4.72) is called the *Newton backward divided difference formula.* It is used to derive a more commonly applied formula known as the *Newton backward-difference formula.*

Since we know by definition of the divided difference that

$$f[x_{n-1}, x_n] = \frac{f(x_n) - f(x_{n-1})}{x_n - x_{n-1}} \quad (4.73)$$

or

$$\bigtriangledown f(x_n) = f(x_n) - f(x_{n-1}) = hf[x_{n-1}, x_n],$$

can also be written as

$$f[x_{n-1}, x_n] = f[x_n, x_{n-1}] = \frac{1}{h} \bigtriangledown f(x_n).$$

Similarly,

$$f[x_n, x_{n-1}, x_{n-2}] = \frac{1}{2h^2} \bigtriangledown^2 f(x_n)$$

and, in general

$$f[x_n, x_{n-1}, \ldots, x_{n-k}] = \frac{1}{k!h^k} \bigtriangledown^k f(x_n). \quad (4.74)$$

Then by using these notations, we can write (4.72) as

$$\begin{aligned} p_n(x) &= f(x_0) + S \bigtriangledown f(x_n) + \frac{S(S+1)}{2} \bigtriangledown^2 f(x_n) \\ &+ \cdots + \frac{S(S+1)\cdots(S+n-1)}{n!} \bigtriangledown^n f(x_n). \end{aligned} \quad (4.75)$$

In terms of binomial coefficient, we have

$$p_n(x) = \sum_{k=0}^{n}(-1)^k \binom{-S}{k} \nabla^k f(x_n), \tag{4.76}$$

where

$$\binom{-S}{k} = \frac{-S(-S-1)\cdots(-S-k+1)}{k!}$$

or

$$\binom{-S}{k} = (-1)^k \frac{S(S+1)\cdots(S+k-1)}{k!}. \tag{4.77}$$

Then formula (4.76) is known as the *Newton backward difference formula.*

Note that for the case of the Newton Backward difference formula, it is necessary that the value of S is negative. Positive values of S would correspond to the process of extrapolation for the estimation of values of the function $f(x)$ at point beyond the end of the table.

The error term for the Newton backward-difference polynomial can be obtained from general error term (4.24), which is

$$f(x) - p_n(x) = \frac{f^{(n+1)}(\eta(x))}{(n+1)!}(x-x_n)(x-x_{n-1})\cdots(x-x_0), \tag{4.78}$$

where $\eta(x)$ lies between x and x_n. Substituting (4.71) into (4.78) gives

$$f(x) - p_n(x) = \frac{1}{(n+1)!}(S+n)(S+n-1)\cdots Sh^{n+1} f^{(n+1)}(\eta(x)), \tag{4.79}$$

which can be written as

$$f(x) - p_n(x) = \binom{S+n}{(n+1)!} h^{n+1} f^{(n+1)}(\eta(x)). \tag{4.80}$$

Then formula (4.80) can be obtained from (4.75) by the following replacement in the *(n + 1)st* term:

$$\nabla^{n+1} f(x_n) \longrightarrow h^{n+1} f^{(n+1)}(\eta(x)).$$

Example 4.18 *Use the Newton backward-difference interpolation formula to estimate the value of $f(4.9)$ using the data given in Example 4.17.*

Solution. *In this case we set $x_n = x_4 = 5,\ h = 1$, and $x = 4.9$, then*

$$S = \frac{x - x_4}{h} = \frac{4.9 - 5}{1} = -0.1.$$

since the cubic polynomial $p_3(x)$ interpolates the last four data points, x_4, x_3, x_2, and x_1, the polynomial $p_4(x)$ interpolates all the data points from x_4 to x_0. Firstly, for polynomial $p_3(x)$, we have from (4.75)

$$p_3(x) = f(x_n) + S \bigtriangledown f(x_n) + \frac{S(S+1)}{2} \bigtriangledown^2 f(x_n) + \frac{S(S+1)(S+2)}{6} \bigtriangledown^3 f(x_n).$$

Now substitution of the tabulated backward-differences (underline) *yields the approximation*

$$\begin{aligned} p_3(4.9) &= 0.333 + (-0.1)(-0.067) + [(-0.1)(0.9)(0.033)]/2 \\ &+ [(-0.1)(0.9)(1.9)(-0.034]/6 \\ &= 0.3392 \end{aligned}$$

so the estimate value of $f(4.9)$ is 0.3392. The exact value of the function $f(1.1)$ is 0.3390, and this estimate $p_3(4.9)$ has an error, 0.0002. Now to find the polynomial $p_4(x)$, one can write

$$p_4(x) = p_3(x) + \frac{S(S+1)(S+2)(S+3)}{24} \bigtriangledown^4 f(x_n)$$

and so for $x = 4.9$, we get

$$\begin{aligned} p_4(4.9) &= p_3(4.9) + [(-0.1)(0.9)(1.9)(2.9)(0.065)]/24 \\ &= 0.3392 - 0.0013 = 0.3379. \end{aligned}$$

The estimate value of the function for this case is 0.3379 *and this estimate has an error,* 0.0011.

We know that the Newton forward-difference and the backward-difference polynomials are essential for fitting an approximating polynomial at the beginning and end, respectively, of a set of tabular data. But these formulas are not appropriate for approximating a value, x, that lies near the center of the table. A number of divided difference formulas are available in this situation. These methods are known as *central-difference formulas.*

4.3.3 central-differences

The first-order central-difference for the *kth* data point is defined as

$$\delta f(x_{k+1/2}) = \delta f_{k+1/2} = f_{k+1} - f_k, \quad k = 0, 1, \ldots, n-1$$

and

$$\delta f_k = f_{k+1/2} - f_{k-1/2}.$$

For the second-order central-difference, we have

$$\delta^2 f_k = \delta f_{k+1/2} - \delta f_{k-1/2} = f_{k+1} - 2f_k + f_{k-1}.$$

The *nth*-order central-difference is given by

$$\delta^n f_k = \delta^{n-1}(f_{k+1} - f_k). \tag{4.81}$$

Consequently, a central-difference table can be developed readily. Obviously, additional data points can be included and their corresponding differences computed in the same way as shown in Table 4.11. Note that for a given set of data

Table 4.11: central-difference table for a function.

k	x_k	f_k	$\delta f_{k+1/2}$	$\delta^2 f_k$	$\delta^3 f_{k+1/2}$	$\delta^4 f_k$
0	x_{-2}	f_{-2}				
			$\delta f_{-3/2}$			
1	x_{-1}	f_{-1}		$\delta^2 f_{-1}$		
			$\delta f_{-1/2}$		$\delta^3 f_{-1/2}$	
2	x_0	f_0		$\delta^2 f_0$		$\delta^4 f_0$
			$\delta f_{1/2}$		$\delta^3 f_{1/2}$	
3	x_1	f_1		$\delta^2 f_1$		
			$\delta f_{3/2}$			
4	x_2	f_2				

points, the actual numbers appearing in the forward, backward, and central-difference tables are identical. The three operators $\triangle$, ∇, and δ simply allow us to label the elements of the difference table in different ways. For instance, $\triangle f_0$, ∇f_1, and $\delta f_{1/2}$ are all equal to $f_1 - f_0$.

There are many methods which are essential for fitting an approximating polynomial in the middle of a set of tabular data. Among of them are the *Gauss forward and backward formulas*, *Stirling's formula*, and *Bessel's formula*.

For the case of the *Gauss forward formula*, we denote the base point by x_0. Since we will be interested in points both to the left and the right of this, we will, of course, be dealing with points such as x_{-1} and x_{-2} as well as x_1 and x_2. The idea is to introduce points in the order $x_0, x_1, x_{-1}, x_2, x_{-2}, \ldots$.

Now introducing the notation in the same manner as before, $x = x_0 + Sh$, by using the Newton divided difference formula (4.49), we obtain

$$\begin{aligned} p_n(x) &= p_n(x_0+Sh) = f_0 + S\delta f_{1/2} + \frac{S(S-1)}{2!}\delta^2 f_0 \\ &+ \frac{S(S-1)(S+1)}{3!}\delta^3 f_{1/2} + \frac{S(S-1)(S+1)(S-2)}{4!}\delta^4 f_0 + \cdots \\ &+ \frac{S(S^2-1)(S^2-2^2)\cdots(S^2-(m-1)^2)(S-m)}{(2m)!}\delta^{2m} f_0 \end{aligned} \quad (4.82)$$

or

$$\begin{aligned} p_n(x) &= p_n(x_0+Sh) = f_0 + S\delta f_{1/2} + \frac{S(S-1)}{2!}\delta^2 f_0 \\ &+ \frac{S(S-1)(S+1)}{3!}\delta^3 f_{1/2} + \frac{S(S-1)(S+1)(S-2)}{4!}\delta^4 f_0 + \cdots \\ &+ \frac{S(S^2-1)(S^2-2^2)\cdots(S^2-m^2)}{(2m+1)!}\delta^{2m+1} f_{1/2} \end{aligned} \quad (4.83)$$

depending on whether we finish with an even or odd order difference.

The *Gauss backward formula* uses the data points in the order $x_0, x_{-1}, x_1, x_{-2}, x_2, \ldots$ and in the same way as for the forward formula, gives

$$\begin{aligned} p_n(x) &= p_n(x_0+Sh) = f_0 + S\delta f_{-1/2} + \frac{S(S+1)}{2!}\delta^2 f_0 \\ &+ \frac{S(S+1)(S-1)}{3!}\delta^3 f_{-1/2} \\ &+ \frac{S(S+1)(S-1)(S+2)}{4!}\delta^4 f_0 + \cdots. \end{aligned} \quad (4.84)$$

Bessel's formula is a simple variation of the Gauss forward and Gauss backward formulas. Bessel's polynomial is defined as

$$\begin{aligned} p_n(x) &= p_n(x_0 + Sh) = \frac{(f_1 + f_0)}{2} + (S - 1/2)\delta f_{1/2} \\ &+ \frac{S(S-1)}{2 \times 2!}(\delta^2 f_1 + \delta^2 f_0) + \frac{S(S-1)(S-1/2)}{2 \times 3!}(\delta^3 f_1 + \delta^3 f_0) \\ &+ \frac{S(S-1)(S+1)(S-2)}{2 \times 4!}(\delta^4 f_1 + \delta^4 f_0) + \cdots \\ &+ \frac{S(S^2-1)\cdots(S^2-(m-1)^2)(S-m)}{2 \times (2m)!}(\delta^{2m} f_1 + \delta^{2m} f_0) \end{aligned} \tag{4.85}$$

or

$$\begin{aligned} p_n(x) &= p_n(x_0 + Sh) = \frac{(f_1 + f_0)}{2} + (S - 1/2)\delta f_{1/2} \\ &+ \frac{S(S-1)}{2 \times 2!}(\delta^2 f_1 + \delta^2 f_0) + \frac{S(S-1)(S-1/2)}{2 \times 3!}(\delta^3 f_1 + \delta^3 f_0) \\ &+ \frac{S(S-1)(S+1)(S-2)}{2 \times 4!}(\delta^4 f_1 + \delta^4 f_0) + \cdots \\ &+ \frac{S\cdots(S^2-(m-1)^2)(S-m)(S-1/2)}{2 \times (2m+1)!}(\delta^{2m+1} f_1 + \delta^{2m+1} f_0). \end{aligned} \tag{4.86}$$

It can be shown by direct substitution that the Bessel centered difference polynomials of odd degree are order n and pass exactly through the data points used to construct the differences appearing in the polynomial. The even degree polynomials use data from one additional point.

The other finite difference interpolation scheme we discuss here is that of *Stirling's difference* interpolation formula. This type of polynomial is defined as

$$\begin{aligned} p(S) &= p_n(x_0 + Sh) = f_0 + \frac{S}{2 \times 1!}(\delta f_{-1/2} + \delta f_{1/2})\frac{S^2(S^2-1)}{2 \times 2!}\delta^2 f_0 \\ &+ \frac{S(S^2-1)}{3!}(\delta^3 f_{-1/2} + \delta^3 f_{1/2}) + \frac{S^2(S^2-1)}{4!}\delta^4 f_0 + \cdots. \end{aligned} \tag{4.87}$$

It can be shown by direct substitution that the Stirling centered difference polynomials of even degree are order n and pass exactly through the data points used to construct the differences appearing in the polynomial. The odd degree polynomials use data from one additional point.

Example 4.19 *Consider the following table for* $f(x) = \dfrac{1.5}{x}$:

x	1.0	2.0	3.0	4.0	5.0
$f(x)$	1.5	0.75	0.5	0.375	0.3

(a) Construct the central-difference table.
(b) Use the Gauss forward and the backward difference formulas to estimate $f(3.15)$.
(c) Use the Bessel's difference formulas to estimate $f(3.15)$.
(d) Use the Stirling's difference formulas to estimate $f(3.15)$.

Solution. *Since* $x = 3.15$ *lies between tabular points 3.0 and 4.0, we use* $x_0 = 3.0$ *and therefore,* $S = 0.15$.
(a) The results are listed in Table 4.12.

Table 4.12: central-difference table for Example 4.19.

k	x_k	f_k	$\delta f_{k+1/2}$	$\delta^2 f_k$	$\delta^3 f_{k+1/2}$	$\delta^4 f_k$
-2	1.0	1.5				
			-0.75			
-1	2.0	0.75		0.5		
			-0.25		-0.375	
0	3.0	0.5		0.125		0.3
			-0.125		-0.075	
1	4.0	0.375		0.05		0.05
			-0.075		-0.025	
2	5.0	0.3		0.025		
			-0.05			
3	6.0	0.25				

(b) **Gauss Forward Formula**

Using the values in Table 4.12 and substituting in (4.83), we have

$$\begin{aligned}
p_4(3.15) &= 0.5 + (0.15)(-0.125) + \frac{(0.15)(0.15-1)}{2}(0.125) \\
&+ \frac{(0.15)(0.15-1)(0.15+1)}{6}(-0.075) \\
&+ \frac{(0.15)(0.15-1)(0.15+1)(0.15-2)}{24}(0.3) \\
&= 0.4785.
\end{aligned}$$

Gauss Backward Formula

Similarly, using (4.84), we have

$$\begin{aligned}
p_4(3.15) &= 0.5 + (0.15)(-0.25) + \frac{(0.15)(0.15+1)}{2}(0.125) \\
&+ \frac{(0.15)(0.15+1)(0.15-1)}{6}(-0.375) \\
&+ \frac{(0.15)(0.15+1)(0.15-1)(0.15+2)}{24}(0.3) \\
&= 0.4786.
\end{aligned}$$

(c) **Bessel's Formula**

Using the values in Table 4.12 and substituting in (4.86), we have

$$\begin{aligned}
p_4(3.15) &= \frac{0.5+0.375}{2} + (0.15-1/2)(-0.125) \\
&+ \frac{(0.15)(0.15-1)(0.125+0.5)}{4} \\
&+ \frac{(0.15)(0.15-1)(0.15-1/2)(-0.375-0.075)}{12} \\
&+ \frac{(0.15)(0.15-1)(0.15+1)(0.15-2)(0.3+0.05)}{48} \\
&= 0.4617.
\end{aligned}$$

(d) **Stirling's Formula**

Using the values in Table 4.12 and substituting in (4.87), we have

$$\begin{aligned}
p_4(3.15) &= 0.5 + \frac{(0.15)}{2}(-0.25 - 0.125) + \frac{(0.15)^2(0.15^2-1)}{4}(0.125) \\
&+ \frac{(0.15)(0.15^2-1)}{6}(-0.375 - 0.075) \\
&+ \frac{(0.15)^2(0.15^2-1)}{24}(0.3) \\
&= 0.4770.
\end{aligned}$$

Since the exact value of the given function at $x = 3.15$ *is* 0.4762*, the corresponding actual errors are* $-0.0023, -0.0024, 0.0145$*, and* -0.0008*, respectively, which showed that the Stirling's approximation is better than the other central difference formulas discussed.*

4.4 Interpolation with Spline Functions

In the previous sections we studied the use of interpolation polynomials for approximating the values of the functions on closed intervals. An alternative approach is to divide the interval into a collection of subintervals and construct a different approximating polynomial on each subinterval. Approximation by polynomial of this type is called *piecewise polynomial approximation.* Here, we will discuss some of the examples of piecewise curve fitting techniques; the use of *piecewise linear interpolation* and *piecewise cubic interpolation.*

Definition 4.1 (Spline Function)

Let $a = x_0 < x_1 < x_2 \cdots < x_n = b$*. A function* $s : [a, b] \to \mathbb{R}$ *is a spline or spline function of degree* m *with points* $x_0, x_1, \ldots, x_n$ *if*
1. a function s *is a piecewise polynomial such that, on each subinterval* $[x_k, x_{k+1}]$*,* s *has degree at most* m*.*
2. a function s *is* $m - 1$ *times differentiable everywhere.*

A spline is a flexible drafting device that can be constrained to pass smoothly through a set of plotted data points. Spline functions are a mathematical tool, which is an adaptation of this idea.

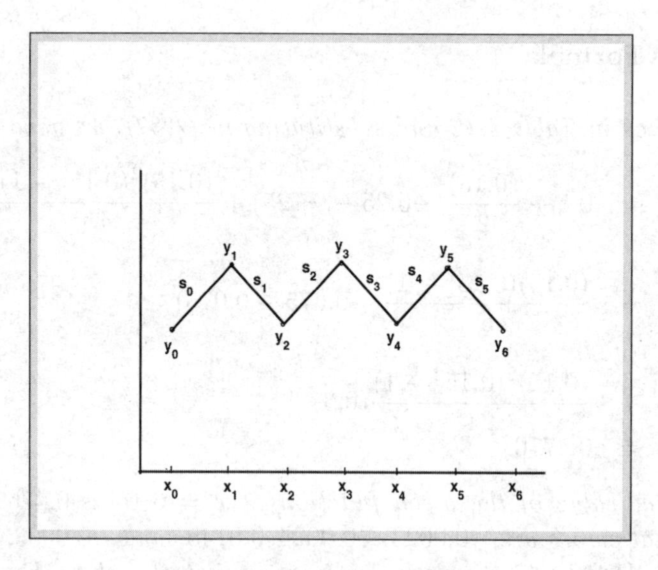

Figure 4.3: Linear spline.

Piecewise Linear Interpolation

Piecewise linear interpolation is the one of the simplest piecewise polynomial interpolations for the approximation of the function called *linear spline.* The linear spline is a continuous function and the basis of it is simply to connect consecutive points with straight lines. Consider a set of seven data points $(x_0, y_0), (x_1, y_1), (x_2, y_2), (x_3, y_3), (x_4, y_4), (x_5, y_5)$, and (x_6, y_6), which define six subintervals. These intervals are denoted as $[x_0, x_1], [x_1, x_2], [x_2, x_3], [x_3, x_4], [x_4, x_5]$, and $[x_5, x_6]$, where $x_0, x_1, x_2, x_3, x_4, x_5$, and x_6 are distinct x-values. If we use a straight line on each subinterval (see Figure 4.3) then we can interpolate the data with a piecewise linear function, where

$$s_k(x) = p_k(x) = \frac{(x - x_{k+1})}{(x_k - x_{k+1})} y_k + \frac{(x - x_k)}{(x_{k+1} - x_k)} y_{k+1}$$

or

$$s_k(x) = y_k + \frac{(y_{k+1} - y_k)}{(x_{k+1} - x_k)} (x - x_k).$$

It gives us

$$s_k(x) = A_k + B_k(x - x_k), \tag{4.88}$$

where the values of the coefficients A_k and B_k are given as

$$A_k = y_k \quad \text{and} \quad B_k = \frac{(y_{k+1} - y_k)}{(x_{k+1} - x_k)}. \tag{4.89}$$

Note that the linear spline must be continuous at given points $x_0, x_1, \ldots, x_n$ and

$$s(x_k) = f(x_k), \qquad \text{for} \quad k = 0, 1, \ldots, n.$$

Example 4.20 *Find the linear spline, which interpolates the following data*

x	1	2	3	4
$f(x)$	1.0	0.67	0.50	0.40

What is its value at $x = 2.9$?

Solution. *Given $x_0 = 1.0, x_1 = 2.0, x_2 = 3.0, x_3 = 4.0$, using (4.89), we have*

$$A_0 = y_0 = 1.0, \; A_1 = y_1 = 0.67, \; A_2 = y_2 = 0.50, \; A_3 = y_3 = 0.5$$

and

$$B_0 = \frac{(y_1 - y_0)}{(x_1 - x_0)} = \frac{(0.67 - 1.0)}{(2.0 - 1.0)} = -0.33$$

$$B_1 = \frac{(y_2 - y_1)}{(x_2 - x_1)} = \frac{(0.50 - 0.67)}{(3.0 - 2.0)} = -0.17$$

$$B_2 = \frac{(y_3 - y_2)}{(x_3 - x_2)} = \frac{(0.40 - 0.50)}{(4.0 - 3.0)} = -0.10.$$

Now using (4.88), the linear spline for three subintervals is defined as

$$s(x) = \begin{cases} s_0(x) = 1.0 - 0.33(x - 1.0) = 1.33 - 0.33x, & 1 \le x \le 2 \\ s_1(x) = 0.67 - 0.17(x - 2.0) = 1.01 - 0.17x, & 2 \le x \le 3 \\ s_2(x) = 0.50 - 0.10(x - 3.0) = 0.80 - 0.10x, & 3 \le x \le 4 \end{cases}.$$

The value $x = 2.9$ lies on the interval $[2, 3]$, so

$$f(2.9) \approx s_1(2.9) = 1.01 - 0.17(2.9) = 0.517.$$

Using the MATLAB command window, we can reproduce the above results as follows:

```
>> X = [1 2 3 4];
>> Y = [1 0.67 0.50 0.40];
>> x = 2.9;
>> s = LSpline(X,Y,x);
```

Program 4.4

MATLAB m-file for the Linear Spline

```
function LS=LSpline(X,Y,x)
n=length(X);
for i=n-1:-1:1
D = x - X(i);
if (D >= 0); break; end; end
D = x - X(i); if (D < 0); i = 0;
D = x - X(1); end
M = (Y(i + 1) - Y(i))/(X(i + 1) - X(i));
LS = Y(i) + M * D; end
```

Piecewise Cubic Interpolation

It is possible to construct splines of other orders, such as quadratic spline (with continuity of function and the first derivative values only) and quintic spline (with continuity of derivatives up to and including those of fourth-order). But here we discuss only cubic spline, which is by far the most widely used in engineering practice.

Now, we construct piecewise cubic interpolating polynomials, which are continuous in position, slope, and curvature; that is, the function, first derivative and second derivative values are matched at the ends of each adjoining subinterval. Piecewise polynomials of this type are known as *cubic splines.*
The cubic spline approximation applied to n ordered pair of data. We seek $(n-1)$ curves that connect points 0 and 1, 1 and 2, 2 and 3, 3 and 4,..., (n - 1) and n (see Figure 4.4).

Definition 4.2 (Cubic Spline)

Given a function f defined on the interval $[a, b]$ and a set of numbers x_i, called points, with $a = x_0 < x_1 < \cdots < x_{n-1} < x_n = b$, a cubic spline interpolation to f is a function s such that

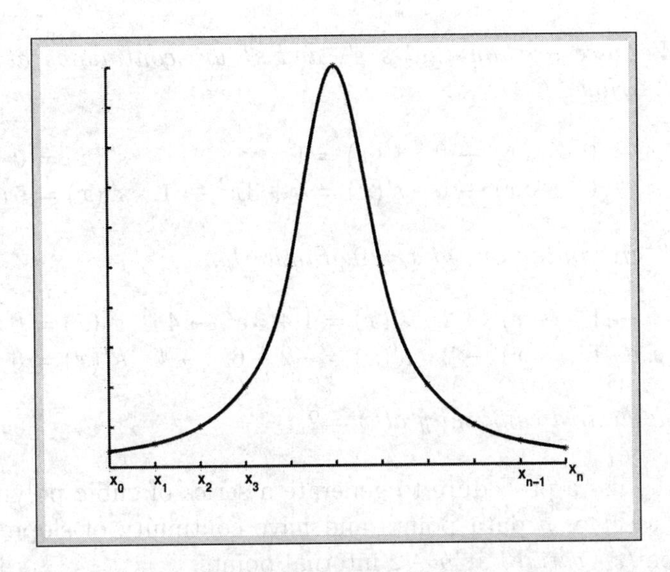

Figure 4.4: Cubic spline.

(a) *s is a cubic polynomial s_i defined on subinterval $[x_i, x_{i+1}]$ for $i = 0, \dots, n-1$.*

(b) $s_i(x_i) = y_i$ *and* $s_i(x_{i+1}) = y_{i+1}, \quad i = 0, 1, \dots, n-1,$ *where* $y_i = f(x_i)$.

(c) $s'_{i-1}(x_i) = s'_i(x_i), \qquad i = 1, 2, \dots, n-1.$

(d) $s''_{i-1}(x_i) = s''_i(x_i), \qquad i = 1, 2, \dots, n-1.$

(e) *Either*

$$s''_0(x_0) = s''_{n-1}(x_n) = 0, \qquad \textit{free boundary conditions}$$

or

$$s'_0(x_0) = y'_0, \quad \textit{and} \quad s'_{n-1}(x_n) = y'_n, \quad \textit{clamped boundary conditions,}$$

where $y'_0 = f'(x_0)$ *and* $y'_n = f'(x_n)$.

Example 4.21 *Show that the following function is a spline of degree three, a cubic spline*

$$s(x) = \begin{cases} x & -1 \le x \le 0 \\ x + x^3 & 0 \le x \le 1 \\ 1 - 2x + 3x^2 & 1 \le x \le 4 \end{cases}.$$

Solution. *We have to show that s, s', and s'' are continuous at the internal points $x = 0, 1$. Now*

$$\begin{array}{llll} As \quad x \to 0^- & s(x) \to 0 & s'(x) = 1 & s''(x) = 0 \\ As \quad x \to 0^+ & s(x) \to 0 & s'(x) = 1 + 3x^2 \to 1 & s''(x) = 6x \to 0 \end{array}$$

and so s, s', s'' are continuous at $x = 0$. Similarly,

$$\begin{array}{llll} As \quad x \to 1^- & s(x) \to 2 & s'(x) = 1 + 3x^2 \to 4 & s''(x) = 6x \to 6 \\ As \quad x \to 1^+ & s(x) \to 2 & s'(x) = -2 + 6x \to 4 & s''(x) = 6 \end{array}$$

establishing the desired continuity at $x = 2$.

Now we will discuss a procedure to generate a series of cubic polynomials that pass through a set of n data points and have continuity of slope and rate of change of slope *(curvature)* at $n - 2$ internal points.

Consider, for the *(i-1)th* interval, and $[x_{i-1}, x_i]$, the polynomial is given as follows:

$$f(x) = A_{i-1} + B_{i-1}(x - x_{i-1}) + C_{i-1}(x - x_{i-1})^2 + D_{i-1}(x - x_{i-1})^3. \qquad (4.90)$$

Since this cubic fits the end points, at one end of the intervals

$$x = x_{i-1}, \qquad f(x_{i-1}) = y_{i-1}$$

and the other end, we get

$$x = x_i, \qquad f(x_i) = y_i.$$

Substituting these conditions in (4.90), we get

$$\begin{aligned} f(x_{i-1}) &= y_{i-1} = A_{i-1} \\ f(x_i) &= y_i = A_{i-1} + B_{i-1}h_{i-1} + C_{i-1}h_{i-1}^2 + D_{i-1}h_{i-1}^3, \end{aligned} \qquad (4.91)$$

where $h_{i-1} = x_i - x_{i-1}$. Since we require the slope and the rate of change of slope or curvature at the end of each polynomial to match that of its neighbor, we must differentiate (4.90) with respect to x twice to obtain the expression for $f'(x)$ and $f''(x)$. Thus,

$$\begin{aligned} f'(x) &= B_{i-1} + 2C_{i-1}(x - x_{i-1}) + 3D_{i-1}(x - x_{i-1})^2 \\ f''(x) &= 2C_{i-1} + 6D_{i-1}(x - x_{i-1}). \end{aligned} \qquad (4.92)$$

If Q_i is the rate of slope at i, then from the expression for $f''(x)$ at $x = x_{i-1}$ and $x = x_i$, we have

$$\begin{aligned} f''(x_{i-1}) &= Q_{i-1} = 2C_{i-1} \\ f''(x_i) &= Q_i = 2C_{i-1} + 6D_{i-1}h_{i-1}. \end{aligned} \tag{4.93}$$

Then (4.90), (4.91), (4.92), and (4.93) can now be solved for the unknowns A_{i-1}, B_{i-1}, C_{i-1}, and D_{i-1} to give

$$\left.\begin{aligned} A_{i-1} &= f_{i-1} \\ B_{i-1} &= \frac{1}{h_{i-1}}(f_i - f_{i-1}) - \frac{h_{i-1}}{6}(Q_i + 2Q_{i-1}) \\ C_{i-1} &= \frac{1}{2}Q_{i-1} \\ D_{i-1} &= \frac{1}{6h_{i-1}}(Q_i - Q_{i-1}) \end{aligned}\right\}. \tag{4.94}$$

It is evident that similar expressions for the remaining intervals can be developed using (4.90). For example, the cubic spline for the *ith* interval is given as

$$f(x) = A_i + B_i(x - x_i) + C_i(x - x_i)^2 + D_i(x - x_i)^3 \tag{4.95}$$

and the unknowns A_i, B_i, C_i, and D_i are evaluated from (4.94) by simply incrementing i by $i + 1$. We now invoke the condition that the slopes of the cubic splines for the *ith* and *(i-1)th* intervals be equal at $x = x_i$. Therefore, for the *(i-1)th* interval, we have

$$f'(x)|_{x=x_i} = B_{i-1} + 2C_{i-1}h_{i-1} + 3D_{i-1}h_{i-1}^2 \tag{4.96}$$

and for the *ith* interval, we differentiate (4.95), and at $x = x_i$, we have

$$f'(x)|_{x=x_i} = B_i. \tag{4.97}$$

One can easily determine B_i from (4.94) by replacing $i - 1$ by i to give

$$f'(x)|_{x=x_i} = B_i = \frac{1}{h_i}(f_{i+1} - f_i) - \frac{h_i}{6}(Q_{i+1} + 2Q_i). \tag{4.98}$$

Now by equating (4.96) and (4.98), and then using (4.92) for the values of the coefficients B_{i-1}, C_{i-1}, and D_{i-1}, we obtain an expression of the form as follows:

$$h_{i-1}Q_{i-1} + 2(h_{i-1} + h_i)Q_i + h_iQ_{i+1} = R_i, \quad i = 2, 3, \ldots, n-1, \tag{4.99}$$

where

$$R_i = \frac{6}{h_{i-1}}y_{i-1} - 6\Big(\frac{1}{h_{i-1}} + \frac{1}{h_i}\Big)y_i + \frac{6}{h_i}y_{i+1}, \quad i = 2, 3, \ldots, n-1. \tag{4.100}$$

Formula (4.99) represents a linear algebraic equation in the unknown second derivatives Q_{i-1}, Q_i, and Q_{i+1} and can be applied to each interval data point where two cubic polynomials join, giving a set of $(n-2)$ equations. However, two further equations are required, in order that we can solve for the n unknown curvatures, and an assumption must be made as to the curvature of the polynomials at each of extreme ends of the complete set of data where there is no neighboring polynomial. This assumption will effect the shape of the polynomial near the end points but the effects will diminish rapidly toward the middle of the data. There are two alternative *end-point conditions*, which are often used and are defined below.

Assumption 4.1 *One possible assumption is that the curvature at these extreme points is zero; that is,*

$$Q_1 = Q_n = 0. \tag{4.101}$$

This is equivalent to assuming the end cubic approach linearity at their extremities. When this free boundary condition (4.101) occurs, the spline is called a natural cubic spline. It is then only necessary to solve the $(n-2)$ equations using (4.99).

Assumption 4.2 *An alternative assumption is using the clamped boundary condition*

$$s'(x_0) = f'(x_0) \quad \text{and} \quad s'(x_n) = f'(x_n). \tag{4.102}$$

This type of spline is known as a clamped cubic spline. It is the best choice if the derivatives of the functions are known. Then one can find the values of curvature at the endpoint by solving the following equations for Q_1 and Q_n as

$$\left.\begin{aligned} Q_1 &= \frac{3}{h_2}(E_1 - s'(x_0)) - \frac{Q_2}{2} \\ Q_n &= \frac{3}{h_{n-1}}(s'(x_n) - E_{n-1}) - \frac{Q_{n-1}}{2} \end{aligned}\right\}, \tag{4.103}$$

where $E_{n-1} = \dfrac{y_n - y_{n-1}}{h_{n-1}}$.

Then using (4.99) and (4.101), or (4.99) and (4.103), we obtain a set of n simultaneous equations, which must be solved to obtain the n unknown curvatures. Once these curvatures coefficients are obtained, the coefficients of the polynomials can be deduced using (4.94).

Irrespective of the assumptions made regarding the curvature of the ends of the data set, it is necessary to solve a set of simultaneous equations which may be large. However, the matrix of coefficients is diagonally dominant and banded so that an iterative method can be used to advantage since the convergence is guaranteed for diagonally dominant systems.

The matrix form of (4.99) is

$$\begin{bmatrix} h_1 & 2(h_1+h_2) & h_2 & & & & \\ & h_2 & 2(h_2+h_3) & h_3 & & \cdots & \\ & & h_3 & 2(h_3+h_4) & h_4 & & \\ & & & \ddots & & & \\ & & & \ddots & & & \\ & & & & h_{n-2} & 2(h_{n-2}+h_{n-1}) & h_{n-1} \end{bmatrix} \begin{bmatrix} Q_2 \\ Q_3 \\ Q_4 \\ \vdots \\ Q_{n-1} \end{bmatrix}$$

$$= \begin{bmatrix} R_2 \\ R_3 \\ R_4 \\ \vdots \\ R_{n-2} \\ R_{n-1} \end{bmatrix}, \qquad (4.104)$$

where the right-hand side $R_i (i = 2, 3, \ldots, n-1)$ is defined by (4.100).

In the above matrix array, there are only $n-2$ equations, but n unknowns. We can eliminate two unknowns Q_1 and Q_n using the above two end condition assumptions. In each case we can reduce the Q vector to $n-2$ elements, and the coefficients matrix becomes square, of size $(n-2) \times (n-2)$. We note that the matrix becomes tridiagonal, and so can be solved speedily by any iterative method for linear systems.

Similarly, one can have the coefficient matrices for above assumptions for the end condition defined in the following.

4.4.1 Natural Cubic Spline

The coefficient matrix for Assumption 4.1 is

$$\begin{bmatrix} 2(h_1+h_2) & h_2 & & & & \\ h_2 & 2(h_2+h_3) & h_3 & & & \\ & h_3 & 2(h_3+h_4) & h_4 & \cdots & \\ & & & \ddots & & \\ & & & & h_{n-2} & 2(h_{n-2}+h_{n-1}) \end{bmatrix}.$$

There exists a unique cubic spline with free boundary conditions $s''(x_1) = 0$ and $s''(x_n) = 0$. Then one has to solve the following linear system

$$\left.\begin{aligned} &2(h_1+h_2)Q_2 + h_2Q_3 = R_2 \\ &h_{i-1}Q_{i-1} + 2(h_{i-1}+h_i)Q_i + h_iQ_{i+1} = R_i, \quad i = 3,4,\ldots,n-2 \\ &h_{n-2}Q_{n-2} + 2(h_{n-2}+h_{n-1})Q_{n-1} = R_{n-1} \end{aligned}\right\}. \quad (4.105)$$

4.4.2 Clamped Spline

The coefficient matrix for assumption 4.2 is

$$\begin{bmatrix} \frac{(3h_1+4h_2)}{2} & h_2 & & & & \\ h_2 & 2(h_2+h_3) & h_3 & & & \\ & h_3 & 2(h_3+h_4) & h_4 & \cdots & \\ & & & \ddots & & \\ & & & & h_{n-2} & \frac{(4h_{n-2}+3h_{n-1})}{2} \end{bmatrix}.$$

There exists a unique cubic spline with first derivative boundary conditions $s'(x_1) = f'(x_1)$ and $s'(x_n) = f'(x_n)$. Then one has to solve the following linear system

$$\left.\begin{aligned} &\frac{(3h_1+4h_2)}{2}Q_2 + h_2Q_3 = R_2 - 3[E_1 - s'(x_1)] \\ &h_{i-1}Q_{i-1} + 2(h_{i-1}+h_i)Q_i + h_iQ_{i+1} = R_i, \quad i = 3,4,\ldots,n-2 \\ &h_{n-2}Q_{n-2} + \frac{(4h_{n-2}+3h_{n-1})}{2}Q_{n-1} = R_{n-1} - 3[s'(x_n) - E_{n-1}] \end{aligned}\right\}. \quad (4.106)$$

Example 4.22 *Find the natural cubic spline that passes through the given data points $(0,1),(1,-2),(2,1)$, and $(3,16)$ with free boundary conditions $s''(0)=0$ and $s''(3)=0$. What is its value at $x=0.9$?*

Solution. *In this case, $n=4$, $h_1=h_2=h_3=1$, and the right-hand side quantities $R_2=36$ and $R_3=72$ can be obtained using (4.100). Then using (4.105), we have to solve the following linear system*

$$4Q_2+Q_3=36$$
$$Q_2+4Q_3=72.$$

The matrix form of this linear system is

$$\begin{bmatrix} 4.0 & 1.0 \\ 1.0 & 4.0 \end{bmatrix} \begin{bmatrix} Q_2 \\ Q_3 \end{bmatrix} = \begin{bmatrix} 36 \\ 72 \end{bmatrix}.$$

One can easily find the solution $Q_2=4.8$ and $Q_3=16.8$. Since $Q_1=s''(0)=0$ and $Q_4=s''(3)=0$, then using (4.94), one can find the spline coefficients as

$$\begin{aligned} A_1 &= y_1 = 1.0 \\ B_1 &= \frac{y_2-y_1}{h_1} - \frac{h_1}{6}(Q_2+2Q_1) = -3.8 \\ C_1 &= \frac{Q_1}{2} = 0 \\ D_1 &= \frac{1}{6h_1}(Q_2-Q_1) = 0.8. \end{aligned}$$

Then the natural spline function $s_0(x)$ that passes through (0,1) and (1,-2) is

$$s_0(x) = 0.8x^3 - 3.8x + 1.0.$$

Similarly, the remaining splines can be determined in the same way and we have

$$s(x) = \begin{cases} s_0(x) = 0.8x^3 - 3.8x + 1.0, & 0 \le x \le 1 \\ s_1(x) = 2.0(x-1)^3 + 2.4(x-1)^2 - 1.4(x-1) - 2.0 & 1 \le x \le 2 \\ s_2(x) = -2.8(x-2)^3 + 8.4(x-2)^2 + 9.4(x-2) + 1.0 & 2 \le x \le 3 \end{cases}.$$

Thus, the value of $s_0(0.9)$ is -1.8368.

Using the MATLAB command window, we can reproduce the above results as follows:

```
>> X = [0 1 2 3];
>> Y = [1 -2 1 16];
>> x = 0.9;
>> s = CubicS(X,Y,x);
```

Program 4.5

MATLAB m-file for the Natural Cubic Spline

```
function s=CubicS(X,Y,x)
N=length(X)-1; P=length(x); h=diff(X);
D=diff(Y)./h; dD3=3*diff(D); a=Y(1:N);
%Generates tridiagonal system
H = diag(2 * (h(2 : N) + h(1 : N - 1)));
for k=1:N-2
H(k, k + 1) = h(k + 1); H(k + 1, k) = h(k + 1); end
c=zeros(1,N+1); %initialize and fixes normal end conditions
c(2 : N) = H\dD3'; b = D-h.*(c(2 : N+1)+2*c(1 : N))/3;
d = (c(2 : N+1)-c(1 : N))./(3*h); %Complete computation
of coefficients
%Begin evaluation of spline at x-values
for i=1:P
k = 1; while(x(i) > X(k + 1))&(k < N); k = k + 1; end
z = x(i) - X(k); s(i) = a(k) + z * (b(k) + z * (c(k) + z *
d(k))); end
```

Example 4.23 *Find the clamped cubic spline that passes through the data points* $(0,1), (1,-2), (2,1)$, *and* $(3,16)$ *with first derivative boundary conditions* $s'(0) = -4$ *and* $s'(3) = 23$. *What is its value at* $x = 0.9$?

Solution. *In this case* $n = 4$, $h_1 = h_2 = h_3 = 1$, *and using (4.106), the right-hand side quantities can be computed as*

$$\begin{aligned} R_2 - 3[E_1 - s'(0)] &= 36 - 3[-3+4] &= 33 \\ R_3 - 3[s'(3) - E_3] &= 72 - 3[23 - 15] &= 48. \end{aligned}$$

Now using (4.106), we have to solve the following linear system

$$\begin{aligned} 3.5Q_2 + Q_3 &= 33 \\ Q_2 + 3.5Q_3 &= 48. \end{aligned}$$

The matrix form of this linear system is

$$\begin{bmatrix} 3.5 & 1.0 \\ 1.0 & 3.5 \end{bmatrix} \begin{bmatrix} Q_2 \\ Q_3 \end{bmatrix} = \begin{bmatrix} 33 \\ 48 \end{bmatrix}.$$

One can easily find the solution of the system as $Q_2 = 6.0$ *and* $Q_3 = 12.0$. *Now to find the value of* Q_1 *and* Q_4, *we use (4.103) as follows:*

$$\begin{aligned} Q_1 &= \frac{3}{h_1}(E_1 - s'(x_1)) - \frac{Q_2}{2} = 3[-3+4] - 3 = 0 \\ Q_4 &= \frac{3}{h_3}(s'(x_4) - E_3) - \frac{Q_3}{2} = 3[-3+4] - 3 = 18. \end{aligned}$$

Now use (4.94) to find the spline coefficients as

$$\begin{aligned} A_1 &= y_1 = 1.0 \\ B_1 &= \frac{y_2 - y_1}{h_1} - \frac{h_1}{6}(Q_2 + 2Q_1) = -4.0 \\ C_1 &= \frac{Q_1}{2} = 0.0 \\ D_1 &= \frac{1}{6h_1}(Q_2 - Q_1) = 1.0. \end{aligned}$$

Then the clamped spline function $s_0(x)$ *that passes through (0,1) and (1,-2) is*

$$s_0(x) = x^3 - 4.0x - 1.0$$

and the remaining splines can be determined the same way. Thus,

$$s(x) = \begin{cases} s_0(x) = x^3 - 4.0x + 1.0, & 0 \le x \le 1 \\ s_1(x) = (x-1)^3 + 3.0(x-1)^2 - (x-1) - 2.0, & 1 \le x \le 2 \\ s_2(x) = (x-2)^3 + 6.0(x-2)^2 + 8.0(x-2) + 1.0, & 2 \le x \le 3. \end{cases}$$

Thus, the value of $s_0(0.9)$ *is* -1.8710.

4.5 Least Squares Approximation

In fitting a curve to given data points, there are two basic approaches. One is to have the graph of the approximating function pass exactly through the given data points. The methods of polynomial interpolation approximation discussed in the previous sections have this special property. If the data values are experimental then they may contain errors or have a limited number of significant digits. In such cases polynomial interpolation methods may yield unsatisfactory results. The second approach, which is discussed here, is usually a more satisfactory one for experimental data, and uses an approximating function that graphs as a smooth curve having the general shapes suggested by the data values but not, in general, passing exactly through all of the data points. Such approach is known as *least squares* data fitting. The least squares method seeks to minimize the sum (over all data points) of the squares of the differences between the function value and the data value. The method is based on results from calculus demonstrating that a function, in this case the total squared error, attains a minimum value when its partial derivatives are zero.

The least squares method of evaluating empirical formulas has been used for many years. In engineering, curve fitting plays an important role in the analysis, interpretation, and correlation of experimental data with mathematical models formulated from fundamental engineering principles.

4.5.1 Linear Least Squares

To introduce the idea of linear least squares approximation, consider the experimental data shown in Figure 4.5.

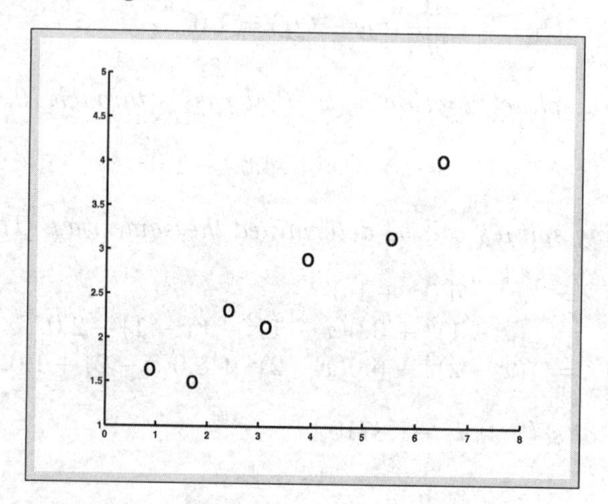

Figure 4.5: Least squares approximation.

A Lagrange interpolation of polynomial of degree six could easily be constructed for this data. However, there is no justification for insisting that the data points be reproduced exactly, and such an approximation may well be very misleading since unwanted oscillations are likely. A more satisfactory approach would be to find a straight line, which passes close to all seven points. One such possibility is shown in Figure 4.6. Here we have to decide what criterion is to be adopted for constructing such an approximation. The most common approach for this curve is known as linear least squares data fitting. The linear least squares approach defines the correct straight line as the one that minimizes the sum of the squares of the distances between the data points and the line. The least squares straight line approximation is an extremely useful and

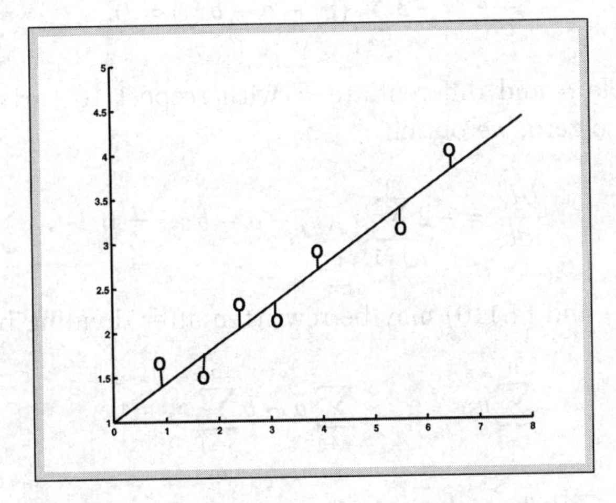

Figure 4.6: Least squares approximation.

common approximate fit. The solution to linear least squares approximation is an important application of the solution of systems of linear equations and leads to other interesting ideas of numerical linear algebra. The least squares approximation is not restricted to a straight line. However, in order to motivate the general case we consider this first. The straight line

$$p_1(x) = a + bx \tag{4.107}$$

should be fitted through the given points $(x_1, y_1), \ldots, (x_n, y_n)$ so that the sum of the squares of the distances of these points from the straight line is minimum, where the distance is measured in the vertical direction (the y-direction). Hence, it will suffice to minimize the function

$$E(a,b) = \sum_{j=1}^{n} (y_j - a - bx_j)^2. \tag{4.108}$$

The minimum of E occurs if the partial derivatives of E with respect to a and b become zero. Note that $\{x_j\}$ and $\{y_j\}$ are constant in (4.108) and unknown parameters a and b are variables. Now differentiate E with respect to variable a by making the other variable b fixed and then put it equal to zero, which gives

$$\frac{\partial E}{\partial a} = -2\sum_{j=1}^{n} (y_j - a - bx_j) = 0. \tag{4.109}$$

Now hold variable a and differentiate E with respect to variable b and then putting it equal to zero, we obtain

$$\frac{\partial E}{\partial b} = -2\sum_{j=1}^{n} x_j(y_j - a - bx_j) = 0. \tag{4.110}$$

Equations (4.109) and (4.110) may be rewritten after dividing by -2 as follows

$$\sum_{j=1}^{n} y_j \quad - \quad \sum_{j=1}^{n} a - b\sum_{j=1}^{n} x_j = 0$$

$$\sum_{j=1}^{n} x_j y_j \quad - \quad a\sum_{j=1}^{n} x_j - b\sum_{j=1}^{n} x_j^2 = 0,$$

which can be arranged to form a 2×2 system that is known as *normal equations*

$$na \quad + \quad b\sum_{j=1}^{n} x_j = \sum_{j=1}^{n} y_j$$

$$a\sum_{j=1}^{n} x_j \quad + \quad b\sum_{j=1}^{n} x_j^2 = \sum_{j=1}^{n} x_j y_j.$$

Now writing in matrix form, we have

$$\begin{pmatrix} n & S1 \\ S1 & S3 \end{pmatrix} \begin{pmatrix} a \\ b \end{pmatrix} = \begin{pmatrix} S2 \\ S4 \end{pmatrix}, \tag{4.111}$$

where

$$
\begin{aligned}
S1 &= \sum x_j \\
S2 &= \sum y_j \\
S3 &= \sum x_j^2 \\
S4 &= \sum x_j y_j.
\end{aligned}
$$

In the foregoing equations the summation is over j from 1 to n.
The solution of system (4.111) can be obtained easily as

$$
a = \frac{S3S2 - S1S4}{nS3 - (S1)^2} \quad \text{and} \quad b = \frac{nS4 - S1S2}{nS3 - (S1)^2}. \tag{4.112}
$$

Formula (4.111) reduces the problem of finding the parameters for least squares linear fit to a simple matrix multiplication.
We shall call a and b the least squares linear parameters for the data and the linear guess function with parameters; that is,

$$
p_1(x) = a + bx
$$

will be called the least squares line (or regression line) for the data.

Example 4.24 *Using the method of least squares, fit a straight line to the four points,* $(1,1), (2,2), (3,2)$, *and* $(4,3)$.

Solution. *The sums required for normal equation (4.111) are easily obtained using the values in Table 4.13. The linear system involving* a *and* b *in (4.111) form*

Table 4.13: Find the coefficients of (4.111).

i	x_i	y_i	x_i^2	$x_i y_i$
1	1.0000	1.0000	1.0000	1.0000
2	2.0000	2.0000	4.0000	4.0000
3	3.0000	2.0000	9.0000	6.0000
4	4.0000	3.0000	16.0000	12.0000
n=4	S1=10	S2=8	S3=30	S4=23

$$
\begin{pmatrix} 4 & 10 \\ 10 & 30 \end{pmatrix} \begin{pmatrix} a \\ b \end{pmatrix} = \begin{pmatrix} 8 \\ 23 \end{pmatrix}.
$$

Then solving the above linear system using LU decomposition by the Cholesky method discussed in Chapter 3, the solution of the linear system is

$$a = 0.5 \quad and \quad b = 0.6.$$

Thus, the least squares line is

$$p_1(x) = 0.5 + 0.6x.$$

Clearly, $p_1(x)$ replaces the tabulated functional relationship given by $y = f(x)$. The original data along with the approximating polynomials are shown graphically in Figure 4.7.

Use the MATLAB command window as follows:

```
>> x = [1 2 3 4];
>> y = [1 2 2 3];
>> [a, b] = linefit(x, y);
```

To plot Figure 4.7 one can use the MATLAB command window as follows:

```
>> xfit = 0 : 0.1 : 5;
>> yfit = 0.6 * xfit + 0.5;
>> plot(x, y,'o', xfit, yfit,'-');
```

Table 4.14 shows the error analysis of the straight line using least squares approximation. Hence, we have

$$E(a, b) = \sum_{i=1}^{4} (y_i - p_1(x_i))^2 = 0.2000.$$

Table 4.14: Error analysis of the linear least squares fit.

i	x_i	y_i	$p_1(x_i)$	$abs(y_i - p_1(x_i))$
1	1.0000	1.0000	1.1000	0.1000
2	2.0000	2.0000	1.7000	0.3000
3	3.0000	2.0000	2.3000	0.3000
4	4.0000	3.0000	2.9000	0.1000

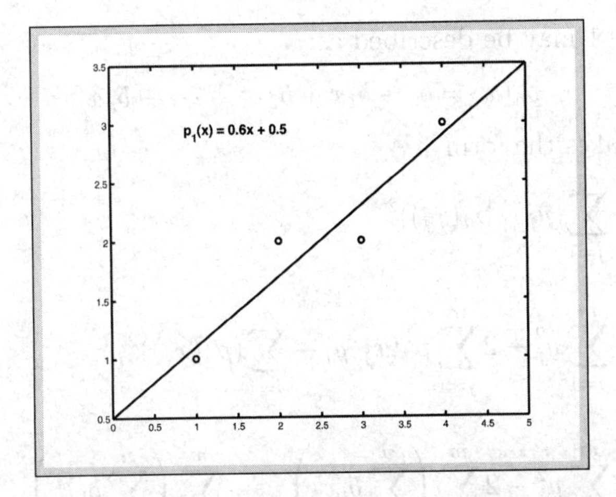

Figure 4.7: Least squares fit of four data points to a line.

Program 4.6

MATLAB m-file for the Linear Least Squares Fit

```
function [a,b]=linefit(x,y)
n=length(x); S1 = sum(x); S2 = sum(y);
S3 = sum(x. * x); S4 = sum(x. * y);
a = (n * S4 - S1 * S2)/(n * S3 - (S1) ^ 2);
b = (S3 * S2 - S4 * S1)/(n * S3 - (S1) ^ 2);
for k = 1 : n
p1 = a + b * x(k);
Error(k) = abs(p1 - y(k)); end
Error = sum(Error. * Error);
```

4.5.2 Polynomial Least Squares

In the previous section we discussed a procedure to derive the equation of a straight line using least squares, which worked very well if the measure data are intrinsically linear. But in many cases, data from experimental results are not linear. Therefore, now we show how to find the least squares parabola and the extension to a polynomial of higher degree is easily made. The general problem of approximating a set of data $\{(x_i, y_i), i = 0, 1, \ldots, m\}$ with a polynomial of degree $n < m-1$ may be described as:

$$p_n(x) = b_0 + b_1 x + b_2 x^2 + \cdots + b_n x^n. \tag{4.113}$$

Then error E takes the form

$$\begin{aligned}
E &= \sum_{j=1}^{m} (y_j - p_n(x_j))^2 \\
&= \sum_{j=1}^{m} y_j^2 - 2\sum_{j=1}^{m} p_n(x_j) y_j + \sum_{j=1}^{m} (p_n(x_j))^2 \\
&= \sum_{j=1}^{m} y_j^2 - 2\sum_{j=1}^{m} \left(\sum_{i=0}^{n} b_i x_j^i\right) y_j + \sum_{j=1}^{m} \left(\sum_{i=0}^{n} b_i x_j^i\right)^2 \\
&= \sum_{j=1}^{m} y_j^2 - 2\sum_{i=0}^{n} b_i \left(\sum_{j=1}^{m} y_j x_j^i\right) + \sum_{i=0}^{n}\sum_{k=0}^{n} b_i b_k \left(\sum_{j=1}^{m} x_j^{i+k}\right).
\end{aligned}$$

Like in the linear least squares, for E to be minimized, it is necessary that $\partial E/\partial b_i = 0$, for each $i = 0, 1, 2, \ldots, n$. Thus, for each i,

$$0 = \frac{\partial E}{\partial b_i} = -2\sum_{j=1}^{m} y_j x_j^i + 2\sum_{k=1}^{n} b_k \sum_{j=1}^{m} x_j^{i+k}. \tag{4.114}$$

This gives (n+1) *normal equations* in the (n+1) unknowns b_i

$$\sum_{k=1}^{n} b_k \sum_{j=1}^{m} x_j^{i+k} = \sum_{j=1}^{m} y_j x_j^i, \quad i = 0, 1, 2, \ldots, n. \tag{4.115}$$

It is helpful to write the equations as follows:

$$\begin{aligned}
b_0 \sum_{j=1}^{m} x_j^0 + b_1 \sum_{j=1}^{m} x_j^1 + b_2 \sum_{j=1}^{m} x_j^2 + \cdots + b_n \sum_{j=1}^{m} x_j^n &= \sum_{j=1}^{m} y_j x_j^0 \\
b_0 \sum_{j=1}^{m} x_j^1 + b_1 \sum_{j=1}^{m} x_j^2 + b_2 \sum_{j=1}^{m} x_j^3 + \cdots + b_n \sum_{j=1}^{m} x_j^{n+1} &= \sum_{j=1}^{m} y_j x_j^1 \\
&\ \vdots \\
b_0 \sum_{j=1}^{m} x_j^n + b_1 \sum_{j=1}^{m} x_j^{n+1} + b_2 \sum_{j=1}^{m} x_j^{n+2} + \cdots + b_n \sum_{j=1}^{m} x_j^{2n} &= \sum_{j=1}^{m} y_j x_j^n.
\end{aligned}$$

Note that the coefficients matrix of this system is symmetric and positive definite. Hence, the normal equations possess a unique solution.

Example 4.25 *Find the least squares polynomial approximation of degree two to the following data:*

x	0	1	2	4	6
y	3	1	0	1	4

Solution. *The coefficient of the least squares polynomial approximation of degree two*

$$p_2(x) = b_0 + b_1 x + b_2 x^2$$

are the solution values b_0, b_1 and b_2 of the linear system

$$\left.\begin{array}{rcl} b_0 m \quad + b_1 \sum x_j^1 + b_2 \sum x_j^2 & = & \sum y_j x_j^0 \\ b_0 \sum x_j^1 + b_1 \sum x_j^2 + b_2 \sum x_j^3 & = & \sum y_j x_j^1 \\ b_0 \sum x_j^2 + b_1 \sum x_j^3 + b_2 \sum x_j^4 & = & \sum y_j x_j^2 \end{array}\right\}. \tag{4.116}$$

The sums required for normal equation (4.116) are easily obtained using the values in Table 4.15.

The linear system involving unknown coefficients $b_0, b_1,$ and b_2 is

$$\begin{array}{rcl} 5b_0 \ + 13b_1 \ + 57b_2 & = & 9 \\ 13b_0 + 57b_1 \ + 289b_2 & = & 29 \\ 57b_0 + 289b_1 + 1569b_2 & = & 161. \end{array}$$

Then solving the above linear system, the solution of the linear system is

$$b_0 = 2.8252, \qquad b_1 = -2.0490, \qquad b_2 = 0.3774.$$

Hence, the parabola equation becomes

$$p_2(x) = 2.8252 - 2.0490x + 0.3774x^2.$$

Table 4.16 shows the error analysis of the parabola using least squares approximation.

Table 4.15: Find the coefficients of (4.116).

i	x_i	y_i	x_i^2	x_i^3	x_i^4	x_iy_i	$x_i^2y_i$
1	0.00	3.00	0.00	0.00	0.00	0.00	0.00
2	1.00	1.00	1.00	1.00	1.00	1.00	1.00
3	2.00	0.00	4.00	8.00	16.00	0.00	0.00
4	4.00	1.00	16.00	64.00	256.00	4.00	16.00
5	6.00	4.00	36.00	216.00	1296.00	24.00	144.00
m=5	13.00	9.00	57.00	289.00	1569.00	29.00	161.00

Table 4.16: Error analysis of the polynomial fit.

i	x_i	y_i	$p_2(x_i)$	$abs(y_i - p_2(x_i))$
1	0.0000	3.0000	2.8252	0.1748
2	1.0000	1.0000	1.1535	0.1535
3	2.0000	0.0000	0.2367	0.2367
4	4.0000	1.0000	0.6674	0.3326
5	6.0000	4.0000	4.1173	0.1173
	13.000	9.0000	9.0001	1.0148

Hence, the error associated with the least squares polynomial approximation of degree two is

$$E(b_0, b_1, b_2) = \sum_{j=1}^{5}(y_i - p(x_i))^2 = 0.2345.$$

Use the MATLAB command window as follows:

```
>> x = [0 1 2 4 6];
>> y = [3 1 0 1 4];
>> n = 2;
>> C = polyfit(x, y, n);
```

Clearly, $p_2(x)$ replaces the tabulated functional relationship given by $y = f(x)$. The original data along with the approximating polynomials are shown graphically in Figure 4.8. To plot Figure 4.8 one can use MATLAB window commands as follows:

```
>> xfit = -1 : 0.1 : 7;
>> yfit = 2.8252 - 2.0490. * xfit + 0.3774. * xfit. * xfit;
>> plot(x, y, 'o', xfit, yfit, '-');
```

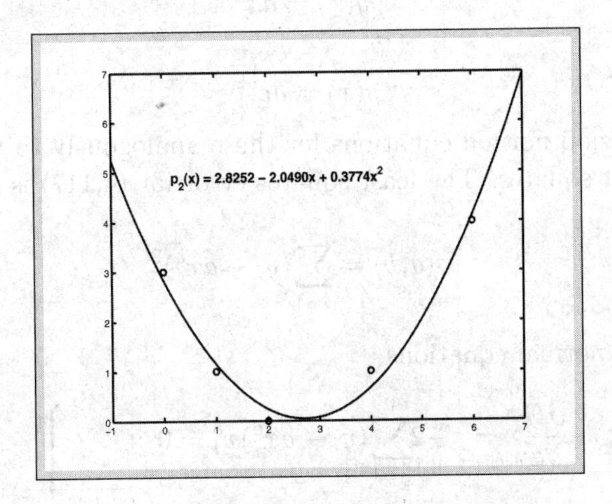

Figure 4.8: Least squares fit of five data points to a parabola.

Program 4.7

MATLAB m-file for the Polynomial Least Square Fit

```
function C=polyfit(x,y,n)
m=length(x); for i = 1 : 2 * n + 1
a(i) = sum(x.^ (i - 1));end
for i = 1 : n + 1 % coefficients of vector b
b(i) = sum(y. * x.^ (i - 1)); end
for i=1:n+1; for j=1:n+1
A(i, j) = a(j + i - 1); end;end
C = A \ b'; % Solving linear system
for k = 1 : m; S = C(1); for i = 2 : m + 1
S = S + C(i). * x(k).^ (i - 1); end;
p2(k) = S; Error(k) = abs(y(k) - p2(k)); end
Error = sum(Error. * Error);
```

4.5.3 Nonlinear Least Squares

Although polynomials are frequently used as the approximating function, they are by no means the only possibility. The most popular forms of nonlinear curves are the exponential forms

$$y(x) = ax^b \tag{4.117}$$

or

$$y(x) = ae^{bx}. \tag{4.118}$$

We can develop normal equations for these analogously to the above development for least squares. The least squares error for (4.117) is given by

$$E(a,b) = \sum_{j=1}^{n} (y_j - ax_j^b)^2 \tag{4.119}$$

with associated normal equations

$$\left.\begin{aligned} \frac{\partial E}{\partial a} &= -2\sum_{j=1}^{n} (y_j - ax_j^b)x_j^b = 0 \\ \frac{\partial E}{\partial b} &= -2\sum_{j=1}^{n} (y_j - ax_j^b)(abx_j^{b-1}) = 0 \end{aligned}\right\}. \tag{4.120}$$

The set of normal equations (4.120) represents the system of two equations in the two unknowns a and b. Such nonlinear simultaneous equations can be solved using Newton's method discussed in Chapter 2. But keep in mind that nonlinear simultaneous equations are more difficult to solve than linear equations. Because of this difficulty, the exponential forms are usually linearized by taking logarithms before determining the required parameters. Therefore, taking logarithms of both sides of (4.117), we get

$$\ln y = \ln a + b \ln x,$$

which may be written as

$$Y = A + BX \tag{4.121}$$

with $A = \ln a$, $B = b$, $X = \ln x$, and $Y = \ln y$. The values of A and B can be chosen to minimize

$$E(A,B) = \sum_{j=1}^{n} (Y_j - (A + BX_j))^2, \tag{4.122}$$

where $X_j = \ln x_j$ and $Y_j = \ln y_j$. After we differentiate E with respect to A and B and then put the results equal to zero, we get the normal equations in linear form as

$$nA \quad + \quad B\sum_{j=1}^{n} X_j = \sum_{j=1}^{n} Y_j$$

$$A\sum_{j=1}^{n} X_j \; + \; B\sum_{j=1}^{n} X_j^2 = \sum_{j=1}^{n} X_j Y_j.$$

Then writing the above equations in matrix form, we have

$$\begin{pmatrix} n & S1 \\ S1 & S3 \end{pmatrix} \begin{pmatrix} A \\ B \end{pmatrix} = \begin{pmatrix} S2 \\ S4 \end{pmatrix}, \tag{4.123}$$

where

$$\begin{aligned} S1 &= \sum X_j \\ S2 &= \sum Y_j \\ S3 &= \sum X_j^2 \\ S4 &= \sum X_j Y_j. \end{aligned}$$

In the foregoing equations the summation is over j from 1 to n. The solution of the above system can be obtained easily as

$$\left.\begin{aligned} A &= \frac{S3S2 - S1S4}{nS3 - (S1)^2} \\ B &= \frac{nS4 - S1S2}{nS3 - (S1)^2} \end{aligned}\right\}. \tag{4.124}$$

Now the data set may be transformed to $(\ln x_j, \ln y_j)$ and determining a and b is a linear least squares problem. The values of unknowns a and b are deduced from relations

$$a = e^A \quad \text{and} \quad b = B. \tag{4.125}$$

Thus, nonlinear guess functions with parameters a and b

$$y(x) = ax^b$$

will be called the nonlinear least squares approximation for the data.

Example 4.26 *Find the best fit of the form $y = ax^b$ using the following data:*

x	1	2	4	10
y	2.87	4.51	6.11	9.43

Solution. *The sums required for normal equation (4.123) are easily obtained using the values in Table 4.17. The linear system involving A and B in (4.123)*

Table 4.17: Find the coefficients of (4.124).

i	X_i	Y_i	X_i^2	X_iY_i
1	0.0000	1.0543	0.0000	0.0000
2	0.6932	1.5063	0.4805	1.0442
3	1.3863	1.8099	1.9218	2.5091
4	2.3026	2.2439	5.3020	5.1668
n=4	S1=4.3821	S2=6.6144	S3=7.7043	S4=8.7201

form is

$$\begin{pmatrix} 4 & 4.3821 \\ 4.3821 & 7.7043 \end{pmatrix} \begin{pmatrix} A \\ B \end{pmatrix} = \begin{pmatrix} 6.6144 \\ 8.7201 \end{pmatrix}.$$

Solving the above linear system, the solution of the linear system is

$$A = 1.0975 \quad and \quad B = 0.5076.$$

Using these values of A and B in (4.125), we have the values of the parameters a and b as

$$a = e^A = 2.9969 \quad and \quad b = B = 0.5076.$$

Hence, the best nonlinear fit is

$$y(x) = 2.9969x^{0.5076}.$$

Program 4.8

MATLAB m-file for the Nonlinear Least Square Fit

```
function [A,B]=exp1fit(x,y) % Least square fit y = ax^b
%Transform the data from (x,y) to (X,Y), X = log(x), Y = log(y);
n = length(x); X = log(x); Y = log(y);
S1 = sum(X); S2 = sum(Y); S3 = sum(X. * X); S4 = sum(W. * Z);
B = (n * S4 - S1 * S2)/(n * S3 - (S1) ^ 2);
A = (S3 * S2 - S4 * S1)/(n * S3 - (S1)^ 2);
b = B; a = exp(A);
for k=1:n
y = a * X(k).^ b; Error(k) = abs(y(k) - y); end
Error = sum(Error. * Error);
```

Use the MATLAB command window as follows:

```
>> x = [1 2 4 10];
>> y = [2.87 4.51 6.11 9.43];
>> [A, B] = exp1fit(x, y);
```

Clearly, $y(x)$ replaces the tabulated functional relationship given by $y = f(x)$. The original data along with the approximating polynomials are shown graphically in Figure 4.9. To plot Figure 4.9 one can use the MATLAB window command as follows:

```
>> xfit = 0 : 0.1 : 11;
>> yfit = 2.9969 * xfit. ^ 0.5076;
>> plot(x, y, 'o', xfit, yfit, '-');
```

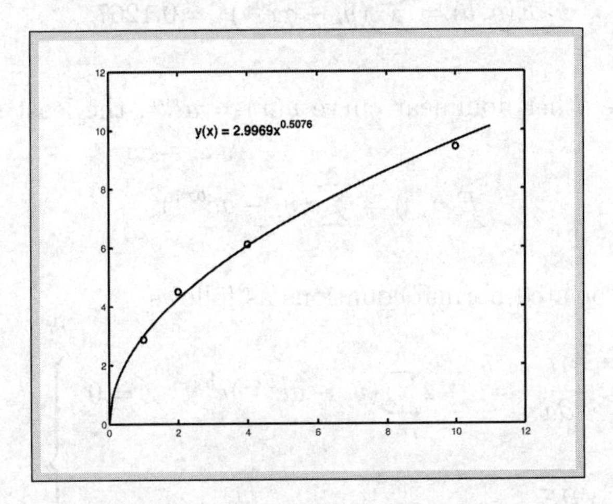

Figure 4.9: Nonlinear least squares fit.

Table 4.18 shows the error analysis of the nonlinear least squares approximation.

Table 4.18: Error analysis of nonlinear fit.

i	x_i	y_i	$y(x_i)$	$abs(y_i - y(x_i))$
1	1.0000	2.870	2.9969	0.1269
2	2.000	4.510	4.2605	0.2495
3	4.000	6.110	6.0569	0.0531
4	10.000	9.430	9.6435	0.2135

Hence, the error associated with the nonlinear least squares approximation is

$$E(a,b) = \sum_{i=1}^{4}(y_i - ax_i^{bx})^2 = 0.1267.$$

Similarly, for the other nonlinear curve $y(x) = ae^{bx}$, the least squares error is defined as

$$E(a,b) = \sum_{j=1}^{n}(y_j - ae^{bx_j})^2 \tag{4.126}$$

and gives the associated normal equations as follows

$$\left.\begin{array}{rcl} \dfrac{\partial E}{\partial a} & = & -2\sum_{j=1}^{n}(y_j - ae^{bx_j})e^{bx_j} = 0 \\ \dfrac{\partial E}{\partial a} & = & -2\sum_{j=1}^{n}(y_j - ae^{bx_j})ax_je^{bx_j} = 0 \end{array}\right\}. \tag{4.127}$$

Then the set of normal equations (4.127) represents the nonlinear simultaneous system. Once again, to make this exponential form as a linearized form, we take the logarithms of both sides of (4.118) and get

$$\ln y = \ln a + bx,$$

which may be written as

$$Y = A + BX \tag{4.128}$$

with $A = \ln a$, $B = b$, $X = x$, and $Y = \ln y$. The values of A and B can be chosen to minimize

$$E(A,B) = \sum_{j=1}^{n} (Y_j - (A + BX_j))^2, \tag{4.129}$$

where $X_j = x_j$ and $Y_j = \ln y_j$. Solve the linear normal equations of the form

$$\left.\begin{array}{rcl} nA & + & B\sum_{j=1}^{n} X_j = \sum_{j=1}^{n} Y_j \\ A\sum_{j=1}^{n} X_j & + & B\sum_{j=1}^{n} X_j^2 = \sum_{j=1}^{n} X_j Y_j \end{array}\right\} \tag{4.130}$$

to get the values of A and B. Now the data set may be transformed to $(x_j, \ln y_j)$ and determining a and b is a linear least squares problem. The values of unknowns a and b are deduced from the relations

$$a = e^A \quad \text{and} \quad b = B. \tag{4.131}$$

Thus, nonlinear guess function with parameters a and b

$$y(x) = ae^{bx}$$

will be called the nonlinear least squares approximation for the data.

Example 4.27 *Find the best-fit of the form $y = ae^{bx}$ using following data:*

x	0	0.25	0.4	0.5
y	9.532	7.983	4.826	5.503

Solution. *The sums required for normal equation (4.130) are easily obtained using the values in Table 4.19.*

The linear system involving unknown coefficients A and B is

$$\begin{array}{rcrcl} 4A & + & 1.1500B & = & 7.6112 \\ 1.1500A & + & 0.4725B & = & 2.0016. \end{array}$$

Solving the above linear system, the solution of the linear system is

$$A = 2.2811 \quad \text{and} \quad B = -1.3157.$$

Table 4.19: Find the coefficients of (4.130).

i	X_i	Y_i	X_i^2	X_iY_i
1	0.0000	2.2546	0.0000	0.0000
2	0.2500	2.0773	0.0625	0.5193
3	0.4000	1.5740	0.1600	0.6296
4	0.5000	1.7053	0.2500	0.8527
n=4	1.1500	7.6112	0.4725	2.0016

Using these values in (4.131), we have the values of the unknown parameters a and b as follows

$$a = e^A = 9.7874 \quad and \quad b = B = -1.3157.$$

Hence, the best nonlinear fit is

$$y(x) = 9.7874x^{-1.3157}.$$

Use the following MATLAB command window as follows:

```
>> x = [0 025 0.4 0.5];
>> y = [9.532 7.983 4.826 5.503];
>> [A, B] = exp2fit(x, y);
```

Clearly, $y(x)$ replaces the tabulated functional relationship given by $y = f(x)$. The original data along with the approximating polynomials are shown graphically in Figure 4.10. To plot Figure 4.10 use the MATLAB command window as follows:

```
>> xfit = -0.1 : 0.1 : 0.6;
>> yfit = 9.7874 * exp(-1.3157. * xfit);
>> plot(x, y,' o', xfit, yfit,' -');
```

Note that the values of a and b calculated for the linearized problem will not necessarily be the same as the values obtained for the original least squares problem. In this example, the nonlinear system becomes:

$$9.532 - a + 7.983e^{0.25b} - ae^{0.5b} + 4.826e^{0.4b} - ae^{0.8b} + 5.503e^{0.5b} - ae^{b} = 0$$

$$1.996e^{0.25b} - 0.25a^2e^{0.5b} + 1.930e^{0.4b} - 0.4a^2e^{0.8b} + 2.752ae^{0.5b} - 0.5a^2e^{b} = 0.$$

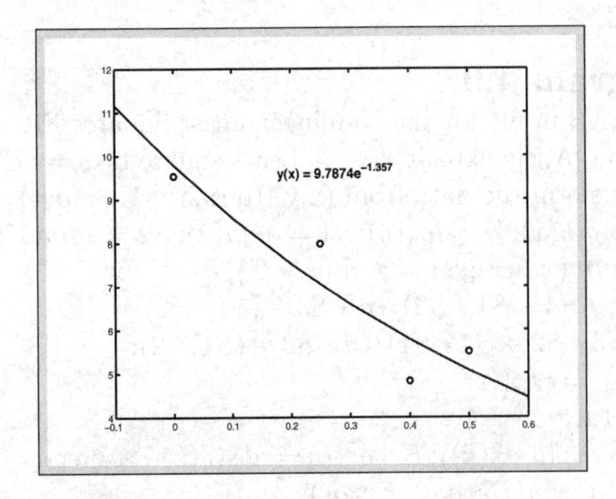

Figure 4.10: Nonlinear least squares fit.

Now Newton's method for nonlinear systems (Chapter 2) can be applied to this system and we get the values of a and b as follows

$$a = 9.731 \qquad \text{and} \qquad b = -1.265.$$

Table 4.20 shows the error analysis of the nonlinear least squares approximation.

Table 4.20: Error analysis of nonlinear fit.

i	x_i	y_i	$y(x_i)$	$abs(y_i - y(x_i))$
1	0.000	9.532	9.7872	0.2552
2	0.250	7.983	7.0439	0.9391
3	0.400	4.826	5.7823	0.9563
4	0.500	5.503	5.0695	0.4335

Hence, the error associated with the nonlinear least squares approximation is

$$E(a,b) = \sum_{i=1}^{4}(y_i - ae^{bx_i})^2 = 2.0496.$$

Program 4.9

MATLAB m-file for the Nonlinear Least Squares Fit

```
function [A,B]=exp2fit(x,y) % Least square fit y = ae^(bx)
% Transform the data from (x,y) to (x,Y), Y = log(Y);
n = length(x); Y = log(y); S1 = sum(x); S2 = sum(Y);
S3 = sum(x.*x); S4 = sum(x.*Y);
B = (n*S4 - S1*S2)/(n*S3 - (S1)^2);
A = (S3*S2 - S4*S1)/(n*S3 - (S1)^2);
b = B; a = exp(A)
for k=1:n
y = a*exp(b*x(k)); Error(k) = abs(y(k) - y); end
Error = sum(Error.*Error);
```

4.5.4 Least Squares Plane

Many problems arise in engineering and science where the dependent variable is a function of two or more variables. For example, $z = f(x, y)$ is a two-variables function. Consider the least squares plane

$$z = ax + by + c, \tag{4.132}$$

where the n points $(x_1, y_1, z_1), \ldots, (x_n, y_n, z_n)$ are obtained by minimizing

$$E(a, b, c) = \sum_{j=1}^{n} (z_j - ax_j - by_j - c)^2. \tag{4.133}$$

The function $E(a, b, c)$ is minimum when

$$\begin{aligned}
\frac{\partial E}{\partial a} &= 2\sum_{j=1}^{n} (z_j - ax_j - bx_j - c)(-x_j) = 0, \\
\frac{\partial E}{\partial b} &= 2\sum_{j=1}^{n} (z_j - ax_j - bx_j - c)(-y_j) = 0, \\
\frac{\partial E}{\partial c} &= 2\sum_{j=1}^{n} (z_j - ax_j - bx_j - c)(-1) = 0.
\end{aligned}$$

Dividing by two and rearranging gives the normal equations:

$$\left.\begin{array}{lllll} (\sum x_j^2)a & + (\sum x_j y_j)b & + (\sum x_j)c & = & \sum z_j x_j \\ (\sum x_j y_j)a & + (\sum y_j^2)b & + (\sum y_j)c & = & \sum z_j y_j \\ (\sum x_j)a & + (\sum y_j)b & + nc & = & \sum z_j \end{array}\right\}. \tag{4.134}$$

The above linear system can be solved for the unknowns a, b, and c.

Example 4.28 *Find the least squares plane $z = ax + by + c$ using the following data:*

x	1	1	2	2	2
y	1	2	1	2	3
z	7	9	10	11	12

Solution. *The sums required for normal equation (4.134) are easily obtained using the values in Table 4.21.*

Linear system (4.134) involving the unknown coefficients a, b, and c is

Table 4.21: Find the coefficients of (4.134).

i	x_i	y_i	z_i	x_i^2	$x_i y_i$	y_i^2	$x_i z_i$	$y_i z_i$
1	1.000	1.000	7.000	1.000	1.000	1.000	7.000	7.000
2	1.000	2.000	9.000	1.000	2.000	4.000	9.000	18.000
3	2.000	1.000	10.000	4.000	2.000	1.000	20.000	10.000
4	2.000	2.000	11.000	4.000	4.000	4.000	22.000	22.000
5	2.000	3.000	12.000	4.000	6.000	9.000	24.000	36.000
n=5	8.000	9.000	49.000	14.000	15.000	19.000	82.000	93.000

$$\begin{aligned} 14a + 15b + 8c &= 82 \\ 15a + 19b + 9c &= 93 \\ 8a + 9b + 5c &= 49. \end{aligned}$$

Then solving the above linear system, the solution of the linear system is

$$a = 2.400, \quad b = 1.200, \quad c = 3.800.$$

Hence, the least squares plane fit is

$$z = 2.400x + 1.200y + 3.800.$$

Use the MATLAB command window as follows:

$$>> x = [1\ 1\ 2\ 2\ 2];$$
$$>> y = [1\ 2\ 1\ 2\ 3];$$
$$>> z = [7\ 9\ 10\ 11\ 12];$$
$$>> sol = planefit(x, y, z);$$

Table 4.22 shows the error analysis of the least squares plane approximation.

Table 4.22: Error analysis of the plane fit.

i	x_i	y_i	z_i	$z(x_i, y_i)$	$abs(z_i - z)$
1	1.0000	1.0000	7.0000	7.4000	0.4000
2	1.0000	2.0000	9.0000	8.6000	0.4000
3	2.0000	1.0000	10.0000	9.8000	0.2000
4	2.0000	2.0000	11.0000	11.0000	0.0000
5	2.0000	3.0000	12.0000	12.2000	0.2000

Hence, the error associated with the least squares plane approximation is

$$E(a, b, c) = \sum_{i=1}^{4} (z_i - ax_i + by_i + c)^2 = 0.4000.$$

Program 4.10

MATLAB m-file for the Least Squares Plane Fit

```
function Sol=planefit(x,y,z)
n = length(x); S1 = sum(x); S2 = sum(y); S3 = sum(z);
S4 = sum(x. * x); S5 = sum(x. * y); S6 = sum(y. * y);
S7 = sum(z. * x); S8 = sum(z. * y);
A = [S4 S5 S1; S5 S6 S2; S1 S2 n]; B = [S7 S8 S3]'; C = A \ B;
for k=1:n
z = C(1) * x(k) + C(2) * y(k) + C(3);
Error(k) = abs(z(k) - z); end
Error = sum(Error. * Error);
```

4.5.5 Overdetermined Linear Systems

In Chapter 3, we discussed methods for computing solution $\mathbf{x}$ to a linear system $A\mathbf{x} = \mathbf{b}$ when coefficient matrix A is square (number of rows and columns are equal). For square matrix A, usually the linear system has a unique solution. Now we consider linear systems where the coefficient matrix is *rectangular* (number of rows and columns are not equal). If A has m rows and n columns, then $\mathbf{x}$ is a vector with n components and $\mathbf{b}$ is a vector with m components. If the number of rows are greater than the number of columns ($m > n$), then the linear system is called an *overdetermined system.* Typically, an overdetermined system has no solution. This type of system generally arises when dealing with experimental data. It is also common in optimization-related problems.

Consider the following linear system

$$\left.\begin{array}{cccc} a_{11}x_1 & + & a_{12}x_2 & = & b_1 \\ a_{21}x_1 & + & a_{22}x_2 & = & b_2 \\ a_{31}x_1 & + & a_{32}x_2 & = & b_3 \end{array}\right\}. \tag{4.135}$$

Obviously, it is impossible to find a solution that can satisfy all of the equations unless two of three equations are dependent. That is, if only two out of the three equations are unique, then a solution is possible. Otherwise, our best hope is to find a solution that minimizes the error. Such a solution is called its least squares solution. Here, we discuss the method for finding the least squares solution to the overdetermined system.

In the least squares method, $\mathbf{x}$ is chosen so that the Euclidean norm of residual $\mathbf{r} = \mathbf{b} - A\mathbf{x}$ is as small as possible. The residual corresponding to system (4.135) is

$$\mathbf{r} = \begin{pmatrix} b_1 - a_{11}x_1 - a_{12}x_2 \\ b_2 - a_{21}x_1 - a_{22}x_2 \\ b_3 - a_{31}x_1 - a_{32}x_2 \end{pmatrix}.$$

The l_2-norm of the residual is the square root of the sum of each component squared:

$$\|\mathbf{r}\|_2 = \sqrt{r_1^2 + r_2^2 + r_3^2}.$$

Since minimizing $\|\mathbf{r}\|_2$ is equivalent to minimizing $(\|\mathbf{r}\|_2)^2$, the least squares solution to (4.135) is values for x_1 and x_2, which minimize the expression

$$(b_1 - a_{11}x_1 - a_{12}x_2)^2 + (b_2 - a_{21}x_1 - a_{22}x_2)^2 + (b_3 - a_{31}x_1 - a_{32}x_2)^2. \tag{4.136}$$

The minimizing of x_1 and x_2 are found by differentiating (4.136) with respect to x_1 and x_2 and setting the derivatives to zero. Then solving for x_1 and x_2, we will obtain the least squares solution $\mathbf{x} = [x_1, x_2]^T$ to system (4.135).

For a general overdetermined linear system $A\mathbf{x} = \mathbf{b}$, the residual is $\mathbf{r} = \mathbf{b} - A\mathbf{x}$ and the l_2-norm of the residual is the square root of $\mathbf{r}^T\mathbf{r}$. The least squares solution to the linear system minimizes

$$\mathbf{r^T r} = (\|\mathbf{r}\|_2)^2 = (\mathbf{b} - A\mathbf{x})^T(\mathbf{b} - A\mathbf{x}). \tag{4.137}$$

Equation (4.137) attains minimum when the partial derivative with respect to each of the variables $x_1, x_2, \ldots, x_n$ is zero. Since

$$\mathbf{r^T r} = r_1^2 + r_2^2 + \cdots + r_m^2 \tag{4.138}$$

the *ith* component of the residual $\mathbf{r}$ is

$$r_i = b_i - a_{i1}x_1 - a_{i2}x_2 - \cdots - a_{in}x_n.$$

Thus, the partial derivative of $\mathbf{r}^T\mathbf{r}$ with respect to x_j is given by

$$\frac{\partial}{\partial x_j}\mathbf{r^T r} = -2r_1a_{1j} - 2r_2a_{2j} - \cdots - 2r_ma_{mj}. \tag{4.139}$$

From the right side of (4.139), we see that the partial derivative of $\mathbf{r}^T\mathbf{r}$ with respect to x_j is -2 times the product between the *jth* column of A and $\mathbf{r}$. Note that the *jth* column of A is the *jth* row of A^T. Since the *jth* component of $A^T\mathbf{r}$ is equal to the jth column of A times $\mathbf{r}$, the partial derivative of $\mathbf{r}^T\mathbf{r}$ with respect to x_j is the *jth* component of the vector $-2A^T\mathbf{r}$. The l_2-norm of the residual is minimized at that point $\mathbf{x}$ where all the partial derivatives vanish; that is,

$$\frac{\partial}{\partial x_1}\mathbf{r^T r} = \frac{\partial}{\partial x_2}\mathbf{r^T r} = \cdots = \frac{\partial}{\partial x_n}\mathbf{r^T r} = 0. \tag{4.140}$$

Since each of these partial derivatives is -2 times the corresponding component of $A^T\mathbf{r}$, we conclude that

$$A^T\mathbf{r} = 0. \tag{4.141}$$

Replacing $\mathbf{r}$ by $\mathbf{b} - A\mathbf{x}$, gives

$$A^T(\mathbf{b} - A\mathbf{x}) = 0 \tag{4.142}$$

or

$$A^TA\mathbf{x} = A^T\mathbf{b}, \tag{4.143}$$

which is called the *normal equation*.

Any $\mathbf{x}$ that minimizes the l_2-norm of the residual $\mathbf{r} = \mathbf{b} - A\mathbf{x}$ is a solution to normal equation (4.143). Conversely, any solution to normal equation (4.143) is a least squares solution to the overdetermined linear system.

Example 4.29 *Solve the following overdetermined linear system of three equations in two unknowns:*

$$\begin{array}{rcrcr} 2x_1 & + & 3x_2 & = & 10 \\ 4x_1 & - & 7x_2 & = & 8 \\ 5x_1 & + & 9x_2 & = & 14 \end{array}$$

Solution. *The matrix form of the given system is*

$$\begin{pmatrix} 2 & 3 \\ 4 & -7 \\ 5 & 9 \end{pmatrix} \begin{pmatrix} x_1 \\ x_2 \end{pmatrix} = \begin{pmatrix} 10 \\ 8 \\ 14 \end{pmatrix}.$$

Using normal equation (4.143), we obtain

$$\begin{pmatrix} 2 & 4 & 5 \\ 3 & -7 & 9 \end{pmatrix} \begin{pmatrix} 2 & 3 \\ 4 & -7 \\ 5 & 9 \end{pmatrix} \begin{pmatrix} x_1 \\ x_2 \end{pmatrix} = \begin{pmatrix} 2 & 4 & 5 \\ 3 & -7 & 9 \end{pmatrix} \begin{pmatrix} 10 \\ 8 \\ 14 \end{pmatrix},$$

which reduces the given system as

$$\begin{pmatrix} 45 & 23 \\ 23 & 139 \end{pmatrix} \begin{pmatrix} x_1 \\ x_2 \end{pmatrix} = \begin{pmatrix} 122 \\ 100 \end{pmatrix}.$$

Solving the above linear system, the values of unknowns are

$$x_1 = 2.5599 \quad \textit{and} \quad x_2 = 0.2959,$$

which is called the least squares solution of the given overdetermined system.

Using the MATLAB command window, the above result can be reproduced as follows:

```
>> A = [2 3; 4 - 7; 5 9];
>> b = [10; 8; 14];
>> x = overd(A, b);
```

Program 4.11

MATLAB m-file for Overdetermined Linear Systems

```
function sol=overd(A,b)
x = (A' * A) \ (A' * b); % Solve the normal equations
sol=x;
```

4.5.6 Least Squares with QR Decomposition

The least squares solutions discussed previously suffer from a frequent problem. The matrix A^TA of the normal equation is usually ill-conditioned; therefore, a small numerical error in performing the Gauss elimination will result in a large error in the least squares.

Usually, the Gauss elimination for A^TA of size $n \geq 5$ does not yield any good approximate solutions. It turns out that the QR decomposition of A (discussed in Chapter 7) yields a more reliable way of computing the least squares approximation of the linear system $A\mathbf{x} = \mathbf{b}$. The idea behind this approach is that because orthogonal matrices preserve length, they should preserve the length of the error as well.

Let A have linearly independent columns and let $A = QR$ be a QR decomposition. In this decomposition, we express a matrix as the product of an orthogonal matrix Q and an upper triangular matrix R.

For $\hat{\mathbf{x}}$, a least squares solution of $A\mathbf{x} = \mathbf{b}$, we have

$$\begin{aligned} A^TA\hat{\mathbf{x}} &= A^T\mathbf{b} \\ (QR)^T(QR)\hat{\mathbf{x}} &= (QR)^T\mathbf{b} \\ R^TQ^TQR\hat{\mathbf{x}} &= R^TQ^T\mathbf{b} \\ R^TR\hat{\mathbf{x}} &= R^TQ^T\mathbf{b} \qquad \text{(because } Q^TQ = \mathbf{I}\text{).} \end{aligned}$$

Since R is invertible, so is R^T, and hence

$$R\hat{\mathbf{x}} = Q^T\mathbf{b}$$

or equivalently,

$$\hat{\mathbf{x}} = R^{-1}Q^T\mathbf{b}.$$

Since R is an upper triangular, in practice it is easier to solve $R\hat{\mathbf{x}} = Q^T\mathbf{b}$ directly (using backward substitution) than to invert R and compute $R^{-1}Q^T\mathbf{b}$.

Theorem 4.5 *If A is an $m \times n$ matrix with linearly independent columns and if $A = QR$ is a QR decomposition, then the unique least squares solutions $\hat{\mathbf{x}}$ of $A\mathbf{x} = \mathbf{b}$ is, theoretically, given by*

$$\hat{\mathbf{x}} = R^{-1}Q^T\mathbf{b} \tag{4.144}$$

and it is usually computed by solving the system

$$R\hat{\mathbf{x}} = Q^T\mathbf{b}. \tag{4.145}$$

Example 4.30 *A QR decomposition of A is given. Use it to find a least squares solution of the linear system $A\mathbf{x} = \mathbf{b}$, where*

$$A = \begin{pmatrix} 2 & 2 & 6 \\ 1 & 4 & -3 \\ 2 & -4 & 9 \end{pmatrix}, \quad \mathbf{b} = \begin{pmatrix} 1 \\ -1 \\ 4 \end{pmatrix}$$

and

$$Q = \begin{pmatrix} \frac{2}{3} & \frac{1}{3} & -\frac{2}{3} \\ \frac{1}{3} & \frac{2}{3} & \frac{2}{3} \\ -\frac{2}{3} & -\frac{2}{3} & \frac{1}{3} \end{pmatrix}, \quad R = \begin{pmatrix} 3 & 0 & 9 \\ 0 & 6 & -6 \\ 0 & 0 & -3 \end{pmatrix}.$$

Solution. *For the right-hand side of (4.145), we obtain*

$$Q^T\mathbf{b} = \begin{pmatrix} \frac{2}{3} & \frac{1}{3} & -\frac{2}{3} \\ \frac{1}{3} & \frac{2}{3} & -\frac{2}{3} \\ -\frac{2}{3} & \frac{2}{3} & \frac{1}{3} \end{pmatrix} \begin{pmatrix} 1 \\ -1 \\ 4 \end{pmatrix} = \begin{pmatrix} 3 \\ -3 \\ 0 \end{pmatrix}.$$

Hence, (4.145) can be written as

$$\begin{pmatrix} 3 & 0 & 9 \\ 0 & 6 & -6 \\ 0 & 0 & -3 \end{pmatrix} \begin{pmatrix} x_1 \\ x_2 \\ x_3 \end{pmatrix} = \begin{pmatrix} 3 \\ -3 \\ 0 \end{pmatrix}$$

or

$$\begin{array}{rcrcl} 3x_1 & & + & 9x_3 & = & 3 \\ & 6x_2 & - & 6x_3 & = & -3 \\ & & - & 3x_3 & = & 0. \end{array}$$

Now using backward substitution, we obtain

$$\hat{\mathbf{x}} = [x_1, x_2, x_3]^T = [1, -1/2, 0]^T,$$

which is called the least squares solution of the given system.

So we conclude that

$$R\hat{\mathbf{x}} = Q^T\mathbf{b}$$

satisfies the solution of $A^TA\hat{\mathbf{x}} = A^T\mathbf{b}$, but because, in general, R is not even square, we cannot use multiplication by $(R^T)^{-1}$ to arrive at this conclusion. In fact, it is not true in general that the solution of

$$R\hat{\mathbf{x}} = Q^T\mathbf{b}$$

even exists; after all, $A\mathbf{x} = \mathbf{b}$ is equivalent to $QR\mathbf{x} = \mathbf{b}$, that is, to $R\mathbf{x} = Q^T\mathbf{b}$, so $R\mathbf{x} = Q^T\mathbf{b}$ can have an actual solution $\mathbf{x}$ only if $A\mathbf{x} = \mathbf{b}$ does. However, we are getting close to finding the least squares solution. Here, we need to find a way to simplify the expression

$$R^TR\hat{\mathbf{x}} = R^TQ^T\mathbf{b}. \tag{4.146}$$

Matrix R is upper triangular, and because we have restricted ourselves with the case $m \geq n$, we may write the $m \times n$ matrix R as

$$R = \begin{pmatrix} R_1 \\ 0 \end{pmatrix} \tag{4.147}$$

in partitioned (block) form, where R_1 is an upper triangular $n \times n$ matrix and 0 represents an $(m-n) \times n$ zero matrix. Since $rank(R) = n$, so that R_1 is nonsingular. Hence, every diagonal element of R_1 must be nonzero. Now we may rewrite

$$R^TR\hat{\mathbf{x}} = R^TQ^T\mathbf{b}$$

as

$$\begin{aligned}
\begin{pmatrix} R_1 \\ 0 \end{pmatrix}^T \begin{pmatrix} R_1 \\ 0 \end{pmatrix} \hat{\mathbf{x}} &= \begin{pmatrix} R_1 \\ 0 \end{pmatrix}^T Q^T\mathbf{b} \\
\begin{pmatrix} R_1^T & 0^T \end{pmatrix} \begin{pmatrix} R_1 \\ 0 \end{pmatrix} \hat{\mathbf{x}} &= \begin{pmatrix} R_1^T & 0^T \end{pmatrix} Q^T\mathbf{b} \\
R_1^TR_1\hat{\mathbf{x}} &= \begin{pmatrix} R_1^T & 0^T \end{pmatrix} (Q^T\mathbf{b}).
\end{aligned}$$

Note that multiplying by the block 0^T (an $n \times (m-n)$ zero matrix) on the right-hand side simply means that the last $(m-n)$ components of $Q^T\mathbf{b}$ do not affect the computation. Since R_1 is nonsingular, then we have

$$\begin{aligned}
R_1\hat{\mathbf{x}} &= (R_1^T)^{-1} \begin{pmatrix} R_1^T & 0^T \end{pmatrix} (Q^T\mathbf{b}) \\
&= \begin{pmatrix} \mathbf{I}_n & 0^T \end{pmatrix} (Q^T\mathbf{b}).
\end{aligned}$$

The left-hand side is $R_1\hat{\mathbf{x}}$, is $(n\times n)\times(n\times 1)\longrightarrow n\times 1$, and the right-hand side is $(n\times(n+(m-n))\times(m\times m)\times(m\times 1)\longrightarrow n\times 1$. If we define the vector $\mathbf{q}$ to be equal to the first n components of $Q^T\mathbf{b}$, then this becomes

$$R_1\hat{\mathbf{x}} = \mathbf{q}, \tag{4.148}$$

which is a square linear system involving a nonsingular upper triangular $n\times n$ matrix. So (4.148) is called the *least squares solution* of the overdetermined system $A\mathbf{x}=\mathbf{b}$ with QR decomposition by backward substitution, where $A=QR$ is the QR decomposition of A and $\mathbf{q}$ is essentially $Q^T\mathbf{b}$.

Note that the last $(m-n)$ columns of Q are not needed to solve the least squares solution of the linear system with QR decomposition. The block-matrix representation of Q corresponding to R (by (4.147)) is

$$Q = [Q_1, Q_2],$$

where Q_1 is the matrix composed of the first m columns of Q and Q_2 is a matrix composed of the remaining columns of Q. Note that only the first n columns of Q are needed to create A. Using the coefficients in R, we can save effort and memory in the process of creating the QR decomposition. The so-called short QR decomposition of A is

$$A = Q_1R_1. \tag{4.149}$$

The only difference between the full QR decomposition and the short decomposition is that the full QR decomposition contains additional $(m-n)$ columns of Q.

Example 4.31 *Find the least solution of the following linear system,* $A\mathbf{x}=\mathbf{b}$, *using* QR *decomposition, where*

$$A=\begin{pmatrix} 2 & 1\\ 1 & 0\\ 3 & 1\end{pmatrix},\quad \mathbf{x}=\begin{pmatrix} x_1\\ x_2\end{pmatrix},\quad \mathbf{b}=\begin{pmatrix} 1.9\\ 0.9\\ 2.8\end{pmatrix}.$$

Solution. *First, we find the* QR *decomposition, and we get*

$$Q=\begin{pmatrix} -0.5345 & 0.6172 & -0.5774\\ -0.2673 & -0.7715 & -0.5774\\ -0.8018 & -0.1543 & 0.5774\end{pmatrix}\quad \textit{and}\quad R=\begin{pmatrix} -3.7417 & -1.3363\\ 0 & 0.4629\\ 0 & 0\end{pmatrix}$$

and

$$Q^T\mathbf{b} = \begin{pmatrix} -0.5345 & -0.2673 & -0.8018 \\ 0.6172 & -0.7715 & -0.1543 \\ -0.5774 & -0.5774 & 0.5774 \end{pmatrix} \begin{pmatrix} 1.9 \\ 0.9 \\ 2.8 \end{pmatrix} = \begin{pmatrix} -3.5011 \\ 0.0463 \\ 0.0000 \end{pmatrix}$$

so that

$$R_1 = \begin{pmatrix} -3.7417 & -1.3363 \\ 0 & 0.4629 \end{pmatrix} \quad \textit{and} \quad \mathbf{q} = \begin{pmatrix} -3.5011 \\ 0.0463 \end{pmatrix}.$$

Hence, we must solve (4.148); that is,

$$R_1\hat{\mathbf{x}} = \mathbf{q},$$

that is

$$\begin{pmatrix} -3.7417 & -1.3363 \\ 0 & 0.4629 \end{pmatrix} \begin{pmatrix} x_1 \\ x_2 \end{pmatrix} = \begin{pmatrix} -3.5011 \\ 0.0463 \end{pmatrix}.$$

Using backward substitution, we obtain

$$\hat{\mathbf{x}} = [x_1, x_2]^T = [0.9000, 0.1000]^T$$

the least solution of the given system.

The MATLAB built-in *qr* function returns the QR decomposition of a matrix. There are two ways of calling *qr*, which are:

```
>> [Q, R] = qr(A);
>> [Q1, R1] = qr(A, 0);
```

where Q and Q_1 are orthogonal matrices and R and R_1 are upper triangular matrices. The above first form returns the full QR decomposition (that is, if A is $(m \times n)$, then Q is $(m \times m)$, and R is $(m \times n)$. The second form returns the short QR decomposition, where Q_1 and R_1 are the matrices in (4.149).

In Example 4.31, we apply the full QR decomposition of A using the first form of the built-in *qr* function as follows:

```
>> A = [2 1; 1 0; 3 1];
>> [Q, R] = qr(A);
```

The short QR decomposition of A can be obtained using the second form of built-in *qr* function as follows:

```
>> A = [2 1;1 0;3 1];
>> [Q1, R1] = qr(A, 0);
```

```
Q1 =
   -0.5345    0.6172
   -0.2673  - 0.7715
   -0.8018  - 0.1543
R1 =
   -3.7417  - 1.3363
         0    0.4629
```

As expected, Q_1 and the first two columns of Q are identical, as are R_1 and the first two rows of R. The short QR decomposition of A possesses all the necessary information in columns of Q_1 and R_1 to reconstruct A.

4.5.7 Least Squares with Singular Value Decomposition

One of the advantages of singular value decomposition (SVD) is that we can efficiently compute the least squares solution. Consider the problem of finding the least squares solution of the overdetermined linear system $A\mathbf{x} = \mathbf{b}$. We discussed previously that the least squares solution of $A\mathbf{x} = \mathbf{b}$ is the solution of $A^TA\hat{\mathbf{x}} = A^T\mathbf{b}$; that is, the solution of

$$
\begin{aligned}
(UDV^T)^TUDV^T\hat{\mathbf{x}} &= (UDV^T)^T\mathbf{b}\\
VD^TU^TUDV^T\hat{\mathbf{x}} &= VD^TU^T\mathbf{b}\\
VD^TDV^T\hat{\mathbf{x}} &= VD^TU^T\mathbf{b}\\
V^TVD^TDV^T\hat{\mathbf{x}} &= V^TVD^TU^T\mathbf{b}\\
D^TDV^T\hat{\mathbf{x}} &= D^TU^T\mathbf{b}\\
DV^T\hat{\mathbf{x}} &= U^T\mathbf{b}\\
V^T\hat{\mathbf{x}} &= D^{-1}U^T\mathbf{b}\\
\hat{\mathbf{x}} &= VD^{-1}U^T\mathbf{b}.
\end{aligned}
$$

This is the same formal solution that we found for the linear system $A\mathbf{x} = \mathbf{b}$ (see Chapter 7), but recall that A is no longer a square matrix.

Note that in exact arithmetic, the solution to a least squares problem via normal equations, QR and SVD, is exactly the same. The main difference between these two approaches is the numerical stability of the methods. To find the least squares solution of the overdetermined linear system with SVD, we will find D_1 as

$$D = \begin{pmatrix} D_1 \\ 0 \end{pmatrix}$$

in partitioned (block) form, where D_1 is an $n \times n$ matrix and 0 represents an $(m-n) \times n$ zero matrix. If we define the right-hand vector $\mathbf{q}$ to be equal to the first n components of $U^T\mathbf{b}$, then the least squares solution of the overdetermined linear system is to solve the following system

$$D_1 V^T \hat{\mathbf{x}} = \mathbf{q} \tag{4.150}$$

or

$$\hat{\mathbf{x}} = V D_1^{-1} \mathbf{q}. \tag{4.151}$$

Example 4.32 *Find the least solution of the following linear system $A\mathbf{x} = \mathbf{b}$ using singular value decomposition, where*

$$A = \begin{pmatrix} 1 & 1 \\ 0 & 1 \\ 1 & 0 \end{pmatrix}, \quad \mathbf{x} = \begin{pmatrix} x_1 \\ x_2 \end{pmatrix}, \quad \mathbf{b} = \begin{pmatrix} 1 \\ 1 \\ 1 \end{pmatrix}.$$

Solution. *First, we find the singular value decomposition of the given matrix. The first step is to find the eigenvalues of the following matrix*

$$A^T A = \begin{pmatrix} 1 & 0 & 1 \\ 1 & 1 & 0 \end{pmatrix} \begin{pmatrix} 1 & 1 \\ 0 & 1 \\ 1 & 0 \end{pmatrix} = \begin{pmatrix} 2 & 1 \\ 1 & 2 \end{pmatrix}.$$

The characteristic polynomial of $A^T A$ is

$$p(\lambda) = \lambda^2 - 4\lambda + 3 = (\lambda - 3)(\lambda - 1) = 0,$$

which gives

$$\lambda_1 = 3, \quad \lambda_2 = 1.$$

The eigenvalues of $A^T A$ and the corresponding eigenvectors are

$$\begin{pmatrix} 1 \\ 1 \end{pmatrix} \quad \textit{and} \quad \begin{pmatrix} -1 \\ 1 \end{pmatrix}.$$

These vectors are orthogonal, so we normalize them to obtain

$$\mathbf{v}_1 = \begin{pmatrix} \frac{\sqrt{2}}{2} \\ \frac{\sqrt{2}}{2} \end{pmatrix} \quad \textit{and} \quad \mathbf{v}_2 = \begin{pmatrix} -\frac{\sqrt{2}}{2} \\ \frac{\sqrt{2}}{2} \end{pmatrix}.$$

The singular values of A are

$$\sigma_1 = \sqrt{\lambda_1} = \sqrt{3} \quad \textit{and} \quad \sigma_2 = \sqrt{\lambda_2} = \sqrt{1} = 1.$$

Thus,

$$V = \begin{pmatrix} \frac{\sqrt{2}}{2} & -\frac{\sqrt{2}}{2} \\ \frac{\sqrt{2}}{2} & \frac{\sqrt{2}}{2} \end{pmatrix} = \begin{pmatrix} 0.7071 & -0.7071 \\ 0.7071 & 0.7071 \end{pmatrix}$$

and

$$D = \begin{pmatrix} \sqrt{3} & 0 \\ 0 & 1 \\ 0 & 0 \end{pmatrix} = \begin{pmatrix} 1.7321 & 0 \\ 0 & 1.0000 \\ 0 & 0 \end{pmatrix}.$$

To find U, we first compute

$$\mathbf{u}_1 = \frac{1}{\sigma_1} A\mathbf{v}_1 = \frac{\sqrt{3}}{3} \begin{pmatrix} 1 & 1 \\ 0 & 1 \\ 1 & 0 \end{pmatrix} \begin{pmatrix} \frac{\sqrt{2}}{2} \\ \frac{\sqrt{6}}{2} \end{pmatrix} = \begin{pmatrix} \frac{\sqrt{6}}{3} \\ \frac{\sqrt{6}}{6} \\ \frac{\sqrt{6}}{2} \end{pmatrix}$$

and similarly

$$\mathbf{u}_2 = \frac{1}{\sigma_2} A\mathbf{v}_2 = \begin{pmatrix} 0 \\ \frac{\sqrt{2}}{2} \\ -\frac{\sqrt{2}}{2} \end{pmatrix}.$$

These are two of the three column vectors of U and already form an orthonormal basis for $\mathbb{R}^2$. Now to find the third column vector, $\mathbf{u}_3$ of U, we will look for a unit vector $\mathbf{u}_3$ that is orthogonal to

$$\sqrt{6}\mathbf{u}_1 = \begin{pmatrix} 2 \\ 1 \\ 1 \end{pmatrix} \quad \textit{and} \quad \sqrt{2}\mathbf{u}_2 = \begin{pmatrix} 0 \\ 1 \\ -1 \end{pmatrix}.$$

To satisfy these two orthogonality conditions, the vector $\mathbf{u}_3$ must be a solution of the homogeneous linear system

$$\begin{pmatrix} 2 & 1 & 1 \\ 0 & 1 & -1 \end{pmatrix} \begin{pmatrix} x_1 \\ x_2 \\ x_3 \end{pmatrix} = \begin{pmatrix} 0 \\ 0 \\ 0 \end{pmatrix},$$

which gives the general solution of the system

$$\begin{pmatrix} x_1 \\ x_2 \\ x_3 \end{pmatrix} = \alpha \begin{pmatrix} -1 \\ 1 \\ 1 \end{pmatrix}, \quad \alpha \in \mathbb{R}.$$

Normalizing the vector on the right-hand side gives

$$\mathbf{u}_3 = \begin{pmatrix} -\dfrac{1}{\sqrt{3}} \\ \dfrac{1}{\sqrt{3}} \\ -\dfrac{1}{\sqrt{3}} \end{pmatrix}.$$

So we have

$$U = \begin{pmatrix} \dfrac{\sqrt{6}}{3} & 0 & -\dfrac{1}{\sqrt{3}} \\ \dfrac{\sqrt{6}}{6} & \dfrac{\sqrt{2}}{2} & \dfrac{1}{\sqrt{3}} \\ \dfrac{\sqrt{6}}{2} & -\dfrac{\sqrt{2}}{2} & \dfrac{1}{\sqrt{3}} \end{pmatrix} = \begin{pmatrix} 0.8165 & 0.0000 & -0.5774 \\ 0.4082 & 0.7071 & 0.5774 \\ 0.4082 & -0.7071 & 0.5774 \end{pmatrix}.$$

This yields SVD

$$A = \begin{pmatrix} \frac{\sqrt{6}}{3} & 0 & -\frac{1}{\sqrt{3}} \\ \frac{\sqrt{6}}{6} & \frac{\sqrt{2}}{2} & \frac{1}{\sqrt{3}} \\ \frac{\sqrt{6}}{2} & -\frac{\sqrt{2}}{2} & \frac{1}{\sqrt{3}} \end{pmatrix} \begin{pmatrix} \sqrt{3} & 0 \\ 0 & 1 \\ 0 & 0 \end{pmatrix} \begin{pmatrix} \frac{\sqrt{2}}{2} & \frac{\sqrt{2}}{2} \\ \frac{\sqrt{2}}{2} & -\frac{\sqrt{2}}{2} \end{pmatrix}.$$

Hence,

$$D_1 = \begin{pmatrix} 1.7321 & 0 \\ 0 & 1.0000 \end{pmatrix} \quad \textit{and} \quad D_1^{-1} = \begin{pmatrix} 0.5774 & 0 \\ 0 & 1.0000 \end{pmatrix}.$$

Also,

$$U^T\mathbf{b} = \begin{pmatrix} 0.8165 & 0.4082 & 0.4082 \\ 0.0000 & 0.7071 & -0.7071 \\ -0.5774 & 0.5774 & 0.5774 \end{pmatrix} \begin{pmatrix} 1 \\ 1 \\ 1 \end{pmatrix} = \begin{pmatrix} 1.6330 \\ 0.0000 \\ 0.5774 \end{pmatrix}$$

and from it we obtain

$$\mathbf{q} = \begin{pmatrix} 1.6330 \\ 0.0000 \end{pmatrix}.$$

Thus, we must solve (4.151); that is,

$$\hat{\mathbf{x}} = VD_1^{-1}\mathbf{q},$$

which gives

$$\begin{pmatrix} x_1 \\ x_2 \end{pmatrix} = \begin{pmatrix} 0.7071 & -0.7071 \\ 0.7071 & 0.7071 \end{pmatrix} \begin{pmatrix} 0.5774 & 0 \\ 0 & 1.0000 \end{pmatrix} \begin{pmatrix} 1.6330 \\ 0.0000 \end{pmatrix} = \begin{pmatrix} 0.6667 \\ 0.6667 \end{pmatrix},$$

which is the least squares solution of the given system.

Like QR decomposition, the MATLAB built-in *svd* function returns the SVD of a matrix. There are two ways of calling *svd*, which are:

```
>> [U, D, V] = svd(A);
>> [U1, D1, V] = svd(A, 0);
```

Here, A is any matrix, D is diagonal matrix having singular values of A in the diagonal, and U and V are orthogonal matrices. The first form returns the full SVD decomposition and the second form returns the short SVD. The second decomposition is useful when A is an $m \times n$ matrix with $m > n$. The second form of SVD decomposition gives U_1 the first n columns of U and square $(n \times n)$ D_1. When $m > n$, then the full SVD of A gives a D matrix with only zeros in the last $(m - n)$ rows. Note that there is no change in V in both forms.

In Example 4.32, we apply the full SVD of A using the first form of the built-in SVD function as follows:

```
>> A = [1 1;0 1;1 0];
>> [U,D,V] = svd(A);
```

The short SVD of A can be obtained using the second form of the built-in *svd* function as follows:

```
>> A = [1 1;0 1;1 0];
>> [U1,D1,V] = svd(A,0);
```

```
U1 =
     0.8165      0.0000
     0.4082      0.7071
     0.4082    - 0.7071
D1 =
     1.7321      0
     0           1.0000
V =
     0.7071    - 0.7071
     0.7071      0.7071
```

As expected, U_1 and the first two columns of U are identical, as are D_1 and the first two rows of D (and no change in V in both forms). The short SVD of A possesses all the necessary information in columns of U_1 and D_1 (with V also) to reconstruct A.

Note that when m and n are similar in size, SVD is significantly more expansive to compute than QR decomposition. If m and n are equal, then solving a least squares problem by SVD is an order of magnitude more costly than using QR decomposition. So for least squares problems it is generally advisable to use QR decomposition. When a least squares problem is known to be a difficult one, using SVD is probably justified.

4.6 Summary

Procedures for developing approximating polynomials for discrete data were discussed in this chapter. Firstly, we discussed the Lagrange and the Newton divided differences polynomials and both yielded the same interpolation for a given set of n data pairs $(x, f(x))$. The pairs are not required to be ordered, nor is the independent variable required to be equally spaced. The dependent variable is approximated as a single-valued function. The Lagrange polynomial works well for small data points. The Newton divided difference polynomial is generally more efficient than the Lagrange polynomial, and it can be adjusted easily for additional data. For efficient implementation of divided difference interpolation, we used Aitken's method, which is designed for the easy evaluation of the polynomial. For equally spaced data, polynomials based on differences are easy to use and recommended. We discussed the forward-difference, the backward-difference and the central difference formulas. The Newton forward- and backward-difference polynomials are simple to fit and evaluate as compared to the central-difference formulas. The Newton forward- and backward-difference polynomials are used to develop procedures for numerical differentiation and integration. We also discussed piecewise linear and cubic splines. Cubic spline approximations are patched among ordered data that maintain continuity and smoothness. They are more powerful than polynomial interpolation for a large number of data points because high degree polynomials may exhibit strong oscillatory behavior.

Finally, procedures for developing least squares approximation for discrete data were also discussed. Least squares approximations are useful for large sets of data and sets of rough data. Least squares polynomial approximation is straight-forward for one independent variable and for two independent variables. The least squares normal equations corresponding to polynomial approximating functions are linear, which leads to very efficient solution procedures. For nonlinear approximating functions, the least squares normal equations are nonlinear, which leads to complicated solution procedures. In the last section we discussed the least squares solution to the overdetermined linear system and also found the least squares solution of linear systems with QR decomposition and singular value decomposition.

4.7 Exercises

1. Find the second-degree Taylor polynomial for the function $f(x) = (e^x + 2)^{1/2}$ expanded about $x_0 = 0$. Then use it to approximate $f(1)$.

2. Use the third-degree Taylor polynomial about $x_0 = 1$ to approximate the function $f(x) = x^2 + 4e^x$ at $x = 1$ and $x = 2$. Find the error bounds for these approximations and compare your results.

3. Find the third-degree Taylor polynomial for the function $f(x) = \ln(x^2 + 2x + 2)$ expanded about $x_0 = 1$. Use this polynomial to approximate $f(1.1)$. Compute the actual error of the problem.

4. Find the fourth-degree Taylor polynomial for the function $f(x) = (x^3 + 1)^{-1}$ expanded about $x_0 = 0$, and use it to approximate $f(0.2)$. Also, find a bound for the error in this approximation.

5. Use the Lagrange interpolation formula based on the points $x_0 = 0, x_1 = 1, x_2 = 2.5$ to find the equation of the quadratic polynomial to approximate $f(x) = \dfrac{2}{x+2}$ at $x = 2.3$.

6. Let $f(x) = \cos(x\pi/4)$, where x is in radian. Use the quadratic Lagrange interpolation formula based on the points $x_0 = 0, x_1 = 1, x_2 = 2$, and $x_3 = 4$ to find the polynomial $p_2(x)$ to approximate the function $f(x)$ at $x = 0.5$ and $x = 3.5$.

7. Let $f(x) = x + 2\ln(x + 2)$. Use the quadratic Lagrange interpolation formula based on the points $x_0 = 0, x_1 = 1, x_2 = 2$, and $x_3 = 3$ to approximate $f(0.5)$ and $f(2.8)$. Also, compute the error bounds for your approximations.

8. Consider the function $f(x) = e^{x^2}$ and $x = 0, 0.25, 0.5, 1$. Then use the suitable Lagrange interpolating polynomial to approximate $f(0.75)$. Also, compute an error bound for your approximation.

9. Let $f(x) = x^4 - 2x + 1$. Use the cubic Lagrange interpolation formula based on the points $x_0 = -1, x_1 = 0, x_2 = 2$, and $x_3 = 3$ to find the polynomial $p_3(x)$ to approximate the function $f(x)$ at $x = 1.1$. Also, compute an error bound for your approximation.

10. Construct Lagrange interpolation polynomials for the following functions and compute the error bounds for the approximations:

(a) $f(x) = x + 2^{x+1}$, $x_0 = 0, x_1 = 1, x_2 = 2.5, x_3 = 3$.
(b) $f(x) = 3x^3 + 2x^2 + 1$, $x_0 = 1, x_1 = 2, x_2 = 3$.
(c) $f(x) = \cos x - \sin x$, $x_0 = 0, x_1 = 0.25, x_2 = 0.5, x_3 = 1$.

11. Consider the following table:

x	0	1	2	3
$f(x)$	2.0	3.72	8.39	21.06

(a) Construct a divided difference table for the tabulated function.
(b) Compute the Newton interpolating polynomials $p_2(x)$ and $p_3(x)$ at $x = 2.2$.

12. Consider the following table:

x	1	2	3	4	5
$f(x)$	3.60	1.80	1.20	0.90	0.72

(a) Construct a divided difference table for the tabulated function.
(b) Compute the Newton interpolating polynomials $p_3(x)$ and $p_4(x)$ at $x = 2.5, 3.5$.

13. Consider the following table of $f(x) = \sqrt{x}$:

x	4	5	6	7	8
$f(x)$	2.0000	2.2361	2.4495	2.6458	2.8284

(a) Construct the divided difference table for the tabulated function.
(b) Find the Newton interpolating polynomials $p_3(x)$ and $p_4(x)$ at $x = 5.9$.
(c) Compute error bounds for your approximations in part *(b)*.

14. Let $f(x) = e^x sinx$, with $x_0 = 0, x_1 = 2, x_2 = 2.5, x_3 = 4$, and $x_4 = 4.5$. Then
(a) Construct the divided-difference table for the given data points.
(b) Find the Newton divided difference polynomials $p_2(x), p_3(x)$, and $p_4(x)$ at $x = 2.4$.
(c) Compute error bounds for your approximations in part *(b)*.
(d) Compute the actual error.

15. Show that if x_0, x_1, and x_2 are distinct then

$$f[x_0, x_1, x_2] = f[x_1, x_2, x_0] = f[x_2, x_0, x_1].$$

16. The divided difference form of the interpolating polynomial $p_3(x)$ is

$$\begin{aligned} p_3(x) &= f[x_0] + (x - x_0)f[x_0, x_1] + (x - x_0)(x - x_1)f[x_0, x_1, x_2, x_0] \\ &+ (x - x_0)(x - x_1)(x - x_2)f[x_0, x_1, x_2, x_3]. \end{aligned}$$

By expressing these divided differences in terms of the function values $f(x_i)(i = 0, 1, 2, 3)$, verify that $p_3(x)$ does pass through the points $(x_i, f(x_i))(i = 0, 1, 2, 3)$.

17. Let $f(x) = x^2 + e^x$ and $x_0 = 0, x_1 = 1$. Use the divided differences to find the value of the second divided difference $f[x_0, x_1, x_0]$.

18. Let $f(x) = x + \dfrac{1}{x^2}$, with points $x_0 = 1,\ x_1 = 1.5,\ x_2 = 2.5$, and $x_3 = 3$. Use Aitken's method to find the estimate value of $f(2.9)$.

19. Consider the following table:

x	1.0	1.1	1.2	1.3	1.4	1.5
$f(x)$	0.8415	0.8912	0.9320	0.9636	0.9854	0.9975

Use Aitken's method to find the estimate value of $f(1.21)$.

20. Let $f(x) = \dfrac{e^x}{2x + 1}$, with points $x_0 = 0,\ x_1 = 1,\ x_2 = 2$, and $x_3 = 3$. Use Aitken's method to find the estimate value of $f(2.5)$.

21. Consider the following table for function $f(x) = \sin\theta$

x	45°	50°	55°	60°
$f(x)$	0.7071	0.7660	0.8192	0.8660

Use Newton's forward interpolation formula to find the value of $\sin 52^o$.

22. Consider function $f(x) = x + 2^{x+1}$. Then

(a) Construct the forward-difference table for the values of the function for

$x_0 = -1, x_1 = 0, x_2 = 1, x_3 = 2$, and $x_4 = 3$.

(b) Use Newton forward-difference formulas to find $p_2(x)$ and $p_3(x)$.

(c) Evaluate the polynomials in part *(b)* at $x = -0.5$.

(d) Compute the actual error.

23. Consider the following table:

x	0.0	0.25	0.5	0.75	1.0
$f(x)$	0.9162	0.8109	0.6931	0.5596	4.055

(a) Construct the central-difference table for the values of the function.
(b) Derive the Newton forward polynomial passing through the data points $x = 0.25, 0.5, 0.75$, and evaluate at $x = 0.6$.
(c) Estimate the error of the interpolation formula at $x = 0.6$.

24. Consider the following table:

x	1.0	1.1	1.2	1.3	1.4
$f(x)$	0.1839	0.16664	0.1506	0.1363	0.1233

(a) Construct the forward-difference table for the values of the function.
(b) Construct the cubic interpolating polynomial in the forward-difference form and estimate $f(1.25)$.
(c) Estimate the error of the interpolation formula at $x = 1.25$.

25. Use Newton's backward-difference formula to find the value of $f(11.8)$, using the following data:

x	2	4	6	8	10	12	14
$f(x)$	23	93	259	569	1071	1813	2843

26. Consider the function $f(x) = x^2 e^x + 3x + 1$. Then
(a) Construct Newton's backward-difference table for the values of the function for $x_0 = 1.1, x_1 = 1.3, x_2 = 1.5$, and $x_3 = 1.7, x_4 = 1.9$.
(b) Find the Newton backward-difference for interpolating polynomials $p_2(x)$ and $p_3(x)$.
(c) Evaluate the polynomials in part (b) at $x = 1.4$.
(d) Compute the actual error.

27. Use the difference table for the values of the function $f(x) = log_{10}(x)$, for $x = 1, 1.1, \ldots, 1.6$ to
(a) Estimate the value of $log_{10}(1.05)$, using Newton's forward-difference formula.
(b) Estimate $log_{10}(1.55)$, using Newton's backward-difference formula.

28. Given the following data, approximate the functional value at $x = 1.7$ using Newton forward-difference and backward-difference interpolating formulas:

x	1.0	1.5	2.0	2.5	3.0
$f(x)$	0.0	0.75	2.0	3.75	6.0

29. Express the following differences in terms of function values:

$$(a)\quad \Delta^4 f_i, \qquad (b)\quad \nabla^3 f_i, \text{ and} \qquad (c)\quad \delta^2 f_i.$$

30. Show that

$$\Delta\nabla f_i = \nabla\Delta f_i = \delta^2 f_i.$$

31. Use the Gauss forward central-difference formula to find the value of $f(41)$ with the help of, following data:

x	30	35	40	45	50
$f(x)$	3678.2	2995.1	24001	1876.2	1416.3

32. Use the Gauss forward central-difference formula to find the value of $f(0.0341)$ using the following data:

x	0.01	0.02	0.03	0.04	0.05
$f(x)$	98.4342	48.4392	31.7775	23.4492	18.4542

33. Given $f(x) = \sqrt{x}$ and the corresponding table

x	30	35	40	45	50
$f(x)$	3678.2	2995.1	24001	1876.2	1416.3

use the Gauss backward central-difference formula to find the value of $f(12516)$. Compute the actual error.

34. Consider the following table:

x	0.0	0.2	0.4	0.6	0.8	1.0	1.2
$f(x)$	1.0	0.98	0.92	0.83	0.70	0.54	0.36

(a) Construct the central-difference table for the values of the function.
(b) Use the Stirling central-difference formula to find $p_2(x)$ and $p_3(x)$.
(c) Evaluate the polynomials in part *(b)* at $x = 1.4$ to approximate $\cos(0.65)$.
(d) Compute the actual error.

35. Use the Stirling central-difference formula to find the value of $f(1.62)$ using the following table:

x	1.2	1.4	1.6	1.8	2.0
$f(x)$	5.6464	6.4422	7.1736	7.8333	8.4147

36. Construct the difference table for the values of function $f(x) = e^{2x} sin 2x$, for $x_0 = 1.0, x_1 = 1.1, x_2 = 1.2, x_3 = 1.3$, and $x_4 = 1.4$. Use the Gauss forward-difference formula to estimate $f(1.01)$, the Gauss backward-difference formula to estimate $f(1.39)$, and the Bessel's central-difference formula to estimate $f(1.25)$.

37. Use the Bessel's central-difference-formula to find the value of $f(1.22)$ using the following table:

x	0.5	1.0	1.5	2.0	2.5	3
$f(x)$	0.1915	0.3413	0.4332	0.4773	0.4938	0.4987

38. Construct the difference table for the values of $f(x) = \ln(x-1)^2$, for $x_0 = 2.0, x_1 = 2.1, x_2 = 2.2, x_3 = 2.3, x_4 = 2.4, x_5 = 2.5$, and $x_6 = 2.6$. Use the Gauss forward-difference formula to estimate $f(2.05)$, the Gauss backward difference formula to estimate $f(2.55)$, and Bessel's and Stirling's central-difference-formulas to estimate $f(2.35)$.

39. Find the linear spline, which interpolates the data:

$$(0, 3.5), (1, 3.9), (2, 4.7), (3, 5.8).$$

What are its values at $x = 0.55, 1.15$ and 2.5?

40. Which of the following functions are linear splines?

$$(a) \qquad s(x) = \begin{cases} x, & 0 \le x \le 1 \\ 2x-1, & 1 \le x \le 2 \\ x+2, & 2 \le x \le 4. \end{cases}$$

$$(b) \qquad s(x) = \begin{cases} 2-x, & 0 \le x \le 1 \\ 2x-1, & 1 \le x \le 2 \\ x+1, & 2 \le x \le 4. \end{cases}$$

41. Find the linear spline, which interpolates the data:

$$(0,0),(0.2,0.18),(0.3,0.26),(0.5,0.41).$$

What are its values at $x = 0.15, 0.25,$ and 0.45?

42. Find the linear splines, which interpolate the following data:

$$(0,0),(1,1),(16,2),(81,3).$$

Compare the interpolated values at $x = 0.5, 11.5,$ and 30.5 to $f(x) = \sqrt[4]{x}$.

43. Find the linear splines, which interpolate the following data:

$$(0,1),(3,2),(8,3),(15,4).$$

Compare interpolated values at $x = 2.5, 5.5$ and 10.5 to $f(x) = \sqrt{x+1}$.

44. Find the linear splines, which interpolate the following data:

$$(0,1),(2,0.9976),(3,0.9945), \quad \text{and} \quad (4,0.9903).$$

Compare interpolated values at $x = 1.5, 2.5$ and 3.5 to $f(x) = \cos(2x)$.

45. Is the following function a cubic spline?

$$s(x) = \begin{cases} x^3 + 8x + 3 & 1 \le x \le 4 \\ x^2 + 12x & 4 \le x \le 5 \\ (x-5)^3 - 6(x-5)^2 + 2 & 5 \le x \le 6 \end{cases}.$$

46. Is the following function a cubic spline?

$$s(x) = \begin{cases} 3x^2 + 3x + 1 & -2 \le x \le -1 \\ -x^3 & -1 \le x \le 1 \\ x^3 - 6x^2 + 6x - 2 & 1 \le x \le 2 \end{cases}.$$

47. Construct the natural cubic spline for the following data:

(a) $(-2,-8),(-1.5,-5.25),(-1,-3),(1,1)$.
(b) $(0,1),(1,0.67),(2,0.5),(3,0.4)$.
(c) $(0,-4),(1,-3),(1.5,-0.25),(2,4)$.
(d) $(0,1),(1,5.437),(2,22.167),(3,80.342)$.

48. Construct the natural cubic spline for the following data:

(a) $(0,4),(1,1),(2,2),(3,3)$.
(b) $(0,3),(1,4),(2,9),(3,14)$.
(c) $(0,7),(1,2),(2,0),(3,-5)$.
(d) $(0,1),(1,4),(2,0),(3,-2)$.

49. Construct the clamped cubic spline using the following data and end conditions:

(a) $(1,0),(2,0.69),(3,1.10),(4,1.39), \quad s'(1)=1,\ s'(4)=0.25$.
(b) $(0,1),(0.2,1.2),(0.4,1.5),(0.6,1.8), \quad s'(0)=1, s'(0.6)=1.75$.
(c) $(0,1),(1,3),(2,11),(3,31), \quad s'(0)=1, s'(3)=6$.
(d) $(0,-4),(1,-2),(2,8),(3,32), \quad s'(0)=0,\ s'(3)=31$.

50. Construct the clamped cubic spline using the following data and end conditions:

(a) $(2,0),(3,1),(4,2),(5,2), \quad s'(2)=-0.4,\ s'(5)=-1.3$.
(b) $(0,0),(1,4),(2,8),(3,9), \quad s'(0)=1, s'(3)=-1$.
(c) $(0,0),(1,2),(2,3),(3,2), \quad s'(0)=1.1, s'(3)=-0.6$.
(d) $(1,4),(2,6),(3,8),(4,10), \quad s'(1)=0,\ s'(4)=15$.

51. The data in Problem 47 were generated using the following functions. For the given value of x, use the constructed splines in Problem 47 to approximate $f(x)$ and also calculate the actual error:

(a) $f(x)=2x-x^2$; approximate $f(-1.7), f'(-1.7)$.
(b) $f(x)=\dfrac{1}{(x+2)}$; approximate $f(1.5), f'(1.5)$.
(c) $f(x)=3x^2-2x-4$; approximate $f(1.1), f'(1.1)$.
(d) $f(x)=(x+1)e^x$; approximate $f(2.7), f'(2.7)$.

52. Fit a natural cubic spline to $f(x)=(x+1)/(2+x^2)$ on the interval $[0,2]$. Use five equispaced points on the function $[x=0(0.5)2]$.

53. Find the least squares line fit $y=ax+b$ for the following data:

(a) $(-2,1),(-1,2),(0,3),(1,4)$.
(b) $(0,0),(2,1.8),(4,3.4),(6,4.6)$.

(c) $(2,0),(3,4),(4,10),(5,16)$.
(d) $(3,1.6),(4,2.4),(5,2.9),(6,3.4),(8,4.6)$.
(e) $(-4,1.2),(-2,2.8),(0,6.2),(2,7.8),(4,13.2)$.

54. Repeat Problem 53 to find the least squares parabolic fit $y = a + bx + cx^2$.

55. Find the least squares parabolic fit $y = a + bx + cx^2$ for the following data:

(a) $(-1,0),(0,-2),(0.5,-1),(1,0)$.
(b) $(-3,15),(-1,5),(1,1),(3,5)$.
(c) $(-3,-1),(-1,25),(1,25),(3,1)$.
(d) $(0,3),(1,1),(2,0),(4,1),(6,4)$.
(e) $(-2,10),(-1,1),(0,0),(1,2),(2,9)$.

56. Repeat Problem 55 to find the best fit of the form $y = ax^b$.

57. Find the best fit of the form $y = ae^{bx}$ for the following data:

(a) $(5,5.7),(10,7.5),(15,8.9),(20,9.9)$.
(b) $(-1,0.1),(1,2.3),(2,10),(3,45)$.
(c) $(3,4),(4,10),(5,16),(6,20)$.
(d) $(-2,1),(-1,2),(0,3),(1,3),92,4)$.
(e) $(-1,6.62),(0,3.94),(1,2.17),(2,1.35),(3,0.89)$.

58. Find the plane $z = ax + by + c$ that best fits the following data:

(a) $(0,1,2),(1,0,3),(2,1,4),(0,2,1)$.
(b) $(2,1,-1),(2,2,0),(1,1,1),(1,2,3)$.
(c) $(1,-1,1),(2,0,-1),(2,1,-2),(0,1,0)$.
(d) $(1,1,0),(2,1,1),(1,2,1),(2,2,2)$.

59. Find the least squares planes for the following data:

$$(3,1,-3),(2,1,-1),(2,2,0),(1,1,1),(1,2,3).$$

60. Find the least squares planes for the following data:

(a) $(1,2,4),(2,3,6),(1,1,8),(2,2,2),(2,1,9)$.
(b) $(1,7,1),(2,5,6),(3,1,-2),(2,1,0),(3,3,3)$.
(c) $(5,4,3),(3,7,9),(3,2,3),(4,4,4),(5,7,8)$.

61. Solve the following overdetermined linear systems:

(a)

$$\begin{aligned} 2x_1 + x_2 &= 4 \\ 3x_1 + 2x_2 &= 1 \\ x_1 + 3x_2 &= 4. \end{aligned}$$

(b)

$$\begin{aligned} x_1 + 3x_2 &= 4 \\ 3x_1 + 5x_2 &= 11 \\ x_1 + 6x_2 &= 12. \end{aligned}$$

(c)

$$\begin{aligned} x_1 + 3x_2 + 2x_3 &= 3 \\ 2x_1 + 5x_2 + 6x_3 &= 8 \\ x_1 + x_2 + x_3 &= -2 \\ 4x_1 + 3x_2 + 4x_3 &= 1. \end{aligned}$$

62. Solve the following overdetermined linear systems:

(a)

$$\begin{aligned} 4x_1 - 2x_2 &= 7 \\ 5x_1 + 11x_2 &= -4 \\ 7x_1 - 12x_2 &= 14. \end{aligned}$$

(b)

$$\begin{aligned} x_1 + 5x_2 + 9x_3 &= 34 \\ 2x_1 - x_2 + 3x_3 &= 45 \\ -3x_1 + x_2 - x_3 &= 2 \\ 3x_1 - 14x_2 + 4x_3 &= -9. \end{aligned}$$

(c)

$$\begin{aligned} 2x_1 + 5x_2 + 2x_3 + x_4 &= 2 \\ x_1 + 4x_2 + 5x_3 + x_4 &= 12 \\ 2x_1 + 3x_2 + 2x_3 - 2x_4 &= 3 \\ 2x_1 - 9x_2 + 11x_3 + 7x_4 &= 10 \\ x_1 + 5x_2 - 8x_3 + 13x_4 &= 14. \end{aligned}$$

63. A QR decomposition of A is given. Use it to find a least squares solution

of $A\mathbf{x} = \mathbf{b}$, where

$$A = \begin{pmatrix} 0 & 3 \\ 0 & 4 \\ 5 & 10 \end{pmatrix}, \quad \mathbf{b} = \begin{pmatrix} 3 \\ 0 \\ -4 \end{pmatrix}$$

$$Q = \begin{pmatrix} 0 & -0.6000 & -0.8000 \\ 0 & -0.8000 & 0.6000 \\ -1.0000 & 0 & 0 \end{pmatrix}, \quad R = \begin{pmatrix} 5 & -10 \\ 0 & -5 \\ 0 & 0 \end{pmatrix}.$$

64. A QR decomposition of A is given. Use it to find a least squares solution of $A\mathbf{x} = \mathbf{b}$, where

$$A = \begin{pmatrix} 1 & 0 \\ 2 & -1 \\ -1 & 1 \end{pmatrix}, \quad \mathbf{b} = \begin{pmatrix} 1 \\ 1 \\ 1 \end{pmatrix}$$

$$Q = \begin{pmatrix} -0.4082 & 0.7071 & -0.5774 \\ -0.8165 & -0.0000 & 0.5774 \\ 0.4082 & 0.7071 & 0.5774 \end{pmatrix}, \quad R = \begin{pmatrix} -2.4495 & 1.2247 \\ 0 & 0.7071 \\ 0 & 0 \end{pmatrix}.$$

65. A QR decomposition of A is given. Use it to find a least squares solution of $A\mathbf{x} = \mathbf{b}$, where

$$A = \begin{pmatrix} 2 & 1 \\ 2 & 0 \\ 1 & 1 \end{pmatrix}, \quad \mathbf{b} = \begin{pmatrix} 2 \\ 3 \\ -1 \end{pmatrix}$$

$$Q = \begin{pmatrix} -0.6667 & 0.3333 & -0.6667 \\ -0.6667 & -0.6667 & 0.3333 \\ -0.3333 & 0.6667 & 0.6667 \end{pmatrix}, \quad R = \begin{pmatrix} -3.0000 & -1.0000 \\ 0 & 1.0000 \\ 0 & 0 \end{pmatrix}.$$

66. A QR decomposition of A is given. Use it to find a least squares solution of $A\mathbf{x} = \mathbf{b}$, where

$$A = \begin{pmatrix} 1 & 1 \\ -1 & 1 \\ 1 & 1 \\ 1 & -1 \end{pmatrix}, \quad \mathbf{b} = \begin{pmatrix} 1 \\ 2 \\ 4 \\ -1 \end{pmatrix}$$

$$Q = \begin{pmatrix} -0.5000 & -0.5000 & -0.7000 & -0.1000 \\ 0.5000 & -0.5000 & -0.1000 & 0.7000 \\ -0.5000 & -0.5000 & 0.7000 & 0.1000 \\ -0.5000 & 0.5000 & -0.1000 & 0.7000 \end{pmatrix}, \quad R = \begin{pmatrix} -2 & 0 \\ 0 & -2 \\ 0 & 0 \\ 0 & 0 \end{pmatrix}.$$

67. A QR decomposition of A is given. Use it to find a least squares solution of $A\mathbf{x} = \mathbf{b}$, where

$$A = \begin{pmatrix} 1 & 2 \\ 1 & -1 \end{pmatrix}, \qquad \mathbf{b} = \begin{pmatrix} 1 \\ 1 \end{pmatrix}$$

$$Q = \begin{pmatrix} 0.7071 & -0.7071 \\ 0.7071 & 0.7071 \end{pmatrix}, \quad R = \begin{pmatrix} 1.4142 & 0.7071 \\ 0.0000 & -2.1213 \end{pmatrix}.$$

68. A QR decomposition of A is given. Use it to find a least squares solution of $A\mathbf{x} = \mathbf{b}$, where

$$A = \begin{pmatrix} 1 & 2 \\ -1 & 3 \end{pmatrix}, \qquad \mathbf{b} = \begin{pmatrix} 2 \\ 1 \end{pmatrix}$$

$$Q = \begin{pmatrix} 0.7071 & 0.7071 \\ -0.7071 & 0.7071 \end{pmatrix}, \quad R = \begin{pmatrix} 1.4142 & -0.7071 \\ 0.0000 & 3.5356 \end{pmatrix}.$$

69. A QR decomposition of A is given. Use it to find a least squares solution of $A\mathbf{x} = \mathbf{b}$, where

$$A = \begin{pmatrix} 1 & 0 \\ 1 & 1 \end{pmatrix}, \qquad \mathbf{b} = \begin{pmatrix} 1 \\ -1 \end{pmatrix}$$

$$Q = \begin{pmatrix} 0.7071 & -0.7071 \\ 0.7071 & 0.7071 \end{pmatrix}, \quad R = \begin{pmatrix} 1.4142 & 0.7071 \\ 0.0000 & 0.7071 \end{pmatrix}.$$

70. A QR decomposition of A is given. Use it to find a least squares solution of $A\mathbf{x} = \mathbf{b}$, where

$$A = \begin{pmatrix} 1 & 3 & 0 \\ 0 & -1 & 8 \\ 1 & 2 & 4 \end{pmatrix}, \qquad \mathbf{b} = \begin{pmatrix} 1 \\ 2 \\ 3 \end{pmatrix}$$

$$Q = \begin{pmatrix} 0.7071 & -0.4082 & -0.5774 \\ 0 & 0.8165 & -0.5774 \\ 0.7071 & 0.4082 & 0.5774 \end{pmatrix}, \quad R = \begin{pmatrix} 1.4142 & 3.5355 & 2.8284 \\ 0 & -1.2247 & 8.1650 \\ 0 & 0 & -2.3094 \end{pmatrix}.$$

71. A QR decomposition of A is given. Use it to find a least squares solution of $A\mathbf{x} = \mathbf{b}$, where

$$A = \begin{pmatrix} 1 & 0 & 2 \\ -1 & 2 & 0 \\ -1 & -2 & 2 \end{pmatrix}, \qquad \mathbf{b} = \begin{pmatrix} 1 \\ 1 \\ 1 \end{pmatrix}$$

$$Q = \begin{pmatrix} 0.7071 & -0.4082 & -0.5774 \\ 0 & 0.8165 & -0.5774 \\ 0.7071 & 0.4082 & 0.5774 \end{pmatrix}, \quad R = \begin{pmatrix} 1.4142 & 3.5355 & 2.8284 \\ 0 & -1.2247 & 8.1650 \\ 0 & 0 & -2.3094 \end{pmatrix}.$$

72. Find the least solution each of the following linear systems, $A\mathbf{x} = \mathbf{b}$, using QR decomposition:

(a)

$$A = \begin{pmatrix} 5 & 3 \\ 1 & 3 \\ 2 & 1 \end{pmatrix}, \quad \mathbf{x} = \begin{pmatrix} x_1 \\ x_2 \end{pmatrix}, \quad \mathbf{b} = \begin{pmatrix} 4.9 \\ 0.8 \\ 1.7 \end{pmatrix}.$$

(b)

$$A = \begin{pmatrix} 1 & -1 & 4 \\ 0 & 2 & 1 \\ 1 & 1 & 0 \\ 2 & -1 & 1 \end{pmatrix}, \quad \mathbf{x} = \begin{pmatrix} x_1 \\ x_2 \\ x_3 \end{pmatrix}, \quad \mathbf{b} = \begin{pmatrix} 1.1 \\ 0.2 \\ 0.9 \\ 1.7 \end{pmatrix}.$$

(c)

$$A = \begin{pmatrix} 1 & -1 & 1 \\ -1 & 4 & 2 \\ -2 & 1 & 2 \\ 1 & 4 & 2 \end{pmatrix}, \quad \mathbf{x} = \begin{pmatrix} x_1 \\ x_2 \\ x_3 \end{pmatrix}, \quad \mathbf{b} = \begin{pmatrix} 0.7 \\ -0.8 \\ -1.5 \\ 1.02 \end{pmatrix}.$$

(d)

$$A = \begin{pmatrix} 3 & 2 & 1 \\ 1 & 2 & 2 \\ 1 & 0 & -1 \\ 2 & 1 & -2 \end{pmatrix}, \quad \mathbf{x} = \begin{pmatrix} x_1 \\ x_2 \\ x_3 \end{pmatrix}, \quad \mathbf{b} = \begin{pmatrix} 2.5 \\ 1.1 \\ 0.8 \\ 1.9 \end{pmatrix}.$$

73. Find the least solution each of the following linear systems, $A\mathbf{x} = \mathbf{b}$, using QR decomposition:

(a)

$$A = \begin{pmatrix} 2 & 1 \\ 1 & 1 \\ 2 & 2 \end{pmatrix}, \quad \mathbf{x} = \begin{pmatrix} x_1 \\ x_2 \end{pmatrix}, \quad \mathbf{b} = \begin{pmatrix} 1 \\ 1 \\ 1 \end{pmatrix}.$$

(b)

$$A = \begin{pmatrix} 1 & -1 \\ 0 & 2 \\ 1 & 1 \end{pmatrix}, \quad \mathbf{x} = \begin{pmatrix} x_1 \\ x_2 \end{pmatrix}, \quad \mathbf{b} = \begin{pmatrix} 0.1 \\ 1.7 \\ 0.9 \end{pmatrix}.$$

(c)

$$A = \begin{pmatrix} 3 & -1 \\ -1 & 4 \\ -2 & 1 \end{pmatrix}, \quad \mathbf{x} = \begin{pmatrix} x_1 \\ x_2 \end{pmatrix}, \quad \mathbf{b} = \begin{pmatrix} 2.7 \\ -0.8 \\ -1.5 \end{pmatrix}.$$

(d)

$$A = \begin{pmatrix} 4 & 2 \\ 1 & 2 \\ 1 & 0 \end{pmatrix}, \quad \mathbf{x} = \begin{pmatrix} x_1 \\ x_2 \end{pmatrix}, \quad \mathbf{b} = \begin{pmatrix} 3.5 \\ 1.1 \\ 0.8 \end{pmatrix}.$$

74. Solve Problem 73 using singular value decomposition.

75. Find the least solution each of the following linear systems, $A\mathbf{x} = \mathbf{b}$, using singular value decomposition:

(a)

$$A = \begin{pmatrix} -2 & 2 \\ -1 & 1 \\ 2 & 2 \end{pmatrix}, \quad \mathbf{x} = \begin{pmatrix} x_1 \\ x_2 \end{pmatrix}, \quad \mathbf{b} = \begin{pmatrix} -1.8 \\ -0.9 \\ 1.9 \end{pmatrix}.$$

(b)

$$A = \begin{pmatrix} 1 & -1 \\ 1 & 2 \\ 1 & 1 \end{pmatrix}, \quad \mathbf{x} = \begin{pmatrix} x_1 \\ x_2 \end{pmatrix}, \quad \mathbf{b} = \begin{pmatrix} 1.1 \\ 0.7 \\ 0.9 \end{pmatrix}.$$

(c)

$$A = \begin{pmatrix} 1 & 0 \\ 1 & 1 \\ -1 & 1 \end{pmatrix}, \quad \mathbf{x} = \begin{pmatrix} x_1 \\ x_2 \end{pmatrix}, \quad \mathbf{b} = \begin{pmatrix} 0.9 \\ 0.85 \\ -0.9 \end{pmatrix}.$$

76. Solve Problem 72 using singular value decomposition.

Chapter 5

Differentiation and Integration

5.1 Introduction

In this chapter we deal with techniques for approximating numerically the two fundamental operations of calculus, differentiation, and integration. Both of these problems may be approached in the same way. Although both numerical differentiation and numerical integration formulas will be discussed, it should be noted that numerical differentiation is inherently much less accurate than numerical integration, and its application is generally avoided whenever possible. Nevertheless, it has been used successfully in certain applications.

Engineers are frequently confronted with the problem of differentiating functions that are defined in tabular or graphical form rather than as explicit functions. The interpretation of experimentally obtained data is a good example of this. A similar situation involves the integration of functions, which have explicit forms that are difficult or impossible to integrate in terms of elementary functions. Graphical techniques, employing the construction of tangents to curves and the estimation of areas under curves, are commonly used in solving such problems, when great accuracy is not a prerequisite for results. However, there are occasions when a higher degree of accuracy is desired, and, for these, various numerical methods are available.

5.2 Numerical Differentiation

Firstly, we discuss the numerical process for approximating the derivative of the function $f(x)$ at the given point. A function $f(x)$, known either explicitly or as a set of data points, is replaced by a simpler function. A polynomial $p(x)$ is the obvious choice of approximating function, since the operation of differentiation is then easily performed. The polynomial $p(x)$ is differentiated to obtain $p'(x)$, which is taken as an approximation of $f'(x)$ for any numerical value of x. Geometrically, this is equivalent to replacing the slope of $f(x)$, at x, by that of $p(x)$. Here, numerical differentiation is derived by differentiating interpolating polynomials.

We now turn our attention to the numerical process for approximating the derivative of a function $f(x)$ at x; that is,

$$f'(x) = \lim_{h \to 0} \frac{f(x+h) - f(x)}{h}, \quad \text{provided the limit exits.} \tag{5.1}$$

In principle, it is always possible to determine an analytic form (5.1) of a derivative for a given function. In some cases, however, the analytic form is very complicated, and a numerical approximation of the derivative may be sufficient for our purpose.

Formula (5.1) provides an obvious way to get an approximation to $f'(x)$; simply compute

$$D_h f(x) = \frac{f(x+h) - f(x)}{h} \tag{5.2}$$

for small values of stepsize h is called the numerical differentiation formula in (5.1).

Numerical differentiation is a much less satisfactory process because the seemingly obvious approximations are not always as good as they seem. Therefore, this process should, for this reason, be avoided if at all possible. We study it mainly as a means of solving differential equations by various numerical methods, based on the approximations we shall obtain for the derivatives of a function. We study it also, because it often happens that the thing we want to differentiate is not a known function. We may, for instance, be given a table of speeds of a body observed at certain times, and wish to estimate its acceleration at these times.

Numerical differentiation is useful in estimating the derivative of a function when either function $f(x)$ is difficult to differentiate easily, or it is not known as an explicit expression in x but the values of the function are described only in terms of tabulated data. Generally, it is considered that numerical differentiation is basically an *unstable process*, which means that small errors made in the initial computations may cause greatly magnified errors in the final result. In fact, we may not always expect reasonable results even when the original data are known to be more accurate.

Here, we shall derive some formulas for estimating derivatives but we should avoid as far as possible numerically calculating derivatives higher than the first, as the error in their evaluation increases with their orders. In spite of some inherent shortcomings, numerical differentiation is important to derive formulas for solving integrals and the numerical solution of both ordinary and partial differential equations.

There are three different approaches for deriving numerical differentiation formulas. The first approach is based on the Taylor expansion of a function about a point; the second is to use difference operators; and the third approach to numerical differentiation is to fit a curve with a simple form to a function, and then to differentiate the curve-fit function. For example, the polynomial interpolation or spline methods of Chapter 4 can be used to fit a curve to tabulated data for a function and the resulting polynomial or spline can then be differentiated. When a function is represented by a table of values, the most obvious approach is to differentiate the Lagrange interpolation formula

$$f(x) = p_n(x) + \frac{f^{(n+1)}(\eta(x))}{(n+1)!} \prod_{i=0}^{n} (x - x_i), \tag{5.3}$$

where the first term $p_n(x)$ of the right-hand side is the Lagrange interpolating polynomial of degree n and the second term is its error term.

It is interesting to note that the process of numerical differentiation may be less satisfactory than interpolation. The closeness of the ordinates of $f(x)$ and $p_n(x)$ on the interval of interest does not guarantee the closeness of their respective derivatives. Note that the derivation and analysis of formulas for numerical differentiation considerably simplifies when the data is equally spaced. It will be assumed, therefore, that the points x_i are given by $x_i = x_0 + ih, (i = 0, 1, \ldots, n)$ for some fixed tabular interval h.

5.3 Numerical Differentiation Formulas

5.3.1 First Derivatives Formulas

To obtain the general formula for approximation of the first derivative of a function $f(x)$, we consider that $\{x_0, x_1, \ldots, x_n\}$ are $(n+1)$ distinct equally spaced points in some interval I and function $f(x)$ is continuous and its $(n+1)th$ derivatives exist in the given interval; that is, $f \in C^{n+1}(I)$. Then by differentiating (5.3) with respect to x and at $x = x_k$, we have

$$f'(x_k) = \sum_{i=0}^{n} f(x_i) L_i'(x_k) + \frac{f^{(n+1)}(\eta(x_k))}{(n+1)!} \prod_{\substack{i=0 \\ i \neq k}}^{n} (x_k - x_i). \tag{5.4}$$

Formula (5.4) is called the *(n+1)-point formula* to approximate $f'(x_k)$. From this formula we can obtain many numerical differentiation formulas but here we shall discuss only three formulas to approximate (5.1) at the given point $x = x_k$. The first one is called the *two-point formula*, which we can get from (5.4) by taking $n = 1$ and $k = 0$. The second numerical differentiation formula is called the *three-point formula*, which can be obtained from (5.4) when $n = 2$ and $k = 0, 1, 2$. Finally, we will discuss the *five-point formula* to approximate (5.1) using (5.4) when $n = 4$ and $k = 0, 1, 2, 3, 4$.

Two-point Formula

Consider two distinct points x_0 and x_1, then, to find the approximation of (5.1), the first derivative of a function at a given point, take $x_0 \in (a, b)$, where $f \in C^2[a, b]$ and that $x_1 = x_0 + h$ for some $h \neq 0$ that is sufficiently small to ensure that $x_1 \in [a, b]$. Consider the linear Lagrange interpolating polynomial $p_1(x)$, which interpolates $f(x)$ at the given points is

$$f(x) \approx p_1(x) = \left(\frac{x - x_1}{x_0 - x_1} \right) f(x_0) + \left(\frac{x - x_0}{x_1 - x_0} \right) f(x_1). \tag{5.5}$$

By taking the derivative of (5.5) with respect to x and at $x = x_0$, we obtain

$$f'(x)|_{x=x_0} \approx p_1'(x)|_{x=x_0} = -\frac{f(x_0)}{x_0 - x_1} + \frac{f(x_1)}{x_1 - x_0}.$$

Simplifying the above expression, we have

$$f'(x_0) \approx -\frac{f(x_0)}{h} + \frac{f(x_0 + h)}{h},$$

which can be written as

$$f'(x_0) \approx \frac{f(x_0+h)-f(x_0)}{h} = D_h f(x_0). \tag{5.6}$$

It is called the *two-point formula* for smaller values of h. For $h > 0$, sometimes formula (5.6) is also called the *two-point forward-difference formula* because it involves only differences of a function values forward from $f(x_0)$. The two-point forward-difference formula has a simple geometric interpretation as the slope of the forward secant line, as shown in Figure 5.1.

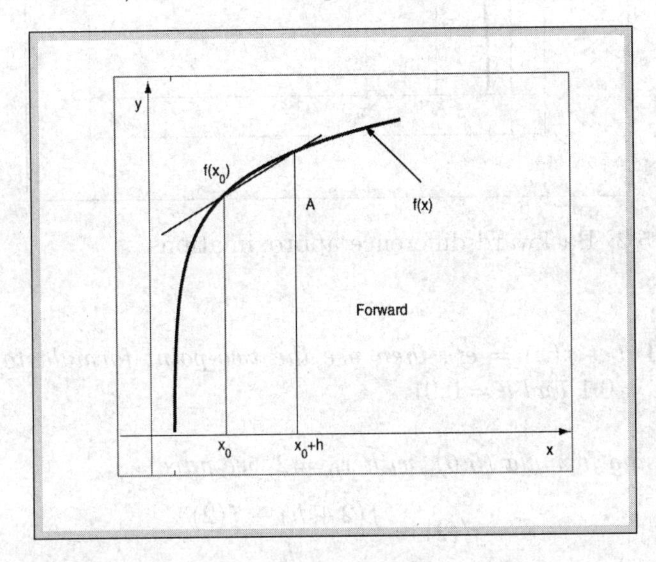

Figure 5.1: Forward-difference approximations.

If $h < 0$, then formula (5.6) is also called the *two-point backward-difference formula*, which can be written as

$$f'(x_0) \approx \frac{f(x_0)-f(x_0-h)}{h}. \tag{5.7}$$

In this case, a value of x behind the point of interest is used. Formula (5.7) is useful in cases where the independent variable represents time. If x_0 denotes the present time, the backward-difference formula uses only present and past samples, it does not rely on future data samples that may not yet be available in a real-time application.

The geometric interpretation of the two-point backward-difference formula, as the slope of the backward secant line, is shown in Figure 5.2.

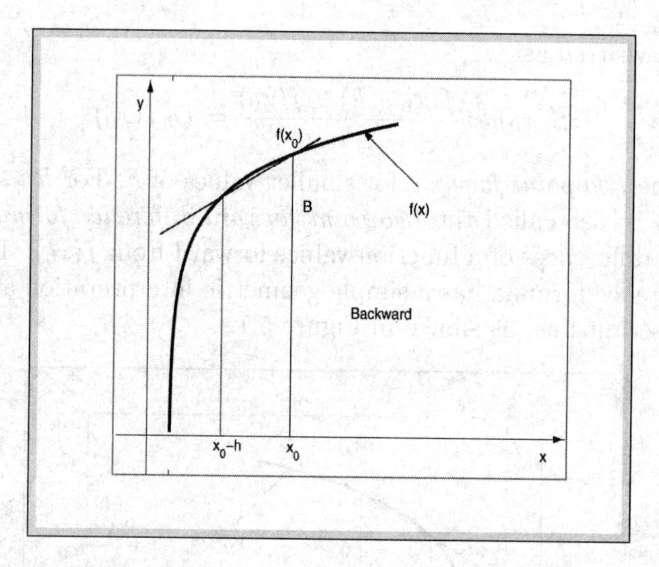

Figure 5.2: Backward-difference approximations.

Example 5.1 *Let $f(x) = e^x$, then use the two-point formula to approximate $f'(2)$, when $h = 0.1$ and $h = 0.01$.*

Solution. *Using formula (5.6), with $x_0 = 2$, we have*

$$f'(2) \approx \frac{f(2+h) - f(2)}{h}.$$

Then for $h = 0.1$, we get

$$\begin{aligned} f'(2) &\approx \frac{f(2.1) - f(2)}{0.1} \\ &\approx \frac{e^{2.1} - e^2}{0.1} = 7.7712. \end{aligned}$$

Similarly, by using $h = 0.01$, we obtain

$$f'(2) \approx \frac{(e^{2.01} - e^2)}{0.01} = 7.4262.$$

Since the exact solution of $f'(2) = e^2$ is 7.3891, the corresponding actual errors with $h = 0.1$ and $h = 0.01$ are -0.3821 and -0.0371, respectively. This shows that the approximation obtained with $h = 0.01$ is better than the approximation with $h = 0.1$.

The above results can be easily achieved with MATLAB commands as follows:

```
>> x0 = 2.0;
>> h = 10.^ -(1 : 2);
>> df = (exp(x0 + h) - exp(x))./h;
```

Similarly, using formula (5.7), with $x_0 = 2$, we have

$$f'(2) \approx \frac{f(2) - f(2-h)}{h}$$

then for $h = 0.1$, we have

$$f'(2) \approx \frac{f(2) - f(1.9)}{0.1} = \frac{e^2 - e^{1.9}}{0.1} = 7.0316.$$

For $h = 0.01$, we have

$$f'(2) \approx \frac{e^2 - e^{1.99}}{0.01} = 7.3522.$$

The corresponding actual errors with $h = 0.1$ and $h = 0.01$ are 0.3575 and 0.0369, respectively, which show that the approximation with $h = 0.01$ is better than the approximation with $h = 0.1$. Note that both errors for $h = 0.1$ and $h = 0.01$, using the backward-difference formula, is better than the forward-difference formula for the same values of h.

Error Formula

Formula (5.6) is not very useful, therefore, let us attempt to find the error involved in our first numerical differentiation formula (5.6). Consider the error term for the linear Lagrange polynomial, which can be written as

$$f(x) - p_1(x) = \frac{f''(\eta(x))}{2!} \prod_{i=0}^{1} (x - x_i)$$

for some unknown point $\eta(x) \in (x_0, x_1)$. By taking the derivative of the above equation with respect to x and at $x = x_0$, we have

$$\begin{aligned} f'(x_0) \quad - \quad p_1'(x_0) &= \left(\frac{d}{dx} f''(\eta(x)) \Big|_{x=x_0} \right) \frac{(x - x_0)(x - x_1)}{2} \\ &+ \quad \frac{f''(\eta(x_0))}{2} \left(\frac{d}{dx}(x^2 - x(x_0 + h) - x x_0 + x_0(x_0 + h)) \Big|_{x=x_0} \right). \end{aligned}$$

Since $\frac{d}{dx}f''(\eta(x)) = 0$ only if $x = x_0$, the error in the forward-difference formula (5.6) is

$$E_F(f,h) = f'(x_0) - D_h f(x_0) = -\frac{h}{2}f''(\eta(x)), \quad \text{where} \quad \eta(x) \in (x_0, x_1), \quad (5.8)$$

which is called the *error formula* of two-point formula (5.6). Hence, formula (5.6) can be written as

$$f'(x_0) = \frac{f(x_0+h) - f(x_0)}{h} - \frac{h}{2}f''(\eta(x)), \quad \text{where} \quad \eta \in (x_0, x_1). \quad (5.9)$$

Formula (5.9) is more useful than formula (5.6) because now on a large class of function, an error term is available along with the basic numerical formula. Note that the error term in (5.9) has two parts; a power of h and a factor involving some higher-order derivative of $f(x)$, which gives us an indication of the class of function to which the error estimate is applicable. The h term in the error makes the entire expression converge to zero as h approaches zero. The rapidity of this convergence will depend on the power of h. These remarks apply to many error estimates in numerical analysis. There will usually be a power of h and a factor telling us to what smoothness class of the function must belong so that the estimate is valid.

Note that formula (5.9) may also be derived from Taylor's theorem. Expansion of function $f(x_1)$ about x_0 as far as term involving h^2 gives

$$f(x_1) = f(x_0) + hf'(x_0) + \frac{h^2}{2!}f''(\eta(x)). \quad (5.10)$$

From this the result follows by subtracting $f(x_0)$ from both sides and dividing both sides by h and putting $x_1 = x_0 + h$.

Note that for a linear function, $f(x) = ax + b$, approximate formula (5.6) is exact; that is, it yields the correct value of the first derivative of the function $f(x)$ for any nonzero value of h.

Example 5.2 *Let $f(x) = x^2 \cos x$. Then*
(a) compute the approximate value of $f'(x)$ at $x = 1$, taking $h = 0.1$ using (5.6).
(b) compute the error bound for your approximation using formula (5.8).
(c) compute the absolute error.
(d) the best maximum value of stepsize h required to obtain the approximate value of $f'(1)$. Correct to two decimal places.

Solution. *(a) Given* $x_0 = 1, h = 0.1$, *by using formula (5.6), we have*

$$f'(1) \approx \frac{f(1+0.1)-f(1)}{0.1} = \frac{f(1.1)-f(1)}{0.1} = D_h f(1).$$

Thus,

$$\begin{aligned} f'(1) &\approx \frac{(1.1)^2\cos(1.1)-(1)^2\cos(1)}{0.1} \\ &\approx \frac{0.5489-0.5403}{0.1} = 0.0860 = D_h f(1), \end{aligned}$$

which is the required approximation of $f'(x)$ *at* $x = 1$.

(b) To find the error bound, we use formula (5.8), which gives

$$E_F(f,h) = -\frac{0.1}{2}f''(\eta(x)), \quad where \quad \eta(x) \in (1, 1.1)$$

or

$$|E_F(f,h)| = \left|-\frac{0.1}{2}\right||f''(\eta(x))|, \quad for \quad \eta \in (1, 1.1).$$

The second derivative $f''(x)$ *of the function can be found as*

$$f(x) = x^2\cos x, \quad gives \quad f''(x) = (2-x^2)\cos x - 4x\sin x.$$

The value of the second derivative $f''(\eta(x))$ *cannot be computed exactly because* $\eta(x)$ *is not known. But one can bound the error by computing the largest possible value for* $|f''(\eta(x))|$. *So bound* $|f''|$ *on* $[1, 1.1]$ *can be obtained*

$$M = \max_{1\leq x\leq 1.1}|(2-x^2)\cos x - 4x\sin x| = 3.5630$$

at $x = 1.1$. *Since* $|f''(\eta(x))| \leq M$, *for* $h = 0.1$, *we have*

$$|E_F(f,h)| \leq \frac{0.1}{2}M = 0.05(3.5630) = 0.1782,$$

which is the possible maximum error in our approximation.

(c) Since the exact value of the derivative $f'(1)$ *is* 0.2392, *the absolute error* $|E|$ *can be computed as follows:*

$$|E| = |f'(1) - D_h f(1)| = |0.2392 - 0.0860| = 0.1532.$$

(d) Since the given accuracy required is 10^{-2},

$$|E_F(f,h)| = |-\frac{h}{2}f''(\eta(x))| \leq 10^{-2}$$

for $\eta(x) \in (1, 1.1)$. *This gives*

$$\frac{h}{2}M \leq 10^{-2}$$

and solving for h, we obtain

$$h \leq \frac{(2 \times 10^{-2})}{M}.$$

Using $M = 3.5630$, *we obtain*

$$h \leq \frac{2}{356.3000} = 0.0056,$$

which is the best maximum value of h to get the required accuracy.

The truncation error in the approximation of (5.9) is roughly proportional to stepsize h used in its computation. The situation is made worse by the fact that the round-off error in computing the approximate derivative (5.6) is roughly proportionate to $\frac{1}{h}$. The overall error, therefore, is of the form

$$E = ch + \frac{\delta}{h},$$

where c and δ are constants. This places serve restriction on the accuracy that can be achieved with this formula.

Now we discuss a little more about the role of round-off errors in numerical differentiation. Consider formula (5.6), which is

$$f'(x_0) \approx \frac{f(x_0 + h) - f(x_0)}{h} = D_h f(x_0).$$

If h is small, then we can reasonably assume that $f(x_0)$ and $f(x_0{+}h)$ have similar magnitude and, therefore, similar round-off errors. Let the actual function values used in the computation be denoted by $\tilde{f}_0$ and $\tilde{f}_1$ with

$$f(x_i) - \tilde{f}_i = \epsilon_i, \qquad \text{for} \quad i = 0, 1$$

the errors in the function values. Thus, the actual quantity calculated is

$$\tilde{D}_h f(x_0) = \frac{\tilde{f}_1 - \tilde{f}_0}{h}.$$

For the error in this quantity, replacing $\tilde{f}_i$ by $f(x_i) - \epsilon_i$, for $i = 0, 1$, we obtain

$$\begin{aligned} f'(x_0) - \tilde{D}_h f(x_0) &= f'(x_0) - \frac{(f(x_1) - \epsilon_1) - (f(x_0) - \epsilon_0)}{h} \\ &= f'(x_0) - \frac{f(x_1) - f(x_0)}{h} + \frac{\epsilon_1 - \epsilon_0}{h}. \end{aligned}$$

Then the overall error is given by

$$|f'(x_0) - \tilde{D}_h f(x_0)| \leq \left| f'(x_0) - \frac{f(x_1) - f(x_0)}{h} \right| + \left| \frac{\epsilon_1 - \epsilon_0}{h} \right|$$

or

$$|f'(x_0) - \tilde{D}_h f(x_0)| \leq \left| \frac{h}{2} f''(\eta(x_0)) \right| + \frac{2\delta}{h}.$$

The errors ϵ_0, ϵ_1 are generally random on some interval $[-\delta, \delta]$.

It is the second term on the right side of this error bound which leads to the growth of error as $h \to 0$. If

$$|f(x_i) - \tilde{f}_i| \leq \frac{1}{2} \times 10^{-t} = \delta$$

and t is the required decimal digits of accuracy, then the maximum rounding error in the two-point formula is $\dfrac{10^{-t}}{h}$. While the truncation error $\left|\dfrac{h}{2} f''(\eta(x_0))\right|$ decreases with h, the rounding error increases. The total error, $E(h)$, therefore, has a minimum with respect to h. If

$$E(h) = E_{trunc} + E_{round} = \frac{h}{2} M + \frac{10^{-t}}{h},$$

where $M = \max\limits_{x_0 \leq x \leq x_1} |f''(\eta(x_0))|$, then

$$\frac{dE}{dh} = \frac{M}{2} - \frac{10^{-t}}{h^2}.$$

A minimum of $E(h)$ satisfies the equation $\dfrac{dE}{dh} = 0$; that is,

$$\frac{dE}{dh} = \frac{M}{2} - \frac{10^{-t}}{h^2} = 0$$

Solving for h, we obtain

$$h = h_{opt} = \sqrt{\frac{2}{M} \times 10^{-t}},$$

which gives the optimal value for h. Thus, the minimum error is

$$E(h_{opt}) = \frac{M}{2}\sqrt{\frac{2}{M} \times 10^{-t}} + \frac{10^{-t}}{\sqrt{\frac{2}{M} \times 10^{-t}}}$$

and simplifying it gives

$$E(h_{opt}) = \sqrt{2M \times 10^{-t}}.$$

Example 5.3 *Consider* $f(x) = x^2 \cos x$ *and* $x_0 = 1$. *To show the effect of round-off error, the values* $\tilde{f}_i$ *are obtained by rounding* $f(x_i)$ *to seven significant digits. Compute the total error for* $h = 0.1$ *and find the optimum* h.

Solution. *Given*

$$|\epsilon_i| \leq \frac{1}{2} \times 10^{-7} = \delta, \quad \text{and} \quad h = 0.1,$$

to calculate the total error, we use

$$E(h) = \frac{h}{2}M + \frac{10^{-t}}{h},$$

where

$$M = \max_{1 \leq x \leq 1.1} |(2 - x^2)\cos x - 4x \sin x| = 3.5630.$$

Then

$$E(h) = \frac{0.1}{2}(3.5630) + \frac{10^{-7}}{0.1}$$

or

$$E(h) = 0.17815 + 0.000001 = 0.178151.$$

Now to find the optimum h, *we use*

$$h = h_{opt} = \sqrt{\frac{2}{M} \times 10^{-t}}$$

and it gives

$$h = h_{opt} = \sqrt{\frac{2}{3.5630} \times 10^{-7}} = 0.00024,$$

which is the smallest value of h, below which the total error will begin to increase.

Note that for

$$\begin{array}{ll} h = 0.00024, & E(h) = 0.000844 \\ h = 0.00015, & E(h) = 0.000934 \\ h = 0.00001, & E(h) = 0.010018. \end{array}$$

A similar effect is present for all numerical differentiation formulas.

Numerical differentiation formulas are often judged by the power of h in the error term. Since h is always small, the higher power of h involved will give a better approximation. In this assessment, formula (5.9) acts poorly, as the error term involved is h of power one. Superior formulas can be obtained by deriving some useful three-point formulas together with error terms, which involve h^2 using formula (5.4) by taking $n = 2$.

Three-point Formula

Consider the quadratic Lagrange interpolating polynomial $p_2(x)$ to the three distinct equally spaced points x_0, x_1, and x_2, with $x_1 = x_0 + h$ and $x_2 = x_0 + 2h$. For smaller value h, we have

$$\begin{aligned} f(x) \approx p_2(x) &= \frac{(x - x_1)(x - x_2)}{(x_0 - x_1)(x_0 - x_2)} f(x_0) + \frac{(x - x_0)(x - x_2)}{(x_1 - x_0)(x_1 - x_2)} f(x_1) \\ &+ \frac{(x - x_0)(x - x_1)}{(x_2 - x_0)(x_2 - x_1)} f(x_2). \end{aligned}$$

Now take the derivative of the above expression with respect to x and then take $x = x_k$. For $k = 0, 1, 2$, we have

$$\begin{aligned} f'(x_k) &\approx \frac{(2x_k - x_1 - x_2)}{(x_0 - x_1)(x_0 - x_2)} f(x_0) + \frac{(2x_k - x_0 - x_2)}{(x_1 - x_0)(x_1 - x_2)} f(x_1) \\ &+ \frac{(2x_k - x_0 - x_1)}{(x_2 - x_0)(x_2 - x_1)} f(x_2). \end{aligned} \tag{5.11}$$

Three different numerical differentiation formulas can be obtained from (5.11) by putting $x_k = x_0$, or $x_k = x_1$, or $x_k = x_2$, which are used to find the approximation of the first derivative of a function defined by formula (5.1) at the given point. Firstly, we take $x_k = x_1$, then formula (5.11) becomes

$$\begin{aligned} f'(x_1) &\approx \frac{(2x_1 - x_1 - x_2)}{(x_0 - x_1)(x_0 - x_2)} f(x_0) + \frac{(2x_1 - x_0 - x_2)}{(x_1 - x_0)(x_1 - x_2)} f(x_1) \\ &+ \frac{(2x_1 - x_0 - x_1)}{(x_2 - x_0)(x_2 - x_1)} f(x_2). \end{aligned}$$

After simplifying, we get

$$f'(x_1) \approx \frac{-h}{2h^2} f(x_0) + 0 + \frac{h}{2h^2} f(x_2),$$

which is equal to

$$f'(x_1) \approx \frac{f(x_1 + h) - f(x_1 - h)}{2h} = D_h f(x_1). \tag{5.12}$$

It is called the *three-point central-difference formula* for finding the approximation of the first derivative of a function at the given point x_1.

Note that the formulation of formula (5.12) uses data points that are centered about the point of interest, x_1, even though it does not appear in the right side of (5.12).

The geometric interpretation of the central-difference formula is shown in Figure 5.3.

Error Formula

Formula (5.12) is not very useful, therefore, let us attempt to find the error involved in formula (5.12) for numerical differentiation. Consider the error term for the quadratic Lagrange polynomial, which can be written as

$$f(x) - p_2(x) = \frac{f'''(\eta(x))}{3!} \prod_{i=0}^{2} (x - x_i)$$

for some unknown point $\eta(x) \in (x_0, x_2)$. By taking the derivative of the above equation with respect to x and then taking $x = x_1$, we have

$$\begin{aligned} f'(x_1) - p_2'(x_1) &= \left(\frac{d}{dx} f'''(\eta(x)) \Big|_{x=x_1} \right) \frac{(x - x_0)(x - x_1)(x - x_2)}{6} \\ &+ \frac{f'''(\eta(x_1))}{6} \left((x - x_1)(x - x_2) + (x - x_0)(x - x_2) + (x - x_0)(x - x_1) \Big|_{x=x_1} \right). \end{aligned}$$

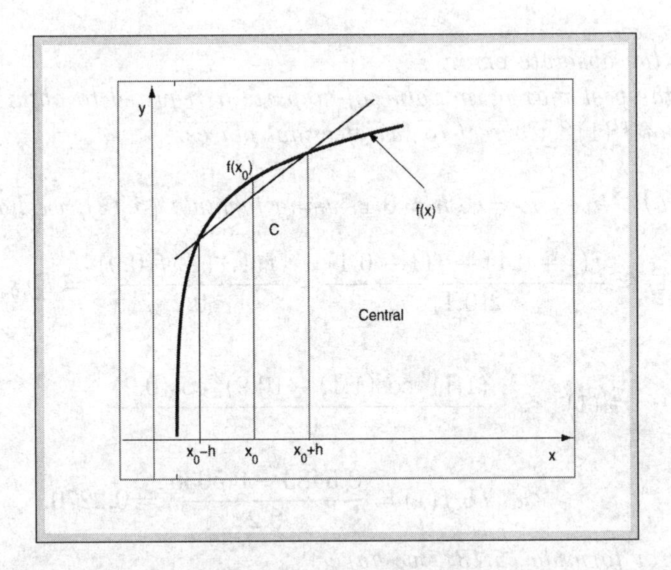

Figure 5.3: Central-difference approximations.

Since $\dfrac{d}{dx} f'''(\eta(x)) = 0$ only if $x = x_1$, the error formula of central-difference formula (5.12) can be written as

$$E_C(f, h) = f'(x_1) - D_h f(x_1) = -\frac{h^2}{6} f'''(\eta(x_1)), \tag{5.13}$$

where $\eta(x_1) \in (x_1 - h, x_1 + h)$. Hence, formula (5.12) can be written as

$$f'(x_1) = \frac{f(x_1 + h) - f(x_1 - h)}{2h} - \frac{h^2}{6} f'''(\eta(x_1)), \tag{5.14}$$

where $\eta(x_1) \in (x_1 - h, x_1 + h)$. Formula (5.14) is more useful than formula (5.12) because now on a large class of function, an error term is available along with the basic numerical formula.
Note that for a quadratic function, $f(x) = ax^2 + bx + c$, approximate formula (5.12) is exact; that is, it yields the correct value of the first derivative of a function $f(x)$ for any nonzero value of h.

Example 5.4 *Let* $f(x) = x^2 \cos x$. *Then*
(a) compute the approximate value of $f'(x)$ *at* $x = 1$, *taking* $h = 0.1$ *using (5.12).*
(b) compute the error bound for your approximation using (5.13).

(c) compute the absolute error.
(d) what is the best maximum value of stepsize h required to obtain the approximate value of $f'(1)$? Correct to two decimal places.

Solution. *(a) Given $x_1 = 1, h = 0.1$, using formula (5.12), we have*

$$f'(1) \approx \frac{f(1+0.1) - f(1-0.1)}{2(0.1)} = \frac{f(1.1) - f(0.9)}{0.2} = D_h f(1).$$

Then

$$\begin{aligned} f'(1) &\approx \frac{(1.1)^2 \cos(1.1) - (0.9)^2 \cos(0.9)}{0.2} \\ &\approx D_h f(1) = \frac{0.5489 - 0.5035}{0.2} = 0.2270. \end{aligned}$$

(b) Using error formula (5.13), we have

$$E_C(f,h) = -\frac{(0.1)^2}{6} f'''(\eta(x_1)), \quad \text{for} \quad \eta(x_1) \in (0.9, 1.1)$$

or

$$|E_C(f,h)| = \left| -\frac{(0.1)^2}{6} \right| |f'''(\eta(x_1))|, \quad \text{for} \quad \eta(x_1) \in (0.9, 1.1)$$

since

$$f'''(\eta(x_1)) = -6\eta(x_1)\cos\eta(x_1) - (6 - \eta(x_1)^2)\sin\eta(x_1).$$

This formula cannot be computed exactly because $\eta(x_1)$ is not known. But one can bound the error by computing the largest possible value for $|f'''(\eta(x_1))|$. So bound $|f'''|$ on $[0.9, 1.1]$ is

$$M = \max_{0.9 \leq x \leq 1.1} |-6x\cos x - (6 - x^2)\sin x| = 7.4222$$

at $x = 0.9$. Thus, for $|f'''(\eta(x_1))| \leq M$ and $h = 0.1$, which gives

$$|E_C(f,h)| \leq \frac{0.01}{6} M = \frac{0.01}{6}(7.4222) = 0.0124,$$

which is the possible maximum error in our approximation.

(c) Since the exact value of the derivative $f'(1)$ is 0.2391, the absolute error $|E|$ can be computed as follows

$$|E| = |f'(1) - D_h f(1)| = |0.2391 - 0.2270| = 0.0121.$$

(d) Since the given accuracy required is 10^{-2},

$$|E_C(f,h)| = \left| -\frac{h^2}{6} f'''(\eta(x_1)) \right| \leq 10^{-2}$$

for $\eta(x_1) \in (0.9, 1.1)$. *Then*

$$\frac{h^2}{6} M \leq 10^{-2}.$$

Solving for h and taking $M = 0.0121$, *we obtain*

$$h^2 \leq \frac{6}{742.22} = 0.01.$$

So the best maximum value of h is 0.1.

To get above results using MATLAB commands, we do the following:

```
>> x0 = 1.0;
>> h = 0.1;
>> df = (x0 + h).^2. * cos(x0 + h) - (x0 - h).^2
   . * cos(x0 - h)./(2. * h);
```

Note that formula (5.14) may also be derived from Taylor's theorem. The second degree Taylor's expansion $f(x)$ about x_1, for $f(x_1+h)$ and $f(x_1-h)$ gives

$$f(x_1+h) = f(x_1) + hf'(x_1) + \frac{h^2}{2!} f''(x_1) + \frac{h^3}{3!} f'''(\eta_1(x))$$

and

$$f(x_1-h) = f(x_1) - hf'(x_1) + \frac{h^2}{2!} f''(x_1) - \frac{h^3}{3!} f'''(\eta_2(x)).$$

Subtracting above two equations, the result is

$$f(x_1+h) - f(x_1-h) = 2hf'(x_1) + \frac{h^3}{3!} \left[f'''(\eta_1(x)) + f'''(\eta_2(x)) \right].$$

Since $f'''(x)$ is continuous, the intermediate value theorem can be used to find a value of $\eta(x)$ so that

$$\frac{f'''(\eta_1(x)) + f'''(\eta_2(x))}{2} = f'''(\eta(x)),$$

which is required formula (5.14).

Similarly, the two other three-point formulas can be obtained by taking $x_k = x_0$ and $x_k = x_2$ in formula (5.11). Firstly, by taking $x_k = x_0$ in formula (5.11) and then after simplifying, we have

$$f'(x_0) \approx \frac{-3f(x_0) + 4f(x_0+h) - f(x_0+2h)}{2h} = D_h f(x_0), \tag{5.15}$$

which is called the *three-point forward-difference formula*, which is used to approximate formula (5.1) at the given point $x = x_0$. The error term of this approximation formula can be obtained in a similar way as for the central-difference formula and it is

$$E_F(f,h) = \frac{h^2}{3} f'''(\eta(x_0)), \tag{5.16}$$

where $\eta(x_0) \in (x_0, x_0+2h)$. Similarly, taking $x_k = x_2$ in formula (5.11), after simplifying, we obtain

$$f'(x_2) \approx \frac{f(x_2-2h) - 4f(x_2-h) + 3f(x_2)}{2h} = D_h f(x_2), \tag{5.17}$$

which is called the *three-point backward-difference formula*, which is use to approximate formula (5.1) at the given point $x = x_2$. It has the error term of the form

$$E_B(f,h) = \frac{h^2}{3} f'''(\eta(x_2)), \tag{5.18}$$

where $\eta(x_2) \in (x_2 - 2h, x_2)$.

Note that backward-difference formula (5.17) can be obtained from the forward-difference formula by replacing h with $-h$. Also, note that the error in (5.12) is approximately half the error in (5.15) and (5.17). This is reasonable since in using central-difference formula (5.12) data is being examined on both sides of point x_1, and for others in (5.15) and (5.17) only on one side. Note that in using the central-difference formula, a function $f(x)$ needs to be evaluated at only two points, whereas in using the other two formulas, we need the values of a function at three points. The approximations in using formulas (5.15) and (5.17) are useful near the ends of the required interval, since the information about a function outside the interval may not be available. Thus, central-difference formula (5.12) is superior to both forward-difference formula (5.15) and backward-difference formula (5.17). The central-difference represents the average of the forward-difference and the backward-difference.

Example 5.5 *Use three-point formulas (5.12), (5.15), and (5.17) to approximate the first derivative of the function $f(x) = e^x$ at $x = 2$, take $h = 0.1$. Also,* *compute the error bound for each approximation.*

Solution. *Given* $f(x) = e^x$ *and* $h = 0.1$, *then*

Central-difference formula:

$$f'(2) \approx \frac{(f(2.1) - f(1.9))}{2h} = \frac{(e^{2.1} - e^{1.9})}{0.2} = 7.4014.$$

Forward-difference formula:

$$f'(2) \approx \frac{-3f(2) + 4f(2.1) - f(2.2)}{2h}$$

$$\approx \frac{-3e^2 + 4e^{2.1} - e^{2.2}}{0.2} = 7.3625.$$

Backward-difference formula:

$$f'(2) \approx \frac{f(1.8) - 4f(1.9) + 3f(2)}{2h}$$

$$\approx \frac{e^{1.8} - 4e^{1.9} + 3e^2}{0.2} = 7.3662.$$

Because the exact solution of the first derivative of the given function at $x = 2$ *is* 7.3891, *so the corresponding actual errors are,* $-0.0123, 0.0266$, *and* 0.0229, *respectively. This shows that the approximate solution we found using the central-difference formula is closer to the exact solution compared with the other two difference formulas.*
The error bounds for the approximations found (5.12), (5.15), and (5.17) are as follows:

Central-difference formula

$$E_C(f, h) = -\frac{h^2}{6} f'''(\eta(x_1))$$

or can be written as

$$|E_C(f, h)| \leq \frac{h^2}{6} |f'''(\eta(x_1))|.$$

Taking $|f'''(\eta(x_1))| \leq M = \max_{1.9 \leq x \leq 2.1} |e^x| = e^{2.1}$ *and* $h = 0.1$, *we obtain*

$$|E_C(f, h)| \leq \frac{(0.1)^2}{6} e^{2.1} = 0.0136.$$

Forward-difference formula

$$E_F(f,h) = \frac{h^2}{3} f'''(\eta(x_0))$$

or can be written as

$$|E_F(f,h)| \leq \frac{h^2}{3} |f'''(\eta(x_0))|.$$

Taking $|f'''(\eta(x_0))| \leq M = \max_{2 \leq x \leq 2.2} |e^x| = e^{2.2}$ *and* $h = 0.1$, *we obtain*

$$|E_F(f,h)| \leq \frac{(0.1)^2}{3} e^{2.2} = 0.0301.$$

Backward difference formula

$$E_B(f,h) = \frac{h^2}{3} f'''(\eta(x_2))$$

or can be written as

$$|E_B(f,h)| \leq \frac{h^2}{3} |f'''(\eta(x_2))|.$$

Taking $|f'''(\eta(x_2))| \leq M = \max_{1.8 \leq x \leq 2} |e^x| = e^2$ *and* $h = 0.1$, *we obtain*

$$|E_B(f,h)| \leq \frac{(0.1)^2}{3} e^2 = 0.0246.$$

Five-point Formula

Similarly, the other superior numerical differentiation formula to approximate (5.1) is called the *five-point formula*, which involves evaluating a function at two or more points compared to formula (5.12) whose error term involves h^4. This central-difference formula can be obtained from formula (5.4) by taking $n = 4$ and $x_k = x_1$, which gives the approximate formula

$$f'(x_1) \approx \frac{f(x_1 - 2h) - 8f(x_1 - h) + 8f(x_1 + h) - f(x_1 + 2h)}{12h} = D_h f(x_1). \tag{5.19}$$

The error term of formula (5.19) can be written as follows

$$E_C(f,h) = \frac{h^4}{30} f^{(5)}(\eta(x_1)), \tag{5.20}$$

where unknown point $\eta(x_1) \in (x_1 - 2h, x_1 + 2h)$.

Five-point formula (5.19) can also be obtained by expanding the function $f(x)$ in Taylor's theorem of order four about x_1, then by evaluating a $f(x)$ at $x_1 + 2h$ and $x_1 - 2h$.

Example 5.6 *Use five-point formula (5.19) to approximate the first derivative of the function $f(x) = e^x$ at $x = 2$, take $h = 0.1$. Also, compute the error bound for the approximation.*

Solution. *Given $f(x) = e^x$, $x_1 = 2$, and $h = 0.1$, then central-difference formula (5.19) can be written as*

$$f'(2) \approx \frac{f(2-2(0.1)) - 8f(2-0.1) + 8f(2+0.1) - f(2+2(0.1))}{12(0.1)}$$

and it gives

$$f'(2) \approx D_h f(2) = \frac{e^{1.8} - 8e^{1.9} + 8e^{2.1} - e^{2.2}}{1.2} = 7.3890,$$

which gives a more accurate approximation than the three-point formula. The actual error in this formula is approximately 0.000025, *which shows that it is superior than the other differentiation formulas we discussed.*

Now to compute the error term for our approximation, we use formula (5.20), which is

$$E_C(f,h) = \frac{h^4}{30} f^{(5)}(\eta(x_1))$$

or it can be written as

$$|E_C(f,h)| = \left|\frac{h^4}{30}\right| |f^{(5)}(\eta(x_1))|.$$

Since $f^{(5)}(\eta(x_1)) = e^{\eta(x_1)}$ and $\eta(x_1)$ is any unknown point on the interval $(1.8, 2.2)$, *we find bound $|f^{(5)}|$ on* $[1.8, 2.2]$, *which is*

$$M = \max_{1.8 \le 2.2} |f^{(5)}(x)| = e^{2.2} = 9.0250.$$

Since $|f^{(5)}(\eta(x_1))| \le M$,

$$|E_C(f,h)| \le \frac{(0.1)^4}{30} M = \frac{(0.1)^4}{30}(9.0250) = 0.00003,$$

which is the possible maximum error in our approximation using central-difference formula (5.19).

5.3.2 Second Derivatives Formulas

It is also possible to estimate second and higher order derivatives numerically. Formulas for higher derivatives can be found by differentiating the interpolating polynomial repeatedly or using Taylor's theorem. Since the *two-point*, *three-point*, and *five-point* formulas for the approximation of the first derivative were derived by differentiating the Lagrange interpolation polynomials for $f(x)$, the derivation of the higher-order can be tedious. Therefore, we shall use here Taylor's theorem for finding the *three-point* and the *five-point* central-difference formulas for finding approximation of the second derivative $f''(x)$ of a function at given point $x = x_1$. The process used to obtain numerical formulas for first and second derivatives can be readily extended to third- and higher-order derivatives.

Three-point Formula

To find the three-point central-difference formula for the approximation of the second derivative of a function at a given point, we use the third-order Taylor's theorem by expanding a function $f(x)$ about a point x_1 and evaluate at $x_1 + h$ and $x_1 - h$. Then

$$f(x_1 + h) = f(x_1) + hf'(x_1) + \frac{1}{2}h^2 f''(x_1) + \frac{1}{6}h^3 f'''(x_1) + \frac{1}{24}h^4 f^{(4)}(\eta_1(x))$$

and

$$f(x_1 - h) = f(x_1) - hf'(x_1) + \frac{1}{2}h^2 f''(x_1) - \frac{1}{6}h^3 f'''(x_1) + \frac{1}{24}h^4 f^{(4)}(\eta_2(x)),$$

where $(x_1 - h) < \eta_2(x) < x_1 < \eta_1(x) < (x_1 + h)$.
By adding these equations and simplifying, we have

$$f(x_1 + h) + f(x_1 - h) = 2f(x_1) + h^2 f''(x_1) + \frac{(f^{(4)}(\eta_1(x)) + f^{(4)}(\eta_2(x)))}{24} h^4.$$

Solving this equation for $f''(x_1)$, we obtain

$$f''(x_1) = \frac{f(x_1 - h) - 2f(x_1) + f(x_1 + h)}{h^2} - \frac{h^4}{24}\left[f^{(4)}(\eta_1(x)) + f^{(4)}(\eta_2(x))\right].$$

If $f^{(4)}$ is continuous on $[x_1 - h, x_1 + h]$, then using the Intermediate Value Theorem, the above equation can be written as

$$f''(x_1) = \frac{f(x_1 - h) - 2f(x_1) + f(x_1 + h)}{h^2} - \frac{h^4}{12} f^{(4)}(\eta(x_1)).$$

Then the following formula

$$f''(x_1) \approx \frac{f(x_1 - h) - 2f(x_1) + f(x_1 + h)}{h^2} = D_h^2 f(x_1) \tag{5.21}$$

is called the *three-point central-difference formula* for the approximation of the second derivative of a function $f(x)$ at the given point $x = x_1$.

Example 5.7 *Let* $f(x) = x \ln x + x$ *and* $x = 0.9, 1.3, 2.1, 2.5, 3.2$. *Then find the approximate value of* $\frac{1}{x}$ *at* $x = 1.9$. *Also, compute the absolute error.*

Solution. *Given* $f(x) = x \ln x + x$, *then one can easily find the second derivative of the function as*

$$f'(x) = \ln x + 2 \quad \textit{and} \quad f''(x) = \frac{1}{x}.$$

To find the approximation of $f''(x) = \frac{1}{x}$ *at given point* $x_1 = 1.9$, *we use three-point formula (5.21)*

$$f''(x_1) \approx \frac{f(x_1 + h) - 2f(x_1) + f(x_1 - h)}{h^2} = D_h^2 f(x_1).$$

Taking the three points $1.3, 1.9$, *and* 2.5 *(equally spaced), giving* $h = 0.6$, *we have*

$$\begin{aligned} f''(1.9) &\approx \frac{f(2.5) - 2f(1.9) + f(1.3)}{0.36} \\ &\approx \frac{((2.5 \ln 2.5 + 2.5) - 2(1.9 \ln 1.9 + 1.9) + (1.3 \ln 1.3 + 1.3))}{0.36} \\ &\approx \frac{4.7907 - 6.2391 + 1.6411}{0.36} = 0.5353 = D_h^2 f(1.9). \end{aligned}$$

Since the exact value of $f''(1.9)$ *is* $\frac{1}{1.9} = 0.5263$, *the absolute error* $|E|$ *can be computed as follows:*

$$|E| = |f''(1.9) - D_h^2 f(1.9)| = |0.5263 - 0.5353| = 0.009.$$

Note that the *error term* of three-point central-difference formula (5.21) for the approximation of the second derivative of function $f(x)$ at given point $x = x_1$ is of the form

$$E_C(f, h) = -\frac{h^2}{12} f^{(4)}(\eta(x_1)) \tag{5.22}$$

for some unknown point $\eta(x_1) \in (x_1 - h, x_1 + h)$.

Note that for a cubic function, $f(x) = ax^3 + bx^2 + cx + d$, difference formula (5.21) is exact; that is, it yields the correct value of $f''(x)$ for any nonzero value of stepsize h.

Example 5.8 *Let $f(x) = x^2 \cos x$. Then*
(a) compute the approximate value of $f''(x)$ at $x = 1$, taking $h = 0.1$ using (5.21).
(b) compute the error bound for your approximation using (5.22).
(c) compute the absolute error.
(d) what is the best maximum value of stepsize h required to obtain the approximate value of $f''(1)$ within the accuracy 10^{-2}?

Solution. *(a) Given $x_1 = 1, h = 0.1$, then formula (5.21) becomes*

$$f''(1) \approx \frac{f(1+0.1) - 2f(1) + f(1-0.1)}{(0.1)^2} = D_h^2 f(1)$$

or

$$\begin{aligned} f''(1) &\approx \frac{f(1.1) - 2f(1) + f(0.9)}{0.01} \\ &\approx \frac{(1.1)^2\cos(1.1) - 2\cos(1) + (0.9)^2\cos(0.9)}{0.01} \\ &\approx \frac{0.5489 - 1.0806 + 0.5035}{0.01} = -2.8200 = D_h^2 f(1). \end{aligned}$$

(b) To compute the error bound for our approximation in part (a), we use formula (5.22) and have

$$E_C(f,h) = -\frac{h^2}{12} f^{(4)}(\eta(x_1)), \quad \text{for} \quad \eta(x_1) \in (0.9, 1.1)$$

or

$$|E_C(f,h)| = \left|-\frac{h^2}{12}\right| |f^{(4)}(\eta(x_1))|, \quad \text{for} \quad \eta(x_1) \in (0.9, 1.1).$$

The fourth derivative of the given function at $\eta(x_1)$ is

$$f^{(4)}(\eta(x_1)) = (-12 + \eta(x_1)^2)\cos\eta(x_1) + 8\eta(x_1)\sin\eta(x_1)$$

and it cannot be computed exactly because $\eta(x_1)$ *is not known. But one can bound the error by computing the largest possible value for* $|f^{(4)}(\eta(x_1))|$. *So bound* $|f^{(4)}|$ *on the interval* $(0.9, 1.1)$ *is*

$$M = \max_{0.9 \leq x \leq 1.1} |(-12 + x^2)\cos x + 8x \sin x| = 2.9483$$

at $x = 1.1$. *Thus, for* $|f^{(4)}(\eta(x))| \leq M$, *we have*

$$|E_C(f, h)| \leq \frac{h^2}{12} M.$$

Taking $M = 2.9483$ *and* $h = 0.1$, *we obtain*

$$|E_C(f, h)| \leq \frac{0.01}{12}(2.9483) = 0.0025,$$

which is the possible maximum error in our approximation.

(c) Since the exact value of $f''(1)$ *is*

$$f''(1) = (2 - 1^2)\cos 1 - 4(1)\sin 1 = -2.8256$$

the absolute error $|E|$ *can be computed as follows:*

$$|E| = |f''(1) - D_h^2 f(1)| = |-2.8256 + 2.8200| = 0.0056.$$

(d) Since the given accuracy required is 10^{-2},

$$|E_C(f, h)| = \left| -\frac{h^2}{12} f^{(4)}(\eta(x_1)) \right| \leq 10^{-2}$$

for $\eta(x_1) \in (0.9, 1.1)$. *Then for* $|f^{(4)}(\eta(x_1))| \leq M$, *we have*

$$\frac{h^2}{12} M \leq 10^{-2}.$$

Solving for h^2, *we obtain*

$$h^2 \leq \frac{(12 \times 10^{-2})}{M} = \frac{(12 \times 10^{-2})}{2.9483} = 0.0407$$

and it gives the value of h *as*

$$h \leq 0.2018.$$

Thus, the best maximum value of h *is* 0.2.

Using the following MATLAB commands, we can easily achieve the above results:

```
>> x0 = 1.0;
>> h = 0.1;
>> ddf = ((x0 + h).^ 2. * cos(x0 + h) - 2. * x0. * cos(x0)+
      (x0 - h).^ 2. * cos(x0 - h))./(h.^ 2)
```

The central-difference formula is probably the most frequently used approximation for derivatives. Many real problems are modeled by second-order differential equations, involving either ordinary or partial derivatives. These equations cannot be solved analytically. To solve the equations numerically requires the replacement of second-order derivatives by difference formula (5.21).

Example 5.9 *The function $f(x)$ satisfies the equation $f''(x) = x^2 f(x)$ and the conditions $f(0) = 1,\ f(0.2) = 3$. Use the central-difference formula for $f''(x)$ and the stepsize $h = 0.1$ to estimate the value of $f(0.1)$.*

Solution. *Given $x_1 = 0.1$ and $h = 0.1$, using the central-difference formula for the second derivative of a function*

$$f''(x_1) \approx \frac{f(x_1 - h) - 2f(x_1) + f(x_1 + h)}{h^2}$$

we obtain

$$f''(0.1) = (0.1)^2 f(0.1) \approx \frac{f(0) - 2f(0.1) + f(0.2)}{0.01},$$

which is equal to

$$\begin{aligned}(0.01)f(0.1) &\approx \frac{1 - 2f(0.1) + 3}{0.01} \\ (0.0001)f(0.1) &\approx (4 - 2f(0.1)).\end{aligned}$$

Solving for $f(0.1)$, we obtain

$$f(0.1) \approx \frac{4}{2 + 0.0001} = 1.9999.$$

Five-point Formula

Similarly, there is another formula called the *five-point formula* for the approximation of the second derivative of a function at given point x_1 that involves evaluating the function at two more points, and whose error term involves h^4. This approximation formula can be written as

$$f'(x_1) \approx \frac{-f(x_1 - 2h) + 16f(x_1 - h) - 30f(x_1) + 16f(x_1 + h) - f(x_1 + 2h)}{12h^2} \tag{5.23}$$

and it has the *error term* of the form

$$E_C(f, h) = f'(x_1) - D_h^2 f(x_1) = \frac{h^4}{90} f^{(6)}(\eta(x_1)), \tag{5.24}$$

where unknown point $\eta(x_1) \in (x_1 - 2h, x_1 + 2h)$.

Note that the tendency for round-off errors to dominate the computation of higher-order derivatives is increased, since the reciprocal power of h in the round-off error estimate rises with the order of the derivative.

Example 5.10 *Use three-point formula (5.21) and five-point formula (5.23) to approximate the second derivative of the function $f(x) = e^x$ at $x = 2$, take $h = 0.1$. Also, compute the actual error and error bound for both approximations.*

Solution. *Given $f(x) = e^x$, $x_1 = 2$ and $h = 0.1$, using three-point formula (5.21), we have*

$$\begin{aligned} f''(x_1) &\approx \frac{f(x_1 - h) - 2f(x_1) + f(x_1 + h)}{h^2} \\ f''(2) &\approx \frac{f(1.9) - 2f(2) + f(2.1)}{0.01} \\ &\approx \frac{e^{1.9} - 2e^2 + e^{2.1}}{0.01} = 7.3952 = D_h^2 f(2). \end{aligned}$$

Similarly, using five-point difference formula (5.23), we have

$$\begin{aligned} f''(2) &\approx \frac{-f(1.8) + 16f(1.9) - 30f(2) + 16f(2.1) - f(2.2)}{12(0.01)} = D_h^2 f(2) \\ &\approx \frac{e^{1.8} + 16e^{1.9} - 30e^2 + 16e^{2.1} - e^{2.2}}{0.12} = 7.3891. \end{aligned}$$

Since the exact solution of the second derivative of function at $x = 2$ *is* 7.3891, *the corresponding actual errors for the three-point formula and the five-point formula are* −0.0061 *and* 0.0000, *respectively. This shows that the five-point formula is correct to four decimal places and clearly the best one.*

Now to compute the error bound for our approximation obtained using three-point formula (5.21), we use its error term formula (5.22), which is

$$E_C(f, h) = -\frac{h^2}{12} f^{(4)}(\eta(x_1)), \quad \textit{for} \quad \eta(x_1) \in (1.9, 2.1)$$

or

$$|E_C(f, h)| = \left| -\frac{h^2}{12} \right| |f^{(4)}(\eta(x_1))|, \quad \textit{for} \quad \eta(x_1) \in (1.9, 2.1).$$

The fourth derivative of the function is e^x. *So bound* $|f^{(4)}|$ *on* $[1.9, 2.1]$ *can be found as*

$$M = \max_{1.9 \le x \le 2.1} |e^x| = 8.1662$$

at $x = 2.1$. *Thus, for* $|f^{(4)}(\eta(x_1))| \le M$, *we have*

$$|E_C(f, h)| \le \frac{h^2}{12} M = 0.0068.$$

Taking $M = 8.1662$ *and* $h = 0.1$, *we get*

$$|E_C(f, h)| \le \frac{0.01}{12}(8.1662) = 0.0068,$$

which is the possible maximum error in our approximation using the tree-point formula. To compute the error bound for the approximation using five-point formula (5.23), we use

$$E_C(f, h) = \frac{h^4}{90} f^{(6)}(\eta(x_1)), \qquad \eta(x_1) \in (1.8, 2.2).$$

The sixth derivative of the function is e^x. *So bound* $|f^{(6)}|$ *on* $[1.8, 2.2]$ *can be obtained*

$$M = \max_{1.8 \le x \le 2.2} |e^x| = 9.0250$$

when $x = 2.2$. *Thus, for* $M = 9.0250$ *and* $h = 0.1$, *we get*

$$|E_C(f, h)| \le \frac{0.0001}{90}(9.0250) = 0.00001,$$

which is the possible maximum error in the approximation using the five-point formula and this error is much smaller than the error got in the three-point central-difference formula.

5.4 Formulas for Computing Derivatives

For convenience, we collected some useful central-difference, forward-difference, and backward-difference, formulas for computing different orders derivatives.

5.4.1 Central Difference Formulas

Central-difference formula (5.12) for first the derivative $f'(x_1)$ of a function required that a function be computed at points that lies on both sides of x_1. The Taylor series can be used to obtain central-difference formulas for higher derivatives. The most usable are those of order $O(h^2)$ and $O(h^4)$ and are given as follows:

$$
\begin{aligned}
f'(x_0) &= \frac{f_1 - f_{-1}}{2h} + O(h^2) \\
f'(x_0) &= \frac{-f_2 + 8f_1 - 8f_{-1} + f_{-2}}{12h} + O(h^4) \\
f''(x_0) &= \frac{f_1 - 2f_0 + f_{-1}}{h^2} + O(h^2) \\
f''(x_0) &= \frac{-f_2 + 16f_1 - 30f_0 + 16f_{-1} - f_{-2}}{12h^2} + O(h^4) \\
f'''(x_0) &= \frac{f_2 - 2f_1 + 2f_{-1} - f_{-2}}{2h^3} + O(h^2) \\
f'''(x_0) &= \frac{-f_3 + 8f_2 - 13f_1 + 13f_{-1} - 8f_{-2} + f_{-3}}{8h^3} + O(h^4) \\
f^{(4)}(x_0) &= \frac{f_2 - 4f_1 + 6f_0 - 4f_{-1} + f_{-2}}{h^4} + O(h^2) \\
f^{(4)}(x_0) &= \frac{-f_3 + 12f_2 - 39f_1 + 56f_0 - 39f_{-1} + 12f_{-2} - f_{-3}}{6h^4} + O(h^4).
\end{aligned}
$$

5.4.2 Forward- and Backward-difference Formulas

If a function cannot be evaluated at points that lie on both sides of x_0, then the central-difference formula cannot be used to approximate the derivatives of a function. When a function can be evaluated at equally spaced points that lie to the right (or left) of point x_0, then forward- (or backward-) difference formulas can be used. These formulas can be derived using the Taylor series, Lagrange interpolating polynomials, or Newton interpolating polynomials. Some of them are mostly usable to find derivatives of a function and are as follows:

$$
\begin{aligned}
f'(x_0) &= \frac{-3f_0 + 4f_1 - f_2}{2h} + O(h^2)\\
f'(x_0) &= \frac{3f_0 - 4f_{-1} + f_{-2}}{2h} + O(h^2)\\
f''(x_0) &= \frac{2f_0 - 5f_1 + 4f_2 - f_3}{h^2} + O(h^2)\\
f''(x_0) &= \frac{2f_0 - 5f_{-1} + 4f_{-2} - f_{-3}}{h^2} + O(h^2)\\
f'''(x_0) &= \frac{-5f_0 + 18f_1 - 24f_2 + 14f_3 - 3f_4}{2h^3} + O(h^2)\\
f'''(x_0) &= \frac{5f_0 - 18f_{-1} + 24f_{-2} - 14f_{-3} + 3f_{-4}}{2h^3} + O(h^2)\\
f^{(4)}(x_0) &= \frac{3f_0 - 14f_1 + 26f_2 - 24f_3 + 11f_4 - 2f_5}{h^4} + O(h^2)\\
f^{(4)}(x_0) &= \frac{3f_0 - 14f_{-1} + 26f_{-2} - 24f_{-3} + 11f_{-4} - 2f_{-5}}{h^4} + O(h^2).
\end{aligned}
$$

5.5 Numerical Integration

Numerical integration has a history extending back to the invention of calculus and before. It is used to integrate tabulated functions or to integrate functions whose integrals are either impossible or very difficult to obtain analytically. Even when analytical integration is easy, numerical integration may save time and effort if only the numerical value of the integral is desired. Consequently, numerical methods of integration represent a natural alternative whenever conventional methods fail to yield a solution.

Now for numerical integration, we wish to find an approximation to the *definite integral*

$$I(f) = \int_a^b f(x)dx \tag{5.25}$$

assuming that $f(x)$ is integrable. If $f(x) \geq 0$ on the given interval $[a, b]$, then geometrically integral (5.25) is equivalent to replacing the *area* under the graph of $f(x)$, the x-axis, and between the ordinates $x = a$ and $x = b$.

The fundamental theorem of calculus shows that integration is the inverse process to differentiation. If we can find a function $F(x)$, called the antiderivative of $f(x)$, that is, $F'(x) = f(x)$, then we can evaluate integral (5.25) using the relation

$$I(f) = \int_a^b f(x)dx = F(b) - F(a). \tag{5.26}$$

Sometimes considerable skill is required to obtain $F(x)$, perhaps by making a change of variable or integrating by parts. But in many cases $F(x)$ cannot be found by elementary methods. In such cases the computation of integral (5.25) by means of formula (5.26) may be either too difficult or practically impossible. Even if $F(x)$ can be found, it may still be more convenient to use a numerical method to estimate integral (5.25) if the evaluation of $F(x)$ required a great deal of computation. Moreover, in practical applications, function $f(x)$ is often given in tabular form and then the entire concept of antiderivative is meaningless. An obvious approach is to replace function $f(x)$ in integral (5.25) by an approximating polynomial $p(x)$; that is,

$$I(f) = \int_a^b f(x)dx \approx \int_a^b p(x)dx.$$

Numerical integration formulas are derived by integrating interpolation polynomials. Therefore, different interpolation formulas will lead to different numerical integration methods.

Definite integral (5.25) may be interpreted as the area under the curve of $y = f(x)$ from a to b as shown in Figure 5.4.

It should be noted that any area beneath the x-axis is counted as negative. Many numerical methods for integration are based on using this interpretation of the integral to derive approximations to it by dividing the interval $[a, b]$ into a number of smaller subintervals. By making simple approximations to the curve $y = f(x)$ in the small subinterval, its area may be obtained and on summing all the contributions we obtain an approximation to an integral on the interval $[a, b]$. Variations of this technique are derived by taking groups of subintervals and fitting different degree polynomials as approximations for each of these groups. The lead of accuracy obtained is dependent on the number of intervals used and the nature of the approximation function.

There are several methods available in the literature for numerical integration but the most commonly used methods may be classified into two groups:

(a) The Newton-Cotes formulas that employ functional values at equally spaced data points.

(b) The Gaussian quadrature formulas that employ unequally spaced data points determined by certain properties of orthogonal polynomials.

Firstly, we shall discuss the Newton-Cotes formulas, which have two different types called the *closed Newton-Cotes* formulas and the *open Newton-Cotes*

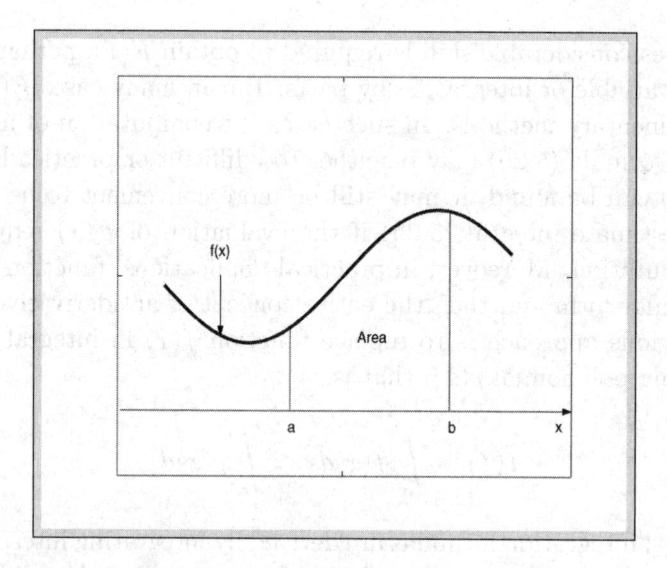

Figure 5.4: Definite integral for $f(x)$.

formulas. We shall discuss in some detail the two mostly usable formulas called the *Trapezoidal rule* and the *Simpson's rule*, which can be derived by integrating Lagrange interpolating polynomials of degree one and two respectively. The use of the closed Newton-Cotes and other integration formulas of order higher than the Simpson's rule is seldom necessary in most engineering applications and can be used for those cases where extremely high accuracy is required.

5.6 Newton-Cotes Formulas

The usual strategy in developing formulas for numerical integration is similar to that for numerical differentiation. We pass a polynomial through points of a function and then integrate this polynomial approximation to a function. This allows us to integrate a function known only as a table of values. Some common formulas based on polynomial interpolation are referred to as Newton-Cotes formulas.

An $(n+1)$-point Newton-Cotes formula for approximating definite integral (5.25) is obtained by replacing the integrand $f(x)$ by the *nth*-degree Lagrange polynomial that interpolates the values of $f(x)$ at equally spaced data points

$$a = x_0 < x_1 < \ldots < x_n = b.$$

Note that if end-points a and b of the given interval $[a, b]$ are in the set of interpolating points; then the Newton-Cotes formulas are called *closed*; otherwise, they are said to be *open*.

5.6.1 Closed Newton-Cotes Formulas

An $(n+1)$-point closed Newton-Cotes formula uses points $x_i = x_0 + ih$, for $i = 0, 1, 2, \ldots, n$, where $x_0 = a$, $x_n = b$ and $h = \dfrac{b-a}{n}$ has the form (see Figure 5.5)

$$\int_a^b f(x)dx = \int_{x_0}^{x_n} f(x)dx \approx \sum_{i=0}^{n} a_i f(x_i), \tag{5.27}$$

where

$$a_i = \int_{x_0}^{x_n} L_i(x)dx = \int_{x_0}^{x_n} \prod_{\substack{j=0 \\ j \neq i}}^{n} \frac{(x - x_j)}{(x_i - x_j)} dx. \tag{5.28}$$

The following theorem describes the error analysis associated with closed Newton-Cotes formulas.

Theorem 5.1 (Closed Newton-Cotes Formulas)

Suppose that $\sum_{i=0}^{n} a_i f(x_i)$ *denotes the* $(n+1)$*-point closed Newton-Cotes formula with* $x_0 = a, x_n = b$ *and* $h = (b-a)/n$. *There exists* $\eta(x) \in (a, b)$ *for which*

$$\int_a^b f(x)dx = \sum_{i=0}^{n} a_i f(x_i) + \frac{h^{n+3} f^{(n+2)}(\eta(x))}{(n+2)!} \int_0^n t^2(t-1)\cdots(t-n)dt \tag{5.29}$$

if n *is even and* $f \in C^{n+2}[a, b]$. *For* $f \in C^{n+1}[a, b]$, *and* n *is odd, then*

$$\int_a^b f(x)dx = \sum_{i=0}^{n} a_i f(x_i) + \frac{h^{n+2} f^{(n+1)}(\eta(x))}{(n+1)!} \int_0^n t(t-1)\cdots(t-n)dt. \tag{5.30}$$

Different numerical integration formulas can be obtained using formulas (5.29) and (5.30) to approximate definite integral (5.25). Using formula (5.30) for $n = 1$, we have a well-known numerical integration formula called the *Trapezoidal rule.* Similarly, using formula (5.29) for $n = 2$, we have one of the best integration rules called the *Simpson's rule.* We shall discuss the formulation of

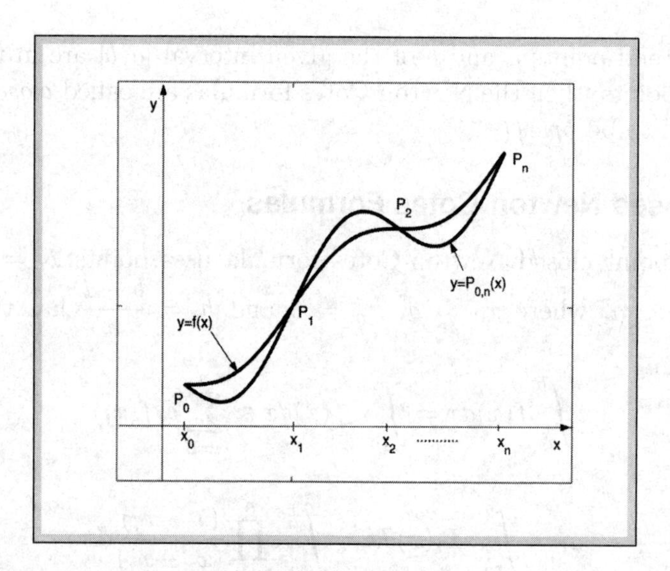

Figure 5.5: Closed Newton-Cotes approximation

both these rules and also discuss their error terms. Later, we shall also consider some other closed Newton-Cotes formulas.

Trapezoidal Rule

It is one of the oldest and best numerical methods for approximating the definite integral (5.25). It is based on approximating a function in each subinterval by a straight line.

To derive the Trapezoidal rule for one strip (one interval), let us consider the first degree Lagrange interpolating polynomial with equally spaced data points; that is, $x_0 = a, x_1 = b$ and $h = x_1 - x_0$, then

$$f(x) \approx p_1(x) = \left(\frac{x - x_1}{x_0 - x_1}\right) f(x_0) + \left(\frac{x - x_0}{x_1 - x_0}\right) f(x_1). \tag{5.31}$$

Taking the integral on both sides of (5.31) with respect to x between limits x_0 and x_1, we have

$$\int_{x_0}^{x_1} f(x)dx \approx \frac{f(x_0)}{x_0 - x_1} \int_{x_0}^{x_1} (x - x_1)dx + \frac{f(x_1)}{x_1 - x_0} \int_{x_0}^{x_1} (x - x_0)dx,$$

which implies that

$$\int_{x_0}^{x_1} f(x)dx \approx \frac{(x_1 - x_0)}{2}[f(x_0) + f(x_1)]$$

or

$$\int_a^b f(x)dx \approx T_1(f) = \frac{h}{2}[f(x_0) + f(x_1)]. \tag{5.32}$$

Then $T_1(f)$ is called the *simple Trapezoidal rule* or the Trapezoidal rule for one trapezoid or one strip and can be used for the approximation of definite integral (5.25). The reason for calling this formula the Trapezoidal rule is that when $f(x)$ is a function with positive values, integral (5.25) is approximated by the area in the trapezoid (see Figure 5.6).

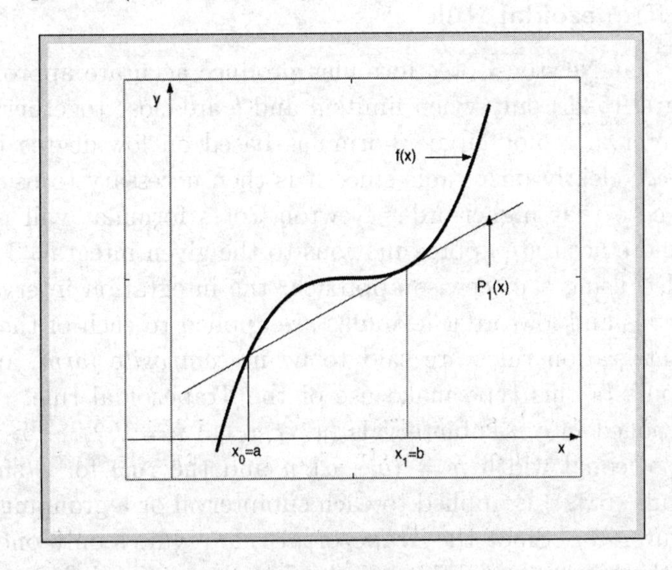

Figure 5.6: Simple Trapezoidal rule.

Example 5.11 *Approximate the following integral*

$$\int_1^2 \frac{1}{x+1}dx$$

using the Trapezoidal rule and compute the absolute error.

Solution. *Since* $f(x) = \dfrac{1}{x+1}$ *and* $h = 2 - 1 = 1$, *using the simple Trapezoidal rule (5.32), we have*

$$T_1(f) = \frac{1}{2}[f(1) + f(2)] = 0.4167.$$

The exact solution of the given integral is

$$I(f) = \ln 3 - \ln 2 = 0.4055.$$

So the absolute error is given as

$$|E_{T_1}(f)| = |I(f) - T_1(f)| = |0.4055 - 0.4167| = 0.0112.$$

Composite Trapezoidal Rule

It is evident that Newton-Cotes formulas produce accurate approximations to definite integral (5.25) only when limits a and b are close together; that is, the integration interval is not large. Formulas based on low-degree interpolating polynomials are clearly unsuitable since it is then necessary to use large values of h. Also, note that higher-order Newton-Cotes formulas will not necessarily produce more accurate approximations to the given integral. This difficulty can be avoided using a piecewise approach; the integration interval is divided into subintervals and low-order formulas are applied to each of these. The corresponding integration rules are said to be in *composite form*, and the most suitable formula of this type make use of the Trapezoidal rule. The interval $[a, b]$ is partitioned into n subintervals (x_{i-1}, x_i), $i = 1, 2, \ldots, n$ with $a = x_0$ and $b = x_n$ of equal width $h = (b - a)/n$ and the rule for a single interval (the simple rule (5.32)) is applied to each subinterval or a grouping of subintervals (see Figure 5.7). Since the Trapezoidal rule requires only one interval for application, there is no restriction on the integer n. We define the composite Trapezoidal rule in the form of the following theorem.

Theorem 5.2 (Composite Trapezoidal Rule)

Let $f \in C^2[a, b]$, n *may be odd or even,* $h = (b - a)/n$, *and* $x_i = a + ih$ *for each* $i = 0, 1, 2, \ldots, n$. *Then the* **composite Trapezoidal rule** *for* n *subintervals can be written as*

$$\int_{a=x_0}^{b=x_n} f(x)dx \approx T_n(f) = \frac{h}{2}\left[f(a) + 2\sum_{i=1}^{n-1} f(x_i) + f(b)\right]. \tag{5.33}$$

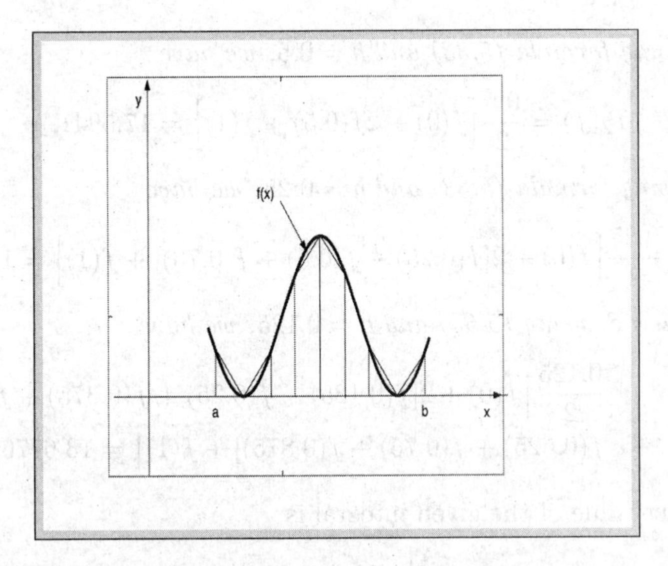

Figure 5.7: Composite Trapezoidal rule.

Proof. *Since for the composite form of the Trapezoidal rule the interval is divided into n equal subintervals of width h so that* $h = \dfrac{b-a}{n}$ *and (n+1) distinct points* $a = x_0 < x_1 < x_2 \ldots < x_n = b$, *then we have*

$$\int_a^b f(x)dx = \int_{x_0}^{x_1} f(x)dx + \int_{x_1}^{x_2} f(x)dx + \cdots + \int_{x_{n-1}}^{x_n} f(x)dx.$$

Applying Trapezoidal rule (5.32) for one strip to each of these integrals, we have

$$\int_a^b f(x)dx \approx \frac{h}{2}[f(x_0) + f(x_1)] + \frac{h}{2}[f(x_1) + f(x_2)] + \cdots + \frac{h}{2}[f(x_{n-1}) + f(x_n)].$$

Note that each of the interior points is counted twice and therefore has a coefficient of two whereas the endpoints are counted once and therefore have a coefficient of one.

Example 5.12 *Evaluate the integral* $\int_0^1 e^{4x}dx$ *using the Trapezoidal rule with* $n = 1, 2, 4, 8$. *Also compute the corresponding actual errors.*

Solution. *For n = 1, we use formula (5.32) for* $h = 1$, *as follows*

$$T_1(f) = \frac{1}{2}\Big[f(0) + f(1)\Big] = 27.7991.$$

For $n = 2$, using formula (5.33) and $h = 0.5$, we have

$$T_2(f) = \frac{0.5}{2}\Big[f(0) + 2f(0.5) + f(1)\Big] = 17.5941.$$

For $n = 4$, using formula (5.33) and $h = 0.25$, we have

$$T_4(f) = \frac{0.25}{2}\Big[f(0) + 2[f(0.25) + f(0.5) + f(0.75)] + f(1)\Big] = 14.4980.$$

Finally, for $n = 8$, using (5.33) and $h = 0.125$, we have

$$\begin{aligned} T_8(f) &= \frac{0.125}{2}\Big[f(0) + 2[f(0.125) + f(0.25) + f(0.375) + f(0.5) \\ &+ f(0.625) + f(0.75) + f(0.875)] + f(1)\Big] = 13.6776, \end{aligned}$$

Since the exact value of the given integral is

$$I(f) = \frac{1}{4}[e^4 - 1] = 13.4000.$$

the corresponding actual errors are $-14.3991, -4.1941, -1.0980$, and -0.2776, respectively, which decrease by a factor of about *four* at each stage.

Note that the Trapezoidal rule of integration involves no restriction relative to number of data points involved. This is not the case with the elaborate methods yet to be discussed. This is one reason it is one of the favored numerical integration methods used in mathematics and engineering.

Error Formulas for the Trapezoidal Rule

Now we discuss the error formula for the Trapezoidal rule. This formula will lead to a better understanding of the method, showing both weaknesses and strengths, and it will allow improvements of the method. We discuss the error for simple Trapezoidal rule (5.32) in the from of the following theorem and then we use it to define the error for composite Trapezoidal rule (5.33).

Theorem 5.3 (Error Formula for the Simple Trapezoidal Rule)

Let $f \in C^2[a, b]$ and $h = (b - a)$. The local error that simple Trapezoidal rule (5.32) makes in estimating definite integral (5.25) is

$$E_{T_1}(f) = -\frac{h^3}{12} f''(\eta(x)), \tag{5.34}$$

where $\eta(x) \in (a, b)$.

Proof. Consider two points $a = x_0 < x_1 = b$ with $h = x_1 - x_0$. From the linear Lagrange interpolation formula with the error terms, we have

$$f(x) = p_1(x) + \frac{f''(\eta(x)(x))}{2!} \prod_{i=0}^{1} (x - x_i) \tag{5.35}$$

and by integrating (5.35) with respect to x and between x_0 and x_1, we have

$$\int_{x_0}^{x_1} f(x)dx = \int_{x_0}^{x_1} p_1(x)dx + \frac{1}{2}\int_{x_0}^{x_1} f''(\eta(x)) \prod_{i=0}^{1} (x - x_i)dx. \tag{5.36}$$

The error term for the Trapezoidal rule of one strip can be obtained as follows:

$$E_{T_1}(f) = \frac{1}{2}\int_{x_0}^{x_1} f''(\eta(x))(x - x_0)(x - x_1)dx.$$

Note that $f''(\eta(x))$ is a continuous function of x and the term $(x - x_0)(x - x_1)$ is negative on (a, b), therefore, using the Mean Value Theorem for integrals, we have

$$E_{T_1}(f) = \frac{f''(\eta(x))}{2} \int_{x_0}^{x_1} (x - x_0)(x - x_1)dx.$$

Now to solve the integral on the right side of the above equation, we use the change of variable

$$x - x_0 = uh, \; x - x_1 = (u - 1)h, \quad \text{and} \quad dx = hdu$$

then we have

$$\begin{aligned} E_{T_1}(f) &= \frac{f''(\eta(x))}{2} \int_0^1 huh(u-1)hdu \\ &= \frac{f''(\eta(x))h^3}{2} \int_0^1 (u^2 - u)du. \end{aligned}$$

Thus,

$$E_{T_1}(f) = -\frac{h^3 f''(\eta(x)))}{12} \tag{5.37}$$

for some $\eta(x) \in (a, b)$. Then formula (5.36) can be written as

$$\int_a^b f(x)dx = \frac{h}{2}[f(x_0) + f(x_1)] - \frac{h^3}{12} f''(\eta(x)) \tag{5.38}$$

for $\eta(x) \in (a, b)$, which is the simple Trapezoidal rule with its error term.

Formula (5.37) indicates that the local error of the Trapezoidal rule is proportional to second derivative f''. So, if the Trapezoidal rule is used to integrate each of $f(x) = 1, x, x^2, x^3, \ldots$, then the results have no error for $f(x) = 1$ and $f(x) = x$ but there are errors for x^2 and higher powers of x.

Example 5.13 *Compute the local error for Trapezoidal rule (5.32) using the integral*

$$\int_1^2 \frac{1}{x+1}\,dx.$$

Solution. *Given* $f(x) = \dfrac{1}{x+1}$ *and* $[a, b] = [1, 2]$, *then the second derivative of the function is*

$$f''(x) = \frac{2}{(x+1)^3},$$

since the error formula for the Trapezoidal rule is

$$E_{T_1}(f) = -\frac{h^3}{12} f''(\eta(x)), \quad \textit{where} \quad \eta(x) \in (1, 2).$$

This formula cannot be computed exactly because $\eta(x)$ *is not known. But one can bound the error by computing the largest possible value for* $|f''(\eta(x))|$. *Bound* $|f''(\eta(x))|$ *on* $[1, 2]$ *is*

$$M = \max_{1 \le x \le 2} \left| \frac{2}{(x+1)^3} \right| = 0.25.$$

Then, for $|f''(\eta(x))| \le M$, *we have*

$$|E_{T_1}(f)| \le \frac{h^3}{12} M.$$

Taking $M = 0.25$ *and* $h = 1$, *we get*

$$|E_{T_1}(f)| \le \frac{0.25}{12} = 0.0208.$$

Comparing this result with the actual error -0.0112, this bound is about two times the actual error.

Error Formula for the Composite Trapezoidal Rule

The *global error* of Trapezoidal rule (5.33) equals the sum of n local errors of Trapezoidal rule (5.32); that is,

$$E_{T_n}(f) = -\frac{h^3}{12}f''(\eta_1(x)) - \frac{h^3}{12}f''(\eta_2(x)) - \cdots - \frac{h^3}{12}f''(\eta_n(x)),$$

which can also written as

$$E_{T_n}(f) = -\frac{h^3}{12}\sum_{i=1}^{n} f''(\eta_i(x)), \quad \text{for} \quad \eta_i(x) \in (x_{i-1}, x_i)$$

or

$$E_{T_n}(f) = -\frac{h^3}{12}nf''(\eta(x)),$$

where $f''(\eta(x))$ is the average of the n individual values of the second derivative. Since $n = \dfrac{b-a}{h}$, thus the global error in composite Trapezoidal rule (5.33) is

$$E_{T_n}(f) = -\frac{h^2}{12}(b-a)f''(\eta(x)), \quad \eta(x) \in (a,b). \tag{5.39}$$

Hence,

$$\int_a^b f(x)dx = \frac{h}{2}\left[f(a) + 2\sum_{i=1}^{n-1} f(x_i) + f(b)\right] - \frac{h^2}{12}(b-a)f''(\eta(x)) \tag{5.40}$$

for $\eta(x) \in (a,b)$ is the composite Trapezoidal rule with its error term.

Note that whereas simple Trapezoidal rule (5.32) has a truncation error of order h^3, composite Trapezoidal rule (5.33) has an error of order h^2. This means that when h is halved and the number of subintervals is doubled, the error decreases by a factor of approximately four (assuming that $f''(\eta(x))$ remains fairly constant throughout $[a,b]$). Of course, it is also possible to express the truncation error in terms of n rather than h. Since $h = \dfrac{b-a}{n}$, it follows that global truncation error (5.39) is of order $O(n^2)$.

Example 5.14 *Consider the integral* $I(f) = \int_1^2 \ln(x+1)dx; \qquad n = 6.$

(a) Compute the approximation of the integral using the composite Trapezoidal rule.

(b) Compute the error bound for your approximation using formula (5.39).
(c) Compute the absolute error.
(d) How many subintervals approximate the given integral to an accuracy of at least 5×10^{-4} using the composite Trapezoidal rule?

Solution. *(a) Given $f(x) = \ln(x+1), n = 6$, and so $h = \dfrac{2-1}{6} = \dfrac{1}{6}$, then composite Trapezoidal rule (5.33) for $n = 6$, can be written as*

$$
\begin{aligned}
T_6(f) &= \frac{1/6}{2}\Big[\ln(1+1) + 2\Big(\ln\Big(\frac{7}{6}+1\Big) + \ln\Big(\frac{8}{6}+1\Big) + \ln\Big(\frac{9}{6}+1\Big) \\
&+ \ln\Big(\frac{10}{6}+1\Big) + \ln\Big(\frac{11}{6}+1\Big)\Big) + \ln(2+1)\Big]
\end{aligned}
$$

$$T_6(f) = \frac{1}{12}[0.6932 + 2(4.5591) + 1.0986].$$

Hence,

$$\int_1^2 \ln(x+1)dx \approx T_6(f) = 0.9092.$$

(b) The second derivative of the function can be obtained as

$$f'(x) = \frac{1}{(x+1)} \quad \text{and} \quad f''(x) = \frac{-1}{(x+1)^2}.$$

Since $\eta(x)$ is unknown point in $(1,2)$, the bound $|f''|$ on $[1,2]$ is

$$M = \max_{1 \le x \le 2} |f''(x)| = \left|\frac{-1}{(x+1)^2}\right| = 0.25.$$

Thus, error formula (5.39) becomes

$$|E_{T_6}(f)| \le \frac{(1/6)^2}{12}(0.25) = 0.0006,$$

which is the possible maximum error in our approximation in part (a).
(c) The absolute error $|E|$ in our approximation is given as

$$|E| = |(3\ln 3 - 2\ln 2 - 1) - T_6(f)| = |0.9095 - 0.9092| = 0.0003.$$

(d) To find the minimum subintervals for the given accuracy, we use formula (5.39) such that

$$|E_{T_n}(f)| \le \frac{|-(b-a)^3|}{12n^2} M \le 10^{-4},$$

where $h = (b - a)/n$. *Since* $M = 0.25$, *then solving for* n^2, *we obtain*

$$n^2 \geq 208.3333, \quad gives \quad n \geq 14.4338.$$

Hence, to get the required accuracy, we need 15 subintervals.

To use MATLAB commands for the composite Trapezoidal rule, first we define a function m-file as fn.m for the function as follows:

```
function y = fn(x)
y = log(x + 1);
>> T6 = TrapezoidalR('fn', 1, 2, 6)
T6 =
    0.9092
```

Note that error formula (5.39) of the composite Trapezoidal rule can only be used to bound the error because $f''(\eta(x))$ is the unknown. This will be improved by the error formula

$$\hat{E}_{T_n}(f) \approx -\frac{h^2}{12}\left[f'(b) - f'(a)\right], \tag{5.41}$$

which is called the *asymptotic error* of the composite Trapezoidal rule. It will be very easy to compute if the first derivative $f'(x)$ can be computable. Error formula (5.41) implies that the convergence of the composite Trapezoidal rule will be very rapid when $f'(b) = f'(a)$. So using error formula (5.41), we have

$$I(f) - T_n(f) \approx -\frac{h^2}{12}\left[f'(b) - f'(a)\right]$$

or

$$I(f) \approx T_n(f) - \frac{h^2}{12}\left[f'(b) - f'(a)\right] = CT_n(f), \tag{5.42}$$

which is called the *corrected Trapezoidal rule* and it is usually more accurate than error formula (5.33).

Example 5.15 *Compute the asymptotic error of the Trapezoidal rule for* $n = 6$ *using the integral* $I(f) = \int_1^2 \ln(x + 1)dx$.

Solution. *Given* $f(x) = \ln(x + 1)$, $n = 6$, *we have the first derivative of the function as* $f'(x) = \dfrac{1}{x+1}$ *and* $h = \dfrac{1}{6}$. *Then using error formula (5.41), we obtain*

$$\hat{E}_{T_6}(f) \approx -\frac{(1/6)^2}{12}\left[\frac{1}{3} - \frac{1}{2}\right] = 0.00038,$$

which is very close to the true error 0.0003.

Program 5.1

MATLAB m-file for the Composite Trapezoidal Rule

```
function TN=TrapezoidR(fn,a,b,n); h=(b-a)/n;
s=(feval(fn,a)+feval(fn,b))/2;
for k=1:n-1
x = a + h * k
s=s+feval(fn,x); end
TN = s * h;
```

Suppose that, due to rounding, the f_i are in error by at most $\frac{1}{2} \times 10^{-t}$. Then we see from (5.33) that the error in the composite Trapezoidal rule due to rounding is not greater than

$$\frac{1}{2}h[1+2+2+\cdots+2+1] \times 10^{-t} = nh(\frac{1}{2} \times 10^{-t}) = \frac{1}{2}(b-a) \times 10^{-t}.$$

Thus, rounding errors do not seriously affect the accuracy of this quadrature rule. This is generally true of numerical integration unlike numerical differentiation, as we saw earlier in the chapter.

Simpson's Rule

The Trapezoidal rule approximates the area under a curve by the area of the trapezoid formed by connecting two points on the curve by a straight line. Simpson's rule gives a more accurate approximation since it consists of connecting three points on the curve by a second-degree parabola and the area under the parabola to obtain the approximate area under the curve (see Figure 5.8).

Let us consider the second-degree Lagrange interpolating polynomial, with equally spaced base points, that is, $x_0 = a, x_1 = a+h$, and $x_2 = a+2h$, with $h = (b-a)/2$, then

$$\begin{aligned} f(x) \approx p_2(x) &= \frac{(x-x_1)(x-x_2)}{(x_0-x_1)(x_0-x_2)}f(x_0) + \frac{(x-x_0)(x-x_2)}{(x_1-x_0)(x_1-x_2)}f(x_1). \\ &+ \frac{(x-x_0)(x-x_1)}{(x_2-x_0)(x_2-x_1)}f(x_2) \end{aligned}$$

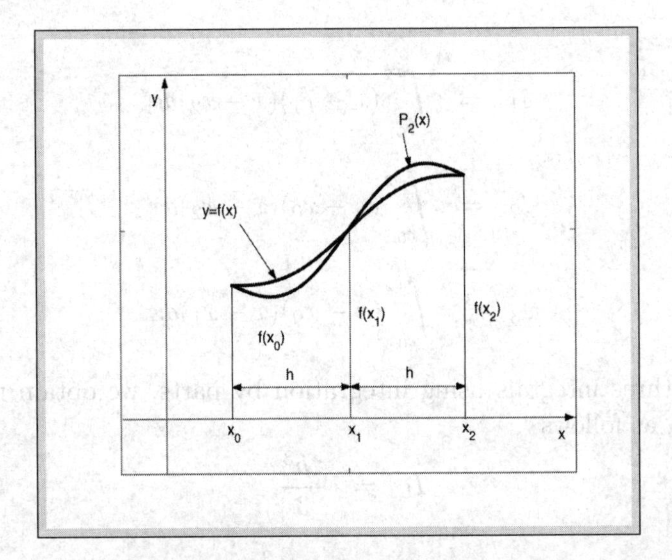

Figure 5.8: Simple Simpson's rule.

Taking the integral on both sides of the above equation with respect to x between limits x_0 and x_2, we have

$$\begin{aligned}\int_{x_0}^{x_2} f(x)dx &\approx \frac{f(x_0)}{(x_0-x_1)(x_0-x_2)}\int_{x_0}^{x_2}(x-x_1)(x-x_2)dx \\ &+ \frac{f(x_1)}{(x_1-x_0)(x_1-x_2)}\int_{x_0}^{x_2}(x-x_0)(x-x_2)dx \\ &+ \frac{f(x_2)}{(x_2-x_0)(x_2-x_1)}\int_{x_0}^{x_2}(x-x_0)(x-x_1)dx,\end{aligned}$$

which implies that

$$\int_a^b f(x)dx \approx \frac{f(x_0)}{2h^2}I_1 + \frac{f(x_1)}{-h^2}I_2 + \frac{f(x_2)}{2h^2}I_3,$$

where

$$I_1 = \int_{x_0}^{x_2} (x - x_1)(x - x_2)dx$$

$$I_2 = \int_{x_0}^{x_2} (x - x_0)(x - x_2)dx$$

$$I_3 = \int_{x_0}^{x_2} (x - x_0)(x - x_1)dx.$$

Solving the three integrals using integration by parts, we obtain the values of I_1, I_2, and I_3 as follows

$$I_1 = \frac{2h^3}{3},$$

$$I_2 = -\frac{4h^3}{3},$$

$$I_3 = \frac{2h^3}{3}.$$

Using these values, we have

$$\int_a^b f(x)dx \approx \frac{f(x_0)}{2h^2}\left(\frac{2h^3}{3}\right) + \frac{f(x_1)}{-h^2}\left(\frac{-4h^3}{3}\right) + \frac{f(x_2)}{2h^2}\left(\frac{2h^3}{3}\right).$$

Simplifying gives

$$\int_a^b f(x)dx \approx S_2(f) = \frac{h}{3}[f(x_0) + 4f(x_1) + f(x_2)], \tag{5.43}$$

which is called the *simple Simpson's rule* or *Simpson's rule for two strips.*

Example 5.16 *Approximate the following integral*

$$\int_1^2 \frac{1}{x+1}dx$$

using simple Simpson's rule. Compute the actual error.

Solution. *Since* $f(x) = \dfrac{1}{x+1}$ *and* $h = (2-1)/2 = 0.5$, *using Simpson's rule (5.43), we have*

$$\begin{aligned} S_2(f) &= \frac{0.5}{3}\Big[f(1) + 4f(1.5) + f(2)\Big] \\ &= (0.1667)[0.5 + 1.6 + 0.3333]. \\ &= 0.4056 \end{aligned}$$

Hence,

$$\int_1^2 \frac{1}{x+1}dx \approx S_2(f) = 0.4056.$$

Since the exact solution of the given integral is 0.4055, *the actual error is*

$$E_{S_2} = I(f) - S_2(f) = -0.0001.$$

The error in Simpson's rule is much smaller than for the Trapezoidal rule by a factor of about 123, *a significant increase in accuracy.*

Composite Simpson's Rule

Just as with simple Trapezoidal rule (5.32), simple Simpson's rule (5.43) can be improved by dividing the integration interval $[a, b]$ into a number of subintervals of equal width h where $h = \dfrac{b-a}{n}$, since simple Simpson's rule (5.43) requires a interval consisting of three points (pair of strips). In practice, we usually take more than three points and add the separate results for the different pairs of strips (see Figure 5.9). Since simple Simpson's rule requires a pair of strips for application, there is restriction on integer n, which must be even. We define the *composite Simpson's rule* in the form of the following theorem.

Theorem 5.4 (Composite Simpson's Rule)

Let $f \in C^4[a,b]$, n *be even*, $h = (b-a)/n$, *and* $x_i = a + ih$ *for each* $i = 0, 1, 2, \ldots, n$. *Then the composite Simpson's rule for* n *subintervals can be written as*

$$\int_a^b f(x)dx \approx S_n(f) = \frac{h}{3}\left[f(a) + 2\sum_{i=1}^{n/2-1} f(x_{2i}) + 4\sum_{i=1}^{n/2} f(x_{2i-1}) + f(b)\right]. \tag{5.44}$$

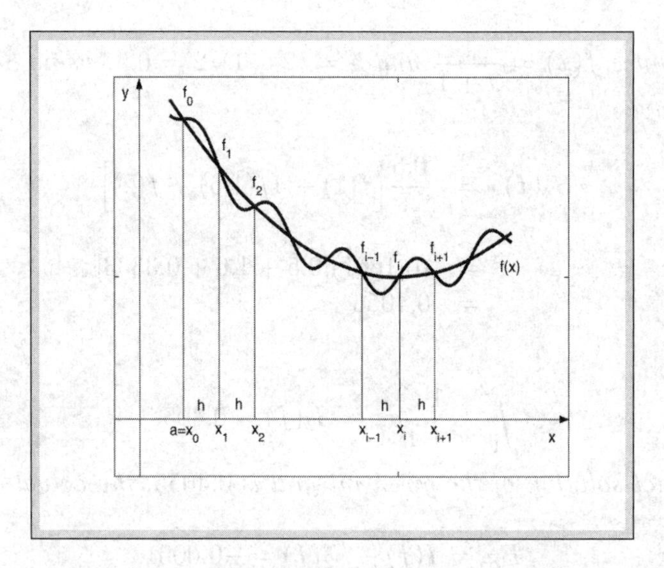

Figure 5.9: Composite Simpson's Rule.

Proof. *Since for the composite form of Simpson's rule, the interval is divided into n equal subintervals of width h so that $h = \dfrac{b-a}{n}$, for this rule to work, n must be an even number and the total number of $(n+1)$ distinct points $a = x_0 < x_1 < x_2 \ldots < x_n = b$ should be odd. The total integral can be represented as*

$$\int_{x_0}^{x_n} f(x)dx = \int_{x_0}^{x_2} f(x)dx + \int_{x_2}^{x_4} f(x)dx + \cdots + \int_{x_{n-2}}^{x_n} f(x)dx.$$

Substitute simple Simpson's rule (5.43) for the individual integral, which yields

$$\begin{aligned} \int_{x_0}^{x_n} f(x)dx \quad &\approx \quad \frac{h}{3}[f(x_0) + 4f(x_1) + f(x_2)] + \frac{h}{3}[f(x_2) + 4f(x_3) + f(x_4)] \\ &+ \quad \cdots + \frac{h}{3}[f(x_{n-2}) + 4f(x_{n-1}) + f(x_n)]. \end{aligned}$$

To avoid the repetition of terms, we summed them. Note that each of the odd interior points is counted as four and therefore has a coefficient of four, whereas each of the even interior points is counted as two and therefore has a coefficient of two. The endpoints are counted once and therefore have a coefficient one.

Example 5.17 *Suppose that $f(1) = 0.5, f(1.5) = 1.5, f(1.25) + f(1.75) = \alpha$, and $f(2) = 2$. Find the approximate value of α if the composite Simpson's rule* *with* $n = 4$ *gives the value* 2.25 *for the integral* $\int_1^2 f(x)\, dx$.

Solution. *Given* $n = 4$, $h = \dfrac{2-1}{4} = 0.25$. *Using composite formula (5.44) for* $n = 4$, *we have*

$$\int_1^2 f(x)\, dx \approx \frac{0.25}{3}\Big[f(1) + 4[f(1.5) + f(1.75)] + 2f(1.25) + f(2)\Big].$$

Now using the given values, we obtain

$$2.25 \approx 0.0833[0.5 + 4(\alpha) + 2(1.5) + 2]$$

or

$$2.25 - 0.4582 \approx 0.3332\alpha,$$

which gives $\quad \alpha \approx 5.3776.$

Example 5.18 *Evaluate the integral* $\int_0^1 e^{4x} dx$ *using Simpson's rule with* $n = 2, 4, 8$. *Also, compute the corresponding actual errors.*

Solution. *For* $n = 2$, *using formula (5.43) and* $h = 0.5$, *we have*

$$S_2(f) = \frac{0.5}{3}\Big[f(0) + 4f(0.5) + f(1)\Big] = 14.1924.$$

For $n = 4$, *using formula (5.44) and* $h = 0.25$, *we have*

$$S_4(f) = \frac{0.25}{3}\Big[f(0) + 4[f(0.25) + f(0.75)] + 2f(0.5) + f(1)\Big] = 13.4659.$$

For $n = 8$, *using formula (5.44) and* $h = 0.125$, *we have*

$$\begin{aligned} S_8(f) = \frac{0.125}{3}\Big[f(0) &+ 4[f(0.125) + f(0.375) + f(0.625) + f(0.875)] \\ &+ 2[f(0.25) + f(0.5) + f(0.75)] + f(1)\Big] = 13.4041. \end{aligned}$$

Note that the exact value of the given integral is 13.39995, *and so the corresponding errors are* $0.79245, 0.06595$, *and* 0.00411, *respectively, which decrease by a factor of about 16 at each stage.*

To use MATLAB commands for the composite Simpson's rule, first we define a function m-file as fn.m for the function as follows:

```
function y = fn(x)
y = exp(4 * x);
>> S2 = SimpsonR('fn', 0, 1, 2)
S2 =
    14.1924
```

.

Program 5.2

MATLAB m-file for the Composite Simpson's Rule

```
function SN=SimpsonR(fn,a,b,n);
h=(b-a)/n;
s=feval(fn,a)+feval(fn,b);
for k=1:2:n-1
s = s + 4 * feval(fn, a + k * h); end
for k=2:2:n-2
s = s + 2 * feval(fn, a + k * h); end
SN = (s * h)/3;
```

One can note that these results are more accurate than those obtained in Example 5.12 using the Trapezoidal rule (for same number of function evaluations). The main disadvantage of Simpson's rule is that it can only be used when the given interval $[a, b]$ is divided into an even number of subintervals.

Error Formulas for Simpson's Rule

Now we discuss the local error formula for Simpson's rule in the form of the following theorem.

Theorem 5.5 (Error Formula for Simple Simpson's Rule)

Let $f \in C^4[a, b]$ and $h = (b-a)/2$. The local error that Simpson's rule makes in estimating definite integral (5.25) is

$$E_{S_2}(f) = -\frac{h^5}{90} f^{(4)}(\eta(x)), \tag{5.45}$$

where $\eta(x) \in (a, b)$.

Proof. Consider three equally spaced points: $a = x_0 < x_1 < x_2 = b$. From the quadratic Lagrange interpolation formula with error terms, we have

$$f(x) = p_2(x) + \frac{f'''(\eta(x))}{3!}\prod_{i=0}^{2}(x - x_i) \tag{5.46}$$

and by integrating (5.46) with respect to x, we have

$$\int_a^b f(x)dx = \int_a^b p_2(x)dx + \frac{1}{6}\int_a^b f'''(\eta(x)(x))\prod_{i=0}^{2}(x - x_i)dx. \tag{5.47}$$

The second term on the right hand-side of (5.47)

$$E_{S_2}(f) = \frac{1}{6}\int_a^b f'''(\eta(x)(x))(x - x_0)(x - x_1)(x - x_2)dx \tag{5.48}$$

is called the error term of Simpson's rule for $n = 2$.

In this way it provides only an $O(h^4)$ error term, involving $f'''(\eta(x))$. By approaching the problem in another way, a higher-order term involving $f^{(4)}(\eta(x))$ can be obtained.

Consider the two intervals from x_{i-1} to x_i and x_i to x_{i+1}, with $h = (x_i - x_{i-1})$, and also, assume that $F(x)$ is the indefinite integral of a function $f(x)$, which we are trying to integrate. Then the exact value of the integral from x_{i-1} to x_{i+1} is

$$I_i(f) = \int_{x_{i-1}}^{x_{i+1}} f(x)dx = F(x_{i+1}) - F(x_{i-1}). \tag{5.49}$$

The approximated value calculated using Simpson's rule is

$$S_{2i}(f) = \frac{h}{3}[f(x_{i-1}) + 4f(x_i) + f(x_{i+1})]. \tag{5.50}$$

Then the error defined in using Simpson's rule on these two intervals is

$$E_i(f) = I_i(f) - S_{2i}(f). \tag{5.51}$$

Expanding $f(x)$ about $x = x_i$ to get $f(x_{i-1})$ in terms of a function and derivatives at $x = x_i$ using the Taylor's series, we have

$$\begin{aligned} f(x_{i-1}) &= f(x_i) + (x_{i-1} - x_i)f'(x_i) + \frac{(x_{i-1} - x_i)^2}{2!}f''(x_i) \\ &+ \frac{(x_{i-1} - x_i)^3}{3!}f'''(x_i) + \frac{(x_{i-1} - x_i)^4}{4!}f^{(4)}(x_i) + \cdots. \end{aligned} \tag{5.52}$$

Since we know that $h = (x_i - x_{i-1})$, or $-h = (x_{i-1} - x_i)$, therefore,

$$\begin{aligned} f(x_{i-1}) &= f(x_i) - hf'(x_i) + \frac{h^2}{2!}f''(x_i) \\ &- \frac{h^3}{3!}f'''(x_i) + \frac{h^4}{4!}f^{(4)}(x_i) - \cdots . \end{aligned} \quad (5.53)$$

In a similar way, we expand $f(x_{i+1})$ about x_i, to get

$$\begin{aligned} f(x_{i+1}) &= f(x_i) + (x_{i+1} - x_i)f'(x_i) + \frac{(x_{i+1} - x_i)^2}{2!}f''(x_i) \\ &+ \frac{(x_{i+1} - x_i)^3}{3!}f'''(x_i) + \frac{(x_{i+1} - x_i)^4}{4!}f^{(4)}(x_i) + \cdots \end{aligned} \quad (5.54)$$

or

$$\begin{aligned} f(x_{i+1}) &= f(x_i) + hf'(x_i) + \frac{h^2}{2!}f''(x_i) \\ &+ \frac{h^3}{3!}f'''(x_i) + \frac{h^4}{4!}f^{(4)}(x_i) + \cdots . \end{aligned} \quad (5.55)$$

So by using (5.53) and (5.55), we get (5.50) of the form

$$\begin{aligned} S_{2i}(f) &= \frac{h}{3}\Big[\{f(x_i) - hf'(x_i) + \frac{h^2}{2!}f''(x_i) - \frac{h^3}{3!}f'''(x_i) + \frac{h^4}{4!}f^{(4)}(x_i) - \cdots\} \\ &+ \{4f(x_i)\} \\ &+ \{f(x_i) + hf'(x_i) + \frac{h^2}{2!}f''(x_i) + \frac{h^3}{3!}f'''(x_i) + \frac{h^4}{4!}f^{(4)}(x_i) + \cdots\}\Big], \end{aligned}$$

which implies that

$$\begin{aligned} S_{2i}(f) &= \frac{h}{3}\left[6f(x_i) + \frac{2h^2}{2!}f''(x_i) + \frac{2h^4}{4!}f^{(4)}(x_i) + \cdots\right] \\ &= 2hf(x_i) + 2f''(x_i)\frac{h^3}{3!} + 2f^{(4)}(x_i)\frac{h^5}{3(4!)} + \cdots . \end{aligned} \quad (5.56)$$

Now we use the same procedure for $F(x)$ as we used for $f(x)$ to get

$$\begin{aligned} F(x_{i-1}) &= F(x_i) - hF'(x_i) + \frac{h^2}{2!}F''(x_i) - \frac{h^3}{3!}F'''(x_i) \\ &+ \frac{h^4}{4!}F^{(4)}(x_i) - \frac{h^5}{5!}F^{(5)}(x_i) + \cdots \end{aligned} \quad (5.57)$$

and

$$\begin{aligned} F(x_{i+1}) &= F(x_i) + hF'(x_i) + \frac{h^2}{2!}F''(x_i) + \frac{h^3}{3!}F'''(x_i) \\ &+ \frac{h^4}{4!}F^{(4)}(x_i) + \frac{h^5}{5!}F^{(5)}(x_i) + \cdots. \end{aligned} \tag{5.58}$$

Then using (5.57) and (5.58), we got (5.49) of the form

$$I_i(f) = 2hF'(x_i) + 2F'''(x_i)\frac{h^3}{3!} + 2F^{(5)}(x_i)\frac{h^5}{5!} + \cdots. \tag{5.59}$$

But we know that

$$\begin{aligned} F'(x) &= f(x) \\ F''(x) &= f'(x) \\ F'''(x) &= f''(x) \\ F^{(4)}(x) &= f'''(x) \\ F^{(5)}(x) &= f^{(4)}(x), \end{aligned}$$

therefore, (5.59) can be written as

$$I_i(f) = 2hf(x_i) + 2f''(x_i)\frac{h^3}{3!} + 2f^{(4)}(x_i)\frac{h^5}{5!} + \cdots. \tag{5.60}$$

So using (5.56) and (5.60), we get (5.51) of the form

$$\begin{aligned} E_i(f) &= -\left[2hf(x_i) + 2f''(x_i)\frac{h^3}{3!} + 2f^{(4)}(x_i)\frac{h^5}{3(4!)} + \cdots\right] \\ &+ \left[2hf(x_i) + 2f''(x_i)\frac{h^3}{3!} + 2f^{(4)}(x_i)\frac{h^5}{5!} + \cdots\right] \\ &= -2f^{(4)}(x_i)\frac{h^5}{3(4!)} + 2f^{(4)}(x_i)\frac{h^5}{5!} \\ &+ \text{higher terms in } h^7 \ + \cdots. \end{aligned} \tag{5.61}$$

Assuming that h is small, we may neglect the terms in h^7 and above, and get the approximate error as follows:

$$\begin{aligned} E_i(f) &\approx f^{(4)}(x_i)h^5\left[\frac{2}{5!} - \frac{2}{3(4!)}\right] \\ &\approx -\frac{1}{90}f^{(4)}(x_i)h^5, \end{aligned} \tag{5.62}$$

which is the desired local error for Simpson's rule.

Since (5.62) indicates that the error of Simpson's rule is proportional to the fourth derivative $f^{(4)}$, if Simpson's rule is used to integrate $f(x) = 1, x, x^2$, and x^3, then the results have no error.

In more general terms, the Newton-Cotes closed formula of odd order n is exact if the integrand is a polynomial of order n or less, whereas that of an even n is exact when the integrand is a polynomial of order $n+1$ or less.

Example 5.19 *Compute the local error for Simpson's rule using the following integral*

$$\int_1^2 \frac{1}{x+1} dx.$$

Solution. *Given* $f(x) = \dfrac{1}{x+1}$, *and* $[a, b] = [1, 2]$, *then the fourth derivative of the function can be obtained as*

$$f' = \frac{-1}{(x+1)^2}, f'' = \frac{2}{(x+1)^3}, f''' = \frac{-6}{(x+1)^4}, f^{(4)} = \frac{24}{(x+1)^5}.$$

Since the error formula for Simpson's rule is

$$E_{S_2}(f) = -\frac{h^5}{90} f^{(4)}(\eta(x)), \quad \textit{where} \quad \eta(x) \in (1, 2)$$

or

$$|E_{S_2}(f)| = \left| -\frac{h^5}{90} \right| \left| f^{(4)}(\eta(x)) \right|, \quad \textit{for} \quad \eta(x) \in (1, 2),$$

this formula cannot be computed exactly because $\eta(x)$ *is not known. But one can bound the error by computing the largest possible value for* $|f^{(4)}|$. *Bound* $|f^{(4)}|$ *on* $[1, 2]$ *is*

$$M = \max_{1 \le x \le 2} = \left| \frac{24}{(x+1)^5} \right| = 0.75.$$

Then for $|f^{(4)}(\eta(x))| \le M$, *we have*

$$|E_{S_2}(f)| \le \frac{h^5}{90} M.$$

Taking $M = 0.75$ *and* $h = 0.5$, *we get*

$$|E_{S_2}(f)| \le \frac{(0.03125)}{90}(0.75) = 0.0003.$$

Comparing this result with the actual error -0.0001, this bound is about three times the actual error.

Error Formula for the Composite Simpson's Rule

Since composite Simpson's rule (5.44) requires that the given interval $[a, b]$ is divided into even number of subintervals and each application of the simple Simpson's rule requires two subintervals, the global error of composite Simpson's rule (5.44) is the sum of $\frac{n}{2}$ local truncation error of the simple Simpson's rule with $n = \frac{b-a}{h}$; that is,

$$E_{S_n}(f) = -\frac{h^5}{90}f^{(4)}(\eta_1(x)) - \frac{h^5}{90}f^{(4)}(\eta_2(x)) - \cdots - \frac{h^5}{90}f^{(4)}(\eta_{n/2}(x)),$$

which implies that

$$E_{S_n}(f) = -\frac{h^5}{90}(\frac{n}{2})\left[\frac{\sum_{i=1}^{n/2} f^{(4)}(\eta_i(x))}{n/2}\right].$$

Thus, using the Intermediate Value Theorem, we have

$$E_{S_n}(f) = -\frac{(b-a)}{180}h^4 f^{(4)}(\eta(x)) \tag{5.63}$$

for $\eta(x) \in (a, b)$, and $nh = b - a$. Then formula (5.63) is known as the *global error* of Simpson's rule.

Note that the truncation error in the composite Simpson's rule is of order h^4. This means that when h is halved and the number of subintervals is doubled the error decreases by a factor of approximately 16, considerably better than the composite Trapezoidal rule.

Example 5.20 *Consider the integral* $I(f) = \int_1^2 \ln(x+1)dx; \qquad n = 6.$

(a) Find the approximation of the given integral using the composite Simpson's rule.
(b) Compute the error bound for the approximation using formula (5.63).
(c) Compute the absolute error.
(d) How many subintervals approximate the given integral to an accuracy of at least 5×10^{-4} *using the composite Simpson's rule?*

Solution. *(a) Given* $f(x) = \ln(x+1), n = 6$, *and so* $h = \frac{2-1}{6} = \frac{1}{6}$, *then composite Simpson's rule (5.44) for* $n = 6$, *can be written as*

$$\begin{aligned} S_6(f) &= \frac{1/6}{3}\Big[\ln(1+1) + 4(\ln\Big(\frac{7}{6}+1\Big) + \ln\Big(\frac{9}{6}+1\Big) + \ln\Big(\frac{11}{6}+1\Big))\Big] \\ &+ \Big[2\Big(\ln\Big(\frac{8}{6}+1\Big) + \ln\Big(\frac{10}{6}+1\Big)\Big) + \ln(2+1)\Big] \end{aligned}$$

$$S_6(f) = \frac{1}{18}\Big[0.6932 + 4(2.7309) + 2(1.8281) + 1.0986\Big].$$

Hence,

$$\int_1^2 \ln(x+1)dx \approx S_6(f) = 0.9095.$$

(b) Since the fourth derivative of the function is

$$f^{(4)}(x) = \frac{-6}{(x+1)^4},$$

and since $\eta(x)$ *is an unknown point in* $(1,2)$, *the bound* $|f^{(4)}|$ *on* $[1,2]$ *is*

$$M = \max_{1 \leq x \leq 2} |f^{(4)}(x)| = \left|\frac{-6}{(x+1)^4}\right| = 6/16 = 0.375.$$

Thus, error formula (5.63) becomes

$$|E_{T_6}(f)| \leq \frac{(1/6)^4}{180}(0.375) = 0.000002,$$

which is the possible maximum error in our approximation in part (a).
(c) The absolute error $|E|$ *in our approximation is given as*

$$|E| = |3\ln 3 - 2\ln 2 - 1 - S_6(f)| == 0.0000003.$$

(d) To find the minimum subintervals for the given accuracy, we use error formula (5.63), which is

$$|E_{S_n}(f)| \leq \frac{(b-a)^5}{180n^4}M \leq 10^{-4}.$$

Since we know $M = 0.375$, *then we have*

$$n^4 \geq 20.83333, \quad \textit{gives} \quad n \geq 2.136435032.$$

Hence, to get the required accuracy, we need four subintervals (because n should be even) that ensures the stipulated accuracy.

To find the approximate value of the given integral within the given accuracy, we use the MATLAB command, which gives us the approximate solution and the number of subintervals n. First, we define m-file as fn.m for the function, so after finding the value of M, it simply computes n using (5.63) and then calls the previously defined SimpsonR function and we have the results:

```
function y = fn(x)
y = log(x + 1);
>> k = ErrorSR('fn', 0, 1, 0.375, 1e - 4)
n =
    4
S =
    0.9095
```

Like the Trapezoidal rule, error bound formula (5.63) of the composite Simpson's rule can be estimated by the error formula

$$\hat{E}_{S_n}(f) \approx -\frac{h^4}{180}\left[f'''(b) - f'''(a)\right], \qquad (5.64)$$

which is called the *asymptotic error* of the composite Simpson's rule. This error is closer to the true error compared to error bound formula (5.63). It will be very easy to compute if the third derivative $f'''(x)$ can be computable. Error formula (5.64) implies that the convergence of the composite Simpson's rule will be very rapid when $f'''(b) = f'''(a)$. Thus,

$$I(f) - S_n(f) \approx -\frac{h^4}{180}\left[f'''(b) - f'''(a)\right]$$

or

$$I(f) \approx S_n(f) - \frac{h^4}{180}\left[f'''(b) - f'''(a)\right] = CS_n(f), \qquad (5.65)$$

which is called the *corrected Simpson's rule* and it is usually more accurate than error formula (5.44).

Example 5.21 *Compute the asymptotic error of Simpson's rule for $n = 6$ using the following integral* $\int_1^2 \ln(x + 1)dx$.

Solution. *Given $f(x) = \ln(x + 1)$ and $n = 6$, we have $f'''(x) = \dfrac{2}{(x+1)^3}$ and $h = \dfrac{1}{6}$. Then from error formula (5.64), we obtain*

$$\hat{E}_{S_6}(f) \approx -\frac{(1/6)^4}{180}\left[\frac{2}{27} - \frac{1}{4}\right] = 0.00000075,$$

which is very close to the true error 0.0000003 *compared to the error bound* 0.000002 *obtained in Example 5.20.*

Program 5.3

MATLAB m-file for computing Error term of the Composite Simpson's Rule

```
function k=ErrorSR(fn,a,b,M,eps)
% M is a bound for the fourth derivative of fn on [a,b]
% eps is the required accuracy
L = abs(b - a); n = ceil(L * sqrt(sqrt(L * M/180/eps)));
if mod(n, 2) == 1; n = n + 1; end
k=SimpsonR(fn,a,b,n);
```

Higher-order Formulas

Similarly, the other higher-order closed Newton-Cotes formulas with their error terms that arise from approximating $y = f(x)$ with polynomial of degree $n = 3, 4, 5, 6$ on $[a, b]$ using formulas (5.29) and (5.30) are summarized as follows:

Simpson's 3/8 Rule

For $n = 3$, formula (5.30) becomes

$$\int_{x_0}^{x_3} f(x)dx = \frac{3h}{8}\Big(f_0 + 3f_1 + 3f_2 + f_3\Big) - \frac{3h^5}{80} f^{(4)}(\eta(x))$$

for $\eta(x) \in (x_0, x_3)$.

Boole's Rule

For $n = 4$, formula (5.29) becomes

$$\int_{x_0}^{x_4} f(x)dx = \frac{2h}{45}\Big(7f_0 + 32f_1 + 12f_2 + 32f_3 + 7f_4\Big) - \frac{8h^7}{945} f^{(6)}(\eta(x))$$

for $\eta(x) \in (x_0, x_4)$.

Six-points Rule

For $n = 5$, formula (5.30) becomes

$$\int_{x_0}^{x_5} f(x)dx = \frac{5h}{288}\Big(19f_0 + 75f_1 + 50f_2 + 50f_3 + 75f_4 + 19f_5\Big) - \frac{275h^7}{12096}f^{(6)}(\eta(x))$$

for $\eta(x) \in (x_0, x_5)$.

Weddle's Rule

For $n = 6$, formula (5.29) becomes

$$\int_{x_0}^{x_6} f(x)dx = \frac{3h}{10}\Big(f_0 + 5f_1 + f_2 + 6f_3 + f_4 + 5f_5 + f_6\Big) - \frac{h^7}{140}f^{(6)}(\eta(x))$$

for $\eta(x) \in (x_0, x_6)$.

Example 5.22 *Evaluate the integral* $\int_0^{\pi/2} \cos(x)dx$ *using the closed Newton-Cotes formulas for* $n = 1, 2, 3, 4, 5, 6$.

Solution. *Given* $f(x) = \cos(x)$, *then*

Trapezoidal Rule ($n = 1$):

$$I_T = \frac{h}{2}\Big[f(x_0) + f(x_1)\Big] = \frac{(\pi/2)}{2}\Big[\cos(0) + \cos(\pi/2)\Big] = \frac{\pi}{4} = 0.7854.$$

Simpson's Rule ($n = 2$):

$$\begin{aligned} I_S &= \frac{h}{3}\Big[f(x_0) + 4(f(x_1) + f(x_2)\Big] = \frac{(\pi/4)}{3}\Big[\cos(0) + 4\cos(\pi/4) + \cos(\pi/2)\Big] \\ &= \frac{\pi}{12}\Big[1 + 4(0.7071) + 0\Big] = 1.0023. \end{aligned}$$

Simpson's 3/8 Rule ($n = 3$):

$$\begin{aligned} I_{SR} &= \frac{3h}{8}\Big[f(x_0) + 3f(x_1) + 3f(x_2) + f(x_3)\Big] \\ &= \frac{(3\pi/6)}{8}\Big[\cos(0) + 3\cos(\pi/6) + 3\cos(\pi/3) + \cos(\pi/2)\Big] \\ &= \frac{\pi}{16}\Big[1 + 3(0.8660) + 3(0.5) + 0\Big] = 1.001001. \end{aligned}$$

Boole's Rule $(n = 4)$:

$$
\begin{aligned}
I_B &= \frac{2h}{45}\Big[7f(x_0) + 32f(x_1) + 12f(x_2) + 32f(x_3) + 7f(x_4)\Big] \\
&= \frac{\pi}{180}\Big[7\cos(0) + 32\cos(\pi/8) + 12\cos(\pi/4) + 32\cos(33\pi/8) + 7\cos(\pi/2)\Big] \\
&= \frac{\pi}{180}\Big[7(1) + 32(0.9239) + 12(0.7071) + 32(0.3827) + 7(0)\Big] = 0.999989
\end{aligned}
$$

$n = 5$:

$$
\begin{aligned}
I_{N5} &= \frac{5h}{288}\Big[19(f(x_0) + f(x_5)) + 75(f(x_1) + f(x_4)) + 50(f(x_2) + f(x_3))\Big] \\
&= \frac{\pi}{576}\Big[19(f(0) + f(\pi/2)) + 75(f(\pi/10) + f(2\pi/5)) + 50(f(\pi/5) + f(3\pi/10))\Big] \\
&= \frac{\pi}{576}\Big[19(\cos(0) + \cos(\pi/2)) + 75(\cos(\pi/10) \\
&+ \cos(2\pi/5)) + 50(\cos(\pi/5) + \cos(3\pi/10))\Big] \\
&= \frac{\pi}{576}\Big[19(1 + 0) + 75(0.9511 + 0.3090) + 50(0.8090 + 0.5878)\Big] = 0.9999995.
\end{aligned}
$$

Weddle's Rule $(n = 6)$:

$$
\begin{aligned}
I_W &= \frac{3h}{10}\Big[(f(x_0) + f(x_2) + f(x_4) + f(x_6)) + 5(f(x_1) + f(x_5)) + 6f(x_3)\Big] \\
&= \frac{\pi}{40}\Big[(f(0) + f(\pi/6) + f(\pi/3) + f(\pi/2)) \\
&+ 5(f(\pi/12) + f(5\pi/12)) + 6f(\pi/4)\Big] \\
&= \frac{\pi}{40}\Big[(\cos(0) + \cos(\pi/6) + \cos(\pi/3) + \cos(\pi/2)) \\
&+ 5(\cos(\pi/12) + \cos(5\pi/12)) + 6\cos(\pi/4)\Big] \\
&= \frac{\pi}{40}\Big[(1 + 0.8660 + 0.5 + 0) + 5(0.9659 + 0.2588) + 6(0.7071)\Big] = 0.9999996.
\end{aligned}
$$

Since the exact value of the given integral is 1.0, the corresponding actual errors are $0.214602, -0.00228, -0.001015, 0.000011, 0.0000005,$ *and* 0.0000004, *respec-*

tively, therefore, Weddle's approximation is better than the rest of the closed methods discussed.

5.6.2 Open Newton-Cotes Formulas

When using the open Newton-Cotes formulas, the data points $x_i = x_0 + ih$ are used for each $i = 0, 1, 2, \ldots, n$, where $x_0 = a + h$ and $h = (b - a)/(n + 2)$. This implies that $x_n = b - h$; so let endpoint $x_{-1} = a$ and $x_{n+1} = b$. Open formulas contain all the points used for approximating within the open interval (a, b). Thus, the formulas become

$$\int_a^b f(x)dx = \int_{x_{-1}}^{x_{n+1}} f(x)dx \approx \sum_{i=0}^{n} a_i f(x_i), \tag{5.66}$$

where

$$a_i = \int_{x_{-1}}^{x_{n+1}} L_i(x)dx. \tag{5.67}$$

The following theorem describes the error analysis associated with the open Newton-Cotes formulas.

Theorem 5.6 (Open Newton-Cotes formulas)

Suppose that $\sum_{i=0}^{n} a_i f(x_i)$ *denotes an* $(n+1)$*-point open Newton-Cotes formula with* $x_{-1} = a$ *and* $x_{n+1} = b$ *and* $h = (b - a)/(n + 2)$. *There exists* $\eta(x) \in (a, b)$ *for which*

$$\int_a^b f(x)dx = \sum_{i=0}^{n} a_i f(x_i) + \frac{h^{n+3} f^{(n+2)}(\eta(x))}{(n+2)!} \int_{-1}^{n+1} t^2(t-1)\cdots(t-n)dt \tag{5.68}$$

when n *is even and* $f \in C^{n+2}[a, b]$. *If* n *is odd and* $f \in C^{n+1}[a, b]$, *then*

$$\int_a^b f(x)dx = \sum_{i=0}^{n} a_i f(x_i) + \frac{h^{n+2} f^{(n+1)}(\eta(x))}{(n+1)!} \int_{-1}^{n+1} t(t-1)\cdots(t-n)dt. \tag{5.69}$$

The open Newton-Cotes formulas with their error terms that arise from approximating $y = f(x)$ with polynomial of degree $n = 0, 1, 2, 3, 4, 5$ on (a, b) are summarized as follows:

Mid-point Rule

If $n = 0$, then

$$\int_{x_{-1}}^{x_1} f(x)dx = 2hf_0 + \frac{h^3}{3}f''(\eta(x))$$

for $\eta(x) \in (x_{-1}, x_1)$.

Two-point Rule

If $n = 1$, then

$$\int_{x_{-1}}^{x_2} f(x)dx = \frac{3h}{2}\Big(f_0 + f_1\Big) + \frac{3h^3}{4}f''(\eta(x))$$

for $\eta(x) \in (x_{-1}, x_2)$.

Milne's Rule

If $n = 2$, then

$$\int_{x_{-1}}^{x_3} f(x)dx = \frac{4h}{3}\Big(2f_0 - f_1 + 2f_2\Big) + \frac{14h^5}{45}f^{(4)}(\eta(x))$$

for $\eta(x) \in (x_{-1}, x_3)$.

Four-point Rule

If $n = 3$, then

$$\int_{x_{-1}}^{x_4} f(x)dx = \frac{5h}{24}\Big(11f_0 + f_1 + f_2 + 11f_3\Big) + \frac{95h^5}{144}f^{(4)}(\eta(x))$$

for $\eta(x) \in (x_{-1}, x_4)$.

Five-point Rule

If $n = 4$, then

$$\int_{x_{-1}}^{x_5} f(x)dx = \frac{6h}{20}\Big(11f_0 - 14f_1 + 26f_2 - 14f_3 + 11f_4\Big) + \frac{41h^7}{140} f^{(6)}(\eta(x))$$

for $\eta(x) \in (x_{-1}, x_5)$.

Six-point Rule

If $n = 5$, then

$$\int_{x_{-1}}^{x_6} f(x)dx = \frac{7h}{1440}\Big[611f_0 - 453(f_1+f_4) + 562(f_2+f_3) + 611f_5\Big] + \frac{5257}{8640} h^7 f^{(6)}(\eta(x))$$

for $\eta(x) \in (x_{-1}, x_6)$.

Note that for each n, the error terms of closed and open formulas are of the same order but with the closed formulas having the smaller multiplicative constant. For example, with $n = 2$, the error term of the closed and open formulas are

$$-\frac{h^5}{90} f^{(4)}(\eta(x)) \quad \text{and} \quad \frac{14h^5}{45} f^{(4)}(\eta(x)), \tag{5.70}$$

respectively. However, the values of number $\eta(x)$ are different for these two terms, and it mostly happen that these numbers are such that the error in the closed formula is smaller than that of the corresponding open formulas. To show this, now we solve the Example 5.22 using open formulas for $n = 0, 1, 2, 3, 4, 5$.

Example 5.23 *Evaluate the integral* $\int_0^{\pi/2} \cos(x)dx$, *using the above open Newton-Cotes formulas for* $n = 0, 1, 2, 3, 4, 5$.

Solution. *Since* $f(x) = \cos(x)$, *therefore,*

n = 0:

$$I_{M0} = 2hf(x_0) = 2(\frac{\pi}{4})\cos(\pi/4) = 1.110721$$

n = 1:

$$I_{M1} = \frac{3h}{2}\Big[f(x_0) + f(x_1)\Big] = \frac{3\pi}{12}\Big[f(\pi/6) + f(\pi/3)\Big]$$

$$= \frac{\pi}{4}\Big[\cos(\pi/6) + \cos(\pi/3)\Big] = 1.072874$$

n = 2:

$$
\begin{aligned}
I_{M2} &= \frac{4h}{3}[2f(x_0) - f(x_1) + 2f(x_2)] \\
&= \frac{4}{3}(\frac{\pi}{8})[2f(\pi/8) - f(\pi/4) + 2f(3\pi/8)] \\
&= \frac{\pi}{6}[2\cos(\pi/8) - \cos(\pi/4) + 2\cos(3\pi/8)] = 0.997984
\end{aligned}
$$

n = 3:

$$
\begin{aligned}
I_{M3} &= \frac{5h}{24}[11f(x_0) + f(x_1) + f(x_2) + 11f(x_3)] \\
&= \frac{5}{24}(\frac{\pi}{10})[11f(\pi/10) + f(\pi/5) + f(3\pi/10) + 11f(2\pi/5)] \\
&= \frac{\pi}{48}[11\cos(\pi/10) + \cos(\pi/5) + \cos(3\pi/10) + 11\cos(2\pi/5)] = 0.998609
\end{aligned}
$$

n = 4:

$$
\begin{aligned}
I_{M4} &= \frac{6h}{20}[11(f(x_0) + f(x_4)) - 14(f(x_1) + f(x_3)) + 26f(x_2)] \\
&= \frac{\pi}{40}[11(f(\pi/12) + f(5\pi/12)) - 14(f(\pi/6) + f(\pi/3)) + 26f(\pi/4)] \\
&= \frac{\pi}{40}[11(0.9659258 + 0.258819) - 14(0.8660254 + 0.5) + 26(0.7071067)] \\
&= 1.0000171
\end{aligned}
$$

n = 5:

$$
\begin{aligned}
I_{M5} &= \frac{7h}{1440}[611(f(x_0)+f(x_5)) - 453(f(x_1)+f(x_4)) + 562(f(x_2)+f(x_3))] \\
&= \frac{\pi}{2880}[611(f(\pi/14)+f(3\pi/7)) - 453(f(\pi/7)+f(5\pi/14)) \\
&+ 562(f(3\pi/14)+f(2\pi/7))] \\
&= \frac{\pi}{2880}[611(\cos(\pi/14)+\cos(3\pi/7)) - 453(\cos(\pi/7)+\cos(5\pi/14)) \\
&+ 562(\cos(3\pi/14)+\cos(2\pi/7))] \\
&= \frac{\pi}{2880}[611(0.9749279+0.2225209) - 453(0.9009688+0.4338837) \\
&+ 562(0.7818314+0.6234898)] = 1.0000121
\end{aligned}
$$

Note that the exact value of the given integral is 1.0, and so the corresponding errors are $-0.110721, -0.072874, 0.002016, 0.001391, -0.0000171$, and -0.0000121, respectively.

It is noted that the comparison of the above results using open formulas with those of Example 5.22 using closed formulas shows that for this particular problem the approximation obtained from the closed formulas are more accurate than the results obtained from the corresponding open formulas.

Theorem 5.7 (Composite Midpoint Rule)

Let $f \in C^2[a,b]$, n be even, $h = (b-a)/(n+2)$, and $x_i = a + (i+1)h$ for each $i = -1, 0, 1, 2, \ldots, n+1$. Then the composite midpoint rule for $[(n/2)+1]$ subintervals with its error term can be written as

$$
\int_a^b f(x)dx = 2h\sum_{i=0}^{n/2} f(x_{2i}) + \frac{(b-a)}{6}h^2 f''(\eta(x)), \quad \eta 9x) \in (a,b). \tag{5.71}
$$

Example 5.24 *Evaluate the integral $\int_0^1 e^{4x}dx$ using the midpoint ruler with $n = 0, 2, 4$.*

Solution. *For $n = 0$, we have*

$$\begin{aligned} I_{M0} &= 2hf(x_0) \\ &= 2(0.5)f(0.5) = 1(e^2) = 7.389056. \end{aligned}$$

For $n = 2$, we have

$$\begin{aligned} I_{M2} &= 2h[f(x_0) + f(x_2)] \\ &= 2(0.25)[f(0.25) + f(0.75)] = (0.5)[e^1 + e^3] = 11.401909. \end{aligned}$$

For $n = 4$, we have

$$\begin{aligned} I_{M4} &= 2h[f(x_0) + f(x_2) + f(x_4)] \\ &= 2(1/6)[f(1/6) + f(1/2) + f(5/6)] = (0.5)[e^{2/3} + e^2 + e^{10/3}]. \\ &= 12.456138 \end{aligned}$$

Note that the exact value of the given integral is 13.39995, and so the corresponding errors are, 6.010894, 1.998041, and 0.943812, respectively.

5.7 Repeated Use of the Trapezoidal Rule

The most difficult problem in numerical integration lies in choosing the right number of intervals. It may not be sensible to start solving the problem with a large number of intervals and then hope for the best. It will not only result in inconvenience but also in the wasting of computer time. In many cases, accuracy can be improved by the subdivision of intervals rather than by using high-order Newton-Cotes formulas.

Here, and in the very next section, we shall show how to tackle this problem in a systematic and efficient manner. Suppose we wish to evaluate integral

$$I(f) = \int_a^b f(x)dx$$

and let I_n be the approximation to I obtained using the Trapezoidal rule with n strips. Our problem is deciding how large n should be so that I_n approximates I to the required accuracy. One possible approach would be to evaluate in turn $I_1, I_2, I_3, I_4, \ldots$ until two successive values agree to the desired accuracy, but this would obviously be very laborious.

A much more sensible approach is to evaluate in turn $I_1, I_2, I_4, I_8, I_{16}, \ldots$ The function values needed for I_n are also needed for I_{2n} so doubling the number of strips at each stage. This will, in theory, converge to the true value I and the process is terminated when two consecutive approximations agree to the desired accuracy. The procedure actually used is as follows (with h = b - a):

(a) Let $T_1 = \frac{1}{2}[f(a) + f(b)]$; then $I_1 = hT_1$.

(b) Let $T_2 = T_1 + [f(a + \frac{1}{2}h)]$; then $I_2 = \frac{1}{2}hT_2$.

(c) Let $T_4 = T_2 + [f(a + \frac{1}{4}h)] + [f(a + \frac{3}{4}h)]$; then $I_4 = \frac{1}{4}hT_4$.

(d) Let $T_8 = T_4 + [f(a + \frac{1}{8}h)] + [f(a + \frac{3}{8}h)] + [f(a + \frac{5}{8}h)] + [f(a + \frac{7}{8}h)]$; then $I_8 = \frac{1}{8}hT_8$; and so on.

Because of the relatively large error of the Trapezoidal rule, it could hardly be described as an efficient approach. If we ask for a maximum error of 0.00000001, the method goes as far as $n = 1024$, and it is time consuming.

Example 5.25 *Approximate the following integral using the repeated Trapezoidal rule with* $n = 8$:

$$\int_1^2 e^{-x/2} dx$$

Solution. *Table 5.1 shows the approximate results. So after 256 intervals, the value of the given integral is 0.477302589.*

Table 5.1: Solution of Example 5.25.

Number of Intervals	Estimates of the integral
1	0.487205051
2	0.479785802
4	0.477923763
8	0.477457799
16	0.477341280
32	0.477312148
64	0.477304865
128	0.477303044
256	0.477302589

To get the above results using MATLAB commands for the repeated Trapezoidal rule, first we define a function m-file as fn.m for the function as follows:

```
function y = fn(x)
y = exp(-x/2)
```

then use the m-file ReaptedTR.m and the following commands:

```
>> n = 8;
>> T = ReaptedTR('fn', 1, 2, n);
```

Program 5.4

MATLAB m-file for the Repeated Trapezoidal Rule

```
function T=ReaptedTR(fn,a,b,n)
n=1;h=b-a;
T(1) = (b - a) * (feval(fn, a) + feval(fn, b))/2;
for j=1:n
n = 2 * n; h = h/2; Sum = 0;
for p=1:n/2
x = a + h * (2 * p - 1); Sum=Sum+feval(fn,x); end
T(j + 1) = T(j)/2 + h * Sum; end
```

5.8 Romberg Integration

The procedure of the last section can be improved enormously by the technique called *Romberg integration.* Let a function be known at equally spaced intervals, or when explicitly known so that it can be computed as desired. Consider the following data:

x	x_0	x_1	x_2	$\cdots$	x_n
$f(x)$	f_0	f_1	f_2	$\cdots$	f_n

The first step in Romberg's method is to define a series of sums: $I_{11}, I_{12}, I_{13}, \ldots,$ where

$$\begin{aligned}
I_{11} &= (f_0 + f_n)/2; \ h = (b-a)/n, \quad \text{where} \quad n = 1 \\
I_{12} &= [I_{11} + f(a + h/2)] \\
I_{13} &= [I_{12} + f(a + h/4) + f(a + 3h/4)] \\
I_{14} &= [I_{13} + f(a + h/8) + f(a + 3h/8) + f(a + 5h/8) + f(a + 7h/8)].
\end{aligned}$$

From these sums, various values $T_{11}, T_{12}, T_{13}, T_{14}, \ldots$ (same as the Trapezoidal rule) are computed using the following relations:

$$\begin{aligned}
T_{11} &= hI_{11} \\
T_{12} &= (h/2)I_{12} \\
T_{13} &= (h/4)I_{13} \\
T_{14} &= (h/8)I_{14}.
\end{aligned}$$

With the values of $T_{11}, T_{12}, T_{13}, \ldots$, we compute the first-order Romberg integration (same as Simpson's rule) as follows:

$$\begin{aligned} T_{22} &= T_{12} + 1/3(T_{12} - T_{11}) \\ T_{23} &= T_{13} + 1/3(T_{13} - T_{12}) \\ T_{24} &= T_{14} + 1/3(T_{14} - T_{13}). \end{aligned}$$

With the values of $T_{22}, T_{23}, T_{24}, \ldots$, we compute the second-order Romberg

Table 5.2: Third-order Romberg table.

Integration sums	Trapezoidal Rule	Simpson's Rule	Boole's Rule	Third Order
I_{11}	T_{11}			
I_{12}	T_{12}	T_{22}		
I_{13}	T_{13}	T_{23}	T_{33}	
I_{14}	T_{14}	T_{24}	T_{34}	T_{44}

integration (the same as Boole's rule) as follows:

$$\begin{aligned} T_{33} &= T_{23} + 1/15(T_{23} - T_{22}) \\ T_{34} &= T_{24} + 1/15(T_{24} - T_{23}). \end{aligned}$$

With the values of $T_{33}, T_{34}, \ldots$, we compute the third-order Romberg integration as follows:

$$T_{44} = T_{34} + 1/63(T_{34} - T_{33}).$$

The general formula to calculate various values as shown in Romberg's Table 5.2 is

$$T_{j+1,k+1} = T_{j,k+1} + \{1/(4^j - 1)\}[T_{j,k+1} - T_{jk}]. \tag{5.72}$$

The procedure continues until the difference between two successive values is less than the required accuracy. Note that in each column, the bottom most number is the most accurate number.

The table is constructed row by row. The new Trapezoidal rule estimate $I_{1,k}$ is appended to the first column and the previously computed values in the *(k - 1)th* row are used to calculate $T_{1,k}, T_{2,k}, \ldots, T_{k,k}$.

Example 5.26 *Use the Romberg method with* $n = 4$ *to find approximations for the following integral*

$$\int_0^4 sin(x)dx = -\cos(4) + \cos(0) \approx 1.65364362.$$

Solution. *The solution of the given problem is listed in Table 5.3.*

Table 5.3: Romberg table for Example 5.26.

Intervals sums	Trapezoidal Rule	Simpson's Rule	Boole's Ruler	Third Order
1	-1.51360499			
2	1.061792336	1.92025814		
4	1.51348717	1.66405211	1.64697171	
8	1.61904831	1.65423535	1.65358090	1.65368581

To get the above results using MATLAB commands for the Romberg method, first we define a function m-file as fn.m for the function and m-file Romberg.m as follows:

```
function y = fn(x)
y = sin(x);
>> T = RombergM('fn', 0, 4, 4)
```

Program 5.5

MATLAB m-file for Romberg's Method

```
function T=RombergM(fn,a,b,n)
n=1; h=b-a;
T(1,1)=(b - a)*(feval(fn,a)+feval(fn,b))/2;
for i=2:n
h=h/2;Sum=0;
for p=1:n
x=a+h*(2*p-1); Sum=Sum+feval(fn,x); end
T(i,1) = T(i-1,1)/2 + h*Sum; n = 2*n;
for k=1:i-1
T(i,k+1)=(4 ^ k*T(i,k) - T(i-1,k))/(4 ^ k-1); end
```

5.9 Gaussian Quadratures

In the previous sections the common feature of numerical methods considered is that integrand $f(x)$ was evaluated at equal intervals within the range of integration. In contrast, here we discuss Gaussian integration, which requires the evaluation of an integrand at specified but unequal intervals.

The Gauss (or Gauss-Legendre) quadratures are numerical integration methods using Legendre points (roots of Legendre polynomials). The Gauss quadratures cannot be used to integrate a function given in tabular form with equal intervals because the Legendre points are not equally spaced, but they are rather suitable for integrating analytical functions. The advantage of Gauss quadratures is that their accuracy is significantly higher than Newton-Cotes formulas.

The general form of the rule is

$$\int_{-1}^{1} f(x)dx = \sum_{i=1}^{n} \alpha_i f(x_i). \tag{5.73}$$

The parameters α_i and x_i are chosen so that for a given n the rule is exact for polynomials, up to and including degree $(2n-1)$. The polynomial with the highest degree has $2n$ coefficients. This value matches the number of parameters x_1 through x_n and α_1 through α_n that are to be selected.

An example of the selection process is shown with n equal to 2. All polynomials up to degree three must be exactly represented by (5.73), by taking an integrand function $f(x)$ in (5.73) as x^k to form the four equations:

$$\int_{-1}^{1} x^k dx = \alpha_1(x_1)^k + \alpha_2(x_2)^k = \begin{cases} 0, & k = \text{odd} \\ \dfrac{2}{k+1}, & k = \text{even} \end{cases} \quad k = 1,2,3,4 \quad .$$

The solution of the following four equations

$$\begin{aligned} 2 &= \alpha_1 + \alpha_2 \\ 0 &= \alpha_1 x_1 + \alpha_2 x_2 \\ 2/3 &= \alpha_1 x_1^2 + \alpha_2 x_2^2 \\ 0 &= \alpha_1 x_1^3 + \alpha_2 x_2^3 \end{aligned}$$

is as follows

$$\alpha_1 = \alpha_2 = 1 \quad \text{and} \quad x_2 = -x_1 = \frac{1}{\sqrt{3}}.$$

The derivation of this integrating formula is the simplest member of the Gauss quadratures. In general the points x_1 through x_n for Gauss quadratures are the roots of the Legendre polynomial $p_n(x)$, which is defined by

$$\begin{aligned} p_0(x) &= 1; \quad p_1(x) = x; \\ p_n(x) &= [(2n-1)xp_{n-1}(x) - (n-1)p_{n-2}(x)]/(n+1); \quad n = 2, 3, \ldots \end{aligned} \tag{5.74}$$

The values of α_1 through α_n can obtained from

$$\alpha_i = 2[1 - x_i^2]/[np_{n-1}(x_i)]^2, \quad i = 1, 2, \ldots, n \tag{5.75}$$

Table 5.4 has been produced for the values of the weights α_i and x_i (the Gauss points) for various values of n (number of the Gauss points). The $\pm$ signs in the table mean that the x values of the Gauss points appear in a pair, one of which is plus and the other minus.

Table 5.4: Gauss Points for $n = 2, 3, 4, 5, 6, 8, 10, 12$.

Gauss points & weights		
n	$\pm x_i$	α_i
2	0.5773503	1.0000000
3	0.0000000	0.8888889
	0.7745967	0.5555556
4	0.3399810	0.6521452
	0.8611363	0.3478549
5	0.0000000	0.5688889
	0.5384693	0.4786287
	0.9061799	0.2369269
6	0.2386192	0.4679139
	0.6612094	0.3607616
	0.9324695	0.1713245
8	0.1834346	0.3626838
	0.5255324	0.3137067
	0.7966665	0.2223810
	0.9602899	0.1012285
10	0.1488743	0.2955242
	0.4333954	0.2692667
	0.6794096	0.2190864
	0.8650634	0.1494514
	0.9739065	0.0666713
12	0.1252334	0.2491471
	0.3678315	0.2334925
	0.5873180	0.2031674
	0.7699027	0.1600783
	0.4091173	0.1069393
	0.9815606	0.0471753

The Gauss integration formula may be applied to any arbitrary interval $[a, b]$ with the transformation

$$x = \frac{2z - a - b}{b - a},$$

where z is the original coordinate in $a \leq z \leq b$, and x is the normalized coordinate in $-1 \leq x \leq 1$. The transformation from x to z is

$$z = \frac{(b-a)x + a + b}{2}. \quad (5.76)$$

Using this transformation, integral (5.73) may be written as

$$\int_a^b f(z)dz = \int_{-1}^{1} f(z)(dz/dx)dx \approx \frac{b-a}{2} \sum_{k=1}^{n} \alpha_k f(z_k), \quad (5.77)$$

where $dz/dx = (b-a)/2$ is used. The values of z_k are obtained by substituting x in (5.76) by the Gauss points, namely

$$z_k = \frac{(b-a)x_k + a + b}{2}.$$

For example, suppose $n = 2$, $a = 0$, and $b = 2$. Since the Gauss points x_k for $n = 2$ on the normalized coordinate x $-1 \leq x \leq 1$ are ± 0.57735, the corresponding points on z are

$$z_1 = \frac{1}{2}[(2-0)(-0.57735) + 0 + 2] = 0.42265$$

$$z_2 = \frac{1}{2}[(2-0)(0.57735) + 0 + 2] = 1.57735.$$

The derivative is

$$dz/dx = (b-a)/2 = 1.$$

So the Gauss quadrature becomes

$$\int_0^2 f(z)dz = \int_{-1}^{1} f(z)(dz/dx)dx \approx (1)[(1)f(0.42265) + (1)f(1.57735)]. \quad (5.78)$$

Example 5.27 *Find the four-term Gauss-Legendre approximations to the following integral*

$$\int_{-1}^{1} \frac{1}{1+x^2} dx.$$

Solution. *Since* $f(x) = \dfrac{1}{1+x^2}$, $n = 4$, $a = -1$, *and* $b = 1$, *then from (5.77), we have*

$$\begin{array}{rcl} \displaystyle\int_{-1}^{1} \frac{1}{1+x^2} dx & = & \alpha_1 f(x_1) + \alpha_2 f(x_2) + \alpha_3 f(x_3) + \alpha_4 f(x_4) \\ & = & 0.34785485 f(0.861136312) + 0.652145155 f(0.339981044) \\ & + & 0.652145155 f(-0.339981044) + 0.34785485 f(0.861136312) \\ & = & 1.5686276 \end{array}.$$

For comparison, we give the true value of the given integral and the other estimates of integrals with the four-points quadrature.

$$\begin{array}{lrcl} \textit{True value:} & I & = & 1.5707963 \\ \textit{Trapezoidal rule:} & I & \approx & 1.550000 \\ \textit{Simpson's rule:} & I & \approx & 1.566667 \\ \textit{Romberg integration:} & I & \approx & 1.565588 \end{array}$$

In this example we note that the estimate from the Gaussian quadrature is better than the estimates from any of the other three quadratures for the same number of control points.

Example 5.28 *Find the two-terms and the four-terms Gauss-Legendre approximations to the integral* $\int_0^2 x e^{x^2} dx$.

Solution. *Firstly, take* $n = 2$, $a = 0$, *and* $b = 2$, *then using (5.78), we have*

$$\begin{array}{rcl} \displaystyle\int_0^2 z e^{z^2} dz & \approx & (1)[(1)(0.42265 e^{(0.42265)^2}) + (1)(1.57735 e^{(1.577735)^2})] \\ & \approx & 19.49277 \end{array}.$$

For $n = 4$, *the Gauss points* x_k *on the normalized coordinate* x, $-1 \leq x \leq 1$ *are* ± 0.33998 *and* ± 0.86114, *so the corresponding points of* z *are calculated as follows:*

$$z_1 = \frac{1}{2}[(2-0)(-0.33998) + 0 + 2] = 0.66002,$$

$$z_2 = \frac{1}{2}[(2-0)(0.33998) + 0 + 2] = 1.33998,$$

$$z_3 = \frac{1}{2}[(2-0)(-0.86114)+0+2] = 0.13886,$$

$$z_4 = \frac{1}{2}[(2-0)(0.86114)+0+2] = 1.86114.$$

The derivative is

$$dz/dx = (b-a)/2 = 1.$$

So the Gauss quadrature becomes

$$\begin{aligned}\int_0^2 ze^{z^2}dz &\approx (1)[0.65215f(0.66002)+0.65215f(1.33998)\\ &+ 0.34786f(0.133886)+0.34786f(1.86114)]\\ &\approx [0.66578+5.26310+0.04924+20.67812]\\ &\approx 26.65624.\end{aligned}$$

The exact solution of the given problem is 26.799075. *Compared with the two-terms and four-terms Gauss quadrature, the actual errors are* 7.306305 *and* 0.142835, *respectively, showing that approximation by four-terms is better.*

To get the above results using the MATLAB command for the Gauss-Legendre formula, first we define a function m-file as fn.m and QaussianQ. m file as follows:

```
function y = fn(x)
y = sin(x);
>> GQ = QaussianQ('fn', 0, 4, 4)
```

Note that the Gauss-Legendre n-points formula is exact for polynomials of degree $\leq 2n-1$, which is about twice that of n-points Newton-Cotes formulas.

Program 5.6

MATLAB m-file for the Gauss-Legendre quadrature

```
function GQ=GaussianQ(fn,a,b,n)
if (n == 2)
z(1) = 0.5774; z(2) = - z(1);
alpha(1) = 1;alpha(2) = 1; end
if (n == 3)
z(1) = 0.0000; z(2) = 0.7746; z(3) = -z(2);
alpha(1) = 0.8889;alpha(2) = 0.5556;
alpha(3)= alpha(2); end
if (n == 4)
z(1) = 0.3340; z(2) = -z(1); z(3) = 0.8611; z(4) = -z(3);
alpha(1) = 0.6522;alpha(2) = alpha(1);
alpha(3)= 0.3479;alpha(4)=alpha(3); end
if (n == 5)
z(1) = 0.0000; z(2) = 0.5385;z(3)=-z(2);
z(4) = 0.9062; z(5) = -z(4);
alpha(1) = 0.5689;alpha(2) = 0.4786;alpha(3)=alpha(2);
alpha(4)= 0.2369;alpha(5)=alpha(4); end sum = 0;
for i=1:n
x = ((b - a) * z(i) + a + b)/2; gn=feval(fn,x);
sum = sum + alpha(i) * gn; end; GQ = sum * (b - a)/2;
```

5.10 Summary

In the first part of the chapter we discussed several methods for approximating the derivatives of the function. The finite difference method is a differentiation technique that uses numerical differences of function values to estimate derivatives. It is assumed that the step size h between the data points is constant. If n is the number of points, then truncation error for the kth derivative is of order $O(h^{n-k})$. We used forward-difference, backward-difference, and central-difference methods for the approximation of the first and the second derivatives of the function at the given point. The forward-difference methods use points in front of the evaluation point, the backward-difference methods use points in back of the evaluation point, and the central-difference methods use points that are symmetrically distributed about the evaluation point. A central-difference approximation of the first derivative requires only $(n-1)$ function values because the coefficient of the central point is always zero. Numerical differentiation is an inherently ill-conditioned process, which tends to amplify additive high-frequency noise that may be present in the data points.

The theme of the second part of the chapter concerned the development of simple numerical methods for estimating the value of the integral $\int_a^b f(x)\,dx$. For approximating the integral of $f(x)$ between a and b, we used Newton-Cotes integration techniques. Newton-Cotes formulas are based on approximating the integrand, $f(x)$, with an interpolating polynomial. The Trapezoidal rule uses piecewise linear interpolation and has a truncation error of order $O(h^2)$. Improved performance is obtained with Simpson's 1/3 rule, which uses piecewise-quadratic interpolation and has a truncation error of order $O(h^4)$. We also used some higher-order closed Newton-Cotes formulas for the approximation of the given definite integral. The Trapezoidal and Simpson's rules are closed integration formulas, which require the values of $f(x)$ at the two endpoints. In contrast, the midpoint rule is an open integration formula, which does not require $f(a)$ and $f(b)$. Open formulas are useful when $f(x)$ has a singularity at one or both ends of the integration interval. The midpoint rule uses piecewise constant interpolation and has a truncation error of order $O(h^2)$. In general, if the integrand is approximated with a polynomial of degree m, the truncation error will be of order $O(h^{m+1})$ when m is odd and $O(h^{m+2})$ when m is even. Simpson's 1/3 rule and the midpoint rule require that the number of integration intervals, n, be even. In the last section, we also discussed alternative integration methods of higher accuracy called the repeated Trapezoidal rule, Romberg integration method, and the Gauss-Legendre formulas.

5.11 Exercises

1. Let $f(x) = (x-1)e^x$ and take $h = 0.01$.
   (a) Calculate approximation to $f'(2.3)$ using the two-point forward-difference formula. Also, compute the actual error and an error bounds for your approximation.
   (b) Solve part *(a)* using the two-point backward-difference formula.

2. Solve Problem 1 for $f(x) = (x^2 + x + 1)e^{2x}$ with $h = 0.05$.

3. Using the following data

$$(1.2, 11.6), (1.29, 13.8), (1.3, 14), (1.31, 14.3), (1.4, 16.8)$$

and compute the best approximations of $f'(1.3)$ using the two-point forward-difference and backward-difference formulas.

4. Let $f(x) = \sin(x+1), h = 0.1$. Compute the approximations of $f'(\frac{\pi}{4})$ using the two-point forward-difference and backward-difference formulas. Compute the actual errors and also find error bounds using the error formulas.

5. Use the three-point central-difference formula to compute the approximate value for $f'(5)$ with $f(x) = (x^2+1)\ln x$ and $h = 0.05$. Compute the actual error and the error bounds for your approximation.

6. Use the three-point central-difference formula to compute the approximate value for $f'(2)$ with $f(x) = e^{x/2} + 2\cos x$ and $h = 0.01$. Compute the actual error and the error bounds for your approximation.

7. Solve Problem 3 to find the best approximation of $f'(1.3)$ using the three-point forward-difference and the backward-difference formulas.

8. Using the following data

$$(1.0, 2.0), (1.5, 1.94), (2.0, 2.25), (3.0, 3.11)$$

find the best approximate values for $f'(1.5)$, $f'(1.0)$, and $f'(3.0)$ using suitable three-point formulas.

9. Use all three-point formulas to compute the approximate value for $f'(2)$ for the derivative of $f(x) = e^{x/2} + x^3$, taking $h = 0.1$. Also, compute the actual errors and error bounds for your approximation.

10. Use all three-point formulas to compute the approximate value for $f'(2.2)$ for the derivative of $f(x) = x^2 e^x - x + 1$, taking $h = 0.2$. Also, compute the actual errors and error bounds for your approximation.

11. Use the most accurate formula to determine approximations that will complete the following table:

x	f(x)	$f'(x)$
2.1	-1.709847	
2.2	-1.373823	
2.3	-1.11921	
2.4	-0.916014	

12. The data in Problem 11 were taken from function $f(x) = \tan(x)$. Compute the actual errors in Problem 11 and also find the error bounds using the error formulas.

13. Use the most accurate formula to determine approximations that will complete the following table:

x	f(x)	$f'(x)$
8.1	16.94410	
8.3	17.56492	
8.5	18.19056	
8.7	18.82091	

14. The data in Problem 13 were taken from function $f(x) = x \ln x$. Compute the actual errors in Problem 13 and also find the error bounds using the error formulas.

15. Let $f(x) = x + \ln(x+2)$, with $h = 0.1$. Use the five-point formula to approximate $f'(2)$. Find the error bound for your approximation and compare the actual error to the bound.

16. Consider the following data

$$(1.1, 0.607), (1.2, 0.549), (1.4, 0.497), (1.5, 0.449), (1.7, 0.427), (1.8, 0.387).$$

Use the five-point formulas to compute the best approximations for $f'(1.3)$ and $f'(1.6)$.

17. Let $f(x) = x^2+1$, with $h = 0.2$. Use the five-point formula to approximate $f'(1.8)$.

18. Let $f(x) = e^{-2x}$, with points $x = 0.25, 0.5, 0.75, 1.25, 1.50$. Use the three-point central-difference formula and the five-point formula to approximate $f'(1.0)$. Also, compute the error bounds for your approximations.

19. Let $f(x) = e^{x^2}$. Use the five-point formula to compute the best approximations for $\dfrac{d^2 f}{dx^2}$ at 0.5 and $h = 0.1$. Find bound on the error for your approximation.

20. Let $f(x) = x + \ln(x+2)$, with $h = 0.1$. Use the three-point formula to approximate $f''(2)$. Find the error bounds for your approximation and compare the actual error to the bound.

21. Let $f(x) = x \sin x$, with $h = 0.1$. Use the five-point formula to approximate $f''(1)$. Find the error bounds for your approximation.

22. Consider the following data

$$(0.2, 0.39), (0.4, 1.08), (0.6, 1.49), (0.8, 1.78), (1, 2), (1.2, 2.18), (1.4, 2.34).$$

Use the five-point formula to compute the best approximation for $f''(0.6)$ and $f''(1.0)$. The data in this problem were taken from function $f(x) = \ln x + 2$. Compute the actual errors and also find the error bounds of your approximations.

23. Let $f(x) = 1/x$, with $h = 0.2$. Use the five-point formula to approximate $f''(2)$. Find the error bounds for your approximation.

24. For the three-point central-difference formula for $f'(x)$, perform an error analysis to show that the optimum stepsize h_{opt} is given by

$$h = h_{opt} = \sqrt[3]{\frac{3}{2M}} \times 10^{-t},$$

where $M = \max|f'''(x)|$.

25. Approximate the integral $\int_0^2 x^2 e^{-x^2} dx$ using suitable composite rules with $n = 4$ and $n = 6$.

26. The following values of a function $f(x) = \tan x / x$ are:

x	f(x)	x	f(x)
1.00	1.5574	1.40	4.1342
1.10	1.7862	1.50	9.4009
1.20	2.1435	1.60	-21.3953
1.30	2.7709		

Find $\int_{1.0}^{1.6} f(x)dx$ using the Trapezoidal rule with $h = 0.1$.

27. Use a suitable composite integration formula to approximate the integral $\int_0^1 \frac{dx}{2e^x - 1}$, with $n = 5$.

28. Use a suitable composite integration formula for the approximation of the integral $\int_1^2 \frac{dx}{3-x}$, with $n = 5$. Compute an upper bound for your approximation.

29. Use the composite Trapezoidal rule for the approximation of the integral $\int_1^3 \frac{dx}{7-2x}$ with $h = 0.5$. Also, compute an error term.

30. Find the stepsize h so that the absolute value of the error for the composite Trapezoidal rule is less than 5×10^{-4} when it is used to approximate the integral $\int_2^7 \frac{dx}{x}$.

31. Estimate the integral $\int_{-1}^1 \frac{dx}{1+x^2}$ using Simpson's rules with $n = 8$.

32. Repeat Problem 26 using the composite Simpson's rule.

33. Repeat Problem 25 using the composite Simpson's rule.

34. Use the composite Trapezoidal and the Simpson's rules to approximate the integral $\int_1^2 \frac{dx}{4x+1}$ such that the error does not exceed 10^{-2}.

35. Evaluate $\int_0^1 e^{x^2} dx$ by Simpson's rule choosing h small enough to guarantee five decimal accuracy. How large can h be?

36. Use a suitable composite integration rule to find the best approximate value of the integral $\int_1^2 \sqrt{1+\sin x}\, dx$, with $h = 0.1$. Estimate the error bound.

37. Evaluate the integral $\int_1^2 \ln x\, dx$ using closed Newton-Cotes formulas for $n = 3$ and $n = 4$. Also, compute the error bounds for your approximations.

38. Evaluate the following integrals using closed Newton-Cotes formulas for $n = 4, 5$, and 6. Also, compute the error bounds for your approximations:

$$(a) \int_0^1 (x + e^{2x})\, dx \quad (b) \int_0^{\pi/2} \cos x\, dx \quad (c) \int_2^3 (x + \ln x)\, dx.$$

39. Evaluate the integral $\int_0^2 \sqrt{x+1}\,dx$ using Weddle's rule and compute the error bound a for your approximation.

40. Evaluate the following integrals using closed Newton-Cotes formulas for $n = 1, 2, 3, 4, 5$, and 6. Also, compute the error bounds for your approximations:

$$(a)\ \int_0^1 (x^2+\sqrt{x+1})\,dx \quad (b)\ \int_0^{\pi/2} (\sin x - x)\,dx \quad (c)\ \int_1^2 (\sqrt{x} + \ln x)\,dx.$$

41. Evaluate the integral $\int_0^1 \sqrt{x+4}\,dx$ using open Newton-Cotes formulas for $n = 0, 1$ and $n = 2$.

42. Evaluate the following integrals using open Newton-Cotes formulas for $n = 0, 1, 2, 3, 4$, and 5. Also, compute the error bounds for your approximations:

$$(a)\ \int_1^2 (x + e^x)\,dx \quad (b)\ \int_0^{\pi} (x + \sin x)\,dx \quad (c)\ \int_2^3 \ln(x^2+4)\,dx.$$

43. Evaluate the integral $\int_1^2 \ln(x+1)\,dx$ using the four-point open Newton-Cotes formula and compute the error bound for your approximation.

44. Repeat Problem 38 using the open Newton-Cotes formulas for $n = 0, 1, 2, 3, 4$, and 5.

45. Evaluate the integral $\int_0^1 (x^2 + e^x)\,dx$ using the composite midpoint rule with $n = 4$. Compute the absolute error and the error bounds for your approximation.

46. Evaluate each of the following integrals using the composite midpoint rule with $n = 4$ and compute the error bounds for your approximation:

$$(a)\ \int_1^2 \ln(x+3)\,dx \quad (b)\ \int_0^{\pi} (x^2+x)\sin x\,dx \quad (c)\ \int_0^1 (\sqrt{x+1} + e^{3x})\,dx.$$

47. Approximate each of the following integrals using the repeated Trapezoidal ruled with $n = 4$:

$$(a)\ \int_2^3 (x\ln(x+1)+\cos x)\,dx \quad (b)\ \int_0^{\pi} x\tan x\,dx \quad (c)\ \int_1^2 \ln(x+1)\,dx.$$

48. Approximate each of the following integrals using the repeated Trapezoidal rule with $n = 4$:

$$(a) \int_1^2 \sqrt{x^2+2x+1}\,dx \quad (b) \int_0^2 (x^2 \sin x + 1)\,dx \quad (c) \int_2^3 \frac{\sqrt{x}+5x}{x-1}\,dx.$$

49. Approximate each of the following integrals using Romberg's method with $n = 4$:

$$(a) \int_1^2 \sqrt{x^2+4}\,dx \quad (b) \int_0^{\pi} x\cos x\,dx \quad (c) \int_0^1 \frac{1+x}{1+x^2}\,dx.$$

50. Evaluate the following integrals using Romberg's method with $n = 4$:

$$(a) \int_0^1 e^{-x^2}\,dx \quad (b) \int_0^2 \sqrt{x+5}\,dx \quad (c) \int_0^1 \frac{dx}{1+x^2}dx.$$

51. Approximate each of the following integral using Romberg's method with $n = 4$:

$$(a) \int_1^2 \ln(x^2+1)\,dx \quad (b) \int_1^2 (x+\sin x)\,dx \quad (c) \int_2^3 \frac{(\sqrt{x}+x)}{1+x}\,dx.$$

52. Find the two-terms and four-terms Gauss-Legendre approximations to each of the following integrals and compare the approximation with the exact answer:

$$(a) \int_1^2 (x+\ln(x+1))\,dx \quad (b) \int_0^1 (x+\cos x)\,dx \quad (c) \int_2^3 \frac{1}{5+x}\,dx.$$

53. Find the four-terms Gauss-Legendre approximations to the integral $\displaystyle\int_0^2 \frac{1}{1+x}\,dx$. Compare the approximations with the exact answer.

54. Find the six-terms and the eight-terms Gauss-Legendre approximations to the integral $\displaystyle\int_{-1}^1 \frac{1}{1+x^2}dx$. Compare the approximations with the exact answer and the approximations given by the Trapezoidal and Simpson's rules.

55. Find the two-terms and four-terms Gauss-Legendre approximations to each of the following integrals and compare the approximations with the exact answer:

$$(a)\ \int_0^{\pi/2} \cos x\, dx \quad (b)\ \int_0^2 (e^x + \tan x))\, dx \quad (c)\ \int_0^1 (x + e^x)\, dx.$$

56. Find the four-terms and six-terms Gauss-Legendre approximations to each of the following integrals and compare the approximations with the exact answer:

$$(a)\ \int_0^{\pi/3} \sqrt{\sin x}\, dx \quad (b)\ \int_0^2 \frac{1}{1+x^2}\, dx \quad (c)\ \int_0^1 e^{-x^2}\, dx.$$

Chapter 6

Ordinary Differential Equations

6.1 Introduction

The differential equations are of fundamental importance in engineering mathematics because many physical laws of biology, chemistry, ecology, economics, business, and relations appear mathematically in the form of such equations.

In this chapter firstly we discuss first-order ordinary differential equations and sets of simultaneous first-order differential equations, since the *nth*-order differential equation may be solved by transforming it to a set of n-simultaneous first-order differential equations. All the specified conditions are on the same endpoints. These are *initial-value problems*. Many numerical methods are discussed for the approximate solutions of such initial value problems. In the end of the chapter we show how to approximate the solution to *boundary-value problems*, differential equations with conditions imposed at different points. For first-order differential equations, only one condition is specified, so there is no distinction between initial value and boundary value problems. We will also consider second-order equations with two boundary values.

Definition 6.1 (Differential Equation)

A differential equation is an equation involving functions and their derivatives. For example, the following equations

$$(a)\quad \frac{dy}{dx} = 3x \qquad (b)\quad \frac{d^2y}{dx^2} + 4\frac{dy}{dx} + y = 0$$

$$(c)\quad \frac{dy}{dx} = x^2 + y^2 \qquad (d)\quad \left(\frac{d^3y}{dx^3}\right)^2 - 5\frac{d^2y}{dx^2} + 2y = 5$$

are differential equations.

For the sake of completeness, we shall define some of the standard terms for differential equations.

Definition 6.2 (Dependent Variable)

The variable that has been differentiated. For example, in each of the above differential equations (a)-(d), y is the dependent variable.

Definition 6.3 (Independent Variable)

The variable with respect to which the differentiation is performed. For example, in each of the above differential equations (a)-(d), x is the independent variable.

Definition 6.4 (Order of Differential Equation)

The order of the differential equation is the order of the highest derivative involved. For example, differential equations (a) and (c) above are of the first-order since the highest derivatives that appear are of first-order, whereas differential equations (b) and (d) are respectively, second-order and third-order.

Definition 6.5 (Degree of Differential Equation)

The degree of the differential equation is the power to which the highest-order derivative is raised. For example, differential equations (a)-(c) are of degree one while differential equation (d) is of degree two.

Definition 6.6 (Linear Differential Equation)

An differential equation is linear if

(1) *The dependent variable y and all its derivatives are of the first degree; that is, the power of each term involving y or its derivatives is one.*

(2) *Each coefficient depends only on the independent variable x or the constant.*

For example, the above differential equations *(a)* and *(b)* are linear differential equations while the differential equations *(c)* and *(d)* are nonlinear differential equations.

Definition 6.7 (Initial Conditions)

When all of the conditions are given at the starting value of independent variable x to solve a given differential equation this condition is called an initial condition. When the conditions are given at the endpoints of x-values, the conditions are called boundary conditions.

6.1.1 Classification of Differential Equations

There are two major types of differential equations called *ordinary differential equations (ODE)* and *partial differential equations (PDE)*. If an equation contains only ordinary derivatives of one or more dependent variables, with respect to a single independent variable, it is then said to be an *ordinary differential equation*. For example, all the differential equations *(a)-(d)* are ordinary differential equations because there is only one independent variable called x.

An equation involving partial derivatives of one or more dependent variables of two or more independent variables is called a *partial differential equation*. For example, the following differential equation

$$\frac{\partial^2 y}{\partial x^2} = c\frac{\partial^2 y}{\partial t^2}$$

is the partial differential because it involves two independent variables, x and t.

Although partial differential equations are very useful and important, their study demands a good foundation in the theory of ordinary differential equations. Consequently, in this chapter in the discussion that follows we shall confine our attention to ordinary differential equations.

As a mathematical form, the ordinary differential equation is a very useful tool. It is used in modeling of a wide variety of physical phenomena; that is, chemical reactions, satellite orbits, vibrating or oscillating systems, electrical networks, and so on. In many cases, the independent variable represents time so that a differential equation describes change, with respect to time, in the system being modeled. The solution of a differential equation will be the representation of the state of the system at any point in time and one can use it to study the behavior of the system. Consequently, the problem of finding the solution of a differential equation plays an important role in scientific research.

The *solution* of a differential equation is the function which satisfies the differential equation. In solving a differential equation analytically, one usually computes a general solution containing an arbitrary constant. The simplest form of the differential equation is

$$y' = f(x) \tag{6.1}$$

with $f(x)$ as the given function. The general solution of this equation is

$$y(x) = \int f(x)dx + C, \tag{6.2}$$

where C is an arbitrary constant. For example, the differential equation of the form

$$y' = \cos x \tag{6.3}$$

has a general solution of the form

$$y(x) = \sin x + C. \tag{6.4}$$

The more general equation is

$$y' = f(x, y(x)). \tag{6.5}$$

Since the general solution of a differential equation depends on an arbitrary constant C, this constant can be calculated by specifying the value of the function $y(x)$ at a particular point x_0

$$y(x_0) = y_0.$$

The point x_0 is called the initial point, and the number y_0 is called the initial value. We call the problem of solving

$$y' = \frac{dy}{dx} = f(x, y); \quad x_0 \leq x \leq x_n, \quad y(x_0) = y_0 \tag{6.6}$$

the *initial-value problem (IVP)*. For example, for finding the solution of differential equation (6.3) satisfying $y(\pi) = 1$, we have the value of the constant $C = 1$, so (6.4) becomes

$$y(x) = \sin x + 1$$

and it is called the particular solution of differential equation (6.3), or the solution of the initial-value problem

$$y' = \cos x, \qquad y(\pi) = 1.$$

The main concern of this chapter is approximating the solution to problem (6.6). Initial-value problems are problems in which the value of the dependent variable y is known at a point x_0. Such a large number of methods are available to handle problems of this type that one may have difficulty in deciding which to use. Solving initial-value problems numerically we will assume that the solution is being sought on a given finite interval $x_0 \leq x \leq x_n$ with $h = (b-a)/n$, where $x_0 = a, x_n = b$ and n are the number of subintervals. In this chapter the most widely used numerical methods are discussed in some detail to find the solutions of initial-value problems. If the analytical process of finding exact solution $y(x)$ is not feasible, it is still useful to know whether a solution exists and unique using numerical methods. To make precise the preceding discussion, we give the following theorem, which gives a sufficient condition for the existence and uniqueness of initial-value problem (6.6).

Theorem 6.1 (Existence and Uniqueness Theorem)

Let $f(x, y)$ and $\dfrac{\partial f}{\partial y}$ be continuous functions of x and y at all points (x, y) in some neighborhood of the initial point (x_0, y_0). Then there is a unique function $y(x)$ defined on some interval $[x_0 - \epsilon, x_0 + \epsilon]$ and satisfying

$$y'(x) = f(x, y(x)), \quad y(x_0) = y_0, \quad x \in [x_0 - \epsilon, x_0 + \epsilon], \quad \epsilon > 0. \tag{6.7}$$

For example, the initial-value problem

$$y'(x) = 2xy^2, \qquad y(0) = 1$$

has a unique solution

$$y(x) = \frac{1}{1 - x^2}, \quad -1 < x < 1$$

because both functions

$$f(x, y) = 2xy^2, \qquad \frac{\partial f}{\partial y} = 4xy$$

are continuous for all (x, y). Note that this example also shows that the continuity of the function $f(x, y)$ and $\frac{\partial f}{\partial y}$ for all (x, y) does not imply the existence of a $y(x)$ that is continuous for all x.

6.2 Numerical Methods for Solving IVP

By a numerical method for solving the initial-value problem (6.6) shows a procedure for finding the approximate values $y_0, y_1, \ldots, y_n$ of the exact solution $y(x)$ at the given points $x_0 < x_1 < \ldots < x_n$. We will let y_i denote the numerical value obtained as an approximation to the exact solution $y(x_i)$, with $x_i = x_0 + ih$ for $i = 0, 1, \ldots, n$, where h (constant) is the size of the interval. Numerical methods for differential equations are of great importance to the engineer and physicist because practical problems often lead to differential equations that cannot be solved by any analytical method or to equations for which the solutions in terms of formulas are so complicated they are often calculated as a table of values by applying a numerical method to such an equation. Two different types of numerical methods are available to solve initial-value problem (6.6). These are called *single-step* and *multi-steps* methods. The methods discussed will vary in complexity, since in general, the greater the accuracy of a method, the greater its complexity. We shall discuss many numerical methods for solving the approximate solution of initial-value problems (6.6) and the error analysis of each of the methods is explained in detail. Firstly, we discuss the single-step methods for solving problem (6.6).

6.3 Single-step Methods for IVP

This type of method refers to estimate $y'(x)$ from the initial condition $y(x_0) = y_0$ and $y'_0 = f(x_0, y_0)$ from (6.6) and proceeds step-wise. In the first step we compute an approximate value y_1 of the solution $y(x)$ at $x = x_1 = x_0 + h$. In the second step we compute an approximate value y_2 of that solution at $x = x_2 = x_0 + 2h$ and so on. Although these methods generally use functional evaluation information for x_i and x_{i+1}, they do not retain that information for direct use in future approximations. All the information used by these methods is consequently obtained within the interval over which the solution is being approximated. Among them we will discuss here Euler's method, Taylor's method of higher-orders, and the Runge-Kutta method of different orders.

6.3.1 Euler's Method

One of the simplest and most straight-forward numerical methods for solving first-order ordinary differential equations of form (6.6) is called *Euler's Method.* This method is not an efficient numerical method and so is seldom used, but it is relatively easy to analyze and many of the ideas involved in the numerical solution of differential equations are introduced most simply with it.

In principle, Euler's method uses the forward-difference formula approximation of $y'(x)$, which we discussed in Chapter 5. That is,

$$y' = \frac{dy}{dx} \approx \frac{y(x_{i+1}) - y(x_i)}{h}, \tag{6.8}$$

where h is the stepsize and is equal to $x_{i+1} - x_i$. Given that $\dfrac{dy}{dx} = f(x, y)$ and the initial conditions $x = x_0$, $y(x) = y(x_0)$, we have

$$\frac{y(x_1) - y(x_0)}{h} \approx f(x_0, y(x_0)),$$

which implies that

$$y(x_1) \approx y(x_0) + hf(x_0, y(x_0)),$$

which shows that $y(x_1)$ is approximately given by $y(x_0) + hf(x_0, y(x_0))$. We can now use this approximation for $y(x_1)$ to estimate $y(x_2)$; that is,

$$y(x_2) \approx y(x_1) + hf(x_1, y(x_1))$$

and so on. In general,

$$y(x_{i+1)} \approx y(x_i) + hf(x_i, y(x_i)), \quad i = 0, 1, \ldots, n-1.$$

Taking $y_i \approx y(x_i)$, for each $i = 1, 2, \ldots, n$, we have

$$y_{i+1} = yx_i + hf(x_i, y_i), \quad i = 0, 1, \ldots, n-1. \tag{6.9}$$

This simple integration strategy is known as Euler's method, or the *Euler-Cauchy method.* It is called an explicit method because the value of $y(x)$ at the next step is calculated only from the value of $y(x)$ at the previous step.

Given the approximate formula, one can solve for y_{i+1} in terms of x_i, y_i, and $f(x_i, y_i)$, all of which are known. Note that formula (6.9) can be derived, using the Taylor series expansion of unknown solution $y(x)$ to problem (6.6) about the point $x = x_i$, for each $i = 0, 1, \ldots, n-1$

$$y(x_{i+1}) = y(x_i) + (x_{i+1} - x_i)y'(x_i) + \frac{(x_{i+1} - x_i)^2}{2!}y''(\eta_i)$$

or

$$y(x_{i+1}) = y(x_i) + hy'(x_i) + \frac{h^2}{2!}y''(\eta_i), \tag{6.10}$$

where unknown point η_i lies on the interval (x_i, x_{i+1}). For the smaller value of stepsize h, the higher power h^2 will be very small and may be neglected. Using $f(x_i, y_i)$ to evaluate $y'(x_i)$ and $y_i \approx y(x_i)$, we have formula (6.9). Geometric interpretation of the method is shown by Figure 6.1.

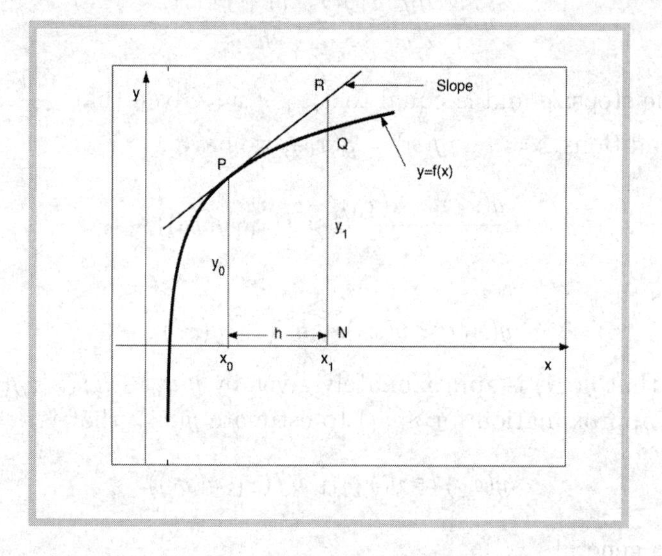

Figure 6.1: Geometric interpretation of Euler's method.

Example 6.1 *Use Euler's method to find the approximate value of* $y(1)$ *for the given initial-value problem*

$$y' = xy + x, \qquad 0 \le x \le 1, \quad y(0) = 0, \quad \textit{with} \quad h = 0.1,\ 0.2.$$

Compare your approximate solutions with the exact solution $y(x) = -1 + e^{x^2/2}$.

Solution. *Since* $f(x, y) = xy + x$, *and* $x_0 = 0$, $y_0 = 0$, *then*

$$y_{i+1} = y_i + hf(x_i, y_i), \quad \text{for} \quad i = 0, 1, \ldots, 9.$$

Then for $h = 0.1$ *and taking* $i = 0$, *we have*

$$y_1 = y_0 + hf(x_0, y_0) = y_0 + h(x_0 y_0 + x_0) = 0 + (0.1)[(0)(0) + (0)],$$

which gives

$$y_1 = 0.0000.$$

In a similar way, we have other approximations by taking $x_i = x_{i-1} + h, i = 1, 2, \ldots, 9$, *as follows*

$$\begin{array}{rclclcl}
y_2 & = & y_1 + hf(x_1, y_1) & = & y_1 + h(x_1 y_1 + x_1) & = & 0.0100 \\
y_3 & = & y_2 + hf(x_2, y_2) & = & y_2 + h(x_2 y_2 + x_2) & = & 0.0302 \\
y_4 & = & y_3 + hf(x_3, y_3) & = & y_3 + h(x_3 y_3 + x_3) & = & 0.0611 \\
y_5 & = & y_4 + hf(x_4, y_4) & = & y_4 + h(x_4 y_4 + x_4) & = & 0.1036 \\
y_6 & = & y_5 + hf(x_5, y_5) & = & y_5 + h(x_5 y_5 + x_5) & = & 0.1587 \\
y_7 & = & y_6 + hf(x_6, y_6) & = & y_6 + h(x_6 y_6 + x_6) & = & 0.2283 \\
y_8 & = & y_7 + hf(x_7, y_7) & = & y_7 + h(x_7 y_7 + x_7) & = & 0.3142 \\
y_9 & = & y_8 + hf(x_8, y_8) & = & y_8 + h(x_8 y_8 + x_8) & = & 0.4194 \\
y_{10} & = & y_9 + hf(x_9, y_9) & = & y_9 + h(x_9 y_9 + x_9) & = & 0.5471
\end{array}$$

with possible error

$$|y(1) - y_{10}| = |0.6487 - 0.5471| = 0.1016.$$

Similarly, the approximations for $h = 0.2$ *give*

$$\begin{array}{rclclcl}
y_1 & = & y_0 + hf(x_0, y_0) & = & y_0 + h(x_0 y_0 + x_0) & = & 0.0000 \\
y_2 & = & y_1 + hf(x_1, y_1) & = & y_1 + h(x_1 y_1 + x_1) & = & 0.0400 \\
y_3 & = & y_2 + hf(x_2, y_2) & = & y_2 + h(x_2 y_2 + x_2) & = & 0.1232 \\
y_4 & = & y_3 + hf(x_3, y_3) & = & y_3 + h(x_3 y_3 + x_3) & = & 0.2580 \\
y_5 & = & y_4 + hf(x_4, y_4) & = & y_4 + h(x_4 y_4 + x_4) & = & 0.4592
\end{array}$$

with possible error

$$|y(1) - y_5| = |0.6487 - 0.4592| = 0.1895.$$

It showed that the result for $h = 0.1$ is better than $h = 0.2$ and for both cases the approximation is not even correct to one decimal place. Clearly, the results using this method are inferior to those we will obtain in the coming methods. However, the accuracy of Euler's method could be considerably improved using a smaller value of h than 0.1.

Note that before calling MATLAB function Euler1, which is defined below, we must define the MATLAB function fun1 as follows:

```
function f = fun1(x, y)
f = x * y + y;
```

Given Euler1.m, and fun1.m the results obtained manually in the preceding example are reproduced with the following MATLAB commands:

```
>> [x', y'] = Euler1('fun1', 0, 1, 0, 10);
>> [x', y'] = Euler1('fun1', 0, 1, 0, 5);
```

The same results are obtained with the following statements that define the MATLAB command as an inline function object:

```
>> sol = inline('x * y + x', 'x', 'y')
>> [x', y'] = Euler1(sol, 0, 1, 0, 10);
>> disp([x', y'])
```

6.3.2 Analysis of Euler's Method

Example 6.1 demonstrated that the error in applying Euler's method is reduced when h is reduced. The question of how well Euler's method for solving initial-value problem (6.6) works is closely related to the truncation error of the method. There are two types of such errors, *local* and *global* truncation errors.

In the case of local truncation error one considers the size of the error made during one step and for global truncation error, one can consider the errors in the entire interval $x_0 \leq x \leq x_n$ over which the solution is sought.

We turn to the Taylor series to find an expression that represents the error, we have

$$y(x+h) = y(x) + hy'(x) + \frac{h^2}{2!}y''(\eta(x))$$

for unknown $\eta(x) \in [x, x+h]$.
If $y(x_{i+1})$ is the true value of $y(x)$, then the Taylor series expression at x_i is

$$y(x_{i+1}) = y(x_i) + hf(x_i, y_i) + \frac{h^2}{2!}y''(\eta(x_i)), \quad \eta(x_i) \in (x_i, x_{i+1})$$

as $y' = f(x_i, y_i)$. Euler's formula uses the recurrence relation

$$y_{i+1} = y(x_i) + hf(x_i, y_i)$$

to estimate y_{i+1} assuming that $y(x_i)$ is the true solution. The error in y_{i+1} is given by $y_{i+1} - y(x_{i+1})$, which can be written as

$$y_{i+1} - y(x_{i+1}) = (y(x_i) + hf(x_i, y_i)) - (y(x_i) + hf(x_i, y_i) + \frac{h^2}{2!}y''(\eta(x_i))),$$

which gives

$$y_{i+1} - y(x_{i+1}) = -\frac{h^2}{2!}y''(\eta(x_i)), \quad i = 0, 1, \ldots, n-1. \tag{6.11}$$

We call term $-\frac{h^2}{2}y''(\eta(x_i))$ the *local* truncation error for Euler's method. It is of order h^2.

Note that this error term only applies in the region (x_i, x_{i+1}), hence it is only the error in estimating y_{i+1} from $y(x_i)$. It does not take into account the compounded error from previous estimates. If we assume that the error is increasing linearly with n, then the error will be proportional to nh^2, but n is dependent on h as $h = \frac{x_n - x_0}{n}$, so the error will be proportional to

$$\frac{x_n - x_0}{h}h^2 = (x_n - x_0)h,$$

which is order h. This error is called the *global truncation* error.
The analysis above leads to important theorem in the analysis of numerical methods.

Theorem 6.2 *For the differential equations* $\frac{dy}{dx} = f(x, y)$, *if the leading term in the local truncation error involves* h^{p+1}, *for some integer* p, *then the global error, for small* h *is of order* h^p; *that is*

$$y_{i+1} - y(x_{i+1}) \approx ch^p,$$

where c *does not depend on stepsize* h.

Note that Euler's method is called the first-order method because of its local truncation error given by formula (6.11), since this arises on each application of the method. Thus, in generating the solution point (x_k, y_k) the truncation error appears k times, once for each application of the method.

Program 6.1

MATLAB m-file for Euler's Method

```
function sol=Euler1(fun1,a,b,y0,n)
h=(b-a)/n; x=a+(0:n)*h; y(1)=y0;
for k = 1 : n
y(k+1) = y(k) + h * feval(fun1, x(k), y(k));
end; sol=[x',y'];
```

Since Euler's method is an iterative method it may converge or diverge. If divergence occurs, then the procedure should be terminated because there may be no solution.

6.3.3 Higher-order Taylor Methods

The basis for many numerical techniques, finding the approximate solution of the initial-value problem can be dependent on the Taylor's series, as we used this series in the previous section in finding Euler's method, which is also called Taylor's method of order one. One can, of course, develop Taylor's method for higher-order to obtain better accuracy, and in general, one can expect that the higher the order of the method, the greater the accuracy for a given stepsize. Taylor's method is relatively easy to use; however; the necessity of calculating the higher derivatives makes Taylor's method completely unsuitable. Nevertheless, it is of great theoretical interest because most practical methods attempt to achieve the same accuracy as Taylor's method of a given order without the disadvantage of having to calculate higher derivatives. Assuming that the solution $y(x)$ of initial-value problem (6.6) has $(n+1)$ continuous derivatives and expanding $y(x)$ in terms of its *nth* degree Taylor polynomial about x_i, we get

$$\begin{aligned} y(x_{i+1}) &= y(x_i) + hy'(x_i) + \frac{h^2}{2!}y''(x_i) + \cdots \\ &+ \frac{h^n}{n!}y^{(n)}(x_i) + \frac{h^{n+1}}{(n+1)!}y^{(n+1)}(\eta(x_i)) \end{aligned} \tag{6.12}$$

for some $\eta(x_i) \in (x_i, x_{i+1})$. The derivatives in this expansion are not known explicitly since the solution is not known. However, if f is sufficiently differentiable, it can be obtained by taking the total derivative of (6.6) with respect to

x, keeping in mind that f is an implicit function of y. Thus,

$$\begin{aligned} y' &= f(x,y) = f \\ y'' &= f' = f_x + f_y f \\ y''' &= f'' = f_{xx} + 2f_{xy}f + f_{yy}f^2 + f_x f_y + f_y^2 f. \\ &\vdots \end{aligned} \tag{6.13}$$

Continuing in this manner, we can express any derivative of y in terms of $f(x,y)$ and its partial derivatives. It is already clear, however, that unless $f(x,y)$ is a very simple function, the higher total derivatives become increasingly complex. Now substituting these results into (6.12), gives

$$\begin{aligned} y(x_{i+1}) &= y(x_i) + hf(x_i, y(x_i)) + \frac{h^2}{2!}f'(x_i, y(x_i)) + \cdots \\ &+ \frac{h^n}{n!}f^{(n-1)}(x_i, y(x_i)) + \frac{h^{n+1}}{(n+1)!}f^{(n)}(\eta(x_i), y(\eta(x_i))). \end{aligned} \tag{6.14}$$

By taking $y_i \approx y(x_i)$, the approximation to the exact solution at x_i, for each $i = 0, 1, 2, \ldots, n-1$, we have

$$y_{i+1} = y_i + hf(x_i, y_i) + \frac{h^2}{2!}f'(x_i, y_i) + \cdots + \frac{h^n}{n!}f^{(n-1)}(x_i, y_i). \tag{6.15}$$

This formula is called Taylor's method of order n. The last term of (6.14), called the remainder, shows that the local error of Taylor's method of order n is

$$E = \frac{h^{n+1}}{(n+1)!}f^{(n)}(\eta_i, y(\eta(x_i))) = \frac{h^{n+1}}{(n+1)!}y^{(n+1)}(\eta(x_i)) \tag{6.16}$$

for some $x_i < \eta(x_i) < x_{i+1}$.

Example 6.2 *Use Taylor's method of order two to find the approximate value of $y(1)$ for the given initial-value problem*

$$y' = xy + x, \quad 0 \leq x \leq 1, \quad y(0) = 0, \quad \textit{with} \quad h = 0.2.$$

Compare your approximate solution with the exact solution $y(x) = -1 + e^{x^2/2}$.

Solution. *Since $f(x,y) = xy + x$, and $x_0 = 0$, $y_0 = 0$, then*

$$y_{i+1} = y_i + hf(x_i, y_i) + \frac{h^2}{2}f'(x_i, y_i), \quad \textit{for} \quad i = 0, 1, 2, 3, 4,$$

where $f'(x_i, y_i) = y_i + x_i^2 y_i + x_i^2 + 1$. *Then for* $i = 0$, *we have*

$$\begin{aligned} y_1 &= y_0 + h(x_0 y_0 + x_0) + \frac{h^2}{2}(y_0 + x_0^2 y_0 + x_0^2 + 1) \\ &= 0 + (0.2)(0) + (0.02)(1) = 0.0200 \end{aligned}$$

and in a similar way, we have for $i = 1, 2, 3, 4$, *as follows*

$$\begin{aligned} y_2 &= y_1 + h(x_1 y_1 + x_1) + \frac{h^2}{2}(y_1 + x_1^2 y_1 + x_1^2 + 1) = 0.0820 \\ y_3 &= y_2 + h(x_2 y_2 + x_2) + \frac{h^2}{2}(y_2 + x_2^2 y_2 + x_2^2 + 1) = 0.1937 \\ y_4 &= y_3 + h(x_3 y_3 + x_3) + \frac{h^2}{2}(y_3 + x_3^2 y_3 + x_3^2 + 1) = 0.3694 \\ y_5 &= y_4 + h(x_4 y_4 + x_4) + \frac{h^2}{2}(y_4 + x_4^2 y_4 + x_4^2 + 1) = 0.6334 \end{aligned}$$

with possible error

$$|y(1) - y_5| = |0.6487 - 0.6334| = 0.0153.$$

We showed that the result is entirely correct to one decimal place. Clearly, the result using this method is better than Euler's method and it could be considerably improved using a smaller value of h than 0.2.

Note that before calling the MATLAB functions *tayl1* and *fun1*, we must define MATLAB function *dfun1* as follows:

```
function f = dfun1(x, y)
f = y + x.^2*y + x.^2 + 1;
```

Given *tayl1.m, fun1.m,* and *dfun1.m,* the results obtained manually in the preceding example are reproduced with the following MATLAB commands:

```
>> [x', y'] = tayl1('fun1','dfun1',0,1,0,5);
>> disp([x', y'])
```

Program 6.2
MATLAB m-file for Taylor's Method of Order 2

```
function sol=tayl1(fun1,dfun1,a,b,y0,n)
h=(b-a)/n; x = a + (0 : n) * h; y(1)=y0;
for k=1:n
y(k + 1) = y(k) + h * feval(fun1, x(k), y(k))
        + (h.^ 2*feval(dfun1, x(k), y(k)))/2;
end; sol = [x', y'];
```

In using Taylor's method, we replace the infinite Taylor series for $f(x+h)$ by a partial sum. The *local truncation* error is inherent in any algorithm that we might choose.

If we retain the term up to and including h^n in the series, then the local truncation error is the sum of all the remaining terms that we do not include by Taylor's method. These terms can be compressed into a single term of the form

$$\frac{h^{n+1}}{(n+1)!}\, f^{(n+1)}(\eta(x), y(\eta(x)))$$

for some unknown point $\eta(x)$. We say that the local truncation error is order h^{n+1}. An error of this sort is present in each step of the numerical solution. The accumulation of all the many local truncation error gives rise to the *global truncation* error, which must be of order h^n because the number of steps necessary to reach an arbitrary point x, having started at x_0, is $\dfrac{x - x_0}{h}$, choosing n large so that this error is small.

6.3.4 Runge-Kutta Methods

Euler's method is not very useful in practical problems because it requires a very small stepsize for reasonable accuracy, and Taylor's method of higher-order is difficult to use because it needs to obtain higher total derivatives of $y(x)$. An important group of methods, which allow us to obtain greater accuracy at each step and yet require only an initial value of $y(x)$ to be given with the differential equation are called Runge-Kutta methods. Runge-Kutta methods attempt to obtain greater accuracy, and at the same time avoid the need of higher derivatives by evaluating the function $f(x, y)$ at selected points on each subintervals. These methods can be used to generate not only starting values but, in fact, whole solutions. They are self-starting and easy to program for a computer. We shall begin by showing how to derive the simplest formulas in this class. These are of the form

$$y_{i+1} = y_i + (w_1 k_1 + w_2 k_2), \tag{6.17}$$

where

$$\begin{aligned} k_1 &= hf(x_i, y_i) \\ k_2 &= hf(x_i + ah, y_i + bk_1). \end{aligned}$$

The parameters w_1, w_2, a, and b are chosen in order to make formula (6.17) as accurate as possible; that is, to make the order of accuracy as large as possible. To this end, we substitute exact value $y(x), y(x_{i+1})$ by the local solution into formula (6.17) and expand about point x_i. The parameters are then chosen to make the resulting expansion agree as much as possible with the Taylor series for $y(x_{i+1})$ about x_i. Upon substituting into (6.17), we first expand $y(x_{i+1})$ in the Taylor series through terms of order h^3, and obtain

$$y(x_{i+1}) = y(x_i) + hy'(x_i) + \frac{h^2}{2!}y''(x_i) + \frac{h^3}{3!}y'''(x_i) + \cdots. \tag{6.18}$$

Since

$$\begin{aligned} y' &= f(x, y) \\ y'' &= f'(x_i, y_i) = (f_x + f_y f)_i \\ y''' &= f''(x_i, y_i) = (f_{xx} + 2f_{xy}f + f_{yy}f^2 + f_x f_y + f_y^2 f)_i + O(h^4). \end{aligned} \tag{6.19}$$

So

$$\begin{aligned} y(x_{i+1}) &= y(x_i) + hf(x_i, y_i) + \frac{h^2}{2!}(f_x + ff_y)_i \\ &+ \frac{h^3}{3!}(f_{xx} + 2f_{xy}f + f_{yy}f^2 + f_x f_y + f_y^2 f)_i + O(h^4), \end{aligned} \tag{6.20}$$

where the subscripts on f denote partial derivatives with respect to the indicated variables, and the subscript i means that all functions involved are to be evaluated at (x_i, y_i). Now using Taylor's expansion for functions of two variables, we find that

$$\begin{aligned} k_2 &= hf(x_i + ah, y_i + bk_1) \\ &= h[f + h(af_x + bff_y) \\ &+ \frac{h^2}{2}(a^2 f_{xx} + 2abff_{xy} + b^2 f^2 f_{yy}) + O(h^4)]_i. \end{aligned} \tag{6.21}$$

Now we substitute this expression for k_2 into (6.17), which gives

$$\begin{aligned} y_{i+1} &= y_i + h[w_1 f(x_i, y_i) + w_2 f(x_i + ah, y_i + bk_1)] \\ &= y_i + h[(w_1 + w_2) f]_i + h^2 w_2 [(af_x + bff_y)]_i \\ &+ \frac{h^3}{2} w_2 [a^2 f_{xx} + 2abff_{xy} + b^2 f^2 f_{yy}]_i + O(h^4). \end{aligned} \quad (6.22)$$

Comparing (6.20) and (6.22), we see that to make the corresponding powers of h and h^2 agree, we must have

$$\begin{aligned} w_1 + w_2 &= 1 \\ a = \frac{1}{2w_2} &= b. \end{aligned}$$

This is a system of two nonlinear equations in the four unknowns a, b, w_1, and w_2 and its solution can be written in the form

$$b = a = \frac{1}{2w_2}, \quad w_1 = 1 - w_2. \quad (6.23)$$

There are many solutions to (6.23) depending on the choices of w_2. These choices lead to the numerical method that has an order of two and some of them correspond to some of the standard numerical integration formulas. Taking the first choice when $w_2 = 1/2$, we have

$$y_{i+1} = y_i + \frac{h}{2}[f(x_i, y_i) + f(x_{i+1}, y_i + hf(x_i, y_i))]. \quad (6.24)$$

It can be written in a standard form as

$$y_{i+1} = y_i + \frac{h}{2}[k_1 + k_2] \quad (6.25)$$

and

$$\begin{aligned} k_1 &= f(x_i, y_i) \\ k_2 &= f(x_{i+1}, y_i + hk_1) \end{aligned}$$

for each $i = 0, 1, \ldots n-1$. Then relation (6.25) is called the *Runge-Kutta method of order two*, which is also known as *Modified Euler's method.* This method corresponds to using the Trapezoidal rule to estimate the integral where a preliminary (full) Euler step is taken to obtain the (approximate) value at x_{i+1}. Geometric interpretation of the method is shown by Figure 6.2. The local error of this formula is, however, of order h^3, whereas that of Euler's method is h^2. We can, therefore, expect to be able to use a large stepsize with this formula.

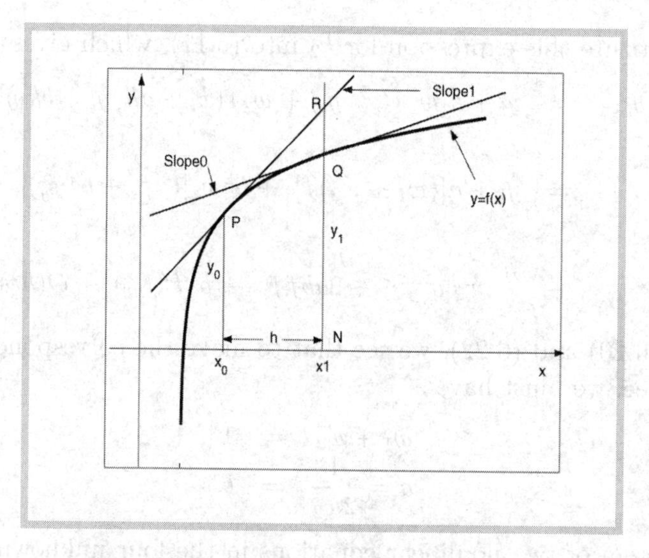

Figure 6.2: Geometric interpretation of Modified Euler's method.

Example 6.3 *Use Modified Euler's method to find the approximate value of $y(1)$ for the given initial-value problem*

$$y' = xy + x, \quad 0 \leq x \leq 1, \quad y(0) = 0, \quad \textit{with} \quad h = 0.2.$$

Compare your approximate solution with the exact solution $y(x) = -1 + e^{x^2/2}$.

Solution. *Since $f(x, y) = xy + x$, and $x_0 = 0$, $y_0 = 0$, then for $i = 0$, we have*

$$\begin{aligned} k_1 &= f(x_0, y_0) = (x_0 y_0 + x_0) = 0.0000 \\ k_2 &= f(x_1, y_0 + hk_1) = (x_1(y_0 + hk_1) + x_1) = (0 + 0.2) \\ &= 0.2000 \end{aligned}$$

and using these values, we have

$$y_1 = y_0 + \frac{h}{2}[k_1 + k_2] = 0 + 0.1(0 + 0.2000) = 0.0200.$$

Continuing in this manner, we have

$$\begin{aligned} k_1 &= 0.01204, & k_2 &= 0.4243, & \textit{then} \quad y_2 &= 0.0828 \\ k_1 &= 0.4331, & k_2 &= 0.7017, & \textit{then} \quad y_3 &= 0.1963 \\ k_1 &= 0.7178, & k_2 &= 1.0719, & \textit{then} \quad y_4 &= 0.3753 \\ k_1 &= 1.1002, & k_2 &= 1.5953, & \textit{then} \quad y_5 &= 0.6449 \end{aligned}$$

with possible error

$$|y(1) - y_5| = |0.6487 - 0.6449| = 0.0039.$$

The results Example 6.3 can be obtained using the following MATLAB command as follows:

```
>> sol = mod1('fun1', 0, 1, 0, 5);
```

So there is a significant improvement in accuracy of this method compared with Euler's method but the problem of accuracy still remains, however, since error will be accurate from step to step. In particular, since the function $f(x, y)$ calculated repeatedly from values of $y(x)$, which include the accumulated error, these errors may grow in an unpredictable way.

The modified Euler's method is classified as a predictor-corrector method. This means that in the case of the modified Euler's method the initial-value problem is given by the formula

$$y_{i+1}^{(k)} = y_i + hf(x_i, y_i), \tag{6.26}$$

which is called the predictor and this is corrected by the repeated application of the formula

$$y_{i+1}^{(k+1)} = y_i + \frac{h}{2}\left[f(x_i, y_i) + f(x_{i+1}, y_{i+1}^{(k)})\right], \quad k = 0, 1, 2, \ldots \tag{6.27}$$

for each $i = 0, 1, \ldots, n-1$. This is called the *corrector*. There are many predictor-corrector formulas and some provide much greater accuracy than the modified Euler's method. These methods, however, require accurate estimates for a number of initial values of $y(x)$ before they can be used. We shall discuss some of those predictor-corrector formulas later in the chapter.

Program 6.3

MATLAB m-file for the Modified Euler's Method

```
function sol=mod1(fun1,a,b,y0,n)
h = (b - a)/n; x = a + (0 : n) * h; y(1) = y0;
for k = 1 : n
k1 = feval(fun1, x(k), y(k));
k2 = feval(fun1, x(k) + h, y(k) + h * k1);
y(k + 1) = y(k) + h * (k1 + k2)/2;
end; sol = [x', y'];
```

Heun's Method

There are also other formulas using the different values of w_2. For example, when $w_2 = 3/4$, we have

$$y_{i+1} = y_i + \frac{h}{4}\Big[f(x_i, y_i) + 3f(x_i + \frac{2h}{3}, y_i + \frac{2h}{3}f(x_i, y_i)\Big] \tag{6.28}$$

or, written in standard form as

$$y_{i+1} = y_i + \frac{h}{4}\Big[k_1 + 3k_2\Big], \tag{6.29}$$

where

$$\begin{aligned} k_1 &= f(x_i, y_i) \\ k_2 &= f(x_i + \frac{2h}{3}, y_i + \frac{2h}{3}k_1) \end{aligned}$$

for each $i = 0, 1, \ldots, n-1$. This method does not correspond to any of the standard numerical integration formulas.

Program 6.4

MATLAB m-file for the Heun Method

```
function sol=heun(fun1,a,b,y0,n)
h = (b - a)/n; x = a + (0 : n) * h; y(1) = y0;
for k = 1 : n
k1 = feval(fun1, x(k), y(k));
k2 = feval(fun1, x(k)+2*h/3, y(k)+(2*h*k1)/3);
y(k + 1) = y(k) + h * (k1 + 3 * k2)/4;
end; sol = [x', y'];
```

Midpoint Method

For $w_2 = 1$, we obtain another method of order two, which is written as follows:

$$y_{i+1} = y_i + hf\Big[x_i + \frac{h}{2}, y_i + \frac{h}{2}f(x_i, y_i)\Big] \tag{6.30}$$

or, written in standard form as

$$y_{i+1} = y_i + hk_2, \tag{6.31}$$

where

$$\begin{aligned} k_1 &= f(x_i, y_i) \\ k_2 &= f(x_i + \frac{h}{2}, y_i + \frac{h}{2}k_1) \end{aligned}$$

for each $i = 0, 1, \ldots, n-1$. This method, which is also called the *corrected Euler*, is essentially the application of the midpoint rule for integration of $f(x, y)$ over the current step with the (approximate) value at the midpoint being obtained by a preliminary (half-) Euler step.

Program 6.5

MATLAB m-file for the Midpoint Method

```
function sol=mid(fun1,a,b,y0,n)
h=(b-a)/n; x = a + (0 : n) * h; y(1) = y0;
for k = 1 : n
k1 = feval(fun1, x(k), y(k));
k2 = feval(fun1, x(k) + h/2, y(k) + h * k1/2);
y(k + 1) = y(k) + h * k2;
end; sol = [x', y'];
```

The last two methods are also classified as the Runge-Kutta method of order two because the truncation error per step is of order h^3.

Example 6.4 *Use Heun's method and the Midpoint method to approximate the solution for the following initial-value problem*

$$y' = xy + x, \quad 0 \leq x \leq 1, \quad y(0) = 0, \quad \textit{with} \quad h = 0.2.$$

Compare your approximate solution with the exact solution $y(x) = -1 + e^{x^2/2}$.

Solution. *Since* $f(x, y) = xy + x$ *and* $x_0 = 0$, $y_0 = 0$, *then for* $i = 0$, *we can show*

Heun's Method.

For $i = 0$, *we have*

$$\begin{aligned} k_1 &= (x_0 y_0 + x_0) = 0.0000 \\ k_2 &= [(x_0 + \tfrac{2}{3}h)(y_0 + \tfrac{2}{3}hk_1) + (x_0 + \tfrac{2}{3}h)] = 0.1333. \end{aligned}$$

Hence,

$$y_1 = y_0 + \frac{h}{4}\Big[k_1 + 3k_2\Big] = 0.0200.$$

Continuing in this manner, we have

$$\begin{array}{lllllll} k_1 & = & 0.2040, & k_2 = & 0.3491, & then & y_2 = & 0.0826 \\ k_1 & = & 0.4330, & k_2 = & 0.6082, & then & y_3 = & 0.1954 \\ k_1 & = & 0.7173, & k_2 = & 0.9468, & then & y_4 = & 0.3733 \\ k_1 & = & 1.0987, & k_2 = & 1.4185, & then & y_5 = & 0.6410 \end{array}$$

with possible error

$$|y(1) - y_5| = |0.6487 - 0.6410| = 0.0077.$$

Midpoint Method

For $i = 0$, we have

$$\begin{array}{lll} k_1 & = & (x_0y_0 + x_0) = 0.0000 \\ k_2 & = & [(x_0 + \frac{1}{2}h)(y_0 + \frac{h}{2}k_1) + (x_0 + \frac{1}{2}h)] = 0.1000. \end{array}$$

Hence,

$$y_1 = y_0 + hk_2 = 0 + 0.2(0.1000) = 0.0200.$$

Continuing in this manner, we have

$$\begin{array}{lllllll} k_1 & = & 0.2040, & k_2 = & 0.3121, & then & y_2 = & 0.0824 \\ k_1 & = & 0.4330, & k_2 = & 0.5629, & then & y_3 = & 0.1950 \\ k_1 & = & 0.7170, & k_2 = & 0.8867, & then & y_4 = & 0.3723 \\ k_1 & = & 1.0979, & k_2 = & 1.3339, & then & y_5 = & 0.6391 \end{array}$$

with

$$|y(1) - y_5| = |0.6487 - 0.6391| = 0.0096$$

possible error.

All three methods are special cases of the Runge-Kutta methods and are also called *two-stage methods* because they use two evaluations, k_1 and k_2, of the derivation $f(x, y)$. The names given to various two-stage methods are not unique; indeed, a particular method may have more than one name, and the same name may be applied to different methods. The modified Euler's method is the same as the predictor-corrector method and also the modified Euler's method is sometimes called Heun's method.

The last two Runge-Kutta methods are also easily programmed. The following MATLAB commands produce the same results obtained in Example 6.4:

```
>> sol = heun('fun1',0,1,0,5);
>> sol = mid('fun1',0,1,0,5);
>> disp(sol)
```

Formulas of the Runge-Kutta method for higher-order (more accurate) can be derived by the procedure used above. The general equation of such formulas is

$$y_{i+1} = y_i + (w_1 k_1 + w_2 k_2 + \cdots + w_s k_s), \tag{6.32}$$

where

$$k_i = hf(x_i + a_j h, y_i + b_{j,1}k_1 + \cdots + b_{j,j-1}k_{j-1}), \quad 1 \leq j \leq s$$

where the parameters a_j and $b_{j,j-1}$ are chosen to give as high order accuracy as possible. However, the derivations become exceedingly complicated. In the following we consider some more higher-order Runge-Kutta methods, which have been used practically with success.

6.3.5 Third-order Runge-Kutta Method

The third-order Runge-Kutta method that is more accurate than the second-order Runge-Kutta method may be derived in a similar way as order two methods. The following version of this method, which is mostly used for solving initial-value problems is written as

$$y_{i+1} = y_i + \frac{h}{6}\Big[k_1 + 4k_2 + k_3\Big], \tag{6.33}$$

where

$$\begin{aligned} k_1 &= f(x_i, y_i) \\ k_2 &= f(x_i + \frac{h}{2}, y_i + \frac{h}{2}k_1) \\ k_3 &= f(x_i + h, y_i - hk_1 + 2hk_2) \end{aligned}$$

for each $i = 0, 1, \ldots, n-1$.

Example 6.5 *Use the Runge-Kutta-method of order three to approximate the solution for the following initial-value problem*

$$y' = xy + x, \quad 0 \leq x \leq 1, \quad y(0) = 0, \quad \textit{with} \quad h = 0.2.$$

Compare your approximate solution with the exact solution $y(x) = -1 + e^{x^2/2}$.

Solution. *We set* $f(x, y) = xy + x$*, and* $x_0 = 0$*,* $y_0 = 0$*, and* $h = 0.2$*, then the first calculated value of* y *can be obtained by taking* $i = 0$ *in (6.33) as follows:*

$$\begin{aligned} k_1 &= (x_0 y_0 + x_0) = 0.0000 \\ k_2 &= [(x_0 + \tfrac{1}{2}h)(y_0 + \tfrac{h}{2}k_1) + (x_0 + \tfrac{1}{2}h)] = 0.1000 \\ k_3 &= [(x_0 + h)(y_0 - hk_1 + 2hk_2) + (x_0 + h)] = 0.2080. \end{aligned}$$

Hence,

$$y_1 = y_0 + \frac{h}{6}\Big[k_1 + 4k_2 + k_3\Big] = 0.0203.$$

Similarly, the other computed values of y*s are as follows:*

$$\begin{array}{lllll} k_1 = 0.2041, & k_2 = 0.3122, & k_3 = 0.4417, & \textit{then} & y_2 = 0.0834 \\ k_1 = 0.4334, & k_2 = 0.5634, & k_3 = 0.7333, & \textit{then} & y_3 = 0.1974 \\ k_1 = 0.7185, & k_2 = 0.8885, & k_3 = 1.1273, & \textit{then} & y_4 = 0.3774 \\ k_1 = 1.1019, & k_2 = 1.3388, & k_3 = 1.6926, & \textit{then} & y_5 = 0.6491 \end{array}$$

with possible error

$$|y(1) - y_5| = |0.6487 - 0.6491| = 0.0004.$$

There is another version of this method, which is written as

$$y_{i+1} = y_i + \frac{h}{9}\Big[2k_1 + 3k_2 + 4k_3\Big], \tag{6.34}$$

where

$$\begin{aligned} k_1 &= f(x_i, y_i) \\ k_2 &= f(x_i + \frac{1}{2}h, y_i + \frac{h}{2}k_1) \\ k_3 &= f(x_i + \frac{3}{4}h, y_i + \frac{3h}{4}k_2) \end{aligned}$$

for each $i = 0, 1, \ldots, n - 1$.

6.3.6 Fourth-order Runge-Kutta Method

The most popular and most commonly used formula of Runge-Kutta methods type are the fourth-order formulas. The fourth-order Runge-Kutta method is accurate to the fourth-order term of the Taylor expansion, so the local error is proportional to h^5. The following version of this method, which is mostly used for solving initial-value problems is based on Simpson's 1/3 rule and is known as the *classic method*, written as

$$y_{i+1} = y_i + \frac{h}{6}\Big[k_1 + 2k_2 + 2k_3 + k_4\Big], \tag{6.35}$$

where

$$\begin{aligned} k_1 &= f(x_i, y_i) \\ k_2 &= f(x_i + \frac{h}{2}, y_i + \frac{h}{2}k_1) \\ k_3 &= f(x_i + \frac{h}{2}, y_i + \frac{h}{2}k_2) \\ k_4 &= f(x_i + h, y_i + hk_3) \end{aligned}$$

for each $i = 0, 1, \ldots, n-1$.

Example 6.6 *Use the Runge-Kutta method of order four to approximate the solution for the following initial-value problem*

$$y' = xy + x, \quad 0 \le x \le 1, \quad y(0) = 0, \quad \textit{with} \quad h = 0.2.$$

Compare your approximate solution with the exact solution $y(x) = -1 + e^{x^2/2}$.

Solution. *We set* $f(x, y) = xy + x$ *and* $x_0 = 0,\ y_0 = 0$, *and* $h = 0.2$, *then for* $i = 0$, *we have first calculated the value of* y *as follows:*

$$\begin{aligned} k_1 &= (x_0y_0 + x_0) = 0.0000 \\ k_2 &= [(x_0 + \tfrac{1}{2}h)(y_0 + \tfrac{h}{2}k_1) + (x_0 + \tfrac{h}{2})] = 0.1000 \\ k_3 &= [(x_0 + \tfrac{h}{2})(y_0 + \tfrac{h}{2}k_2) + (x_0 + \tfrac{1}{2}h)] = 0.1010 \\ k_4 &= [(x_0 + h)(y_0 + hk_3) + (x_0 + h)] = 0.2040. \end{aligned}$$

Hence,

$$y_1 = y_0 + \frac{h}{6}\Big[k_1 + 2k_2 + 2k_3 + k_4\Big] = 0.0202.$$

Similarly, the other computed values of ks are as follows:

$$\begin{array}{llllllll} k_1 &=& 0.2040, & k_2 = & 0.3122, & k_3 = & 0.3154, & k_4 = \; 0.4333 \\ k_1 &=& 0.4333, & k_2 = & 0.5633, & k_3 = & 0.5698, & k_4 = \; 0.7183 \\ k_1 &=& 0.7183, & k_2 = & 0.8883, & k_3 = & 0.9002, & k_4 = \; 1.1018 \\ k_1 &=& 1.1017, & k_2 = & 1.3386, & k_3 = & 1.3599, & k_4 = \; 1.6491 \end{array}$$

and the remaining calculated values of ys are given as:

$$y_2 = 0.0833, \quad y_3 = 0.1972, \quad y_4 = 0.3771, \quad y_5 = 0.6487$$

with possible error

$$|y(1) - y_5| = |0.6487 - 0.6487| = 0.0000.$$

We showed that the result is entirely correct to four decimal places. Clearly, the results using this method are superior to those we obtained in the preceding methods. However, the accuracy of this method could be considerably improved using a smaller value of h than 0.2. The Runge-Kutta method of order four is programmed using the following MATLAB commands, which produce the same results as obtained in Example 6.6:

```
>> sol = RK4('fun1',0,1,0,5);
> disp(sol)
```

Program 6.6

MATLAB m-file for Runge-Kutta Method of Order four

```
function sol=RK4(fun1,a,b,y0,n)
h=(b-a)/n; x = a + (0 : n) * h; y(1) = y0;
for k = 1 : n
k1 = feval(fun1, x(k), y(k));
k2 = feval(fun1, x(k) + h/2, y(k) + h * k1/2);
k3 = feval(fun1, x(k) + h/2, y(k) + h * k2/2);
k4 = feval(fun1, x(k) + h, y(k) + h * k3);
y(k + 1) = y(k) + h * (k1 + 2 * (k2 + k3) + k4)/6;
end; sol = [x', y'];
```

The local error term for fourth-order Runge-Kutta is $O(h^5)$; the global error would be about $O(h^4)$. It is computationally more efficient than the modified Euler's method because, while four evaluations of the function are required per step rather than two, the steps can be many-fold larger for the same accuracy. It is easy to see why the Runge-Kutta method is so popular. Since going from second to fourth order was beneficial, we may wonder why we should use still a higher order of formula. Higher-order (fifth, sixth, etc.) Runge-Kutta formulas have been developed and can be used to advantage in determining a suitable size h, as we shall see later. There are other versions of this method. One of them is based on Simpson's 3/8 rule and is written as

$$y_{i+1} = y_i + \frac{h}{8}\Big[k_1 + 3k_2 + 3k_3 + k_4\Big], \tag{6.36}$$

where

$$\begin{aligned} k_1 &= f(x_i, y_i) \\ k_2 &= f(x_i + \frac{h}{3}, y_i + \frac{h}{3}k_1) \\ k_3 &= f(x_i + \frac{2h}{3}, y_i + \frac{h}{3}k_1 + hk_2)) \\ k_4 &= f(x_i + h, y_i + h(k_1 - k_2 + k_3)) \end{aligned}$$

for each $i = 0, 1, \ldots, n-1$, and the other variation of the Runge-Kutta method of order four is due to the Gill and is called *Runge-Kutta-Gill Method*, which is written as

$$y_{i+1} = y_i + \frac{h}{6}\Big[k_1 + (2-\sqrt{2})k_2 + (2+\sqrt{2})k_3 + k_4\Big], \tag{6.37}$$

where

$$\begin{aligned} k_1 &= f(x_i, y_i) \\ k_2 &= f(x_i + \frac{h}{2}, y_i + \frac{h}{2}k_1) \\ k_3 &= f(x_i + \frac{h}{2}, y_i + \frac{(\sqrt{2}-1)}{2}hk_1 + \frac{(2-\sqrt{2})}{2}hk_2) \\ k_4 &= f(x_i + h, y_i - \frac{\sqrt{2}}{2}hk_2 + (1 + \frac{\sqrt{2}}{2})hk_3) \end{aligned}$$

for each $i = 0, 1, \ldots, n-1$.

A number of other higher-order forms of the Runge-Kutta method have been derived which have particularly advantageous properties. The equations for these methods and their important features are as follows.

6.3.7 Fifth-order Runge-Kutta Method

The fifth-order Runge-Kutta method is written as

$$y_{i+1} = y_i + \frac{h}{8}\Big[\frac{k_1}{3} + \frac{5}{6}k_4 + \frac{27}{7}k_5 + \frac{125}{42}k_6\Big], \tag{6.38}$$

where

$$\begin{aligned}
k_1 &= f(x_i, y_i) \\
k_2 &= f(x_i + \frac{h}{2}, y_i + \frac{h}{2}k_1) \\
k_3 &= f(x_i + \frac{h}{2}, y_i + \frac{h}{4}(k_1 + k_2)) \\
k_4 &= f(x_i + h, y_i + h(2k_3 - k_2)) \\
k_5 &= f(x_i + \frac{2h}{3}, y_i + \frac{h}{27}(7k_1 + 10k_2 + k_3) + 2hk_4) \\
k_6 &= f(x_i + \frac{h}{5}, y_i + \frac{h}{625}(28k_1 + 546k_3 + 54k_4 - 378k_5) - \frac{1}{5}hk_2)
\end{aligned}$$

for each $i = 0, 1, \ldots, n-1$.

6.3.8 Runge-Kutta-Merson Method

This method has an error term of order h^5, and in addition it allows an estimate of the local truncation error to be obtained at each step in terms of known values. This method is defined as follows:

$$y_{i+1} = y_i + \frac{1}{6}\Big[k_1 + 4k_4 + k_5\Big], \tag{6.39}$$

where

$$
\begin{aligned}
k_1 &= hf(x_i, y_i) \\
k_2 &= hf(x_i + \frac{h}{3}, y_i + \frac{1}{3}k_1) \\
k_3 &= hf(x_i + \frac{h}{3}, y_i + \frac{1}{6}(k_1 + k_2)) \\
k_4 &= hf(x_i + \frac{h}{2}, y_i + \frac{1}{8}(k_1 + 3k_3)) \\
k_5 &= hf(x_i + h, y_i + \frac{1}{2}(k_1 - 3k_3 + 4k_4))
\end{aligned}
$$

for each $i = 0, 1, \ldots, n-1$, and

$$E \approx \frac{1}{30}\left[2k_1 - 9k_3 + 8k_4 - k_5\right]$$

is called the local error term.

Program 6.7

MATLAB m-file for the Runge-Kutta Merson Method

```
function sol=RKM(fun1,a,b,y0,n)
h=(b-a)/n; x = a + (0 : n) * h; y(1) = y0;
for k = 1 : n
k1 = feval(fun1, x(k), y(k));
k2 = feval(fun1, x(k) + h/3, y(k) + h * k1/3);
k3 = feval(fun1, x(k) + h/3, y(k) + h * (k1 + k2)/6);
k4 = feval(fun1, x(k) + h/2, y(k) + h * (k1 + 3 * k3)/8);
k5 = feval(fun1, x(k) + h, y(k) + h * (k1 - 3 * k3 + 4 * k4)/2);
y(k + 1) = y(k) + h * (k1 + 4 * k4 + k5)/6;
end; sol = [x', y'];
```

Example 6.7 *Use the Runge-Kutta-Merson method to approximate the solution for the following initial-value problem*

$$y' = xy + x, \quad 0 \leq x \leq 1, \quad y(0) = 0, \quad \textit{with} \quad h = 0.2.$$

Compare your approximate solution with the exact solution $y(x) = -1 + e^{x^2/2}$.

Solution. *We set $f(x,y) = xy + x$ and $x_0 = 0,\ y_0 = 0$, and $h = 0.2$, then the first calculated value of y can be obtain by taking $i = 0$ in (6.39) as follows:*

$$
\begin{aligned}
k_1 &= (x_0 y_0 + x_0) = 0.0000 \\
k_2 &= [(x_0 + \tfrac{h}{3})(y_0 + \tfrac{h}{3}k_1) + (x_0 + \tfrac{h}{3})] = 0.0667 \\
k_3 &= [(x_0 + \tfrac{h}{3})(y_0 + \tfrac{h}{6}(k_1 + k_2)) + (x_0 + \tfrac{h}{3})] = 0.06668 \\
k_4 &= [(x_0 + \tfrac{h}{2})(y_0 + \tfrac{h}{8}(k_1 + 3k_3)) + (x_0 + \tfrac{h}{2})] = 0.1005 \\
k_5 &= [(x_0 + h)(y_0 + \tfrac{h}{2}(k_1 - 3k_3 + 4k_4)) + (x_0 + h)] = 0.2040.
\end{aligned}
$$

Hence,

$$y_1 = y_0 + \frac{h}{6}\Big[k_1 + 4k_4 + k_5\Big] = 0.0202.$$

Similarly, the other computed values of ks are as follows:

$$
\begin{array}{lllll}
k_1 = 0.2040, & k_2 = 0.2757, & k_3 = 0.2763, & k_4 = 0.3138, & k_5 = 0.4333 \\
k_1 = 0.4333, & k_2 = 0.5190, & k_3 = 0.5203, & k_4 = 0.5666, & k_5 = 0.7183 \\
k_1 = 0.7183, & k_2 = 0.8301, & k_3 = 0.8326, & k_4 = 0.8943, & k_5 = 1.1016 \\
k_1 = 1.1017, & k_2 = 1.2572, & k_3 = 1.2617, & k_4 = 1.3494, & k_5 = 1.6485
\end{array}
$$

and the remaining calculated values of ys are given as:

$$y_2 = 0.083287, \quad y_3 = 0.197216, \quad y_4 = 0.377125, \quad y_5 = 0.648715$$

with possible error

$$|y(1) - y_5| = |0.648721 - 0.648715| = 0.000006.$$

We showed that the result is entirely correct to five decimal places. Clearly, the results using this method are superior to those obtained in the preceding methods. However, the accuracy of this method could be considerably improved using smaller value of h than 0.2.

The following MATLAB commands will give the same results as we obtained in the preceding example of the Runge-Kutta Merson method:

```
>> sol = RKM('fun1',0,1,0,5);
>> disp(sol)
```

6.3.9 Runge-Kutta-Lawson's Fifth-order Method

This method provides higher accuracy at each step, the error being of order h^6. This method is defined as follows:

$$y_{i+1} = y_i + \frac{h}{90}\Big[7k_1 + 32k_3 + 12k_4 + 32k_5 + 7k_6\Big], \tag{6.40}$$

where

$$\begin{aligned}
k_1 &= hf(x_i, y_i) \\
k_2 &= hf(x_i + \frac{h}{2}, y_i + \frac{1}{2}k_1) \\
k_3 &= hf(x_i + \frac{h}{4}, y_i + \frac{1}{16}(3k_1 + k_2)) \\
k_4 &= hf(x_i + \frac{h}{2}, y_i + \frac{1}{2}k_3) \\
k_5 &= hf(x_i + \frac{3h}{4}, y_i - \frac{1}{16}(3k_2 - 6k_3 - 9k_4)) \\
k_6 &= hf(x_i + h, y_i + \frac{1}{7}(k_1 + 4k_2 + 6k_3 - 12k_4 + 8k_5))
\end{aligned}$$

for each $i = 0, 1, \ldots, n-1$.

6.3.10 Runge-Kutta-Butcher Sixth-order Method

This method provides higher accuracy at each step, the error being of order h^6. This method is defined as follows:

$$y_{i+1} = y_i + \Big[\frac{11}{20}k_1 + \frac{27}{40}(k_3 + k_4) - \frac{4}{15}(k_5 + k_6) + \frac{11}{120}k7\Big], \tag{6.41}$$

where

$$
\begin{aligned}
k_1 &= hf(x_i, y_i) \\
k_2 &= hf(x_i + \frac{h}{3}, y_i + \frac{1}{3}k_1) \\
k_3 &= hf(x_i + \frac{2h}{3}, y_i + \frac{2}{3}k_2) \\
k_4 &= hf(x_i + \frac{1}{3}h, y_i - \frac{1}{12}(k_1 + 4k_2 - k_3)) \\
k_5 &= hf(x_i + \frac{h}{2}, y_i - \frac{1}{16}(k_1 - 18k_2 + 3k_3 + 6k_4)) \\
k_6 &= hf(x_i + \frac{h}{2}, y_i + \frac{1}{8}(9k_2 - 3k_3 - 6k_3 + 6k_4 + 4k_5)) \\
k_7 &= hf(x_i + h, y_i + \frac{9}{44}(k_1 - 4k_2 + 7k_3 + 8k_4 - \frac{64}{9}k_6))
\end{aligned}
$$

for each $i = 0, 1, \ldots, n-1$.

6.3.11 Runge-Kutta-Fehlberg Method

One way to guarantee accuracy in the solution of an initial-value problem is to solve the problem twice using stepsize h and $h/2$ and compare answers at data points corresponding to the larger stepsize. But this requires a significant amount of computation for the smaller stepsize and must be repeated if it is determined that the agreement is not enough. For the Runge-Kutta formulas, using the one step, two half steps procedure can be very expansive. Consider, for instance, classic fourth-order formula (6.35). The cost of each step would be 11 function evaluations, four for first value y_{i+1} and seven for second (more accurate) value y^*_{i+1}. A better procedure, known as the Fehlberg's method, is to compute y^*_{i+1} using the Runge-Kutta formula of higher-order accuracy rather than the one used for y_{i+1}. This may seem inefficient but the key is to use a pair of formulas with a common set of $k_i s$. If the answers are in close agreement, the approximation is accepted. If the answers do not agree to a specified accuracy, the stepsize is reduced. If the answer agrees to more significant digits than required, the stepsize is increased. Using the Runge-Kutta of order four, each step requires the use of the following five function values:

$$k_1 = hf(x_i, y_i)$$

$$k_2 = hf(x_i + \frac{h}{4}, y_i + \frac{1}{4}k_1)$$

$$k_3 = hf(x_i + \frac{3h}{8}, y_i + \frac{1}{32}(3k_1 + 9k_2))$$

$$k_4 = hf(x_i + \frac{12h}{13}, y_i + \frac{1}{2197}(1932k_1 - 7200k_2 + 7296k_3))$$

$$k_5 = hf(x_i + h, y_i + \frac{439h}{216}k_1 - 8hk_2 + \frac{3680h}{513}k_3 - \frac{845h}{4104}k_4)$$

for each $i = 0, 1, \ldots, n-1$. Then an approximation to the solution of the initial-value problem is made using the Runge-Kutta of order four:

$$y_{i+1} = y_i + \left[\frac{25}{216}k_1 + \frac{1408}{2565}k_3 - \frac{2197}{4104}k_4 - \frac{1}{5}k_5\right] + O(h^5) \quad (6.42)$$

for each $i = 0, 1, \ldots, n-1$, and the four functions values f_1, f_3, f_4, and f_5 are used. A better value for the solution is determined using the Runge-Kutta method of order five:

$$y^*_{i+1} = y_i + \left[\frac{16}{135}k_1 + \frac{6656}{12825}k_3 + \frac{28561}{36430}k_4 - \frac{9}{50}k_5 + \frac{2}{55}k_6\right] + O(h^6). \quad (6.43)$$

Each step requires the use of the following six function values:

$$k_1 = hf(x_i, y_i)$$

$$k_2 = hf(x_i + \frac{h}{4}, y_i + \frac{1}{4}k_1)$$

$$k_3 = hf(x_i + \frac{3h}{8}, y_i + \frac{1}{32}(3k_1 + 9k_2))$$

$$k_4 = hf(x_i + \frac{12h}{13}, y_i + \frac{1}{2197}(1932k_1 - 7200k_2 + 7296k_3))$$

$$k_5 = hf(x_i + h, y_i + \frac{439h}{216}k_1 - 8hk_2 + \frac{3680h}{513}k_3 - \frac{845h}{4104}k_4)$$

$$k_6 = f(x_i + \frac{h}{2}, y_i - \frac{8h}{27}k_1 + 2hk_2 - \frac{3544h}{2565}k_3 + \frac{1859h}{4104}k_4 - \frac{11h}{40}k_5)$$

for each $i = 0, 1, \ldots, n-1$. The estimated error is the value computed from the fifth-order method minus the value computed from the fourth-order method, which can be expressed directly as

$$E \approx \left[\frac{1}{360}k_1 - \frac{128}{4275}k_3 - \frac{2197}{75240}k_4 + \frac{1}{50}k_5 + \frac{2}{55}k_6\right].$$

Note that despite the complicated appearance, this pair of formulas is considered easy to implement since it can be coded in a straight-forward way. Looking at efficiency, we observe that the formula for y_{i+1} is fourth-order accurate and requires five evaluations, which is more expansive than the fourth-order Runge-Kutta formula for the same order of accuracy. However, y^*_{i+1} can be obtained at the cost of only one more function evaluation (to compute k_6). Therefore, the local error of a step using (6.42) is only six function evaluations compared to 11 for the fourth-order Runge-Kutta method using the half-steps to compute y^*_{i+1}. In addition, since formula (6.43) gives y^*_{i+1} that is significantly more accurate than y_{n+1}, we get a more reliable estimate for error e_{i+1}. Thus, we noted that Fehlberg-type pairs of Runge-Kutta formulas are much more efficient than a single formula of comparable accuracy. For this reason, they are the preferred Runge-Kutta formulas. The optimal step qh size can be determined by multiplying the scalar q times the current step size h. The scalar q is

$$\begin{aligned} q &= \left(\frac{Tol\ h}{2|y^*_{i+1} - y_{i+1}|}\right)^{1/4} \\ &\approx 0.84\left(\frac{Tol\ h}{|y^*_{i+1} - y_{i+1}|}\right)^{1/4}, \end{aligned}$$

where Tol is the specified error central tolerance. The disadvantage of Runge-Kutta methods is that they require a considerable number of function evaluations at each step. In the case of the fourth-order method each step needs four evaluations. In the coming section we shall consider a number of predictor-corrector methods, some of which give the same order truncation errors as Runge-Kutta methods at each step, but may require fewer function evaluations.

6.4 Multi-step Methods for IVP

So far we have discussed the methods for the approximation of initial-value problem (6.6), the Taylor's method of order n and Runge-Kutta types methods. All these methods are called single-step methods because they require information about a solution only at a single point $x = x_i$, from which the methods proceed to obtain the solution $y(x)$ at the very next point $x = x_{i+1}$. Although these methods generally use functional evaluation information at points x_i and x_{i+1}, they don't retain that information for direct use in future approximations. All the information used by these methods is consequently obtained within the interval over which the solution is being approximated. Methods that make use of information about the solution at more than one point are called *multi-step methods.* One class of multi-step methods is based on the principle of numerical integration and this type of multi-step method is discussed here. Let us assume that we have already obtained approximations to y' and y at a number of equally spaced points, say $x_0, x_1, \ldots, n$. Consider the differential equation

$$y' = f(x, y). \tag{6.44}$$

By integrating (6.44) over the interval $[x_i, x_{i+1}]$, we obtain

$$\int_{x_i}^{x_{i+1}} y' dx = \int_{x_i}^{x_{i+1}} f(x, y) dx \tag{6.45}$$

and it gives

$$y(x_{i+1}) = y(x_i) + \int_{x_i}^{x_{i+1}} f(x, y) dx. \tag{6.46}$$

Now we will develop numerical methods for finding the solution $y(x)$ by approximating the integral in (6.46). There are many such methods, and we will consider only the most popular of them, the *Adams methods.* These methods use the information at multiple steps of the solution to obtain the solution at the next x-value. The Adams methods form two main classes, the *explicit Adams-Bashforth methods* and the *implicit Adams-Moulton methods.* These may be combined to form *Adams-Bashforth-Moulton predictor-corrector methods,* which we shall discuss in the next section. The Adams methods are generally more efficient than Runge-Kutta methods, especially if one wishes to find the solution with a high degree of accuracy or if the derivative of $f(x, y)$ is too expansive to evaluate. These multi-step methods can be derived using the Taylor series, or Newton's-backward formula, or by using the Lagrange interpolating polynomial. To approximate the integral

$$\int_{x_i}^{x_{i+1}} P(x) dx, \quad \text{where} \quad P(x) = f(x, y) = y' \tag{6.47}$$

means to approximate $P(x)$ using polynomial interpolation and then integrating the interpolating polynomial.

Consider the first Adams-bashforth method, which is based on linear interpolation. The linear polynomial interpolating $P(x)$ at $\{x_{i-1}, x_i\}$ is

$$p_1(x) = \frac{1}{h}[(x_i - x)P(x_{i-1}) + (x - x_{i-1})P(x_i)],$$

where $h = x_i - x_{i-1}$. Now integrating over the interval $[x_i, x_{i+1}]$, we get

$$\int_{x_i}^{x_{i+1}} P(x)dx \approx \int_{x_i}^{x_{i+1}} p_1(x)dx = \frac{-P(x_{i-1})}{h}\int_{x_i}^{x_{i+1}} (x - x_i)\, dx + \frac{P(x_i)}{h}\int_{x_i}^{x_{i+1}} (x - x_{i-1})\, dx.$$

Solving the above integrals, we obtain

$$\int_{x_i}^{x_{i+1}} P(x)dx = \int_{x_i}^{x_{i+1}} f(x,y)\, dx \approx \frac{3h}{2}P(x_i) - \frac{h}{2}P(x_{i-1}). \tag{6.48}$$

Using relation (6.48) in (6.46), it becomes

$$y_{i+1} = y_i + \frac{h}{2}[3f(x_i, y_i) - f(x_{i-1}, y_{i-1})], \tag{6.49}$$

where $i = 1, 2, \ldots, n-1$. Similarly, the local truncation error of (6.49) can be derived using the error term of the linear interpolation

$$E = \frac{(x - x_{i-1})(x - x_i)}{2!}P''(\eta(x_i)).$$

Then integrating over the interval $[x_i, x_{i+1}]$ and simplifying gives

$$E = \frac{5}{12}h^3 y'''(\eta(x_i)), \quad x_{i-1} < \eta(x_i) < x_{i+1}, \tag{6.50}$$

which is called the error term of (6.49). Relation (6.49) represents the *Adams-bashforth method of order two.* With this method, note that it is necessary to have $i \geq 1$. Both y_0 and y_1 are needed in finding y_2, and y_1 cannot be found from (6.49). The value of y_1 must be obtained by another method. Method (6.49) is an example of a two-step method since the values at x_{i-1} and x_i are needed to find the value at x_{i+1}.

Example 6.8 *Solve the initial-value problem*

$$y' = xy + x, \qquad y(0) = 1, \quad h = 0.2$$

to find the approximate value of $y(1)$ *using the Adams-Bashforth two-step method. Generate the other initial value* y_1 *using the Modified Euler's method and compare your approximate solution with the true solution* $y(x) = -1 + e^{x^2/2}$.

Solution. *We set* $f(x,y) = xy + x$, *and* $x_0 = 0$, $y_0 = 0$, *and* $h = 0.2$. *The unknown initial value can be found using the Modified Euler's method and it gives* $y_1 = 0.0200$. *Then the next approximate solution using the Adam-Bashforth two-step method is*

$$\begin{aligned} y_2 &= y_1 + \frac{h}{2}[3(x_1y_1 + x_1) - (x_0y_0 + x_0)]. \\ &= 0.0200 + 0.1[3(0.2040) - (0.0)] \\ &= 0.0812 \end{aligned}$$

Similarly, the other three computed approximate values are as follows

$$y_3 = 0.1905, \quad y_4 = 0.3616, \quad \textit{and} \quad y_5 = 0.6169$$

with possible error

$$|y(1) - y_5| = |0.6487 - 0.6169| = 0.0318.$$

Note that before calling the MATLAB function *AB2M* which, is defined below, we must define the MATLAB function *fn* as follows:

```
function f = fun1(x, y)
f = x * y + x;
```

Given *AB2M.m* and *fn.m* the results obtained manually in the preceding example are reproduced with the following MATLAB command:

```
>> T = AB2M('fun1', 0, 1, 0, 5);
```

Program 6.8

MATLAB m-file for Adams-Bashforth Two-step Method

```
function T=AB2M(fun1,a,b,y0,n)
h=(b-a)/n; x = a + (0 : n) * h;
y(1)=y0; f(1)=feval(fun1,x(1),y(1));S(1)=y(1);
for i=1:n % Modified Euler's method
k1 = f(i); k2 = feval(fun1, x(i) + h, y(i) + h * k1);
y(i + 1) = y(i) + h * (k1 + k2)/2; S(i + 1) = y(i + 1);
f(i+1)=feval(fun1,x(i+1),y(i+1)); end
for k=2:n % Now use 2-step formula
S1(1)=1;S1(2)=S(2);
y(k + 1) = y(k) + h * (3 * f(k) - f(k - 1))/2;
S1(k+1)=y(k+1);
f(k+1)=feval(fun1,x(k+1),y(k+1));
end; T = [x', S', S1'];
```

Higher-order Adams-bashforth methods can be obtained using higher degree polynomial interpolation in the approximation of integration formula (6.48). The Adams-bashforth methods together with their required starting values and local truncation errors are given as follows. The derivation of these methods is similar to the procedure discussed above.

Third-order Adams-Bashforth (Three-step) Method

$$y_{i+1} = y_i + \frac{h}{12}\Big[23f(x_i, y_i) - 16f(x_{i-1}, y_{i-1}) + 5f(x_{i-2}, y_{i-2})\Big], \tag{6.51}$$

where $i = 2, 3, \ldots, n-1$, and y_0, y_1, y_2 are given. The local truncation error is

$$E = \frac{3}{8}h^3 y^{(4)}(\eta(x_i)), \quad x_{i-2} < \eta(x_i) < x_{i+1}. \tag{6.52}$$

Fourth-order Adams-Bashforth (Four-step) Method

$$y_{i+1} = y_i + \frac{h}{24}\Big[55f(x_i, y_i) - 59f(x_{i-1}, y_{i-1}) + 37f(x_{i-2}, y_{i-2}) - 9f(x_{i-3}, y_{i-3})\Big], \tag{6.53}$$

where $i = 3, 4, \ldots, n-1$, and y_0, y_1, y_2, y_3 are given. The local truncation error is

$$E = \frac{251}{720}h^4 y^{(5)}(\eta(x_i)), \quad x_{i-3} < \eta(x_i) < x_{i+1}. \tag{6.54}$$

Fifth-order Adams-Bashforth (Five-step) Method

$$y_{i+1} = y_i + \frac{h}{720}\Big[1901f(x_i, y_i) - 2774f(x_{i-1}, y_{i-1}) + 2616f(x_{i-2}, y_{i-2}) - 1274f(x_{i-3}, y_{i-3}) + 251f(x_{i-4}, y_{i-4})\Big], \quad (6.55)$$

where $i = 4, 5, \ldots, n-1$, and y_0, y_1, y_2, y_3, y_4 are given. The local truncation error is

$$E = \frac{95}{288}h^5 y^{(6)}(\eta(x_i)), \quad x_{i-4} < \eta(x_i) < x_{i+1}. \quad (6.56)$$

Among the most popular implicit multi-step methods are the Adams-Moulton methods. The development of Adams-Moulton methods is similar to that of the preceding Adams-Bashforth methods. The difference is that the interpolating polynomial now includes the data point $P(x_{i+1})$. As a result, the Adams-Moulton methods are implicit. The order of an Adams-Moulton method is one order higher than the number of steps involved in the method. Some of the more common implicit Adams-Moulton methods are given as follows:

Third-order Adams-Moulton (Two-step) Method

$$y_{i+1} = y_i + \frac{h}{12}\Big[5f(x_{i+1}, y_{i+1}) + 8f(x_i, y_i) - f(x_{i-1}, y_{i-1})\Big], \quad (6.57)$$

where $i = 1, 2, \ldots, n-1$, and y_0, y_1 are given. The local truncation error is

$$E = -\frac{1}{24}h^3 y^{(4)}(\eta(x_i)), \quad x_{i-1} < \eta(x_i) < x_{i+1}. \quad (6.58)$$

Fourth-order Adams-Moulton (Three-step) Method

$$y_{i+1} = y_i + \frac{h}{24}\Big[9f(x_{i+1}, y_{i+1}) + 19f(x_i, y_i) - 5f(x_{i-1}, y_{i-1}) + f(x_{i-2}, y_{i-2})\Big], \quad (6.59)$$

where $i = 2, 3, \ldots, n-1$, and y_0, y_1, y_2 are given. The local truncation error is

$$E = -\frac{19}{720}h^4 y^{(5)}(\eta(x_i)), \quad x_{i-2} < \eta(x_i) < x_{i+1}. \quad (6.60)$$

Fifth-order Adams-Moulton (Four-step) Method

$$y_{i+1} = y_i + \frac{h}{720}\Big[251f(x_{i+1}, y_{i+1}) + 646f(x_i, y_i) - 264f(x_{i-1}, y_{i-1}) + 106f(x_{i-2}, y_{i-2}) - 19f(x_{i-3}, y_{i-3})\Big], \quad (6.61)$$

where $i = 3, 4, \ldots, n-1$, and y_0, y_1, y_2, y_3 are given. The local truncation error is

$$E = -\frac{3}{160} h^5 y^{(6)}(\eta(x_i)), \quad x_{i-3} < \eta(x_i) < x_{i+1}. \tag{6.62}$$

Note that the implicit Adams-Moulton methods give better results than the explicit Adams-Bashforth methods of the same order. But the weakness of the implicit Adams-Moulton methods is that before using it we have to convert the asking method algebraically to an explicit representation for y_{i+1}, and it is not always possible.

Example 6.9 *Solve the initial-value problem*

$$y' = xy + x, \qquad y(0) = 0, \quad h = 0.2$$

to find the approximate value of $y(1)$ using the Adam-Bashforth three-step method and the Adams-Moulton two-step method. Generate the initial values y_1 and y_2 using the exact solution $y(x) = -1 + e^{x^2/2}$ and compare both your approximations solutions with the exact solution.

Solution. *We set $f(x, y) = xy + x$, and $x_0 = 0$, $y_0 = 0$, and $h = 0.2$. The initial values using the exact solution are $y_1 = 0.0202$ and $y_2 = 0.0833$. Firstly, we shall use Adam-Bashforth three-step method (6.51), which gives*

$$\begin{aligned} y_3 &= y_2 + \frac{h}{12}\Big[23(x_2y_2 + x_2) - 16(x_1y_1 + x_1) + 5(x_0y_0 + x_0)\Big] \\ &= 0.0833 + 0.0167[23(0.4333) - 16(0.20404) + 5(0.0)]. \\ &= 0.1952 \end{aligned}$$

Similarly, the other two computed approximated values are

$$y_4 = 0.3719 \quad \textit{and} \quad y_5 = 0.6380$$

with possible error

$$|y(1) - y_5| = |0.6487 - 0.6380| = 0.0107.$$

Now we use Adams-Moulton two-step method (6.57) for the given problem as follows:

$$y_{i+1} = y_i + \frac{h}{12}\Big[5(x_{i+1}y_{i+1} + x_{i+1}) + 8(x_iy_i + x_i) - (x_{i-1}y_{i-1} + x_{i-1})\Big].$$

Then solving for y_{i+1} gives

$$y_{i+1} = \frac{1}{(1 - 0.0833x_{i+1})}\Big[y_i + \frac{h}{12}[5x_{i+1} + 8(x_i y_i + x_i) - (x_{i-1}y_{i-1} + x_{i-1})]\Big],$$

where $i = 1, 2, 3, 4$. Now using $i = 1$, we have the first approximate solution as follows:

$$\begin{aligned} y_2 &= (1.0345)\Big[0.0202 + (0.0167)[2 + 8(0.20404) - (0.0)]\Big] \\ &= (1.0345)[0.0202 + 0.0607]. \\ &= 0.0837 \end{aligned}$$

Similarly, the other three computed approximate values are

$$y_3 = 0.1982, \quad y_4 = 0.3791, \quad \textit{and} \quad y_5 = 0.6524$$

with possible error

$$|y(1) - y_5| = |0.6487 - 0.6524| = 0.0037.$$

It is interesting to compare an m-step Adams-bashforth explicit method to an (m-1)-step Adams-Moulton implicit method. Both involve n evaluations of function f per step, and both have the terms $h^m y^{(m+1)}(\eta(x_i))$ in their local truncation errors. The coefficients of the error terms are, in general, smaller for implicit than the corresponding explicit methods of the same order. Thus, implicit methods have less round-off error than do explicit methods.

A major disadvantage of multi-step formulas is that they are not self-starting. Thus, in the fourth-order Adams-Bashforth method, we must have four successive values of $f(x, y)$ at equally spaced points before the formula can be used. We might, for example, use Taylor's method or one of the Runge-Kutta methods to obtain the starting values. We must also be assured that these starting values are as accurate as necessary for the overall required accuracy. A second disadvantage of fourth step Adams-Bashforth method is that, although the local discretization error is $O(h^5)$, the coefficient in the error term is somewhat larger than for methods of Runge-Kutta type of the same order. Runge-Kutta methods are generally, although not always, more accurate for this reason. On the other hand, multi-step formulas require only one derivative evaluation per step, compared with four evaluations per step with Runge-Kutta methods and are, therefore, considerably faster and require less computational work.

6.5 Predictor-corrector Methods

The multi-step methods of the previous section were derived using polynomials that interpolated at point x_i and at points backward from x_i. These are sometimes formulas known as an *explicit* or *open* type. While the formulas of an *implicit* or *closed* type are derived by basing the interpolating polynomial on the point x_{i+1} as well as on x_i and points backward from x_i, the combination of an explicit and implicit methods is called the *predictor-corrector* method. A predictor-corrector method consists of a predictor step and corrector step in each interval. The predictor estimates the solution for the new point, and the corrector improves its accuracy. Predictor-corrector methods use the solutions for previous points instead of using intermediate points in each interval. The predictor formula is used once in an iteration while the corrector formula is repeated as many times as necessary to obtain the required level of accuracy; that is, until the two successive corrected values in the iterations are in agreement to the required number of decimal places. There are several predictor-corrector methods but we shall discuss here those methods which are more easy to develop and are widely used. These are the *Milne-Simpson method* and the *Adams-Bashforth-Moulton method.*

6.5.1 Milne-Simpson Method

The *Milne-Simpson method* is a multi-step method. For the derivation of this method, we suppose four equally spaced starting values of $y(x)$ are known at points x_i, x_{i-1}, x_{i-2}, and x_{i-3}. Then considering the quadratic Lagrange polynomial interpolating $p_2(x)$ at points x_{i-2}, x_{i-1}, x_i and integrating over $[x_{i-3}, x_{i+1}]$, we obtain

$$\int_{x_{i-3}}^{x_{i+1}} y' dx = \int_{x_{i-3}}^{x_{i+1}} f(x,y) dx \approx \int_{x_{i-3}}^{x_{i+1}} p_2(x) dx. \tag{6.63}$$

Since

$$\begin{aligned}\int_{x_{i-3}}^{x_{i+1}} p_2(x) dx = \frac{f_{i-2}}{2h^2} \int_{x_{i-3}}^{x_{i+1}} (x - x_{i-1})(x - x_i) dx \\ + \frac{f_{i-1}}{-h^2} \int_{x_{i-3}}^{x_{i+1}} (x - x_{i-2})(x - x_{i-2}) dx \\ + \frac{f_i}{2h^2} \int_{x_{i-3}}^{x_{i+1}} (x - x_{i-2})(x - x_{i-1}) dx\end{aligned} \tag{6.64}$$

after evaluating the above integrals, we obtain

$$\int_{x_{i-3}}^{x_{i+1}} p_2(x)dx = \frac{4h}{3}\Big[2f_i - f_{i-1} + 2f_{i-2}\Big].$$

Using the relation in (6.63), we get

$$y_{i+1}^{(0)} \approx y_{i-3} + \frac{4h}{3}\Big[2f_i - f_{i-1} + 2f_{i-2}\Big]. \tag{6.65}$$

The error term of (6.65) can be found by integrating the error term of the polynomial $p_2(x)$ over the interval $[x_{i-3}, x_{i+1}]$ and after simplifying, we obtain

$$E = \frac{28}{90}h^5 y^{(5)}(\eta_1(x)), \tag{6.66}$$

where the unknown point $\eta_1(x)$ is lies between x_{i-2} and x_i. With this value of $y_{i+1}^{(0)}$ one can calculate f_{i+1} reasonably accurately. In Milne's method, we use (6.65) as a predictor formula and then correct with the following formula by taking the polynomial $p_2(x)$, which fits at points x_{i-1}, x_i, and x_{i+1} as

$$\int_{x_{i-1}}^{x_{i+1}} y' dx = \int_{x_{i-1}}^{x_{i+1}} f(x,y)dx \approx \int_{x_{i-1}}^{x_{i+1}} p_2(x)dx. \tag{6.67}$$

Solving this, we have

$$y_{i+1}^{(1)} \approx y_{i-1} + \frac{h}{3}\Big[f_{i+1}^{(0)} + 4f_i + f_{i-1}\Big] \tag{6.68}$$

with its error term

$$E = -\frac{1}{90}h^5 y^{(5)}(\eta_2(x)) \tag{6.69}$$

for each $x_{i-1} < \eta_2(x) < x_{i+1}$.

Formula (6.68) is based on Simpson's 1/3 rule for numerical integration. The values of y_{i+1} from the predictor and the corrector do not agree. The Milne-Simpson method is unstable and is therefore not very popular, although it can produce accurate solutions over a short enough range of x.

Consideration of error terms (6.66) and (6.69) suggests that the true value will usually lie between the two values and closer to the corrector value. While $\eta_1(x)$ and $\eta_2(x)$ are not necessarily the same value, they lie in similar intervals. If one assumes that the values of $y^{(5)}(\eta_1(x))$ and $y^{(5)}(\eta_2(x))$ are equal, the error in the corrector formula is (1/28) times the error in the predictor formula. Hence, the difference between the predictor and the corrector formulas is 29 times the error in the corrected value.

For smaller values of h, the Milne method would appear to give high accuracy at each step. Comparing it with the fourth-order Runge-Kutta method, we see that the order of truncation error is the same. However, the fourth-order Runge-Kutta method requires four function evaluations while the Milne method may, with careful choice of stepsize, require only two for each step. Apart from the need for starting values, the Milne method would appear to be generally superior to the fourth-order Runge-Kutta method. For a true comparison of these methods, however, it is necessary to consider how they behave over a whole range of problems since applying any method to some differential equations results at each step in a growth of error that ultimately swamps the calculation.

Example 6.10 *Use the Milne-Simpson predictor-corrector method to find the approximation of the solution $y(1)$ to the initial-value problem*

$$y' = xy + y, \qquad y(0) = 0, \quad h = 0.2$$

using the starting values, $y_1 = 0.0200$, $y_2 = 0.0828$, and $y_3 = 0.1963$. Compare both your approximate solutions with the true solution $y(x) = -1 + e^{x^2/2}$.

Solution. *Given $f(x, y) = xy + x$, then*

Milne predictor formula

$$y_{i+1}^{(0)} = y_{i-3} + \frac{4h}{3}\Big[2(x_i y_i + x_i) - (x_{i-1}y_{i-1} + x_{i-1}) + 2(x_{i-2}y_{i-2} + x_{i-2})\Big], \quad i = 3, 4.$$

First taking $i = 3$ in the above formula, we obtain

$$\begin{aligned} y_4^{(0)} &= y_0 + \frac{4h}{3}\Big[2(x_3y_3 + x_3) - (x_2y_2 + x_2) + 2(x_1y_1 + x_1)\Big] \\ &= 0.0 + 0.2667[1.4356 - 0.4331 + 0.4080]. \\ &= 0.3762 \end{aligned}$$

The next approximate value can be obtained by taking $n = 4$ in the formula to obtain

$$\begin{aligned} y_5^{(0)} &= y_1 + \frac{4h}{3}\Big[2(x_4y_4^{(0)} + x_4) - (x_3y_3 + x_3) + 2(x_2y_2 + x_2)\Big] \\ &= 0.02 + 0.2667[2.2019 - 0.7178 + 0.8662] \\ &= 0.6468 \end{aligned}$$

with possible error

$$|y(1) - y_5^{(0)}| = |0.6487 - 0.6468| = 0.0019.$$

Simpson corrector formula

$$y_{i+1}^{(1)} = y_{i-1} + \frac{h}{3}\Big[(x_{i+1}y_{i+1}^{(0)} + x_{i+1}) + 4(x_i y_i + x_i) + (x_{i-1}y_{i-1} + x_{i-1})\Big]$$

for $i = 3, 4$. Using $i = 3$ in the above formula, we obtain

$$\begin{aligned} y_4^{(1)} &= y_2 + \frac{h}{3}\Big[(x_4 y_4^{(0)} + x_4) + 4(x_3 y_3 + x_3) + (x_2 y_2 + x_2)\Big] \\ &= 0.0828 + 0.0667[1.1010 + 2.8711 + 0.4331]. \\ &= 0.3767 \end{aligned}$$

For $n = 4$, we obtain

$$\begin{aligned} y_5^{(1)} &= y_3 + \frac{h}{3}[(x_5 y_5^{(0)} + x_5) + 4(x_4 y_4^{(0)} + x_4) + (x_3 y_3 + x_3)] \\ &= 0.1963 + 0.0667[1.6468 + 4.4038 + 0.7178] \\ &= 0.6478 \end{aligned}$$

with possible error

$$|y(1) - y_5^{(1)}| = |0.6487 - 0.6478| = 0.0009.$$

Note that the error obtained by the Simpson corrector method is smaller than that obtained by the Milne predictor method.

The above results can be obtained using MATLAB commands as follows:

```
>> T = MSIMP('fun1', 0, 1, 0, 5);
```

Program 6.9

MATLAB m-file for the Milne-Simpson Method

```
function T=MSIMP(fun1,a,b,y0,n)
h=(b-a)/n; x = a + (0 : n) * h; y(1) = y0;
f(1)=feval(fun1,x(1),y(1)); S(1)=y(1);
for k = 1 : n %Modified Euler method for starting y values
k1 =feval(fun1,x(k),y(k));
k2 = feval(fun1, x(k) + h, y(k) + h * k1);
y(k + 1) = y(k) + h * (k1 + k2)/2; S(k + 1) = y(k + 1);
f(k+1)=feval(fun1,x(k+1),S(k+1)); end
for k=4:n
S1(1)=S(1);S1(2)=S(2);S1(3)=S(3);S1(4)=S(4);
f1=feval(fun1,x(k),y(k));
f2=feval(fun1,x(k-1),y(k-1));
f3=feval(fun1,x(k-2),y(k-2));
P = y(k-3)+(4*h/3)*(2*f1-f2+2*f3); % Predictor
S1(k+1)=P; f4 = feval(fun1, x(k + 1), S1(k + 1));
S2(1)=S(1);S2(2)=S(2);S2(3)=S(3);S2(4)=S(4);
y(k+1) = y(k-1)+h*(f4+4*f1+f2)/3 % Corrector
S2(k+1)=y(k+1); end; T = [x', S', S1', S2'];
```

6.5.2 Adams-Bashforth-Moulton Method

In order to take advantage of the beneficial properties of implicit methods while avoiding the difficulties inherent in solving the implicit equation, an explicit and implicit method can be combined. The explicit method is used to predict a value of y_{i+1}. Then this value is used in the right-hand side of the implicit method, which produces an improved or corrected value of y_{i+1}. So combining an Adams-Bashforth method (as a predictor) with the corresponding order Adams-Moulton method (as a corrector) gives a predictor-corrector method of the same order. Here, we use the fourth-order Adams-Bashforth method

$$y_{i+1}^{(0)} = y_i + \frac{h}{24}\Big[55f_i - 59f_{i-1} + 37f_{i-2} - 9f_{i-3}\Big] \tag{6.70}$$

as a predictor and the fourth-order Adams-Moulton method

$$y_{i+1}^{(1)} = y_i + \frac{h}{24}\Big[9f_{i+1}^{(0)} + 19f_i - 5f_{i-1} + f_{i-2}\Big] \tag{6.71}$$

as the corrector. The procedure is started when $i = 3$ and consequently the method requires the four starting values y_3, y_2, y_1, and y_0. Corrector formula (6.71) is iterated until convergence is achieved. It can be shown that each step of this procedure has a truncation error of order h^5 and that the method is generally more stable than the Milne-Simpson method. The local truncation error for predictor formula (6.70) is

$$E_p = \frac{251}{720} h^5 y^{(5)}(\eta_1(x)) \tag{6.72}$$

and the error of corrector formula (6.71) is

$$E_c = -\frac{19}{720} h^5 y^{(5)}(\eta_2(x)). \tag{6.73}$$

Let $y_{i+1}^{(0)}$ represent the value of y_{i+1} obtained from (6.70) and $y_{i+1}^{(1)}$ the results obtained with one iteration of (6.71). If the values of f are assumed to be exact at all points up to and including x_i and if $y(x_{i+1})$ represents the exact value of y at $x = x_{i+1}$, then from (6.72) and (6.73) we obtain the error estimates

$$y(x_{n+1}) - y_{i+1}^{(0)} = \frac{251}{720} h^5 y^{(5)}(\eta_1(x)) \tag{6.74}$$

and

$$y(x_{i+1}) - y_{i+1}^{(1)} = -\frac{19}{720} h^5 y^{(5)}(\eta_2(x)). \tag{6.75}$$

In general, $\eta_1(x) \neq \eta_2(x)$. However, if we assume that over the interval of interest $y^{(5)}$ is approximately constant, then on subtracting (6.73) from (6.72), we obtain the following estimate for the derivative $y^{(5)}(\eta(x))$ of the form

$$y_{i+1}^{(1)} - y_{i+1}^{(0)} = \left(\frac{19}{270} h^5 y^{(5)}(\eta(x)) + \frac{251}{720} h^5 y^{(5)}(\eta(x)) \right)$$

or

$$h^5 y^{(5)} = \frac{720}{270} \left(y_{i+1}^{(1)} - y_{i+1}^{(0)} \right). \tag{6.76}$$

Substituting this into (6.75), we find that

$$y(x_{i+1}) - y_{i+1}^{(1)} = -\frac{19}{270} \left(y_{i+1}^{(1)} - y_{i+1}^{(0)} \right)$$

or

$$y(x_{i+1}) - y_{i+1}^{(1)} \approx -\frac{1}{14} (y_{i+1}^{(1)} - y_{i+1}^{(0)}). \tag{6.77}$$

Thus, the error of the corrected value is approximately $(-1/14)$ of the difference between the corrected and the predicted values. It is advisable to use the corrected formula only once. If the accuracy determined by formula (6.71) is not sufficient, it is better to reduce the stepsize than to correct more than once. Note that the local truncation error involved with the Milne-Simpson predictor-corrector method is generally smaller than that of the Adams-Bashforth-Moulton method but the Milne's method has limited use because of problems of stability, which do not occur with the Adams technique.

Example 6.11 *Use the Adams-Bashforth-Moulton predictor-corrector method to find the approximation of the solution $y(1)$ to the initial-value problem*

$$y' = xy + x, \qquad y(0) = 0, \quad h = 0.2$$

with starting values obtained by the Modified Euler's method. Compare both your approximate solutions with the true solution $y(x) = -1 + e^{x^2/2}$.

Solution. *Given $f(x,y) = xy + x$, then using the Modified Euler's method, we have*

$$y_1 = 0.0200, \quad y_2 = 0.0828, \quad \textit{and} \quad y_3 = 0.1963.$$

Adams-Bashforth Predictor Formula

$$\begin{aligned} y_{i+1}^{(0)} &= y_i + \frac{h}{24}\Big[[55(x_i y_i + x_i) - 59(x_{i-1}y_{i-1} + x_{i-1})] \\ &+ [37(x_{i-2}y_{i-2} + x_{i-2}) - 9(x_{i-3}y_{i-3} + x_{i-3})]\Big] \end{aligned}$$

for $i = 3, 4$. Let $i = 3$, we get

$$\begin{aligned} y_4^{(0)} &= y_3 + \frac{h}{24}\Big[[55(x_3y_3 + x_3) - 59(x_2y_2 + x_2)] \\ &+ [37(x_1y_1 + x_1) - 9(x_0y_0 + x_0)]\Big] \\ &= 0.1963 + 0.00833[39.4790 - 25.5541 + 7.5480 - 0] \\ &= 0.3752 \end{aligned}$$

and for $i = 4$, we obtain

$$\begin{aligned} y_5^{(0)} &= y_4 + \frac{h}{24}\Big[[55(x_4 y_4^{(0)} + x_4) - 59(x_3 y_3 + x_3)] \\ &+ [37(x_2 y_2 + x_2) - 9(x_1 y_1 + x_1)]\Big] \\ &= 0.6460 \end{aligned}$$

with possible error

$$|y(1) - y_5| = |0.6487 - 0.6460| = 0.0027.$$

Adams-Moulton Corrector Formula

$$\begin{aligned} y_{i+1}^{(1)} &= y_i + \frac{h}{24}\Big[[9(x_{i+1} y_{i+1} + x_{i+1}) + 19(x_i y_i + x_i)] \\ &- [5(x_{i-1} y_{i-1} + x_{i-1}) - (x_{i-2} y_{i-2} + x_{i-2})]\Big] \end{aligned}$$

for $i = 3, 4$. By taking $i = 3$, we obtain

$$\begin{aligned} y_4^{(1)} &= y_3 + \frac{h}{24}\Big[[9(x_4 y_4^{(0)} + x_4) + 19(x_3 y_3 + x_3)] \\ &- [5(x_2 y_2 + x_2) - (x_1 y_1 + x_1)]\Big] \\ &= 0.1963 + 0.00833[9.9015 + 13.6378 - 2.1656 + 0.2040] \\ &= 0.3761. \end{aligned}$$

Now for $i = 4$, we obtain

$$\begin{aligned} y_5^{(1)} &= y_4^{(0)} + \frac{h}{24}\Big[[9(x_5 y_5 + x_5) + 19(x_4 y_4 + x_4)] \\ &- [5(x_3 y_3 + x_3) - (x_2 y_2 + x_2)]\Big] \\ &= 0.6476 \end{aligned}$$

with possible error

$$|y(1) - y_5| = |0.6487 - 0.6476| = 0.0011.$$

The actual error obtained by the Milne-Simpson predictor-corrector method is smaller than that of the Adams-Bashforth-Moulton method for the same problem.

Use the following MATLAB command to get the above results as follows:

$>> T = ABMM('fun1', 0, 1, 0, 5);$.

Program 6.10

MATLAB m-file for the Adams-Bashforth-Moulton Method

```
function T=ABMM(fun1,a,b,y0,n)
h=(b-a)/n; x = a + (0 : n) * h; y(1) = y0;
f(1)=feval(fun1,x(1),y(1)); S(1)=y(1);
for k = 1 : n %Modified Euler method for starting y values
k1 =feval(fun1,x(k),y(k));
k2 = feval(fun1, x(k) + h, y(k) + h * k1);
y(k + 1) = y(k) + h * (k1 + k2)/2; S(k + 1) = y(k + 1);
f(k+1)=feval(fun1,x(k+1),S(k+1)); end
for k=4:n
S1(1)=S(1);S1(2)=S(2);S1(3)=S(3);S1(4)=S(4);
f1=feval(fun1,x(k),y(k));
f2=feval(fun1,x(k-1),y(k-1));
f3=feval(fun1,x(k-2),y(k-2));
f4=feval(fun1,x(k-3),y(k-3));
P = y(k) + h * (55 * f1 - 59 * f2 + 37 * f3 - 9 * f4)/24; % Predictor
S1(k+1)=P; f4 = feval(fun1, x(k + 1), S1(k + 1));
S2(1)=S(1);S2(2)=S(2);S2(3)=S(3);S2(4)=S(4);
y(k + 1) = y(k) + h * (9 * f5 + 19 * f1 - 5 * f2 + f3)/24 % Corrector
S2(k+1)=y(k+1); end; T = [x', S', S1', S2'];
```

The predictor-corrector methods are widely used for solving initial-value problem (6.6). An advantage of the predictor-corrector methods is their efficiency; they use information from previous steps. Indeed, the function $f(x, y)$ is evaluated only twice in each step regardless of the order of the predictor-corrector methods, whereas the fourth-order Runge-Kutta method evaluates $f(x, y)$ four times in each interval. Another advantage is that the local error can be detected at each step with little additional effort. On the other hand, there are some disadvantages of these types of methods. Firstly, they cannot be self-started because they need previous points. Until the solutions for enough points are determined, another method such as the fourth-order Runge-Kutta method must be used. The other disadvantage is that because previous points are used, changing the interval size in the middle of the solution is not easy. Although the predictor-corrector formulas may be derived on non-uniformly points, the coefficients of the formulas change for each interval, so programming becomes very cumbersome. Also, the predictor-corrector method cannot be used if y' becomes discontinuous. This can happen when one of the coefficients of the differential equation changes discontinuously in the middle of the domain.

6.6 Systems of Simultaneous ODE

So far in this chapter we have considered only single first-order ordinary differential equations, but many practical problems in engineering and science require the solution of a system of simultaneous first-order ordinary differential equations rather than a single equation. Generally, an *nth*-order system of a first-order initial-value problem has the form

$$\begin{aligned} y_1' &= f_1(x, y_1, y_2, \ldots, y_n) \\ y_2' &= f_2(x, y_1, y_2, \ldots, y_n) \\ y_3' &= f_3(x, y_1, y_2, \ldots, y_n) \\ &\vdots \\ y_n' &= f_n(x, y_1, y_2, \ldots, y_n) \end{aligned} \tag{6.78}$$

for $a = x_0 \leq x \leq x_n = b$, with the initial conditions

$$\begin{aligned} y_1(x_0) &= y_{1,0} \\ y_2(x_0) &= y_{2,0} \\ y_3(x_0) &= y_{3,0} \\ &\vdots \\ y_n(x_0) &= y_{n,0}. \end{aligned} \tag{6.79}$$

The object is to find n functions $y_1, y_2, \ldots, y_n$ that satisfy each of differential equations of system (6.78) as well as initial conditions (6.79).

System (6.78) may be written more concisely as

$$\mathbf{y}' = \mathbf{f}(x, \mathbf{y}) \tag{6.80}$$

where $\mathbf{y}, \mathbf{y}'$, and $\mathbf{f}$ are vectors with components y_i, y_i', and f_i, for $i = 1, 2, \ldots, n$, respectively. If this system is to possess a unique solution, it is necessary to impose an additional condition on $\mathbf{y}$. This usually takes the form

$$\mathbf{y}(x_0) = \mathbf{y_0} \tag{6.81}$$

for a given numbers x_0 and $\mathbf{y_0}$.

The numerical methods available for the initial-value of this kind are essentially the same as those for single first-order differential equation that we have already discussed. The main difference is that, at each step, a vector step must be taken. Since the fourth-order Runge-Kutta method is both easy to program and very efficient, we shall concentrate on its use for solving systems. In vector form, the fourth-order Runge-Kutta method can be written as

$$\mathbf{y_{i+1}} = \mathbf{y_i} + \frac{1}{6}\Big[\mathbf{k_1} + 2\mathbf{k_2} + 2\mathbf{k_3} + \mathbf{k_4}\Big], \tag{6.82}$$

where

$$\left.\begin{aligned} \mathbf{k_1} &= h\mathbf{f}(x_i, \mathbf{y_i}) \\ \mathbf{k_2} &= h\mathbf{f}(x_i + \frac{h}{2}, \mathbf{y_i} + \frac{\mathbf{k_1}}{2}) \\ \mathbf{k_3} &= h\mathbf{f}(x_i + \frac{h}{2}, \mathbf{y_i} + \frac{\mathbf{k_2}}{2}) \\ \mathbf{k_4} &= h\mathbf{f}(x_{i+1}, \mathbf{y_i} + \mathbf{k_3}) \end{aligned}\right\} \tag{6.83}$$

for each $i = 0, 1, \ldots, n-1$. The simplest system of this kind is the pair of equations

$$\left.\begin{aligned} y' &= f(x, y, z) \\ z' &= g(x, y, z) \end{aligned}\right\}. \tag{6.84}$$

In this system, x is the independent variable, and we are required to find y and z as functions of x subject to initial conditions of the form

$$y(x_0) = y_0, \qquad z(x_0) = z_0. \tag{6.85}$$

Using the fourth-order Runge-Kutta method, the following formulas would be applied

$$\begin{aligned} y_{i+1} &= y_i + \frac{1}{6}\Big[k_1 + 2k_2 + 2k_3 + k_4\Big] \\ z_{i+1} &= z_i + \frac{1}{6}\Big[l_1 + 2l_2 + 2l_3 + l_4\Big], \end{aligned} \tag{6.86}$$

where

$$\left.\begin{aligned} k_1 &= hf(x_i, y_i, z_i) \\ k_2 &= hf(x_i + \frac{h}{2}, y_i + \frac{k_1}{2}, z_i + \frac{l_1}{2}) \\ k_3 &= hf(x_i + \frac{h}{2}, y_i + \frac{k_2}{2}, z_i + \frac{l_2}{2}) \\ k_4 &= hf(x_{i+1}, y_i + k_3, z_i + l_3) \end{aligned}\right\} \tag{6.87}$$

and

$$\left.\begin{aligned} l_1 &= hg(x_i, y_i, z_i) \\ l_2 &= hg(x_i + \frac{h}{2}, y_i + \frac{k_1}{2}, z_i + \frac{l_1}{2}) \\ l_3 &= hg(x_i + \frac{h}{2}, y_i + \frac{k_2}{2}, z_i + \frac{l_2}{2}) \\ l_4 &= hg(x_{i+1}, y_i + k_3, z_i + l_3) \end{aligned}\right\} \tag{6.88}$$

for each $i = 0, 1, \ldots, n-1$.

Example 6.12 *Find the approximate solution of the following system*

$$\begin{aligned} y' &= yz^2, & y(0) &= 1 \\ z' &= -z - e^y, & z(0) &= 1 \end{aligned}$$

at $x = 1$, using the fourth-order Runge-Kutta method, with $h = 0.1$.

Solution. *Given $f(x, y, z) = yz^2$ and $g(x, y, z) = -z - e^y$, we can find the approximate values of $y(1)$ and $z(1)$ using (6.86)-(6.88) and taking $i = 0, 1, \ldots, 9$. When $i = 0$, we have*

$$y_1 = y_0 + \frac{1}{6}\Big[k_1 + 2k_2 + 2k_3 + k_4\Big]$$

$$z_1 = z_1 + \frac{1}{6}\Big[l_1 + 2l_2 + 2l_3 + l_4\Big],$$

where

$$k_1 = h(y_0 z_0^2) = 0.1(1) = 0.1$$
$$l_1 = h(-z_0 - e^{y_0}) = 0.1(-1 - e^1) = -0.3718$$

$$k_2 = h[(y_0 + k_1/2)(z_0 + l_1/2)^2] = 0.0696$$
$$l_2 = h[-(z_0 + l_1/2) - e^{(y_0+k_1/2)}] = -0.3672$$

$$k_3 = h[(y_0 + k_2/2)(z_0 + l_2/2)^2] = 0.0690$$
$$l_3 = h[-(z_0 + l_2/2) - e^{(y_0+k_2/2)}] = -0.3630$$

$$k_4 = h[(y_0 + k_3)(z_0 + l_3)^2] = 0.0434$$
$$l_4 = h[-(z_0 + l_3) - e^{(y_0+k_3)}] = -0.3549.$$

Using these values, we obtain

$$y_1 = y_0 + \frac{1}{6}\Big[k_1 + 2k_2 + 2k_3 + k_4\Big]$$

$$= 1 + \frac{1}{6}\Big[0.1 + 2(0.0696 + 0.0690) + 0.0434\Big] = 1.0701$$

and

$$z_1 = z_0 + \frac{1}{6}\Big[l_1 + 2l_2 + 2l_3 + l_4\Big]$$

$$= 1 + \frac{1}{6}\Big[-0.3718 + 2(-0.3672 - 0.3631) - 0.3549\Big] = 0.6355.$$

The first and the remaining values are listed in Table 6.1.

Note that before calling MATLAB function *RK4FS*, we must define MATLAB functions *fn* and *gn* as follows:

```
function f1 = fn(x, y, z)
f1 = y * z.^ 2;
function f2 = gn(x, y, z)
f2 = -z - exp(y);
```
.

Table 6.1: Solution of Example 6.12.

n	x_n	y_n	z_n
0	0.0000	1.0000	1.0000
1	0.1000	1.0701	0.6355
2	0.2000	1.0942	0.2933
3	0.3000	1.0971	-0.0195
4	0.4000	1.1008	-0.3030
5	0.5000	1.1224	-0.5628
6	0.6000	1.1772	-0.8089
7	0.7000	1.2846	-1.0564
8	0.8000	1.4811	-1.3327
8	0.9000	1.8598	-1.7047
10	1.0000	2.8091	-2.4817

Given *RK4FS.m, fn.m,* and *gn.m*, the results obtained manually in the preceding example are reproduced with the following MATLAB command:

```
>> S = [x', y', z'] = RK4FS('fn','gn', 0, 0.2, 1, 1, 2);
```

Program 6.11

MATLAB m-file for System of ODE using
Runge-Kutta Method of Order Four

```
function S=RK4FS(fn,gn,a,b,y0,z0,n)
h=(b-a)/n; x = a + (0 : n) * h; y(1) = y0; z(1) = z0;
for i=1:n
k1 = feval(fn, x(i), y(i), z(i));
l1 = feval(gn, x(i), y(i), z(i));
k2 = feval(fn, x(i) + h/2, y(i) + h * k1/2, z(i) + h * l1/2);
l2 = feval(gn, x(i) + h/2, y(i) + h * k1/2, z(i) + h * l1/2);
k3 = feval(fn, x(i) + h/2, y(i) + h * k2/2, z(i) + h * l2/2);
l3 = feval(gn, x(i) + h/2, y(i) + h * k2/2, z(i) + h * l2/2);
k4 = feval(fn, x(i) + h, y(i) + h * k3, z(i) + h * l3);
l4 = feval(gn, x(i) + h, y(i) + h * k3, z(i) + h * l3);
y(i + 1) = y(i) + h * (k1 + 2 * (k2 + k3) + k4)/6;
z(i + 1) = z(i) + h * (l1 + 2 * (l2 + l3) + l4)/6;
end; S = [x', y', z'];
```

6.7 Higher-order Differential Equations

So far we have considered single first-order differential equations and systems of first-order differential equations but of course we often want to solve equations of higher-order. Higher-order differential equations are an important class of differential equations. Here, we will give a brief introduction of higher-order differential equations. Higher-order differential equations involve higher-order derivatives $y''(x), y'''(x)$, and so on. They arise in mathematical models for problems in physics and engineering.

Consider an *nth*-order differential equation

$$y^{(n)} = f(x, y, y', \ldots, y^{(n-1)}). \tag{6.89}$$

The general solution of this equation has contained n arbitrary constants, and so n differential conditions are needed to seek the solution. To solve the higher-order problems, we consider the following initial conditions

$$y(x_0) = y_0,\ y'(x_0) = y'_0, \ldots, y^{(n-1)} = y_0^{(n-1)} \tag{6.90}$$

for given numbers $x_0, y_0, y'_0, \ldots, y_0^{(n-1)}$.

The general approach to solving any *nth*-order initial-value problem (6.89)-(6.90) is to convert it to an equivalent system of n first-order equations. By introducing new variables

$$y_1 = y, y_2 = y', y_3 = y'', \ldots, y_n = y^{(n-1)} \tag{6.91}$$

and then substituting these values in (6.89), we obtain the first-order system

$$\begin{aligned}
y'_1 &= y_2 \\
y'_2 &= y_3 \\
y'_3 &= y_4 \\
\vdots &= \vdots \\
y'_{n-1} &= y_n \\
y'_n &= f(x, y_1, y_2, \ldots, y_n)
\end{aligned} \tag{6.92}$$

subject to the initial conditions

$$\begin{aligned}
y_1(x_0) &= y_0 \\
y_2(x_0) &= y'_0 \\
\vdots & \quad \vdots \\
y_n(x_0) &= y_0^{(n-1)}.
\end{aligned} \tag{6.93}$$

This is now of the form

$$\mathbf{y}' = \mathbf{f}(x, \mathbf{y}), \quad \mathbf{y}(x_0) = \mathbf{y_0}.$$

Hence, it can be solved using the fourth-order Runge-Kutta method or any other method described earlier in the chapter.

Example 6.13 *Use the fourth-order Runge-Kutta method to solve the second-order differential equation*

$$y'' + 3y' + 2y = 2e^{-3x}, \quad y(0) = 1, \; y'(0) = -2$$

over the interval $0 \leq x \leq 1$*, with* $h = 0.2$.

Solution. *The given second-order differential equation can be converted to the system of first-order equations if we substitute*

$$y' = z(x), \quad \textit{then} \quad y'' = z'(x).$$

Then the given differential equation becomes the system

$$\begin{aligned} y' &= z(x) \\ z' &= f(x, y, z) = -2y + 2e^{-3x} - 3z \end{aligned}$$

with initial conditions

$$y(0) = 1, z(0) = -2.$$

Now use the fourth-order Runge Kutta method (6.86), which will generate two sequences $\{y_k\}$ *and* $\{z_k\}$*. A summary of the calculations is given in Table 6.2.*

Comparing this numerical solution with the true solution $y(1) = 0.2823$*, we got three decimal accuracy and it can be improved by taking a smaller value of* h *than 0.2.*

6.8 Boundary-value Problems

Recall from our discussion at the beginning of the chapter that an ordinary differential equation is accompanied by auxiliary conditions, which are used to evaluate the constants of integration that result during the solution of the equation. For an *nth*-order equation, n conditions are required and if all the conditions are specified at the same value of the independent variable, then we are

Table 6.2: Solution of Example 6.13.

k	x_k	y_k	z_k	$\lvert y(x_k) - y_k \rvert$
0	0.0	1.000000	-2.000000	-0.1814342
1	0.2	0.6975	-1.1250	0.0479
2	0.4	0.5225	-0.6768	0.0123
3	0.6	0.4132	-0.4428	0.0031
4	0.8	0.3383	-0.3181	0.0008
5	1.0	0.2825	-0.2469	0.0002

dealing with the initial-value problem. In contrast, there is another application for which the conditions are not known at a single point, but rather, are known at different values of the independent variable because these values are specified at the extreme points or boundaries of a system. Ordinary differential equations with known conditions at more than one value of the independent variable are known as *boundary-value problems (BVP)*. A variety of significant engineering applications are involved within this type. Here, we shall discuss two numerical methods for the solution of second-order, two-point boundary-value problems. The first method is known as the *shooting method*, which forms a linear combination of the solutions to two initial-value problems (*linear shooting method*) and by converting a boundary-value problem to a sequence of initial-value problems (*nonlinear shooting method*), which can be solved using the methods described earlier in the chapter. The second of these two methods is known as the *finite difference method*, which is based on finite differences and it reduces a boundary-value problem to a system of linear equations, which can be solved using the methods described in Chapter 3.

Consider second-order differential equation

$$y'' = f(x, y, y') \tag{6.94}$$

defined on an interval $[a, b]$. Here, f is a given function of three variables and y is an unknown function of the independent variable x. Two associated conditions are required for (6.94) if the solution y is to be unique. We shall use the simplest type of boundary condition, the Dirichlet boundary condition, in which the value of the function is given at each end of the interval. When one of the conditions is given at $x = a$ and the other one at $x = b$, then the Dirichlet boundary conditions

$$y(a) = \alpha \quad \text{and} \quad y(b) = \beta \tag{6.95}$$

for α and β are known. Thus, for the second-order, two-point boundary-value problem of the form

$$y'' = f(x, y, y'), \qquad \text{subject to} \quad y(a) = \alpha \quad \text{and} \quad y(b) = \beta \tag{6.96}$$

we have to determine the solution $y(x)$ of (6.96) for $x \in [a, b]$. We may use the single-step methods employed for initial-value problems to boundary-value problems. Suppose that we have an estimate λ of $y'(a)$ where $y(x)$ is the solution of boundary-value problem (6.96). Now we solve the initial-value problem

$$y'' = f(x, y, y'), \qquad \text{subject to} \quad y(a) = \alpha \quad \text{and} \quad y'(a) = \lambda \tag{6.97}$$

using any single-step method from $x = a$ to $x = b$. If the corresponding solution happens to satisfy the right-hand boundary condition $y(b) = \beta$, then no further action is necessary since the solutions of the initial-value and the boundary-value problems are same. It is most likely that $y(b) \neq \lambda$, then we adjust λ and repeat the calculation to try a better guess.

6.8.1 The Shooting Method

The idea of this method is analogous to shooting at a target with a rifle. One set of conditions consists of the location of the rifle and the direction in which it is aimed. Another condition is the location of the target. We know the locations of the rifle and the target, but we don't know with certainty the direction in which we should aim. We therefore estimate the direction and then adjust it as needed when we see the results of taking the shots.

The Linear Shooting Method

A linear two-point boundary-value problem can be solved by forming a linear combination of the solutions to two initial-value problems. Consider a linear boundary-value problem

$$y'' = f(x, y, y') = p(x)y' + q(x)y + r(x), \ a \leq x \leq b, \ y(a) = \alpha, \ y(b) = \beta, \tag{6.98}$$

where $p(x), q(x)$, and $r(x)$ are continuous on $[a, b]$ and $q(x) > 0$. Then linear boundary-value problem (6.98) has a unique solution $y = y(x)$ over $[a, b]$. This unique solution can be obtained by forming a linear combination of the solution to the following two second-order initial-value problems

$$y_1'' = p(x)y_1' + q(x)y_1 + r(x), \ a \leq x \leq b, \ y_1(a) = \alpha, \ y_1'(a) = 0 \tag{6.99}$$

and

$$y_2'' = p(x)y_2' + q(x)y_2, \ a \leq x \leq b, \ y_2(a) = 0, \ y_2'(a) = 1. \quad (6.100)$$

Let $y_1(x)$ be the solution of (6.99) and $y_2(x)$ be the solution of (6.100). Then the linear combination

$$y(x) = y_1(x) + wy_2(x)$$

is the solution of problem (6.98) and it can be verified easily as follows

$$\begin{aligned} y'(x) &= y_1'(x) + wy_2'(x) \\ y''(x) &= y_1''(x) + wy_2''(x). \end{aligned}$$

Then using (6.99) and (6.100), we have

$$\begin{aligned} y''(x) &= p(x)y_1' + q(x)y_1 + r(x) + w[p(x)y_2' + q(x)y_2] \\ &= p(x)[y_1' + wy_2'] + q(x)[y_1 + wy_2] + r(x) \\ &= p(x)y' + q(x)y + r(x). \end{aligned}$$

Also,

$$\begin{aligned} y(a) &= y_1(a) + wy_2(a) = \alpha + w(0) = \alpha \\ y(b) &= y_1(b) + wy_2(b) = \beta, \end{aligned}$$

which gives the value of w as

$$w = \frac{\beta - y_1(b)}{y_2(b)} y_2(x).$$

Thus, if $y_2(b) \neq 0$, then

$$y(x) = y_1(x) + \left(\frac{\beta - y_1(b)}{y_2(b)} \right) y_2(x) \quad (6.101)$$

is the unique solution of linear boundary-value problem (6.98).

Example 6.14 *Use the linear shooting method to solve the following linear boundary-value problem*

$$y'' = -xy' + y + 2x + \frac{2}{x}, \quad y(1) = 0, \ y(2) = 4\ln 2, \ h = 0.2.$$

Solution. *To find the solution of the given linear boundary-value problem, firstly, we convert the given problem into two initial-value problems as*

$$y_1'' = -xy_1' + y_1 + 2x + \frac{2}{x}, \quad y_1(1) = 0, \ y_1'(1) = 0$$

and

$$y_2'' = -xy_2' + y_2, \quad y_2(1) = 0, \; y_2'(1) = 1.$$

To convert the above two second-order differential equations into the systems of first order-equations, we substitute

$$y_1' = z_1(x), \quad \textit{then} \quad y_1'' = z_1'(x)$$

and then the first differential equation becomes a system

$$\begin{aligned} y_1' &= z_1 \\ z_1' &= f(x, u, z_1) = -xz_1 + u + 2x + \frac{2}{x} \end{aligned}$$

with

$$u(1) = 0, z_1(1) = 0.$$

Similarly, the second differential equation can be converted as

$$y_2' = z_2(x), \quad \textit{then} \quad y_2'' = z_2'(x)$$

and then the second differential equation becomes a system

$$\begin{aligned} y_2' &= z_2 \\ z_2' &= f(x, v, z_2) = -xz_2 + v \end{aligned}$$

with

$$v(1) = 0, z_2(1) = 1.$$

Solve both above systems using fourth-order Runge Kutta method (6.86) to get solutions $y_1(x)$ and $y_2(x)$. Then using (6.101), we can have the solution of the given linear boundary-value problem. A summary of the calculations is given in Table 6.3. The value listed as u_n approximates $y_1(x)$, the value v_n approximates $y_2(x)$, and y_n approximates

$$y(x_n) = u(x_n) + \left(\frac{4\ln 2 - u(2)}{v(2)} \right) v(x_n).$$

Table 6.3: Shooting method for Example 6.14.

n	x_n	u_n	v_n	y_n	$\lvert y(x_n) - y_n\rvert$
00	1.0	0.0000	0.0000	0.0000	0.0000
01	1.2	0.0750	0.1813	0.4370	0.0006
02	1.4	0.2812	0.3305	0.9423	0.0002
03	1.6	0.5944	0.4548	1.5041	0.0001
04	1.8	0.9945	0.5608	2.1162	0.0002
05	2.0	1.4655	0.6535	2.7726	0.0000

By comparing the numerical solution with the true solution $y(2) = 2.7726$, we have an accuracy of four decimal places. It can be improved by taking a smaller value of h than 0.2.

Program 6.12

MATLAB m-file for the Linear Shooting Method

```
function S=LSOOT(fn,gn,a,b,alpha,beta,n)
h = (b - a)/n; y1 = alpha; y2 = 0; u = 0; v = 0;
for i=1:n; x = a + (i - 1) * h;
k1 = feval(fn, x, u, v); l1 = u;
k2 = feval(fn, x + h/2, u + h * k1/2, v + h * l1/2); l2 =u+h/2*k1;
k3 = feval(fn, x+h/2, u+h*k2/2, v+h*l2/2); l3 = u+h*k2/2;
k4 = feval(fn, x + h, u + h * k3, v + h * l3); l4 = u + h * k3;
u = u + h * (k1 + 2 * (k2 + k3) + k4)/6;
v = v + h * (l1 + 2 * (l2 + l3) + l4)/6; y1(i + 1) = v; end
u=1;v=0; for i=1:n; x = a + (i - 1) * h;
k1 =feval(gn,x,u,v); l1 =u;
k2 = feval(gn, x+h/2, u+h*k1/2, v+h*l1/2); l2 = u+h*k1/2;
k3 = feval(gn, x+h/2, u+h*k2/2, v+h*l2/2); l3 = u+h*k2/2;
k4 = feval(gn, x + h, u + h * k3, v + h * l3); l4 = u + h * k3;
u = u + h * (k1 + 2 * (k2 + k3) + k4)/6;
v = v + h * (l1 + 2 * (l2 + l3) + l4)/6; y2(i + 1) = v; end
for i =1:n+1; x = a + (i - 1) * h;
y = y1(i) + (beta - y1(n + 1))/y2(n + 1) * y2(i); end
```

To get the above results using MATLAB commands we first define the following functions as follows:

```
function f = fn(x)
f = -x * u + v + 2 * x + 2./x;
function f = gn(x)
f = -x * u + v;
```

Then the following single MATLAB command can be used to generate the solution of Example 6.14:

```
>> s = LSOOT('fn','gn',1,2,0,4*log(2),5)
```

6.8.2 The Nonlinear Shooting Method

The idea of the shooting method for nonlinear second-order boundary-value problem (6.96) is similar to the linear case, except that the solution to a nonlinear problem cannot be simply expressed as a linear combination of the solutions to two-point initial-value problems. For a nonlinear boundary-value problem, we have an iterative procedure and we need to find a zero of the function representing the error; that is, the amount by which the solution to the initial-value problem fails to satisfy the boundary condition at $x = b$ or in other words, the amount by which $y(b, \lambda)$ misses the target value β. So the problem of solving two-point boundary-value problem (6.96) can be viewed as that of finding an approximate root λ^* of $F(\lambda)$; that is, solving

$$F(\lambda) = y(b, \lambda) - \beta = 0. \tag{6.102}$$

Once $\lambda^* = y'(a)$ has been found, then the desired solution $y(x)$ is $y(x, \lambda^*)$. The methods considered in Chapter 2 for solving a single nonlinear equation are applicable here. For example, the bisection method, the secant method, and Newton's method are all available. The function F is an expansive one to compute because each value of $F(\lambda)$ is obtained by numerically solving an initial-value problem. Since each evaluation of $F(\lambda)$ required a lot of work, it is important to find the desired root of $F(\lambda)$ using a method that converges rapidly. Although Newton's method was the most rapidly convergent of Chapter 2, it is not suitable for this application because we generally have no analytic expression for $F(\lambda)$ to differentiate. On the other hand, the secant method is ideally suited. We select two initial guesses λ_0, λ_1 for λ and solve initial-value problem (6.97) for each of them. After finding $F(\lambda_0)$ and $F(\lambda_1)$ we can apply the secant method to generate our next guess λ_2 using

$$\lambda_2 = \lambda_1 - \left(\frac{\lambda_1 - \lambda_0}{F(\lambda_1) - F(\lambda_0)} \right) F(\lambda_1).$$

The secant iterative method can then be repeated as usual but the only difference here is you solve a second-order initial-value problem on each iteration obtaining a sequence of values $\lambda_2, \lambda_3, \ldots, \lambda_n$ by

$$\lambda_{i+1} = \lambda_i - \left(\frac{\lambda_i - \lambda_{i-1}}{F(\lambda_i) - F(\lambda_{i-1})} \right) F(\lambda_i), \quad \text{for} \quad i = 1, 2, \ldots, n-1. \qquad (6.103)$$

One should note that in order for the method to converge, a good choice of two initial approximations λ_0 and λ_1 is necessary.

Example 6.15 *Use the nonlinear shooting method to solve the following two-point nonlinear boundary-value problem*

$$y'' = 2y^3, \quad y(1) = 0.25, \; y(2) = 0.2, \; h = 0.1.$$

Compare your results with the exact solution $y(x) = (x+3)^{-1}$.

Solution. *We wish to find an approximate root* λ^* *of*

$$F(\lambda) = y(b, \lambda) - \beta = y(2, \lambda) - 0.2,$$

where $y(x, \lambda)$ *is the solution of the associated IVP*

$$y'' = 2y^3, \quad y(1) = 0.25, \; y'(1) = \lambda, \; h = 0.1,$$

which reduces to the first-order system

$$\begin{array}{rclcrcl} y_1' & = & y_2, & & y_1(1) & = & 0.25 \\ y_2' & = & 2y_1^3, & & y_2(1) & = & \lambda \end{array},$$

where $y_1 = y$ *and* $y_2 = y'$. *For the first initial approximation of* $y'(1)$, *we set*

$$\lambda_0 = \frac{\beta - \alpha}{b - a} = \frac{0.2 - 0.25}{2 - 1} = -0.05$$

and we take $\lambda_1 = -0.5$ *as the second initial approximation for* $y'(1)$. *Using the fourth-order Runge-Kutta with* $h = 0.1$, *we find that*

$$y(2, \lambda_0) = 0.21314938 \quad \textit{and} \quad F(\lambda_0) = 0.01314938$$

and

$$y(2, \lambda_1) = -0.24684241 \quad \textit{and} \quad F(\lambda_1) = -0.44684241.$$

For the third approximation, we can now begin iteration. Taking $\lambda_0 = -0.05$, *and* $\lambda_1 = -0.5$, *and then using secant method (6.103) by taking* $i = 1$, *we have*

$$\begin{aligned} \lambda_2 &= \lambda_1 - \left(\frac{\lambda_1 - \lambda_0}{F(\lambda_1) - F(\lambda_0)}\right) F(\lambda_1) \\ &= -0.24684241 - \frac{(-0.44684241)(-0.5 + 0.05)}{-0.44684241 - 0.01314938} \\ &= -0.06286375. \end{aligned}$$

Once again using the fourth-order Runge Kutta method with $h = 0.1$, *we have*

$$y(2, \lambda_2) = 0.19961790 \quad \text{and} \quad F(\lambda_2) = -0.00038210.$$

The further steps of the secant method are given as follows:

$$\begin{aligned} \lambda_3 &= -0.06248963, \quad y(2, \lambda_3) = 0.20001090, \quad \text{and} \quad F(\lambda_3) = 0.00001090 \\ \lambda_4 &= -0.06250001, \quad y(2, \lambda_4) = 0.20000000, \quad \text{and} \quad F(\lambda_4) = 0.00000000. \end{aligned}$$

After four attempts, the value $\lambda_4 = -0.06250001$ *has been located for the solution* $y(2, \lambda) = 2$ *to within nine decimal places. The intermediate values of* y *associated with value* $\lambda = -0.06250001$ *are given in Table 6.4.*

Table 6.4: The Shooting method for Example 6.15.

n	x_n	y_n	$y(x_n)$	$\|y(x_n) - y_n\|$
00	1.0	0.25000000	0.25000000	0.00000000
01	1.1	0.24390244	0.24390244	2.02E-11
02	1.2	0.23809524	0.23809524	1.47E-10
03	1.3	0.23255814	0.23255814	3.63E-10
04	1.4	0.22727273	0.22727273	6.56E-10
05	1.5	0.22222222	0.22222222	1.02E-09
06	1.6	0.21739130	0.21739130	1.43E-09
07	1.7	0.21276596	0.21276596	1.90E-09
08	1.8	0.20833333	0.20833333	2.42E-09
09	1.9	0.20408163	0.20408163	2.98E-09
10	2.0	0.20000000	0.20000000	3.57E-09

Note that the accuracy obtained by the nonlinear shooting method is limited only by the accuracy of the calculated $y(b, \lambda)$ values. The shooting method can become arduous for higher-order differential equations. Especially with problems of fourth and higher-order, the necessity to assume two or more conditions makes the approach somewhat more difficult. For these reasons, the alternative methods are available, and one of them we discuss next.

6.8.3 The Finite Difference Method

An alternative strategy for solving two-point boundary-value problem (6.96) for the special solution where the differential equation is linear specifically, we consider a boundary value problem of the form

$$y'' = p(x)y' + q(x)y + r(x) \tag{6.104}$$

subject to the boundary conditions

$$y(a) = \alpha \quad \text{and} \quad y(b) = \beta. \tag{6.105}$$

The interval $[a, b]$ over which the solution is sought is divided into N subintervals of length h. We need to compute the values $y_1, y_2, \ldots, y_{n-1}$ of the solution at the intermediate points

$$x_k = x_0 + kh, \qquad k = 1, 2, \ldots, n-1$$

with $x_0 = a, x_n = b$, and $h = \dfrac{x_n - x_0}{n}$.

The basic idea behind the finite difference method is that we use different approximations for the derivatives to obtain a system of linear equations for the y-values. The numerical approximation to the exact solution $y(x_k)$ is denoted by y_k. From (6.105) we may take $y_0 = \alpha$ and $y_n = \beta$.

As in Chapter 5 we used the central difference formula to approximate the derivatives. For sufficiently smooth y, both approximations have second-order accuracy.

$$y'(x_k) \approx \frac{y_{k+1} - y_{k-1}}{2h} \tag{6.106}$$

and

$$y''(x_k) \approx \frac{y_{k+1} - 2y_k + y_{k-1}}{h^2}. \tag{6.107}$$

At any point $x_k (1 \leq k < n)$, using approximations (6.106) and (6.107), differential equation (6.104) becomes

$$\frac{y_{k+1} - 2y_k + y_{k-1}}{h^2} = p_k(\frac{y_{k+1} - y_{k-1}}{2h}) + q_k y_k + r_k, \tag{6.108}$$

where

$$p_k = p(x_k), \quad q_k = q(x_k), \quad \text{and} \quad r_k = r(x_k).$$

After simplifying, we get (6.108) as

$$\left(1 + \frac{hp_k}{2}\right) y_{k-1} - \left(2 + h^2 q_k\right) y_k + \left(1 - \frac{hp_k}{2}\right) y_{k+1} = h^2 r_k, \tag{6.109}$$

which provides a tridiagonal system of linear equations for the unknown values $y_1, y_2, \cdots, y_{n-1}$ with the known values of p_k, q_k and r_k. The first and the last of these equations (for $k = 1$ and $k = n - 1$) must be adjusted to absorb the terms relating to α and β into their respective right-hand sides. The resulting system can be written as

$$T\mathbf{y} = \mathbf{c}, \tag{6.110}$$

where

$$T = \begin{pmatrix} -(2 + h^2 q_1) & (1 - \frac{hp_1}{2}) & 0 & \cdots & 0 \\ (1 + \frac{hp_2}{2}) & -(2 + h^2 q_2) & (1 - \frac{hp_2}{2}) & \cdots & 0 \\ \ddots & \ddots & \ddots & \ddots & \ddots \\ 0 & & & & (1 - \frac{hp_{n-2}}{2}) \\ 0 & 0 & \cdots & (1 + \frac{hp_{n-1}}{2}) & -(2 + h^2 q_{n-1}) \end{pmatrix}$$

and

$$\mathbf{y} = \begin{pmatrix} y_1 \\ y_2 \\ \vdots \\ y_{n-1} \end{pmatrix}, \quad \mathbf{c} = \begin{pmatrix} h^2 r_1 - (1 + \frac{hp_1}{2})\alpha \\ h^2 r_2 \\ \vdots \\ h^2 r_{n-1} - (1 - \frac{hp_{n-1}}{2})\beta \end{pmatrix}.$$

Example 6.16 *Use the finite difference method to solve the following boundary value problem*

$$y'' = -\frac{2y'}{x} + \frac{4}{x}, \qquad y(0.5) = 3, \; y(1) = 3.$$

Solution. *Given* $p(x_k) = \frac{-2}{x_k}$, $q(x_k) = 0$, *and* $r(x_k) = \frac{4}{x_k}$, *using (6.109), we obtain*

$$(1 - \frac{h}{x_k})y_{k-1} - 2y_k + (1 + \frac{h}{x_k})y_{k+1} = \frac{4h^2}{x_k}.$$

For $n = 5$, *with* $h = 0.1$ *gives the tridiagonal system*

$$\begin{pmatrix} -2 & \frac{7}{6} & 0 & 0 \\ \frac{6}{7} & -2 & \frac{8}{7} & 0 \\ 0 & \frac{8}{9} & -2 & \frac{9}{8} \\ 0 & 0 & \frac{8}{9} & -2 \end{pmatrix} \begin{pmatrix} y_1 \\ y_2 \\ y_3 \\ y_4 \end{pmatrix} = \begin{pmatrix} -\frac{37}{18} \\ \frac{16}{49} \\ \frac{1}{4} \\ -\frac{254}{81} \end{pmatrix}.$$

Then solve the above system using LU *decomposition (with* $l_{ii} = 1$*), which we discussed in Chapter 3, we obtain*
$y_1 = 2.8667$, $y_2 = 2.8286$, $y_3 = 2.8500$, $y_4 = 2.9111$.

To get above results using MATLAB commands we first define the following functions as follows:

```
function f = fp(x)
f = -2./x;
function f = fq(x)
f = 0;
function f = fr(x)
f = 4./x;
```

Then the following single MATLAB command can be used to generate the solution of Example 6.16 as:

```
>> s = FDM('fp','fq','fr',0.5,1,3,3,5)
```

Program 6.13

MATLAB m-file for the Finite Difference Method

```
function S=FDM(fp,fq,fr,a,b,alpha,beta,n);
h=(b-a)/n;
for k=1:n-1; x = a + k * h;
if(k~=n-1); U(k) = 1 - h/2 * feval(fp, x);end
D(k) = -(2 + h^2*feval(fq, x));
if(k~=1); L(k - 1) = 1 + h/2 * feval(fp, x); end; end
c(1) = h^2*feval(fr, a+h) - (1+h/2*feval(fp, a+h))*alpha;
c(n-1) = h^2*feval(fr, b-h) - (1-h/2*feval(fp, b-h))*beta;
for k=2:n-2; x = a + k * h; c(k) = h^2 * feval(fr, x); end
for k=2:n-1; mult=L(k-1)/D(k-1);
D(k) = D(k) - mult * U(k - 1); c(k) = c(k) - mult * c(k - 1); end
y(n)=beta; y(n-1)=c(n-1)/D(n-1);
for k=n-2:-1:1; y(k) = (c(k) - U(k) * y(k + 1))/D(k); end
```

We observed that system (6.110) is tridiagonal and therefore speedy to solve and also economical of memory space to store the coefficients. This will be true even if the value of n is very large (or h is very small) because we only use y_{k-1}, y_k, and y_{k+1} in any equation to replace y' and y''. This is one reason the finite difference method is widely used to solve second-order linear boundary-value problems.

So if a two-point boundary-value problem is linear, then the finite difference method yields a linear system with a diagonally dominant, banded coefficient matrix that can be solved easily by the Gaussian elimination method. But if the boundary-value problem is nonlinear, then the finite difference method yields a nonlinear system that can be solved by the methods discussed in Chapter 2. Whether a boundary-value problem is linear or nonlinear, the solution obtained by the finite difference method reflects the truncation error of the finite difference approximations of y' and y''; the accuracy should be assessed by comparing the calculated y_k values to those obtained with n doubled.

6.9 Summary

In this chapter we described many numerical methods for solving first-order ordinary differential equations and systems of first-order ordinary differential equations. The performance of these methods is problem dependent and consequently they should be used with care. Euler's method is the simplest and least accurate of a family of techniques called single-step methods. For getting satisfactory results by this method the stepsize should be small. But it requires many iterations of the method and can also lead to round-off error accumulation. Euler's method does not require the computation of higher-order derivatives, which can be a problem with higher-order methods and is thus quite simple to apply. Generally, Euler's method is a very poor method and is not recommended.

Taylor's methods are excellent when higher-order derivatives can be found. The main disadvantage of these methods is that higher-order derivatives may be quite complex and difficult to find. Since these methods are based on derivatives rather than differences, round-off error tends to be less of a problem. We also discussed Runge-Kutta methods of different order, but of order four is most recommended for the solution of initial-value problems and also for the starting values of the multi-step methods.

We also discussed multi-step methods for the solution of initial-value problems. The Adams multi-step method requires only two function evaluations per step, but is not a self-starting method because it requires four initial values. These initial values can be obtained using Runge Kutta methods. The multi-step methods are most effective when the right-hand side function $f(x, y)$ is smooth. We also discussed predictor-corrector methods. A predictor-corrector method consists of a predictor and a corrector. The Adams-bashforth-Moulton and the Milne-Simpson predictor-corrector methods were discussed. The second-order Adams-bashforth-Moulton predictor-corrector is identical to the second-order Runge-Kutta method. The third and fourth-order predictor-corrector methods cannot be self-started. However, once they are started, their computational efficiency is higher than that of the fourth-order Runge-Kutta method. The error check in each interval is easier than for the fourth-order Runge-Kutta method. Thus, for solving initial-value problems the fourth-order Runge-Kutta method and predictor-corrector methods are most popular and widely used, although no one method will perform uniformly better than another method on all problems. For example, the Runge-Kutta methods have an important advantage: they are self-starting. In addition, they are more stable, provide good accuracy, and as a computer program, occupy a relatively small amount of core storage. They provide, however, no estimate of the accuracy being achieved, so the user has no way of knowing whether stepsize h being used is adequate. One can, of course, run the same problem with several values of h and estimate the accuracy by comparing the results, but this is expansive in machine time. The second major disadvantage of the fourth-order Runge-Kutta method is that it requires four derivative evaluations per step, compared with only two using the fourth-order predictor-corrector methods. On some problems Runge-Kutta methods require almost twice as much computing time. Predictor-corrector methods provide an automatic error estimate at each step, thus allowing the program to select an optimum value of h for a required accuracy. They are also fast since they require only two derivative evaluations peer step. On the other hand, predictor-corrector methods have the disadvantage that they require special techniques for starting and for doubling and halving the stepsize and they may be subject to numerical instability. We also discussed the solutions of the systems of simultaneous differential equations using the Runge-Kutta method of order four and the solution of higher-order ordinary differential equations. In the last section we discussed two basic numerical methods (the Shooting method and the finite difference method) for the solution of boundary-value problems of second-order ordinary differential equations.

6.10 Exercises

1. Find the general solution of the differential equation

$$y' = \frac{x}{y}.$$

2. Put the following differential equations into a form for numerical solution by Euler's method.

(a) $y + 2yy' - y' = 0.$

(b) $\ln y' = x^2 - y^2.$

(c) $y' - x^2 y' = y.$

3. Solve the following initial-value problems using Euler's method.

(a) $y' = y + x^2, \quad x = 0(0.2)1, \quad y(0) = 1.$

(b) $y' = (x-1)(x+y+1), \quad x = 0(0.2)1, \quad y(0) = 1.$

(c) $y' = y + \sin(x), \quad x = 0.2(0.01)0.25, \quad y(0.2) = 1.5.$

4. Solve the following initial-value problems and compare the numerical solutions obtained with Euler's method using the values of $h = 0.1$ and $h = 0.2$. Compare the results to the actual values.

(a) $y' = 1 + x^2, \quad 0 \le x \le 1, \quad y(0) = 0, \quad y(x) = \tan x.$

(b) $y' = 2(y + 1), \quad 0 \le x \le 1, \quad y(0) = 2, \quad y(x) = e^{2x} - 1.$

(c) $y' = 2(y - 1)^2, \ 0 \le x \le 1, \ y(0) = 2, \ y(x) = 2(1 - x)/(1 - 2x).$

5. Solve the following initial-value problems using Taylor's method of order two.

(a) $y' = 2x^2 - y, \quad x = 0(0.2)1, \quad y(0) = -1.$

(b) $y' = 3x^2 y, \quad x = 0(0.2)1, \quad y(0) = 1.$

(c) $y' = x/y - x, \quad x = 0(0.2)1, \ y(0) = 2.$

6. Solve the initial-value problems using Taylor's method of order three of Problem 3.

7. Solve the following initial-value problems using the Modified Euler's method.

(a) $y' = y^2 x^2, \quad x = 1(0.2)2, \quad y(1) = -1.$

(b) $y' = x - y/2x, \quad x = 1(0.02)1.10, \quad y(1) = 0.25.$

(c) $y' = 1/y^2 - yx, \quad x = 1(0.2)2, \quad y(1) = 1.$

8. Solve the following initial-value problems and compare the numerical solutions obtained with the Modified Euler's method using the values of $h = 0.05$ and $h = 0.1$ and compare the results with the actual values.

(a) $y' = x + \dfrac{3y}{x}, \quad 1 \le x \le 2, \quad y(0) = 0, \quad y(x) = x^3 - x^2.$

(b) $y' = \sqrt{y}, \quad 0 \le x \le 1, \quad y(0) = 1, \quad y(x) = 1/4(x + 2)^2.$

(c) $y' = 4 - 3y, \quad 0 \le x \le 1, \quad y(0) = 5, \quad y(x) = 4/3 + 11/3e^{-3x}.$

9. Solve the following initial-value problems using Heun's method and the Midpoint method.

(a) $y' = (x + 1)y, \quad x = 0.5(0.2)1.5, \ y(0.5) = 1.$

(b) $y' = -xy^2, \quad x = 0(0.2)1, \quad y(0) = 2.$

(c) $y' = x^2 + y^2, \quad x = 1(0.2)2, \ y(1) = -1.$

10. Solve the initial-value problems in Problem 8 and compare the numerical solutions obtained with Heun's method and the Midpoint method.

11. Solve the following initial-value problems using the fourth-order Runge-Kutta formula using $h = 0.2$

(a) $y' = 1 + \frac{y}{x}, \quad 1 \leq x \leq 2 \quad y((1) = 1.$

(b) $y' = y \tan x, \quad 0 \leq x \leq 1, \quad y(0) = 2.$

(c) $y' = (1-x)y^2 - y, \quad 1 \leq x \leq 2 \quad y(0) = 1.$

12. Solve the following initial-value problems and compare the numerical solutions obtained with the fourth-order Runge-Kutta formula and the fourth-order Taylor's method using the values of $h = 0.1$ and $h = 0.2$, over interval $[a, b]$.

(a) $y' = 4 - 3y, \quad [0, 1], \quad y(0) = 5, \quad y(x) = 4/3 + 11/3e^{-3x}.$

(b) $y' = (2-x)y, \quad [2, 3], \quad y(2) = 1, \quad y(x) = e^{-1/2(x-2)^2}.$

(c) $y' = \frac{1}{1+x^2} - 2y^2, \quad [0, 1], \; y(0) = 0, \quad y(x) = \frac{x}{1+x^2}.$

13. Solve Problem 11 using the Runge-Kutta Merson method.

14. Use the Runge-Kutta- Fehlberg method to approximate the solutions to the following initial-value problems.

(a) $y' = 1 - y, \; [0, 1], \; y(0) = 0, \; Tol = 10^{-6}, hmax = 0.2.$

(b) $y' = 1 + y^2, \; [0, \frac{\pi}{4}], \; y(0) = 0, \; Tol = 10^{-4}, hmax = 0.2.$

(c) $y' = 2(\frac{y}{x} + x^2 e^x, \; [1, 2], \; y(1) = 0, \; Tol = 10^{-4}, hmax = 0.2.$

15. Use the Adams-Bashforth three-step method to approximate the solutions to the following initial-value problems. In each case use exact starting values and compare the results to the actual values, use $h = 0.1$ each case.

$$
\begin{aligned}
(a) \; y' &= 1 + x^2, \; 0 \leq x \leq 1, \; y(0) = 0; \; y(x) = \tan x. \\
(b) \; y' &= 2(y+1), \; 0 \leq x \leq 1, \; y(0) = 2; \; y(x) = e^{2x} - 1. \\
(c) \; y' &= x + \frac{3y}{x}, \; 1 \leq x \leq 2, \; y(0) = 0; \; y(x) = x^3 - x^2. \\
(d) \; y' &= (1+x)^{-1} y, \; 0 \leq x \leq 1, \; y(0) = 1; \; y(x) = 1 + x.
\end{aligned}
$$

16. Use all the Adams-Moulton methods to approximate the solutions to Problem 15. In each case use exact starting values and explicitly solve for y_{n+1}. Compare the results to the actual values.

17. Use the Milne-Simpson predictor-corrector method to approximate the solutions to the following initial-value problems. Use starting values obtained from the modified Euler's method and compare the results to the actual values, with $h = 0.2$ each case.

$$
\begin{array}{llcl}
(a) & y' & = & 4 - 3y, \ 0 \leq x \leq 1, \ y(0) = 5; \ y(x) = \frac{4}{3} + \frac{11}{3}e^{-3x}. \\
(b) & y' & = & y \tan x, \ 0 \leq x \leq 1, \ y(0) = 2; \ y(x) = 2/\cos x. \\
(c) & y' & = & (2 - x)y, \ 2 \leq x \leq 3, \ y(2) = 1; \ y(x) = exp[-0.5(x-2)^2]. \\
(d) & y' & = & (1 - x)y^2 - y, \ 0 \leq x \leq 1, \ y(0) = 1; \ y(x) = 1/(e^x - x).
\end{array}
$$

18. Use the Milne-Simpson predictor-corrector method to approximate solutions to Problem 15.

19. Use the Adams-Bashforth-Moulton predictor-corrector method to approximate solutions to Problem 17.

20. Use the Adams-Bashforth-Moulton predictor-corrector method to approximate solutions to Problem 15.

21. Use the Runge-Kutta method for a system of differential equations to approximate solutions of the following systems of first-order differential equations, using $h = 0.2$ each case.

(a)

$$
\begin{aligned}
y' &= 1 + 2e^{-z}, \quad 1 \leq x \leq 2, \quad y(1) = -1 \\
z' &= x - y, \quad 1 \leq x \leq 2, \quad z(1) = 0.
\end{aligned}
$$

(b)

$$
\begin{aligned}
y' &= 1, \quad 0 \leq x \leq 1, \quad y(0) = 1 \\
z' &= 3y^2, \quad 0 \leq x \leq 1, \quad z(0) = 0.
\end{aligned}
$$

(c)

$$
\begin{aligned}
y' &= 4y + z, \quad 0 \leq x \leq 1, \quad y(0) = 1 \\
z' &= -2y + z, \quad 0 \leq x \leq 1, \quad z(0) = 0.
\end{aligned}
$$

22. Use the Runge-Kutta method for a system of differential equations to approximate solutions of the following systems of first-order differential equations, using $h = 0.2$ on the interval $[0, 1]$ each case and compare the approximate solutions with the given exact solutions.

(a)

$$\begin{aligned} y' &= 3z^3, \quad y(0) = 1, \quad y(x) = e^{3x} \\ z' &= z, \quad z(0) = 1, \quad z(x) = e^{x}. \end{aligned}$$

(b)

$$\begin{aligned} y' &= 2y + z - e^{2x}, \; y(0) = 1, \; y(x) = e^{x} + e^{-x} + \cos x + \sin x \\ z' &= y + 2z, \quad z(0) = -1, \; z(x) = e^{x} + e^{-x} - \cos x - \sin x. \end{aligned}$$

(c)

$$\begin{aligned} y' &= 2y + 3z, \quad y(0) = 2, \; y(x) = 3e^{4x} - e^{-x} \\ z' &= 2y + z, \quad z(0) = 3, \; z(x) = 2e^{4x} + e^{-x}. \end{aligned}$$

23. Express the following second-order ordinary differential equations and then solve the pair of first-order ordinary differential equations using the Runge-Kutta method of order four, with $h = 0.2$.

(a) $y'' + xy' - 3y = x^2, \quad y(0) = 3, \quad y'(0) = -6, \quad 0 \leq x \leq 1.$

(b) $y'' - \cos(x - y) + y^2, \quad y(0) = 1, \; y'(0) = 0, \quad 0 \leq x \leq 1.$

(c) $xy'' - y' - 8x^3y^3 = 0, \quad y(1) = 0.5, \; y'(1) = -0.5, \quad 1 \leq x \leq 2.$

24. Use the nonlinear shooting method to approximate the solutions of the following boundary-value problems within accuracy 10^{-4} using $h = 0.2$:

(a) $y'' = 2y^3 - 6y - 2x^3, \quad y(1) = 2, \quad y(2) = 2.5, \quad 1 \leq x \leq 2.$

(b) $y'' = x(y')^2, \quad y(0) = \pi/2, \; y(2) = \pi/4, \quad 0 \leq x \leq 2.$

25. Approximate the solutions of the following boundary-value problems using the linear shooting method using $h = 0.2$:

    $(a)\ y'' = 2y' + y + 4, \quad y(0) = 0, \quad y(1) = 1, \quad 0 \le x \le 1.$

    $(b)\ y'' = xy' + (1 + x)y + 5, \quad y(1) = 1,\ y(2) = 3, \quad 1 \le x \le 2.$

26. Use the linear finite difference method to approximate the solutions of following boundary-value problems using $h = 0.1$:

    $(a)\ y'' = -3y' + 2y + 2x + 3, \quad y(0) = 2, \quad y(1) = 1, \quad 0 \le x \le 1.$

    $(b)\ y'' = 3xy' - 9y + 3x^2 + 2, \quad y(1) = 1,\ y(2) = -2, \quad 1 \le x \le 2.$

Chapter 7

Eigenvalues and Eigenvectors

7.1 Introduction

In this chapter we describe numerical methods for solving eigenvalue problems, which arise in many branches of science and engineering and seem to be a very fundamental part of the structure of the universe. Eigenvalue problems are important in a less direct manner in numerical applications. For example, to discover the condition factor in the solution of a set of linear algebraic equations involves finding the ratio of the largest to the smallest eigenvalue values of the underlying matrix. Also, the eigenvalue problem is involved when establishing the stiffness of ordinary differential equation problems. In solving eigenvalue problems, mainly we are concerned with the task of finding the values of parameter λ and vector $\mathbf{x}$, which satisfy a set of equations of the form

$$A\mathbf{x} = \lambda\mathbf{x}. \tag{7.1}$$

Linear equation (7.1) represents the eigenvalue problem where A is an $n \times n$ coefficient matrix, also called the system matrix, $\mathbf{x}$ is an unknown column vector, and λ is an unknown scalar. If the set of equations has a zero on the right-hand side, then a very important special case arises. For such case, one solution of (7.1) for a real square matrix A is the trivial solution $\mathbf{x} = \mathbf{0}$. However, there is a set of values for the parameter λ for which non-trivial solutions for the vector $\mathbf{x}$ exist. These non-trivial solutions are called eigenvectors, and characteristic vectors or latent vectors of a matrix A and the corresponding values of the parameter λ are called eigenvalues, characteristic values or latent roots of A. The set of all eigenvalues of A is called the *spectrum* of A. Eigenvalues may be real or complex, distinct or multiple. From (7.1), we deduce

$$A\mathbf{x} = \lambda\mathbf{I}\mathbf{x},$$

which gives

$$(A - \lambda\mathbf{I})\mathbf{x} = \mathbf{0}, \tag{7.2}$$

where $\mathbf{I}$ is an $n \times n$ identity matrix. Matrix $(A - \lambda\mathbf{I})$ appears as follows

$$\begin{pmatrix} (a_{11}-\lambda) & a_{12} & \cdots & a_{1n} \\ a_{21} & (a_{22}-\lambda) & \cdots & a_{2n} \\ \vdots & \vdots & \ddots & \vdots \\ a_{n1} & a_{n2} & \cdots & (a_{nn}-\lambda) \end{pmatrix}$$

and the result of the multiplication of (7.2) is a set of homogeneous equations of the form

$$\begin{array}{ccccccccc} (a_{11}-\lambda)x_1 & + & a_{12}x_2 & + & \cdots & + & a_{1n}x_n & = & 0 \\ a_{21}x_1 & + & (a_{22}-\lambda)x_2 & + & \cdots & + & a_{2n}x_n & = & 0 \\ \vdots & & \vdots & & \ddots & & \vdots & & \vdots \\ a_{n1}x_1 & + & a_{n2}x_2 & + & \cdots & + & (a_{nn}-\lambda)x_n & = & 0. \end{array} \tag{7.3}$$

Then using Cramer's rule, we see that the determinant of the denominator, namely, the determinant of matrix of system (7.3) must vanish if there is to be a non-trivial solution; that is, a solution other than $\mathbf{x} = \mathbf{0}$. Geometrically, $A\mathbf{x} = \lambda\mathbf{x}$ says, that under transformation by A, eigenvectors experience only changes in magnitude or sign; the orientation of $A\mathbf{x}$ in $\mathbb{R}^n$ is the same as that of $\mathbf{x}$. The eigenvalue λ is simply the amount of "stretch" or "shrink" to which the eigenvector $\mathbf{x}$ is subjected when transformed by A (see Figure 7.1).

For most matrices, there is no doubt that the eigenvalue problem is computationally more difficult than the linear system $A\mathbf{x} = \mathbf{b}$. With a linear system a finite number of elimination steps produce the exact answer in a finite time. In the case of eigenvalue, no such steps and no such formula can exist. The

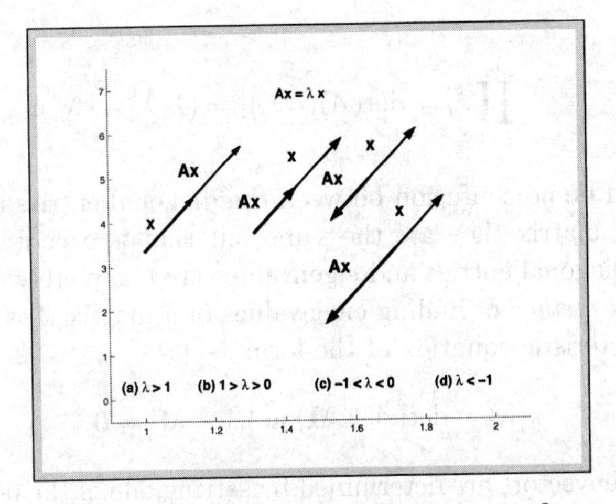

Figure 7.1: Shows the situation in $\mathbb{R}^2$.

characteristic polynomial of 5×5 matrix is a quintic and it is proved there can be no algebraic form for the roots of a fifth degree polynomial. Although, there are a few simple checks on the eigenvalues, after they have been computed, and we mention two of them here:

1. The *sum* of n eigenvalues of matrix A equals the sum of n diagonal entries; that is,

$$\sum_{i=1}^{n} \lambda_i = \sum_{i=1}^{n} a_{ii} = (a_{11} + a_{22} + \cdots + a_{nn}).$$

This sum is known as the *trace* of A.

Program 7.1
MATLAB m-file for finding trace of the matrix

```
function [trc]=trac(A)
n=max(size(A)); trc=0;
for i=1:n; for k=1:n
if i==k tracc=A(i,k); trc=trc+tracc; else trc=trc;
end; end; end
```

2. The *product* of the n eigenvalues of a matrix A equals the determinant of

A; that is

$$\prod_{i=1}^{n} \lambda_i = \det(A) = |A| = (\lambda_1 \lambda_2 \cdots \lambda_n).$$

There should be no confusion between the diagonal entries and eigenvalues. For a triangular matrix they are the same but that is exceptional. Normally, the pivots and diagonal entries and eigenvalues are completely different.

The *classical method* of finding eigenvalues of a matrix A is to estimate the roots of a characteristic equation of the form

$$p(\lambda) = \det(A - \lambda \mathbf{I}) = |A - \lambda \mathbf{I}| = \mathbf{0}. \tag{7.4}$$

Then the eigenvectors are determined by setting one of the nonzero elements of $\mathbf{x}$ to unity and calculating the remaining elements by equating coefficients in relation (7.2).

It should be noted that the system matrix A of (7.1) may be real and symmetric, or real and nonsymmetric, or complex with symmetric real and skew symmetric imaginary parts. These different types of a matrix A are explained as follows:

1. If the given matrix A is a *real symmetric matrix*, then the eigenvalues of A are real but not necessarily positive, and the corresponding eigenvectors are also real. Also, if λ_i, $\mathbf{x}_i$ and λ_j, $\mathbf{x}_j$ satisfy eigenvalue problem (7.1) and λ_i and λ_j are distinct then

$$\mathbf{x}_i^T \mathbf{x}_j = 0, \qquad i \neq j \tag{7.5}$$

and

$$\mathbf{x}_i^T A \mathbf{x}_j = 0, \qquad i \neq j. \tag{7.6}$$

Equations (7.5) and (7.6) represent the *orthogonality* relationships. Note that if $i = j$, then in general $\mathbf{x}_i^T \mathbf{x}_i$ and $\mathbf{x}_i^T A \mathbf{x}_i$ are not zero. Recalling that $\mathbf{x}_i$ includes an arbitrary scaling factor, then the product $\mathbf{x}_i^T \mathbf{x}_i$ must also be arbitrary. However, if the arbitrary scaling factor is adjusted so that

$$\mathbf{x}_i^T \mathbf{x}_j = 1 \tag{7.7}$$

then

$$\mathbf{x}_i^T A \mathbf{x}_j = \lambda_i \tag{7.8}$$

and the eigenvectors are known to be *normalized*.

Sometimes the eigenvalues are not distinct and the eigenvectors associated with these equal or repeated eigenvalues are not, of necessity, orthogonal. If $\lambda_i = \lambda_j$ and the other eigenvalues, λ_k, are distinct then

$$\mathbf{x}_i^T \mathbf{x}_k = 0, \quad k = 1, 2, \cdots, n \quad k \neq i, \quad k \neq j \tag{7.9}$$

and

$$x_j^T x_k = 0, \quad k = 1, 2, \cdots, n \quad k \neq i, \quad k \neq j. \tag{7.10}$$

When $\lambda_i = \lambda_j$, the eigenvectors $\mathbf{x}_i$ and $\mathbf{x}_j$ are not unique and a linear combination of them, that is, $a\mathbf{x}_i + b\mathbf{x}_j$, where a and b are arbitrary constants, also satisfy the eigenvalue problems. One important result is that *a symmetric matrix of order n has always n distinct eigenvectors even if some of the eigenvalues are repeated.*

2. If a given A is *real nonsymmetric matrix,* then a pair of related eigenvalue problems can arise as follows:

$$A\mathbf{x} = \lambda \mathbf{x} \tag{7.11}$$

and

$$A^T \mathbf{y} = \beta \mathbf{y}. \tag{7.12}$$

By taking the transpose of (7.12), we have

$$\mathbf{y}^T A = \beta \mathbf{y}^T. \tag{7.13}$$

Vectors $\mathbf{x}$ and $\mathbf{y}$ are called the right-hand and left-hand vectors of A, respectively. The eigenvalues of A and A^T are identical, that is, $\lambda_i = \beta_i$, but the eigenvectors $\mathbf{x}$ and $\mathbf{y}$ will, in general, differ from each other. The eigenvalues and eigenvectors of a nonsymmetric real matrix are either real or pairs of complex conjugates. If $\lambda_i, \mathbf{x}_i, \mathbf{y}_i$, and $\lambda_j, \mathbf{x}_j, \mathbf{y}_j$ are solutions that satisfy the eigenvalue problems of (7.11) and (7.12) and λ_i and λ_j are distinct then

$$\mathbf{y}_j^T \mathbf{x}_i = 0, \quad i \neq j \tag{7.14}$$

and

$$\mathbf{y}_j^T A\mathbf{x}_i = 0, \quad i \neq j. \tag{7.15}$$

Equations (7.14) and (7.15) are called *bi-orthogonal* relationships. Note that if, in these equations, $i = j$, then in general $\mathbf{y}_i^T \mathbf{x}_i$ and $\mathbf{y}_i^T A\mathbf{x}_i$ are not zero. Eigenvectors $\mathbf{x}_i$ and $\mathbf{y}_i$ include arbitrary scaling factors and so the product of these vectors will also be arbitrary. However, if the vectors are adjusted so that

$$\mathbf{y}_i^T \mathbf{x}_i = 1 \tag{7.16}$$

then

$$\mathbf{y}_i^T A \mathbf{x}_i = \lambda_i. \tag{7.17}$$

We cannot, in these circumstances, describe neither $\mathbf{x}_i$ nor $\mathbf{y}_i$ as normalized; the vectors still include arbitrary scaling factors, only their product is uniquely chosen. If for a nonsymmetric matrix $\lambda_i = \lambda_j$ and the remaining eigenvalues λ_k, are distinct then

$$\mathbf{y}_j^T \mathbf{x}_k = 0, \quad \mathbf{y}_i^T \mathbf{x}_k = 0, \qquad k = 1, 2, \cdots, n, \quad k \neq i, \quad k \neq j \tag{7.18}$$

and

$$\mathbf{x}_j^T \mathbf{x}_k = 0, \quad \mathbf{y}_k^T \mathbf{x}_j = 0, \qquad k = 1, 2, \cdots, n, \quad k \neq i, \quad k \neq j. \tag{7.19}$$

For certain matrices with repeated eigenvalues, the eigenvectors may also be repeated. Consequently, for an *nth*-order matrix of this type we may have less than n distinct eigenvectors. This type of matrix is called *deficient*.

3. Let us consider the case when given A is a *complex matrix*. The properties of one particular complex matrix is an *Hermitian matrix*, which is defined as

$$H = A + \mathbf{i}B, \tag{7.20}$$

where A and B are real matrices such that $A = A^T$ and $B = -B^T$. Hence, A is symmetric and B is skew symmetric with zero terms on the leading diagonal. Thus, by definition of an Hermitian matrix, H, has a symmetric real part and a skew symmetric imaginary part making H equal to the transpose of its complex conjugate, denoted by H^*. Consider now the eigenvalue problem

$$H\mathbf{x} = \lambda \mathbf{x}. \tag{7.21}$$

If $\lambda_i, \mathbf{x}_i$ are solutions of (7.21), then $\mathbf{x}_i$ is complex but λ_i is real. Also, if $\lambda_i, \mathbf{x}_i$ and $\lambda_j, \mathbf{x}_j$ satisfy the eigenvalue problem (7.21) and λ_i and λ_j are distinct then

$$\mathbf{x}_i^* \mathbf{x}_j = 0, \qquad i \neq j, \tag{7.22}$$

and

$$\mathbf{x}_i^* H \mathbf{x}_j = 0, \qquad i \neq j \tag{7.23}$$

where $\mathbf{x}_i^*$ is the transpose of the complex conjugate of $\mathbf{x}_i$. As before, $\mathbf{x}_i$ includes an arbitrary scaling factor and the product $\mathbf{x}_i^* \mathbf{x}_i$ must also be arbitrary. However, if the arbitrary scaling factor is adjusted so that

$$\mathbf{x}_i^* \mathbf{x}_i = 1 \tag{7.24}$$

then

$$\mathbf{x}_i^* H \mathbf{x}_i = \lambda_i \tag{7.25}$$

and the eigenvectors are then said to be normalized.

A large number of numerical techniques have been developed to solve eigenvalue problems. Before discussing all these numerical techniques, we shall start with a hand calculation, mainly to reinforce the definition and solve the following examples.

Example 7.1 *Find the eigenvalues and eigenvectors of the following matrix:*

$$A = \begin{pmatrix} 3 & 0 & 1 \\ 0 & -3 & 0 \\ 1 & 0 & 3 \end{pmatrix}.$$

Solution. *Firstly, we shall find the eigenvalues of the given matrix A. From (7.2), we have*

$$\begin{pmatrix} 3 & 0 & 1 \\ 0 & -3 & 0 \\ 1 & 0 & 3 \end{pmatrix} \begin{pmatrix} x_1 \\ x_2 \\ x_3 \end{pmatrix} = \begin{pmatrix} 0 \\ 0 \\ 0 \end{pmatrix}.$$

For non-trivial solutions, using (7.4), we get

$$\begin{vmatrix} 3-\lambda & 0 & 1 \\ 0 & -3-\lambda & 0 \\ 1 & 0 & 3-\lambda \end{vmatrix} = 0,$$

which gives a characteristic equation of the form

$$\lambda^3 - 3\lambda^2 - 10\lambda + 24 = 0,$$

which factorizes to

$$(\lambda + 3)(\lambda - 2)(\lambda - 4) = 0$$

and gives the eigenvalues 4, -3, and 2 of the given matrix A. One can note that the sum of these eigenvalues is 3, and this agrees with the trace of A.

The elements of vector P are arranged in decreasing power of λ.

After finding the eigenvalues of matrix A, we turn to the problem of finding the corresponding eigenvectors. The eigenvectors of A corresponding to the eigenvalues λ are the nonzero vectors $\mathbf{x}$ that satisfy (7.2). Equivalently, the eigenvectors corresponding to λ are the nonzero vectors in the solution space of (7.2). We call this solution space the eigenspace of A corresponding to λ.

Now to find the eigenvectors of the given matrix A corresponding to each of these eigenvalues, we substitute each of these three eigenvalues in (7.2). When $\lambda_1 = 4$, we have

$$\begin{pmatrix} -1 & 0 & 1 \\ 0 & -7 & 0 \\ 1 & 0 & -1 \end{pmatrix} \begin{pmatrix} x_1 \\ x_2 \\ x_3 \end{pmatrix} = \begin{pmatrix} 0 \\ 0 \\ 0 \end{pmatrix},$$

which implies that

$$\begin{array}{rcrcrcl} -x_1 & + & 0x_2 & + & x_3 & = 0 & \Rightarrow \quad x_1 = x_3 \\ 0x_1 & - & 7x_2 & + & 0x_3 & = 0 & \Rightarrow \quad x_2 = 0 \\ x_1 & + & 0x_2 & - & x_3 & = 0 & \Rightarrow \quad x_1 = x_3. \end{array}$$

Solving this system, we got $x_1 = x_3 = \infty$ and $x_2 = 0$. Hence, eigenvector $\mathbf{x}_1$ corresponds to the first eigenvalue 4 by choosing $x_3 = 1$, is

$$\mathbf{x}_1 = \alpha[1, 0, 1]^T, \quad where \quad \alpha \in \mathbb{R}, \quad \alpha \neq 0.$$

When $\lambda_2 = -3$, we have

$$\begin{pmatrix} 6 & 0 & 1 \\ 0 & 0 & 0 \\ 1 & 0 & 6 \end{pmatrix} \begin{pmatrix} x_1 \\ x_2 \\ x_3 \end{pmatrix} = \begin{pmatrix} 0 \\ 0 \\ 0 \end{pmatrix},$$

which implies that

$$\begin{array}{rcrcrcl} 6x_1 & + & 0x_2 & + & x_3 & = 0 & \Rightarrow \quad x_1 = -\frac{1}{6}x_3 \\ 0x_1 & + & 0x_2 & + & 0x_3 & = 0 & \Rightarrow \quad x_2 = \infty \\ x_1 & + & 0x_2 & + & 6x_3 & = 0 & \Rightarrow \quad x_1 = -6x_3, \end{array}$$

which gives the solution $x_1 = -6x_3 = 0$ and $x_2 = \infty$. Hence, eigenvector $\mathbf{x}_2$ corresponding to the second eigenvalue, -3, by choosing $x_2 = 1$, is

$$\mathbf{x}_2 = \alpha[0, 1, 0]^T, \quad where \quad \alpha \in \mathbb{R}, \quad \alpha \neq 0.$$

Finally, when $\lambda_3 = 2$, *we have*

$$\begin{pmatrix} 1 & 0 & 1 \\ 0 & -5 & 0 \\ 1 & 0 & 1 \end{pmatrix} \begin{pmatrix} x_1 \\ x_2 \\ x_3 \end{pmatrix} = \begin{pmatrix} 0 \\ 0 \\ 0 \end{pmatrix},$$

which implies that

$$\begin{array}{rcrcrl} x_1 & + & 0x_2 & + & x_3 & = 0 \\ 0x_1 & - & 5x_2 & + & 0x_3 & = 0 \\ x_1 & + & 0x_2 & + & x_3 & = 0 \end{array} \qquad \begin{array}{ccl} \Rightarrow & x_1 = & -x_3 \\ \Rightarrow & x_2 = & 0 \\ \Rightarrow & x_1 = & -x_3 \end{array}$$

and gives, $x_1 = -x_3 = \infty$ *and* $x_2 = 0$. *Hence, eigenvector* $\mathbf{x}_3$ *corresponding to the third eigenvalue, 2, by choosing* $x_1 = 1$, *is*

$$\mathbf{x}_3 = \alpha[1, 0, -1]^T, \quad \textit{where} \quad \alpha \in \mathbb{R}, \quad \alpha \neq 0.$$

MATLAB can handle eigenvalues, eigenvectors, and the characteristic polynomial. The built-*poly* function in MATLAB computes the characteristic polynomial of a matrix:

```
>> A = [3 0 1;0 -3 0;1 0 3];
>> P = poly(A);
```

The elements of vector P are arranged in decreasing power of x. To solve the characteristic equation (in order to obtain the eigenvalues of A), ask for the roots of P:

```
>> roots(P);
```

If all we require are the eigenvalues of A, we can use the MATLAB command *eig*, which is the basic eigenvalue and eigenvector routine. The command

```
>> d = eig(A);
```

returns a vector containing all the eigenvalues of a matrix A. If the eigenvectors are also wanted, the syntax

```
>> [X, D] = eig(A);
```

will return a matrix X whose columns are eigenvectors of A corresponding to the eigenvalues in the diagonal matrix D.

To get the results of Example 7.1, we use the MATLAB command window as follows:

```
>> A = [3 0 1;0 -3 0;1 0 3];
>> P = poly(A);
>> [X, D] = eig(A);
>> λ = diag(D);
>> x1 = X(:,1); x2 = X(:,2); x3 = X(:,3);
```

.

Example 7.2 *Find the eigenvalues and eigenvectors of the following matrix*

$$A = \begin{pmatrix} 1 & 2 & 2 \\ 0 & 3 & 3 \\ -1 & 1 & 1 \end{pmatrix}.$$

Solution. *From (7.2), we have*

$$\begin{pmatrix} 1 & 2 & 2 \\ 0 & 3 & 3 \\ -1 & 1 & 1 \end{pmatrix} \begin{pmatrix} x_1 \\ x_2 \\ x_3 \end{pmatrix} = \begin{pmatrix} 0 \\ 0 \\ 0 \end{pmatrix}.$$

For non-trivial solutions, using (7.4), we get

$$\begin{vmatrix} 1-\lambda & 2 & 2 \\ 0 & 3-\lambda & 3 \\ -1 & 1 & 1-\lambda \end{vmatrix} = 0,$$

which gives a characteristic equation of the form

$$-\lambda^3 + 5\lambda^2 - 6\lambda = 0,$$

which factorizes to

$$\lambda(\lambda - 2)(\lambda - 3) = 0$$

and gives the eigenvalues 0, 2, and 3 of the given matrix A. One can note that the sum of these three eigenvalues is 5, and this agrees with the trace of A.

To find the eigenvectors corresponding to each of these eigenvalues, we substitute each of the three eigenvalues of A in (7.2). When $\lambda = 0$, we have

$$\begin{pmatrix} 1 & 2 & 2 \\ 0 & 3 & 3 \\ -1 & 1 & 1 \end{pmatrix} \begin{pmatrix} x_1 \\ x_2 \\ x_3 \end{pmatrix} = \begin{pmatrix} 0 \\ 0 \\ 0 \end{pmatrix}.$$

The augmented matrix form of the system is

$$\begin{pmatrix} 1 & 2 & 2 & 0 \\ 0 & 3 & 3 & 0 \\ -1 & 1 & 1 & 0 \end{pmatrix},$$

which can be reduced to

$$\begin{pmatrix} 1 & 2 & 2 & 0 \\ 0 & 1 & 1 & 0 \\ 0 & 0 & 0 & 0 \end{pmatrix}.$$

Thus, the components of an eigenvector must satisfy the relation

$$\begin{array}{rcrcrcl} x_1 & + & 2x_2 & + & 2x_3 & = & 0 \\ 0x_1 & + & x_2 & + & x_3 & = & 0 \\ 0x_1 & + & 0x_2 & + & 0x_3 & = & 0. \end{array}$$

This system has an infinite set of solutions. Arbitrarily, we choose $x_2 = 1$, then x_3 can be equal to -1, whence $x_1 = 0$. This gives solutions of the first eigenvector of the form $\mathbf{x}_1 = \alpha[0, 1, -1]^T$, with $\alpha \in \mathbb{R}$ and $\alpha \neq 0$. Thus, $\mathbf{x}_1 = \alpha[0, 1, -1]^T$ as the most general eigenvector corresponding to eigenvalue 0.
A similar procedure can be applied to the other two eigenvalues. The result is that we have two other eigenvectors $\mathbf{x}_2 = \alpha[4, 3, -1]^T$ and $\mathbf{x}_3 = \alpha[1, 1, 0]^T$ corresponding to the eigenvalues 2 and 3, respectively.

Example 7.3 *Find the eigenvalues and eigenvectors of the following matrix:*

$$A = \begin{pmatrix} 3 & 2 & -1 \\ 2 & 6 & -2 \\ -1 & -2 & 3 \end{pmatrix}.$$

Solution. *From (7.2), we have*

$$\begin{pmatrix} 3 & 2 & -1 \\ 2 & 6 & -2 \\ -1 & -2 & 3 \end{pmatrix} \begin{pmatrix} x_1 \\ x_2 \\ x_3 \end{pmatrix} = \begin{pmatrix} 0 \\ 0 \\ 0 \end{pmatrix}.$$

For non-trivial solutions, using (7.4), we get

$$\begin{vmatrix} 3-\lambda & 2 & -1 \\ 2 & 6-\lambda & -2 \\ -1 & -2 & 3-\lambda \end{vmatrix} = 0,$$

which gives a characteristic equation

$$\lambda^3 - 12\lambda^2 + 36\lambda - 32 = 0,$$

which factorizes to

$$(\lambda - 2)^2(\lambda - 8) = 0$$

and gives the eigenvalue 2 of multiplicity 2 and eigenvalue 8 of multiplicity 1, and the sum of these three eigenvalues is 12, which agrees with the trace of A. When $\lambda = 2$, we have

$$(A - 2I) = \begin{pmatrix} 1 & 2 & -1 \\ 2 & 4 & -2 \\ -1 & -2 & 1 \end{pmatrix}$$

and so from (7.2), we have

$$\begin{array}{rcrcrcl} x_1 & + & 2x_2 & - & x_3 & = & 0 \\ 2x_1 & + & 4x_2 & - & 2x_3 & = & 0 \\ -x_1 & - & 2x_2 & + & x_3 & = & 0. \end{array}$$

Let $x_2 = s$ and $x_3 = t$, then the solution to this system is

$$\begin{pmatrix} x_1 \\ x_2 \\ x_3 \end{pmatrix} = \begin{pmatrix} -2s + t \\ s \\ t \end{pmatrix} = s \begin{pmatrix} -2 \\ 1 \\ 0 \end{pmatrix} + t \begin{pmatrix} 1 \\ 0 \\ 1 \end{pmatrix}.$$

So the two eigenvectors of the given matrix A are $\mathbf{x}_1 = \alpha[-2, 1, 0]^T$ and $\mathbf{x}_2 = \alpha[1, 0, 1]^T$, corresponding to eigenvalue 2, with $s, t \in \mathbb{R}$, and $s, t \neq 0$. Similarly, we can find the third eigenvector, $\mathbf{x}_3 = \alpha[0.5, 1, -0.5]^T$ of A corresponding to the other eigenvalue 8.

Note that in all three examples and any other example, there is always an infinite number of choices for each eigenvector. We arbitrarily choose a simple one by setting one or more of the elements x_is equal to a convenient number. Here we have set one of the element x_is equal to 1.

7.2 Linear Algebra and Eigenvalues Problems

The solution of many physical problems requires the calculation of eigenvalues and the corresponding eigenvectors of a matrix associated with a linear system of equations, since matrix A of order n has n not necessary distinct, eigenvalues, which are the roots of characteristic equation (7.4). Theoretically, the eigenvalues of A can be obtained by finding the n roots of a characteristic polynomial $p(\lambda)$ and then the associated linear system can be solved to determine the corresponding eigenvectors. For a large value of n, the polynomial $p(\lambda)$ is difficult to

obtain except for small values of n. So it is necessary to construct approximation techniques for finding the eigenvalues of A.
Before discussing such approximation techniques for finding eigenvalues and eigenvectors of a given matrix A, we need some definitions and results from linear algebra.

Definition 7.1 (Basis of Vector)

Let V be a vector space. A finite set $S = \{\mathbf{v}_1, \mathbf{v}_2, \ldots, \mathbf{v}_n\}$ of vectors in V is a basis for V if and only if any vector $\mathbf{v}$ in V can be written, in a unique way, as a linear combination of the vectors in S; that is, if and only if any vector $\mathbf{v}$ has the form

$$\mathbf{v} = k_1\mathbf{v}_1 + k_2\mathbf{v}_2 + \cdots + k_n\mathbf{v}_n \tag{7.26}$$

for one and only one set of real numbers $k_1, k_2, \ldots, k_n$.

Definition 7.2 (Linear Independent Vectors)

Vectors $\mathbf{v}_1, \mathbf{v}_2, \ldots, \mathbf{v}_n$ are said to be linear independent if whenever

$$k_1\mathbf{v}_1 + k_2\mathbf{v}_2 + \cdots + k_n\mathbf{v}_n = \mathbf{0} \tag{7.27}$$

then all of the coefficients $k_1, k_2, \ldots, k_n$ must equal to zero; that is,

$$k_1 = k_2 = \cdots = k_n = 0. \tag{7.28}$$

If vectors $\mathbf{v}_1, \mathbf{v}_2, \ldots, \mathbf{v}_n$ are not linear independent, then we say that they are linear dependent. In other words, vectors $\mathbf{v}_1, \mathbf{v}_2, \ldots, \mathbf{v}_n$ are linear dependent if and only if there exists numbers $k_1, k_2, \ldots, k_n$, not all zero, for which

$$k_1\mathbf{v}_1 + k_2\mathbf{v}_2 + \cdots + k_n\mathbf{v}_n = \mathbf{0}.$$

Sometimes we say that the set $\{\mathbf{v}_1, \mathbf{v}_2, \ldots, \mathbf{v}_n\}$ is linearly independent (or linearly dependent) instead of saying that vectors $\mathbf{v}_1, \mathbf{v}_2, \ldots, \mathbf{v}_n$ are linearly independent (or linearly dependent).

Example 7.4 *Let us consider vectors $\mathbf{v}_1 = (1, 2)$ and $\mathbf{v}_2 = (-1, 1)$ in $\mathbb{R}^2$. To show that the vectors are linearly independent, we write*

$$\begin{aligned} k_1v_1 + k_2v_2 &= \mathbf{0} \\ k_1(1,2) + k_2(-1,1) &= (0,0) \\ (k_1 - k_2, 2k_1 + k_2) &= (0,0) \end{aligned}$$

showing that

$$\begin{aligned} k_1 - k_2 &= 0 \\ 2k_1 + k_2 &= 0 \end{aligned}$$

the only solution to the above system is a trivial solution; that is, $k_1 = k_2 = 0$. Thus, the vectors are linearly independent.

Example 7.5 *Consider vectors $\mathbf{v}_1 = (1, 2, 1)$, $\mathbf{v}_2 = (1, 3, -2)$, and $\mathbf{v}_3 = (0, 1, -3)$ in $\mathbb{R}^3$. If k_1, k_2, and k_3 are numbers with*

$$k_1\mathbf{v}_1 + k_2\mathbf{v}_2 + k_3\mathbf{v}_3 = \mathbf{0},$$

which is equivalent to

$$k_1(1, 2, 1) + k_2(1, 3, -2) + k_3(0, 1, -3) = (0, 0, 0),$$

thus we have system

$$\begin{array}{rcrcrcl} k_1 & + & k_2 & + & 0k_3 & = & 0 \\ 2k_1 & + & 3k_2 & + & k_3 & = & 0 \\ k_1 & - & 2k_2 & - & 3k_3 & = & 0. \end{array}$$

This system has infinitely many solutions, one of which is $k_1 = 1$, $k_2 = -1$, and $k_3 = 1$. So

$$\mathbf{v}_1 - \mathbf{v}_2 + \mathbf{v}_3 = \mathbf{0}.$$

Thus, vectors $\mathbf{v}_1, \mathbf{v}_2$, and $\mathbf{v}_3$ are linearly dependent.

Theorem 7.1 *If $\{\mathbf{v}_1, \mathbf{v}_2, \ldots, \mathbf{v}_n\}$ is a set of n linearly independent vectors in $\mathbb{R}^n$, then any vector $\mathbf{x} \in \mathbb{R}^n$ can be written uniquely as*

$$\mathbf{x} = k_1\mathbf{v}_1 + k_2\mathbf{v}_2 + \cdots + k_n\mathbf{v}_n \tag{7.29}$$

for some collection of constants $k_1, k_2, \ldots, k_n$.

Note that any collection of n linearly independent vectors in $\mathbb{R}^n$ is a basis for $\mathbb{R}^n$.

Theorem 7.2 *If A is an $n \times n$ matrix and $\lambda_1, \ldots, \lambda_n$ are distinct eigenvalues of A with associated eigenvectors $\mathbf{v}_1, \ldots, \mathbf{v}_n$, then the set $\{\mathbf{v}_1, \ldots, \mathbf{v}_n\}$ is linearly independent.*

Definition 7.3 (Orthogonal Vectors)

A set of vectors $\{\mathbf{v}_1, \mathbf{v}_2, \ldots, \mathbf{v}_n\}$ *is called orthogonal if*

$$\mathbf{v}_i^T \mathbf{v}_j = 0, \qquad \textit{for all} \qquad i \neq j. \tag{7.30}$$

If, in addition,

$$\mathbf{v}_i^T \mathbf{v}_i = 1, \qquad \textit{for all} \quad i = 1, 2, \ldots, n \tag{7.31}$$

then the set is called orthonormal.

Theorem 7.3 *An orthogonal set of vectors that does not contain zero vectors is linearly independent.*

The proof of this theorem is beyond the scope of this text and will be omitted. However, the result is extremely important and can be easily understood and used. We illustrate this result by considering the following matrix:

$$A = \begin{pmatrix} 6 & -2 & 2 \\ -2 & 5 & 0 \\ 2 & 0 & 7 \end{pmatrix},$$

which has the eigenvalues 3, 6, and 9. The corresponding eigenvectors of A are $[2, 2, -1]^T$, $[-1, 2, 2]^T$, and $[2, -1, 2]^T$ and they form an orthogonal set. To show that the vectors are linearly independent, we write

$$k_1\mathbf{v}_1 + k_2\mathbf{v}_2 + k_3\mathbf{v}_3 = \mathbf{0}$$

then the equation

$$k_1(2, 2, -1) + k_2(-1, 2, 2) + k_3(2, -1, 2) = (0, 0, 0),$$

which leads to the homogeneous system of three equations in three unknown, k_1, k_2, and k_3:

$$\begin{array}{rcrcrcl} 2k_1 & - & k_2 & + & 2k_3 & = & 0 \\ 2k_1 & + & 2k_2 & - & k_3 & = & 0 \\ -k_1 & + & 2k_2 & + & 2k_3 & = & 0. \end{array}$$

Thus, the vectors will be linearly independent if and only if above system has a trivial solution. By writing the above system as an augmented matrix form and then row-reducing, it gives:

$$\left(\begin{array}{rrr|r} 2 & -1 & 2 & 0 \\ 2 & 2 & -1 & 0 \\ -1 & 2 & 2 & 0 \end{array}\right) \longrightarrow \left(\begin{array}{rrr|r} 1 & 0 & 0 & 0 \\ 0 & 1 & 0 & 0 \\ 0 & 0 & 1 & 0 \end{array}\right),$$

which gives, $k_1 = 0$, $k_2 = 0$, and $k_3 = 0$. Hence, the given vectors are linearly independent.

7.3 Diagonalization of Matrices

Of special importance for the study of eigenvalues are diagonal matrices. These will be denoted by

$$D = \begin{pmatrix} \lambda_1 & 0 & 0 & \cdots & 0 \\ 0 & \lambda_2 & 0 & \cdots & 0 \\ \vdots & \vdots & \vdots & \ddots & \vdots \\ 0 & 0 & 0 & \cdots & \lambda_n \end{pmatrix}$$

and are called *spectral matrices*; that is, all the diagonal elements of D are the eigenvalues of A. This simple but useful result makes it desirable to find ways to transform a general $n \times n$ matrix A into a diagonal matrix having the same eigenvalues. Unfortunately, the elementary operations that can be used to reduce $A \rightarrow D$ are not suitable because the scale and subtract operations alter eigenvalues. Here, what we needed are *similarity transformations*. Similarity transformations occur frequently in the context of relating coordinate systems.

Definition 7.4 (Similar Matrix)

Let A and B be square matrices of the same size. A matrix B is said to be similar to A if there exists a nonsingular matrix Q such that $B = Q^{-1}AQ$. The transformation of a matrix A into matrix B in this manner is called a **similarity transformation**.

Example 7.6 *Consider the following matrices A and Q, Q is nonsingular. Use the similarity transformation $Q^{-1}AQ$ to transform A into a matrix B.*

$$A = \begin{pmatrix} 0 & 0 & -2 \\ 1 & 2 & 1 \\ 1 & 0 & 3 \end{pmatrix}, \qquad Q = \begin{pmatrix} -1 & 0 & -2 \\ 0 & 1 & 1 \\ 1 & 0 & 1 \end{pmatrix}.$$

Solution. *Let*

$$\begin{aligned} B = Q^{-1}AQ &= \begin{pmatrix} -1 & 0 & -2 \\ 0 & 1 & 1 \\ 1 & 0 & 1 \end{pmatrix}^{-1} \begin{pmatrix} 0 & 0 & -2 \\ 1 & 2 & 1 \\ 1 & 0 & 3 \end{pmatrix} \begin{pmatrix} -1 & 0 & -2 \\ 0 & 1 & 1 \\ 1 & 0 & 1 \end{pmatrix} \\ &= \begin{pmatrix} 1 & 0 & 2 \\ 1 & 1 & 1 \\ -1 & 0 & -1 \end{pmatrix} \begin{pmatrix} 0 & 0 & -2 \\ 1 & 2 & 1 \\ 1 & 0 & 3 \end{pmatrix} \begin{pmatrix} -1 & 0 & -2 \\ 0 & 1 & 1 \\ 1 & 0 & 1 \end{pmatrix} \\ &= \begin{pmatrix} 2 & 0 & 0 \\ 0 & 2 & 0 \\ 0 & 0 & 1 \end{pmatrix}. \end{aligned}$$

In Example 7.6, matrix A is transformed into diagonal matrix B. Not every square matrix can be "diagonalized" in this manner. Here, we will discuss conditions under which a matrix can be diagonalized and when it can, ways of constructing an approximate transforming matrix Q. We will find that eigenvalues and eigenvectors play a key role in this discussion.

Theorem 7.4 *Similar matrices have the same eigenvalues.*

Proof. *Let A and B be similar matrices. Hence, there exists a matrix Q such that $B = Q^{-1}AQ$. The characteristic polynomial of B is $|B - \lambda\mathbf{I}|$. Substituting for B and using the multiplicative properties of determinants, we get*

$$\begin{aligned} |B - \lambda\mathbf{I}| &= |Q^{-1}AQ - \lambda\mathbf{I}| = |Q^{-1}(A - \lambda\mathbf{I})Q| \\ &= |Q^{-1}||A - \lambda\mathbf{I}||Q| = |(A - \lambda\mathbf{I})|Q^{-1}||Q| \\ &= |A - \lambda\mathbf{I}||Q^{-1}Q| = |A - \lambda\mathbf{I}||\mathbf{I}| \\ &= |A - \lambda\mathbf{I}|. \end{aligned}$$

The characteristic polynomials of A and B are identical. This means that their eigenvalues are the same.

Definition 7.5 (Diagonalizable Matrix)

A square matrix A is called diagonalizable if there exists an invertible matrix Q such that

$$D = Q^{-1}AQ \tag{7.32}$$

is a diagonal matrix. Note that all the diagonal elements of it are eigenvalues of A, and an invertible matrix Q can be written as

$$Q = \Big(\mathbf{x}_1|\mathbf{x}_2|\cdots|\mathbf{x}_n\Big)$$

and is called a model matrix because its columns contain, $\mathbf{x}_1, \mathbf{x}_2, \ldots, \mathbf{x}_n$, which are the eigenvectors of A corresponding to eigenvalues $\lambda_1, \ldots, \lambda_n$.

Theorem 7.5 *Any matrix having linearly independent eigenvectors corresponding to distinct and real eigenvalues is diagonalizable; that is*

$$Q^{-1}AQ = D,$$

where D is a diagonal matrix and Q is an invertible matrix.

Proof. *Let $\lambda_1, \ldots, \lambda_n$ be eigenvalues of a matrix A, with corresponding linearly independent eigenvectors $\mathbf{x}_1, \ldots, \mathbf{x}_n$. Let Q be the matrix having $\mathbf{x}_1, \ldots, \mathbf{x}_n$ as column vectors; that is,*

$$Q = (\mathbf{x}_1 \cdots \mathbf{x}_n).$$

Since $A\mathbf{x}_1 = \lambda\mathbf{x}_1, \ldots, A\mathbf{x}_n = \lambda_n\mathbf{x}_n$, matrix multiplication in terms of columns gives

$$\begin{aligned} AQ &= (A\mathbf{x}_1 \cdots A\mathbf{x}_n) \\ &= (\lambda_1\mathbf{x}_1 \cdots \lambda_n\mathbf{x}_n) \\ &= (\mathbf{x}_1 \cdots \mathbf{x}_n)\begin{pmatrix} \lambda_1 & & 0 \\ & \ddots & \\ 0 & & \lambda_n \end{pmatrix} \\ &= Q\begin{pmatrix} \lambda_1 & & 0 \\ & \ddots & \\ 0 & & \lambda_n \end{pmatrix}. \end{aligned}$$

Since the columns of Q are linearly independent, Q is invertible. Thus,

$$Q^{-1}AQ = \begin{pmatrix} \lambda_1 & & 0 \\ & \ddots & \\ 0 & & \lambda_n \end{pmatrix} = D.$$

Therefore, if a square matrix A has n linearly independent eigenvectors, these eigenvectors can be used as the columns of a matrix Q that diagonalizes A. The diagonal matrix has the eigenvalues of A as diagonal elements.
Note that the converse of the above theorem also exists; that is, if A is diagonalizable, then it has n linearly independent eigenvectors.

Example 7.7 *Consider the following matrix*

$$A = \begin{pmatrix} 0 & 0 & 1 \\ 3 & 7 & -9 \\ 0 & 2 & -1 \end{pmatrix},$$

which has a characteristic equation

$$\lambda^3 - 6\lambda^2 + 11\lambda - 6 = 0$$

that cubic factorizes to give

$$(\lambda - 1)(\lambda - 2)(\lambda - 3) = 0.$$

The eigenvalues of A, therefore, are 1, 2, and 3, with the sum of 6, which agrees with the trace of A. Corresponding to these eigenvalues, the eigenvectors of A are $\mathbf{x}_1 = [1, 1, 1]^T$, $\mathbf{x}_2 = [1, 3, 2]^T$, and $\mathbf{x}_3 = [1, 6, 3]^T$. Thus, the nonsingular matrix Q is given by

$$Q = \begin{pmatrix} 1 & 1 & 1 \\ 1 & 3 & 6 \\ 1 & 2 & 3 \end{pmatrix}$$

and the inverse of this matrix is given by

$$Q^{-1} = \begin{pmatrix} 3 & 1 & -3 \\ -3 & -2 & 5 \\ 1 & 1 & -2 \end{pmatrix}.$$

Thus,

$$Q^{-1}AQ = \begin{pmatrix} 3 & 1 & -3 \\ -3 & -2 & 5 \\ 1 & 1 & -2 \end{pmatrix} \begin{pmatrix} 0 & 0 & 1 \\ 3 & 7 & -9 \\ 0 & 2 & -1 \end{pmatrix} \begin{pmatrix} 1 & 1 & 1 \\ 1 & 3 & 6 \\ 1 & 2 & 3 \end{pmatrix},$$

which implies that

$$Q^{-1}AQ = \begin{pmatrix} 1 & 0 & 0 \\ 0 & 2 & 0 \\ 0 & 0 & 3 \end{pmatrix} = D.$$

The above results can be obtained using the MATLAB command window as follows:

```
>> A = [0 0 1; 3 7 − 9; 0 2 − 1];
>> P = poly(A);
>> [X, D] = eig(A);
>> eigenvalues = diag(D);
>> x1 = X(:, 1); x2 = X(:, 2); x3 = X(:, 3);
>> Q = [x1 x2 x3];
>> D = inv(Q) ∗ A ∗ Q;
```

It is possible that independent eigenvectors may exist even though the eigenvalues are not distinct though no theorem exists to show under what conditions they do so. The following example shows the situation that can arise.

Example 7.8 *Consider the following matrix:*

$$A = \begin{pmatrix} 2 & 1 & 1 \\ 2 & 3 & 2 \\ 3 & 3 & 4 \end{pmatrix},$$

which has a characteristic equation

$$\lambda^3 - 9\lambda^2 + 15\lambda - 7 = 0$$

and it can be easily factorized to give

$$(\lambda - 7)(\lambda - 1)^2 = 0.$$

The eigenvalues of A are 7 of multiplicity one, and 1 of multiplicity two. The eigenvectors corresponding to these eigenvalues are $\mathbf{x}_1 = [1, 2, 3]^T$, $\mathbf{x}_2 = [1, 0, -1]^T$, and $\mathbf{x}_3 = [0, 1, -1]^T$. Thus, nonsingular matrix Q is given by

$$Q = \begin{pmatrix} 1 & 1 & 0 \\ 2 & 0 & 1 \\ 3 & -1 & -1 \end{pmatrix}$$

and the inverse of this matrix is

$$Q^{-1} = \frac{1}{6}\begin{pmatrix} 1 & -1 & -1 \\ -5 & 1 & 1 \\ 2 & -4 & 2 \end{pmatrix}.$$

Thus,

$$Q^{-1}AQ = \begin{pmatrix} 7 & 0 & 0 \\ 0 & 1 & 0 \\ 0 & 0 & 1 \end{pmatrix} = D.$$

Computing Powers of a Matrix

There are numerous problems in applied mathematics that require the computation of higher powers of a square matrix. Now we shall show how a diagonalization can be used to simplify such computations for diagonalizable matrices.

If A is a square matrix and Q is an invertible matrix, then

$$(Q^{-1}AQ)^2 = Q^{-1}AQQ^{-1}AQ = Q^{-1}AIAQ = Q^{-1}A^2Q.$$

More generally, for any positive integer k, we have

$$(Q^{-1}AQ)^k = Q^{-1}A^kQ.$$

It follows from this equation that if A is diagonalizable and $Q^{-1}AQ = D$ is a diagonal matrix, then

$$Q^{-1}A^kQ = (Q^{-1}AQ)^k = D^k.$$

Solving this equation for A^k yields

$$A^k = QD^kQ^{-1}. \tag{7.33}$$

Therefore, in order to compute the *kth* power of A, all we need to do is compute the *kth* power of a diagonal matrix D and then form matrices Q and Q^{-1} as indicated in (7.33). Taking the *kth* power of a diagonal matrix is easy, for it simply amounts to taking the *kth* power of each of the entries on the main diagonal.

Example 7.9 *Consider the following matrix:*

$$A = \begin{pmatrix} 1 & 1 & -4 \\ 2 & 0 & -4 \\ -1 & 1 & -2 \end{pmatrix},$$

which has a characteristic equation

$$\lambda^3 + \lambda^2 - 4\lambda - 4 = 0,$$

which factorizes to

$$(\lambda + 1)(\lambda + 2)(\lambda - 2) = 0$$

and gives eigenvalues 2, -2, and -1 of the given matrix A with corresponding eigenvectors $[1, 2, 1]^T, [1, 1, 1]^T$, *and* $[1, 1, 0]^T$. *Then the factorization*

$$A = QDQ^{-1}$$

becomes

$$\begin{pmatrix} 1 & 1 & -4 \\ 2 & 0 & -4 \\ -1 & 1 & -2 \end{pmatrix} = \begin{pmatrix} 1 & 1 & 1 \\ 2 & 1 & 1 \\ 1 & 1 & 0 \end{pmatrix} \begin{pmatrix} -1 & 0 & 0 \\ 0 & -2 & 0 \\ 0 & 0 & 2 \end{pmatrix} \begin{pmatrix} -1 & 1 & 0 \\ 1 & -1 & 1 \\ 1 & 0 & -1 \end{pmatrix}$$

and from (7.33), we have

$$A^k = \begin{pmatrix} 1 & 1 & 1 \\ 2 & 1 & 1 \\ 1 & 1 & 0 \end{pmatrix} \begin{pmatrix} (-1)^k & 0 & 0 \\ 0 & (-2)^k & 0 \\ 0 & 0 & (2)^k \end{pmatrix} \begin{pmatrix} -1 & 1 & 0 \\ 1 & -1 & 1 \\ 1 & 0 & -1 \end{pmatrix},$$

which implies that

$$A^k = \begin{pmatrix} -(-1)^k + (-2)^k + 2^k & (-1)^k - (-2)^k & (-2)^k - 2^k \\ -2(-1)^k + (-2)^k + 2^k & 2(-1)^k - (-2)^k & (-2)^k - 2^k \\ -(-1)^k + (-2)^k & (-1)^k - (-2)^k & (-2)^k \end{pmatrix}.$$

For this formula, we can easily compute any power of a given matrix A. For example, if $k = 10$, *then*

$$A^{10} = \begin{pmatrix} 2047 & -1023 & 0 \\ 2046 & -1022 & 0 \\ 1023 & -1023 & 1024 \end{pmatrix}.$$

The above results can be obtained using the MATLAB command window as follows:

```
>> A = [1 1 - 4; 2 0 - 4; -1 1 - 2];
>> P = poly(A);
>> [X, D] = eig(A);
>> eigenvalues = diag(D);
>> x1 = X(:, 1); x2 = X(:, 2); x3 = X(:, 3);
>> Q = [x1 x2 x3];
>> A10 = Q * D^ 10*inv(Q);
```

Example 7.10 *Show that the following matrix A is not diagonalizable.*

$$A = \begin{pmatrix} 5 & -3 \\ 3 & -1 \end{pmatrix}.$$

Solution. *To compute the eigenvalues and corresponding eigenvectors of given matrix A, we have the characteristic equation of the form*

$$\begin{aligned} |A - \lambda \mathbf{I}| &= 0 \\ \lambda^2 - 4\lambda + 4 &= 0, \end{aligned}$$

which factorizes to

$$(\lambda - 2)(\lambda - 2) = 0$$

and gives repeated eigenvalues 2 and 2 of the given matrix A. To find the corresponding eigenvectors, we solve (7.2) for $\lambda = 2$, *and we get*

$$\begin{pmatrix} 3 & -3 \\ 3 & -3 \end{pmatrix} \begin{pmatrix} x_1 \\ x_2 \end{pmatrix} = \mathbf{0}.$$

Solving the above homogeneous system gives $3x_1 - 3x_2 = 0$ *and we have* $x_1 = x_2 = \alpha$. *Thus, the eigenvectors are nonzero vectors of the form*

$$\alpha \begin{pmatrix} 1 \\ 1 \end{pmatrix}.$$

The eigenspace is a one-dimensional space. Since A is the 2×2 *matrix, it does not have two linearly independent eigenvectors. Thus, given matrix A is not diagonalizable.*

Definition 7.6 (Orthogonal Matrix)

An orthogonal matrix is a square matrix whose inverse can be determined by transposing it; that is,

$$A^{-1} = A^T.$$

Such matrices do occur in some engineering problems. The matrix used to obtain rotation of coordinates about the origin of a Cartesian system is one example of an orthogonal matrix. For example, consider the square matrix

$$A = \begin{pmatrix} 0.6 & -0.8 \\ 0.8 & 0.6 \end{pmatrix}.$$

One can easily verify that the given matrix A is orthogonal because

$$A^{-1} = \begin{pmatrix} 0.6 & 0.8 \\ -0.8 & 0.6 \end{pmatrix} = A^T.$$

Orthogonal Diagonalization

Let Q be an orthogonal matrix; that is, $Q^{-1} = Q^T$. Thus, if such a matrix is used in a similarity transformation, the transformation becomes $D = Q^T AQ$. This type of similarity transformation is very easy to calculate; Its inverse is simply its transpose. There is, therefore, considerable advantage in searching for situations where a reduction to a diagonal matrix using an orthogonal matrix is possible.

Definition 7.7 (Orthogonally Diagonalizable Matrix)

A square matrix A is said to be orthogonally diagonalizable if there exists an orthogonal matrix Q such that

$$D = Q^{-1}AQ = Q^T AQ$$

is a diagonal matrix.

The following theorem tells us that the set of orthogonally diagonalizable matrices is in fact the set of symmetric matrices.

Theorem 7.6 *A square matrix A is orthogonally diagonalizable if it is a symmetric matrix.*

Proof. *Suppose that a matrix A is orthogonally diagonalizable then there exists an orthogonal matrix Q such that*

$$D = Q^T AQ.$$

Therefore,

$$A = QDQ^T.$$

Taking its transpose gives

$$A^T = (QDQ^T)^T = (Q^T)^T(QD)^T = QDQ^T = A.$$

Thus, A is symmetric.

The converse of this theorem is also true but it is beyond the scope of the text and will be omitted.

Symmetric Matrices

Now our next goal is to devise a procedure for orthogonally diagonalizing a symmetric matrix, but before we can do so, we need an important theorem about eigenvalues and eigenvectors of symmetric matrices.

Theorem 7.7 *If A is a symmetric matrix, then:*

(a) the eigenvalues of A are all real numbers.

(b) eigenvectors from distinct eigenvalues are orthogonal.

Diagonalization of a Symmetric Matrices

As a consequence of the proceeding theorem we obtain the following procedure for orthogonally diagonalizing a symmetric matrix.

1. Find a basis for each eigenspace of A.

2. Find an orthonormal basis for each eigenspace.

3. Form the matrix Q whose columns are these orthonormal vectors.

4. The matrix $D = Q^T AQ$ will be a diagonal matrix.

Example 7.11 *Consider the following matrix:*

$$A = \begin{pmatrix} 3 & -1 & 0 \\ -1 & 2 & -1 \\ 0 & -1 & 3 \end{pmatrix},$$

which has a characteristic equation

$$\lambda^3 - 8\lambda^2 + 19\lambda - 12 = 0$$

and it gives the eigenvalues 1, 3, and 4 for the given matrix A. Corresponding to these eigenvalues are eigenvectors of A, $\mathbf{x}_1 = [1, 2, 1]^T$, $\mathbf{x}_2 = [1, 0, -1]^T$, and $\mathbf{x}_3 = [1, -1, 1]^T$, and they form the orthogonal set. Note that the following *vectors*

$$\mathbf{u}_1 = \frac{\mathbf{x1}}{\|\mathbf{x}_1\|_2} = \tfrac{1}{\sqrt{6}}[1,2,1]^T,$$

$$\mathbf{u}_2 = \frac{\mathbf{x}_2}{\|\mathbf{x}_2\|_2} = \tfrac{1}{\sqrt{2}}[1,0,-1]^T,$$

$$\mathbf{u}_3 = \frac{\mathbf{x}_3}{\|\mathbf{x}_3\|_2} = \tfrac{1}{\sqrt{3}}[1,-1,1]^T$$

form an orthonormal set, since they inherit orthogonality from $\mathbf{x}_1, \mathbf{x}_2,$ *and* $\mathbf{x}_3,$ *and in addition*

$$\|\mathbf{u}_1\|_2 = \|\mathbf{u}_2\|_2 = \|\mathbf{u}_3\|_2 = 1.$$

Then an orthogonal matrix Q formed from an orthonormal set of vectors as

$$Q = \begin{pmatrix} \frac{1}{\sqrt{6}} & \frac{1}{\sqrt{2}} & \frac{1}{\sqrt{3}} \\ \frac{2}{\sqrt{6}} & 0 & -\frac{1}{\sqrt{3}} \\ \frac{1}{\sqrt{6}} & -\frac{1}{\sqrt{2}} & \frac{1}{\sqrt{3}} \end{pmatrix}$$

and

$$Q^T AQ = \begin{pmatrix} \frac{1}{\sqrt{6}} & \frac{2}{\sqrt{6}} & \frac{1}{\sqrt{6}} \\ \frac{1}{\sqrt{2}} & 0 & -\frac{1}{\sqrt{2}} \\ \frac{1}{\sqrt{3}} & -\frac{1}{\sqrt{3}} & \frac{1}{\sqrt{3}} \end{pmatrix} \begin{pmatrix} 3 & -1 & 0 \\ -1 & 2 & -1 \\ 0 & -1 & 3 \end{pmatrix} \begin{pmatrix} \frac{1}{\sqrt{6}} & \frac{1}{\sqrt{2}} & \frac{1}{\sqrt{3}} \\ \frac{2}{\sqrt{6}} & 0 & -\frac{1}{\sqrt{3}} \\ \frac{1}{\sqrt{6}} & -\frac{1}{\sqrt{2}} & \frac{1}{\sqrt{3}} \end{pmatrix},$$

which implies that

$$Q^T AQ = \begin{pmatrix} 1 & 0 & 0 \\ 0 & 3 & 0 \\ 0 & 0 & 4 \end{pmatrix} = D.$$

Note that the eigenvalues $1, 3,$ and 4 of the matrix A are real and its eigenvectors form an orthonormal set, since they inherit orthogonally from $\mathbf{x}_1, \mathbf{x}_2,$ and $\mathbf{x}_3,$

which satisfy the proceeding theorem.

The results of Example 7.11 can be obtained using the MATLAB command window as follows:

```
>> A = [3 -1 0; -1 2 -1; 0 -1 3];
>> P = poly(A);
>> [X, D] = eig(A);
>> λ = diag(D);
>> x1 = X(:,1); x2 = X(:,2); x3 = X(:,3);
>> u1 = x1/norm(x1); u2 = x2/norm(x2); u3 = x3/norm(x3);
>> Q = [u1 u2 u3];
>> D = Q' * A * Q;
```

In the following section we shall discuss some extremely important properties of the eigenvalue problem. Before this we discuss some special matrices.

Definition 7.8 (Conjugate of Matrix)

If the entries of an $n \times n$ matrix A are complex numbers, we can write

$$A = (a_{ij}) = (b_{ij} + \mathbf{i}c_{ij}),$$

where b_{ij} and c_{ij} are real numbers. The conjugate of a matrix A is matrix

$$\bar{A} = (\bar{a}_{ij}) = (b_{ij} - \mathbf{i}c_{ij}).$$

For example, the conjugate of the following matrix

$$A = \begin{pmatrix} 2 & \pi & \mathbf{i} \\ 3 & 7 & 0 \\ 4 & 1-\mathbf{i} & 4 \end{pmatrix} \quad is \quad \bar{A} = \begin{pmatrix} 2 & \pi & -\mathbf{i} \\ 3 & 7 & 0 \\ 4 & 1+\mathbf{i} & 4 \end{pmatrix}.$$

Definition 7.9 (Hermitian Matrix)

Hermitian matrix is a square matrix $A = (a_{ij})$ that is equal to its conjugate transpose

$$A = A^* = \bar{A}^T;$$

that is, whenever $a_{ij} = \bar{a}_{ji}$. This is the complex analog of symmetry. For example, the following matrix A is Hermitian if it has the form

$$A = \begin{pmatrix} a & b + \mathbf{i}c \\ b - \mathbf{i}c & d \end{pmatrix},$$

where a, b, c, and d are real. An Hermitian matrix may be or may not be symmetric. For example, the following matrices

$$A = \begin{pmatrix} 1 & 2+4\mathrm{i} & 1-3\mathrm{i} \\ 2-4\mathrm{i} & 3 & 8+6\mathrm{i} \\ 1+3\mathrm{i} & 8-6\mathrm{i} & 5 \end{pmatrix} \quad \text{and} \quad B = \begin{pmatrix} 1 & 2+4\mathrm{i} & 1-3\mathrm{i} \\ 2+4\mathrm{i} & 3 & 8+6\mathrm{i} \\ 1-3\mathrm{i} & 8+6\mathrm{i} & 5 \end{pmatrix}$$

are Hermitian where the matrix A is symmetric but the matrix B is not. Note that:
1. every diagonal matrix is Hermitian if and only if it is real.
2. the square matrix A is said to be a skew Hermitian when

$$A = -A^* = -\bar{A}^T;$$

that is, whenever $a_{ij} = -\bar{a}_{ji}$. This is the complex analog of skew symmetry. For example, the following matrix A is skew Hermitian if it has the form

$$A = \begin{pmatrix} 0 & 1+\mathrm{i} \\ -1+\mathrm{i} & \mathrm{i} \end{pmatrix}.$$

Definition 7.10 (Unitary Matrix)

Let $A = (a_{ij})$ be the square matrix, then if

$$AA^* = A^*A = \mathbf{I}_n,$$

where $\mathbf{I}_n$ is an $n \times n$ identity matrix, then A is called the unitary matrix. For example, for any real number θ, the following matrix

$$A = \begin{pmatrix} \cos\theta & -\sin\theta \\ \sin\theta & \cos\theta \end{pmatrix}$$

is unitary.
Note that:
1. the identity matrix is unitary.
2. the inverse of a unitary matrix is unitary.
3. a product of unitary matrix is unitary.
4. a real matrix A is unitary if and only if $A^T = A^{-1}$.
5. a square matrix A is unitarily similar to the square matrix B if and only if there is an unitary matrix Q of same size as A and B such that

$$A = QBQ^{-1}.$$

A square matrix A is *unitarily diagonalizable* if and only if it is unitarily similar to a diagonal matrix.

Theorem 7.8 *A square matrix A is unitarily diagonalizable if and only if there is a unitary matrix Q of the same size whose columns are eigenvectors of A.*

Definition 7.11 (Normal Matrix)

A square matrix A is normal if and only if it commutes with its conjugate transpose; that is,

$$AA^* = A^*A$$

For example, if a and b are real numbers, then the following matrix

$$A = \begin{pmatrix} a & b \\ -b & a \end{pmatrix}$$

is normal because

$$AA^* = \begin{pmatrix} a^2+b^2 & 0 \\ 0 & a^2+b^2 \end{pmatrix} = A^*A.$$

However, its eigenvalues are $a \pm ib$. Note that all the Hermitian, skew Hermitian, and unitary matrices are normal matrices.

7.4 Basic Properties of Eigenvalue Problems

1. A square matrix A is *singular* if and only if at least one of its eigenvalue is zero. It can be easily prove, since for $\lambda = 0$, we have (7.4) of the form

$$|A - \lambda I| = |A| = 0.$$

Example 7.12 *Consider the following matrix:*

$$A = \begin{pmatrix} 3 & 1 & 0 \\ 2 & -1 & -1 \\ 4 & 3 & 1 \end{pmatrix}.$$

Then the characteristic equation of A takes the form

$$-\lambda^3 + 3\lambda^2 = 0.$$

By solving this cubic equation, the eigenvalues of A are 0, 0, and 3. Hence, the given matrix is singular because two of its eigenvalues are zero.

2. The eigenvalues of a matrix A and its transpose A^T are identical.

   It is well known that the determinant of a matrix A and of its A^T are the same. Therefore, they must have the same characteristic equation and the same eigenvalues.

Example 7.13 *Consider a matrix A and its transpose matrix A^T as*

$$A = \begin{pmatrix} 1 & 1 & 0 \\ 3 & 0 & 3 \\ 2 & -1 & 3 \end{pmatrix} \quad \textit{and} \quad A^T = \begin{pmatrix} 1 & 3 & 2 \\ 1 & 0 & -1 \\ 0 & 3 & 3 \end{pmatrix}.$$

The characteristic equations of A and A^T are the same, which is

$$-\lambda^3 + 4\lambda^2 - 3\lambda = 0.$$

Solving this cubic polynomial equation, we have the eigenvalues $0, 1,$ *and* 3 *for matrix A and its transpose A^T.*

3. The eigenvalues of an inverse matrix A^{-1}, provided that A^{-1} exists, are the inverses of the eigenvalues of A.

   To prove this, let us consider λ is an eigenvalue of A and using (7.4), gives

$$|A - \lambda I| = |A - \lambda A A^{-1}| = |A(I - \lambda A^{-1})| = |\lambda||A||\frac{1}{\lambda}I - A^{-1}| = 0.$$

   Since matrix A is nonsingular, $|A| \neq 0$, and also, $\lambda \neq 0$. Hence,

$$\left|\frac{1}{\lambda}I - A^{-1}\right| = 0,$$

   which shows that $\frac{1}{\lambda}$ is an eigenvalue of matrix A^{-1}.

Example 7.14 *Consider matrix A and its inverse matrix A^{-1} as*

$$A = \begin{pmatrix} 3 & 0 & 1 \\ 0 & -3 & 0 \\ 1 & 0 & 3 \end{pmatrix} \quad \textit{and} \quad A^{-1} = \begin{pmatrix} \frac{3}{8} & 0 & -\frac{1}{8} \\ 0 & -\frac{1}{3} & 0 \\ -\frac{1}{8} & 0 & \frac{3}{8} \end{pmatrix}.$$

Then a characteristic equation of A has the form

$$\lambda^3 - 3\lambda^2 - 10\lambda + 24 = 0,$$

which gives the eigenvalues 4, -3, and 2 of A. Also, the characteristic equation of A^{-1} *is*

$$\lambda^3 - \frac{5}{12}\lambda^2 - \frac{1}{8}\lambda + \frac{1}{24} = 0$$

and it gives the eigenvalues,

$$\frac{1}{4}, -\frac{1}{3}, \quad and \quad \frac{1}{2}$$

which are reciprocal to the eigenvalues $4, -3,$ *and* 2 *of matrix* A.

4. The eigenvalues of A^k (k an integer) are eigenvalues of A raised to the kth power. To prove this, consider the characteristic equation of matrix A

$$|A - \lambda I| = 0,$$

which can be also written as

$$0 = |A - \lambda I| = |A - \lambda I||A + \lambda I| = |(A - \lambda I)(A + \lambda I)| = |A^2 - \lambda^2 I|.$$

Example 7.15 *Consider the following matrix:*

$$A = \begin{pmatrix} 1 & 1 & 0 \\ 3 & 0 & 3 \\ 2 & -1 & 3 \end{pmatrix},$$

which has eigenvalues, 0, 1, and 3. Now

$$AA = A^2 = \begin{pmatrix} 4 & 1 & 3 \\ 9 & 0 & 9 \\ 5 & -1 & 6 \end{pmatrix}$$

has the characteristic equation of the form

$$\lambda^3 - 10\lambda^2 + 9\lambda = 0.$$

Solving this cubic equation, the eigenvalues of A^2 *are, 0, 1, and 9, which are double the eigenvalues* $0, 1,$ *and* 3 *of* A.

5. The eigenvalues of a diagonal matrix or a triangular (upper or lower) matrix are their diagonal elements.

Example 7.16 *Consider the following matrices:*

$$A = \begin{pmatrix} 1 & 0 & 0 \\ 0 & 2 & 0 \\ 0 & 0 & 3 \end{pmatrix} \quad \textit{and} \quad B = \begin{pmatrix} 2 & 2 & 3 \\ 0 & 3 & 3 \\ 0 & 0 & 4 \end{pmatrix}.$$

Since the characteristic equation of A is

$$\lambda^3 - 6\lambda^2 + 11\lambda - 6 = (1-\lambda)(2-\lambda)(3-\lambda) = 0$$

and it gives the eigenvalues $1, 2,$ and $3,$ which are the diagonal elements of the given matrix A, similarly, the characteristic equation of B is

$$\lambda^3 - 9\lambda^2 + 26\lambda - 24 = (2-\lambda)(3-\lambda)(4-\lambda) = 0$$

and it gives the eigenvalues $2, 3,$ and $4,$ which are the diagonal elements of given matrix B. Hence, the eigenvalues of a diagonal matrix A and the upper-triangular matrix B are their diagonal elements.

6. Every square matrix satisfies its own characteristic equation.

This is a well-known theorem called the *Cayley-Hamilton Theorem.* If a characteristic equation of A is

$$\lambda^n + \alpha_{n-1}\lambda^{n-1} + \alpha_{n-2}\lambda^{n-2} + \cdots + \alpha_1\lambda + \alpha_0 = 0,$$

the matrix itself satisfies the same equation, namely,

$$A^n + \alpha_{n-1}A^{n-1} + \alpha_{n-2}A^{n-2} + \cdots + \alpha_1 A + \alpha_0 = 0. \tag{7.34}$$

Multiplying each term in (7.34) by A^{-1}, when A^{-1} exists and thus $\alpha_0 \neq 0$, gives an important relationship for the inverse of a matrix:

$$A^{n-1} + \alpha_{n-1}A^{n-2} + \alpha_{n-2}A^{n-3} + \cdots + \alpha_1 I + \alpha_0 A^{-1} = 0$$

or

$$A^{-1} = -\frac{1}{\alpha_0}[A^{n-1} + \alpha_{n-1}A^{n-2} + \alpha_{n-2}A^{n-3} + \cdots + \alpha_1 I].$$

Example 7.17 *Consider the following square matrix:*

$$A = \begin{pmatrix} 2 & 1 & 2 \\ 0 & 2 & 3 \\ 0 & 0 & 5 \end{pmatrix},$$

which has a characteristic equation of the form

$$p(\lambda) = \lambda^3 - 9\lambda^2 + 24\lambda - 20 = 0$$

and one can write

$$p(A) = A^3 - 9A^2 + 24A - 20I = 0.$$

Then the inverse of A can be obtained as follows:

$$A^2 - 9A + 24I - 20A^{-1} = 0,$$

which gives

$$A^{-1} = \frac{1}{20}[A^2 - 9A + 24I].$$

Computing the right-hand side, we have

$$A^{-1} = \frac{1}{20}\begin{pmatrix} 10 & -5 & -1 \\ 0 & 10 & -6 \\ 0 & 0 & 4 \end{pmatrix}.$$

Similarly, one can also find the higher power of the given matrix A. For example, one can compute the value of matrix A^5 by solving the following expression

$$A^5 = 9A^4 - 24A^3 + 20A^2$$

and it gives

$$A^5 = \begin{pmatrix} 32 & 80 & 3013 \\ 0 & 32 & 3093 \\ 0 & 0 & 3125 \end{pmatrix}.$$

To find the coefficients of a characteristic equation and the inverse of matrix A by the Cayley-Hamilton theorem using MATLAB commands, we do the following:

$$>> A = [2\ 1\ 2; 0\ 2\ 3; 0\ 0\ 5];$$
$$c = CAYHAM(A);$$

Program 7.2

MATLAB m-file for using the Cayley-Hamilton Theorem

```
function [c,Ainv]= CAYHAM(A)
n=max(size(A));
for i=1:n; for j=1:n
I(i,j)=0; I(i,i)=1; end; end
AA=A; AAA=A; c=[1];
for k=1:n; traceA=0;
for g=1:n % Loop to find the trace of matrix A.
traceA=traceA+A(g,g); end
cc = -1/k * traceA; % To find coefficients of the polynomial.
c = [c, cc]; if k < n;
for i=1:n; for j=1:n; b(i, j) = A(i, j) + cc * I(i, j); end; end
for i=1:n; for j=1:n; s=0;
for m=1:n; ss = AA(i, m) * b(m, j); s=s+ss; end;
A(i,j)=s; end; end; end; end
for i=1:n; for j=1:n
su1(i, j) = c(n) * I(i, j) + c(n - 1) * AA(i, j); su2(i,j)=0; end; end
if n > 2
for z=2:n-1; for i=1:n; for j=1:n; s=0;
for m=1:n; ss = AAA(i, m) * AA(m, j); s = s + ss; end
am(i,j)=s; end; end; AAA=am;
for i=1:n; for j=1:n
su2(i, j) = su2(i, j) + c(n - z) * AAA(i, j); end; end;end;end
for i=1:n; for j=1:n
su(i,j)=su1(i,j)+su2(i,j); Ainv(i, j) = -1/c(n + 1) * su(i, j);
end;end
```

7. The eigenvectors of A^{-1} are the same as the eigenvectors of A.

   Let $\mathbf{x}$ be an eigenvector of A, which satisfies the following equation

$$A\mathbf{x} = \lambda\mathbf{x}$$

   then

$$\frac{1}{\lambda}A^{-1}A\mathbf{x} = A^{-1}\mathbf{x},$$

hence,

$$A^{-1}\mathbf{x} = \frac{1}{\lambda}\mathbf{x},$$

which shows that $\mathbf{x}$ is also an eigenvector of A^{-1}.

8. The eigenvectors of matrix (kA) are identical to the eigenvectors of A, for any scalar k.

Since the eigenvalues of (kA) are k times the eigenvalues of A, if, $A\mathbf{x} = \lambda\mathbf{x}$, then

$$(kA)\mathbf{x} = (k\lambda)\mathbf{x}.$$

9. A symmetric matrix A is positive-definite if and only if all the eigenvalues of A are positive.

Example 7.18 *Consider the following matrix:*

$$A = \begin{pmatrix} 2 & 0 & 1 \\ 0 & 2 & 0 \\ 1 & 0 & 2 \end{pmatrix},$$

which has the characteristic equation of the form

$$\lambda^3 - 6\lambda^2 + 11\lambda - 6 = 0$$

and it gives the eigenvalues 3, 2, and 1 of A. Since all the eigenvalues of the matrix A are positive, A is positive-definite.

10. For any $n \times n$ matrix A, we have

$$\left.\begin{array}{lcl} B_0 & = & I \\ A_k & = & AB_{k-1} \\ c_k & = & -\frac{1}{k}\mathrm{tr}(A_k) \\ B_k & = & A_k + c_k I \end{array}\right\} \qquad k = 1, 2, \ldots, n,$$

where $tr(A)$ is a trace of a matrix A. Then a characteristic polynomial of A is

$$p(\lambda) = |A - \lambda I| = \lambda^n + c_1\lambda^{n-1} + \cdots + c_{n-1}\lambda + c_n.$$

If $c_n \neq 0$, then the inverse of A can be obtained as follows:

$$A^{-1} = -\frac{1}{c_n}B_{n-1}.$$

This is called the *Sourian-Frame theorem.*

Example 7.19 *Find a characteristic polynomial of the following matrix:*

$$A = \begin{pmatrix} 5 & -2 & -4 \\ -2 & 2 & 2 \\ -4 & 2 & 5 \end{pmatrix}$$

and then find A^{-1} using the Sourian-Frame theorem.

Solution. *Since the given matrix is of size 3×3, a possible form of a characteristic polynomial will be as follows:*

$$p(\lambda) = \lambda^3 + c_1\lambda^2 + c_2\lambda + c_3.$$

The values of the coefficients c_1, c_2, and c_3 of the above characteristic polynomial can be computed as:

$$A_1 = AB_0 = AI = A$$

so

$$c_1 = -\frac{1}{1}tr(A_1) = -12.$$

Now

$$B_1 = A_1 + c_1 I = A_1 - 12I = \begin{pmatrix} -7 & -2 & -4 \\ -2 & -10 & 2 \\ -4 & 2 & -7 \end{pmatrix}$$

and

$$A_2 = AB_1 = \begin{pmatrix} -15 & 2 & -4 \\ 2 & -12 & -2 \\ 4 & -2 & -15 \end{pmatrix}$$

so

$$c_2 = -\frac{1}{2}tr(A_2) = 21.$$

Now

$$B_2 = A_2 + c_2 I = A_2 + 21I = \begin{pmatrix} 6 & 2 & 4 \\ 2 & 9 & -2 \\ 4 & -2 & 6 \end{pmatrix}$$

and

$$A_3 = AB_2 = \begin{pmatrix} 10 & 0 & 0 \\ 0 & 10 & 0 \\ 0 & 0 & 10 \end{pmatrix}$$

so

$$c_3 = -\frac{1}{3} tr(A_3) = -10.$$

Thus,

$$p(\lambda) = \lambda^3 - 12\lambda^2 + 21\lambda - 10$$

and the inverse of given matrix A is

$$A^{-1} = -\frac{1}{c_3} B_2 = \frac{1}{10} \begin{pmatrix} 6 & 2 & 4 \\ 2 & 9 & -2 \\ 4 & -2 & 6 \end{pmatrix}.$$

The above results can be obtained using the MATLAB the command window as follows:

```
>> A = [5 -2 -4; -2 2 2; -4 2 5];
>> [c, Ainv] = SOURIANF(A);
```

Program 7.3

MATLAB m-file for using the Sourian-Frame theorem

```
function [c,Ainv]=SOURIANF(A)
[n,n]=size(A);b=eye(n);
for k=1:n; for i=1:n; for j=1:n;
sum=0; for p=1:n;
sum=sum+A(i,p)*b(p,j);end;
d(i,j)=sum;end;end; a=d;
sum=0; for i=1:n; sum=sum+a(i,i);end
if i==j; b(i,j)=a(i,j)+c(k); else
b(i,j)=a(i,j);end;end
c(k) = -1/k * sum; for i=1:n; for j=1:n;
Ainv(i, j) = -1/c(k) * b(i, j);
for i=1:n; for j=1:n;
ai(i,j)=-1/c(n+1)*b(i,j); end; end;end;end
disp('Coefficients of the polynomial') c
disp('Inverse of the matrix A=Aniv') Ainv
```

11. If a characteristic equation of an $n \times n$ matrix A is

$$\lambda^n + \alpha_{n-1}\lambda^{n-1} + \alpha_{n-2}\lambda^{n-2} + \alpha_{n-3}\lambda^{n-3} + \cdots + \alpha_1\lambda + \alpha_0 = 0$$

the values of the coefficients of a characteristic polynomial are then found from the following sequence of computations:

$$\begin{aligned} \alpha_{n-1} &= -\mathrm{tr}(A) \\ \alpha_{n-2} &= -\frac{1}{2}[\alpha_{n-1}\mathrm{tr}(A) + \mathrm{tr}(A^2)] \\ \alpha_{n-3} &= -\frac{1}{3}[\alpha_{n-2}\mathrm{tr}(A) + \alpha_{n-1}\mathrm{tr}(A^2) + \mathrm{tr}(A^3)] \\ \vdots \quad & \quad \vdots \\ \alpha_0 &= -\frac{1}{n}[\alpha_1\mathrm{tr}(A) + \alpha_2\mathrm{tr}(A^2) + \cdots + \mathrm{tr}(A^n)]. \end{aligned}$$

This formula is called *Bocher's formula*, which can be used to find coefficients of a characteristic equation of a square matrix.

Example 7.20 *Find the characteristic equation of the following matrix using Bocher's formula:*

$$A = \begin{pmatrix} 1 & 1 & 0 \\ 3 & 0 & 3 \\ 2 & -1 & 3 \end{pmatrix}$$

Solution. *Since the size of the given matrix is* 3×3, *we have to find the coefficients* α_2, α_1, *and* α_0 *of the following characteristic equation*

$$\lambda^3 + \alpha_2\lambda^2 + \alpha_1\lambda + \alpha_0 = 0,$$

where

$$\begin{aligned} \alpha_2 &= -tr(A) \\ \alpha_1 &= -\frac{1}{2}[\alpha_2 tr(A) + tr(A^2)] \\ \alpha_0 &= -\frac{1}{3}[\alpha_1 tr(A) + \alpha_2 tr(A^2) + tr(A^3)]. \end{aligned}$$

In order to find the values of the above coefficients, we must compute the powers of matrix A as follows

$$A^2 = \begin{pmatrix} 4 & 1 & 3 \\ 9 & 0 & 9 \\ 5 & -1 & 6 \end{pmatrix} \quad \textit{and} \quad A^3 = \begin{pmatrix} 13 & 1 & 12 \\ 27 & 0 & 27 \\ 14 & -1 & 15 \end{pmatrix}.$$

Using these matrices, we can find the coefficients of the characteristic equation as

$$\begin{aligned}\alpha_2 &= -(4) = -4\\ \alpha_1 &= -\frac{1}{2}[-4(4)+10] = 3\\ \alpha_0 &= -\frac{1}{3}[3(4)+(-4)(10)+28] = 0.\end{aligned}$$

Hence, the characteristic equation of A is

$$\lambda^3 - 4\lambda^2 + 3\lambda = 0.$$

To find the coefficients of the characteristic equation by the Bocher's theorem using MATLAB commands we do as follows:

```
>> A = [1 1 0; 3 0 3; 2 -1 3];
c = BOCHER(A);
```

Program 7.4

MATLAB m-file for using Bocher's Theorem

```
function c=BOCHER(A)
[n,n]=size(A);
for i=1:n; for j=1:n; I(i,i)=1; end; end
a(n-1)=-trace(A);
for i=2:n; s=0; p=1;
for k=i-1:-1:1
s = s + a(n - k) * trace(A^ p); p = p + 1; end
if i~=n; a(n - i) = (-1/i) * (s + trace(A^ i)); else
ao = (-1/i) * (s + trace(A^ i)); end; end
coeff=[ao  a]; if ao~=0; s=A ^ (n-1);
for i=1:n-2
s = s + a(n - i) * A^ (n - (i + 1)); end
s = s + A(1) * I; Ainv = -s/ao; else; end
```

12. For an $n \times n$ matrix A with the characteristic equation

$$|A - \lambda I| = \lambda^n + \alpha_{n-1}\lambda^{n-1} + \alpha_{n-2}\lambda^{n-2} + \cdots + \alpha_1\lambda + \alpha_0 = 0.$$

The unknown coefficients can be computed as follows:

$$\begin{aligned}\alpha_{n-1} &= -\mathrm{tr}(AD_1)\\ \alpha_{n-2} &= -\frac{1}{2}(AD_2)\\ \alpha_{n-3} &= -\frac{1}{3}(AD_3)\\ \vdots & \qquad \vdots\\ \alpha_0 &= -\frac{1}{n}(AD_n),\end{aligned}$$

where

$$\begin{aligned}D_1 &= I\\ D_2 &= AD_1 + \alpha_{n-1}I = A + \alpha_{n-1}I\\ D_3 &= AD_2 + \alpha_{n-2}I = A^2 + \alpha_{n-1}A + \alpha_{n-2}I\\ &\vdots\\ D_n &= AD_{n-1} + \alpha_1 I = A^{n-1} + \alpha_{n-1}A^{n-2} + \cdots + \alpha_2 A + \alpha_1 I\end{aligned}$$

and also

$$D_{n+1} = AD_n + \alpha_0 I = 0.$$

Then the determinant of A is

$$\det(A) = |A| = (-1)^n \alpha_0,$$

the adjoint of A is

$$adj(A) = (-1)^{n+1} D_n,$$

and the inverse of A is

$$A^{-1} = -\frac{1}{\alpha_0} D_n.$$

Note that a singular matrix is indicated by $\alpha_0 = 0$.

This result is known as the *Faddeev-Leverrier method.* This method which is recursive, yields, a characteristic equation for a square matrix, adjoint of a matrix and its inverse (if it exists). The determinant of a matrix, being the negative of the last coefficient in the characteristic equation, is also computed.

Example 7.21 *Find the characteristic equation, determinant, adjoint, and inverse of the following matrix using the Faddeev-Leverrier method:*

$$A = \begin{pmatrix} 2 & 2 & -1 \\ -1 & 0 & 4 \\ 3 & -1 & -3 \end{pmatrix}.$$

Solution. *Since the given matrix is of order* 3×3, *the possible characteristic equation will be of the form*

$$|A - \lambda I| = \lambda^3 + \alpha_2 \lambda^2 + \alpha_1 \lambda + \alpha_0 = 0.$$

The values of the unknown coefficients, α_2, α_1, *and* α_0 *can be computed as follows:*

$$\alpha_2 = -tr(AD_1) = -tr(A) = 1$$

and

$$D_2 = AD_1 + \alpha_2 I = A + I = \begin{pmatrix} 3 & 2 & -1 \\ -1 & 1 & 4 \\ 3 & -1 & -2 \end{pmatrix}.$$

Also,

$$AD_2 = \begin{pmatrix} 1 & 7 & 8 \\ 9 & -6 & -7 \\ 1 & 8 & -1 \end{pmatrix}$$

so

$$\alpha_1 = -\frac{1}{2} tr(AD_2) = -\frac{1}{2}(-6) = 3.$$

Similarly, we have

$$D_3 = AD_2 + \alpha_1 I = AD_2 + 3I = \begin{pmatrix} 4 & 7 & 8 \\ 9 & -3 & -7 \\ 1 & 8 & 2 \end{pmatrix}$$

and

$$AD_3 = \begin{pmatrix} 25 & 0 & 0 \\ 0 & 25 & 0 \\ 0 & 0 & 25 \end{pmatrix}.$$

It gives

$$\alpha_0 = -\frac{1}{3} tr(AD_3) = -\frac{1}{3}(25) = -25,$$

which showed that the given matrix is nonsingular. Hence, the characteristic equation is

$$|A - \lambda I| = \lambda^3 + \lambda^2 + 3\lambda - 25 = 0.$$

Thus, the determinant of A is

$$|A| = (-1)^3(-25) = 25$$

and the adjoint of A is

$$adj(A) = (-1)^4(D_3) = D_3 = \begin{pmatrix} 4 & 7 & 8 \\ 9 & -3 & -7 \\ 1 & 8 & 2 \end{pmatrix}.$$

Finally, the inverse of A is

$$A^{-1} = -\frac{1}{\alpha_0} D_3 = \frac{1}{25} \begin{pmatrix} 4 & 7 & 8 \\ 9 & -3 & -7 \\ 1 & 8 & 2 \end{pmatrix}.$$

13. All the eigenvalues of a Hermitian matrix are real.

To prove this, consider (7.1), which is

$$A\mathbf{x} = \lambda\mathbf{x}$$

and it implies that

$$\mathbf{x}^* A\mathbf{x} = \lambda\mathbf{x}^*\mathbf{x}.$$

Since A is Hermitian, that is, $A = A^*$ and $\mathbf{x}^* A\mathbf{x}$ is a scalar,

$$\mathbf{x}^* A\mathbf{x} = \mathbf{x}^* A^*\mathbf{x} = (\mathbf{x}^* A\mathbf{x})^* = \overline{\mathbf{x}A\mathbf{x}}.$$

Thus, the scalar is equal to its own conjugate and, hence, real. Therefore, λ is real.

Example 7.22 *Consider the Hermitian matrix*

$$A = \begin{pmatrix} 2 & -i & 0 \\ i & 2 & 0 \\ 0 & 0 & 3 \end{pmatrix},$$

which has a characteristic equation

$$\lambda^3 - 7\lambda^2 + 15\lambda - 9 = 0$$

and it gives real eigenvalues 1, 3, and 3 for given matrix A.

14. A matrix that is unitarily similar to a Hermitian matrix is itself Hermitian. To prove this, assume that $B^* = B$ and $A = QBQ^{-1}$, where $Q^{-1} = Q^*$. Then

$$\begin{aligned} A^* &= (QBQ^{-1})^* \\ &= (QBQ^*)^* \\ &= Q^{**}B^*Q^* \\ &= QBQ^{-1} = A. \end{aligned}$$

This shows that matrix A is Hermitian.

15. If A is a Hermitian matrix with distinct eigenvalues, then it is unitary similar to a diagonal matrix; that is, there exists a unitary matrix μ such that

$$\mu^T A\mu = D$$

is a diagonal matrix.

Example 7.23 *Consider the following matrix*

$$A = \begin{pmatrix} 11 & 2 & 8 \\ 2 & 2 & -10 \\ 8 & -10 & 5 \end{pmatrix},$$

which has the characteristic equation

$$\lambda^3 - 18\lambda^2 - 81\lambda + 1458 = 0.$$

It can be easily factorized to give

$$(\lambda + 9)(\lambda - 9)(\lambda - 18) = 0$$

and the eigenvalues of A are, -9, 9, and 18. The eigenvectors corresponding to these eigenvalues are

$$\mathbf{x}_1 = [-1, 2, 2]^T, \quad \mathbf{x}_2 = [2, 2, -1]^T, \quad and \quad \mathbf{x}_3 = [2, -1, 2]^T$$

and they form an orthogonal set. Note that the following vectors

$$\begin{aligned} \mathbf{u}_1 &= \frac{\mathbf{x}_1}{\|\mathbf{x}_1\|_2} = \tfrac{1}{3}[-1, 2, 2]^T, \\ \mathbf{u}_2 &= \frac{\mathbf{x}_2}{\|\mathbf{x}_2\|_2} = \tfrac{1}{3}[2, 2, -1]^T, \\ \mathbf{u}_3 &= \frac{\mathbf{x}_3}{\|\mathbf{x}_3\|_2} = \tfrac{1}{3}[2, -1, 2]^T \end{aligned}$$

form an orthonormal set, since they inherit orthogonality from $\mathbf{x}_1, \mathbf{x}_2$, *and* $\mathbf{x}_3$, *and in addition*

$$\|\mathbf{u}_1\|_2 = \|\mathbf{u}_2\|_2 = \|\mathbf{u}_3\|_2 = 1.$$

Thus, the unitary matrix μ *is given by*

$$\mu = \frac{1}{3}\begin{pmatrix} -1 & 2 & 2 \\ 2 & 2 & -1 \\ 2 & -1 & 2 \end{pmatrix}$$

and

$$\mu^T A \mu = \frac{1}{9}\begin{pmatrix} -1 & 2 & 2 \\ 2 & 2 & -1 \\ 2 & -1 & 2 \end{pmatrix}\begin{pmatrix} 11 & 2 & 8 \\ 2 & 2 & -10 \\ 8 & -10 & 5 \end{pmatrix}\begin{pmatrix} -1 & 2 & 2 \\ 2 & 2 & -1 \\ 2 & -1 & 2 \end{pmatrix},$$

which implies that

$$\mu^T A \mu = \begin{pmatrix} -9 & 0 & 0 \\ 0 & 9 & 0 \\ 0 & 0 & 18 \end{pmatrix} = D.$$

16. A matrix Q is *unitary* if and only if its conjugate transpose is its inverse:

$$Q^* = Q^{-1}.$$

Note that a real matrix Q is unitary if and only if $Q^T = Q^{-1}$. For any square matrix A, there is a unitary matrix μ such that

$$\mu^{-1} A \mu = T$$

an upper-triangular matrix whose diagonal entries consists of eigenvalues of A. It is a well-known lemma called *Shur's lemma.*

Example 7.24 *Consider the following matrix:*

$$A = \begin{pmatrix} 2 & -1 \\ 1 & 0 \end{pmatrix},$$

which has an eigenvalue 1 of multiplicity 2. The eigenvector corresponding to this eigenvalue is $[1, 1]^T$. *Thus, the first column of unitary matrix* μ *is*

$[\frac{1}{\sqrt{2}}, \frac{1}{\sqrt{2}}]^T$ *and the other column is orthogonal to it; that is,* $[\frac{1}{\sqrt{2}}, -\frac{1}{\sqrt{2}}]^T$. *So*

$$\mu^{-1}A\mu = \begin{pmatrix} \frac{1}{\sqrt{2}} & \frac{1}{\sqrt{2}} \\ \frac{1}{\sqrt{2}} & -\frac{1}{\sqrt{2}} \end{pmatrix} \begin{pmatrix} 2 & -1 \\ 1 & 0 \end{pmatrix} \begin{pmatrix} \frac{1}{\sqrt{2}} & \frac{1}{\sqrt{2}} \\ \frac{1}{\sqrt{2}} & -\frac{1}{\sqrt{2}} \end{pmatrix},$$

which gives

$$\mu^{-1}A\mu = \begin{pmatrix} 1 & 2 \\ 0 & 1 \end{pmatrix} = T.$$

17. A matrix that is unitarily similar to a normal matrix is itself normal. Assume that $BB^* = B^*B$ and $A = QBQ^{-1}$, where $Q^{-1} = Q^*$. Then

$$\begin{aligned} A^* &= (QBQ^{-1})^* \\ &= (QBQ^*)^* \\ &= Q^{**}B^*Q^* \\ &= QB^*Q^{-1}. \end{aligned}$$

So

$$\begin{aligned} AA^* &= (QBQ^{-1})(QB^*Q^{-1}) \\ &= (QBB^*Q^{-1}) \\ &= (QB^*BQ^{-1}) \\ &= (QB^*Q^{-1}QBQ^{-1}) \\ &= (QB^*Q^{-1})(QBQ^{-1}) = A^*A. \end{aligned}$$

This shows that matrix A is normal.

18. The value of the exponential of matrix A can be calculated from

$$e^A = Q(exp\Lambda)Q^{-1},$$

where $exp\Lambda$ is a diagonal matrix whose elements are the exponential of successive eigenvalues and Q is a matrix of the eigenvectors of A.

Example 7.25 *Consider the following matrix*

$$A = \begin{pmatrix} 0.1 & 0.1 \\ 0.0 & 0.2 \end{pmatrix}.$$

In order to find the value of the exponential of given matrix A, we have to find the eigenvalues of A and also, the eigenvectors of A. The eigenvalues *of A are 0.1 and 0.2, and the corresponding eigenvectors are $[1,0]^T$ and $[1,1]^T$. Then matrix Q is*

$$Q = \begin{pmatrix} 1 & 1 \\ 0 & 1 \end{pmatrix}.$$

Its inverse can be found as

$$Q^{-1} = \begin{pmatrix} 1 & -1 \\ 0 & 1 \end{pmatrix}.$$

Thus,

$$e^A = Q(exp\Lambda)Q^{-1} = \begin{pmatrix} 1 & 1 \\ 0 & 1 \end{pmatrix} \begin{pmatrix} e^{0.1} & 0 \\ 0 & e^{0.2} \end{pmatrix} \begin{pmatrix} 1 & -1 \\ 0 & 1 \end{pmatrix},$$

which gives

$$e^A = \begin{pmatrix} 1.105 & 0.116 \\ 0.000 & 1.221 \end{pmatrix}.$$

In the following section we shall discuss some very important results concerned with eigenvalue problems. The proofs of all the results are beyond the text and will be omitted. However, they are very easily understood and can be used. We can show these results by using the different matrices.

7.5 Some Important Results of Eigenvalue Problems

1. If A is a Hermitian matrix then

$$\|A\|_2 = \sqrt{\rho(A^H A)} = \rho(A).$$

Example 7.26 *Consider the Hermitian matrix*

$$A = \begin{pmatrix} 0 & \bar{\alpha} \\ \alpha & 0 \end{pmatrix}.$$

Then the characteristic equation of A is

$$\lambda^2 - \alpha\bar{\alpha} = \lambda^2 - |\alpha|^2 = 0,$$

which implies that

$$\lambda = |\alpha| = \rho(A).$$

Also,

$$A^H A = \begin{pmatrix} 0 & \bar{\alpha} \\ \alpha & 0 \end{pmatrix} \begin{pmatrix} 0 & \bar{\alpha} \\ \alpha & 0 \end{pmatrix} = \begin{pmatrix} \alpha\bar{\alpha} & 0 \\ 0 & \alpha\bar{\alpha} \end{pmatrix}$$

and the characteristic equation of $A^H A$ is

$$(\alpha\bar{\alpha} - \lambda)^2 = 0,$$

which implies that

$$\|A\|_2 = [\rho(A^H A)]^{1/2} = |\alpha| = \rho(A).$$

2. For an arbitrary nonsingular matrix A

$$\frac{1}{\rho(A^{-1})} \geq \min_i \left(|a_{ii}| - \sum_{j=1}^{n} |a_{ij}| \right), \quad j \neq i. \tag{7.35}$$

Example 7.27 *Consider the following matrix*

$$A = \begin{pmatrix} 0 & \sqrt{3} \\ \sqrt{3} & 2 \end{pmatrix}.$$

To satisfy relation (7.35), firstly, we compute the inverse of given matrix A, which is

$$A^{-1} = \begin{pmatrix} \frac{2}{3} & \frac{\sqrt{3}}{3} \\ \frac{\sqrt{3}}{3} & 0 \end{pmatrix}.$$

Now to find the eigenvalues of the above inverse matrix, we solve the characteristic equation of the form as follows

$$\det(A^{-1} - \lambda I) = \begin{vmatrix} \frac{2}{3} - \lambda & \frac{\sqrt{3}}{3} \\ \frac{\sqrt{3}}{3} & -\lambda \end{vmatrix} = 3\lambda^2 - 2\lambda - 1 = 0,$$

which gives eigenvalues $-1, \frac{1}{3}$ of A^{-1}. Hence, the spectral radius of matrix A^{-1} is

$$\rho(A^{-1}) = \max\{|-1|, |\tfrac{1}{3}|\} = 1.$$

Thus,

$$1 > \min\{-1.7321, -0.2680\},$$

which satisfies relation (7.35).

3. Let A be a symmetric matrix with $\|.\| = \|.\|_2$, then

$$\text{cond}A = \frac{\text{largest } |\lambda_i|}{smallest\ |\lambda_i|}.$$

It is well-known theorem called the *spectral radius theorem*, which shows that, for a symmetric matrix A, ill-conditioning corresponds to A having eigenvalues of both large and small magnitude. It is most commonly used to define the condition number of a matrix. A matrix is ill conditioned if its condition number is large, as we discussed in Chapter 3. Strictly speaking a matrix has many condition numbers, and the word "large" is not itself well-defined in this context. To have an idea of what "large" means is to deal with the Hilbert matrix. For example, in dealing with a 3×3 Hilbert matrix, we have:

```
>> A = hilb(3);
A =
     1.0000  0.5000  0.3333
     0.5000  0.3333  0.2500
     0.3333  0.2500  0.2000
```

and one can find the condition number of the above Hilbert matrix as follows:

```
>> cond(A)
ans =
     524.0568
```

By adapting the above result, we can easily confirm that the condition numbers of Hilbert matrices increase rapidly as the size of the matrices increases. Large Hilbert matrices are, therefore, considered extremely ill-conditioned.

Example 7.28 *Find the conditioning of the following matrix:*

$$A = \begin{pmatrix} 0 & \sqrt{3} \\ \sqrt{3} & 2 \end{pmatrix}.$$

Solution. *Since*

$$\det(A - \lambda I) = \begin{vmatrix} -\lambda & \sqrt{3} \\ \sqrt{3} & 2-\lambda \end{vmatrix} = 0,$$

it gives a characteristic equation

$$\lambda^2 - 2\lambda - 3 = 0.$$

Solving the above equation gives the solution 3 *and* −1, *which are the eigenvalues of matrix A. Thus, the largest eigenvalue of A is* 3, *and the smallest one is* 1. *Hence, the condition number of matrix A is*

$$cond\ A = \frac{3}{1} = 3.$$

Since 3 is of the order of magnitude of 1, A is well conditioned.

4. Let A be a nonsymmetric matrix A, with $\|.\| = \|.\|_2$, then

$$cond\ A = \left(\frac{\text{largest eigenvalue } |\lambda_i| \text{ of } A^T A}{\text{smallest eigenvalue} |\lambda_i| \text{ of } A^T A} \right)^{1/2}.$$

Example 7.29 *Find the conditioning of the following matrix:*

$$A = \begin{pmatrix} 4 & 5 \\ 3 & 2 \end{pmatrix}.$$

Solution. *Since*

$$det(A^T A - \lambda I) = \begin{vmatrix} 25-\lambda & 26 \\ 26 & 29-\lambda \end{vmatrix} = 0$$

solving the above equation gives

$$\lambda^2 - 54\lambda + 49 = 0.$$

Solutions 53.08 *and* 0.92 *of the above equation are called the eigenvalues of matrix $A^T A$. Thus, the conditioning of the given matrix can be obtained as*

$$condA = \left(\frac{53.08}{0.92}\right)^{1/2} \approx 7.6,$$

which shows that given matrix A is not ill-conditioned.

7.6 Numerical Methods for Eigenvalue Problems

Now we shall discuss some numerical methods for finding approximations of eigenvalues and the corresponding eigenvectors of the given matrix.

One class of techniques, called iterative methods, can be used to find some or all of the eigenvalues and eigenvectors of the given matrix. They start with an arbitrary approximation to one of the eigenvectors and successively improve this until the required accuracy is obtained. Among them is the power method of the inverse iteration, which is used to find all of the eigenvectors of a matrix from known approximations to its eigenvalues.

The other class of techniques, which can only be applied to symmetric matrices such as the Jacobi methods, the Given's methods, and the Householder's methods, which reduce a given symmetric matrix to a special form whose eigenvalues are readily computed. For general matrices (symmetric or nonsymmetric matrices), the QR method and the LR method are the most widely used techniques for solving eigenvalue problems. Most of these procedures make use of a series of similarity transformations.

7.7 Vector Iterative Methods for Eigenvalues

So far as we have discussed classical methods for evaluating eigenvalues and eigenvectors for different matrices. It is evident that these methods became impractical as the matrices involved became large. Consequently, iterative methods are used for that purpose, such as power methods. These methods are an easy means to compute eigenvalues and eigenvectors of a given matrix.

Power methods include three versions. First, there is the *regular power method* or *simple iteration* based on the power of a matrix. Second, there is the *inverse power method*, which is based on the inverse power of a matrix. Third, there is the *shifted inverse power method* in which the given matrix A is replaced by $(A - \mu\mathbf{I})$ for any given scalar μ. In the following section, we discuss all of these methods in some detail.

7.7.1 Power Method

The basic power method can be used to compute the eigenvalue of the largest modules and the corresponding eigenvector of a general matrix. The eigenvalue of largest magnitude is often called the *dominant eigenvalue*. The implication of the power method is that if we assume a vector $\mathbf{x}_k$, then a new vector $\mathbf{x}_{k+1}$ can be calculated. The new vector is normalized by factoring its largest coefficient. This coefficient is then taken as a first approximation to the largest eigenvalue and the resulting vector represents the first approximation to the corresponding eigenvector. This process is continued by substituting the new eigenvector and determining a second approximation until the desired accuracy is achieved.

Consider an $n \times n$ matrix A, then the eigenvalues and eigenvectors satisfy

$$A\mathbf{v}_i = \lambda_i \mathbf{v}_i, \tag{7.36}$$

where λ_i is the *ith* eigenvalue and $\mathbf{v}_i$ is the corresponding *ith* eigenvector of A. The power method can be used on both symmetric and nonsymmetric matrices. If A is a symmetric matrix, then all the eigenvalues are real. If A is a nonsymmetric, then there is a possibility that there is not a single real dominant eigenvalue but a complex conjugate pair. Under these conditions the power method does not converge. We assume that the largest eigenvalue is real and not repeated and that eigenvalues are numbered in increasing order; that is,

$$|\lambda_1| > |\lambda_2| \geq |\lambda_3| \cdots \geq |\lambda_{n-1}| \geq |\lambda_n|. \tag{7.37}$$

The power method starts with an initial guess for the eigenvector $\mathbf{x}_0$, which can be any nonzero vector. The power method is defined by the iteration

$$\mathbf{x}_{k+1} = A\mathbf{x}_k, \qquad \text{for} \quad k = 0, 1, 2, \ldots \tag{7.38}$$

and it gives

$$\begin{aligned} \mathbf{x}_1 &= A\mathbf{x}_0 \\ \mathbf{x}_2 &= A\mathbf{x}_1 = A^2\mathbf{x}_0 \\ \mathbf{x}_3 &= A\mathbf{x}_2 = A^3\mathbf{x}_0. \\ &\vdots \end{aligned}$$

Thus,

$$\mathbf{x}_k = A^k \mathbf{x}_0, \qquad \text{for} \quad k = 1, 2, \ldots.$$

The vector $\mathbf{x}_0$ is an unknown linear combination of all eigenvectors of the system provided they are linearly independent. Thus,

$$\mathbf{x}_0 = \alpha_1 \mathbf{v}_1 + \alpha_2 \mathbf{v}_2 + \cdots + \alpha_n \mathbf{v}_n.$$

Let

$$\begin{aligned} \mathbf{x}_1 = A\mathbf{x}_0 &= A(\alpha_1 \mathbf{v}_1 + \alpha_2 \mathbf{v}_2 + \cdots + \alpha_n \mathbf{v}_n) \\ &= \alpha_1 A\mathbf{v}_1 + \alpha_2 A\mathbf{v}_2 + \cdots + \alpha_n A\mathbf{v}_n \\ &= \alpha_1 \lambda_1 \mathbf{v}_1 + \alpha_2 \lambda_2 \mathbf{v}_2 + \cdots + \alpha_n \lambda_n \mathbf{v}_n \end{aligned}$$

since, from the definition of an eigenvector, $A\mathbf{v}_i = \lambda_i \mathbf{v}_i$. Similarly,

$$\begin{aligned}\mathbf{x}_2 = A\mathbf{x}_1 &= A(\alpha_1\lambda_1\mathbf{v}_1 + \alpha_2\lambda_2\mathbf{v}_2 + \cdots + \alpha_n\lambda_n\mathbf{v}_n)\\ &= \alpha_1\lambda_1 A\mathbf{v}_1 + \alpha_2\lambda_2 A\mathbf{v}_2 + \cdots + \alpha_n\lambda_n A\mathbf{v}_n\\ &= \alpha_1\lambda_1^2\mathbf{v}_1 + \alpha_2\lambda_2^2\mathbf{v}_2 + \cdots + \alpha_n\lambda_n^2\mathbf{v}_n.\end{aligned}$$

Continuing in this way gives

$$\mathbf{x}_k = \alpha_1\lambda_1^k\mathbf{v}_1 + \alpha_2\lambda_2^k\mathbf{v}_2 + \cdots + \alpha_n\lambda_n^k\mathbf{v}_n,$$

which may be written as

$$\mathbf{x}_k = A^k\mathbf{x}_0 = \lambda_1^k\left[\alpha_1\mathbf{v}_1 + \alpha_2\left(\frac{\lambda_2}{\lambda_1}\right)^k\mathbf{v}_2 + \alpha_3\left(\frac{\lambda_3}{\lambda_1}\right)^k\mathbf{v}_3 + \cdots + \alpha_n\left(\frac{\lambda_n}{\lambda_1}\right)^k\mathbf{v}_n\right]. \tag{7.39}$$

All of the terms except the first in relation (7.39) converge to the zero vector as $k \to \infty$, since $|\lambda_1| > |\lambda_i|$ for $i \neq 1$. Hence,

$$\mathbf{x}_k \approx (\lambda_1^k\alpha_1)\mathbf{v}_1, \quad \text{for large } k, \quad \text{provided that} \quad \alpha_1 \neq 0.$$

Since $\lambda_1^k\alpha_1\mathbf{v}_1$ is a scalar multiple of $\mathbf{v}_1$, $\mathbf{x}_k = A^k\mathbf{x}_0$ will approach an eigenvector for the dominant eigenvalue λ_1; that is,

$$A\mathbf{x}_k \approx \lambda_1\mathbf{x}_k$$

if $\mathbf{x}_k$ is scaled so that its dominant component is 1, then dominant component of $A\mathbf{x}_k) \approx \lambda_1\times$ dominant component of $\mathbf{x}_k) = \lambda_1 \times 1 = \lambda_1$.

The rate of convergence of the power method is primarily dependent on distribution of eigenvalues; the smaller the ratio $\left|\frac{\lambda_i}{\lambda_1}\right|$, (for $i = 2, 3, \ldots, n$), the faster the convergence; in particular, this rate depends on the ratio $\left|\frac{\lambda_2}{\lambda_1}\right|$. The number of iterations required to get a desired degree of convergence depends on both the rate of convergence and how large λ_1 is compared to the other λ_i, the latter depending, in turn, on the choice of initial approximation $\mathbf{x}_0$.

Example 7.30 *Find the first five iterations obtained by the power method applied to the following matrix using the initial approximation* $\mathbf{x}_0 = [1, 1, 1]^T$*:*

$$A = \begin{pmatrix} 1 & 1 & 2 \\ 1 & 2 & 1 \\ 1 & 1 & 0 \end{pmatrix}.$$

Solution. *Starting with an initial vector* $\mathbf{x}_0 = [1,1,1]^T$, *we have*

$$A\mathbf{x}_0 = \begin{pmatrix} 1 & 1 & 2 \\ 1 & 2 & 1 \\ 1 & 1 & 0 \end{pmatrix} \begin{pmatrix} 1 \\ 1 \\ 1 \end{pmatrix} = \begin{pmatrix} 4.0000 \\ 4.0000 \\ 2.0000 \end{pmatrix},$$

which gives

$$A\mathbf{x}_0 = \lambda_1 \mathbf{x}_1 = 4.0000 \begin{pmatrix} 1.0000 \\ 1.0000 \\ 0.5000 \end{pmatrix}.$$

Similarly, the other possible iterations are as follows:

$$\begin{aligned} A\mathbf{x}_1 &= \begin{pmatrix} 1 & 1 & 2 \\ 1 & 2 & 1 \\ 1 & 1 & 0 \end{pmatrix} \begin{pmatrix} 1.0000 \\ 1.0000 \\ 0.5000 \end{pmatrix} = \begin{pmatrix} 3.0000 \\ 3.5000 \\ 2.0000 \end{pmatrix} \\ &= 3.5000 \begin{pmatrix} 0.8571 \\ 1.0000 \\ 0.5714 \end{pmatrix} = \lambda_2 \mathbf{x}_2 \end{aligned}$$

$$\begin{aligned} A\mathbf{x}_2 &= \begin{pmatrix} 1 & 1 & 2 \\ 1 & 2 & 1 \\ 1 & 1 & 0 \end{pmatrix} \begin{pmatrix} 0.8571 \\ 1.0000 \\ 0.5714 \end{pmatrix} = \begin{pmatrix} 3.0000 \\ 3.4286 \\ 1.8571 \end{pmatrix} \\ &= 3.4286 \begin{pmatrix} 0.8750 \\ 1.0000 \\ 0.5417 \end{pmatrix} = \lambda_3 \mathbf{x}_3 \end{aligned}$$

$$\begin{aligned} A\mathbf{x}_3 &= \begin{pmatrix} 1 & 1 & 2 \\ 1 & 2 & 1 \\ 1 & 1 & 0 \end{pmatrix} \begin{pmatrix} 0.8750 \\ 1.0000 \\ 0.5417 \end{pmatrix} = \begin{pmatrix} 2.9583 \\ 3.4167 \\ 1.8750 \end{pmatrix} \\ &= 3.4167 \begin{pmatrix} 0.8659 \\ 1.0000 \\ 0.5488 \end{pmatrix} = \lambda_4 \mathbf{x}_4 \end{aligned}$$

$$A\mathbf{x}_4 = \begin{pmatrix} 1 & 1 & 2 \\ 1 & 2 & 1 \\ 1 & 1 & 0 \end{pmatrix} \begin{pmatrix} 0.8659 \\ 1.0000 \\ 0.5488 \end{pmatrix} = \begin{pmatrix} 2.9634 \\ 3.4146 \\ 1.8659 \end{pmatrix}$$

$$= 3.4146 \begin{pmatrix} 0.8679 \\ 1.0000 \\ 0.5464 \end{pmatrix} = \lambda_5 \mathbf{x}_5$$

since the eigenvalues of given matrix A are $3.4142, 0.5858$ *and* -1.0000. *Hence, the approximation of the dominant eigenvalue after the five iterations is* $\lambda_5 = 3.4146$ *and the corresponding eigenvector is* $[0.8679, 1.0000, 0.5464]^T$.

To get the above results using the MATLAB command window, we do the following:

```
>> A = [1 1 2; 1 2 1; 1 1 0];
>> X = [1 1 1]';
>> sol = POWERM(A, X);
```

Program 7.5

MATLAB m-file for the Power method

```
function sol=POWERM(A,X)
[n,n]=size(A); X = A * X;
r(1)=abs(X(1,1)); m=r(1);
for i=1:n
if abs(r(1)) < abs(X(i,1))
r(1)=X(i,1); end end
r(1)=X(i,1); X = X/r(1); dd=1; k=1;
while dd > 1e - 2  X = A * X; r(k + 1) = (X(1, 1));
for i=1:n
if abs(r(k + 1)) < abs(X(i, 1))
r(k+1)=X(i,1); end end
```

The power method has the disadvantage that it is unknown at the outset whether or not a matrix has a single dominant eigenvalue. Nor it is known how an initial vector $\mathbf{x}_0$ should be chosen so as to ensure that its representation in terms of the eigenvectors of a matrix will contain a nonzero contribution from the eigenvector associated with the dominant eigenvalue, should it exist.

Note that the dominant eigenvalue λ of a matrix can also be obtained from two successive iterations, by dividing the corresponding elements of vectors $\mathbf{x}_n$ and $\mathbf{x}_{n-1}$.

Example 7.31 *Find the dominant eigenvalue of the following matrix:*

$$A = \begin{pmatrix} 4 & 1 & 3 \\ 9 & 0 & 9 \\ 5 & -1 & 6 \end{pmatrix}.$$

Solution. *Let us consider an arbitrary vector* $\mathbf{x}_0 = [1, 0, 0]^T$. *Then*

$$\mathbf{x}_1 = A\mathbf{x}_0 = \begin{pmatrix} 4 & 1 & 3 \\ 9 & 0 & 9 \\ 5 & -1 & 6 \end{pmatrix} \begin{pmatrix} 1 \\ 0 \\ 0 \end{pmatrix} = \begin{pmatrix} 4 \\ 9 \\ 5 \end{pmatrix}$$

$$\mathbf{x}_2 = A\mathbf{x}_1 = \begin{pmatrix} 4 & 1 & 3 \\ 9 & 0 & 9 \\ 5 & -1 & 6 \end{pmatrix} \begin{pmatrix} 4 \\ 9 \\ 5 \end{pmatrix} = \begin{pmatrix} 40 \\ 81 \\ 41 \end{pmatrix}$$

$$\mathbf{x}_3 = A\mathbf{x}_2 = \begin{pmatrix} 4 & 1 & 3 \\ 9 & 0 & 9 \\ 5 & -1 & 6 \end{pmatrix} \begin{pmatrix} 40 \\ 81 \\ 41 \end{pmatrix} = \begin{pmatrix} 364 \\ 729 \\ 365 \end{pmatrix}.$$

Then the dominant eigenvalue can be obtained as

$$\lambda_1 \approx \frac{\mathbf{x}_3}{\mathbf{x}_2} \approx \frac{364}{40} \approx \frac{729}{81} \approx \frac{365}{41} \approx 9.$$

7.7.2 Inverse Power Method

The power method can be modified by replacing the given matrix A by its inverse matrix A^{-1} and is called the *inverse power method.* Since we know that the eigenvalues of A^{-1} are reciprocals of A, the power method applied to A^{-1} will find the smallest eigenvalue of A. Thus, the smallest (or least) value of eigenvalues for A will become the maximum value for A^{-1}. Of course, we must assume that the smallest eigenvalue of A is real and not repeated, otherwise, the method does not work.

In this method the solution procedure is a little more involved than that discussed for finding the largest eigenvalue of the given matrix. Fortunately, it is just as straight-forward. Consider

$$A\mathbf{x} = \lambda\mathbf{x} \tag{7.40}$$

multiplying by A^{-1}, we have

$$A^{-1}A\mathbf{x} = \lambda A^{-1}\mathbf{x}$$

or

$$A^{-1}\mathbf{x} = \frac{1}{\lambda}\mathbf{x}. \tag{7.41}$$

The solution procedure is initiated by starting with an initial guess for the vector $\mathbf{x}_i$ and improving the solution by getting new vector $\mathbf{x}_{i+1}$, and so on until the vector $\mathbf{x}_i$ is approximately equal to $\mathbf{x}_{i+1}$.

Example 7.32 *Use the inverse power method to find the first seven approximations of the least dominant eigenvalue and the corresponding eigenvector of the following matrix using an initial approximation* $\mathbf{x}_0 = [0, 1, 2]^T$.

$$A = \begin{pmatrix} 3 & 0 & 1 \\ 0 & -3 & 0 \\ 1 & 0 & 3 \end{pmatrix}$$

Solution. *The inverse of given matrix A is*

$$A^{-1} = \begin{pmatrix} \frac{3}{8} & 0 & -\frac{1}{8} \\ 0 & -\frac{1}{3} & 0 \\ -\frac{1}{8} & 0 & \frac{3}{8} \end{pmatrix}.$$

Start with the given initial vector $\mathbf{x}_0 = [0, 1, 2]^T$, *we have*

$$A^{-1}\mathbf{x}_0 = \begin{pmatrix} \frac{3}{8} & 0 & -\frac{1}{8} \\ 0 & -\frac{1}{3} & 0 \\ -\frac{1}{8} & 0 & \frac{3}{8} \end{pmatrix} \begin{pmatrix} 0 \\ 1 \\ 2 \end{pmatrix} = \begin{pmatrix} -0.2500 \\ -0.3333 \\ 0.7500 \end{pmatrix}$$

$$= 0.75 \begin{pmatrix} -0.3333 \\ -0.4444 \\ 1.0000 \end{pmatrix} = \lambda_1 \mathbf{x}_1.$$

Similarly, the other possible iterations are as follows:

$$A^{-1}\mathbf{x}_1 = \begin{pmatrix} \frac{3}{8} & 0 & -\frac{1}{8} \\ 0 & -\frac{1}{3} & 0 \\ -\frac{1}{8} & 0 & \frac{3}{8} \end{pmatrix} \begin{pmatrix} -0.3333 \\ -0.4444 \\ 1.0000 \end{pmatrix} = \begin{pmatrix} -0.2500 \\ 0.1481 \\ 0.4167 \end{pmatrix}$$

$$= 0.4167 \begin{pmatrix} -0.6000 \\ 0.3558 \\ 1.0000 \end{pmatrix} = \lambda_2 \mathbf{x}_2$$

$$A^{-1}\mathbf{x}_2 = \begin{pmatrix} \frac{3}{8} & 0 & -\frac{1}{8} \\ 0 & -\frac{1}{3} & 0 \\ -\frac{1}{8} & 0 & \frac{3}{8} \end{pmatrix} \begin{pmatrix} -0.6000 \\ 0.3558 \\ 1.0000 \end{pmatrix} = \begin{pmatrix} -0.3500 \\ -0.1185 \\ 0.4500 \end{pmatrix}$$

$$= 0.4500 \begin{pmatrix} -0.7778 \\ -0.2634 \\ 1.0000 \end{pmatrix} = \lambda_3 \mathbf{x}_3$$

$$A^{-1}\mathbf{x}_3 = \begin{pmatrix} \frac{3}{8} & 0 & -\frac{1}{8} \\ 0 & -\frac{1}{3} & 0 \\ -\frac{1}{8} & 0 & \frac{3}{8} \end{pmatrix} \begin{pmatrix} -0.7778 \\ -0.2634 \\ 1.0000 \end{pmatrix} = \begin{pmatrix} -0.4167 \\ 0.0878 \\ 0.4722 \end{pmatrix}$$

$$= 0.4722 \begin{pmatrix} -0.8824 \\ 0.1859 \\ 1.0000 \end{pmatrix} = \lambda_4 \mathbf{x}_4$$

$$A^{-1}\mathbf{x}_4 = \begin{pmatrix} \frac{3}{8} & 0 & -\frac{1}{8} \\ 0 & -\frac{1}{3} & 0 \\ -\frac{1}{8} & 0 & \frac{3}{8} \end{pmatrix} \begin{pmatrix} -0.8824 \\ 0.1859 \\ 1.0000 \end{pmatrix} = \begin{pmatrix} -0.4559 \\ -0.0620 \\ 0.4853 \end{pmatrix}$$

$$= 0.4853 \begin{pmatrix} -0.9394 \\ -0.1277 \\ 1.0000 \end{pmatrix} = \lambda_5 \mathbf{x}_5$$

$$A^{-1}\mathbf{x}_5 = \begin{pmatrix} \frac{3}{8} & 0 & -\frac{1}{8} \\ 0 & -\frac{1}{3} & 0 \\ -\frac{1}{8} & 0 & \frac{3}{8} \end{pmatrix} \begin{pmatrix} -0.9394 \\ -0.1277 \\ 1.0000 \end{pmatrix} = \begin{pmatrix} -0.4773 \\ -0.0426 \\ 0.4924 \end{pmatrix}$$

$$= 0.4924 \begin{pmatrix} -0.9692 \\ -0.0864 \\ 1.0000 \end{pmatrix} = \lambda_6 \mathbf{x}_6$$

$$A^{-1}\mathbf{x}_6 = \begin{pmatrix} \frac{3}{8} & 0 & -\frac{1}{8} \\ 0 & -\frac{1}{3} & 0 \\ -\frac{1}{8} & 0 & \frac{3}{8} \end{pmatrix} \begin{pmatrix} -0.9692 \\ 0.0864 \\ 1.0000 \end{pmatrix} = \begin{pmatrix} -0.4885 \\ -0.0288 \\ 0.4962 \end{pmatrix}$$

$$= 0.4962 \begin{pmatrix} -0.9845 \\ -0.0581 \\ 1.0000 \end{pmatrix} = \lambda_7 \mathbf{x}_7$$

since the eigenvalues of given matrix A are $-3.0000, 2.0000$, and 4.0000. Hence, the dominant eigenvalue of A^{-1} after the seven iterations is $\lambda_7 = 0.4962$ and is converging to $\frac{1}{2}$ and so the smallest dominant eigenvalue of given matrix A is the reciprocal of dominant eigenvalue $\frac{1}{2}$ of matrix A^{-1}; that is, 2 and the corresponding eigenvector is $[-0.9845, -0.0581, 1.0000]^T$.

To get the above results using the MATLAB command window, we do:

```
>> A = [3 0 1; 0 -3 0; 1 0 3];
>> x = [0 1 2]';
>> Tol = 1e-2;
>> sol = INVERSEPM(A, x, Tol);
```

Program 7.6

MATLAB m-file for Using the Inverse Power method

```
function sol=INVERSEPM(A,x,Tol)
[n,n]=size(A); for i=1:n  for j=1:n
if i == j  id(i,j)=1;
else  id(i,j)=0; end end end
for i=1:n  for j=1:n
ar(i,j)=A(i,j); ar(i,j+n)=id(i,j); end end
for p=1:n  for i=1:n
if i ~= p  m(i,p)=ar(i,p)/ar(p,p);
for j = p : 2 * n
ar(i,j) = ar(i,j) - m(i,p) * ar(p,j); end end end end
for i=1:n  for j=1 : 2 * n
ad(i,j) = ar(i,j)/ar(i,i); end end
for i=1:n  for j=1:n
b(i,j)=ad(i,j+n); end end
t=1;m=1; k=0;  while t > Tol  k=k+1;
for i=1:n  sum=0;  for j=1:n
sum=sum+b(i,j) * x(j,k);  end  x(i,k+1)=sum;  end
m(k+1)=x(1,k+1);
for i=2:n  if x(i,k+1)>m(k+1)
m(k+1)=x(i,k+1); end end
for i=1:n  x(i,k + 1) = x(i,k + 1)/m(k + 1); end
t=abs(m(k+1)-m(k)); end
m(k);lum=1/m(k);eigvec=x(:,k);iterat=k
```

7.7.3 Shifted Inverse Power Method

Another modification of the power method consists of replacing the given matrix A by $(A - \mu I)$ for any scalar μ; that is,

$$(A - \mu I)\mathbf{x} = (\lambda - \mu)\mathbf{x} \tag{7.42}$$

and it follows that eigenvalues of $(A - \mu I)$ are the same as those of A except that they have all been shifted by an amount μ. The eigenvectors remain unaffected by the shift.

The shifted inverse power method is to apply the power method to the system

$$(A - \mu I)^{-1}\mathbf{x} = \frac{1}{(\lambda - \mu)}\mathbf{x}. \tag{7.43}$$

Thus, the iteration of $(A-\mu I)^{-1}$ leads to the largest value of $\frac{1}{(\lambda-\mu)}$; that is, the smallest value of $(\lambda-\mu)$. The smallest value of $(\lambda-\mu)$ implies that the value of λ determined will be the value closest to μ. Thus, by a suitable choice of μ we have a procedure for finding subdominant eigensolutions. So $(A-\mu I)^{-1}$ has the same eigenvectors as A but with eigenvalues $\frac{1}{(\lambda-\mu)}$.

In practice, the inverse of $(A-\mu I)$ is never actually computed, especially if the given matrix A is a large sparse matrix. It is computationally more efficient if $(A-\mu I)$ is decomposed into the product of a lower-triangular matrix L and an upper-triangular matrix U. If $\mathbf{u}_s$ is an initial vector for the solution of (7.43) then

$$(A-\mu I)^{-1}\mathbf{u}_s=\mathbf{v}_s \tag{7.44}$$

and

$$\mathbf{u}_{s+1}=\mathbf{v}_s/max(\mathbf{v}_s). \tag{7.45}$$

By rearranging (7.44), we obtain

$$\begin{aligned}\mathbf{u}_s &= (A-\mu I)\mathbf{v}_s.\\ &= LU\mathbf{v}_s\end{aligned}$$

Let

$$U\mathbf{v}_s=\mathbf{z} \tag{7.46}$$

then

$$L\mathbf{z}=\mathbf{u}_s. \tag{7.47}$$

By using an initial value, we can find $\mathbf{z}$ from (7.47) by applying forward substitution and knowing $\mathbf{z}$ we can find $\mathbf{v}_s$ from (7.46) by applying backward substitution. The new estimate for the vector $\mathbf{u}_{s+1}$ can then be found from (7.45). The iteration is terminated when $\mathbf{u}_{s+1}$ is sufficiently close to $\mathbf{u}_s$ and it can be shown when convergence is completed.

Let λ_μ be an eigenvalue of A nearest μ, then

$$\lambda_\mu=\frac{1}{\text{dominant eigenvalue of }(A-\mu I)^{-1}}+\mu. \tag{7.48}$$

The shifted inverse power method uses the power method as a basis but gives faster convergence. Convergence is to the eigenvalue λ, which is close to μ, and if this eigenvalue is extremely close to μ, the rate of convergence will be very rapid. Inverse iteration, therefore, provides a means of determining an eigenvector of a matrix for which the corresponding eigenvalue has already been determined to moderate accuracy by alternative methods, such as the QR method or the Strum sequence iteration, which we will discuss later in the chapter.

When inverse iteration is used to determine eigenvectors corresponding to known eigenvalues, the matrix to be inverted, even if symmetric, will not normally be positive definite, and if nonsymmetric, will not normally be diagonally dominant. The computation of an eigenvector corresponding to a complex conjugate eigenvalue by inverse iteration is more difficult than for a real eigenvalue.

Example 7.33 *Use the shifted inverse power method to find the first five approximations of the eigenvalue nearest $\mu = 6$ of the following matrix using initial approximation $\mathbf{x}_0 = [1,1]^T$:*

$$A = \begin{pmatrix} 4 & 2 \\ 3 & 5 \end{pmatrix}.$$

Solution. *Consider*

$$B = (A - 6I) = \begin{pmatrix} -2 & 2 \\ 3 & -1 \end{pmatrix}.$$

The inverse of B is

$$B^{-1} = (A - 3I)^{-1} = \begin{pmatrix} \frac{1}{4} & \frac{1}{2} \\ \frac{3}{4} & \frac{1}{2} \end{pmatrix}.$$

Now applying the power method, we obtain the following iterations:

$$\begin{aligned} B^{-1}\mathbf{x}_0 &= \begin{pmatrix} \frac{1}{4} & \frac{1}{2} \\ \frac{3}{4} & \frac{1}{2} \end{pmatrix} \begin{pmatrix} 1 \\ 1 \end{pmatrix} = \begin{pmatrix} 0.7500 \\ 1.2500 \end{pmatrix} \\ &= 1.2500 \begin{pmatrix} 0.6000 \\ 1.0000 \end{pmatrix} = \lambda_1 \mathbf{x}_1. \end{aligned}$$

Similarly, other approximations can be computed as

$$B^{-1}\mathbf{x}_1 = \begin{pmatrix} \frac{1}{4} & \frac{1}{2} \\ \frac{3}{4} & \frac{1}{2} \end{pmatrix} \begin{pmatrix} 0.6000 \\ 1.0000 \end{pmatrix} = \begin{pmatrix} 0.6500 \\ 0.9500 \end{pmatrix}$$

$$= 0.9500 \begin{pmatrix} 0.6842 \\ 1.0000 \end{pmatrix} = \lambda_2 \mathbf{x}_2$$

$$B^{-1}\mathbf{x}_2 = \begin{pmatrix} \frac{1}{4} & \frac{1}{2} \\ \frac{3}{4} & \frac{1}{2} \end{pmatrix} \begin{pmatrix} 0.6842 \\ 1.0000 \end{pmatrix} = \begin{pmatrix} 0.6711 \\ 1.0132 \end{pmatrix}$$

$$= 1.0132 \begin{pmatrix} 0.6623 \\ 1.0000 \end{pmatrix} = \lambda_3 \mathbf{x}_3$$

$$B^{-1}\mathbf{x}_3 = \begin{pmatrix} \frac{1}{4} & \frac{1}{2} \\ \frac{3}{4} & \frac{1}{2} \end{pmatrix} \begin{pmatrix} 0.6623 \\ 1.0000 \end{pmatrix} = \begin{pmatrix} 0.6656 \\ 0.9968 \end{pmatrix}$$

$$= 0.9968 \begin{pmatrix} 0.6678 \\ 1.0000 \end{pmatrix} = \lambda_4 \mathbf{x}_4$$

$$B^{-1}\mathbf{x}_4 = \begin{pmatrix} \frac{1}{4} & \frac{1}{2} \\ \frac{3}{4} & \frac{1}{2} \end{pmatrix} \begin{pmatrix} 0.6678 \\ 1.0000 \end{pmatrix} = \begin{pmatrix} 0.6669 \\ 1.0008 \end{pmatrix}$$

$$= 1.0008 \begin{pmatrix} 0.6664 \\ 1.0000 \end{pmatrix} = \lambda_5 \mathbf{x}_5.$$

Thus, the fifth approximation of the dominant eigenvalue of the matrix $B^{-1} = (A - 3I)^{-1}$ is $\lambda_5 = 1.0008$ and it is converging to 1 with the eigenvector $[1.0000, 0.7000]^T$. Hence, the eigenvalue λ_μ of A nearest to $\mu = 6$ is

$$\lambda_\mu = \frac{1}{1} + 6 = 7.$$

To get the above results using the MATLAB command window, we do the following:

```
>> A = [4 2; 3 5];
>> mu = 6;
>> x = [1 1]';
>> Tol = 0.005;
>> sol = ShiptedIPM(A, x, mu, Tol);
```

Program 7.7
MATLAB m-file for using the Shifted Inverse Power Method

```
function [k,eigenvalue,eigenvector]=ShiftedIPM
(A, x, mu, Tol)
[n,n]=size(A); ID=zeros(n,n);
for i=1:n  for j=1:n
if i == j  ID(i,j)=1; end end end
B = A - mu * ID; R = 1; k = 1; C = inv(B);
x = C * x; m(1) = (x(1, 1));
for i=1:n  if abs(m(1))< abs(x(i,1))
m(1)=x(i,1); end end mm=m(1); x = x/m(1);
while R > Tol  x = C * x; m(k+1)=(x(1,1));
for i=1:n  if abs(m(k+1))< abs(x(i,1))
m(k+1)=x(i,1); end end; mm=m(k+1)
x = x/m(k + 1); R = abs(m(k + 1) - m(k)); k=k+1;end
number of Iterations = k; eigenvalue=1/mm+mu; eigen-
vector=x
```

7.8 Location of Eigenvalues

Here, we discuss two well-known theorems, which are some of the more important among the many theorems, which deal with the location of eigenvalues of both symmetric and nonsymmetric matrices; that is, the location of zeros of the characteristic polynomial. The eigenvalues of a nonsymmetric matrix could, of course, be complex, in which case the theorems give us a means of locating these numbers in the complex plane. The theorems can be used also to estimate the magnitude of the largest and smallest eigenvalues in magnitude, and thus to estimate the *spectral radius* $\rho(A)$ of A and the *condition number* of A. Such estimates can be used to generate initial approximations to be used in iterative methods for determining eigenvalues.

7.8.1 Gerschgorin Circles Theorem

Let A be an $n \times n$ matrix and R_i denote the circles in the complex plane C with centre a_{ii} and radius $\sum_{\substack{j=1 \\ j \neq i}}^{n} |a_{i,j}|$; that is,

$$R_i = \{z \in C : \quad |z - a_{ii}| \leq \sum_{\substack{j=1 \\ j \neq i}}^{n} |a_{ij}|\}, \quad i = 1, 2, \cdots, n, \tag{7.49}$$

where the variable z is complex valued.

The eigenvalues of A are contained within $R = \cup_{i=1}^{n} R_i$ and the union of any k of these circles that do not intersect the remaining $(n-k)$ must contain precisely k (counting multiplication) of the eigenvalues.

Example 7.34 *Consider the following matrix:*

$$A = \begin{pmatrix} 10 & 1 & 1 & 2 \\ 1 & 5 & 1 & 0 \\ 1 & 1 & -5 & 0 \\ 2 & 0 & 0 & -10 \end{pmatrix},$$

which is symmetric and has only real eigenvalues. The Gerschgorin circles associated with A are given by

$$\begin{aligned} R_1 &= \{z \in C|\ |z - 10| \leq 4\} \\ R_2 &= \{z \in C|\ |z - 5|\ \leq 2\} \\ R_3 &= \{z \in C|\ |z + 5|\ \leq 2\} \\ R_4 &= \{z \in C|\ |z + 10| \leq 2\}. \end{aligned}$$

These circles are illustrated in Figure 7.2, and the Gerschgorin's theorem indicates that the eigenvalues of A lie inside the circles. The circles about -10 and -5 each must contain an eigenvalue. The other eigenvalues must lie on the interval $[3, 14]$. *Using the shifted inverse power method with $\epsilon = 0.000005$ with initial approximations of* $10, 5, -5,$ *and* -10 *leads to approximations of*

$$\begin{aligned} \lambda_1 &= 10.4698, \qquad \lambda_2 = 4.8803 \\ \lambda_3 &= -5.1497, \qquad \lambda_4 = -10.2004, \end{aligned}$$

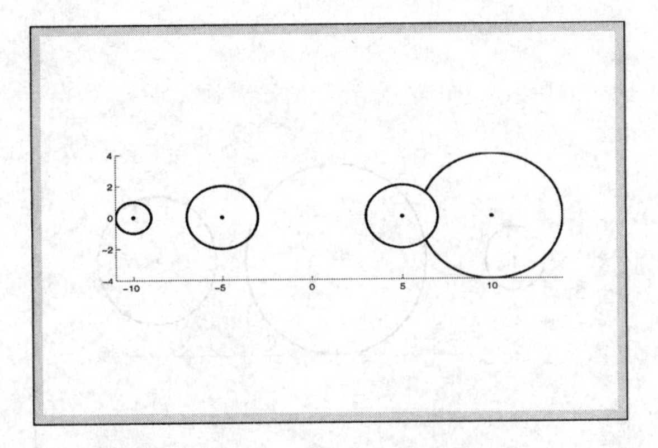

Figure 7.2: Circles for Example 7.34.

respectively. The number of iterations require range from 9 to 13.

Example 7.35 *Consider the following matrix:*

$$A = \begin{pmatrix} 1 & 2 & -1 \\ 2 & 7 & 0 \\ -1 & 0 & -5 \end{pmatrix},$$

which is symmetric and so has only real eigenvalues. The Gerschgorin circles are

$$\begin{array}{lrl} C_1 : & |z-1| & \leq 3 \\ C_2 : & |z-7| & \leq 2 \\ C_3 : & |z+5| & \leq 1. \end{array}$$

These circles are illustrated in Figure 7.3, and the Gerschgorin's theorem indicates that the eigenvalues of A lie inside the circles.

Then by the Gerschgorin theorem, any eigenvalues of A must lie on one of the three intervals $[-2,4],[5,9]$, and $[-6,-4]$, since the eigenvalues of A are 0, 5, and 9. Hence, $\lambda_1 = 0$ lies on circle C_1, $\lambda_2 = 5$ and $\lambda_3 = 9$ lie on circle C_2.

7.8.2 Rayleigh Quotient

The shifted inverse power method requires the input of an initial approximation μ for the eigenvalue λ of a matrix A. It can be obtained by the Rayleigh quotient as

$$\mu = \frac{\mathbf{x}^T A \mathbf{x}}{\mathbf{x}^T \mathbf{x}}. \tag{7.50}$$

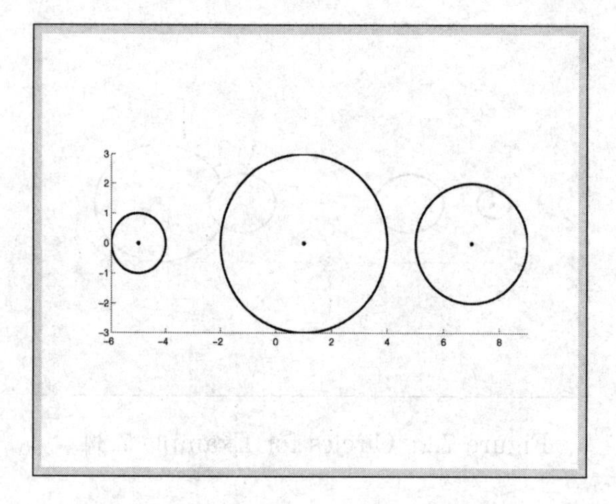

Figure 7.3: Circles for Example 7.35.

The maximum eigenvalue λ_1 can be obtained when $\mathbf{x}$ is the corresponding vector as

$$\lambda_1 = \max_{\mathbf{x} \neq 0} \frac{\mathbf{x}^T A \mathbf{x}}{\mathbf{x}^T \mathbf{x}}. \tag{7.51}$$

Theorem 7.9 (Rayleigh Quotient Theorem)

If the eigenvalues of a real symmetric matrix A are

$$\lambda_1 \geq \lambda_2 \geq \lambda_3 \cdots \geq \lambda_n \tag{7.52}$$

and if $\mathbf{x}$ is any nonzero vector, then

$$\lambda_n \leq \frac{\mathbf{x}^T A \mathbf{x}}{\mathbf{x}^T \mathbf{x}} \leq \lambda_1. \tag{7.53}$$

Example 7.36 *Consider the symmetric matrix*

$$A = \begin{pmatrix} 2 & -1 & 3 \\ -1 & 1 & -2 \\ 3 & -2 & 1 \end{pmatrix}$$

and the vector $\mathbf{x}$ as

$$\mathbf{x} = \begin{pmatrix} 1 \\ 1 \\ 1 \end{pmatrix}.$$

Then

$$\mathbf{x}^T\mathbf{x} = \begin{pmatrix} 1 & 1 & 1 \end{pmatrix} \begin{pmatrix} 1 \\ 1 \\ 1 \end{pmatrix} = 3$$

and

$$\begin{aligned} \mathbf{x}^T A\mathbf{x} &= \begin{pmatrix} 1 & 1 & 1 \end{pmatrix} \begin{pmatrix} 2 & -1 & 3 \\ -1 & 1 & -2 \\ 3 & -2 & 1 \end{pmatrix} \begin{pmatrix} 1 \\ 1 \\ 1 \end{pmatrix}. \\ &= \begin{pmatrix} 1 & 1 & 1 \end{pmatrix} \begin{pmatrix} 4 \\ -2 \\ 2 \end{pmatrix} = 4 \end{aligned}$$

Thus,

$$\lambda_3 \leq \frac{4}{3} \leq \lambda_1.$$

If μ is close to an eigenvalue λ_1, then convergence will be quite rapid.

7.9 Intermediate Eigenvalues

Once the largest eigenvalue is determined, then there is a method to obtain the approximations to the other possible eigenvalues of a matrix. This method is called *matrix deflation* and it is applicable to both symmetrical and nonsymmetrical coefficients matrices. The deflation method involves forming a new matrix B whose eigenvalues are the same as those of A, except that the dominant eigenvalue of A is replaced by the eigenvalue zero in B.

It is evident that this process can be continued until all eigenvalues have been extracted. Although this method shows promise, it does have a significant drawback. That is, at each iteration performed in deflating the original matrix, any errors in the computed eigenvalues and eigenvectors will be passed on to the next eigenvectors. This could result in serious inaccuracy especially when dealing with large eigenvalue problems. This is precisely why this method is generally used for small eigenvalue problems.

The following preliminary results are essential when using this technique.

Theorem 7.10 *If a matrix A has eigenvalues λ_i corresponding to eigenvectors $\mathbf{x}_i$, then $Q^{-1}AQ$ has the same eigenvalues as A but with eigenvectors $Q^{-1}\mathbf{x}_i$ for any nonsingular matrix Q.*

Theorem 7.11 *Let*

$$B = \begin{pmatrix} \lambda_1 & a_{12} & a_{13} & \cdots & a_{1n} \\ 0 & c_{22} & c_{23} & \cdots & c_{2n} \\ 0 & c_{32} & c_{33} & \cdots & c_{3n} \\ \vdots & \vdots & \vdots & \cdots & \vdots \\ 0 & c_{n2} & c_{n3} & \cdots & c_{nn} \end{pmatrix} \tag{7.54}$$

and let C be an $(n-1) \times (n-1)$ matrix obtained by deleting the first row and first column of matrix B. Matrix B has eigenvalues λ_1 together with the $(n-1)$ eigenvalues of C. Moreover, if $(\beta_2, \beta_3, \ldots, \beta_n)^T$ is an eigenvector of C with eigenvalue $\mu \neq \lambda_1$, then the corresponding eigenvector of B is $(\beta_1, \beta_2, \ldots, \beta_n)^T$ with

$$\beta_1 = \frac{\sum_{j=2}^{n} a_{1j}\beta_j}{\mu - \lambda_1}. \tag{7.55}$$

Note that eigenvectors $\mathbf{x}_i$ of A can be recovered by pre-multiplication by Q.

Example 7.37 *Consider the following matrix*

$$A = \begin{pmatrix} 10 & -6 & -4 \\ -6 & 11 & 2 \\ -4 & 2 & 6 \end{pmatrix},$$

which has dominant eigenvalue $\lambda_1 = 18$ with the corresponding eigenvector $\mathbf{x}_1 = [1, -1, -\frac{1}{2}]^T$. Use the deflation method to find the other eigenvalues and eigenvectors of A.

Solution. *The transformation matrix is given as*

$$Q = \begin{pmatrix} 1 & 0 & 0 \\ -1 & 1 & 0 \\ -\frac{1}{2} & 0 & 1 \end{pmatrix}.$$

Then

$$B = Q^{-1}AQ = \begin{pmatrix} 1 & 0 & 0 \\ 1 & 1 & 0 \\ \frac{1}{2} & 0 & 1 \end{pmatrix} \begin{pmatrix} 10 & -6 & -4 \\ -6 & 11 & 2 \\ -4 & 2 & 6 \end{pmatrix} \begin{pmatrix} 1 & 0 & 0 \\ -1 & 1 & 0 \\ -\frac{1}{2} & 0 & 1 \end{pmatrix}.$$

After simplifying, we get

$$B = \begin{pmatrix} 18 & -6 & -4 \\ 0 & 5 & -2 \\ 0 & -1 & 4 \end{pmatrix}.$$

So the deflated matrix is

$$C = \begin{pmatrix} 5 & -2 \\ -1 & 4 \end{pmatrix}.$$

Now we can easily find the eigenvalues of C, which are 6 and 3 with corresponding eigenvectors $[1, -\frac{1}{2}]^T$ and $[1, 1]^T$, respectively. Thus, the other two eigenvalues of A are 6 and 3. Now we calculate the eigenvectors of A corresponding to these two eigenvalues. First, we calculate the eigenvectors of B corresponding to $\lambda = 6$ from the following system

$$\begin{pmatrix} 18 & -6 & -4 \\ 0 & 5 & -2 \\ 0 & -1 & 4 \end{pmatrix} \begin{pmatrix} \beta_1 \\ 1 \\ -\dfrac{1}{2} \end{pmatrix} = 6 \begin{pmatrix} \beta_1 \\ 1 \\ -\dfrac{1}{2} \end{pmatrix}.$$

Then by solving the above system, we have

$$18\beta_1 - 4 = 6\beta_1,$$

which gives $\beta_1 = \frac{1}{3}$. Similarly, we can find value of β_1 corresponding to $\lambda = 3$ by using the following system

$$\begin{pmatrix} 18 & -6 & -4 \\ 0 & 5 & -2 \\ 0 & -1 & 4 \end{pmatrix} \begin{pmatrix} \beta_1 \\ 1 \\ 1 \end{pmatrix} = 3 \begin{pmatrix} \beta_1 \\ 1 \\ 1 \end{pmatrix},$$

which gives $\beta_1 = \dfrac{2}{3}$. Thus, the eigenvectors of B are $\mathbf{v}_1 = [\frac{1}{3}, 1, -\frac{1}{2}]^T$ and $\mathbf{v}_2 = [\frac{2}{3}, 1, 1]^T$.

Now we find the eigenvectors of the original matrix A, which can be obtained by pre-multiplying the vectors of B by nonsingular matrix Q. First, the second eigenvector of A can be found as

$$\mathbf{x}_2 = Q\mathbf{v}_1 = \begin{pmatrix} 1 & 0 & 0 \\ -1 & 1 & 0 \\ -\dfrac{1}{2} & 0 & 1 \end{pmatrix} \begin{pmatrix} \dfrac{1}{3} \\ 1 \\ -\dfrac{1}{2} \end{pmatrix} = \begin{pmatrix} \dfrac{1}{3} \\ \dfrac{2}{3} \\ -\dfrac{2}{3} \end{pmatrix}$$

or, equivalently, $\mathbf{x}_2 = [\frac{1}{2}, 1, -1]^T$. *Similarly, the third eigenvector of the given matrix* A *can be computed as*

$$\mathbf{x}_3 = Q\mathbf{v}_2 = \begin{pmatrix} 1 & 0 & 0 \\ -1 & 1 & 0 \\ -\frac{1}{2} & 0 & 1 \end{pmatrix} \begin{pmatrix} \frac{2}{3} \\ 1 \\ 1 \end{pmatrix} = \begin{pmatrix} \frac{2}{3} \\ \frac{1}{3} \\ \frac{2}{3} \end{pmatrix}$$

or, equivalently, $\mathbf{x}_3 = [1, \frac{1}{2}, 1]^T$.

Note that in this example the deflated matrix C is nonsymmetric even though the original matrix A is symmetric. We deduce that the property of symmetry is not preserved in the deflation process. Also, note that the method of deflation fails whenever the first element of given vector $\mathbf{x}_1$ is zero, since $\mathbf{x}_1$ cannot then be scaled so that this number is one.

The above results can be reproduced using the following MATLAB commands:

```
>> A = [10 - 6 - 4; -6 11 2; -4 2 6];
>> Lamda = 18;
>> XA = [1 - 1 - 0.5]';
>> [Lamda, X] = DEFLATION(A, Lamda, XA);
```

Program 7.8

MATLAB m-file for Using the Deflation Method

```
function [Lamda,X]=DEFLATION(A,Lamda,XA)
[n,n]=size(A); Q=eye(n);
Q(:,1) = XA(:,1); B = inv(Q) * A * Q; c=B(2:n,2:n);
[xv, ev] = eig(c,'nobalance');
for i=1:n-1
b = -(B(1,2 : n) * xv(:,i))/(Lamda - ev(i,i));
Xb(:,i) = [bxv(:,i)']'; XA(:,i + 1) = Q * Xb(:,i); end
Lamda=[Lamda;diag(ev)]; Lamda; XA; end
```

7.10 Eigenvalues of Symmetric Matrices

In the previous section we discussed power methods for finding individual eigenvalues. The regular power method can be used to find the distinct eigenvalue with the largest magnitude; that is, the dominant eigenvalue and the inverse power method can find the eigenvalue called the smallest eigenvalue and the shifted inverse power method can find the subdominant eigenvalues. In this section we develop some methods to find all eigenvalues of a given matrix. The basic approach is to find a sequence of similarity transformations that transform the original matrix into a simple form. Clearly, the best form for the transformed matrix would be a diagonal one but this is not always possible, since some transformed matrix would be a tridiagonal one. Furthermore, these techniques are generally limited to symmetrical matrices with real coefficients.

Before we discuss these methods, we define some special matrices, which are very useful in discussing these methods.

Definition 7.12 (Orthogonally Similar Matrix)

A matrix A is said to be orthogonally similar to a matrix B if there is an orthogonal matrix Q for which

$$A = QBQ^T. \tag{7.56}$$

If A is a symmetric and $B = Q^{-1}AQ$, then

$$B^T = (Q^TAQ)^T = Q^TAQ = B.$$

Thus, similarity transformations on symmetric matrices that use orthogonal matrices produce matrices which are again symmetric.

Definition 7.13 (Rotation Matrix)

A rotation matrix Q is an orthogonal matrix that differs from an identity matrix in at most four elements. These four elements at the vertices of the rectangle have been replaced by $\cos\theta$, $-\sin\theta$, $\sin\theta$, and $\cos\theta$ in the positions pp, pq, qp, and qq, respectively. For example, the following matrix:

$$B = \begin{pmatrix} 1 & 0 & 0 & 0 & 0 \\ 0 & \cos\theta & 0 & -\sin\theta & 0 \\ 0 & 0 & 1 & 0 & 0 \\ 0 & \sin\theta & 0 & \cos\theta & 0 \\ 0 & 0 & 0 & 0 & 1 \end{pmatrix} \tag{7.57}$$

is the rotation matrix, where $p = 2$ and $q = 4$. Note that a rotation matrix is also an orthogonal matrix; that is, $B^T B = I$.

7.10.1 Jacobi Method

This method can be used to find all the eigenvalues and eigenvectors of a symmetric matrix by performing a series of similarity transformations. The Jacobi method permits the transformation of a symmetric matrix into a diagonal one having the same eigenvalues as the original matrix. This can be done by eliminating each off-diagonal elements in a systematic way. The method requires an infinite number of iterations to produce the diagonal form. This is because the reduction of a given element to zero in a matrix will most likely introduce a nonzero element into a previous zero coefficient. Hence, the method can be viewed as an iterative procedure that can approach a diagonal form using a finite number of steps. The implication is that the off-diagonal coefficients will be close to zero rather than exactly equal to zero.

Consider the eigenvalue problem

$$A\mathbf{v} = \lambda\mathbf{v}, \tag{7.58}$$

where A is a symmetric matrix of order $n \times n$ and let the solution of (7.58) give the eigenvalues $\lambda_1, \ldots, \lambda_n$ and corresponding eigenvectors $\mathbf{v}_1, \ldots, \mathbf{v}_n$ of A, since the eigenvectors of a symmetric matrix are orthogonal; that is,

$$\mathbf{v}^T = \mathbf{v}^{-1}. \tag{7.59}$$

So by using (7.59), we can write (7.58) as follows

$$\mathbf{v}^T A\mathbf{v} = \lambda. \tag{7.60}$$

The basic procedure for the Jacobi method is as follows:
Assume that

$$\begin{aligned}
A_1 &= Q_1^T A Q_1 \\
A_2 &= Q_2^T A_1 Q_2 = Q_2^T Q_1^T A Q_1 Q_2 \\
A_3 &= Q_3^T A_2 Q_3 = Q_3^T Q_2^T Q_1^T A Q_1 Q_2 Q_3 \\
&\vdots \\
A_k &= Q_k^T \cdots Q_1^T A Q_1 \cdots Q_k.
\end{aligned}$$

We see that as $k \to \infty$, then

$$A_k \to \lambda \quad \text{and} \quad Q_1 Q_2 \cdots Q_k \to \mathbf{v}. \tag{7.61}$$

Matrix $Q_i (i = 1, 2, \ldots, k)$ is a rotation matrix, which is constructed in such a way that off-diagonal coefficients in matrix A_k is reduced to zero. In other words, in a rotation matrix

$$Q_k = \begin{pmatrix} 1 & & & & & \\ & \ddots & & & & \\ & & cos\theta & & -sin\theta & \\ & & & & & \\ & & sin\theta & & cos\theta & \\ & & & & \ddots & \\ & & & & & 1 \end{pmatrix}$$

the value of θ is selected in such a way that the a_{pq} coefficient in A_k is reduced to zero; that is

$$\tan 2\theta = \frac{2a_{pq}}{a_{pp} - a_{qq}}. \tag{7.62}$$

Theoretically there are an infinite number of θ values corresponding to the infinite matrices A_k. However, as θ approaches zero, a rotation matrices Q_k become identity matrix and no further transformations are required.

There are three strategies for annihilating off-diagonals. The first one is called the *serial method*, which selects the elements in row order; that is, in the positions $(1,2), \ldots, (1,n)$; $(2,3), \ldots, (2,n)$; $\ldots$; $(n-1,n)$ in turn, which is then repeated. The second method is called the *natural method*, which searches through all of the off-diagonals and annihilates the elements of the largest modules at each stage. Although this method converges faster than the serial method, it is not recommended for large values of n, since the actual search procedure itself can be extremely time consuming. The third method is known as the *threshold serial method* in which the off-diagonals are cycled in row order as in the serial method, omitting transformations on any element whose magnitude is below some threshold value. This value is usually decreased after each cycle. The advantage of this approach is that zeros are only created in positions where it is worthwhile to do so, without the need for a lengthy search. Here, we shall use only the natural method for annihilating the off-diagonal elements.

Theorem 7.12 *Consider a matrix A and a rotation matrix Q as*

$$A = \begin{pmatrix} a_{11} & a_{12} \\ a_{12} & a_{22} \end{pmatrix} \quad and \quad Q = \begin{pmatrix} p_{11} & p_{12} \\ p_{21} & p_{22} \end{pmatrix}.$$

Then there exists θ such that

1. $Q^T Q = \mathbf{I}$
2. $Q^T A Q = D$

where $\mathbf{I}$ is an identity matrix and D is a diagonal matrix, and its diagonal elements, λ_1 and λ_2, are the eigenvalues of A.

Proof. To convert the given matrix A into a diagonal matrix D, we have to make off-diagonal element a_{12} of A to zero; that is, $p = 1$ and $q = 2$. Consider $p_{11} = \cos\theta = p_{22}$ and $p_{12} = -p_{21} = \sin\theta$, then the matrix Q has a form

$$Q = \begin{pmatrix} \cos\theta & -\sin\theta \\ \sin\theta & \cos\theta \end{pmatrix}.$$

The corresponding matrix A_1 can be constructed as

$$A_1 = Q_1^T A Q_1$$

or

$$\begin{pmatrix} a_{11}^* & a_{12}^* \\ a_{12}^* & a_{22}^* \end{pmatrix} = \begin{pmatrix} \cos\theta & \sin\theta \\ -\sin\theta & \cos\theta \end{pmatrix} \begin{pmatrix} a_{11} & a_{12} \\ a_{21} & a_{22} \end{pmatrix} \begin{pmatrix} \cos\theta & -\sin\theta \\ \sin\theta & \cos\theta \end{pmatrix}.$$

Since our task is to reduce a_{12}^* to zero, carrying out the multiplication on the right-hand side and using matrix equality gives

$$a_{12}^* = 0 = -(\sin\theta\cos\theta)a_{11} + (\cos^2\theta)a_{12} - (\sin^2\theta)a_{12} + (\cos\theta\sin\theta)a_{22}.$$

Simplifying and rearranging gives

$$\frac{\sin\theta\cos\theta}{\cos^2\theta - \sin^2\theta} = \frac{a_{12}}{a_{11} - a_{22}}$$

$$\frac{\sin 2\theta}{2\cos 2\theta} = \frac{a_{12}}{a_{11} - a_{22}}$$

$$\frac{\sin 2\theta}{\cos 2\theta} = \frac{2a_{12}}{a_{11} - a_{22}}$$

or more simplified

$$\tan 2\theta = \frac{2a_{12}}{a_{11} - a_{22}}, \qquad a_{11} \neq a_{22}.$$

Note that if $a_{11} = a_{22}$, this implies that $\theta = \frac{\pi}{4}$. We found that for a 2×2 matrix, it required only one iteration to convert given matrix A to a diagonal matrix D.

Similarly, for a higher-order matrix, a diagonal matrix D can be obtained by a number of such multiplications; that is

$$Q_k^T Q_{k-1}^T \cdots Q_1^T A Q_1 \cdots Q_{k-1} Q_k = D.$$

The diagonal elements of D are all the eigenvalues λ of A and the corresponding eigenvectors $\mathbf{v}$ of A can be obtained as

$$Q_1 Q_2 \cdots Q_k = \mathbf{v}.$$

Example 7.38 *Use the Jacobi method to find the eigenvalues and the eigenvectors of the following matrix*

$$A = \begin{pmatrix} 3.0 & 0.01 & 0.02 \\ 0.01 & 2.0 & 0.1 \\ 0.02 & 0.1 & 1.0 \end{pmatrix}.$$

Solution. *The largest off-diagonal entry of given matrix A is $a_{23} = 0.1$, so we begin by reducing element a_{23} to zero. Since $p = 2$ and $q = 3$, the first orthogonal transformation matrix has the form*

$$Q_1 = \begin{pmatrix} 1 & 0 & 0 \\ 0 & c & -s \\ 0 & s & c \end{pmatrix}.$$

The values of $c = \cos\theta$ and $s = \sin\theta$ can be obtained as follows

$$\theta = \frac{1}{2}\arctan\left(\frac{2a_{23}}{a_{22} - a_{33}}\right) = \frac{1}{2}\arctan\left(\frac{2(0.1)}{2-1}\right) \approx 6.2833$$

$$\cos\theta \approx 0.9951 \quad \textit{and} \quad \sin\theta \approx 0.0985.$$

Then

$$Q_1 = \begin{pmatrix} 1 & 0 & 0 \\ 0 & 0.9951 & -0.0985 \\ 0 & 0.0985 & 0.9951 \end{pmatrix} \quad \textit{and} \quad Q_1^T = \begin{pmatrix} 1 & 0 & 0 \\ 0 & 0.9951 & 0.0985 \\ 0 & -0.0985 & 0.9951 \end{pmatrix}$$

and

$$A_1 = Q_1^T A Q_1 = \begin{pmatrix} 3.0 & 0.0119 & 0.0189 \\ 0.0119 & 2.0099 & 0 \\ 0.0189 & 0 & 0.9901 \end{pmatrix}.$$

Note that the rotation makes a_{32} *and* a_{23} *zero, increasing slightly* a_{21} *and* a_{12}*, and decreasing the second dominant off-diagonal entries* a_{13} *and* a_{31}*.*

Now the largest off-diagonal element of matrix A_1 *is* $a_{13} = 0.0189$*, so to make this position zero, we consider the second orthogonal matrix of the form*

$$Q_2 = \begin{pmatrix} c & 0 & -s \\ 0 & 1 & 0 \\ s & 0 & c \end{pmatrix}$$

and the values of c and s can be obtained as follows

$$\theta = \frac{1}{2}\arctan\left(\frac{2a_{13}}{a_{11} - a_{33}}\right) = \frac{1}{2}\arctan\left(\frac{2(0.0189)}{3 - 0.9901}\right) \approx 0.5984$$

$$\cos\theta \approx 0.9999 \quad \textit{and} \quad \sin\theta \approx 0.0094.$$

Then

$$Q_2 = \begin{pmatrix} 0.9999 & 0 & -0.0094 \\ 0 & 1 & 0 \\ 0.0094 & 0 & 0.9999 \end{pmatrix} \quad \textit{and} \quad Q_2^T = \begin{pmatrix} 0.9999 & 0 & 0.0094 \\ 0 & 1 & 0 \\ -0.0094 & 0 & 0.9999 \end{pmatrix}.$$

Hence,

$$A_2 = Q_2^T A_1 Q_2 = Q_2^T Q_1^T A Q_1 Q_2 = \begin{pmatrix} 3.0002 & 0.0119 & 0 \\ 0.0119 & 2.0099 & -0.0001 \\ 0 & -0.0001 & 0.9899 \end{pmatrix}.$$

Similarly, to make off-diagonal element $a_{12} = 0.0119$ *of matrix* A_2 *zero, we consider the third orthogonal matrix of the form*

$$Q_3 = \begin{pmatrix} c & -s & 0 \\ s & c & 0 \\ 0 & 0 & 1 \end{pmatrix}$$

and

$$\theta = \frac{1}{2}\arctan\left(\frac{2a_{12}}{a_{11} - a_{22}}\right) = \frac{1}{2}\arctan\left(\frac{2(0.0119)}{3.0002 - 2.0099}\right) \approx 0.7638$$

$$\cos\theta \approx 0.9999 \quad \textit{and} \quad \sin\theta \approx 0.0120.$$

Then

$$Q_3 = \begin{pmatrix} 0.9999 & -0.0120 & 0 \\ 0.0120 & 0.9999 & 0 \\ 0 & 0 & 1 \end{pmatrix} \quad and \quad Q_3^T = \begin{pmatrix} 0.9999 & 0.0120 & 0 \\ -0.0120 & 0.9999 & 0 \\ 0 & 0 & 1 \end{pmatrix}.$$

Program 7.9

MATLAB m-file for the Jacobi Method

```
function sol=JOBM(A)
[n,n]=size(A); QQ=[ ];
for u = 1 : .5 * n * (n - 1); for i=1:n; for j=1:n
if (j > i); aa(i,j)=A(i,j); else; aa(i,j)=0;
end; end; end
aa=abs(aa); mm=max(aa); m=max(mm);
[i,j]=find(aa==m); i=i(1); j=j(1);
t = .5 * atan(2 * A(i, j)/(A(i, i) - A(j, j))); c=cos(t); s=sin(t);
for ii=1:n; for jj=1:n; Q(ii,jj)=0.0;
if (ii==jj); Q(ii,jj)=1.0; end; end; end
Q(i,i)=c; Q(i,j)=-s; Q(j,i)=s; Q(j,j)=c;
for i1=1:n; for j1=1:n;
QT(i1,j1)=Q(j1,i1); end; end
for i2=1:n; for j2=1:n; s=0;
for k = 1 : n; ss = QT(i2, k) * A(k, j2);
s=s+ss; end; QTA(i2,j2)=s; end; end
for i3=1:n; for j3=1:n; s=0;
for k=1:n; ss = QTA(i3, k) * Q(k, j3); s=s+ss; end
A(i3,j3)=s; end; end; QQ=[QQ,Q]; end; D=A
y=[]; for k=1:n; yy=A(k,k); y=[y;yy]; end; eigvals=y
x=Q; if (n > 2) % Compute eigenvectors
x(1:n,1:n)=QQ(1:n,1:n);
for c = n + 1 : n : n * n; xx(1:n,1:n)=QQ(1:n,c:n+c-1);
for i=1:n; for j=1:n; s=0;
for k=1:n; ss = x(i, k) * xx(k, j); s=s+ss; end
x1(i,j)=s; end; end; x=x1; end; end
```

Hence,

$$A_3 = Q_3^T Q_2^T Q_1^T A Q_1 Q_2 Q_3 = \begin{pmatrix} 3.0003 & 0 & -1.35E-6 \\ 0 & 2.00 & -1.122E-4 \\ -1.35E-6 & -1.122E-4 & 0.9899 \end{pmatrix},$$

which gives diagonal matrix D, and its diagonal elements are converging to $3, 2$, and 1, which are the eigenvalues of the original matrix A. The corresponding eigenvectors can be computed as follows:

$$\mathbf{v} = Q_1 Q_2 Q_3 = \begin{pmatrix} 0.9998 & -0.0121 & -0.0094 \\ 0.0111 & 0.9951 & -0.0985 \\ 0.0106 & 0.0984 & 0.9951 \end{pmatrix}.$$

To reproduce the above results using the Jacobi method by the MATLAB command window, we do the following:

```
>> A = [3 0.01 0.02; 0.01 2 0.1; 0.02 0.1 1];
>> sol = JOBM(A);
```

7.10.2 Sturm Sequence Iteration

When a symmetric matrix is tridiagonal, the eigenvalues of a tridiagonal matrix can be computed to any specified precision using a simple method called the *Sturm sequence iteration.* In the coming sections we will discuss two methods, which will convert the given symmetric matrices into a symmetrical tridiagonal form using similarity transformations. The Sturm sequence iteration below can, therefore, be used in the calculation of eigenvalues of any symmetric tridiagonal matrix. Consider a symmetric tridiagonal matrix of order 4×4 as

$$A = \begin{pmatrix} a_1 & b_2 & 0 & 0 \\ b_2 & a_2 & b_3 & 0 \\ 0 & b_3 & a_3 & b_4 \\ 0 & 0 & b_4 & a_4 \end{pmatrix}$$

and assume that $b_i \neq 0$, for each $i = 2, 3, 4$. Then one can define the characteristic polynomial of given matrix A as

$$f_4(\lambda) = det(A - \lambda I), \tag{7.63}$$

which is equivalent to

$$f_4(\lambda) = \begin{vmatrix} a_1 - \lambda & b_2 & 0 & 0 \\ b_2 & a_2 - \lambda & b_3 & 0 \\ 0 & b_3 & a_3 - \lambda & b_4 \\ 0 & 0 & b_4 & a_4 - \lambda \end{vmatrix}.$$

We expand by minors in the last row as

$$f_4(\lambda) = (a_4 - \lambda)\begin{vmatrix} a_1 - \lambda & b_2 & 0 \\ b_2 & a_2 - \lambda & b_3 \\ 0 & b_3 & a_3 - \lambda \end{vmatrix} - b_4 \begin{vmatrix} a_1 - \lambda & b_2 & 0 \\ b_2 & a_2 - \lambda & 0 \\ 0 & b_3 & b_4 \end{vmatrix}$$

or

$$f_4(\lambda) = (a_4 - \lambda) f_3(\lambda) - b_4^2 f_2(\lambda). \tag{7.64}$$

Recurrence relation (7.64) is true for a matrix of any order say $r \times r$; that is,

$$f_r(\lambda) = (a_r - \lambda) f_{r-1}(\lambda) - b_r^2 f_{r-2}(\lambda) \tag{7.65}$$

provided that we define $f_0(\lambda) = 1$ and evaluate $f_1(\lambda) = a_1 - \lambda$.

The sequence $\{f_0, f_1, \ldots, f_r, \ldots\}$ is known as the Sturm sequence. So starting with $f_0(\lambda) = 1$, we can eventually find a characteristic polynomial of A using

$$f_n(\lambda) = 0. \tag{7.66}$$

Example 7.39 *Use the Sturm sequence iteration to find the eigenvalues of the following symmetric tridiagonal matrix*

$$A = \begin{pmatrix} 1 & 2 & 0 & 0 \\ 2 & 4 & 1 & 0 \\ 0 & 1 & 5 & -1 \\ 0 & 0 & -1 & 3 \end{pmatrix}.$$

Solution. *We compute the Sturm sequences as follows*

$$\begin{aligned} f_0(\lambda) &= 1 \\ f_1(\lambda) &= (a_1 - \lambda). \\ &= 1 - \lambda \end{aligned}$$

The second sequence is

$$\begin{aligned} f_2(\lambda) &= (a_2 - \lambda) f_1(\lambda) - b_2^2 f_0(\lambda) \\ &= (4 - \lambda)(1 - \lambda) - 4(1) \\ &= \lambda^2 - 5\lambda \end{aligned}$$

and the third sequence is

$$\begin{aligned} f_3(\lambda) &= (a_3 - \lambda) f_2(\lambda) - b_3^2 f_1(\lambda) \\ &= (5 - \lambda)(\lambda^2 - 5\lambda) - (1)^2(1 - \lambda). \\ &= -\lambda^3 + 10\lambda^2 - 24\lambda - 1 \end{aligned}$$

Finally, the fourth sequence is

$$\begin{aligned} f_4(\lambda) &= (a_4 - \lambda)f_3(\lambda) - b_4^2 f_2(\lambda) \\ &= (3 - \lambda)(-\lambda^3 + 10\lambda^2 - 24\lambda - 1) - (-1)^2(\lambda^2 - 5\lambda). \\ &= \lambda^4 - 13\lambda^3 + 53\lambda^2 - 66\lambda - 3 \end{aligned}$$

Thus,

$$f_4(\lambda) = \lambda^4 - 13\lambda^3 + 53\lambda^2 - 66\lambda - 3 = 0.$$

Program 7.10

MATLAB m-file for the Sturm Sequence method

```
function sol=SturmS(A)
% This evaluates the eigenvalues of a tridiagonal sym-
metric matrix
[n,n] = size(A);
ff(1,:)=[1 0 0 0 0]; ff(2,:)=[A(1,1) -1 0 0 0];
for i=3:n+1; h=[A(i-1,i-1) -1];
ff(i,1) = h(1)*ff(i-1,1)-A(i-1,i-2)^2*ff(i-2,1);
for z=2:n+1
ff(i,z) = h(1)*ff(i-1,z) + h(2)*ff(i-1,z-1) -
A(i-1,i-2)^2*ff(i-2,z); end; end
for i=1:n+1; y(i)=ff(n+1,n+2-i);end; eigval=roots(y)
```

Solving the above equation, we have eigenvalues $6.11, 4.41, 2.54$, and -0.04 of the given symmetric tridiagonal matrix.

To get the above results using the MATLAB command window, we do the following:

```
>> A = [1 2 0 0; 2 4 1 0; 0 1 5 -1; 0 0 -1 3];
>> sol = SturmS(A);
```

Theorem 7.13 *For any real number λ^*, the number of agreements in signs of the successive terms of the Sturm sequence $\{f_0(\lambda^*), f_1(\lambda^*), \ldots, f_n(\lambda^*)\}$ is equal to the number of eigenvalues of tridiagonal matrix A greater than λ^*. The sign of a zero is taken to be opposite to that of the previous term.*

Example 7.40 *Find the number of eigenvalues of the following matrix:*

$$A = \begin{pmatrix} 3 & -1 & 0 \\ -1 & 2 & -1 \\ 0 & -1 & 3 \end{pmatrix}$$

lying on the interval $(0, 4)$.

Solution. *Since the given matrix is of size* 3×3, *we have to compute the Sturm sequences* $f_3(0)$ *and* $f_3(4)$. *Firstly, for* $\lambda^* = 0$, *we have*

$$f_0(0) = 1 \quad \text{and} \quad f_1(0) = (a_1 - 0) = (3 - 0) = 3.$$

Also,

$$\begin{aligned} f_2(0) &= (a_2 - 0)f_1(0) - b_2^2 f_0(0) \\ &= (2)(3) - (-1)^2(1) = 5. \end{aligned}$$

Finally, we have

$$\begin{aligned} f_3(0) &= (a_3 - 0)f_2(0) - b_3^2 f_1(0) \\ &= (3)(5) - (-1)^2(3) = 12, \end{aligned}$$

which have signs $++++$, *with three agreements. So all three eigenvalues are greater than* $\lambda^* = 0$.

Similarly, we can calculate for $\lambda^* = 4$. *The Sturm sequences are as follows:*

$$f_0(4) = 1 \qquad \text{and} \quad f_1(4) = (a_1 - 4) = (3 - 4) = -1.$$

Also,

$$\begin{aligned} f_2(4) &= (a_2 - 4)f_1(4) - b_2^2 f_0(4) \\ &= (2 - 4)(-1) - (-1)^2(1) = 1. \end{aligned}$$

In the last, we have

$$\begin{aligned} f_3(4) &= (a_3 - 4)f_2(4) - b_3^2 f_1(4) \\ &= (3 - 4)(1) - (-1)^2(-1) = 0, \end{aligned}$$

which have signs $+-+-$, with no agreements. So no eigenvalues are greater than $\lambda^* = 4$. Hence, there is exactly three eigenvalues on $[0, 4]$. Furthermore, since $f_3(0) \neq 0$ and also, $f_3(4) = 0$, we deduce that no eigenvalue is exactly equal to zero but one eigenvalue is exactly equal to four, because $f_3(\lambda^*) = \det(A - \lambda^*\mathbf{I})$, the characteristic polynomial of A. Therefore, there are three eigenvalues on half-open interval $(0, 4]$ and two eigenvalues are on open interval $(0, 4)$. Since given matrix A is positive-definite, by well-known result, all eigenvalues of A must be strictly positive. Note that the eigenvalues of given matrix A are $1, 3$, and 4.

Note that if the sign pattern is $+ + + - -$ for a 4×4 matrix for $\lambda = c$, then there are three eigenvalues greater than $\lambda = c$. If the sign pattern is $+ - + - -$ for a 4×4 matrix for $\lambda = c$, then there is one eigenvalue greater than $\lambda = c$. If the sign pattern is $+ - 0 + +$ for a 4×4 matrix for $\lambda = c$, then there are two eigenvalues greater than $\lambda = c$.

7.10.3 Given's Method

This method is also based on similarity transformations of the same type as used for the Jacobi method. The zeros created are retained and the symmetric matrix is reduced into symmetric tridiagonal matrix C rather than a diagonal form using a finite number of orthogonal similarity transformations. The eigenvalues of original matrix A are the same as those of symmetric tridiagonal matrix C. Given's method is generally preferable to the Jacobi method in that it requires a finite number of iterations.

For Given's method, the angle θ is chosen to create zeros, not in the (p, q) and (q, p) positions as in the Jacobi method but in the $(p - 1, q)$ and $(q, p - 1)$ positions. This is because zeros can be created in row order without destroying those previously obtained.

In the first stage of Given's method we annihilate elements along the first row (and by symmetry, down the first column) in positions $(1, 3), \ldots, (1, n)$ using the rotation matrices $Q_{23}, \ldots, Q_{2n}$ in turn. Once a zero has been created in positions $(1, j)$, subsequent transformations use matrices Q_{pq} with $p, q \neq 1, j$ and so zeros are not destroyed. In the second stage we annihilate elements in positions $(2, 4), \ldots, (2, n)$ using $Q_{34}, \ldots, Q_{3n}$. Again, any zeros produced by these transformations are not destroyed as subsequent zeros are created along the second row. Furthermore, zeros previously obtained in the first row are also preserved. The process continues until a zero is created in position $(n - 2, n)$ using Q_{n-1n}. The original matrix can therefore be converted into a symmetric tridiagonal matrix C in exactly

$$(n - 2) + (n - 3) + \cdots + 1 \equiv \frac{1}{2}(n - 1)(n - 2)$$

steps. This method also uses rotation matrices as the Jacobi method does but in the following form

$$\cos\theta = (p-1, p-1), \quad \sin\theta = (p-1, q), \quad -\sin\theta = (q, p-1), \quad \cos\theta = (q, q)$$

and

$$\theta = -\arctan\left(\frac{a_{p-1q}}{a_{p-1p}}\right).$$

We can also find the values of $\cos\theta$ and $\sin\theta$ using

$$\cos\theta = \frac{|a_{p-1p}|}{R} \quad \text{and} \quad \sin\theta = \frac{|a_{p-1q}|}{R},$$

where

$$R = \sqrt{(a_{p-1p})^2 + (a_{p-1q})^2}.$$

Example 7.41 *Use Given's method to reduce the following matrix*

$$A = \begin{pmatrix} 4 & 3 & 2 & 1 \\ 3 & 4 & 3 & 2 \\ 2 & 3 & 4 & 3 \\ 1 & 2 & 3 & 4 \end{pmatrix}$$

to a symmetric tridiagonal form and then find the eigenvalues of A.

Solution.
Step I. *Create a zero in the $(1,3)$ position by using the first orthogonal transformation matrix as*

$$Q_{23} = \begin{pmatrix} 1 & 0 & 0 & 0 \\ 0 & c & s & 0 \\ 0 & -s & c & 0 \\ 0 & 0 & 0 & 1 \end{pmatrix}.$$

To find the value of the $\cos\theta$ and $\sin\theta$, we have

$$\theta = -\arctan\left(\frac{a_{13}}{a_{12}}\right) = -\arctan\left(\frac{2}{3}\right) \approx -0.5880$$

$$\cos\theta = 0.8321, \quad \textit{and} \quad \sin\theta = -0.5547.$$

Then

$$Q_{23} = \begin{pmatrix} 1.0000 & 0 & 0 & 0 \\ 0 & 0.8321 & -0.5547 & 0 \\ 0 & 0.5547 & 0.8321 & 0 \\ 0 & 0 & 0 & 1.0000 \end{pmatrix}$$

and

$$Q_{23}^T = \begin{pmatrix} 1.0000 & 0 & 0 & 0 \\ 0 & 0.8321 & 0.5547 & 0 \\ 0 & -0.5547 & 0.8321 & 0 \\ 0 & 0 & 0 & 1.0000 \end{pmatrix},$$

which gives

$$A_1 = Q_{23}^T A Q_{23} = \begin{pmatrix} 4.0000 & 3.6057 & 0.0001 & 1.0000 \\ 3.6057 & 6.7697 & 1.1541 & 3.3283 \\ 0.0001 & 1.1541 & 1.2309 & 1.3869 \\ 1.0000 & 3.3283 & 1.3869 & 4.0000 \end{pmatrix}.$$

Note that because of the symmetry matrix, the lower part of A_1 is the same as the upper part.

Step II. *Create a zero in the $(1,4)$ position using the second orthogonal transformation matrix as*

$$Q_{24} = \begin{pmatrix} 1 & 0 & 0 & 0 \\ 0 & c & 0 & s \\ 0 & 0 & 1 & 0 \\ 0 & -s & 0 & c \end{pmatrix}$$

and

$$\begin{aligned} \theta &= -\arctan\left(\frac{a_{14}}{a_{12}}\right) = -\arctan\left(\frac{1}{3.6057}\right) \approx -0.2705 \\ \cos\theta &= 0.9636, \quad \textit{and} \quad \sin\theta = -0.2673. \end{aligned}$$

Then

$$Q_{24} = \begin{pmatrix} 1.0000 & 0 & 0 & 0 \\ 0 & 0.9636 & 0 & -0.2673 \\ 0 & 0 & 1.0000 & 0 \\ 0 & 0.2673 & 0 & 0.9636 \end{pmatrix}$$

and

$$Q_{24}^T = \begin{pmatrix} 1.0000 & 0 & 0 & 0 \\ 0 & 0.9636 & 0 & 0.2673 \\ 0 & 0 & 1.0000 & 0 \\ 0 & -0.2673 & 0 & 0.9636 \end{pmatrix},$$

which gives

$$A_2 = Q_{24}^T A_1 Q_{24} = \begin{pmatrix} 4.0000 & 3.7418 & 0.0001 & -0.0002 \\ 3.7418 & 8.2862 & 1.4828 & 2.1392 \\ 0.0001 & 1.4828 & 1.2309 & 1.0279 \\ -0.0002 & 2.1392 & 1.0279 & 2.4832 \end{pmatrix}.$$

Step III. *Create a zero in the* $(2,4)$ *position using the third orthogonal transformation matrix as*

$$Q_{34} = \begin{pmatrix} 1 & 0 & 0 & 0 \\ 0 & 1 & 0 & 0 \\ 0 & 0 & c & s \\ 0 & 0 & -s & c \end{pmatrix}$$

and

$$\begin{aligned} \theta &= -\arctan\left(\frac{a_{24}}{a_{23}}\right) = -\arctan\left(\frac{2.1392}{1.4828}\right) \approx -0.9649 \\ \cos\theta &= 0.5695, \quad \textit{and} \quad \sin\theta = -0.8220. \end{aligned}$$

Then

$$Q_{34} = \begin{pmatrix} 1.0000 & 0 & 0 & 0 \\ 0 & 1.0000 & 0 & 0 \\ 0 & 0 & 0.5695 & -0.8220 \\ 0 & 0 & 0.8220 & 0.5695 \end{pmatrix}$$

and

$$Q_{34}^T = \begin{pmatrix} 1.0000 & 0 & 0 & 0 \\ 0 & 1.0000 & 0 & 0 \\ 0 & 0 & 0.5695 & 0.8220 \\ 0 & 0 & -0.8220 & 0.5695 \end{pmatrix},$$

which gives

$$A_3 = Q_{34}^T A_2 Q_{34} = \begin{pmatrix} 4.0000 & 3.7418 & -0.0001 & -0.0002 \\ 3.7418 & 8.2862 & 2.6029 & -0.0006 \\ -0.0001 & 2.6029 & 3.0395 & 0.2251 \\ -0.0002 & -0.0006 & 0.2251 & 0.6747 \end{pmatrix} = C.$$

Program 7.11

MATLAB m-file for Given's method

```
function sol=Given(A)
[n,n] = size(A);t = 0;  for i=1:n; for j=1:n
if i==j; Q(i,j)=1; else; Q(i,j)=0; end; end;end
for i=1:n-2; for j=i+2:n; t=t+1;
for f=1:n; for g=1:n
Q(f,t*n+g) = Q(f,g); end; end;
theta=atan(A(i,j)/A(i,i+1));
Q(i+1,t*n+i+1) = cos(theta); Q(i+1,t*n+j) = -sin(theta);
Q(j,t*n+i+1) = sin(theta); Q(j,t*n+j) = cos(theta);
for f=1:n; for g=1:n; sum=0; for l=1:n
sum = sum + A(f,l) * Q(l,t*n+g); ;end; aa(f,g)=sum; end; end
for f=1:n; for g=1:n; sum=0; for l=1:n
sum = sum + Q(l,t*n+f) * aa(l,g); end
A(f,g)=sum; end; end; end; end T=A
% Solve the tridiagonal matrix using Sturm sequence method
ff(1,:)=[1  0  0  0  0]; ff(2,:)=[A(1,1)  -1  0  0  0];
for i=3:n+1; h=[A(i-1,i-1)  -1];
ff(i,1) = h(1) * ff(i-1,1) - A(i-1,i-2)^ 2*ff(i-2,1);
for z=2:n+1
ff(i,z) = h(1)*ff(i-1,z)+h(2)*ff(i-1,z-1)-A(i-1,i-2)
^ 2*ff(i-2,z); end;end
for i=1:n+1; y(i) = ff(n+1,n+2-i); end; eigval = roots(y)
```

Using the Sturm sequence iteration, the eigenvalues of symmetric tridiagonal matrix C are $11.0990, 3.4142, 0.9010,$ and 0.5858, which are also the eigenvalues of A.

To get the above results using the MATLAB command window, we do the following:

```
>> A = [4 3 2 1; 3 4 3 2; 2 3 4 3; 1 2 3 4];
>> sol = Given(A);
```

7.10.4 Householder's Method

This method is a variation of Given's method and enables us to reduce a symmetric matrix A to a symmetric tridiagonal matrix form C having the same eigenvalues. It reduces a given matrix into a symmetric tridiagonal form with about half as much computation as Given's method required. This method reduces a whole row and column (except for the tridiagonal elements) to zero. It should be noted that the symmetric tridiagonal matrix form by Given's method and Householder's method may be different but the eigenvalues will be same.

Definition 7.14 (Householder Matrix)

A Householder matrix H_w is a matrix of the form

$$H_w = I - 2\mathbf{w}\mathbf{w}^T = I - \left(\frac{2}{\mathbf{w}^T\mathbf{w}}\right)\mathbf{w}\mathbf{w}^T,$$

where I is an $n \times n$ identity matrix and $\mathbf{w}$ is some $n \times 1$ vector satisfying

$$\mathbf{w}^T\mathbf{w} = \sum_{k=1}^{n} w_k^2 = 1$$

that is, the vector $\mathbf{w}$ has unit length.

It is easy to verify that Householder matrix H_w is symmetric, that is

$$H_w = \begin{pmatrix} 1 - 2w_1^2 & -2w_1w_2 & \cdots & -2w_1w_n \\ -2w_1w_2 & 1 - 2w_2^2 & \cdots & -2w_2w_n \\ \vdots & \vdots & \vdots & \vdots \\ -2w_1w_n & -2w_2w_n & \cdots & 1 - 2w_n^2 \end{pmatrix} = H_w^T$$

and is orthogonal, that is

$$\begin{aligned} H_w^2 &= (I - 2\mathbf{w}\mathbf{w}^T)(I - 2\mathbf{w}\mathbf{w}^T) \\ &= I - 4\mathbf{w}\mathbf{w}^T + 4\mathbf{w}\mathbf{w}^T\mathbf{w}\mathbf{w}^T \\ &= I - 4\mathbf{w}\mathbf{w}^T + 4\mathbf{w}\mathbf{w}^T \qquad (\text{since } \mathbf{w}^T\mathbf{w} = 1). \\ &= I \end{aligned}$$

Thus,

$$H_w = H_w^{-1} = H_w^T,$$

which shows that H_w is symmetric. Note that the determinant of Householder matrix H_w is always equal to -1.

Example 7.42 *Consider a vector* $\mathbf{w} = [1, 2]^T$, *then*

$$H_w = I - \frac{2}{5}\mathbf{w}\mathbf{w}^T$$

so

$$H_w = \begin{pmatrix} 1 & 0 \\ 0 & 1 \end{pmatrix} - \frac{2}{5}\begin{pmatrix} 1 & 2 \\ 2 & 4 \end{pmatrix} = \begin{pmatrix} \frac{3}{5} & -\frac{4}{5} \\ -\frac{4}{5} & -\frac{3}{5} \end{pmatrix},$$

which shows that given Householder matrix H_w is symmetric and orthogonal and the determinant of H_w is -1.

Householder matrix H_w corresponding to a given $\mathbf{w}$ may be generated using the MATLAB command window as follows:

```
>> w = [1 2]';
>> w = w/norm(w);
>> Hw = eye(2) - 2 * w * w';
```

The basic steps of Householder's method, which converts the symmetric matrix into a symmetric tridiagonal matrix are as follows:

$$\begin{aligned} A_1 &= A \\ A_2 &= Q_1^T A_1 Q_1 \\ A_3 &= Q_2^T A_2 Q_2 \\ &\vdots \\ A_{k+1} &= Q_k^T A_k Q_k, \end{aligned}$$

where Q_k matrices are Householder transformation matrices and can be constructed as

$$Q_k = I - s_k \mathbf{w}_k \mathbf{w}_k^T$$

and

$$s_k = \frac{2}{\mathbf{w}_k^T \mathbf{w}_k}.$$

The coefficients of a vector $\mathbf{w}_k$ are defined in term of a matrix A as

$$w_{ik} = \begin{cases} 0, & \text{for } i = 1, 2, \ldots, k \\ a_{ik}, & \text{for } i = k+2, k+3, \ldots, n \end{cases}$$

and

$$w_{k+1k} = a_{k+1k} \pm \sqrt{\sum_{i=k+1}^{n} a_{ik}^2}.$$

The positive sign or negative sign of w_{k+1k} can be taken depending on the sign of a coefficient a_{k+1k} of a given matrix A.

Householder's method transforms a given $n \times n$ symmetric matrix to a symmetric tridiagonal matrix in exactly (n - 2) steps. Each step of the method creates zero in a complete row and column. The first step annihilates elements in the positions $(1,3), (1,4), \ldots, (1,n)$ simultaneously. Similarly, step r annihilates elements in positions $(r, r+2), (r, r+3), \ldots, (r, n)$ simultaneously. Once a symmetric tridiagonal form has been achieved, then the eigenvalues of a given matrix can be calculated using the Sturm sequence iteration. After calculating the eigenvalues, the shifted inverse power method can be used to find the eigenvectors of a symmetric tridiagonal matrix and then the eigenvectors of original matrix A can be found by pre-multiplying these eigenvectors (of symmetric tridiagonal matrix) by the product of successive transformation matrices.

Example 7.43 *Reduce the following matrix*

$$A = \begin{pmatrix} 30 & 6 & 5 \\ 6 & 30 & 9 \\ 5 & 9 & 30 \end{pmatrix}$$

to symmetric tridiagonal form using Householder's method, and then approximate the eigenvalues of A.

Solution. *Since the given matrix is of size* 3×3, *only one iteration is required in order to reduce the given symmetric matrix into symmetric tridiagonal form. Thus, for* $k = 1$, *we construct the elements of the vector* $\mathbf{w}_1$ *as follows:*

$$\begin{aligned} w_{11} &= 0 \\ w_{31} &= a_{31} = 5 \\ w_{21} &= a_{21} \pm \sqrt{a_{21}^2 + a_{31}^2} = 6 \pm \sqrt{6^2 + 5^2} = 6 \pm 7.81. \end{aligned}$$

Since the given coefficient a_{21} *is a positive, positive sign must be used for* w_{21}*; that is*

$$w_{21} = 13.81.$$

Therefore, the vector $\mathbf{w}_1$ *is now determined to be*

$$\mathbf{w}_1 = [0, 13.81, 5]^T$$

and

$$s_1 = \frac{2}{(0)^2 + (13.81)^2 + (5)^2} = 0.0093.$$

Thus, the first transformation matrix Q_1 *for the first iteration is*

$$Q_1 = \begin{pmatrix} 1 & 0 & 0 \\ 0 & 1 & 0 \\ 0 & 0 & 1 \end{pmatrix} - 0.009 \begin{pmatrix} 0 \\ 13.81 \\ 5 \end{pmatrix} \begin{pmatrix} 0 & 13.81 & 5 \end{pmatrix}$$

and it gives

$$Q_1 = \begin{pmatrix} 1 & 0 & 0 \\ 0 & -0.7682 & -0.6402 \\ 0 & -0.6402 & 0.7682 \end{pmatrix}.$$

Therefore,

$$A_2 = Q_1^T A_1 Q_1 = \begin{pmatrix} 30.0 & -7.810 & 0 \\ -7.810 & 38.85 & -1.622 \\ 0 & -1.622 & 21.15 \end{pmatrix},$$

which is the symmetric tridiagonal form. The eigenvalues of this symmetric tridiagonal matrix can be found using the Sturm sequence iteration and it gives the characteristic equation of the form

$$f_3(\lambda) = (21.15 - \lambda)(\lambda^2 - 68.85\lambda + 1104.5) - 2.63(30 - \lambda) = 0.$$

Now solving this characteristic equation, the eigenvalues of the symmetric tridiagonal matrix are $43.493, 25.595$ *and* 20.913*, which are also the eigenvalues of* A*. Once the eigenvalues of* A *are obtained, then the corresponding eigenvectors of* A *can be obtained using the shifted inverse power method.*

To get the above results using the MATLAB command window, we do the following:

```
>> A = [30 6 5; 6 30 9; 5 9 30];
>> sol = HouseHM(A);
```

Program 7.12

MATLAB m-file for the Householder method

```
function sol=HouseHM(A)
[n, n] = size(A); Q = eye(n); for k=1:n-2
alfa = sign(A(k+1, k)) * sqrt(A((k+1) : n, k)' * A((k+1) : n, k));
w = zeros(n, 1);
w(k+1, 1) = A(k+1, k)+alfa; w((k+2) : n, 1) = A((k+2) : n, k);
P = eye(n) - 2 * w * w'/(w' * w); Q = Q * P; A = P * A * P; end
T=A % this is the tridiagonal matrix
% using Sturm sequence method
ff(1,:)=[1 0 0 0 0]; ff(2,:)=[A(1,1) -1 0 0 0];
for i=3:n+1
h = [A(i-1, i-1) - 1]; ff(i, 1) = h(1) * ff(i-1, 1) - A(i-1, i-2)
^ 2*ff(i - 2, 1);
for z=2:n+1
ff(i, z) = h(1) * ff(i-1, z) + h(2) * ff(i-1, z-1) - A(i-1, i-2)^
2*ff(i - 2, z); end; end
for i=1:n+1; y(i)=ff(n+1,n+2-i); end; alfa; u; Q; eig-val=roots(y)
```

7.11 Matrix Decomposition Methods

7.11.1 QR Method

We know that the Jacobi, Given's, and Householder's methods are applicable only to symmetric matrices for finding all the eigenvalues of a matrix A. Now we describe the QR method, which can find all the eigenvalues of a general matrix. In this method we decompose an arbitrary real matrix A into a product QR, where Q is a orthogonal matrix and R is a upper-triangular matrix with nonnegative diagonal elements. Note that when A is nonsingular, then this decomposition is unique.

Starting with $A_1 = A$, the QR method iteratively computes similar matrices A_i, $i = 2, 3, \ldots$, in two stages:

(1) Factor A_i into $Q_i R_i$, that is, $A_i = Q_i R_i$.

(2) Define $A_{i+1} = R_iQ_i$.

Note that from stage *(1)*, we have

$$R_i = Q_i^{-1}A_i$$

using this, stage *(2)* can be written as

$$A_{i+1} = Q_i^{-1}A_iQ_i = Q_i^TA_iQ_i,$$

where all A_i are similar to A and thus have the same eigenvalues. It turns out that in the case where the eigenvalues of A all have a different magnitude,

$$|\lambda_1| > |\lambda_2| > \cdots > |\lambda_n|$$

then the QR iterates an A_i approach to an upper-triangular matrix, and thus the elements of the main diagonal approach the eigenvalues of a given matrix A. When there are distinct eigenvalues of the same size, iterates A_i may not approach a upper-triangular matrix; however, they do approach a matrix that is near enough to an upper-triangular matrix to allow us to find the eigenvalues of A.

If a given matrix A is symmetric and tridiagonal, since the QR transformation preserves symmetry, all subsequent matrices A_i will be symmetric and hence tridiagonal. Thus, the combined method of first reducing a symmetric matrix to a symmetric tridiagonal form by Householder transformations and then applying the QR method is probably the most effective to evaluate all the eigenvalues of a symmetric matrix.

The simplest way of calculating the QR decomposition of an $n \times n$ matrix A is to pre-multiply A by a series of rotation matrices, and the values of p, q, and θ are chosen to annihilate one of the lower-triangular elements. The value of θ, which is chosen to create zero in the (q, p) position is defined as

$$\theta = -\arctan\left(\frac{a_{qp}}{a_{pp}}\right).$$

The first stage of decomposition annihilates the element in position (2,1) using rotation matrix Q_{12}^T. The next two stages annihilate elements in positions (3,1) and (3,2) using the rotation matrices Q_{13}^T and Q_{23}^T, respectively. The process continues in this way, creating zeros in row order, until rotation matrix Q_{n-1n}^T is used to annihilate the element in position (n,n-1). The zeros created are retained in a similar way as in the Given's method and an upper-triangular matrix R is produced after $n(n-1)/2$ pre-multiplications; that is,

$$Q_{n-1n}^T \cdots Q_{13}^T Q_{12}^T A = R,$$

which can be rearranged as

$$A = (Q_{12}Q_{13} \cdots Q_{n-1n})R = QR$$

since $Q_{pq}^T = Q_{pq}^{-1}$.

Example 7.44 *Find the first QR iteration for the following matrix*

$$A = \begin{pmatrix} 1 & 4 & 3 \\ 2 & 3 & 1 \\ 2 & 6 & 5 \end{pmatrix}.$$

Solution. Step I. *Create a zero in the* $(2,1)$ *position using the first orthogonal transformation matrix as*

$$Q_{12} = \begin{pmatrix} c & s & 0 \\ -s & c & 0 \\ 0 & 0 & 1 \end{pmatrix}$$

and

$$\theta = -\arctan\left(\frac{a_{21}}{a_{11}}\right) = -\arctan(2) = -70.4833,$$

$$\cos\theta \approx 0.4472 \quad \text{and} \quad \sin\theta \approx -0.8944.$$

Then

$$Q_{12} = \begin{pmatrix} 0.4472 & -0.8944 & 0 \\ 0.8944 & 0.4472 & 0 \\ 0 & 0 & 1 \end{pmatrix} \quad \text{and} \quad Q_{12}^T = \begin{pmatrix} 0.4472 & -0.8944 & 0 \\ 0.8944 & 0.4472 & 0 \\ 0 & 0 & 1 \end{pmatrix},$$

which gives

$$Q_{12}^T A = \begin{pmatrix} 2.2360 & 4.4720 & 2.2360 \\ 0 & -2.2360 & -2.2360 \\ 2.0000 & 6.0000 & 5.0000 \end{pmatrix}.$$

Step II. *Create a zero in the* $(3,1)$ *position using the second orthogonal transformation matrix as*

$$Q_{13} = \begin{pmatrix} c & 0 & s \\ 0 & 1 & 0 \\ -s & 0 & c \end{pmatrix}$$

with

$$\theta = -\arctan\left(\frac{a_{31}}{a_{11}}\right) = -\arctan\left(\frac{2}{2.2360}\right) \approx -46.4585$$

$$\cos\theta \approx 0.7453, \quad \textit{and} \quad \sin\theta \approx -0.6667.$$

Then

$$Q_{13} = \begin{pmatrix} 0.7453 & 0 & -0.6667 \\ 0 & 1 & 0 \\ 0.6667 & 0 & 0.7453 \end{pmatrix} \quad \textit{and} \quad Q_{13}^T = \begin{pmatrix} 0.7453 & 0 & 0.6667 \\ 0 & 1 & 0 \\ -0.6667 & 0 & 0.7453 \end{pmatrix},$$

which gives

$$Q_{13}^T(Q_{12}^T A) = \begin{pmatrix} 3.0001 & 7.3336 & 5.0002 \\ 0 & -2.2360 & -2.2360 \\ 0.0001 & 1.4909 & 2.2363 \end{pmatrix}.$$

Step III. *Create a zero in the* $(3,2)$ *position using the third orthogonal transformation matrix as*

$$Q_{23} = \begin{pmatrix} 1 & 0 & 0 \\ 0 & c & s \\ 0 & -s & c \end{pmatrix}$$

with

$$\theta = -\arctan\left(\frac{a_{32}}{a_{22}}\right) = -\arctan\left(\frac{1.4909}{-2.2360}\right) \approx 37.4393$$

$$\cos\theta \approx 0.8320, \quad \textit{and} \quad \sin\theta \approx 0.5548.$$

Then

$$Q_{23} = \begin{pmatrix} 1 & 0 & 0 \\ 0 & 0.8320 & 0.5548 \\ 0 & -0.5548 & 0.8320 \end{pmatrix} \quad \textit{and} \quad Q_{23}^T = \begin{pmatrix} 1 & 0 & 0 \\ 0 & 0.8320 & -0.5548 \\ 0 & 0.5548 & 0.8320 \end{pmatrix},$$

which gives

$$R_1 = Q_{23}^T(Q_{13}^T Q_{12}^T A) = \begin{pmatrix} 3 & 7.3333 & 5 \\ 0 & -2.6874 & -3.1009 \\ 0 & 0 & 0.6202 \end{pmatrix},$$

which is the required upper-triangular matrix R_1. Matrix Q_1 can be computed as

$$Q_1 = Q_{12}Q_{13}Q_{23} = \begin{pmatrix} 0.3333 & -0.5788 & -0.7442 \\ 0.6667 & 0.7029 & -0.2481 \\ 0.6667 & -0.4134 & 0.6202 \end{pmatrix}.$$

Hence, the original matrix A can be decomposed as

$$A_1 = Q_1 R_1 = \begin{pmatrix} 0.9999 & 3.9997 & 2.9997 \\ 2.0001 & 3.0001 & 1.0000 \\ 2.0001 & 6.0001 & 5.0001 \end{pmatrix}$$

and the new matrix can be obtained as

$$A_2 = R_1 Q_1 = \begin{pmatrix} 9.2222 & 1.3506 & -0.9509 \\ -3.8589 & -0.6068 & -1.2564 \\ 0.4134 & -0.2564 & 0.3846 \end{pmatrix},$$

which is the required first QR iteration for the given matrix.

Note that if we continue in the same way until we reach 21 iterations, the new matrix A_{21} becomes the upper-triangular matrix

$$A_{21} = R_{20}Q_{20} = \begin{pmatrix} 8.5826 & -4.9070 & -2.1450 \\ 0 & 1 & -1.1491 \\ 0 & 0 & -0.5825 \end{pmatrix}$$

and its diagonal elements are the eigenvalues, $\lambda = 8.5826, 1, -0.5825$, of the given matrix A. Once the eigenvalues have been determined, the corresponding eigenvectors can be computed by the shifted inverse power method.

To get the above results using the MATLAB command window, we do the following:

```
>> A = [1 4 3; 2 3 1; 2 6 5];
>> sol = QRM(A);
```

Program 7.13
MATLAB m-file for the QR method

```
function sol=QRM(A,maxI)
[n,n]=size(A); M=0; for i=1:n; for j=1:n
if j < i; M=M+1;end; end;end;
for i=1:n; I(i,i)=1;end
for k=1:maxI  Q=I; Qs=I; kk=1;
for i=2:n; for j=1:i-1
t = -atan((A(i,j)/A(j,j))); Q(j,j) = cos(t);
Q(j,i) = sin(t); Q(i,j) = -sin(t); Q(i,i) = cos(t); Q;
A = Q' * A; Qs(:,:,kk) = Q; kk = kk + 1; Q = I; end; end;
Q = Qs(:,:,M); forc = M - 1 : -1 : 1
Q = Qs(:,:,c) * Q; end; R = A; Q; A = R * Q; k = 1;
for i=1:n; for j=1:n
if j < i; m(k) = A(i,j); k = k + 1; end; end; end;
m; dd = max(abs(m)); end; for i=1:n; eigvals(i)=A(i,i); end end
```

7.11.2 LR Method

Another method, which is very similar to the QR method is Rutishauser's LR method. This method is based on the decomposition of a matrix A into the product of lower-triangular matrix L (with unit diagonal elements) and upper-triangular matrix R. Starting with $A_1 = A$, the LR method iteratively computes similar matrices A_i, $i = 2, 3, \dots$, in two stages:

(1) Factor A_i into L_iR_i; that is, $A_i = L_iR_i$.

(2) Define $A_{i+1} = R_iL_i$.

Each complete step is a similarity transformation because

$$A_{i+1} = R_iL_i = L_i^{-1}A_iL_i$$

and so all of the matrices A_i have the same eigenvalues. This triangular-decomposition-based method enables us to reduce a given nonsymmetric matrix to an upper-triangular matrix whose diagonal elements are the possible eigenvalues of a given matrix A, in decreasing order of magnitude. The rate at which the lower-triangular elements $a_{jk}^{(i)}$ of A_i converges to zero is of order $\left(\frac{\lambda_j}{\lambda_k}\right)^i$, $j > k$.

This implies, in particular, that the order of convergence of the elements along the first subdiagonal is $\left(\frac{\lambda_{j+1}}{\lambda_j}\right)^i$ and so convergence will be slow whenever two or more real eigenvalues are close together. The situation is more complicated if any of the eigenvalues are complex.

Since we know that the triangular decomposition is not always possible, we will use the decomposition by the use of the partial pivoting. Using it, we start

$$P_i A_i = L_i R_i,$$

where P_i represents the row permutations used in the decomposition. In order to preserve eigenvalues it is necessary to calculate A_{i+1} from

$$A_{i+1} = (R_i P_i) L_i.$$

It is easy to see that this is a similarity transformation because

$$A_{i+1} = (R_i P_i) L_i = L_i^{-1} P_i A_i P_i L_i$$

and $P_i^{-1} = P_i$.

Matrix P_i does not have to computed explicitly; $R_i P_i$ is just a column permutation of R_i using interchanges corresponding to row interchanges used in the decomposition of A_i.

Example 7.45 *Use the LR method to find the eigenvalues of the following matrix*

$$A = \begin{pmatrix} 2 & -2 & 3 \\ 0 & 3 & -2 \\ 0 & -1 & 2 \end{pmatrix}.$$

Solution. *The exact eigenvalues of given matrix A are $\lambda = 1, 2, 4$. The first triangular decomposition of $A = A_1$ produces*

$$L_1 = \begin{pmatrix} 1.0000 & 0 & 0 \\ 0 & 1.0000 & 0 \\ 0 & -0.3333 & 1.0000 \end{pmatrix}, \qquad R_1 = \begin{pmatrix} 2.0000 & -2.0000 & 3.0000 \\ 0 & 3.0000 & -2.0000 \\ 0 & 0 & 1.3333 \end{pmatrix}$$

and no rows are interchanged. Then

$$A_2 = R_1 L_1 = \begin{pmatrix} 2.0000 & -3.0000 & 3.0000 \\ 0 & 3.6667 & -2.0000 \\ 0 & -0.4444 & 1.3333 \end{pmatrix}.$$

The second triangular decomposition of A_2 *produces*

$$L_2 = \begin{pmatrix} 1.0000 & 0 & 0 \\ 0 & 1.0000 & 0 \\ 0 & -0.1212 & 1.0000 \end{pmatrix}, \qquad R_2 = \begin{pmatrix} 2.0000 & -3.0000 & 3.0000 \\ 0 & 3.6667 & -2.0000 \\ 0 & 0 & 1.0909 \end{pmatrix}$$

and again no rows are interchanged. Then

$$A_3 = R_2 L_2 = \begin{pmatrix} 2.0000 & -3.3636 & 3.0000 \\ 0 & 3.9091 & -2.0000 \\ 0 & -0.1322 & 1.0909 \end{pmatrix}.$$

In a similar way, the next matrices in the sequence are

$$A_4 = \begin{pmatrix} 2 & -3.3636 & 3.0000 \\ 0 & 3.9091 & -2.0000 \\ 0 & 0 & 1.0233 \end{pmatrix} \begin{pmatrix} 1 & 0 & 0 \\ 0 & 1.0000 & 0 \\ 0 & -0.0338 & 1 \end{pmatrix} = \begin{pmatrix} 2 & -3.4651 & 3.0000 \\ 0 & 3.9767 & -2.0000 \\ 0 & -0.0346 & 1.0233 \end{pmatrix},$$

$$A_5 = \begin{pmatrix} 2 & -3.4651 & 3.0000 \\ 0 & 3.9767 & -2.0000 \\ 0 & 0 & 1.0058 \end{pmatrix} \begin{pmatrix} 1 & 0 & 0 \\ 0 & 1.0000 & 0 \\ 0 & -0.0087 & 1 \end{pmatrix} = \begin{pmatrix} 2 & -3.4912 & 3.0000 \\ 0 & 3.9942 & -2.0000 \\ 0 & -0.0088 & 1.0058 \end{pmatrix},$$

$$A_6 = \begin{pmatrix} 2 & -3.4912 & 3.0000 \\ 0 & 3.9942 & -2.0000 \\ 0 & 0 & 1.0015 \end{pmatrix} \begin{pmatrix} 1 & 0 & 0 \\ 0 & 1.0000 & 0 \\ 0 & -0.0022 & 1 \end{pmatrix} = \begin{pmatrix} 2 & -3.4978 & 3.0000 \\ 0 & 3.9985 & -2.0000 \\ 0 & -0.0022 & 1.0015 \end{pmatrix},$$

$$A_7 = \begin{pmatrix} 2 & -3.4978 & 3 \\ 0 & 3.9985 & -2 \\ 0 & 0 & 1 \end{pmatrix} \begin{pmatrix} 1 & 0 & 0 \\ 0 & 1.0000 & 0 \\ 0 & -0.0005 & 1 \end{pmatrix} = \begin{pmatrix} 2 & -3.4995 & 3.0000 \\ 0 & 3.9996 & -2.0000 \\ 0 & -0.0005 & 1.0004 \end{pmatrix},$$

$$A_8 = \begin{pmatrix} 2 & -3.4995 & 3 \\ 0 & 3.9996 & -2 \\ 0 & 0 & 1 \end{pmatrix} \begin{pmatrix} 1 & 0 & 0 \\ 0 & 1.0000 & 0 \\ 0 & -0.0001 & 1 \end{pmatrix} = \begin{pmatrix} 2 & -3.4999 & 3.0000 \\ 0 & 3.9999 & -2.0000 \\ 0 & -0.0001 & 1.0001 \end{pmatrix},$$

$$A_9 = \begin{pmatrix} 2 & -3.4999 & 3 \\ 0 & 3.9999 & -2 \\ 0 & 0 & 1 \end{pmatrix} \begin{pmatrix} 1 & 0 & 0 \\ 0 & 1 & 0 \\ 0 & 0 & 1 \end{pmatrix} = \begin{pmatrix} 2 & -3.5 & 3 \\ 0 & 4 & -2 \\ 0 & 0 & 1 \end{pmatrix}.$$

7.11.3 Upper Hessenberg Form

When employing the QR method or the LR method to find the eigenvalues of a nonsymmetric matrix A, it is preferable to first use similarity transformations to convert A to upper Hessenberg form and then go on to demonstrate its usefulness in the QR and LR methods.

Definition 7.15 *A matrix A is in upper Hessenberg form if*

$$a_{ij} = 0, \quad \textit{for all } i, j \quad \textit{such that } i - j > 1.$$

For example, in the following 4×4 matrix case, the nonzero elements are

$$A = \begin{pmatrix} 3 & 2 & 1 & 5 \\ 4 & 6 & 7 & 3 \\ 0 & 8 & 9 & 5 \\ 0 & 0 & 7 & 8 \end{pmatrix}.$$

Note that one way to characterize upper Hessenberg form is that it is almost triangular. This is important since the eigenvalues of the triangular matrix are the diagonal elements. The upper Hessenberg form of a matrix A can be achieved by a sequence of Householder transformations or the Gaussian elimination procedure. Here, we will use the Gaussian elimination procedure since it is about a factor of 2 more efficient than the Householder method. It is possible to construct matrices for which the Householder reduction, being orthogonal, is stable and elimination is not, but such matrices are extremely rare in practice.

A general $n \times n$ matrix A can be reduced to upper Hessenberg form in exactly $n - 2$ steps.

Consider an 5×5 matrix

$$A = \begin{pmatrix} a_{11} & a_{12} & a_{13} & a_{14} & a_{15} \\ a_{21} & a_{22} & a_{23} & a_{24} & a_{25} \\ a_{31} & a_{32} & a_{33} & a_{34} & a_{35} \\ a_{41} & a_{42} & a_{43} & a_{44} & a_{45} \\ a_{51} & a_{52} & a_{53} & a_{54} & a_{55} \end{pmatrix}.$$

The first step of reducing given matrix $A = A_1$ into upper Hessenberg form is to eliminate the elements in the $(3, 1), (4, 1)$, and $(5, 1)$ positions. It can be done by subtracting multiples $m_{31} = \dfrac{a_{31}}{a_{21}}, m_{41} = \dfrac{a_{41}}{a_{21}}$ and $m_{51} = \dfrac{a_{51}}{a_{21}}$ of row 2 from rows 3, 4, and 5, respectively and considering the matrix

$$M_1 = \begin{pmatrix} 1 & 0 & 0 & 0 & 0 \\ 0 & 1 & 0 & 0 & 0 \\ 0 & m_{31} & 1 & 0 & 0 \\ 0 & m_{41} & 0 & 1 & 0 \\ 0 & m_{51} & 0 & 0 & 1 \end{pmatrix}.$$

Since we wish to carry out a similarity transformation to preserve eigenvalues, it is necessary to find the inverse of matrix M^{-1} and compute

$$A_2 = M_1^{-1} A_1 M_1.$$

The right-hand side multiplication gives us

$$A_2 = \begin{pmatrix} a_{11} & a_{12}^{(2)} & a_{13} & a_{14} & a_{15} \\ a_{21} & a_{22}^{(2)} & a_{23} & a_{24} & a_{25} \\ 0 & a_{32}^{(2)} & a_{33}^{(2)} & a_{34}^{(2)} & a_{35}^{(2)} \\ 0 & a_{42}^{(2)} & a_{43}^{(2)} & a_{44}^{(2)} & a_{45}^{(2)} \\ 0 & a_{52}^{(2)} & a_{53}^{(2)} & a_{54}^{(2)} & a_{55}^{(2)} \end{pmatrix},$$

where $a_{ij}^{(2)}$ denotes the new element in (i, j). In the second step, we eliminate the elements in the $(4, 2)$ and $(5, 2)$ positions. This can be done by subtracting multiples $m_{42} = \frac{a_{42}^{(2)}}{a_{32}^{(2)}}$ and $m_{52} = \frac{a_{52}^{(2)}}{a_{32}^{(2)}}$ of row 3 from rows 4 and 5, respectively, and considering the matrix

$$M_2 = \begin{pmatrix} 1 & 0 & 0 & 0 & 0 \\ 0 & 1 & 0 & 0 & 0 \\ 0 & 0 & 1 & 0 & 0 \\ 0 & 0 & m_{42} & 1 & 0 \\ 0 & 0 & m_{52} & 0 & 1 \end{pmatrix}.$$

Hence,

$$A_3 = M_2^{-1}A_2M_2 = \begin{pmatrix} a_{11} & a_{12}^{(2)} & a_{13}^{(3)} & a_{14} & a_{15} \\ a_{21} & a_{22}^{(2)} & a_{23}^{(3)} & a_{24} & a_{25} \\ 0 & a_{32}^{(2)} & a_{33}^{(3)} & a_{34}^{(2)} & a_{35}^{(2)} \\ 0 & 0 & a_{43}^{(3)} & a_{44}^{(3)} & a_{45}^{(3)} \\ 0 & 0 & a_{53}^{(3)} & a_{54}^{(3)} & a_{55}^{(3)} \end{pmatrix},$$

where $a_{ij}^{(3)}$ denotes the new element in (i, j). In the third step, we eliminate the elements in the $(5, 3)$ position. This can be done by subtracting multiples $m_{53} = \dfrac{a_{53}^{(3)}}{a_{43}^{(3)}}$ of row 4 from row 5 and considering the matrix

$$M_3 = \begin{pmatrix} 1 & 0 & 0 & 0 & 0 \\ 0 & 1 & 0 & 0 & 0 \\ 0 & 0 & 1 & 0 & 0 \\ 0 & 0 & 0 & 1 & 0 \\ 0 & 0 & 0 & m_{53} & 1 \end{pmatrix}.$$

Hence,

$$A_4 = M_3^{-1}A_3M_3 = \begin{pmatrix} a_{11} & a_{12}^{(2)} & a_{13}^{(3)} & a_{14}^{(4)} & a_{15} \\ a_{21} & a_{22}^{(2)} & a_{23}^{(3)} & a_{24}^{(4)} & a_{25} \\ 0 & a_{32}^{(2)} & a_{33}^{(3)} & a_{34}^{(4)} & a_{35}^{(2)} \\ 0 & 0 & a_{43}^{(3)} & a_{44}^{(4)} & a_{45}^{(4)} \\ 0 & 0 & 0 & a_{54}^{(4)} & a_{55}^{(4)} \end{pmatrix},$$

which is in upper Hessenberg form.

Example 7.46 *Use the Gaussian elimination procedure to convert the following* *matrix*

$$A_1 = \begin{pmatrix} 5 & 3 & 6 & 4 & 9 \\ 4 & 6 & 5 & 3 & 4 \\ 4 & 2 & 3 & 1 & 1 \\ 2 & 4 & 6 & 3 & 3 \\ 2 & 5 & 6 & 4 & 7 \end{pmatrix}$$

into upper Hessenberg form.

Solution. *In the first step we eliminate the elements in the* $(3,1),(4,1)$, *and* $(5,1)$ *positions. It can be done by subtracting multiples* $m_{31} = \frac{4}{4} = 1, m_{41} = \frac{2}{4} = 0.5$ *and* $m_{51} = \frac{2}{4} = 0.5$ *of row 2 from rows 3, 4 ,and 5, respectively. Matrices* M_1 *and* M_1^{-1} *are as follows:*

$$M_1 = \begin{pmatrix} 1 & 0 & 0 & 0 & 0 \\ 0 & 1 & 0 & 0 & 0 \\ 0 & -1 & 1 & 0 & 0 \\ 0 & -0.5 & 0 & 1 & 0 \\ 0 & -0.5 & 0 & 0 & 1 \end{pmatrix} \quad \text{and} \quad M_1^{-1} = \begin{pmatrix} 1 & 0 & 0 & 0 & 0 \\ 0 & 1 & 0 & 0 & 0 \\ 0 & 1 & 1 & 0 & 0 \\ 0 & 0.5 & 0 & 1 & 0 \\ 0 & 0.5 & 0 & 0 & 1 \end{pmatrix}.$$

Then the transformation is

$$A_2 = M_1^{-1} A_1 M_1 = \begin{pmatrix} 5 & 15.50 & 6 & 4 & 9 \\ 4 & 14.50 & 5 & 3 & 4 \\ 0 & -8.50 & -2 & -2 & -3 \\ 0 & 5.75 & 3.50 & 1.50 & 1 \\ 0 & 9.25 & 3.50 & 2.50 & 5 \end{pmatrix}.$$

In the second step, we eliminate the elements in the $(4,2)$ *and* $(5,2)$ *positions. This can be done by subtracting multiples* $m_{42} = \frac{5.75}{-8.50} = -0.6765$ *and* $m_{52} = \frac{9.25}{-8.50} = -1.0882$ *of row 3 from rows 4 and 5, respectively. Matrices* M_2 *and* M_2^{-1} *are as follows:*

$$M_2 = \begin{pmatrix} 1 & 0 & 0 & 0 & 0 \\ 0 & 1 & 0 & 0 & 0 \\ 0 & 0 & 1 & 0 & 0 \\ 0 & 0 & 0.6765 & 1 & 0 \\ 0 & 0 & 1.0882 & 0 & 1 \end{pmatrix} \quad \text{and} \quad M_2^{-1} = \begin{pmatrix} 1 & 0 & 0 & 0 & 0 \\ 0 & 1 & 0 & 0 & 0 \\ 0 & 0 & 1 & 0 & 0 \\ 0 & 0 & -0.6765 & 1 & 0 \\ 0 & 0 & -1.0882 & 0 & 1 \end{pmatrix}.$$

Then the transformation is

$$A_3 = M_2^{-1} A_1 M_2 = \begin{pmatrix} 5 & 15.50 & -6.50 & 4 & 9 \\ 4 & 14.50 & -1.3824 & 3 & 4 \\ 0 & -8.50 & 2.6176 & -2 & -3 \\ 0 & 0 & 3.1678 & 0.1471 & -1.0294 \\ 0 & 0 & -0.7837 & 0.3235 & 1.7353 \end{pmatrix}.$$

In the last step, we eliminate the elements in the $(5,3)$ *position. This can be done by subtracting multiples* $m_{53} = \dfrac{-0.7837}{3.1678} = -0.2474$ *of row 4 from row 5. The matrices* M_3 *and* M_3^{-1} *are as follows:*

$$M_3 = \begin{pmatrix} 1 & 0 & 0 & 0 & 0 \\ 0 & 1 & 0 & 0 & 0 \\ 0 & 0 & 1 & 0 & 0 \\ 0 & 0 & 0 & 1 & 0 \\ 0 & 0 & 0 & 0.2474 & 1 \end{pmatrix} \quad \text{and} \quad M_3^{-1} = \begin{pmatrix} 1 & 0 & 0 & 0 & 0 \\ 0 & 1 & 0 & 0 & 0 \\ 0 & 0 & 1 & 0 & 0 \\ 0 & 0 & 0 & 1 & 0 \\ 0 & 0 & 0 & -0.2474 & 1 \end{pmatrix}.$$

Then the transformation is

$$A_4 = \begin{pmatrix} 5 & 15.50 & -6.50 & 1.7733 & 9 \\ 4 & 14.50 & -1.3824 & 2.0104 & 4 \\ 0 & -8.50 & 2.6176 & -1.2578 - 3 & \\ 0 & 0 & 3.1678 & 0.4017 & -1.0294 \\ 0 & 0 & 0 & -0.0064 & 1.4806 \end{pmatrix},$$

which is in the required upper Hessenberg form.

To get the above results using the MATLAB command window, we do the following:

```
>>>> A = [5 3 6 4 9; 4 6 5 3 4; 4 2 3 1 1; 2 4 6 3 3; 2 5 6 4 7];
>> sol = HessenB(A);
```

Program 7.14
MATLAB m-file for the Upper Hessenberg Form
```
function sol=HessenB(A)
n = length(A(1, :)); for i = 1:n-1; m = eye(n);
[w j] = max(abs(A(i + 1 : n, i)));
if j > i + 1;
t = m(i + 1, :); m(i + 1, :) = m(j, :);
m(j, :) = t; A = m * A * m'; end;
m = eye(n); m(i+2 : n, i+1) = -A(i+2 : n, i)/(A(i+
1, i));
mi = m; mi(i + 2 : n, i + 1) = -m(i + 2 : n, i + 1);
A = m * A * mi; mesh(abs(A)); end
```

Note that the above reduction fails if any $a^{(j)}_{j+1,j} = 0$ and, as in Gaussian elimination, is unstable whenever $|m_{ij}| > 1$. Row and column interchanges are used to avoid these difficulties (i.e., the Gaussian elimination with pivoting). At step j, the elements below the diagonal in column j are examined. If the element of largest modulus occurs in row r_j, say, then rows $j+1$ and r_j are interchanged. Here, we perform the transformation

$$A_{j+1} = M_j^{-1}(I^{-1}_{j+1,r_j} A_j I_{j+1,r_j})M_j,$$

where I_{j+1,r_j} denotes a matrix obtained from the identity matrix by interchanging rows $j+1$ and r_j, and the elements of M_j are all less than or equal to one in modulus. Note that

$$I^{-1}_{j+1,r_j} = I_{j+1,r_j}.$$

Example 7.47 *Use Gaussian elimination with pivoting to convert the following matrix*

$$A = \begin{pmatrix} 3 & 2 & 1 & -1 \\ 1 & 4 & 2 & 1 \\ 2 & 2 & 3 & -2 \\ 5 & 1 & 2 & 3 \end{pmatrix}$$

into upper Hessenberg form.

Solution. *The element of the largest modulus below the diagonal occurs in the* *fourth row, so we need to interchange rows 2 and 3 and columns 2 and 3 to get*

$$A_1 = \mathbf{I}_{24} A \mathbf{I}_{24} = \begin{pmatrix} 1 & 0 & 0 & 0 \\ 0 & 0 & 0 & 1 \\ 0 & 0 & 1 & 0 \\ 0 & 1 & 0 & 0 \end{pmatrix} \begin{pmatrix} 3 & 2 & 1 & -1 \\ 1 & 4 & 2 & 1 \\ 2 & 2 & 3 & -2 \\ 5 & 1 & 2 & 3 \end{pmatrix} \begin{pmatrix} 1 & 0 & 0 & 0 \\ 0 & 0 & 0 & 1 \\ 0 & 0 & 1 & 0 \\ 0 & 1 & 0 & 0 \end{pmatrix},$$

which gives

$$A_1 = \begin{pmatrix} 3 & -1 & 1 & 2 \\ 5 & 3 & 2 & 1 \\ 2 & -2 & 3 & 2 \\ 1 & 1 & 2 & 4 \end{pmatrix}.$$

Now we eliminate the elements in the $(3,1)$ *and* $(4,1)$ *positions. It can be done by subtracting multiples* $m_{31} = \frac{2}{5} = 0.4$ *and* $m_{41} = \frac{1}{5} = 0.2$ *of row 2 from rows 3 and 4, respectively. Then the transformation*

$$A_2 = M^{-1} A_1 M = \begin{pmatrix} 1 & 0 & 0 & 0 \\ 0 & 1 & 0 & 0 \\ 0 & -0.4 & 1 & 0 \\ 0 & -0.2 & 0 & 1 \end{pmatrix} \begin{pmatrix} 3 & -1 & 1 & 2 \\ 5 & 3 & 2 & 1 \\ 2 & -2 & 3 & 2 \\ 1 & 1 & 2 & 4 \end{pmatrix} \begin{pmatrix} 1 & 0 & 0 & 0 \\ 0 & 1 & 0 & 0 \\ 0 & 0.4 & 1 & 0 \\ 0 & 0.2 & 0 & 1 \end{pmatrix},$$

which gives

$$A_2 = \begin{pmatrix} 3 & -0.2 & 1 & 2 \\ 5 & 4 & 2 & 1 \\ 0 & -2 & 2.2 & 1.6 \\ 0 & 1.8 & 1.6 & 3.8 \end{pmatrix}.$$

The element of the largest modulus below the diagonal in the second column occurs in the third row, and so there is no need to interchange the row and column. Now we eliminate the elements in the $(4,2)$ *position. This can be done by subtracting multiple* $m_{42} = \frac{1.8}{-2} = -0.9$ *of row 3 from row 4. Then the transformation*

$$A_3 = M_2^{-1} A_2 M_2 = \begin{pmatrix} 1 & 0 & 0 & 0 \\ 0 & 1 & 0 & 0 \\ 0 & 0 & 1 & 0 \\ 0 & 0 & 0.9 & 1 \end{pmatrix} \begin{pmatrix} 3 & -0.2 & 1 & 2 \\ 5 & 4 & 2 & 1 \\ 0 & -2 & 2.2 & 1.6 \\ 0 & 1.8 & 1.6 & 3.8 \end{pmatrix} \begin{pmatrix} 1 & 0 & 0 & 0 \\ 0 & 1 & 0 & 0 \\ 0 & 0 & 1 & 0 \\ 0 & 0 & 0.9 & 1 \end{pmatrix}$$

and it gives

$$A_3 = \begin{pmatrix} 3 & -0.2 & -0.8 & 2 \\ 5 & 4 & 1.1 & 1 \\ 0 & -2 & 0.76 & 1.6 \\ 0 & 0 & -1.136 & 5.24 \end{pmatrix},$$

which is upper Hessenberg form.

Example 7.48 *Convert the following matrix to upper Hessenberg form and then apply the QR method to find its eigenvalues.*

$$A = \begin{pmatrix} 1 & 4 & 3 \\ 2 & 3 & 1 \\ 2 & 6 & 5 \end{pmatrix}$$

Solution. *Since the upper Hessenberg form of the given matrix is*

$$H_1 = \begin{pmatrix} 1 & 7 & 3 \\ 2 & 4 & 1 \\ 0 & 7 & 4 \end{pmatrix},$$

apply the QR method on the upper Hessenberg matrix H_1 and the resulting transformation matrices after iterations $1, 10, 14,$ and 19 are as follows:

$$H_2 = R_1Q1 = \begin{pmatrix} 7.0000 & -1.3460 & 2.0466 \\ -7.4297 & 1.8551 & -5.5934 \\ -0.0000 & -0.2268 & 0.1449 \end{pmatrix},$$

$$H_{10} = R_9Q_9 = \begin{pmatrix} 8.5826 & 4.4029 & 6.8555 \\ -0.0000 & 1.0069 & -1.8962 \\ 0.0000 & 0.0058 & -0.5895 \end{pmatrix},$$

$$H_{14} = R_{13}Q_{13} = \begin{pmatrix} 8.5826 & 4.4248 & 6.8413 \\ -0.0000 & 1.0008 & -1.9013 \\ 0.0000 & 0.0007 & -0.5834 \end{pmatrix},$$

$$H_{19} = R_{18}Q_{18} = \begin{pmatrix} 8.5826 & 4.4279 & 6.8393 \\ -0.0000 & 0.9999 & -1.9020 \\ -0.0000 & -0.0000 & -0.5825 \end{pmatrix}.$$

In this case the QR method converged 19 iterations faster than the QR method applied on the original matrix A in Example 7.44.

Note that the calculation of the QR decomposition is simplified if a given matrix is converted to upper Hessenberg form. So instead of applying the decomposition to the original matrix $A = A_1$, the original matrix is first transformed to the Hessenberg form. When $A_1 = H_1$ is in the upper Hessenberg form, all the subsequent H_i are also in the same form. Unfortunately, although transformation to upper Hessenberg form reduces the number of calculations at each step, the method may still prove to be computationally inefficient if the number of steps required for convergence is too large. Therefore, we use the more efficient process called the *shifting* QR *method.* Here, the iterative procedure

$$\begin{aligned} H_i &= Q_i R_i \\ H_{i+1} &= R_i Q_i \end{aligned}$$

is changed to

$$\begin{aligned} H_i - \mu_i \mathbf{I} &= Q_i R_i \\ H_{i+1} &= R_i Q_i + \mu_i \mathbf{I}. \end{aligned} \tag{7.67}$$

This change is called *shift* because subtracting $\mu_i \mathbf{I}$ from H_i shifts the eigenvalues of the right side by μ_i as well as the eigenvalues of $R_i Q_i$. Adding $\mu_i \mathbf{I}$ in the second equation in (7.67) shifts the eigenvalues of H_{i+1} back to the original values. However, the shifts accelerate convergence of the eigenvalues close to μ_i.

7.11.4 Singular Value Decomposition

We have considered two principal methods for the decomposition of the matrix, QR decomposition and LR decomposition. There is another important method for matrix decomposition called *singular value decomposition (SVD).*

Here, we show that every rectangular real matrix A can be decomposed into a product UDV^T of two orthogonal matrices U and V and a generalized diagonal matrix D. The construction of UDV^T is based on the fact that for all real matrices A a matrix $A^T A$ is symmetric and therefore there exists an orthogonal matrix Q and a diagonal matrix D for which

$$A^T A = QDQ^T.$$

As we know the diagonal entries of D are the eigenvalues of $A^T A$, now we show that they are nonnegative in all cases and that their square roots, called the *singular values* of A, can be used to construct UDV^T.

Singular Values of a Matrix

For any $m \times n$ matrix A, an $n \times n$ matrix A^TA is symmetric and hence can be orthogonally diagonalized. Not only are the eigenvalues of A^TA all real, they are all nonnegative. To show this, let λ be an eigenvalue of A^TA with corresponding unit vector $\mathbf{v}$. Then

$$\begin{aligned} 0 \leq \|A\mathbf{v}\|^2 &= (A\mathbf{v}).(A\mathbf{v}) = (A\mathbf{v})^T A\mathbf{v} = \mathbf{v}^T A^T A\mathbf{v}. \\ &= \mathbf{v}^T \lambda \mathbf{v} = \lambda(\mathbf{v}.\mathbf{v}) = \lambda\|\mathbf{v}\|^2 = \lambda \end{aligned}$$

It therefore makes sense to take (positive) square roots of these eigenvalues.

Definition 7.16 (Singular Values of a Matrix)

If A is an $m \times n$ matrix, the singular values of A are the square roots of the eigenvalues of A^TA and are denoted by $\sigma_1, \ldots, \sigma_n$. It is conventional to arrange the singular values so that $\sigma_1 \geq \sigma_2 \geq \cdots \sigma_n$.

Example 7.49 *Find the singular values of*

$$A = \begin{pmatrix} 1 & 0 & 1 \\ 1 & 1 & 0 \end{pmatrix}.$$

Solution. *Since the singular values of A are the square roots of the eigenvalues of A^TA, we compute*

$$A^TA = \begin{pmatrix} 1 & 1 \\ 0 & 1 \\ 1 & 0 \end{pmatrix} \begin{pmatrix} 1 & 0 & 1 \\ 1 & 1 & 0 \end{pmatrix} = \begin{pmatrix} 2 & 1 & 1 \\ 1 & 1 & 0 \\ 1 & 0 & 1 \end{pmatrix}.$$

The matrix A^TA has eigenvalues $\lambda_1 = 3, \lambda_2 = 1$, and $\lambda_3 = 0$. Consequently, the singular values of A are $\sigma_1 = \sqrt{3} = 1.7321$, $\sigma_2 = \sqrt{1} = 1$, and $\sigma_3 = \sqrt{0} = 0$.

Note that the singular values of A are not the same as its eigenvalues but there is a connection between them if A is a symmetric matrix.

Theorem 7.14 *If $A = A^T$ is a symmetric matrix, then its singular values are the absolute values of its nonzero eigenvalues; that is,*

$$\sigma_i = |\lambda_i| > 0.$$

Theorem 7.15 *The condition number of a nonsingular matrix is the ratio between its largest singular value σ_1 (or dominant singular value) and its smallest singular value σ_n; that is,*

$$K(A) = \frac{\sigma_1}{\sigma_n}.$$

Singular Value Decomposition

The following are some of the *properties* that make singular value decompositions useful:

1. All real matrices have singular value decompositions.
2. A real square matrix is invertible if and only if all its singular values are nonzero.
3. For any $m \times n$ real rectangular matrix A, the number of nonzero singular values of A is equal to the rank of A.
4. If $A = UDV^T$ is a singular value decomposition of an invertible matrix A, then $A^{-1} = VD^{-1}U^T$.
5. For positive definite symmetric matrices, the orthogonal decomposition QDQ^T and the singular value decomposition UDV^T coincide.

Theorem 7.16 (Singular Value Decomposition Theorem)

Every $m \times n$ matrix A can be factored into the product of an $m \times m$ matrix U with orthonormal columns, so $U^TU = \mathbf{I}$, the $m \times n$ diagonal matrix $D = diag(\sigma_1, \ldots, \sigma_r)$ that has the singular values of A as its diagonal entries, and an $n \times n$ matrix V with orthonormal rows, so $V^TV = \mathbf{I}$; that is,

$$A = UDV^T = (\mathbf{u}_1\mathbf{u}_2 \cdots \mathbf{u}_r\mathbf{u}_{r+1} \cdots \mathbf{u}_n)\begin{pmatrix} \sigma_1 & & & & & \mathbf{0} \\ & \sigma_2 & & & & \\ & & \ddots & & & \\ & & & \sigma_r & & \\ & & & & 0 & \\ & & & & & \ddots \\ \mathbf{0} & & & & & 0 \end{pmatrix}\begin{pmatrix} \mathbf{v}_1^T \\ \mathbf{v}_2^T \\ \vdots \\ \mathbf{v}_r^T \\ \mathbf{v}_{r+1}^T \\ \vdots \\ \mathbf{v}_n^T \end{pmatrix}.$$

Note that the columns of U, $\mathbf{u}_1, \mathbf{u}_2, \ldots, \mathbf{u}_r$ are called *left singular vectors* of A, and the columns of V, $\mathbf{v}_1, \mathbf{v}_2, \ldots, \mathbf{v}_r$ are called *right singular vectors* of A. The matrices U and V are not uniquely determined by A, but a matrix D must contain the singular values, $\sigma_1, \sigma_2, \ldots, \sigma_r$, of A.

To construct the orthogonal matrix V, we must find an orthonormal basis $\{\mathbf{v}_1, \mathbf{v}_2, \ldots, \mathbf{v}_n\}$ for $\mathbb{R}^n$ consisting of eigenvectors of an $n \times n$ symmetric matrix $A^T A$. Then

$$V = [\mathbf{v}_1 \mathbf{v}_2 \cdots \mathbf{v}_n]$$

is an orthogonal $n \times n$ matrix.
For the orthogonal matrix U, we first note that $\{A\mathbf{v}_1, A\mathbf{v}_2, \ldots, A\mathbf{v}_n\}$ is an orthogonal set of vectors in $\mathbb{R}^m$. To see this, suppose that $\mathbf{v}_i$ is a eigenvector of $A^T A$ corresponding to a eigenvalue λ_i, then, for $i \neq j$, we have

$$(A\mathbf{v}_i).(A\mathbf{v}_j) = (A\mathbf{v}_i)^T A\mathbf{v}_j = \mathbf{v}_i{}^T A^T A\mathbf{v}_j = \mathbf{v}_i{}^T \lambda_j \mathbf{v}_j = \lambda_j(\mathbf{v}_i.\mathbf{v}_j) = 0$$

since eigenvectors $\mathbf{v}_i$ are orthogonal. Now recall that the singular values satisfy $\sigma_i = \|A\mathbf{v}_i\|$ and that the first r of these are nonzero. Therefore, we can normalize $A\mathbf{v}_1, \ldots, A\mathbf{v}_r$, by setting

$$\mathbf{u}_i = \frac{1}{\sigma_i} A\mathbf{v}_i, \qquad \text{for} \quad i = 1, 2, \ldots, r.$$

This guarantees that $\{\mathbf{u}_1, \mathbf{u}_2, \ldots, \mathbf{u}_r\}$ is an orthonormal set in $\mathbb{R}^m$, but if $r < m$ it will not be a basis for $\mathbb{R}^m$. In this case, we extend the set $\{\mathbf{u}_1, \mathbf{u}_2, \ldots, \mathbf{u}_r\}$ to an orthonormal basis $\{\mathbf{u}_1, \mathbf{u}_2, \ldots, \mathbf{u}_m\}$ for $\mathbb{R}^m$.

Example 7.50 *Find the singular value decomposition of the following matrix:*

$$A = \begin{pmatrix} 1 & 0 & 1 \\ 1 & 1 & 0 \end{pmatrix}.$$

Solution. *We compute*

$$A^T A = \begin{pmatrix} 1 & 1 \\ 0 & 1 \\ 1 & 0 \end{pmatrix} \begin{pmatrix} 1 & 0 & 1 \\ 1 & 1 & 0 \end{pmatrix} = \begin{pmatrix} 2 & 1 & 1 \\ 1 & 1 & 0 \\ 1 & 0 & 1 \end{pmatrix}$$

and find that its eigenvalues are

$$\lambda_1 = 3, \quad \lambda_2 = 1, \quad \lambda_3 = 0$$

with corresponding eigenvectors

$$\begin{pmatrix} 2 \\ 1 \\ 1 \end{pmatrix}, \quad \begin{pmatrix} 0 \\ -1 \\ 1 \end{pmatrix}, \quad \begin{pmatrix} -1 \\ 1 \\ 1 \end{pmatrix}.$$

These vectors are orthogonal, so we normalize them to obtain

$$\mathbf{v}^{(1)} = \begin{pmatrix} \frac{2}{\sqrt{6}} \\ \frac{1}{\sqrt{6}} \\ \frac{1}{\sqrt{6}} \end{pmatrix}, \quad \mathbf{v}^{(2)} = \begin{pmatrix} 0 \\ -\frac{1}{\sqrt{2}} \\ \frac{1}{\sqrt{2}} \end{pmatrix}, \text{ and } \quad \mathbf{v}^{(3)} = \begin{pmatrix} -\frac{1}{\sqrt{3}} \\ \frac{1}{\sqrt{3}} \\ \frac{1}{\sqrt{3}} \end{pmatrix}.$$

The singular values of A are

$$\sigma_1 = \sqrt{\lambda_1} = \sqrt{3}, \quad \sigma_2 = \sqrt{\lambda_2} = \sqrt{1} = 1, \quad \sigma_3 = \sqrt{\lambda_3} = \sqrt{0} = 0.$$

Thus,

$$V = \begin{pmatrix} \frac{2}{\sqrt{6}} & 0 & -\frac{1}{\sqrt{3}} \\ \frac{1}{\sqrt{6}} & -\frac{1}{\sqrt{2}} & \frac{1}{\sqrt{3}} \\ \frac{1}{\sqrt{6}} & \frac{1}{\sqrt{2}} & \frac{1}{\sqrt{3}} \end{pmatrix}, \quad D = \begin{pmatrix} \sqrt{3} & 0 & 0 \\ 0 & 1 & 0 \end{pmatrix}.$$

To find U, we compute

$$\mathbf{u}_1 = \frac{1}{\sigma_1} A\mathbf{v}_1 = \frac{1}{\sqrt{3}} \begin{pmatrix} 1 & 0 & 1 \\ 1 & 1 & 0 \end{pmatrix} \begin{pmatrix} \frac{2}{\sqrt{6}} \\ \frac{1}{\sqrt{6}} \\ \frac{1}{\sqrt{6}} \end{pmatrix} = \begin{pmatrix} \frac{1}{\sqrt{2}} \\ \frac{1}{\sqrt{2}} \end{pmatrix}$$

and

$$\mathbf{u}_2 = \frac{1}{\sigma_2} A\mathbf{v}_2 = \frac{1}{1} \begin{pmatrix} 1 & 0 & 1 \\ 1 & 1 & 0 \end{pmatrix} \begin{pmatrix} 0 \\ -\frac{1}{\sqrt{2}} \\ \frac{1}{\sqrt{2}} \end{pmatrix} = \begin{pmatrix} \frac{1}{\sqrt{2}} \\ -\frac{1}{\sqrt{2}} \end{pmatrix}.$$

These vectors already form an orthonormal basis for $\mathbb{R}^2$, *so we have*

$$U = \begin{pmatrix} \frac{1}{\sqrt{2}} & \frac{1}{\sqrt{2}} \\ \frac{1}{\sqrt{2}} & -\frac{1}{\sqrt{2}} \end{pmatrix}.$$

This yields the SVD

$$A = \begin{pmatrix} \frac{1}{\sqrt{2}} & \frac{1}{\sqrt{2}} \\ \frac{1}{\sqrt{2}} & -\frac{1}{\sqrt{2}} \end{pmatrix} \begin{pmatrix} \sqrt{3} & 0 & 0 \\ 0 & 1 & 0 \end{pmatrix} \begin{pmatrix} \frac{2}{\sqrt{6}} & 0 & -\frac{1}{\sqrt{3}} \\ \frac{1}{\sqrt{6}} & -\frac{1}{\sqrt{2}} & \frac{1}{\sqrt{3}} \\ \frac{1}{\sqrt{6}} & \frac{1}{\sqrt{2}} & \frac{1}{\sqrt{3}} \end{pmatrix}.$$

The MATLAB built-in *svd* function performs the *SVD* of a matrix. Thus, to reproduce the above results using the MATLAB command window, we do the following:

```
>> A = [1 0 1; 1 1 0];
>> [U, D, V] = svd(A);
```

The singular value decomposition occurs in many applications. For example, if we can compute the *SVD* accurately, then we can solve a linear system very efficiently. Since we know that the nonzero singular values of A are the square roots of the nonzero eigenvalues of a matrix AA^T, which are the same as the nonzero eigenvalues of A^TA, there are exactly $r = rank(A)$ positive singular values.

Suppose that A is square and has full rank. Then if $A\mathbf{x} = \mathbf{b}$, we have

$$\begin{aligned} UDV^T\mathbf{x} &= \mathbf{b} \\ U^TUDV^T\mathbf{x} &= U^T\mathbf{b} \\ DV^T\mathbf{x} &= U^T\mathbf{b} \\ V^T\mathbf{x} &= D^{-1}U^T\mathbf{b} \\ VV^T\mathbf{x} &= VD^{-1}U^T\mathbf{b} \\ \mathbf{x} &= VD^{-1}U^T\mathbf{b} \end{aligned}$$

(since $U^TU = 1, VV^T = 1$ by orthogonality).

Example 7.51 *Find the solution of the linear system* $A\mathbf{x} = \mathbf{b}$ *using singular value decomposition, where*

$$A = \begin{pmatrix} -4 & -6 \\ 3 & -8 \end{pmatrix} \quad and \quad \mathbf{b} = \begin{pmatrix} 1 \\ 4 \end{pmatrix}.$$

Solution. *First we have to compute the singular value decomposition of* A. *For this we have to compute*

$$A^T A = \begin{pmatrix} -4 & 3 \\ -6 & -8 \end{pmatrix} \begin{pmatrix} -4 & -6 \\ 3 & -8 \end{pmatrix} = \begin{pmatrix} 25 & 0 \\ 0 & 100 \end{pmatrix}.$$

The characteristic polynomial of $A^T A$ *is*

$$\lambda^2 - 125\lambda + 2500 = (\lambda - 100)(\lambda - 25) = 0$$

and it gives the eigenvalue of $A^T A$

$$\lambda_1 = 100 \quad and \quad \lambda_2 = 25.$$

Corresponding to the eigenvalues λ_1 *and* λ_2, *we can have eigenvectors*

$$\begin{pmatrix} 0 \\ 1 \end{pmatrix} \quad and \quad \begin{pmatrix} 1 \\ 0 \end{pmatrix},$$

respectively. These vectors are orthogonal, so we normalize them to obtain

$$\mathbf{v}_1 = \begin{pmatrix} 0 \\ 1 \end{pmatrix} \quad and \quad \mathbf{v}_2 = \begin{pmatrix} 1 \\ 0 \end{pmatrix}.$$

The singular values of A *are*

$$\sigma_1 = \sqrt{\lambda_1} = \sqrt{100} = 10 \quad and \quad \sigma_2 = \sqrt{\lambda_2} = \sqrt{25} = 5.$$

Thus,

$$V = \begin{pmatrix} 0 & 1 \\ 1 & 0 \end{pmatrix} \quad and \quad D = \begin{pmatrix} 10 & 0 \\ 0 & 5 \end{pmatrix}.$$

To find U, *we compute*

$$\mathbf{u}_1 = \frac{1}{\sigma_1} A\mathbf{v}_1 = \frac{1}{10}\begin{pmatrix} -4 & -6 \\ 3 & -8 \end{pmatrix}\begin{pmatrix} 0 \\ 1 \end{pmatrix} = \begin{pmatrix} -0.6 \\ -0.8 \end{pmatrix}$$

and

$$\mathbf{u}_2 = \frac{1}{\sigma_2} A\mathbf{v}_2 = \frac{1}{5}\begin{pmatrix} -4 & -6 \\ 3 & -8 \end{pmatrix}\begin{pmatrix} 1 \\ 0 \end{pmatrix} = \begin{pmatrix} -0.8 \\ 0.6 \end{pmatrix}.$$

These vectors already form an orthonormal basis for $\mathbb{R}^2$*, so we have*

$$U = \begin{pmatrix} -0.6 & -0.8 \\ -0.8 & 0.6 \end{pmatrix}.$$

This yields the SVD

$$A = \begin{pmatrix} -0.6 & -0.8 \\ -0.8 & 0.6 \end{pmatrix}\begin{pmatrix} 10 & 0 \\ 0 & 5 \end{pmatrix}\begin{pmatrix} 0 & 1 \\ 1 & 0 \end{pmatrix}.$$

Now to find the solution of the given linear system, we solve

$$\mathbf{x} = VD^{-1}U^T\mathbf{b}$$

or

$$\begin{pmatrix} x_1 \\ x_2 \end{pmatrix} = \begin{pmatrix} 0 & 1 \\ 1 & 0 \end{pmatrix}\begin{pmatrix} 0.1 & 0 \\ 0 & 0.2 \end{pmatrix}\begin{pmatrix} -0.6 & -0.8 \\ -0.8 & 0.6 \end{pmatrix} = \begin{pmatrix} -0.04 \\ -0.14 \end{pmatrix}$$

and so

$$x_1 = -0.04 \quad and \quad x_2 = -0.14,$$

which is the solution of the given linear system.

7.12 Summary

In this chapter we discussed the approximation of eigenvalues and eigenvectors. We discussed similar, unitary, and diagonalizable matrices. The set of diagonalizable matrices includes matrices with n distinct eigenvalues and symmetric matrices. Matrices that are not diagonalizable are sometimes referred to as defective matrices.

We discussed the Cayley-Hamilton theorem for finding the power and inverse of a matrix. We also discussed the Sourian-Frame theorem, Bocher's theorem, and the Faddeev-Laverrier theorem for computing the coefficients of the characteristic polynomial $p(\lambda)$ of a matrix A. There are no restrictions on A. In theory, the eigenvalues of A can be obtained by factoring $p(\lambda)$ using polynomial root finding techniques. However, this approach is practical only for relatively small values of n.

We discussed many numerical methods for finding eigenvalues and eigenvectors. Many eigen problems do not require computation of all of the eigenvalues. The power method gives us a mechanism for computing the dominant eigenvalue along with its associated eigenvector for an arbitrary matrix. The convergence rate of the power method is poor when the two largest eigenvalues in magnitude are nearly equal. The technique of shifting the matrix by an amount $(-\mu\mathbf{I})$ can help us to overcome this disadvantage, and it can also be used to find intermediate eigenvalues by the power method. Also, if a matrix A is symmetric, then the power method gives faster convergence to the dominant eigenvalue and associated eigenvector. The inverse power method is used to estimate the least dominant eigenvalue of a nonsingular matrix. The inverse power method is guaranteed to converge if a matrix A is diagonalizable with the single least dominant nonzero eigenvalue. The inverse power method requires more computational effort than the power method because a linear algebraic system must be solved at each iteration. The LU decomposition method (Chapter 3) can be used to efficiently accomplish this task. We also discussed the deflation method to obtain other eigenvalues once the dominant eigenvalue is known and the Gerschgorin Circles theorem, which gives a crude approximation to the location of the eigenvalues of a matrix.

A technique for symmetric matrices, which occurs frequently, is the Jacobi method. It is an iterative method that uses orthogonal similarity transformation based on plane rotations to reduce a matrix to diagonal form with diagonal elements as the eigenvalues of a matrix. The rotation matrices are used at the same time to form a matrix whose columns contain the eigenvectors of the matrix. The disadvantage of this method is that it may take many rotations to converge to a diagonal form. The rate of convergence for this method is increased by first preprocessing a matrix by the Given's method and the Householder transformation. These methods use orthogonal similarity transformations to convert a given symmetric matrix to a symmetric tridiagonal matrix.

In the last section we discussed methods dependent on matrix decomposition. Methods such as the QR method and the LR method can be applied to a general matrix. To improve computational efficiency of these methods, instead of applying the decomposition to a original matrix, a original matrix is first transformed to the upper Hessenberg form. In the last section we also discussed the singular values of a matrix and the singular value decomposition of a matrix.

7.13 Exercises

1. Find the characteristic polynomial, eigenvalues, and eigenvectors of each matrix:

$$(a)\begin{pmatrix} 3 & 2 & 1 \\ 2 & 1 & 3 \\ 1 & 2 & 3 \end{pmatrix} \quad (b)\begin{pmatrix} -2 & 1 & 1 \\ -6 & 1 & 3 \\ -12 & -2 & 8 \end{pmatrix} \quad (c)\begin{pmatrix} 2 & -1 & 1 \\ 1 & 2 & -1 \\ 1 & 1 & 2 \end{pmatrix}$$

$$(d)\begin{pmatrix} 1 & 1 & 1 \\ 1 & 1 & 0 \\ 1 & 0 & 1 \end{pmatrix} \quad (e)\begin{pmatrix} 3 & 2 & -2 \\ -3 & -1 & 3 \\ 1 & 2 & 0 \end{pmatrix} \quad (f)\begin{pmatrix} 4 & 3 & 2 & 1 \\ 3 & 3 & 2 & 1 \\ 2 & 2 & 2 & 1 \\ 1 & 1 & 1 & 1 \end{pmatrix}.$$

2. Determine whether each of the given sets of vectors is linearly dependent or independent.
   (a) $(-3,4,2),\ (7,-1,3),$ and $(1,1,8)$.
   (b) $(1,0,2),\ (2,6,4),$ and $(1,12,2)$.
   (c) $(1,-2,1,1),(3,0,2,-2),(0,4,-1,1),$ and $(5,0,3,-1)$.
   (d) $(3,-2,4,5),(0,2,3,-4),(0,0,2,7),$ and $(0,0,0,4)$.

3. For what values of k are the following vectors in $\mathbb{R}^3$ linearly independent?
   (a) $(-1,0,-1),\ (2,1,2),$ and $(1,1,k)$.
   (b) $(1,2,3),\ (2,-1,4),$ and $(3,k,4)$.
   (c) $(2,k,1),\ (1,0,1),$ and $(0,1,3)$.
   (d) $(k,1/2,1/2),\ (1/2,k,1/2),$ and $(1/2,1/2,k)$.

4. Show that the vectors $(1,a,a^2),(1,b,b^2)$, and $(1,c,c^2)$ are linearly independent if
   $a \neq b, a \neq c$, and $b \neq c$.

5. Determine whether the each of the given matrices is diagonalizable:

$$(a)\begin{pmatrix} 1 & 1 & -4 \\ 2 & 0 & -4 \\ -1 & 1 & -2 \end{pmatrix} \quad (b)\begin{pmatrix} 1 & 1 & 0 \\ 3 & 0 & 3 \\ 2 & -1 & 3 \end{pmatrix} \quad (c)\begin{pmatrix} -46 & 6 & 18 \\ 0 & 2 & 0 \\ -120 & 15 & 47 \end{pmatrix}$$

$$(d)\ \begin{pmatrix} -5 & -25 & 6 \\ 10 & 12 & 9 \\ -3 & -9 & 4 \end{pmatrix} \quad (e)\ \begin{pmatrix} 10 & 11 & 3 \\ -3 & -4 & -3 \\ -8 & -8 & 11 \end{pmatrix} \quad (f)\ \begin{pmatrix} 1 & 0 & 0 & 2 \\ 0 & 1 & 3 & 0 \\ 0 & 0 & 2 & 0 \\ 0 & 0 & 0 & 2 \end{pmatrix}.$$

6. Find an 3×3 nondiagonal matrix whose eigenvalues are -2, -2, and 3 and associated eigenvectors are:

$$\begin{pmatrix} 1 \\ 0 \\ 1 \end{pmatrix}, \quad \begin{pmatrix} 0 \\ 1 \\ 1 \end{pmatrix}, \quad \begin{pmatrix} 1 \\ 1 \\ 1 \end{pmatrix}.$$

7. Find a nonsingular matrix Q such that $Q^{-1}AQ$ is a diagonal matrix using Problem 1.

8. Find the formula for the *kth* power of each matrix considered in Problem 5 and then compute A^5.

9. Show that the following matrices are similar:

$$A = \begin{pmatrix} 1 & 0 & 0 \\ 0 & 1 & 0 \\ 1 & 0 & 1 \end{pmatrix} \quad \text{and} \quad B = \begin{pmatrix} 1 & 0 & 0 \\ 1 & 1 & 0 \\ 0 & 0 & 1 \end{pmatrix}.$$

10. Prove that:
   (a) Matrix A is similar to itself.
   (b) If A is similar to B, then B is also similar to A.
   (c) If A is similar to B, and B is similar to C, then A is similar to C.
   (d) If A is similar to B, then $det(A) = det(B)$.
   (e) If A is similar to B, then A^2 is similar to B^2.
   (f) If A is a noninvertible and B is similar to A, then B is also noninvertible.

11. Find a diagonal matrix that is similar to the following given matrices:

$$(a)\ \begin{pmatrix} 3 & -1 & -1 \\ -12 & 0 & 5 \\ 4 & -2 & -1 \end{pmatrix}, \quad (b)\ \begin{pmatrix} -5 & 0 & 0 \\ -4 & 0 & -4 \\ -7 & -7 & 0 \end{pmatrix}, \quad (c)\ \begin{pmatrix} -3 & 2 & -3 \\ 1 & -2 & -3 \\ 1 & -1 & 3 \end{pmatrix}.$$

12. Show that the each of the given matrices is not diagonalizable:

$$(a)\ \begin{pmatrix} 10 & 11 & 3 \\ -3 & -4 & -3 \\ -8 & -8 & -1 \end{pmatrix}, \quad (b)\ \begin{pmatrix} 3 & 3 & 3 \\ 2 & 2 & 2 \\ 1 & 1 & 1 \end{pmatrix}, \quad (c)\ \begin{pmatrix} 2 & 0 & 0 \\ 3 & 2 & 0 \\ 0 & 0 & 5 \end{pmatrix}.$$

13. Find the orthogonal transformations matrix Q to reduce the following given matrices to diagonal matrices:

$$(a)\begin{pmatrix} 1 & 0 & 0 \\ 0 & 0 & 1 \\ 0 & 1 & 0 \end{pmatrix}, \quad (b)\begin{pmatrix} -1 & 2 & 2 \\ 2 & -1 & 2 \\ 2 & 2 & -1 \end{pmatrix}, \quad (c)\begin{pmatrix} 3 & 2 & 2 \\ 2 & 2 & 0 \\ 2 & 0 & 4 \end{pmatrix}$$

$$(d)\begin{pmatrix} 8 & -1 & 1 \\ -1 & 8 & 1 \\ 1 & 1 & 8 \end{pmatrix}, \quad (e)\begin{pmatrix} 5 & -2 & -4 \\ -2 & 8 & -2 \\ -4 & -2 & 5 \end{pmatrix}, \quad (f)\begin{pmatrix} -2 & 3 & 0 \\ 3 & 4 & 0 \\ 0 & 0 & 2 \end{pmatrix}.$$

14. Find a characteristic polynomial and inverse of each of the matrices considered in Problem 5 using the Cayley-Hamilton theorem.

15. Use the Cayley-Hamilton theorem to compute a characteristic polynomial, power powers A^3, A^4, and inverse matrices A^{-1}, A^{-2} for the each of following the given matrices:

$$(a)\begin{pmatrix} 2 & 3 \\ -4 & 5 \end{pmatrix}, \quad (b)\begin{pmatrix} 2 & -2 & 1 \\ -2 & 4 & 2 \\ 1 & 2 & 3 \end{pmatrix}, \quad (c)\begin{pmatrix} 1 & 1 & 0 \\ -1 & 0 & 1 \\ -2 & 1 & 0 \end{pmatrix}.$$

16. Find a characteristic polynomial and inverse for each of the following matrices using the Sourian-Frame theorem:

$$(a)\begin{pmatrix} 5 & 0 & 0 \\ 2 & 1 & 2 \\ 0 & 1 & 1 \end{pmatrix}, \quad (b)\begin{pmatrix} 2 & 2 & 1 \\ -2 & 1 & 2 \\ 1 & -2 & 2 \end{pmatrix}, \quad (c)\begin{pmatrix} 1 & 1 & -1 \\ 3 & 1 & 0 \\ 1 & -2 & 1 \end{pmatrix}.$$

$$(d)\begin{pmatrix} 1 & 2 & 1 \\ 1 & 2 & 0 \\ 1 & 4 & 2 \end{pmatrix}, \quad (e)\begin{pmatrix} 4 & 0 & 0 & 2 \\ 0 & 3 & 0 & -1 \\ 1 & 0 & 2 & 2 \\ 0 & 0 & 0 & 2 \end{pmatrix}, \quad (f)\begin{pmatrix} 1 & 0 & 2 & -1 & 4 \\ 5 & 3 & -1 & 0 & 1 \\ 8 & 5 & -3 & -1 & 4 \\ 6 & 2 & 0 & 0 & 1 \\ 0 & 1 & 4 & 2 & 0 \end{pmatrix}.$$

17. Find a characteristic polynomial and inverse of each of the given matrices considered in Problem 11 using the Sourian-Frame theorem.

18. Use Bocher's formula to find the coefficients of the characteristic equation of each of the matrices considered in Problem 1.

19. Find a characteristic equation, determinant, adjoint, and inverse of each of the following given matrices using the Faddeev-Leverrer method:

$$(a)\ \begin{pmatrix} 3 & 5 & 0 \\ 4 & -2 & 1 \\ 6 & -3 & 4 \end{pmatrix},\quad (b)\ \begin{pmatrix} 5 & 5 & 5 \\ 2 & 10 & 1 \\ 6 & 3 & -9 \end{pmatrix},\quad (c)\ \begin{pmatrix} 1 & -3 & 0 & -2 \\ 3 & -12 & -2 & -6 \\ -2 & 10 & 2 & 5 \\ -1 & 6 & 1 & 3 \end{pmatrix}.$$

20. Find the exponential of each of the matrices considered in Problem 1.

21. Find the first four iterations of the power method applied to the following matrices:

$$(a)\ \begin{pmatrix} 2 & 3 & 1 \\ 1 & 4 & -1 \\ 3 & 1 & 2 \end{pmatrix}, \qquad \text{start} \quad \mathbf{x}_0 = [0, 1, 1]^T.$$

$$(b)\ \begin{pmatrix} 5 & 4 & 6 \\ 2 & 2 & -3 \\ 3 & 1 & 1 \end{pmatrix}, \qquad \text{start} \quad \mathbf{x}_0 = [1, 1, 1]^T$$

$$(c)\ \begin{pmatrix} 1 & 1 & 1 \\ -2 & 2 & 1 \\ 5 & 1 & 1 \end{pmatrix}, \qquad \text{start} \quad \mathbf{x}_0 = [1, 1, 0]^T.$$

$$(d)\ \begin{pmatrix} 3 & 0 & 0 & 2 \\ 0 & 3 & 0 & -1 \\ 1 & 0 & 2 & 2 \\ 0 & 0 & 4 & 2 \end{pmatrix}, \qquad \text{start} \quad \mathbf{x}_0 = [1, 0, 0, 0]^T.$$

22. Repeat Problem 21 using the inverse power method.

23. Find the first four iterations of the following matrices using the shifted inverse power method:

$$(a)\ \begin{pmatrix} 2 & 3 & 3 \\ 1 & 4 & -1 \\ 3 & 1 & 2 \end{pmatrix}, \qquad \text{start} \quad \mathbf{x}_0 = [0, 1, 1]^T, \quad \mu = 4.5.$$

$$(b)\ \begin{pmatrix} 1 & 1 & -1 \\ 2 & 1 & -3 \\ 2 & -4 & 1 \end{pmatrix}, \qquad \text{start} \quad \mathbf{x}_0 = [1, 1, 1]^T, \quad \mu = 5.$$

$$(c) \quad \begin{pmatrix} 1 & 1 & 1 \\ -2 & 2 & 1 \\ 3 & 3 & 3 \end{pmatrix}, \qquad \text{start} \quad \mathbf{x}_0 = [1,1,0]^T, \quad \mu = 4.$$

$$(d) \quad \begin{pmatrix} 3 & 0 & 3 & 2 \\ 1 & 3 & 0 & -1 \\ 1 & 0 & 2 & 2 \\ 0 & 0 & 0 & 2 \end{pmatrix}, \qquad \text{start} \quad \mathbf{x}_0 = [1,0,0,0]^T, \quad \mu = 3.5.$$

24. Also, solve by using the shifted inverse power method, with $\mathbf{x}^{(0)} = [1,1,1]^t$, (only four iterations):

$$A = \begin{pmatrix} 3 & 0 & 1 \\ 2 & 2 & 2 \\ 4 & 2 & 5 \end{pmatrix}.$$

Also, solve using the shifted inverse power method by taking the initial value of the eigenvalue using the Rayleigh Quotient theorem.

25. Use the Gerschgorin Circles theorem to determine bounds for the eigenvalues of each of the given matrices:

$$(a) \quad \begin{pmatrix} 3 & 2 & 1 \\ 2 & 3 & 0 \\ 1 & 0 & 3 \end{pmatrix}, \quad (b) \quad \begin{pmatrix} 1 & 1 & 1 \\ 1 & 1 & 0 \\ 1 & 0 & 1 \end{pmatrix}, \quad (c) \quad \begin{pmatrix} 2 & -2 & 1 \\ -2 & 1 & 1 \\ 1 & 1 & 2 \end{pmatrix}.$$

26. Consider the following matrix:

$$A = \begin{pmatrix} 1 & 1 & -2 \\ -1 & 2 & 1 \\ 0 & 1 & -1 \end{pmatrix},$$

which has an eigenvalue 2 with an eigenvector $[1,3,1]^T$. Use the deflation method to find the remaining eigenvalues and eigenvectors of A.

27. Consider the following matrix:

$$A = \begin{pmatrix} 2 & -3 & 6 \\ 0 & 3 & -4 \\ 0 & 2 & -3 \end{pmatrix},$$

which has an eigenvalue 2 with an eigenvector $[1,0,0]^T$. Use the deflation method to find the remaining eigenvalues and eigenvectors of A.

28. Consider the following matrix:

$$A = \begin{pmatrix} 8 & -2 & -3 & 1 \\ 7 & -1 & -3 & 1 \\ 6 & -2 & -1 & 1 \\ 5 & -2 & -3 & 4 \end{pmatrix},$$

which has an eigenvalue 4 with an eigenvector $[1, 1, 1, 1]^T$. Use the deflation method to find the remaining eigenvalues and eigenvectors of A.

29. Find the eigenvalues and corresponding eigenvectors of the following matrices using the Jacobi method:

$$(a) \begin{pmatrix} 5 & 3 & 7 \\ 3 & 1 & 6 \\ 7 & 6 & 2 \end{pmatrix}, \quad (b) \begin{pmatrix} 2 & -1.5 & 0 \\ -1.5 & 2 & -0.5 \\ 0 & -0.5 & 2 \end{pmatrix}, \quad (c) \begin{pmatrix} 4 & 6 & 7 \\ 6 & 5 & -3 \\ 7 & -3 & 2 \end{pmatrix}$$

$$(d) \begin{pmatrix} 0.4 & 0.3 & 0.1 \\ 0.3 & 0.5 & 0.2 \\ 0.1 & 0.2 & 0.6 \end{pmatrix}, \quad (e) \begin{pmatrix} 4 & 4 & 4 & 1 \\ 4 & 6 & 1 & 4 \\ 4 & 1 & 6 & 4 \\ 1 & 4 & 4 & 6 \end{pmatrix}, \quad (f) \begin{pmatrix} 2 & -1 & 3 & 2 \\ -1 & 3 & 1 & -2 \\ 3 & 1 & 4 & 1 \\ 2 & -2 & 1 & -3 \end{pmatrix}.$$

30. Use the Sturm sequence iteration to find number of eigenvalues of the following matrices lying on the given intervals (a, b):

$$(a) \begin{pmatrix} 2 & -1 & 0 \\ -1 & 2 & -1 \\ 0 & -1 & 2 \end{pmatrix}, \quad (-1, 3) \quad (b) \begin{pmatrix} 5 & -1 & 0 \\ -1 & 2 & 2 \\ 0 & 2 & 3 \end{pmatrix}, \quad (0, 4).$$

31. Use the Sturm sequence iteration to find eigenvalues of the following matrices:

$$(a) \begin{pmatrix} 1 & 4 & 0 \\ 4 & 1 & 4 \\ 0 & 4 & 1 \end{pmatrix}, \quad (b) \begin{pmatrix} 1 & 2 & 0 \\ 2 & 2 & 4 \\ 0 & 4 & 4 \end{pmatrix}, \quad (c) \begin{pmatrix} 1 & 2 & 0 \\ 2 & 1 & 2 \\ 0 & 2 & 1 \end{pmatrix}$$

32. Use Given's method to convert each matrix considered in Problem 29 into tridiagonal form.

33. Use Given's method to convert each matrix considered in Problem 31 into tridiagonal form and then use the Sturm sequence iteration to find the eigenvalues of each of the matrices.

34. Use Householder's method to convert each matrix considered in Problem 29 into tridiagonal form.

35. Use Householder's method to convert each matrix into tridiagonal form and then use the Sturm sequence iteration to find the eigenvalues of each of the matrices:

$$(a)\ \begin{pmatrix} 2 & 3 & 4 \\ 3 & 4 & 5 \\ 4 & 5 & 6 \end{pmatrix},\quad (b)\ \begin{pmatrix} 5 & -2 & 1 \\ -2 & 7 & 9 \\ 1 & 9 & 8 \end{pmatrix},\quad (c)\ \begin{pmatrix} 4 & -2 & 1 & 4 \\ -2 & 5 & 0 & 3 \\ 1 & 0 & 6 & 2 \\ 4 & 3 & 2 & 7 \end{pmatrix}$$

36. Find the first four QR iterations for each of the following matrices:

$$(a)\ \begin{pmatrix} 1 & 0 & 2 \\ -2 & 1 & 1 \\ -2 & -5 & 1 \end{pmatrix},\quad (b)\ \begin{pmatrix} 2 & -1 & 2 \\ 3 & 1 & 0 \\ 0 & 2 & 1 \end{pmatrix},\quad (c)\ \begin{pmatrix} -21 & -9 & 12 \\ 0 & 6 & 0 \\ -24 & -8 & 15 \end{pmatrix}.$$

37. Find the first 15 QR iterations for each matrix in Problem 31.

38. Find the eigenvalues using the LR method for each of the following matrices:

$$(a)\ \begin{pmatrix} 1 & 2 & 4 \\ 5 & 1 & 1 \\ 2 & 1 & 1 \end{pmatrix},\quad (b)\ \begin{pmatrix} 3 & 3 & 3 \\ 3 & 3 & 3 \\ -3 & -3 & -3 \end{pmatrix},\quad (c)\ \begin{pmatrix} 15 & 13 & 20 \\ -21 & 12 & 15 \\ -8 & -8 & 11 \end{pmatrix}.$$

39. Find the eigenvalues using the LR method for each of the following matrices:

$$(a)\ \begin{pmatrix} 3 & 1 & 1 \\ 2 & 1 & 1 \\ 1 & 1 & 1 \end{pmatrix},\quad (b)\ \begin{pmatrix} 2 & 1 & 2 \\ 3 & 1 & 0 \\ 1 & 2 & 1 \end{pmatrix},\quad (c)\ \begin{pmatrix} 4 & 0 & 1 \\ -2 & 1 & 0 \\ -2 & 0 & 1 \end{pmatrix}.$$

40. Transform each of the following matrices into upper Hessenberg form:

$$(a)\ \begin{pmatrix} 1 & 6 & 4 \\ 5 & 1 & 3 \\ 2 & 4 & 4 \end{pmatrix},\quad (b)\ \begin{pmatrix} 5 & 4 & 3 \\ 2 & 3 & 3 \\ -3 & -3 & 8 \end{pmatrix},\quad (c)\ \begin{pmatrix} 2 & 5 & 2 \\ 11 & 6 & 7 \\ 9 & 15 & 22 \end{pmatrix}$$

$$(d)\ \begin{pmatrix} 2 & -1 & 4 & 2 \\ 3 & 2 & 3 & 2 \\ 1 & 2 & 2 & 2 \\ 2 & -3 & 4 & 4 \end{pmatrix},\quad (e)\ \begin{pmatrix} 2 & 1 & -2 & -3 \\ 2 & 2 & -3 & 2 \\ -3 & -3 & 4 & 5 \\ 7 & 8 & 3 & 2 \end{pmatrix},\quad (f)\ \begin{pmatrix} 9 & 2 & 1 & -2 \\ 2 & 1 & 1 & -5 \\ -2 & 1 & 6 & -2 \\ -2 & -1 & 1 & -3 \end{pmatrix}.$$

41. Transform each of the following matrices into upper Hessenberg form using Gaussian elimination with pivoting. Then use the QR method and the LR method to find their eigenvalues:

$$(a) \begin{pmatrix} 11 & 33 & 45 \\ 12 & 21 & 23 \\ 18 & 22 & 31 \end{pmatrix}, \quad (b) \begin{pmatrix} 4 & 3 & 3 \\ 2 & 5 & 4 \\ -3 & 2 & 1 \end{pmatrix}, \quad (c) \begin{pmatrix} 14 & 22 & 2 & 1 \\ 5 & 1 & 5 & -2 \\ 6 & 1 & 6 & 1 \\ 7 & -2 & 7 & 4 \end{pmatrix}.$$

42. Find the singular values for each of the following matrices:

$$(a) \begin{pmatrix} 2 & 0 & 1 \\ 0 & 2 & 0 \end{pmatrix}, \quad (b) \begin{pmatrix} 3 & 0 & 0 \\ -2 & 3 & -2 \\ 2 & 0 & 5 \end{pmatrix}, \quad (c) \begin{pmatrix} 1 & 0 & 1 \\ 0 & 1 & 0 \\ 0 & 1 & 2 \end{pmatrix}$$

$$(d) \begin{pmatrix} 4 & 0 & 1 \\ 0 & 1 & 0 \\ 2 & 1 & 1 \end{pmatrix}, \quad (e) \begin{pmatrix} 2 & 0 & 1 \\ -4 & 6 & -2 \\ 2 & 0 & 7 \end{pmatrix}, \quad (f) \begin{pmatrix} 2 & 0 & 1 & 2 \\ 0 & 1 & 1 & 3 \\ 0 & 3 & 2 & 1 \\ 1 & 0 & 3 & 1 \end{pmatrix}.$$

43. Show that the singular values of the following matrices are the same as the eigenvalues of the matrices:

$$(a) \begin{pmatrix} 4 & 2 & 1 \\ 2 & 8 & 0 \\ 1 & 0 & 8 \end{pmatrix}, \quad (b) \begin{pmatrix} 3 & 0 & 1 \\ 0 & 5 & 0 \\ 1 & 0 & 5 \end{pmatrix}, \quad (c) \begin{pmatrix} 2 & 0 & 0 \\ 0 & 6 & 0 \\ 0 & 0 & 7 \end{pmatrix}.$$

44. Show that if A is a positive definite matrix, then A has a singular value decomposition of the form QDQ^T.

45. Show that all singular values of an orthogonal matrix are 1.

46. Find an SVD for each of the following matrices:

$$(a) \begin{pmatrix} 0 & -4 \\ -6 & 0 \end{pmatrix}, \quad (b) \begin{pmatrix} 2 & 1 & 0 \\ 1 & 3 & 0 \end{pmatrix}$$

$$(c) \begin{pmatrix} 1 & 0 \\ 1 & 2 \\ -1 & 3 \end{pmatrix}, \quad (d) \begin{pmatrix} 1 & 2 & 1 \\ 1 & 2 & 1 \end{pmatrix}.$$

47. Find an SVD for each of the following matrices:

(a) $\begin{pmatrix} 0 & -2 \\ -3 & 0 \end{pmatrix}$, (b) $\begin{pmatrix} 2 & 0 & 0 \\ 0 & 3 & 0 \end{pmatrix}$

(c) $\begin{pmatrix} 1 & 0 \\ 1 & 1 \\ -1 & 1 \end{pmatrix}$, (d) $\begin{pmatrix} 1 & 1 & 1 \\ 1 & 1 & 1 \end{pmatrix}$.

48. Find an SVD for each of the following matrices:

(a) $\begin{pmatrix} 0 & -2 & 1 \\ -3 & 0 & 2 \\ 0 & 1 & 1 \end{pmatrix}$, (b) $\begin{pmatrix} 2 & -4 & 3 \\ 6 & 6 & 3 \\ -4 & 2 & 4 \end{pmatrix}$

(c) $\begin{pmatrix} 1 & 2 & 0 \\ 1 & 1 & 1 \\ -1 & 1 & 2 \end{pmatrix}$, (d) $\begin{pmatrix} 8 & -3 & 7 & 1 \\ 3 & 11 & 3 & 2 \\ 1 & 2 & 5 & 2 \\ 2 & 0 & 7 & 2 \end{pmatrix}$

(e) $\begin{pmatrix} 2 & 1 & -2 & -13 \\ 11 & 12 & -3 & 12 \\ 3 & 22 & 24 & 15 \\ 7 & 8 & 3 & 2 \end{pmatrix}$, (f) $\begin{pmatrix} 10 & 1 & 0 & -2 \\ 2 & 1 & 1 & -5 \\ 4 & 6 & 7 & -2 \\ 5 & -1 & 1 & -3 \end{pmatrix}$.

49. Find the solution for each of the following linear systems $A\mathbf{x} = \mathbf{b}$ using singular value decomposition:

(a)

$$A = \begin{pmatrix} 1 & -3 \\ 3 & -5 \end{pmatrix}, \quad \mathbf{x} = \begin{pmatrix} x_1 \\ x_2 \end{pmatrix}, \quad \mathbf{b} = \begin{pmatrix} 1 \\ 2 \end{pmatrix}$$

(b)

$$A = \begin{pmatrix} 1 & -1 \\ 1 & 4 \end{pmatrix}, \quad \mathbf{x} = \begin{pmatrix} x_1 \\ x_2 \end{pmatrix}, \quad \mathbf{b} = \begin{pmatrix} 1.1 \\ 0.5 \end{pmatrix}$$

(c)

$$A = \begin{pmatrix} 3 & -1 & 4 \\ -1 & 0 & 1 \\ 4 & 1 & 2 \end{pmatrix}, \quad \mathbf{x} = \begin{pmatrix} x_1 \\ x_2 \\ x_3 \end{pmatrix}, \quad \mathbf{b} = \begin{pmatrix} 1 \\ 2 \\ 3 \end{pmatrix}$$

(d)

$$A = \begin{pmatrix} 4 & 3 & 2 \\ 1 & 2 & -1 \\ 1 & 3 & 2 \end{pmatrix}, \quad \mathbf{x} = \begin{pmatrix} x_1 \\ x_2 \\ x_3 \end{pmatrix}, \quad \mathbf{b} = \begin{pmatrix} 2.5 \\ 1.5 \\ 0.85 \end{pmatrix}.$$

50. Find the solution each of the following linear system $A\mathbf{x} = \mathbf{b}$ using singular value decomposition.

(a)
$$A = \begin{pmatrix} 2 & 2 \\ 1 & 3 \end{pmatrix}, \quad \mathbf{x} = \begin{pmatrix} x_1 \\ x_2 \end{pmatrix}, \quad \mathbf{b} = \begin{pmatrix} 1 \\ 0.9 \end{pmatrix}$$

(b)
$$A = \begin{pmatrix} 1 & 0 \\ 3 & -2 \end{pmatrix}, \quad \mathbf{x} = \begin{pmatrix} x_1 \\ x_2 \end{pmatrix}, \quad \mathbf{b} = \begin{pmatrix} 1 \\ 2 \end{pmatrix}$$

(c)
$$A = \begin{pmatrix} 1 & -1 & 0 \\ 2 & 0 & 1 \\ 3 & 0 & 2 \end{pmatrix}, \quad \mathbf{x} = \begin{pmatrix} x_1 \\ x_2 \\ x_3 \end{pmatrix}, \quad \mathbf{b} = \begin{pmatrix} 1 \\ 1 \\ 1 \end{pmatrix}$$

(d)
$$A = \begin{pmatrix} 1 & 2 & 3 \\ 2 & 1 & 2 \\ 1 & 1 & 1 \end{pmatrix}, \quad \mathbf{x} = \begin{pmatrix} x_1 \\ x_2 \\ x_3 \end{pmatrix}, \quad \mathbf{b} = \begin{pmatrix} 1 \\ 0 \\ 1 \end{pmatrix}.$$

Appendix A

Some Mathematical Preliminaries

Because the number of results and theorems from calculus are frequently used in this book, we collect here a number of these results for ready reference, and to refresh the student's memory.

Open and Closed Intervals

For the open interval $a < x < b$, we use notation (a, b), and for the closed interval $a \leq x \leq b$, we use notation $[a, b]$.

Limits and Continuity

Definition A.1 (Limits)

Let a function f be defined in an open interval and L be a real number. Then for the limit of a function, we write

$$\lim_{x \to 0} f(x) = L$$

if for every $\epsilon > 0$, *there is a* $\delta > 0$ *such that*

$$\text{if} \quad 0 < |x - a| < \delta, \quad \text{then} \quad 0 < |f(x) - L| < \epsilon.$$

Definition A.2 (Continuity)

A function f *is continuous at* $x = a$ *if it satisfies the following three conditions:*

$$\begin{array}{ll} (a) & f(x) \text{ is defined at } x = a. \\ (b) & \lim_{x \to a} f(x) \text{ exists.} \\ (c) & \lim_{x \to a} f(x) = f(a). \end{array}$$

Note that:
1. A polynomial function f *is continuous at each point of the real line.*
2. If a function is continuous for all x*-values in an interval, it is said to be continuous on the interval.*
3. Let $f(x)$ *be continuous on the closed interval* $[a, b]$ *then* $f(x)$ *assumes its maximum and minimum values on* $[a, b]$*; that is, there are real numbers* $x_1, x_2 \in [a, b]$ *such that*

$$f(x_1) \leq f(x) \leq f(x_2)$$

for all $x \in [a, b]$.

Definition A.3 *Suppose that* $\{x_n\}_{n=1}^{\infty}$ *is an infinite sequence. Then the sequence is said to have the limit* L, *and we write*

$$\lim_{n \to \infty} x_n = L$$

if, given any $\epsilon > 0$, *there exists a positive integer* $N = N(\epsilon)$ *such that* $n > N$ *implies that* $|x_n - L| < \epsilon$.

When a sequence has a limit, we say that it is a convergent sequence.

Theorem A.1 *Assume that* $f(x)$ *is defined on the set* S *and* $x_0 \in S$. *The following statements are equivalents:*
(a) The function $f(x)$ *is continuous at* x_0.
(b) If $\lim_{n \to \infty} x_n = x_0$, *then* $\lim_{n \to \infty} f(x_n) = f(x_0)$.

Differentiation

Definition A.4 *The derivative of a function f is the function f' defined by*

$$f'(x) = Df(x) = \frac{df(x)}{dx} = \lim_{h \to 0} \frac{f(x+h) - f(x)}{h}$$

provided the limit exists.

Theorem A.2 *A polynomial function f is differentiable at each point of the real line.*

Theorem A.3 *If a function f is differentiable at $x = a$, then f is continuous at a.*

Some Differentiation Formulas

$$\begin{aligned}
[f(x) + g(x)]' &= f'(x) + g'(x), \qquad \text{(sum rule)} \\
[f(x)g(x)]' &= f'(x)g(x) + f(x)g'(x), \qquad \text{(product rule)} \\
\left[\frac{f(x)}{g(x)}\right]' &= \frac{g(x)f'(x) - f(x)g'(x)}{(g(x))^2}, \qquad \text{(quotent rule)} \\
[f(g(x))]' &= f'(g(x))g'(x) \qquad \text{(chain rule)} \\
\frac{d}{dx}x^n &= nx^{n-1}, \quad n \neq 0, \qquad \text{(power rule)} \\
\frac{d}{dx}\sin(x) &= \cos(x), \qquad \frac{d}{dx}\cos(x) = -\sin(x) \\
\frac{d}{dx}\tan(x) &= \frac{d}{dx}\left(\frac{\sin(x)}{\cos(x)}\right) = \sec^2(x)
\end{aligned}$$

$$\frac{d}{dx}\cot(x) = \frac{d}{dx}\left(\frac{\cos(x)}{\sin(x)}\right) = -\csc^2(x)$$

$$\frac{d}{dx}\csc(x) = \frac{d}{dx}\left(\frac{1}{\sin(x)}\right) = -\csc(x)\cot(x)$$

$$\frac{d}{dx}\sec(x) = \frac{d}{dx}\left(\frac{1}{\cos(x)}\right) = \sec(x)\tan(x)$$

$$\frac{d}{dx}[\sin^{-1}(x)] = \frac{1}{\sqrt{1-x^2}}, \qquad \frac{d}{dx}[\cos^{-1}(x)] = \frac{-1}{\sqrt{1-x^2}}$$

$$\frac{d}{dx}[\tan^{-1}(x)] = \frac{1}{1+x^2}, \qquad \frac{d}{dx}[\cot^{-1}(x)] = \frac{-1}{1+x^2}$$

$$\frac{d}{dx}[\csc^{-1}(x)] = \frac{1}{|x|\sqrt{x^2-1}}, \qquad \frac{d}{dx}[\sec^{-1}(x)] = \frac{-1}{|x|\sqrt{x^2-1}}$$

$$\frac{d}{dx}[e^x] = e^x, \qquad \frac{d}{dx}[a^x] = [\log a]a^x$$

$$\frac{d}{dx}[\ln x] = \frac{1}{x}, \qquad \frac{d}{dx}[\log_a x] = \frac{1}{x(\log a)}$$

Integration Some Integration Formulas

$$\int_a^b [f(x)+g(x)]\,dx = \int_a^b f(x)\,dx + \int_a^b g(x)dx$$

$$\int_a^b cf(x)\,dx = c\int_a^b f(x)\,dx, \qquad c \text{ is constant}$$

$$\int_a^b f(x)\,dx = \int_a^c f(x)\,dx + \int_c^b f(x)dx$$

$$\int_a^b f(x)\,dx = -\int_b^a f(x)\,dx$$

$$\int_a^b f(x)g'(x)\,dx = \Big[f(x)g(x)\Big]_a^b - \int_a^b f'(x)g(x)\,dx, \qquad \text{(integration by parts)}$$

$$\int_a^b f(g(x))g'(x)\,dx = \int_{g(a)}^{g(b)} f(u)\,du$$

$$\int x^n\,dx = \frac{x^{n+1}}{n+1} + C, \quad n \neq -1$$

$$\int \frac{1}{x}\,dx = \ln|x| + C$$

$$\int e^x\,dx = e^x + C, \qquad \int a^x\,dx = \frac{a^x}{\log(a)} + C$$

$$\int \frac{1}{x^2+a^2}\,dx = \frac{1}{a}\tan^{-1}\left(\frac{x}{a}\right) + C$$

$$\int \frac{1}{\sqrt{a^2-x^2}}\,dx = \sin^{-1}\left(\frac{x}{a}\right) + C$$

$$\int \frac{1}{x\sqrt{x^2-a^2}}\,dx = \frac{1}{a}\sec^{-1}\left|\frac{x}{a}\right| + C$$

$$\int \sin(x)\,dx = -\cos(x) + C, \qquad \int \cos(x)\,dx = \sin(x) + C$$

$$\int \tan(x)\,dx = \ln|\sec(x)| + C$$

$$\int \cot(x)\,dx = \ln|\sin(x)| + C$$

$$\int \sec(x)\,dx = \ln|\sec(x) + \tan(x)| + C$$

$$\int \csc(x)\,dx = \ln|\csc(x) - \cot(x)| + C$$

$$\int \sin^2(x)\,dx = \frac{x}{2} - \frac{\sin(2x)}{4} + C, \qquad \left(\sin^2(x) = \frac{1-\cos(2x)}{2}\right)$$

$$\int \cos^2(x)\,dx = \frac{x}{2} + \frac{\sin(2x)}{4} + C, \qquad \left(\cos^2(x) = \frac{1+\cos(2x)}{2}\right)$$

$$\int \tan^2(x)\,dx = \tan(x) - x + C,$$

$$\int \cot^2(x)\,dx = -\cot(x) - x + C,$$

$$\int \sec^n(x)\,dx = \frac{1}{n-1}\sec^{n-2}(x)\tan(x) + \frac{n-2}{n-1}\int \sec^{n-2}(x)\,dx + C,$$

$$\int \csc^n(x)\,dx = -\frac{1}{n-1}\csc^{n-2}(x)\cot(x) + \frac{n-2}{n-1}\int \csc^{n-2}(x)\,dx + C,$$

$$\int \sin^{-1}(x)\,dx = x\sin^{-1}(x) + \sqrt{1-x^2} + C,$$

$$\int \cos^{-1}(x)\,dx = x\cos^{-1}(x) - \sqrt{1-x^2} + C,$$

$$\int \tan^{-1}(x)\,dx = x\tan^{-1}(x) - \frac{1}{2}\ln(1+x^2) + C,$$

Definition A.5 *The definite integral is defined as*

$$\int_a^b f(x)\,dx = \lim_{n\to\infty}\sum_{j=1}^{n} f(p_j)(x_j - x_{j-1})$$

with $a = x_0 < x_1 < \cdots < x_n = b$, p_j *in the interval* $[x_{j-1}, x_j]$, *and* $max(x_j - x_{j-1}) \to 0$ *as* $n \to \infty$. *The above sums are called Riemann sums.*

Theorem A.4 (Fundamental Theorem for Calculus)
Suppose that f *is continuous on the closed interval* $[a, b]$. *Let* $F(x)$ *be an antiderivative of* $f(x)$ *for all* x *in* $[a, b]$. *Then*
Form 1

$$\int_a^b f(x)\,dx = \Big[F(x)\Big]_{x=a}^{x=b} = F(b) - F(a).$$

Form 2

$$\frac{d}{dx}[F(x)] = f(x) = \frac{d}{dx}\int_a^x f(t)\,dt.$$

Quadratic Formula

The familiar solutions of the quadratic equation

$$ax^2 + bx + c = 0, \quad a \neq 0$$

where a, b, and c are real numbers, are given by

$$x = \frac{-b \pm \sqrt{b^2 - 4ac}}{2a}.$$

A less familiar form of the solution is

$$x = \frac{2c}{[-b \pm \sqrt{b^2 - 4ac}]}.$$

Series Expansions for Common Functions

Binomial

$$(x+y)^n = x^n + nx^{n-1}y + \frac{n(n-1)}{2!}x^{n-2}y^2 + \cdots + y^n, \quad y^2 < x^2$$

or

$$(x+y)^n = \sum_{i=0}^{n} \binom{n}{i} x^{n-i}y^i,$$

where

$$\binom{n}{i} = \frac{n!}{i!(n-i)!}.$$

Exponential and Logarithmic

$$e^x = 1 + x + \frac{x^2}{2!} + \frac{x^3}{3!} + \cdots$$

$$\ln x = 2(u + \frac{u^3}{3} + \frac{u^5}{5} + \cdots), \qquad u = (x-1)(x+1)$$

$$\ln(x+1) = x - \frac{x^2}{2} + \frac{x^3}{3} - \frac{x^4}{4} + \cdots, \qquad -1 \le x \le 1$$

Trigonometric

$$\sin x = x - \frac{x^3}{3!} + \frac{x^5}{5!} - \frac{x^7}{7!} + \cdots, \qquad x \text{ in radians}$$

$$\cos x = 1 - \frac{x^2}{2!} + \frac{x^4}{4!} - \frac{x^6}{6!} + \cdots, \qquad x \text{ in radians}$$

$$\tan x = x + \frac{x^3}{3} + \frac{2x^5}{15} + \frac{17x^7}{315} + \cdots, \qquad x \text{ in radians}$$

Partial Derivatives

Suppose that f is a function of several variables, say $x_1, x_2, \ldots, x_n$. It will be convenient to use the vector $\mathbf{x}$ to denote the n-tuple $(x_1, x_2, \ldots, x_n)$. Then $f(x_1, x_2, \ldots, x_n)$ will be abbreviated as $f(\mathbf{x})$

Definition A.6 (Partial Derivative of a Function)

The partial derivative of f with respect to x_j at $\mathbf{x} = (x_1, x_2, \ldots, x_n)$ is the number

$$\frac{\partial f}{\partial x_j}(\mathbf{x}) = \lim_{h \to 0} \frac{f(\mathbf{x} + h\mathbf{u}_j) - f(\mathbf{x})}{h},$$

where

$$\mathbf{x} + h\mathbf{u}_j = (x_1, \ldots, x_{j-1}, x_j + h, x_{j+1}, \ldots, x_n).$$

The number $\dfrac{\partial f}{\partial x_j}(\mathbf{x})$ *is also denoted by* $f_{x_j}(\mathbf{x})$.

Theorem A.5 *If the second partial*

$$f_{x_i x_j} = \frac{\partial}{\partial x_j}(f_{x_i}) \quad \text{and} \quad f_{x_j x_i} = \frac{\partial}{\partial x_i}(f_{x_j})$$

both exist in an open region (of n-space) on which one of them is continuous, then

$$f_{x_i x_j}(\mathbf{x}) = f_{x_j x_i}(\mathbf{x}), \quad \text{for all } \mathbf{x} \text{ in this region.}$$

This result allows us to find higher-order partials of f in any convenient order as long as the result is continuous.

Definition A.7 (Total Differential)

If $f_{x_1}(\mathbf{x}), f_{x_2}(\mathbf{x}), \ldots, f_{x_n}(\mathbf{x})$ all exits, then the total differential of f at x is defined by

$$df(\mathbf{x}) = f_{x_1}(\mathbf{x})dx_1 + f_{x_2}(\mathbf{x})dx_2 + \cdots + f_{x_n}(\mathbf{x})dx_n,$$

where $dx_1, dx_2, \ldots, dx_n$ can be viewed as increments from $x_1, x_2, \ldots, x_n$, respectively. The following result justifies the use of linear approximation

$$f(\mathbf{x} + \mathbf{d}x) = f(x_1 + dx_1, \ldots, x_n + dx_n) \approx f(\mathbf{x}) + df(\mathbf{x})$$

when increment $\mathbf{d}x = (dx_1, \ldots, dx_n)$ *is small in the sense that*

$$\|\mathbf{d}x\| = max\{|dx_1|, \ldots, |dx_n|\} \approx 0.$$

Theorem A.6 (Chain Rule)

If f *is a differentiable function of* $x_1, x_2, \ldots, x_n$ *and each of these is a function of* $u_1, \ldots, u_m$*, then for any* $j = 1, 2, \ldots, m$

$$\frac{\partial f}{\partial u_j} = \frac{\partial f}{\partial x_1} \cdot \frac{\partial x_1}{\partial u_j} + \frac{\partial f}{\partial x_2} \cdot \frac{\partial x_2}{\partial u_j} + \cdots + \frac{\partial f}{\partial x_n} \cdot \frac{\partial x_n}{\partial u_j},$$

where the partials with respect to u_j *are at a fixed* $\mathbf{u} = (u_1, u_2, \ldots, u_m)$ *and the partials with respect to* $x_1, x_2, \ldots, x_n$ *are at* $\mathbf{x} = (x_1, x_2, \ldots, x_n)$ *corresponding to this fixed* $\mathbf{u}$.

Theorem A.7 (Extreme Value Theorem)

If f *is continuous on a closed interval* $[a, b]$*, then* f *assumes a maximum value* M *and a minimum value* m*; that is, there exists points* x_M *and* x_m *in* $[a, b]$ *such that*

$$m = f(x_m) \leq f(x) \leq f(x_M) = M, \qquad \text{for any} \quad x \in [a, b].$$

If, in addition, f *is differentiable on* $[a, b]$*, then numbers* x_m *and* x_M *occur either at endpoints of* $[a, b]$ *or where* f' *is zero.*

Theorem A.8 (Intermediate Value Theorem)

If f *is continuous on a closed interval* $[a, b]$ *and* $m < c < M$*, where* m *and* M *are minimum and maximum values of function, respectively, then there exists at least one point* η *in* $[a, b]$ *such that*

$$f(\eta) = c.$$

Note that if f *is continuous on* $[a, b]$ *and* $f(a)$ *and* $f(b)$ *have opposite sign, then it follows from the Intermediate Value theorem that* f *has at least one root* η *between* a *and* b *such that*

$$f(\eta) = 0.$$

Taken together, the Extreme value Theorem and the Intermediate Value Theorem say that if f *is continuous on a closed interval* $[a, b]$*, then its range is also a closed* *interval, namely* $[m, M]$*; moreover, if* $\eta_1, \eta_2, \ldots, \eta_n$ *lie on the interval* $[a, b]$ *then, since*

$$nm \leq \sum_{k=1}^{n} f(\eta_k) \leq nM$$

there must be a point η *in* $[a, b]$ *such that*

$$f(\eta) = \frac{1}{n}[f(\eta_1) + \cdots + f(\eta_n)].$$

Theorem A.9 (Mean Value Theorem)

If f *is continuous on a closed interval* $[a, b]$ *and differentiable on* (a, b)*, then there exists at least one number* η *between a and b such that*

$$f(b) = f(a) + f'(\eta)(b - a).$$

Suppose that f *is continuous on* $[a, b]$ *and* $f^{(n)}$ *exists on* (a, b) *and it is known that* f *has* $n + 1$ *distinct roots, say* $x_0, x - 1, \ldots, x_n$*, on* $[a, b]$*. Since* $f(\eta_i) = 0$ *for* $i = 0, 1, \ldots, n$*, applying the Mean Value Theorem to the n subintervals*

$$[x_0, x_1],\ [x_1, x_2],\ \ldots,\ [x_{n-1}, x_n]$$

shows that there are points η_i *in* (x_{i-1}, x_i) *such that* $f'(\eta_i) = 0$ *for* $i = 0, 1, \ldots, n$*. In other words,* f' *has n distinct roots on* (a, b)*. Continuing inductively, we see that* $f'', f''', \ldots, f^{(n)}$ *have, respectively,* $n - 1, n - 2, \ldots, 1$ *roots. This proves the following result.*

Theorem A.10 (Generalized Rolle's Theorem)

If f *is continuous on a closed interval* $[a, b]$ *and* $f^{(n)}$ *exists on* (a, b)*, then there exists a point* η *between a and b such that*

$$f^{(n)}(\eta) = 0.$$

Theorem A.11 (Generalized Mean Value Theorem for Integrals)

If $f(x)$ *and* $g(x)$ *are continuous and integrable on a closed interval* $[a, b]$ *and* $g(x)$ *does not change sign on the interval* $[a, b]$*, then there exists at least one point* η *between a and b such that*

$$\int_a^b f(x)g(x)dx = f(\eta) \int_a^b g(x)dx.$$

Theorem A.12 (One-Dimensional Taylor's Theorem)

Let $f(x)$ be a function such that its $(n+1)$ derivative $f^{(n+1)}(x)$ is continuous on interval (a,b). If x and x_1 are any two points in (a,b), then where

$$\begin{aligned} f(x) &= f(x_1) + (x - x_1)\frac{(x-x_1)}{1!}f'(x_1) + \frac{(x-x_1)^2}{2!}f''(x_1) + \cdots \\ &+ \frac{(x-x_1)^n}{n!}f^{(n)}(x_1) + R_{n+1}(x_1;x) \end{aligned}$$

there exists a number η between x and x_1 such that

$$R_{n+1}(x_1;x) = \frac{(x-x_1)^{n+1}}{(n+1)!}f^{(n+1)}(\eta).$$

The infinite series can be obtained by taking the limit as $n \to \infty$ and is called Taylor series for f about x_1. When $x_1 = 0$ in the Taylor series, we get the Maclaurin series.

Theorem A.13 (Two-dimensional Taylor's Theorem)

If $f(x,y)$ and its partial derivatives of order $n+1$ exist and are continuous in a neighborhood of (a,b), which contains the line joining (a,b) to $(a+h, b+k)$, then there is a number $\theta \in (0,1)$ such that

$$\begin{aligned} f(a+h, b &+ k) = f(a,b) + \left(h\frac{\partial}{\partial x}\right) f(a,b) + \frac{1}{2!}\left(h\frac{\partial}{\partial x} + k\frac{\partial}{\partial y}\right)^2 f(a,b) \\ &+ \cdots + \frac{1}{n!}\left(h\frac{\partial}{\partial x} + k\frac{\partial}{\partial y}\right)^n f(a,b) + \frac{1}{(n+1)!}\left(h\frac{\partial}{\partial x} + k\frac{\partial}{\partial y}\right)^2 f(a,b). \end{aligned}$$

Theorem A.14 (Theorems on Polynomials)

Let

$$p(x) = a_n x^n + a_{n-1}x^{n-1} + \cdots + a_1 + a_0$$

be a polynomial of degree n and $a_n \neq 0$. Then the following theorems on the zeros of $p(x)$ holds:

1. *Every polynomial of degree $n \geq 1$ has at least one zero. This zero may be real or complex.*

2. *A polynomial of degree $n \geq 1$ has exactly n zeros provided that a root of multiplicity k is counted k times.*

3. *If two polynomials each of degree $\leq n$ coincide more than n distinct values of x, the polynomials are identical.*

4. *If $x_1, x_2, \ldots, x_n$ are the zeros of polynomial $p(x)$, then $p(x)$ can be expressed uniquely in factored form:*

$$p(x) = a_n(x - x_1)(x - x_2)\cdots(x - x_n).$$

Complex Numbers and Inner Products

In this appendix, firstly, we give a review of complex numbers and how they can be used in linear algebra. This Appendix also devoted to general inner product spaces and to how different notations and processes generalize.

Complex Numbers

Although physical applications ultimately require real answers, complex numbers and complex vector spaces play an extremely useful, if not essential, role in the intervening analysis. Particularly in the description of periodic phenomena, complex numbers, and complex exponentials to help simplify complicated trigonometric formulae.

Complex numbers arise naturally in the course of solving polynomial equations. For example, the solutions of the quadratic equation

$$ax^2 + bx + c = 0,$$

which are given by the quadratic formula

$$x = \frac{-b \pm \sqrt{b^2 - 4ac}}{2a}$$

are complex numbers if $b^2 - 4ac < 0$. To deal with the problem that the equation $x^2 = -1$ has no real solution, mathematicians of the eighteenth century inverted the "imaginary" number

$$\mathbf{i} = \sqrt{-1},$$

which is assumed to have the property

$$\mathbf{i}^2 = (\sqrt{-1})^2 = -1$$

but which otherwise has the algebraic properties of a real number.

A *complex number* z is of the form

$$z = a + \mathbf{i}b, \tag{A.1}$$

where a and b are real numbers, a is called the *real part* of z and is denoted by Re(z), and b is called the *imaginary part* of z and is denoted by Im(z).

We say that two complex numbers $z_1 = a_1 + \mathbf{i}b_1$ and $z_2 = a_2 + \mathbf{i}b_2$ are **equal** if their real and imaginary parts are equal; that is, if

$$a_1 = a_2 \quad \text{and} \quad b_1 = b_2.$$

Note that every real number a is a complex number with its imaginary part zero, $a = a + \mathbf{i}0$.
The complex number $z = 0 + \mathbf{i}0$ corresponds to zero.
If $a = 0$ and $b \neq 0$, then $z = \mathbf{i}b$ is called the imaginary number, or a purely imaginary number.

Geometric Representation of a Complex Number

A complex number $z = a + \mathbf{i}b$ may be regarded as an ordered pair (a, b) of real numbers. This ordered pair of real numbers corresponds to a point in the plane. Such a correspondence naturally suggests that we represent $a + \mathbf{i}b$ as a point in the complex plane (see Figure A.1, where the *horizontal axis* (also called the *real axis*) is used to represent the real part of z and the *vertical axis* (also called the *imaginary axis*) is used to represent the imaginary part of the complex number z).

Operations on Complex Numbers

Complex numbers are added, subtracted, and multiplied in accordance with the standard rules of algebra but $\mathbf{i}^2 = -1$.

If $z_1 = a_1 + \mathbf{i}b_1$ and $z_2 = a_2 + \mathbf{i}b_2$ are two complex numbers, then their *sum* is

$$z_1 + z_2 = (a_1 + a_2) + \mathbf{i}(b_1 + b_2)$$

and their *difference* is

$$z_1 - z_2 = (a_1 - a_2) + \mathbf{i}(b_1 - b_2).$$

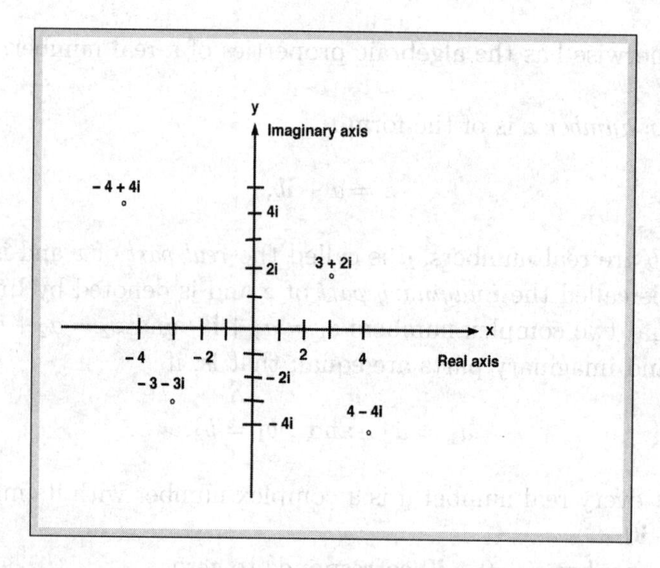

Figure A.1: Geometric representation of a complex number.

The *product* of z_1 and z_2 is

$$z_1 z_2 = (a_1 + \mathbf{i}b_1)(a_2 + \mathbf{i}b_2) = (a_1 a_2 - b_1 b_2) + \mathbf{i}(a_1 b_2 + a_2 b_1).$$

This multiplication formula is obtained by expanding the left side and using the fact that $\mathbf{i}^2 = -1$.

One can multiply a complex number by a real number α according to

$$\alpha z = \alpha a + \mathbf{i}\alpha b.$$

Finally, *division* is obtained in the following manner:

$$\begin{aligned} \frac{z_1}{z_2} &= \frac{a_1 + \mathbf{i}b_1}{a_2 + \mathbf{i}b_2} = \frac{(a_1 + \mathbf{i}b_1)(a_2 - \mathbf{i}b_2)}{(a_2 + \mathbf{i}b_2)(a_2 - \mathbf{i}b_2)} \\ &= \left(\frac{a_1 a_2 + b_1 b_2}{a_2^2 + b_2^2}\right) + \mathbf{i}\left(\frac{b_1 a_2 - a_1 b_2}{a_2^2 + b_2^2}\right). \end{aligned}$$

As important quantity associated with complex number z is its *complex conjugate*, defined by

$$\overline{z} = a - \mathbf{i}b.$$

Note that

$$z\overline{z} = (a + \mathbf{i}b)(a - \mathbf{i}b) = a^2 + b^2$$

is an intrinsically positive quantity (unless $a = b = 0$).

We call $\sqrt{z\overline{z}}$ the *modulus*, or **absolute value**, or the magnitude of z and write

$$|z| = |a + \mathbf{i}b| = \sqrt{z\overline{z}} = a^2 + b^2.$$

This also tells us that

$$\frac{1}{z} = \frac{\overline{z}}{|z|^2}.$$

Note that a complex number cannot be ordered, in the sense that the inequality $z_1 < z_2$ has no meaning. Nevertheless, the absolute values of complex numbers, being real numbers, can be ordered. Thus, for example, $|z| < 1$ means that z is such that $\sqrt{a^2 + b^2} < 1$.

Note that:

1. $\overline{\overline{z}} = z$.

2. $\overline{z_1 + z_2} = \overline{z_1} + \overline{z_2}$.

3. $\overline{z_1 z_2} = \overline{z_1}\ \overline{z_2}$.

4. If $z_2 \neq 0$, then $\overline{\left(\frac{z_1}{z_2}\right)} = \frac{\overline{z_1}}{\overline{z_2}}$.

5. z is real if and only if $\overline{z} = z$.

A *complex vector space* is defined in exactly the same manner as its real counterpart, the only difference being that we replace real scalars by *complex scalars*. The terms complex vector space and real vector space emphasize the set from which the scalars are chosen. The most basic example is the n-dimensional complex vector space $\mathbb{C}^n$ consisting of all column vectors $\mathbf{z} = (z_1, z_2, \ldots, z_n)^n$ that have n complex entries $z_1, z_2, \ldots, z_n$ in $\mathbb{C}^n$. Note that

$$\mathbf{z} \in \mathbb{R}^n \subset \mathbf{C}^n$$

is a real vector if and only if $z = \overline{z}$.

Polar Form of Complex Number

As we have seen, the complex number $z = a + \mathbf{i}b$ can be represented geometrically by point (a, b). This point can also be expressed in terms of *polar coordinates* (r, θ), where $r \geq 0$, as shown in Figure A.2. We have

$$a = r\cos\theta \quad \text{and} \quad b = r\sin\theta$$

so

$$z = a + \mathbf{i}b = r\cos\theta + \mathbf{i}r\sin\theta.$$

Thus, any complex number can be written in the polar form

$$z = r(\cos\theta + \mathbf{i}\sin\theta),$$

where

$$r = |z| = \sqrt{a^2 + b^2} \quad \text{and} \quad \tan\theta = \frac{b}{a}.$$

The angle θ is called an *argument* of z and is denoted by *argz*. Observe that *argz*

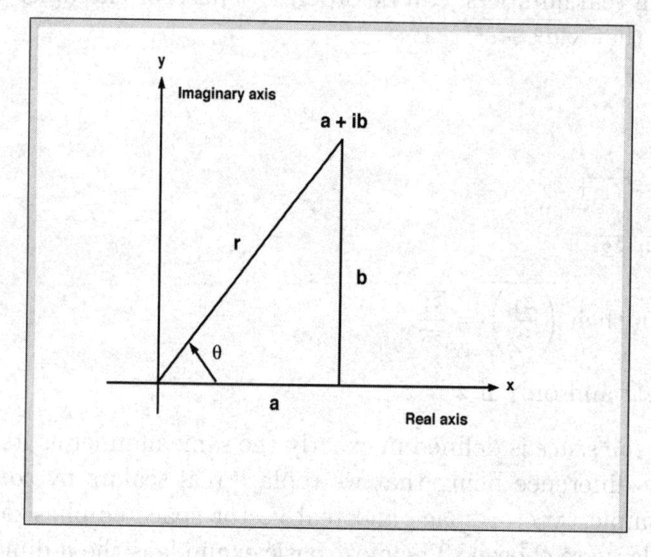

Figure A.2: Polar form of a complex number.

is not unique. Adding or subtracting any integer multiple of 2π gives another argument of z. However, there is only one argument θ that satisfies

$$-\pi < \theta \leq \pi.$$

This is called the *principal argument* of z and is denoted by *Argz*.
Note that:

$$z_1 z_2 = [r_1(\cos\theta_1 + \mathbf{i}\sin\theta_1)][r_2(\cos\theta_2 + \mathbf{i}\sin\theta_2)],$$

which can be written as (after using the trigonometric identities)

$$z_1 z_2 = r_1 r_2[\cos(\theta_1 + \theta_2) + \mathbf{i}\sin(\theta_1 + \theta_2)],$$

which means that to multiply two complex numbers, we multiply their absolute values and add their arguments. Similarly, we can get

$$\frac{z_1}{z_2} = \frac{r_1}{r_2}\left[\cos(\theta_1 - \theta_2) + \mathbf{i}\sin(\theta_1 - \theta_2)\right], \quad z_2 \neq 0, \tag{A.2}$$

which means that to divide two complex numbers, we divide their absolute values and subtract their arguments.

As a special case of (A.2), we obtain a formula for the reciprocal of a complex number in polar form. Setting $z_1 = 1$ (and therefore $\theta_1 = 0$) and $z_2 = z$ (and therefore $\theta_2 = 0$), we obtain:
If $z = r(\cos\theta + \mathbf{i}\sin\theta)$ is nonzero, then

$$\frac{1}{z} = \frac{1}{r}\left(\cos\theta - \mathbf{i}\sin\theta\right).$$

In the following we give some well-known theorems concerned with the polar form of a complex number.

Theorem A.15 (De Moivre's Theorem)

If $z = r(\cos\theta + \mathbf{i}\sin\theta)$ and n is a positive integer, then

$$z^n = r^n(\cos n\theta + \mathbf{i}\sin n\theta).$$

Theorem A.16 (Euler's Formula)

For any real number α,

$$e^{\mathbf{i}\alpha} = \cos\alpha + \mathbf{i}\sin\alpha.$$

Using Euler's formula, we see that the polar form of a complex number can be written more compactly as

$$z = r(\cos\theta + \mathbf{i}\sin\theta) = re^{\mathbf{i}\theta}$$

and $\overline{z} = re^{-\mathbf{i}\theta}$.

Theorem A.17 (Multiplication Rule)

If $z_1 = r_1 e^{\mathbf{i}\theta_1}$ and $z_2 = r_1 e^{\mathbf{i}\theta_2}$ are complex numbers in polar form, then

$$z_1 z_2 = r_1 r_2 e^{\mathbf{i}(\theta_1 + \theta_2)}.$$

In the following we give other important theorems concerned with complex numbers.

Theorem A.18 *If α_1 and α_2 are roots of the quadratic equation*

$$x^2 + ux + v = 0$$

then $\alpha_1 + \alpha_2 = -u$ and $\alpha_1\alpha_2 = v$.

Theorem A.19 (Fundamental Theorem of Algebra)

Every polynomial $f(x)$ of positive degree with complex coefficients has a complex root.

Theorem A.20 *Every complex polynomial of degree $n \geq 1$ has the form*

$$f(x) = u(x - u_1)(x - u_2) \cdots (x - u_n),$$

where $u_1, u_2, \ldots, u_n$ are the roots of $f(x)$ (and need not all be distinct), and u is the coefficient of x^n.

Theorem A.21 *Every polynomial $f(x)$ of positive degree with real coefficients can be factored as a product of linear and irreducible quadratic factors.*

Theorem A.22 (nth Roots of Unity)

If $n \geq 1$ is an integer, the nth roots of unity (that is, the solution to $z^n = 1$) are given by

$$z = e^{2\pi ki}, \quad k = 0, 1, \ldots, n-1.$$

Matrices with Complex Entries

If the entries of a matrix are complex numbers, we can perform the matrix operations of addition, subtraction, multiplication, and scalar multiplication in the same manner as for the case of real matrices. The validity of these operations can be verified using properties of complex arithmetic, just imitating the proofs for real matrices discussed in Chapter 2. For example, consider the following matrices:

$$A = \begin{pmatrix} 1+\mathrm{i} & 3\mathrm{i} \\ 5+\mathrm{i} & 6\mathrm{i} \end{pmatrix}, \quad B = \begin{pmatrix} 2-2\mathrm{i} & 4\mathrm{i} \\ 1+3\mathrm{i} & 2\mathrm{i} \end{pmatrix}, \quad C = \begin{pmatrix} 1+\mathrm{i} & 2+\mathrm{i} \\ 5+\mathrm{i} & 4+5\mathrm{i} \\ 2-4\mathrm{i} & 1+2\mathrm{i} \end{pmatrix}.$$

Then

$$A+B = \begin{pmatrix} 1+\mathrm{i} & 3\mathrm{i} \\ 5+\mathrm{i} & 6\mathrm{i} \end{pmatrix} + \begin{pmatrix} 2-2\mathrm{i} & 4\mathrm{i} \\ 1+3\mathrm{i} & 2\mathrm{i} \end{pmatrix} = \begin{pmatrix} 3-\mathrm{i} & 7\mathrm{i} \\ 6+4\mathrm{i} & 8\mathrm{i} \end{pmatrix}$$

and

$$A-B = \begin{pmatrix} 1+\mathrm{i} & 3\mathrm{i} \\ 5+\mathrm{i} & 6\mathrm{i} \end{pmatrix} - \begin{pmatrix} 2-2\mathrm{i} & 4\mathrm{i} \\ 1+3\mathrm{i} & 2\mathrm{i} \end{pmatrix} = \begin{pmatrix} -1+3\mathrm{i} & -\mathrm{i} \\ 4-2\mathrm{i} & 4\mathrm{i} \end{pmatrix}.$$

Also,

$$CA = \begin{pmatrix} 1+\mathrm{i} & 2+\mathrm{i} \\ 5+\mathrm{i} & 4+5\mathrm{i} \\ 2-4\mathrm{i} & 1+2\mathrm{i} \end{pmatrix} \begin{pmatrix} 1+\mathrm{i} & 3\mathrm{i} \\ 5+\mathrm{i} & 6\mathrm{i} \end{pmatrix} = \begin{pmatrix} 9-9\mathrm{i} & -9+15\mathrm{i} \\ 19+35\mathrm{i} & -30+39\mathrm{i} \\ 9+9\mathrm{i} & 12\mathrm{i} \end{pmatrix}$$

and

$$3\mathrm{i}A = 3\mathrm{i} \begin{pmatrix} 1+\mathrm{i} & 3\mathrm{i} \\ 5+\mathrm{i} & 6\mathrm{i} \end{pmatrix} = \begin{pmatrix} -3+3\mathrm{i} & -9\mathrm{i} \\ -3+15\mathrm{i} & -18 \end{pmatrix}.$$

There are special types of complex matrices, like the Hermitian matrix, unitary matrix, and normal matrix which we discussed in Chapter 7.

Solving Systems with Complex Entries

The results and techniques dealing with the solution of linear systems that we developed in Chapter 3 carry over directly to linear systems with complex coefficients.

For example, the solution of the following linear system

$$\begin{array}{rcrcl} 3\mathrm{i}x_1 & + & 4x_2 & = & 5+15\mathrm{i} \\ 5-\mathrm{i}x_1 & + & 3-4\mathrm{i}x_2 & = & 24+5\mathrm{i} \end{array}$$

can be obtained using the Gauss-Jordan method as follows:

$$[A|\mathbf{b}] = \begin{pmatrix} 3\mathrm{i} & 4 & \vdots & 5+15\mathrm{i} \\ 5-\mathrm{i} & 3-4\mathrm{i} & \vdots & 24+5\mathrm{i} \end{pmatrix} \sim \begin{pmatrix} 3\mathrm{i} & 4 & \vdots & 5+15\mathrm{i} \\ 0 & \frac{13}{3}+\frac{8}{3}\mathrm{i} & \vdots & \frac{2}{3}+\frac{55}{3}\mathrm{i} \end{pmatrix}$$

$$\sim \begin{pmatrix} 3\mathrm{i} & 4 & \vdots & 5+15\mathrm{i} \\ 0 & 1 & \vdots & 2+3\mathrm{i} \end{pmatrix} \sim \begin{pmatrix} 3\mathrm{i} & 0 & \vdots & -3+3\mathrm{i} \\ 0 & 1 & \vdots & 2+3\mathrm{i} \end{pmatrix}$$

$$\sim \begin{pmatrix} 1 & 0 & \vdots & 1+\mathrm{i} \\ 0 & 1 & \vdots & 2+3\mathrm{i} \end{pmatrix}.$$

Thus, the solution to the given system is $x_1 = 1 + \mathbf{i}$ and $x_2 = 2 + 3\mathbf{i}$.

Determinants of Complex Numbers

The definition of a determinant and all its properties derived in Chapter 3 apply to matrices with complex entries. For example, the determinant of the matrix

$$A = \begin{pmatrix} 1+\mathrm{i} & 3\mathrm{i} \\ 5+\mathrm{i} & 6\mathrm{i} \end{pmatrix}$$

can be obtained as

$$|A| = \begin{vmatrix} 1+\mathrm{i} & 3\mathrm{i} \\ 5+\mathrm{i} & 6\mathrm{i} \end{vmatrix} = (6\mathrm{i} + 6\mathrm{i}^2) - (15\mathrm{i} + 3\mathrm{i}^2) = -9\mathrm{i} - 3.$$

Complex Eigenvalues and Eigenvectors

Let A be an $n \times n$ matrix. Complex number λ is an eigenvalue of A if there exists a nonzero vector $\mathbf{x}$ in $\mathbb{C}^n$ such that

$$A\mathbf{x} = \lambda\mathbf{x}. \tag{A.3}$$

Every nonzero vector $\mathbf{x}$ satisfying (A.3) is called an eigenvector of A associated with the eigenvalue λ. Relation (A.3) can be rewritten as

$$(A - \lambda\mathbf{I})\mathbf{x} = \mathbf{0}. \tag{A.4}$$

This homogeneous system has a nonzero solution $\mathbf{x}$ if and only if

$$\det(A - \lambda\mathbf{I}) = 0$$

has a solution. Like in Chapter 5, $\det(A - \lambda\mathbf{I})$ is called the characteristic polynomial of matrix A, which is a complex polynomial of degree n in λ. The eigenvalues of complex matrix A are the complex roots of the characteristic polynomial. For example, let

$$A = \begin{pmatrix} 0 & 1 \\ -1 & 0 \end{pmatrix}$$

then

$$\det(A - \lambda\mathbf{I}) = \lambda^2 + 1 = 0$$

gives eigenvalues $\lambda_1 = \mathbf{i}$ and $\lambda_2 = -\mathbf{i}$ of A. One can easily find eigenvectors

$$\mathbf{x}_1 = [1,\ \mathbf{i}]^T \quad \text{and} \quad \mathbf{x}_2 = [-1,\ \mathbf{i}]^T$$

associated with eigenvalues **i** and **−i**, respectively.

Inner Product Space

Now we study a more advanced topic in linear algebra called inner product spaces. Inner products lie at the heart of linear (and nonlinear) analysis, both in finite-dimensional vector spaces and infinite-dimensional function spaces. It is impossible to overemphasize their importance for both theoretical developments, practical application, and in the design of numerical solution techniques. Inner products are widely used from theoretical analysis to applied signal processing. Here, we discuss the basic properties of inner products and give some important theorems.

Definition A.8 (Inner Product)

An inner product on a vector space V is an operation that assigns to every pair of vectors $\mathbf{u}$ *and* $\mathbf{v}$ *in V a real number* $<\mathbf{u},\mathbf{v}>$ *such that the following properties hold for all vectors* $\mathbf{u},\mathbf{v}$*, and* $\mathbf{w}$ *in V and all scalars* α*:*

1. $<\mathbf{u},\mathbf{v}> = <\mathbf{v},\mathbf{u}>$.

2. $<\mathbf{u},\mathbf{v}+\mathbf{w}> = <\mathbf{u},\mathbf{v}> + <\mathbf{u},\mathbf{w}>$.

3. $<\alpha\mathbf{u},\mathbf{v}> = \alpha<\mathbf{u},\mathbf{v}>$.

4. $<\mathbf{u},\mathbf{u}> \geq 0$ *and* $<\mathbf{u},\mathbf{u}> = 0$ *if and only if* $\mathbf{u} = \mathbf{0}$.

A vector space with an inner product is called an inner product space. The most basic example of an inner product is a familiar dot product

$$<\mathbf{u},\mathbf{v}> = \mathbf{u}.\mathbf{v} = u_1v_1 + u_2v_2 + \cdots + u_nv_n = \sum_{j=1}^{n} u_jv_j$$

between (column) vectors

$$\mathbf{u} = (u_1, u_2, \ldots, u_n)^T \quad \textit{and} \quad \mathbf{v} = (v_1, v_2, \ldots, v_n)^T$$

lying in Euclidean space $\mathbb{R}^n$.

Properties of Inner Products

The following theorem summarizes some additional properties that follow from the definition of inner product.

Theorem A.23 *Let* $\mathbf{u}, \mathbf{v}$, *and* $\mathbf{w}$ *be vectors in an inner product space* V *and let* α *be a scalar:*

1. $< \mathbf{u} + \mathbf{v}, \mathbf{w} > = < \mathbf{u}, \mathbf{w} > + < \mathbf{u}, \mathbf{v} >.$
2. $< \mathbf{u}, \alpha\mathbf{v} > = \alpha < \mathbf{u}, \mathbf{v} >.$
3. $< \mathbf{u}, \mathbf{0} > = < \mathbf{0}, \mathbf{v} > = \mathbf{0}.$

In an inner product space, we can define the *length of a vector*, the distance between vectors, and the orthogonal vectors.

Definition A.9 (Length of a Vector)

Let $\mathbf{v}$ *be a vector in an inner product space* V. *Then, the* **length** *(or* **norm***) of* $\mathbf{v}$ *is defined as:*

$$\|\mathbf{v}\| = \sqrt{< \mathbf{v}, \mathbf{v} >}.$$

Theorem A.24 (Inner Product Norm Theorem)

If V *is a real vector space with an inner product* $< \mathbf{u}, \mathbf{v} >$, *then the function* $\|\mathbf{u}\| = \sqrt{< \mathbf{u}, \mathbf{u} >}$ *is a norm on* V.

Definition A.10 (Distance Between Vectors)

Let $\mathbf{u}$ *and* $\mathbf{v}$ *be vectors in an inner product space* V. *Then, the distance between* $\mathbf{u}$ *and* $\mathbf{v}$ *is defined as:*

$$d(\mathbf{u}, \mathbf{v}) = \|\mathbf{u} - \mathbf{v}\|.$$

Note that:

$$d(\mathbf{u}, \mathbf{0}) = d(\mathbf{0}, \mathbf{u}) = \|\mathbf{u}\|.$$

A vector with norm 1 is called a unit vector. The set S *of all unit vectors is called unit circle or unit sphere* $S = \{\mathbf{u} | \mathbf{u} \in V \text{ and } \|\mathbf{u}\| = 1\}$.

The following theorem summarizes the most important properties of a distance function.

Theorem A.25 *Let* d *be a distance function defined on a normed linear space* V. *The following properties hold for all* $\mathbf{u}, \mathbf{v}$, *and* $\mathbf{w}$ *as vectors in* V.

1. $d(\mathbf{u}, \mathbf{v}) \geq 0$, *and* $d(\mathbf{u}, \mathbf{v}) = 0$, *if and only if* $\mathbf{u} = \mathbf{v}$.

2. $d(\mathbf{u},\mathbf{v}) = d(\mathbf{v},\mathbf{u})$.

3. $d(\mathbf{u},\mathbf{w}) \leq d(\mathbf{u},\mathbf{v}) + d(\mathbf{v},\mathbf{w})$.

Definition A.11 (Orthogonal Vectors)

Let $\mathbf{u}$ *and* $\mathbf{v}$ *be vectors in an inner product space* V. *Then,* $\mathbf{u}$ *and* $\mathbf{v}$ *are orthogonal if*

$$< \mathbf{u},\mathbf{v} > = 0.$$

In the following we give some well-known theorems concerning inner product space.

Theorem A.26 (Pythagoras' Theorem)

Let $\mathbf{u}$ *and* $\mathbf{v}$ *be vectors in an inner product space* V. *Then,* $\mathbf{u}$ *and* $\mathbf{v}$ *are orthogonal if and only if*

$$\|\mathbf{u}+\mathbf{v}\|^2 = \|\mathbf{u}\|^2 + \|\mathbf{v}\|^2.$$

Theorem A.27 (Orthogonality Test for Linear Independence)

Nonzero orthogonal vectors in an inner product space are linearly independent.

Theorem A.28 (Normalization Theorem)

For every nonzero vector $\mathbf{u}$ *in an inner product space* V, *the vector* $\mathbf{v} = \mathbf{u}/\|\mathbf{u}\|$ *is a unit vector.*

Theorem A.29 (Cauchy-Schwarz Inequality)

Let $\mathbf{u}$ *and* $\mathbf{v}$ *be vectors in an inner product space* V. *Then*

$$|< \mathbf{u},\mathbf{v} >| \leq \|\mathbf{u}\|\|\mathbf{v}\|$$

with inequality holding if and only if $\mathbf{u}$ *and* $\mathbf{v}$ *are scalar multiples of each other.*

Theorem A.30 (Triangle Inequality)

Let $\mathbf{u}$ *and* $\mathbf{v}$ *be vectors in an inner product space* V. *Then*

$$\|\mathbf{u}+\mathbf{v}\| \leq \|\mathbf{u}\| + \|\mathbf{v}\|.$$

Theorem A.31 (Parallelogram Law)

Let V be an inner product space. For any vectors $\mathbf{u}$ *and* $\mathbf{v}$ *of* V*, we have*

$$\|\mathbf{u}+\mathbf{v}\|^2 + \|\mathbf{u}-\mathbf{v}\|^2 = 2\|\mathbf{u}\|^2 + 2\|\mathbf{v}\|^2.$$

Theorem A.32 (Polarization Identity)

Let V be an inner product space. For any vectors $\mathbf{u}$ *and* $\mathbf{v}$ *of* V*, we have*

$$< \mathbf{u}, \mathbf{v} > = \frac{1}{4}\|\mathbf{u}+\mathbf{v}\|^2 - \frac{1}{4}\|\mathbf{u}-\mathbf{v}\|^2.$$

Theorem A.33 *Let V be an inner product space. For any vectors* $\mathbf{u}$ *and* $\mathbf{v}$ *of* V*, we have*

$$\|\mathbf{u}+\mathbf{v}\|^2 = \|\mathbf{u}\|^2 + \|\mathbf{v}\|^2 + 2 < \mathbf{u}, \mathbf{v} > .$$

Theorem A.34 *Every inner product on* $\mathbb{R}^n$ *is given by*

$$< \mathbf{u}, \mathbf{v} > = \mathbf{u}^T A \mathbf{v} \quad \textit{for} \quad \mathbf{u}, \mathbf{v} \in \mathbb{R}^n,$$

where A is a symmetric, positive-definite matrix.

Theorem A.35 *All Gram matrices are positive semi-definite. The* **Gram matrix**

$$A = \begin{pmatrix} < \mathbf{u_1}, \mathbf{u_1} > & < \mathbf{u_1}, \mathbf{u_2} > & \cdots & < \mathbf{u_1}, \mathbf{u_n} > \\ < \mathbf{u_2}, \mathbf{u_1} > & < \mathbf{u_2}, \mathbf{u_2} > & \cdots & < \mathbf{u_2}, \mathbf{u_n} > \\ \vdots & \vdots & \ddots & \vdots \\ < \mathbf{u_n}, \mathbf{u_1} > & < \mathbf{u_n}, \mathbf{u_2} > & \cdots & < \mathbf{u_n}, \mathbf{u_n} > \end{pmatrix}$$

(where $\mathbf{u_1}, \mathbf{u_2}, \ldots, \mathbf{u_n}$ *are vectors in the inner product space V) is positive-definite if and only if* $\mathbf{u_1}, \mathbf{u_2}, \ldots, \mathbf{u_n}$ *are linearly independent.*

Complex Inner Products

Certain applications of linear algebra require *complex-valued inner products.*

Definition A.12 (Complex Inner Product)

An inner product on a complex vector space V is a function that associates a complex number $< \mathbf{u}, \mathbf{v} >$ *with each pair of vectors* $\mathbf{u}$ *and* $\mathbf{v}$ *in such a way that the following axioms are satisfied for all vectors* $\mathbf{u}, \mathbf{v}$*, and* $\mathbf{w}$ *in V and all scalars* α*:*

1. $<\mathbf{u}, \mathbf{v}> = <\overline{\mathbf{v}, \mathbf{u}}>$.

2. $<\mathbf{u}+\mathbf{v}, \mathbf{w}> = <\mathbf{u}, \mathbf{w}> + <\mathbf{v}, \mathbf{w}>$.

3. $<\alpha\mathbf{u}, \mathbf{v}> = \alpha <\mathbf{u}, \mathbf{v}>$.

4. $<\mathbf{v}, \mathbf{v}> \geq 0$ *and* $<\mathbf{v}, \mathbf{v}> = 0$ *if and only if* $\mathbf{v} = \mathbf{0}$.

The scalar $<\overline{\mathbf{v}, \mathbf{u}}>$ *is the complex conjugate of* $<\mathbf{v}, \mathbf{u}>$. *Complex inner products are no longer symmetric since* $<\mathbf{v}, \mathbf{u}>$ *is not always equal to its complex conjugate. A complex vector space with an inner product is called a complex inner product space or unitary space.*

The following additional properties follow immediately from the four inner product axioms:

1. $<\mathbf{0}, \mathbf{u}> = <\mathbf{v}, \mathbf{0}> = 0$.

2. $<\mathbf{u}, \mathbf{v}+\mathbf{w}> = <\mathbf{u}, \mathbf{v}> + <\mathbf{u}, \mathbf{w}>$.

3. $<\mathbf{u}, \alpha\mathbf{v}> = \overline{\alpha} <\mathbf{u}, \mathbf{v}>$.

An inner product can then be used to define norm, orthogonality, and distance, as for a real vector space.

Let $\mathbf{u} = (u_1, u_2, \ldots, u_n)$ and $\mathbf{v} = (v_1, v_2, \ldots, v_n)$ be elements of $\mathbb{C}^n$. The most useful inner product for $\mathbb{C}^n$ is

$$<\mathbf{u}, \mathbf{v}> = u_1\overline{v_1} + u_2\overline{v_2} + \cdots + u_n\overline{v_n}.$$

It can be shown that this definition satisfies the inner product axioms for a complex vector space.

This inner product leads to the following definitions of norm, distance, and orthogonality for $\mathbb{C}^n$:

1. $\|\mathbf{u}\| = \sqrt{u_1\overline{u_1} + u_2\overline{u_2} + \cdots + u_n\overline{u_n}}$.

2. $d(\mathbf{u}, \mathbf{v}) = \|\mathbf{u} - \mathbf{v}\|$.

3. $\mathbf{u}$ is orthogonal to $\mathbf{v}$ if $<\mathbf{u}, \mathbf{v}> = 0$.

Appendix B

Introduction to MATLAB

In this appendix we discuss the programming and software package MATLAB. The name 'MATLAB' is an abbreviation for "Matrix Laboratory." MATLAB is an extremely powerful package for numerical computing and programming. In MATLAB we can give direct commands, as on a hand calculator, and we can write programs.

MATLAB software exists as a primary application program and a large library of program modules called the standard toolbox. Most of the numerical methods described in this textbook are implemented in one form or another in the toolbox. The MATLAB toolbox contains an extensive library for solving many practical numerical problems, such as root-finding, interpolation, numerical integration and differentiation, solving systems of linear and nonlinear equations, and solving ordinary differential equations.

The MATLAB package also consists of an extensive library of numerical routines, easily accessed two- and three dimensional graphics, and a high level programming format. The ability to quickly implement and modify programs makes MATLAB an appropriate format for exploring and executing the algorithms in this textbook. MATLAB is a mathematical software package based on matrices. It is a highly optimized and extremely reliable system for numerical linear algebra. Many numerical tasks can be concisely expressed in the language of linear algebra.

1. Some Basic MATLAB Operations

(1) Throughout this discussion we shall use >> to indicate a MATLAB command statement. The command prompt >> may vary from system to system. The command prompt >> is given by the system and you only need to enter the MATLAB command.

(2) It is possible to include comments in the MATLAB workspace. Typing % before a statement indicates a comment statement. Comment statements are not executable. For example:

>> % Finding root of nonlinear equation $f(x) = 0$

(3) To get help on a topic, say, a determinant, enter

>> help determinant

(4) A semicolon placed at the end of an expression suppresses the computer output. For example:

```
>> a = 25;
```

.

Without ; a was displayed.

2. MATLAB Numbers and Numeric Formats

All numerical variables are stored in MATLAB in double-precision, floating-point form. It is possible to force some variables to be other types but not easily and this ability is not needed.

The default output to the screen is to have four digits to the right of the decimal point. To control the formatting of output to the screen, use the command format. The default formatting is obtained using the following command:

```
>> format short
>> pi
ans =
   3.1416
```

.

To obtain the full accuracy available in a number, we can use the following:

```
>> format long
>> pi
ans =
    3.14159265358979
```

.

The other format commands called format short e and format long e will use 'scientific notation' for the output:

```
>> format short e
>> pi
ans =
    3.1416e + 000
```

.

As part of its syntax and semantics, MATLAB provides for exceptional values. Positive infinity is represented by *Inf*, negative infinity by - *Inf*, and not a real number by *NAN*. These exceptional values are carried through the computations in a logically consistent way.

3. Arithmetic Operations

Arithmetic in MATLAB follows all the rules and uses the standard computer symbols for its arithmetic operation signs.

Symbol	**Effect**
$+$	*Addition*
$-$	*Subtraction*
$*$	*Multiplication*
$\backslash$	*Division*
$\wedge$	*Power*
$'$	*Conjugate transpose*
pi, e	*Constants*

In the present context we shall consider these operations as scalar arithmetic operations, which is to say that they operate on 2 numbers in the conventional manner:

```
>> (4 − 2 + 3 * pi)/2
ans =
    5.7124
>> a = 2; b = sin(a);
>> 2 * b^2
ans =
    1.6537
```

.

MATLAB's arithmetic operations are actually much more powerful than this. We shall see just a little of this extra power later.

There are some arithmetic operations which require great care. The order in which multiplication and division operations are specified is especially important. For example:

```
>> a = 2; b = 3; c = 4;
>> a/b * c
```

.

Here the absence of any parentheses results in MATLAB executing the two operations from left-to-right so that:
First a is divided by b, and then: The result is multiplied by c.
The result is therefore:

```
ans = 2.6667
```

.

This arithmetic is equivalent to $\frac{a}{b}c$ or as MATLAB command:

```
>> (a/b) * c
```

.

Similarly, $a/b/c$ yields the same result as $\frac{a/b}{c}$ or $\frac{a}{b/c}$, which could be achieved with the MATLAB command:

```
>> a/(b * c)
```

.

Use parentheses to be sure that MATLAB does what you want.

4. MATLAB Mathematical Functions

All of the standard mathematical functions—often called the elementary functions that we met in our calculus courses are available in MATLAB using their usual mathematical names. The important functions for our purposes are:

Symbol	**Effect**
$abs(x)$	*Absolute value*
$sqrtx$	*Square root*
$\sin(x)$	*Sine function*
$\cos(x)$	*Cosine function*
$\tan(x)$	*Tangent function*
$\log(x)$	*Natural logarithmic function*
$\exp(x)$	*Exponential function*
$atan(x)$	*Inverse tangent function*
$acos(x)$	*Inverse cosine function*
$asin(x)$	*Inverse sine function*
$\cos h(x)$	*Hyperbolic cosine function*
$\sin h(x)$	*Hyperbolic sine function*

Note that the various trigonometric functions expect their argument to be radian (or pure number) form but not in degree form. For example:

```
>> cos(pi/3)
```

gives the output:

```
ans = 0.5
```

As we discussed before, the variables appear to be scalars. In fact, all MATLAB variables are array. An important aspect of MATLAB is that it works very efficiently with arrays and main tasks are best done with arrays.

5. Vectors

In MATLAB the word *vector* can really be interpreted simply as 'list of numbers.' Strictly it could be a list of other objects than numbers but 'list of numbers' will fit our need for now.

There are two basic kinds of MATLAB vectors: Row and column vectors. As the names suggests, a row vector stores its numbers in a long 'horizontal list' such as

$$1, 2, 3, 1.23, -10.3, 1.2,$$

which is a row vector with 6 components. A column vector stores its numbers in a vertical list such as:

$$\begin{matrix} 1 \\ 2 \\ 3 \\ 1.23 \\ -10.3 \\ 2.1 \end{matrix}$$

which is a column vector with six components. In mathematical notation these arrays are usually enclosed in brackets [ ].

There are various convenient forms of these vectors, for allocating values to them and accessing the values that are stored in them. The most basic method of accessing or assigning individual components of a vector is based on using an index, or subscript, which indicates the position of the particular component in the list. MATLAB notation for this subscript is to enclose it in parentheses ( ). For assigning a complete vector in a single statement, we can use the square brackets [ ] notation. For example:

```
>> x = [1, 2, 3.4, 1.23, -10.3, 2.1]
x =
    1.0000 2.0000 3.4000 1.2300  - 10.3000 2.1000
>> x(3) = x(1) + 3 * x(6)
x =
    1.0000 2.0000 7.3000 1.2300  - 10.3000 2.1000
```

Remember that in entering values for a row vector, space could be used in place of commas. For the corresponding column vector simply replace the commas with semicolons. To switch between column and row format for MATLAB vector we use the transpose operator denoted by ′. For example:

$$\boxed{\begin{array}{l} >> x = x' \\ x = \\ \quad\quad 1.0000 \\ \quad\quad 2.0000 \\ \quad\quad 7.3000 \\ \quad\quad 1.2300 \\ \quad -10.3000 \\ \quad\quad 2.1000 \end{array}}.$$

MATLAB has several convenient ways of allocating values to a vector where these values fit a simple pattern.

The *Colon :* has a very special and powerful role in MATLAB. Basically, it allows an easy way to specify a vector of equally spaced numbers. There are two basic forms of the MATLAB colon notation.

The first one is that two arguments are separated by a colon as in:

$$\boxed{>> x = -2:4}.$$

which generates a row vector with first component -2, last one 4, and others spaced at unit intervals.

The second one is that three arguments separated by two colons has the effect of specifying the *starting value : spacing : final value.* For example:

$$\boxed{>> x = -2:0.5:1}.$$

which generates

$$\boxed{\begin{array}{l} x = \\ \quad -2.0 \ \ -1.5 \ \ -1.0 \ \ -0.5 \ \ 0.0 \ \ 0.5 \ \ 1.0 \end{array}}.$$

Also, one can use MATLAB colon notation as follows:

$$\boxed{>> y = x(2:6)},$$

which generates

$$\boxed{\begin{array}{l} y = \\ \quad -1.5 \ \ -1.0 \ \ -0.5 \ \ 0.0 \ \ 0.5 \end{array}}.$$

MATLAB has two other commands for specifying vectors conveniently. The first one is called the *linspace* function, which is used to specify a vector with a given number of equally spaced elements between specified start and finish points. For example:

```
>> x = linspace(0, 1, 10)
x =
    0.000 0.111 0.222 0.333 0.444 0.556 0.667 0.778 0.889 1.000
```

.

Using 10 points results in just 9 steps.

The other command is called the *logspace* function, which is similar to the linspace function, except that it creates elements that are logarithmically equally spaced. The statement:

```
>> lognspace(start value, endvalue, numpoints)
```

will create numpoints elements between $10^{start\ value}$ and $10^{end\ value}$. For example:

```
>> x = lognspace(1, 4, 4)
x =
    10 100 1000 10000
```

.

We can use MATLAB's vectors to generate tables of function values. For example:

```
>> x = linspace(0, 1, 11);
>> y = cos(x);
>> [x', y']
ans =
    0.0000  1.0000
    0.1000  0.9950
    0.2000  0.9801
    0.3000  0.9553
    0.4000  0.9211
    0.5000  0.8776
    0.6000  0.8253
    0.7000  0.7648
    0.8000  0.6967
    0.9000  0.6216
    1.0000  0.5403
```

.

Note the use of the transpose to convert the row vectors to columns, and the separation of these two columns by a comma.
Note also that the standard MATLAB functions are defined to operate on vectors of inputs in an element-by-element manner. The following example illustrates the use of the colon (:) notation and arithmetic within the argument of a function as:

```
>> y = sqrt(4 + 2 * (0 : 0.1 : 1)')
ans =
    2.0000
    2.0494
    2.0976
    2.1448
    2.1909
    2.2361
    2.2804
    2.3238
    2.3664
    2.4083
    2.4495
```

6. Matrices

A matrix is a two-dimensional array of numerical values that obeys the rules of linear algebra as discussed in Chapter 3.

To *enter* a matrix, list all the entries of the matrix with the first row, separating the entries by blank space or commas, separating two rows by a semicolon, and enclosing the list in square brackets. For example to enter 3×4 matrix A, we do the following:

```
>> A = [1 2 3 4; 3 2 1 4; 4 1 2 3]
```

and it will appears as follows:

```
A =
    1 2 3 4
    3 2 1 4
    4 1 2 3
```

There are also other options available when directly defining an array. When wanting to define a column vector, we can use the transpose operation. For example:

```
>> [1 2 5]'
```

results in the column vector:

```
ans =
    1
    2
    5
```

.

The components (entries) of matrices can be manipulated in several ways. For example:

```
>> A = [1 2 3; 4 5 6; 7 8 9];
>> A(2,3)
ans =
     6
```

.

Select a *submatrix* of A as follows:

```
>> A([1 3], [1 3])
ans =
    1  3
    7  9
```

or

```
>> A(1 : 2, 2 : 3)
ans =
    2  3
    5  6
```

.

An individual element or group of elements can be *deleted* from vectors and matrices by assigning these elements to the null (zero) matrix, [ ]. For example:

$$\begin{array}{l}
>> x = [1\ 2\ 3\ 4\ 5];\\
>> x(3) = [\,]\\
x =\\
\quad [1\ 2\ 4\ 5]\\
>> A = [1\ 2\ 3; 4\ 5\ 6; 7\ 8\ 9];\\
>> A(:,1) = [\,]\\
ans =\\
\quad 2\ \ 3\\
\quad 5\ \ 6\\
\quad 8\ \ 9
\end{array}.$$

To *interchange* the two rows of a given matrix A, we type the following:

$$\boxed{>> B = A([\text{new order of rows separating the entries by commas}],:)}.$$

For example, if matrix A has three rows and we want to change rows 1 and 3, type:

$$\boxed{>> B = A([3,2,1],:)}.$$

For example:

$$\begin{array}{l}
>> A = [1\ 2\ 3; 4\ 5\ 6; 7\ 8\ 9]\\
>> B = A([3,2,1],:)\\
B =\\
\quad 7\ \ 8\ \ 9\\
\quad 4\ \ 5\ \ 6\\
\quad 1\ \ 2\ \ 3
\end{array}.$$

Note that the method can be used to change the order of any number of rows.

Similarly, one can interchange the columns easily by typing:

$$\boxed{>> B = A(:,[\text{new order of columns separating the entries by commas}])}.$$

For example, if matrix A has three columns and we want to change column 1 and 3, type:

$$\begin{array}{l}
>> B = A(:,[3,2,1])\\
B =\\
\qquad 3\ \ 2\ \ 1\\
\qquad 6\ \ 5\ \ 4\\
\qquad 9\ \ 8\ \ 7
\end{array}.$$

Note that the method can be used to change the order of any number of columns. In order to *replace* the *kth* row of matrix A set $A(k,:)$ equal to new entries of the row separated by a space and enclosed in square brackets; that is, type:

```
>> A(k,:) = [New entries of kth row]
```

For example, to change second row of a 3×3 matrix A to $[2,2,2]$, type the command:

```
>> A(2,:) = [2 2 2]
```

For example:

```
>> A = [1 2 3; 4 5 6; 7 8 9]
>> A(2,:) = [2 2 2]
A =
    1  2  3
    2  2  2
    7  8  9
```

Similarly, one can replace the *kth* column of a matrix A equal to new entries of the column in square brackets separated by semicolons; that is, type:

```
>> A(:,k) = [New entries of kth column]
```

For example, to change the second column of a 3×3 matrix A to $[2,2,2]'$, type the command:

```
>> A(:,2) = [2 2 2]
```

For example:

```
>> A = [1 2 3; 4 5 6; 7 8 9]
>> A(:,2) = [2; 2; 2]
A =
    1  2  3
    4  2  6
    7  2  9
```

7. Creating Special Matrices

There are several built-in functions for *creating* vectors and matrices.

- Create a zero matrix with m rows and n columns using *zeros* function as follows:

$$>> A = zeros(m, n).$$

or, one can create an $n \times n$ zero matrix as follows:

$$>> A = zeros(n).$$

For example:

```
>> A = zeros(3)
A =
    0 0 0
    0 0 0
    0 0 0
```

- Create an $n \times n$ ones matrix using the *ones* function as follows:

$$>> A = ones(n, n).$$

For example the 3×3 ones matrix:

```
>> A = ones(3,3)
A =
    1 1 1
    1 1 1
    1 1 1
```

Of course, the matrix need not be square:

```
>> A = ones(2,4)
A =
    1 1 1 1
    1 1 1 1
```

Indeed, ones and zeros can be used to create row and column vectors:

```
>> u = ones(1,4)
u =
    1 1 1 1
```

and

```
>> v = ones(1,4)
v =
    1
    1
    1
    1
```

.

- Create an $n \times n$ identity matrix using the *eye* function as follows:

```
>> I = eye(n)
```

.

For example:

```
>> I = eye(3)
I =
    1 0 0
    0 1 0
    0 0 1
```

.

- Create an $n \times n$ *diagonal* matrix using the *diag* function, which either creates a matrix with specified values on the diagonal or it extracts the diagonal entries. Using the diag function, the argument must be a vector:

```
>> v = [4 5 6];
>> A = diag(v)
A =
    4 0 0
    0 5 0
    0 0 6
```

.

or it can be specified directly in the input argument as in:

```
>> A = diag([4 5 6])
```

To extract the diagonal entries of an existing matrix, the same *diag* function is used, but with input being a matrix instead of a vector:

```
>> u = diag(A)
u =
    4
    5
    6
```

- Create the *length* function and *size* function, which are used to determine the number of elements in vectors and matrices. These functions are useful when one is dealing with matrices of unknown or variable size, especially when writing loops. To define length function, type:

```
>> u = 1 : 5
u =
    1  2  3  4  5
```

Then

```
>> n = length(u)
n = 5
```

Now define the size command, which returns two values and has the syntax:

$$[nr, nc] = size(A)],$$

where nr is the number of rows and nc is the number of columns in matrix A. For example:

```
>> A = eye(3,4);
>> [nr, nc] = size(A)
nr = 3
nc = 4
>> B = ones(size(A))
B =
    1  1  1  1
    1  1  1  1
    1  1  1  1
```

- Create a square root of matrix A using the *sqrt* function means to obtain a matrix B with entries square root of entries of matrix A, type:

$$>> B = sqrt(A).$$

For example:

```
>> A = [1 4 5; 2 3 4; 4 7 8];
>> B = sqrt(A)
B =
      1.0000  2.0000  2.2361
      1.4142  1.7321  2.0000
      2.0000  2.6458  2.8284
```

- Create an upper triangular matrix for a given matrix A using the *triu* function as follows:

$$>> U = triu(A).$$

For example:

```
>> A = [1 2 3; 4 5 6; 7 8 9];
>> U = triu(A)
A =
      1  2  3
      0  5  6
      0  0  9
```

Also, one can create an upper triangular matrix from a given matrix A with zero diagonal as:

$$>> W = triu(A, 1)$$

For example:

```
>> A = [1 2 3; 4 5 6; 7 8 9];
>> W = triu(A, 1)
W =
      0  2  3
      0  0  6
      0  0  0
```

- Create a lower triangular matrix A for a given matrix using the *tril* function as follows:

$$\boxed{>> L = tril(A)}.$$

For example:

```
>> A = [1 2 3; 4 5 6; 7 8 9];
>> L = tril(A)
A =
     1  0  0
     4  5  0
     7  8  9
```

.

Also, one can create an lower triangular matrix from a given matrix A with a zero diagonal as follows:

$$\boxed{>> V = tril(A, 1)}.$$

For example:

```
>> A = [1 2 3; 4 5 6; 7 8 9];
>> V = tril(A, 1)
V =
     0  0  0
     4  0  0
     7  8  0
```

.

- Create an $n \times n$ random matrix using the *rand* function as follows:

$$\boxed{>> R = rand(n)}.$$

For example:

```
>> R = rand(3)
R =
     0.6038  0.0153  0.9318
     0.2722  0.7468  0.4660
     0.1988  0.4451  0.4186
```

.

- Create a reshape matrix of matrix A using the *reshape* function as follows:

$$\boxed{>> B = reshape(A, newrows, newcols)}.$$

For example:

```
>> A = [1 2 3; 4 5 6; 7 8 9; 10 11 12]
>> B = reshape(A, 2, 6)
B =
     1   7   2   8   3   9
     4  10   4  11   6  12
```

and

```
>> c = reshape(A, 1, 12)
c =
     1   4   7   10   2   5   8   11   3   6   9   12
```

.

- Create $n \times n$ Hilbert matrix using the *hilb* function as follows:

$$\boxed{>> H = hilb(n)}$$

For example:

```
>> H = hilb(3)
H =
     1.0000   0.5000   0.3333
     0.5000   0.3333   0.2500
     0.3333   0.2500   0.2000
```

.

- Create a Toeplitz matrix with a given column vector C as the first column and a given row vector R as a first row using the *toeplitz* function as follows:

$$\boxed{>> U = toeplitz(C, R)}.$$

8. Matrix Operations

The basic arithmetic operations of addition, subtraction, and multiplication may be applied directly to matrix variables, provided that the particular operation is legal under the rules of linear algebra. When two matrices have the same size, we add and subtract them in the standard way matrices are added and subtracted. For example:

```
>> A = [3 2 -3; 4 5 6; 7 6 7];
>> B = [1 2 3; 4 -2 1; 7 5 -4];
>> C = A + B
C =
     4    4    0
     8    3    7
    14   11    3
```

and the difference of A and B gives:

```
>> D = A - B
D =
     2    0   -6
     0    7    5
     0    1   11
```
.

Matrix multiplication has the standard meaning as well. Given any two compatible matrix variables A and B, the MATLAB expression $A * B$ evaluates the product of A and B as defined by the rules of linear algebra. For example:

```
>> A = [2 3; -1 4];
>> B = [5 -2 1; 3 8 -6];
>> C = A * B
C =
    19   20  -16
     7   34  -25
```
.

Also

```
>> A = [1 2;3 4];
>> B = A';
>> C = 3 * (A * B)^ 3
C =
      13080  29568
      29568  66840
```

Similarly, the two vectors have the same size, they can be added or subtracted from one other. They can be multiplied, or divided, by a scalar, or a scalar can be added to each of their components.

Mathematically the operation of division by a vector does not make sense. To achieve the corresponding componentwise operation, we use the ./ operator. Similarly, multiplication and powers we use .* and .∧, respectively. For example:

```
>> a = [1 2 3];
>> b = [2 -1 4];
>> c = a. * b
c =
      2  -2  12
```

Also

```
>> c = a./b
c =
      0.5  -2.0  0.75
```

and

```
>> c = a.^ 3
c =
      1  8  27
```

Similarly,

```
>> c = 2.^ a
c =
      2  4  8
```

and

```
>> c = b.^ b
c =
    2  1  64
```

Note that these operations apply to matrices as well as vectors. For example:

```
>> A = [1 2 3; 4 5 6; 7 8 9];
>> B = [9 8 7; 6 5 4; 3 2 1];
>> C = A. * B
C =
     9  16  21
    24  25  24
    21  16   9
```

Note that $A. * B$ is not the same as $A * B$:

```
>> C = A.^ 2
C =
     1   4   9
    16  25  36
    49  64  81
```

and

```
>> C = A.^ (1/2)
C =
    1.0000  1.4142  1.7321
    2.0000  2.2361  2.4495
    2.6458  2.8284  3.0000
```

Note that there are no such special operators for addition and subtraction.

9. String and Printing

Strings are matrices with character elements. In more advanced applications such as symbolic computation, string manipulation is a very important topic. For our purposes, however, we shall need only very limited skills in handling strings initially.

One most important use might be to include your name. Strings can be defined in MATLAB by simply enclosing the appropriate string of characters in single quotes such as:

```
>> first =' Rizwan';
>> last =' Butt';
>> name = [first, last]
name =
    Rizwan Butt
```

Since the transpose operator and the string delimiter are the same character (the single quote), creating a single column vector with a direct assignment requires enclosing the string literal in parentheses:

```
>> Last Name = ('Butt')'
Name =
    B
    u
    t
    t
```

String matrices can also be created as follows:

```
>> Name = ['Rizwan';' Butt']
Name =
    Rizwan
    Butt
```

There are two functions for text output called *disp* and *fprintf*. The disp function is suitable for simple printing task. The fprintf function provides fine control over the displayed information as well as the capability of directing the output to a file.
The *disp* function takes only one argument, which may be either a string matrix or a numerical matrix. For example:

```
>> disp('Hello')
Hello
```

and

$$
\begin{array}{l}
>> x = 0 : pi/5 : 2 * pi; \\
>> y = \sin(x); \\
>> disp([x' \ y']) \\
\quad 0.0000 \quad 1.0000 \\
\quad 0.6283 \quad 0.8090 \\
\quad 1.2566 \quad 0.3090 \\
\quad 1.8850 \quad -0.3090 \\
\quad 2.5133 \quad -0.8090 \\
\quad 3.1416 \quad -1.0000 \\
\quad 3.7699 \quad -0.8090 \\
\quad 4.3982 \quad -0.3090 \\
\quad 5.0265 \quad 0.3090 \\
\quad 5.6549 \quad 0.8090 \\
\quad 6.2832 \quad 1.0000
\end{array}.
$$

More complicated strings can be printed using the fprintf function. This is essentially a C programming command, which can be used to obtain a wide range of printing specifications. For example:

$$
\begin{array}{l}
>> fprintf('My\ Name\ is \backslash n\ Rizwan\ Butt\ \backslash n') \\
My\ Name\ is \\
\quad Rizwan\ Butt
\end{array}
$$

where the $\backslash n$ is the newline command.

The sprintf function allows specification of the number of digits in the display, as in:

$$
\begin{array}{l}
>> root2 = fprintf('The\ square\ root\ of\ 2\ is\ \%9.7f', (sqrt(2))) \\
root2 = \\
\quad \text{The square root of 2 is 1.4142136}
\end{array}
$$

or use of exponential format:

$$
\begin{array}{l}
>> root2 = fprintf('The\ square\ root\ of\ 2is\ \%11.5e', (sqrt(2))) \\
root2 = \\
\quad \text{The square root of 2 is } 1.41421e + 000
\end{array}.
$$

10. Solving Linear Systems

MATLAB started as a linear algebra extension of Fortran. Since its early days, MATLAB has been extended beyond its initial purpose, but linear algebra methods are still one of its strongest features. To solve the linear system

$$A\mathbf{x} = \mathbf{b}$$

we can just set

$$>> \mathbf{x} = A \backslash \mathbf{b}$$

with A is a nonsingular matrix. For example:

$$
\begin{array}{l}
>> A = [1\ 1\ 1; 2\ 3\ 1; 1\ -1\ -2];\\
>> \mathbf{b} = [2; 3; -6];\\
>> \mathbf{x} = A \backslash \mathbf{b}\\
\mathbf{x} =\\
\quad -1\\
\quad\ \ 1\\
\quad\ \ 2
\end{array}
$$

There are a small number of functions that should be mentioned.

- Reducing a given matrix A to reduced row echelon form using the *rref* function as:

$$>> rref(A)$$

For example:

$$
\begin{array}{l}
>> A = [1\ 1\ 1; 2\ 3\ 1; 1\ -1\ -2];\\
>> rref(A)\\
ans =\\
\quad 1\ 0\ 0\\
\quad 0\ 1\ 0\\
\quad 0\ 0\ 1
\end{array}
$$

- Finding determinant of a matrix A using the *det* function as:

$$\boxed{>> det(A)}.$$

For example:

$$\boxed{\begin{array}{l} >> A = [1\ 2\ \ -1; 3\ 0\ 1; 4\ 2\ 1]; \\ >> det(A) \\ ans = -6 \end{array}}.$$

- Finding rank of a matrix A using the *rank* function as:

$$\boxed{>> rank(A)}.$$

For example:

$$\boxed{\begin{array}{l} >> A = [1\ 4\ 5; 2\ 3\ 4; 4\ 7\ 8]; \\ >> rank(A) \\ ans = 3 \end{array}}.$$

- Finding the inverse of a nonsingular matrix A using the *inv* function as:

$$\boxed{>> inv(A)}.$$

For example:

$$\boxed{\begin{array}{l} >> A = [1\ 1\ 1; 1\ 2\ 4; 1\ 3\ 9]; \\ >> inv(A) \\ ans = \\ \quad \begin{array}{rrr} 3.0000 & -3.0000 & 1.0000 \\ -2.5000 & 4.0000 & -1.5000 \\ 0.5000 & -1.0000 & 0.5000 \end{array} \end{array}}.$$

- Finding the augmenting matrix $[A\ \mathbf{b}]$, which is the combination of coefficient matrix A and the right-hand side vector b of linear system $A\mathbf{x} = \mathbf{b}$ and saving the answer in matrix C, type:

$$\boxed{>> C = [A\ \mathbf{b}];}.$$

For example:

```
>> A = [1 1 1; 1 2 4; 1 3 9];
>> b = [2; 3; 4];
>> C = [A b]
C =
    1  1  1  2
    1  2  3  3
    1  4  9  4
```
.

- LU decomposition of a matrix A can be computed using the *lu* function as:

```
>> [L, U] = lu(A)
```
.

For example:

```
>> A = [1 4 5; 2 3 4; 4 7 8];
>> [L, U] = lu(A)
L =
    0.2500  1.0000   0.0000
    0.5000  -0.2222  1.0000
    1.0000  0.0000   0.0000
```

and

```
U =
    4.0000  7.0000  8.0000
    0.0000  2.2500  3.0000
    0.0000  0.0000  0.6667
```
.

- Using indirect LU decomposition, one can compute:

```
>> [L, U, P] = lu(A)
```
.

For example:

```
>> A = [1 4 5; 2 3 4; 4 7 8];
>> [L, U, P] = lu(A)
L =
       1     0  0
    0.25     1  0
     0.5  -0.2  1
```

and

$$U = \begin{matrix} 4 & 7 & 8 \\ 0 & 2.25 & 3 \\ 0 & 0 & 0.67 \end{matrix}$$

and

$$P = \begin{matrix} 0 & 0 & 1 \\ 1 & 0 & 0 \\ 0 & 1 & 0 \end{matrix}.$$

- One can compute the various norms of the vectors and matrices using the *norm* function. The expression $norm(A, 2)$ or *norm*(A) gives the Euclidean norm or l_2-norm of A while *norm*(A,Inf) gives the maximum or l_∞-norm. Here, A can be a vector or a matrix. The l_1-norm of a vector or matrix can be obtained by *norm*(A,1). For example, the different norms of the vector can be obtained as:

```
>> a = [6, 7, 8]
V1 >> norm(a)
V1 = 12.8841
V2 >> norm(a, 1)
V2 = 22
V3 >> norm(a, Inf)
V3 = 8
```

Similarly, for finding the different norms of matrix A, type:

```
>> A = [1 1 1; 1 2 4; 1 3 9]
M1 >> norm(A)
M1 = 10.6496
M2 >> norm(A, 1)
M2 = 14
M3 >> norm(A, Inf)
M3 = 13
```

- The condition number of a matrix A can be obtained using the *cond* function as cond(A). This equivalent to $norm(A, Inf) * norm(inv(A), Inf)$. For example:

$$
\boxed{\begin{array}{l}
>> A = [1\ 1\ 1; 1\ 2\ 4; 1\ 3\ 9] \\
>> B = inv(A) = [3\ \ -3\ 1; -2.5\ 4\ \ -1; 0.5\ \ -1\ 0.5] \\
N1 >> norm(A, Inf) \\
N1 = 13 \\
N2 >> norm(B, Inf) \\
N2 = 8
\end{array}}.
$$

Thus, the condition number of a matrix A is computed by cond(A) as:

$$
\boxed{\begin{array}{l}
>> cond(A) = N1 * N2 \\
cond(A) = 104
\end{array}}.
$$

- The root of polynomial $p(x)$ can be obtained using *roots* function as roots(p). For example, if $p(x) = 3x^2 + 5x - 6$ is a polynomial, enter:

$$
\boxed{\begin{array}{l}
>> p = [35 - 6]; \\
>> r = roots(p) \\
r = \\
\quad -2.4748 \\
\quad\ \ 0.8081
\end{array}}.
$$

- Use the *polyvar* function to evaluate a polynomial $p_n(x)$ at a particular point x. For example, to find the polynomial function $p_3(x) = x^3 - 2x + 12$ at given point $x = 1.5$, type:

$$
\boxed{\begin{array}{l}
>> ceof = [1\ 0\ \ -2\ 12]; \\
>> sol = polyvar(coef, 1.5) \\
sol = 12.3750
\end{array}}.
$$

- Creating eigenvalues and eigenvectors of given matrix A using the *eig* function as follows:

$$
\boxed{>> [U, D] = eig(A)}.
$$

Here, U is a matrix with columns as eigenvectors and D is a diagonal matrix with eigenvalues on the diagonal. For example:

```
>> A = [1 1 2; -1 2 1; 0 1 3];
>> [U, D] = eig(A)
U =
      0.4082   -0.5774   0.7071
      0.8165    0.5774   0.0000
     -0.4082   -0.5774   0.7071
```

and

```
D =
     1  0  0
     0  2  0
     0  0  3
```
,

which shows that 1, 2, and 3 are eigenvalues of the given matrix.

11. Graphing in MATLAB

MATLAB can produce two- and three-dimensional plots of curves and surfaces. The *plot* command is used to generate graphs of two-dimensional functions. MATLAB's plot function has the ability to plot many types of 'linear' two-dimensional graphs from data which is stored in vectors or matrices. For producing two-dimensional plots we have to do the following:

- Divide the interval into subintervals of equal width. To do this type:

  ```
  >> x = a : d : b;
  ```
  ,

  where a is the lower limit, d, the width of each subinterval, and b, the upper limit of the interval.

- Enter the expression for y in term of x as:

  ```
  >> y = f(x);
  ```
  .

- Create the plot by typing:

$>> plot(x, y)$.

For example, to graph the function $y = e^x + 10$, type:

$$
\begin{array}{l}
>> x = -2 : 0.1 : 2; \\
>> y = exp(x) + 10; \\
>> plot(x, y)
\end{array}
$$

.

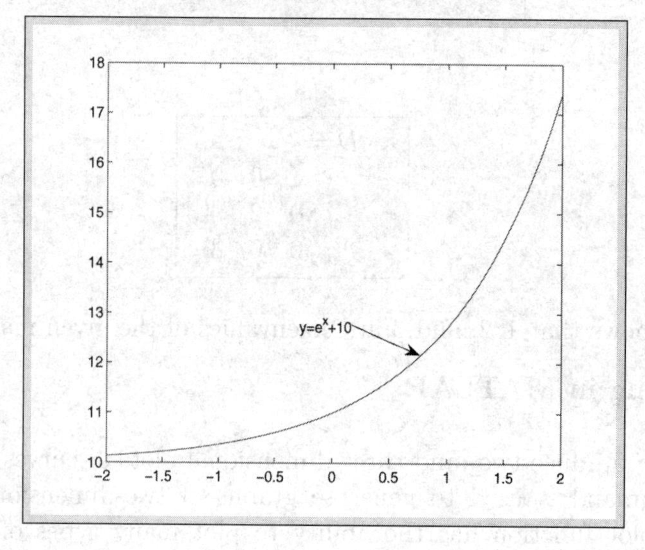

Figure B.1: Graph of $y = e^x + 10$.

By default, the plot function connects the data with a solid line. The markers used for points in a plot may be any of the following:

Symbol	**Effect**
•	*Point*
∘	*Circle*
×	*Cross*
⋆	*Star*

For example, to put a marker for points in the above function plot, type:
we see

$$
\begin{array}{l}
>> x = -2 : 0.1 : 2; \\
>> y = exp(x) + 10; \\
>> plot(x, y,' o')
\end{array}.
$$

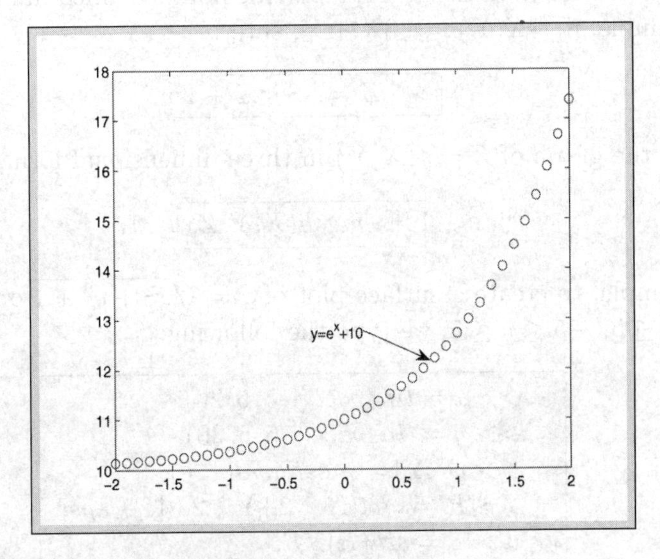

Figure B.2: Graph of $y = e^x + 10$.

Three-dimensional surface plots are obtained by specifying a rectangular subset of the domain of a function with the *meshgrid* command and then using the *mesh* or *surf* commands to obtain a graph. For a three-dimensional graph, we do the following:

For function of two variable $Z = f(X, Y)$ and three-dimensional plots, use the following procedure:

- Define scaling vector for X. For example, to divide interval $[-2, 2]$ for x into subintervals of width 0.1, enter:

$$\boxed{>> x = -2 : 0.1 : 2;}.$$

- Define scaling vectors for Y. In order to use the same scaling for y, enter:

$$\boxed{>> y = x;}.$$

One may, however, use a different scaling for y.

- Create a meshgrid for the x and y axis:

$$\boxed{>> [X, Y] = meshgrid(x, y);}.$$

- Compute the function $Z = f(X, Y)$ at the points defined in first two steps. For example, if $f(X, Y) = -3X + Y$, enter:

$$\boxed{>> Z = -3 * X + Y;}.$$

- To plot the graph of $Z = f(X, Y)$ in three-dimensional form, type:

$$\boxed{>> mesh(X, Y, Z)}.$$

For example, to create a surface plot of $z = \sqrt{x^2 + y^2 + 1}$ on the domain $-5 \leq x \leq 5$, $-5 \leq y \leq 5$, we type the following:

```
>> x = linspace(-5, 5, 20);
>> y = linspace(-5, 5, 20);
>> [X, Y] = meshgrid(x, y);
>> R = sqrt(X.^2+Y.^2+1) + eps;
>> Z = sin(R)./R;
>> surf(X, Y, Z)
```

Adding *eps* (a MATLAB command that returns the smallest floating-point number on your system) avoids the indeterminate 0/0 at the origin.

Subplots

Often, it is of interest to place more than one plot in a single figure window. This is possible with the graphic command called *subplot function*, which is always called with three arguments as in:

$$\boxed{>> subplot(nrows, ncols, thisplot)},$$

where nrows and ncols define a visual matrix of plots to be arranged in a single figure window and *thisplot* indicates the number of the subplot that is being currently drawn. *Thisplot* is an integer that counts across rows and then columns. For a given arrangement of subplots in a figure window, the nrows and ncols arguments do not change. Just before each plot in the matrix is drawn, the subplot function is issued with the appropriate value of *thisplot*. The following figure shows four subplots created with the following statements:

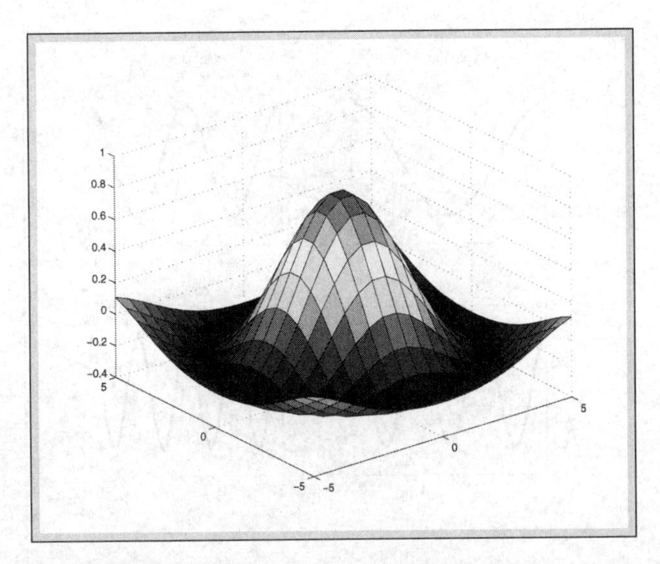

Figure B.3: Surface plot of $z = \sin(\sqrt{x^2 + y^2 + 1})/\sqrt{x^2 + y^2 + 1}$.

```
>> x = linspace(0, 2 * pi);
>> subplot(2, 2, 1);
>> plot(x, cos(x));      axis([0 2 * pi  − 1.5 1.5]);     title('cos(x)');
>> subplot(2, 2, 2);
>> plot(x, cos(2 * x));  axis([0 2 * pi  − 1.5 1.5]);     title('cos(2x)');
>> subplot(2, 2, 3);
>> plot(x, cos(3 * x));  axis([0 2 * pi  − 1.5 1.5]);     title('cos(3x)');
>> subplot(2, 2, 4);
>> plot(x, cos(4 * x));  axis([0 2 * pi  − 1.5 1.5]);     title('cos(4x)');
```

Similarly, one can use the subplots function for creating surface plots using the following command:

```
>> x = linspace(−5, 5, 20);
>> y = linspace(−5, 5, 20);
>> [X, Y] = meshgrid(x, y); Z = 2 + (X.^ 2+Y.^ 2);
>> subplot(2, 2, 1);    mesh(x, y, Z);     title('meshplot');
>> subplot(2, 2, 2);    surf(x, y, Z);     title('surfplot');
>> subplot(2, 2, 3);    surfc(x, y, Z);    title('surfcplot');
>> subplot(2, 2, 4);    surfl(x, y, Z);    title('surflplot');
```

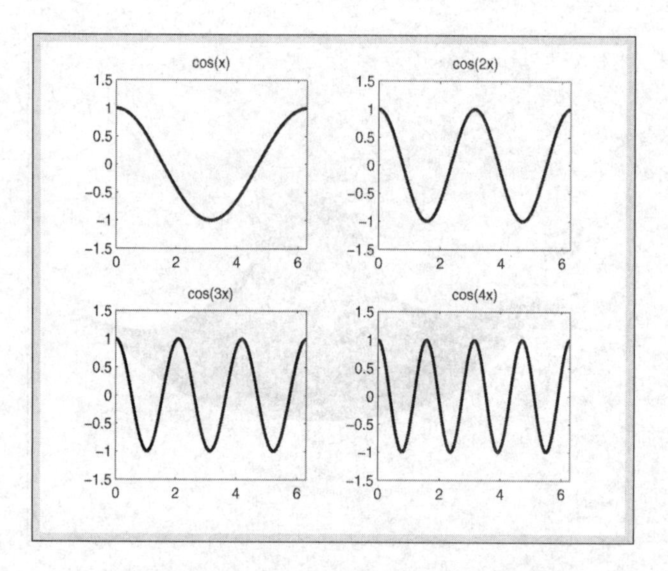

Figure B.4: Four subplots in a figure window.

Programming in MATLAB

Here we discuss the structure and syntax of MATLAB programs. There are many similarities between MATLAB and other high-level languages. The syntax is similar to Fortran, with some ideas borrowed from C. MATLAB has loop and conditional execution constructs.

Several important features of MATLAB differentiate it from other high-level languages. MATLAB programs are tightly integrated into an interactive environment. MATLAB programs are interpreted, not compiled. All MATLAB variables are sophisticated data structures that manifest themselves to the user as matrices. MATLAB automatically manages dynamic memory allocation for matrices, which affords convenience and flexibility in the development of algorithms. MATLAB provides highly optimized, built-in routines for multiplication, adding, and subtracting matrices, along with solving linear systems and computing eigenvalues.

Statements for Control Flow

The commands *for*, *while*, and *if* define decision-making structures called control flow statements for the execution of parts of a script based on various conditions.

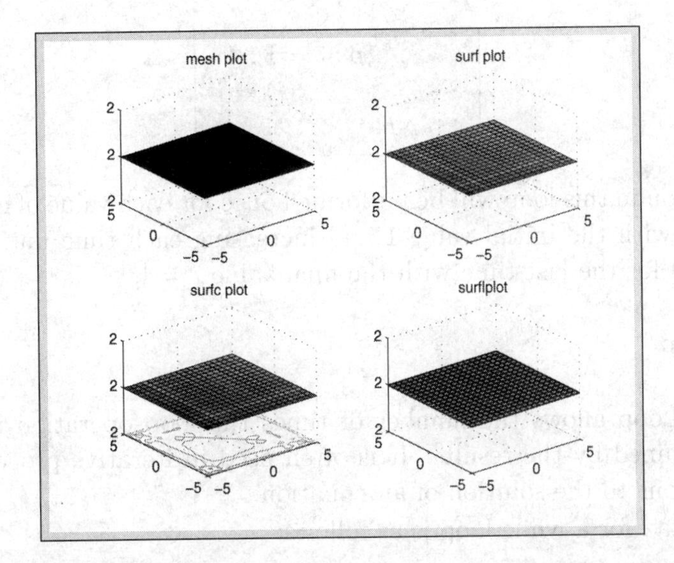

Figure B.5: Four types of surface plots.

Each of three structures is ended by an *end* command. These statements used to control the flow are *called relations.*

The repetitions can be handled in MATLAB using a *for loop* or a *while loop.* The syntax is similar to the syntax of such loops in any programming language. In the following we will discuss such loops.

For Loop

The For Loop enables us to have an operation repeated a specified number of times. This may be required in summing terms of series, or specifying the elements of a nonuniformly spaced vectors such as the first terms of a sequence defined recursively.

The syntax includes a counter variable, initial value of the counter, the final value of the counter, and the action to be performed written in the following format:

$$>> for\ counter\ name = initial\ value : final\ value, action; end.$$

For example, in order to create the 1×4 row vector x with entries according to formula $x(i) = i$; type:

$$\boxed{\begin{array}{l} >> for\ i = 1:4 \\ \qquad\qquad x(i) = i \\ end \end{array}}.$$

The action in this loop will be performed once for each value of counter name i beginning with the initial value 1 and increasing each time until the actions are executed for the last time with the final value $i = 4$.

While Loop

The While Loop allows the number of times the loop operation is performed to be determined by the results. It is often used in iterative processes such as approximations to the solution of an equation.

The syntax for a while loop is as follows:

$$\boxed{>> while\ (condition)\ action; increment\ action; end}.$$

The loop is executed until the condition (the statement in parentheses) is evaluated. Note that the counter variable must be initialized before using the above command and the increment action gives the increment in the counter variable. For example:

$$\boxed{\begin{array}{l} >> x = 1; \\ while\ x > 0.01 \\ \qquad\quad x = x/2; \\ end \\ >> disp(x) \end{array}},$$

which generates:

$$\boxed{\begin{array}{l} x = \\ \qquad 0.5000 \\ \qquad 0.2500 \\ \qquad 0.1250 \\ \qquad 0.0625 \\ \qquad 0.0313 \\ \qquad 0.0156 \\ \qquad 0.0078 \end{array}}.$$

Nested for Loops

In order to have a nest of for loops or while loops, each type of loop must have a separate counter. The syntax for two nested for loops is:

$$
\begin{array}{l}
>> for\ counter1 = initial\ value1 : final\ value1,\\
>> for\ counter2 = initial\ value2 : final\ value2,\\
\quad action;\\
\quad end\\
end
\end{array}
$$

For example, in order to create a 5×4 matrix A by formula $A(i,j) = i + j$, type:

$>> for\ i = 1 : 5,$
for $j = 1 : 4,$
$A(i,j) = i + j;$
end,
end

which generates a matrix of the form:

$$
ans = \begin{array}{cccc}
2 & 3 & 4 & 5\\
3 & 4 & 5 & 6\\
4 & 5 & 6 & 7\\
5 & 6 & 7 & 8\\
6 & 7 & 8 & 9
\end{array}.
$$

if Structure

Finally, we introduce the basic structure of MATLAB's logical branching commands. Frequently in programs we wish the computer to take different actions depending on the value of some variables. Strictly these are logical variables, or, more commonly, logical expressions similar to those we saw in defining while loops.

Two types of decision statements are possible in MATLAB, one-way decision statements and two-way decision statements.

The syntax for the one-way decision statement is:

$$>> if\ (condition), action,\ end$$

in which the statements in the action block are executed only if the condition is satisfied (true). If the condition is not satisfied (false) then the action block is skipped. For example:

$$>> if\ x > 0, disp('x\ is\ positive\ number'); end.$$

For a two-way decision statement we define its syntax as:

$$>> if\ (condition), action,\ else\ action, end$$

in which the first set of instructions in the action block are executed if the condition is satisfied while the second set, the action block, are executed if the condition is not satisfied. For example, if x and y are two numbers and we want to display the value of the number, type:

$$>> if\ (x > y), disp(x), else\ disp(y), end.$$

MATLAB also contains a number of logical and relational operators.

The logical operations include the following:

Symbol	Effect
&	*and*
|	*or*
~	*not*

However, these operators not only apply to side variables but such operators will also work on vectors and matrices when the operation is valid.

The relational operators in MATLAB are:

Symbol	Effect
==	is equal to
<=	is less than or equal
>=	is greater than or equal
~=	is not equal to
<	is less than
>	is greater than

The relational operators are used to compare values or elements of arrays. If the relationship is true, the result is a logical variable whose value is one. Otherwise, the value is zero if the relationship is false.

Defining Function

MATLAB allows us to define their own functions by constructing an M-file in the M-file editor. If the M-file is to be a function M-file, the first word of the file is function, and we must also specify names for the function input and output. The last two of these are purely local variable names.

The first line of the function has the form:

function y = function name(input arguments)

For example, to define the function

$$f(x) = e^x - \frac{2x}{(1 + x^3)}$$

type:

$$\begin{array}{l} functiony = fn1(x) \\ y = exp(x) - 2 * x./(1 + x.\hat{}\ 3); \end{array}$$

Once this function is saved as an M-file named f1.m, we can use the MATLAB command window to compute function at any given point. For example:

$$\begin{array}{l} >> x = (0 : 0.2 : 2)'; \\ >> fx = fn1(x); \\ >> [x', fx'] \end{array}$$

generates the following table:

```
ans =
    0.0000  1.0000
    0.2000  0.8246
    0.4000  0.7399
    0.6000  0.8353
    0.8000  1.1673
    1.0000  1.7183
    1.2000  2.4404
    1.4000  3.3073
    1.6000  4.3251
    1.8000  5.5227
    2.0000  6.9446
```

Conclusion

MATLAB has a wide range of capabilities. In this book, we used only a small portion of its features. We found that MATLAB's command structure is very close to the way one writes algebraic expressions and linear algebra operations. The names of many MATLAB commands closely parallel those of the operations and concepts of linear algebra. We gave descriptions of commands and features of MATLAB that related directly to this course. A more detailed discussion of MATLAB commands can be found in MATLAB user's guide that accompanies the software and in the following books:

Experiments in computational Matrix Algebra by David R. Hill (New York, Random House, 1988).
Linear Algebra LABS with MATLAB, second edition by David R. Hill and David E. Zitaret (Prentice Hall, Inc, 1996).
For a very complete introduction to MATLAB graphics, one can use the following book:
Graphics and GUIs with MATLAB, second edition by P. Marchand (CRC Press, 1999).
There are many websites to aid you in learning about MATLAB and you can locate many of those by using a web search engine (e.g., Google). Alternatively, MATLAB software provides immediate on screen descriptions using the *help* command or, one can contact Mathworks using following address:
www.mathworks.com

Appendix C

Index of MATLAB Programs

In this appendix we list all of the MATLAB functions supplied in this book. These functions are contained in a CD-Rom attached to the cover of the book. The CD-ROM included with Introduction to Numerical Analysis Using MATLAB by Rizwan Butt contains a MATLAB program for each of the methods presented in the book. Every program is illustrated with a sample problem or example that is closely correlates to the text. The programs can easily be modified for other problems by making minor changes. All the programs are designed to run on a minimally configured computer. Minimal hard disk space plus the MATLAB package are all that is needed to run the programs. All the programs are given as ASCII files called M-files with the .m extension. They can be altered using any word processor that creates standard ASCII files. The M-files can run from within MATLAB by entering the name without the .m extension. For example, fixpt.m can be run using fixpt as the command. The files should be placed in the MATLAB\work subdirectory of MATLAB.

MATLAB Function	Definitions	Chapter Two
bisect	Bisection method	program 2.1
falseP	False position method	program 2.2
fixpt	Fixed-point method	program 2.3
newton	Newton's method	program 2.4
secant	Secant method	program 2.5
mnewton1	First modified Newton's method	program 2.6
mnewton2	Second modified Newton's method	program 2.7
AitK	Aitken's Δ^2 method	program 2.8
newton2	Newton's method for system	program 2.9
HornM	Horner's method	program 2.10
Muller	Muller's method	program 2.11
BairSM	Bairstow's method	program 2.12

MATLAB Function	Definitions	Chapter Three
INVMAT	Inverse of a matrix	program 3.1
WP	Gauss elimination without pivoting	program 3.2
PP	Gauss elimination with partial pivoting	program 3.3
TP	Gauss elimination with complete pivoting	program 3.4
GaussJ	Gauss-Jordan method	program 3.5
lu-guass	LU decomposition method	program 3.6
Doll	LU decomposition with Doolittle's method	program 3.7
Crout	LU decomposition with Crout's method	program 3.8
cholesky	LU decomposition with Cholesky method	program 3.9
TridLU	LU decomposition of tridiagonal system	program 3.10
JacobiM	Jacobi iterative method	program 3.11
GaussSM	Gauss-Seidel iterative method	program 3.12
SORM	SOR iterative method	program 3.13
CONJG	Conjugate gradient method	program 3.14
RESID	Calculate residual vector	program 3.15

MATLAB Function	Definitions	Chapter Four
lint	Lagrange interpolation method	program 4.1

divdiff	Divided difference	program 4.2
Ndivf	Newton's interpolation method	program 4.3
LSpline	Linear spline approximation	program 4.4
CubicS	Natural cubic spline approximation	program 4.5
linefit	Linear least square fit	program 4.6
polyfit	Polynomial least square fit	program 4.7
ex1fit	Nonlinear least square fit	program 4.8
ex2fit	Nonlinear least square fit	program 4.9
planefit	Least square plane fit	program 4.10
overd	Overdetermined	program 4.11

MATLAB Function	**Definitions**	**Chapter Five**
TrapezoidR	Trapezoidal rule	program 5.1
SimpsonR	Simpson's rule	program 5.2
ErrorSR	Error term for Simpson's rule	program 5.3
RepeatedTR	Repeated Trapezoidal rule	program 5.4
RombergM	Romberg integration	program 5.5
GaussianQ	Gaussian quadrature	program 5.6

MATLAB Function	**Definitions**	**Chapter Six**
Euler1	Euler's method	program 6.1
tayl1	Taylor's method	program 6.2
mod1	Modified Euler's method	program 6.3
heun	Heun's method	program 6.4
mid	Midpoint method	program 6.5
Rk4	Runge-Kutta method of order 4	program 6.6
RKM	Runge-Kutta-Merson method	program 6.7
Ab2M	Adams-Bashforth two-step method	program 6.8
MSIMP	Milne-Simpson method	program 6.9
ABMM	Adam-Bashforth Moulton method	program 6.10
RK4FS	Runge-Kutta method of order 4 for system	program 6.11
Lsoot	Linear shooting method	program 6.12
FDM	Finite difference method	program 6.13

MATLAB Function	Definitions	Chapter Seven
trac	Trace of a matrix	program 7.1
CAYHAM	Cayley-Hamilton theorem	program 7.2
SOURIAN	Sourian frame theorem	program 7.3
BOCHER	Bocher's theorem	program 7.4
POWERM	Power method	program 7.5
INVERSEPM	Inverse power method	program 7.6
ShiftedIPM	Shifted inverse power method	program 7.7
DEFLATION	Deflation method	program 7.8
JOBM	Jacobi method for eigenvalues	program 7.9
sturmS	Sturm sequence method	program 7.10
Given	Given's method	program 7.11
HouseHM	Householder method	program 7.12
QRM	QR method	program 7.13
HessenB	Upper Hessenberg form	program 7.14

MATLAB Built-in Functions

Listed below are some of the MATLAB built-in-functions grouped by subject areas:

Built-in Function	Definitions
abs	absolute value
cos	cosine function
sin	sine function
tan	tangent function
cosh	cosine hyperbolic function
sinh	sine hyperbolic function
tanh	tangent hyperboic function
acos	inverse cosine function
asin	inverse sine function
atan	inverse tangent function
erf	error function $erf(x) = (2/\sqrt{\pi}) \int_0^x e^{-t^2}\, dt$
exp	exponential function
expm	matrix exponential

Built-in Function	Definitions
log	natural logarithm
log10	common logarithm
sqrt	calculate square root
sqrm	calculate square root of a matrix
sort	arrange elements in asending order
std	calculate standard deviation
mean	calculate mean value
median	calculate median value
sum	calculate sum of elements
angle	calculate phase angle
fix	round toward zero
floor	round toward - ∞
ceil	round toward ∞
sign	signum function
round	round to nearest integer
dot	dot product of two vectors
cross	cross product of two vectors
dist	distance between two points
frac	return the rational approximation
max	return maximum value
min	return minimum value
factorial	factorial function
rref	reduce echelon form
zeros	generates matrix of all zeros
ones	generates matrix of all ones
eye	generates identity matrix
hilb	generates a Hilbert matrix
reshape	rearrange a matrix
det	calculate determinant of a matrix
eig	calculate eigenvalues of a matrix
rank	calculate rank of a matrix
norm(v)	calculate the Euclidean norm of a vector v
norm(v,inf)	calculate maximum norm of vector v
cond(A,2)	calculate condition number of matrix using Euclidean norm

Built-in Function	Definitions
cond(A,inf)	calculate condition number of matrix using maximum norm
toeplitz	creates a Toeplitz matrix
inv	finding inverse of a matrix
pinv	finding pseudoinverse matrix
diag	create a diagonal matrix
length	number of elements in a vector
size	size of an array
qr	create QR-decomposition of a matrix
svd	calculate singular value decomposition of a matrix
polyval	calculate the value of a polynomial
roots	calculate the roots of a polynomial
conv	multiplies two polynomials
deconv	divide two polynomials
polyder	calculate derivative of a polynomial
polyint	calculate integral of a polynomial
polyfit	calculate coefficients of a polynomial
fzero	solve an equation with one variable
quad	integrate a function
linspace	create equally spaced vector
logspace	create logarithically spaced elements
axis	sets limts to axes
plot	create a plot
pie	create a pie plot
polar	create a polar plot
hist	create a histogram
bar	create a vertical bar plot
barh	create a horizontal bar plot
fplot	plot a function
bar3	create a vertical 3-D bar plot
contour	create a 2-D contour plot
contour3	create a 3-D contour plot
cylinder	create a cylinder
mesh	create a mesh plot

Built-in Function	Definitions
meshc	create a mesh and a contour plot
surf	create a surface plot
surfc	create a surface and a contour plot
surfl	create a surface plot with lighting
sphere	create a sphere
subplot	create multiple plot on one page
title	add a title to a plot
xlabel	add label to x-axis
ylabel	add label to y-axis
grid	add grid to a plot

Appendix D

Symbolic Computation

In this appendix we discuss symbolic computation, which is an important and complementary aspect of computing. As we have noted, MATLAB uses floating-point arithmetic for its calculations. But one can also do exact arithmetic with symbolic expression. Here, we will give many examples to get the exact arithmetic.

Many applications in mathematics, science, and engineering require symbolic operations, which are mathematical operations with expressions that contain symbolic variables. Symbolic variables are such variables that don't have specific numerical values when the operation is executed. The result of such operations is also mathematical expression in terms of the symbolic variables. Symbolic operations can be performed by MATLAB when the Symbolic Math Toolbox is installed. The Symbolic Math Toolbox is included in the student version of the software and can be added to the standard program. The Symbolic Math Toolbox is a collection of MATLAB functions that are used for execution of symbolic operations. The commands and functions for the symbolic operations have the same style and syntax as those for the numerical operations.

Symbolic computations are performed by computer programs such as Derive®, Maple®, and Mathematica®. MATLAB also supports symbolic computation through the Symbolic Math Toolbox, which uses the symbolic routines of Maple. T, check if the Symbolic Math Toolbox is installed, one can type:

```
>> ver
```

In response, MATLAB displays information about the version that is used as well as a list of the toolboxes that are installed. Using MATLAB Symbolic Math Toolbox, we can carry out algebraic or symbolic calculations such as factoring polynomials or solving algebraic equations. For example, to add three numbers $\frac{3}{4}$, $\frac{1}{4}$, and $\frac{5}{4}$ symbolically, we do the following:

```
>> sym('3/4') + sym('1/4') + sym('5/4')
ans =
    9/4
```

Symbolic computations can be performed without the approximations that are necessary for numerical calculations. For example, to evaluate $\sqrt{5}\sqrt{5} - 5$ symbolically, we do:

```
>> sym(sqrt(5)) * sym(sqrt(5)) - 5
ans =
    0
```

But when we do the same calculation numerically, we have:

```
>> sqrt(5) * sqrt(5) - 5
ans =
    8.8818e - 016
```

In general, numerical results are obtained much more quickly with a numerical computation than with numerical evaluation of a symbolic calculation. To perform symbolic computations, we must use *syms* to declare the variables we plan to use as symbolic variables. For example, the quadratic formula can be defined in terms of a symbolic expression by the following kind of commands:

```
>> syms x a b c
>> sym(sqrt(a * x^ 2+b * x + c))
ans =
    (a * x^ 2+b * x + c) ^ (1/2)
```

Some Important Symbolic Commands

The collect command:

This command collects the terms in the expression that have the variable with the same power. In the new expression the terms will be ordered in decreasing order of power. The form of this command is:

$$>> collect(f)$$

or

$$>> collect(f, var_name).$$

For example, if $f = (2x^2 + y^2)(x + y^2 + 3)$, then we use the following commands:

```
>> syms x y
>> f = (2*x^2+y^2)*(x+y^2+3)
>> collect(f)
ans =
    2*x^3+(2*y^2+6)*x^2+y^2*x+y^2*(y^2+3)
```

But if we take y as a symbolic variable, then we do the following:

```
>> syms x y
>> f = (2*x^2+y^2)*(x+y^2+3)
>> collect(f,y)
ans =
    y^4+(2*x^2+x+3)*y^2+2*x^2*(x+3)
```

The factor command:

This command changes an expression that is a polynomial to be a product of polynomials of lower degree. The form of this command is:

$$>> factor(f).$$

For example, if $f = x^3 - 3x^2 - 4x + 12$, then use the following commands:

```
>> syms x
>> f = x^3-3*x^2-4*x+12
>> factor(f)
ans =
    (x-2)*(x-3)*(x+2)
```

The expand command:

This command multiplies the expressions. The form of this command is:

```
>> expand(f)
```

For example, if $f = (x^3 - 3x^2 - 4x + 12)(x - 3)^2$, then use the following commands:

```
>> syms x
>> f = (x^3-3*x^2-4*x+12)*(x-3)^3
>> expand(f)
ans =
    x^6-12*x^5+50*x^4-60*x^3-135*x^2+432*x-324
```

The simplify command:

This command is used to generate a simpler form of the expression. The form of this command is:

```
>> simplify(f)
```

For example, if $f = (x^3 - 3x^2 - 4x + 12)/(x - 3)^2$, then use the following commands:

```
>> syms x
>> f = (x^3-3*x^2-4*x+12)/(x-3)^3
>> simplify(f)
ans =
    (x^2-4)/(x-3)^2
```

The simple command:

This command finds a form of the expression with the fewest number of characters. The form of this command is:

$$>> simple(f)$$.

For example, if $f = (\cos x \cos y + \sin x \sin y)$, then using the *simplify* command, we get:

```
>> syms x y
>> f = (cos(x) * cos(y) + sin(x) * sin(y))
>> simplify(f)
ans =
    cos(x) * cos(y) + sin(x) * sin(y)
```

But if we use the *simple* command, we get:

```
>> syms x y
>> f = (cos(x) * cos(y) + sin(x) * sin(y))
>> simple(f)
ans =
    cos(x - y)
```

The pretty command:

This command displays a symbolic expression in a format in which expressions are generally typed. The form of this command is:

$$>> pretty(f)$$.

For example, if $f = \sqrt{x^3 - 3x^2 - 4x + 12}$, then use the following commands:

```
>> syms x
>> f = sqrt(x^3-3*x^2-4*x+12)
>> pretty(f)
ans =
    (x^3 - 3x^2 - 4x + 12)^1/2
```

The findsym command:

To determine what symbolic variables are used in an expression, we use the *findsym* command. For example, given the symbolic expressions $f1$ and $f2$ are defined by

```
>> syms a b c x y z
>> f1 = a * x^2+b * x + c
>> f2 = x * y * z
>> findsym(f1)
ans =
   a, b, c
>> findsym(f2)
ans =
   x, y, z
```

.

The subs command:

We can substitute a numerical value for a symbolic variable using the *subs* command. For example, to substitute the value $x = 2$ in the $f = x^3y + 12xy + 12$, we use the following commands:

```
>> syms x y
>> f = x^3*y + 12 * x * y + 12
>> subs(f, 2)
ans =
   32 * y + 12
```

.

Note that if we do not specify a variable to substitute for, MATLAB chooses a default variable according to the following rule. For one-letter variables, MATLAB chooses the letter closest to x in the alphabet. If there are two letters equally close to x, MATLAB chooses the one that comes later in the alphabet. In the preceding function, *subs(f,2)* returns the same answer as *subs(f,x,2)*. One can also use the findsym command to determine the default variable. For example,

```
>> syms u v
>> f = u * v
>> findsym(f, 1)
ans =
   v
```

.

Solving Equations Symbolically

We can find the solutions of certain equations symbolically using the MATLAB command *solve*. For example, to solve the nonlinear equation $x^3 - 2x - 1 = 0$

we define symbolic variable x and expression $f = x^3 - 2x - 1$ with the following commands:

```
>> syms x
>> f = x^3-2*x-1
>> solve(f)
ans =
    [-1]
    [1/2*5^(1/2)+1/2]
    [1/2-1/2*5^(1/2)]
```

Note that equation to be solved is specified as a string; that is, it is surrounded by single quotes. The answer consists of the exact(symbolic) solutions $-1, 1/2 \pm 1/2\sqrt{5}$. To get numerical solutions, type *double*(ans):

```
>> double(ans)
ans =
    -1.0000
    1.6180
    -0.6180
```

or type *vpa*(ans):

```
>> vpa(ans)
ans =
    [-1.]
    [1.618033988749894902525738871190 7]
    [-.6180339887498949025257388711907 0]
```

The command *solve* can also be used to solve polynomial equations of higher degree, as well as many other types of equations. It can also solve equations involving more than one variable. For example, solving two equations $3x+3y=2$ and $x+2y^2=1$, we do the following:

```
>> syms x y
>> [x,y] = solve('3*x+3*y=2','x+2*y^2=1')
x =
    [5/12-1/12*33^(1/2)]
    [5/12+1/12*33^(1/2)]

y =
    [1/4+1/12*33^(1/2)]
    [1/4-1/12*33^(1/2)]
```

.

Note that both solutions can be extracted with $x(1), y(1), x(2)$, and $y(2)$. For example, type:

```
>> x(1)
ans =
    [5/12-1/12*33^(1/2)]
```

and

```
>> y(1)
ans =
    [1/4+1/12*33^(1/2)]
```

.

If we want to solve $x + xy^2 + 3xy = 3$ for y in terms of x, then we have to specify the equation as well as variable y as a string:

```
>> syms x y
>> solve('x+x*y^2+3*x*y=3','y')
ans =
    [1/2/x*(-3*x+(5*x^2+12*x)^(1/2))]
    [1/2/x*(-3*x-(5*x^2+12*x)^(1/2))]
```

.

Calculus

The Symbolic Math Toolbox provides functions to do basic operations of calculus. Here, we describe these functions.

Symbolic Differentiation

This can be performed using the *diff* command as follows:

```
>> diff(f)
```

or

```
>> diff(f,var)
```
,

where the command diff(f, var) is used for differentiation of expression with several symbolic variables. For example, to find the first derivative of $f = x^3 + 3x^2 + 20x - 12$, we use the following commands:

```
>> syms x
>> f = x^3+3*x^2+20*x-12
>> diff(f)
ans =
    3*x^2+6*x+20
```
.

Note that if $f = x^3 + x \ln y + ye^{x^2}$ is taken, then MATLAB differentiates f with respect to x (default symbolic variable) as:

```
>> syms x y
>> f = x^3+x*log(y)+y*exp(x^2)
>> diff(f)
ans =
    3*x^2+log(y)+2*y*x*exp(x^2)
```
.

If we want to differentiate $f = x^3 + x \ln y + ye^{x^2}$ with respect to y, then we use the MATLAB *diff(f, y)* command as:

```
>> syms x y
>> f = x^3+x*log(y)+y*exp(x^2)
>> diff(f,y)
ans =
    x/y+exp(x^2)
```
.

To find the numerical value of the symbolic expression use the MATLAB *subs* command. For example, to find the derivative of $f = x^3 + 3x^2 + 20x - 12$ at $x = 2$, we do the following:

```
>> syms x
>> f = x^3+3*x^2+20*x-12
>> df = diff(f)
>> subs(df,x,2)
ans =
   44
```

We can also find the second and higher derivative of expressions using the following commands:

```
>> diff(f,n)
```

or

```
>> diff(f,var,n)
```

where n is a positive integer. For $n = 2$ and $n = 3$ are the second and third derivatives, respectively. For example, to find the second derivative of $f = x^3 + x \ln y + ye^{x^2}$ with respect to y, we use MATLAB *diff(f, y, 2)* command as:

```
>> syms x y
>> f = x^3+x*log(y)+y*exp(x^2)
>> diff(f,y,2)
ans =
   -x/y^2
```

Symbolic Integration

Integration can be performed symbolically using the *int* command. This command can be used to determine indefinite integrals and definite integrals of expression f. For indefinite integration, we use:

```
>> int(f)
```

or

```
>> int(f,var)
```

If in using *int*(f) command the expression contains one symbolic variable, then integration took place with respect to that variable. But if the expression contains more than one variable, then the integration is performed with respect to the default symbolic variable. For example, to find the indefinite integral (antiderivative) of $f = x^3 + x \ln y + ye^{x^2}$ with respect to y, we use the MATLAB *int(f, y)* command as:

```
>> syms x y
>> f = x^3+x*log(y) + y*exp(x^2)
>> int(f,y)
ans =
    x^3*y + x*y*log(y) - x*y + 1/2*y^2*exp(x^2)
```

Similarly, for the case of the definite integral, we use the following commands:

```
>> int(f,a,b)
```

or

```
>> int(f,var,a,b)
```

where a and b are the limits of integration. Note that limits a and b may be numbers or symbolic variables. For example, to determine the value of $\int_0^1 (x^2 + 3e^x + x \ln y)\, dx$, we use the following commands:

```
>> syms x y
>> f = x^2+3*exp(x) + x*log(y)
>> int(f,0,1)
ans =
    -8/3 + 3*exp(1) + 1/2*log(y)
```

We can also used symbolic integration to evaluate the integral when f has some parameters. For example, to evaluate the $\int_{-\infty}^{\infty} e^{-ax^2}\, dx$, we do the following:

```
>> syms a positve
>> syms x
>> f = exp(-a*x^2)
>> int(f,x,-inf,inf)
ans =
    1/(a)^2(1/2)*pi^2(1/2)
```

Note that if we don't assign a value to a, then MATLAB assumes that a represents a complex number, and therefore gives a complex answer. If a is any real number, then we do the following:

```
>> syms a real
>> syms x
>> f = exp(-a * x^ 2)
>> int(f, x, -inf, inf)
ans =
    PIECEWISE([1/a^ 2*pi^ 2(1/2), signum(a) = 1], [Inf, otherwise])
```

.

Symbolic Limits

The Symbolic Math Toolbox provides the *limit* command, which allows us to obtain the limits of functions directly. For example, to use the definition of the derivative of function

$$f'(x) = \lim_{h \to 0} \frac{f(x+h) - f(x)}{h}, \quad \text{provided limit exists}$$

for finding the derivative of the function $f(x) = x^2$, we use the following commands:

```
>> syms h x
>> f = (x + h)^ 2-x^ 2
>> limit(f/h, h, 0)
ans =
    2 * x
```

.

We can also find one-sided limits with the Symbolic Math Toolbox. To find the limit as x approaches a from the left, we use the commands:

```
>> syms a real
>> syms x
>> limit(f, x, a, 'left')
```

and to find the limit as x approaches a from the right, we use the commands:

```
>> syms a real
>> syms x
>> limit(f, x, a, 'right')
```

.

For example, to find the limit of $\frac{|x-3|}{x-3}$ when x approaches 3, we need to calculate

$$\lim_{x \to 3^-} \frac{|x-3|}{x-3} \quad \text{and} \quad \lim_{x \to 3^+} \frac{|x-3|}{x-3}.$$

Now to calculate the left side limit, we do as follows:

```
>> syms a real
>> syms x
>> a = 3
>> f = abs(x - 3)/(x - 3)
>> limit(f, x, a, 'left')
ans =
   -1
```

and to calculate the right side limit, we use the commands:

```
>> syms a real
>> syms x
>> a = 3
>> f = abs(x - 3)/(x - 3)
>> limit(f, x, a, 'right')
ans =
    1
```

Since the limit from the left does not equal the limit from the right, the limit does not exits. It can be checked using the following commands:

```
>> syms a real
>> syms x
>> a = 3
>> f = abs(x - 3)/x - 3
>> limit(f, x, a)
ans =
    NaN
```

Taylor's Polynomial of a Function

The Symbolic Math Toolbox provides the *taylor* command, which allows us to obtain the analytical expression of Taylor's polynomial of a given function. In particular, having defined in the string function f on which we want to operate *taylor* (f, x, n+1) the associated Taylor polynomial of degree n expanded about $x_0 = 0$. For example, to find the Taylor polynomial of degree three for $f(x) = e^x \sin x$ expanded about $x_0 = 0$, we use the following commands:

```
>> syms x
>> f = exp(x) * sin(x)
>> taylor(f, x, 4)
ans =
    x + x^ 2+1/3 * x^ 3
```

Symbolic Ordinary Differential Equation

Like differentiation and integration, an ordinary differential equation can be solved symbolically using the *dsolve* command. This command can be used to solve a single equation or a system of differential equations. This command can also be used in getting a general solution or particular solution of an ordinary differential equation. For first-order ordinary differential equation, we use:

```
>> dsolve('eq')
```

or

```
>> dsolve('eq','var')
```

For example, in finding the general solution of the ordinary differential equation

$$\frac{dy}{dt} = t + \frac{3y}{t}$$

we use the following commands:

```
>> syms t y
>> f = x + 3 * y/t
>> dsolve('Dy = f')
ans =
    -t^ 2+t^ 3*C1
```

For finding a particular solution of first-order ordinary differential equation, we use the following command:

```
>> dsolve('eq','cond1')
```

For example, in finding the particular solution of the ordinary differential equation

$$\frac{dy}{dt} = t + \frac{3y}{t}$$

with the initial condition $y(1) = 4$, we do the following:

```
>> syms t y
>> f = x + 3 * y/t
>> dsolve('Dy = f','y(1) = 4','t')
ans =
     -t^2+5*t^3
```

Similarly, a higher-order ordinary differential equation can be solved symbolically using the following command:

```
>> dsolve('eq','cond1','cond2',···,'var')
```

For example, the second-order ordinary differential equation

$$\frac{d^2y}{dx^2} - 4\frac{dy}{dx} - 5y = 0, \quad y(1) = 0, \; y'(1) = 2$$

can be solved using the following commands:

```
>> syms x y
>> dsolve('D2y - 4 * Dy - 5 * y = 0','y(1) = 0','Dy(1) = 2','x')
ans =
    1/3/exp(1)^5*exp(5 * x) - 1/3 * exp(1) * exp(-x)
```

Linear Algebra

Consider the following matrix:

$$A = \begin{bmatrix} 3 & 2 \\ x & y \end{bmatrix} \quad \text{and} \quad \mathbf{b} = \begin{bmatrix} 1 \\ y \end{bmatrix}.$$

Since the matrix A is the symbolic expression, we can calculate the determinant and the inverse of A, and also solve the linear system using the vector b:

```
>> syms x y
>> A = [3 2; x y])
>> det(A)
ans =
    3 * y - 2 * x
>> inv(A)
ans =
    [-y/(-3 * y + 2 * x), 2/(-3 * y + 2 * x)]
    [x/(-3 * y + 2 * x), -3/(-3 * y + 2 * x)]
>> b = [1; x]
>> A\b
ans =
    [(2 * x - y/(-3 * y + 2 * x)]
    [-2 * x/(-3 * y + 2 * x)]
```

Eigenvalues and Eigenvectors

To find a characteristic equation of the following matrix

$$A = \begin{bmatrix} 3 & -1 & 0 \\ -1 & 2 & -1 \\ 0 & -1 & 3 \end{bmatrix}$$

we use the following commands:

```
>> A = [3 -1 0; -1 2 -1; 0 -1 3]
>> poly(sym(A))
ans =
    x^3-8 * x^2+19 * x - 12

>> factor(ans)
ans =
    (x - 1) * (x - 3) * (x - 4)
```

We can also get the eigenvalues and eigenvectors of a square matrix A symbolically using the *eig*(sym(A)) command. The form of this command is as follows:

```
>> [X, D] = eig(sym(A))
```

For example, to find eigenvalues and eigenvectors of matrix A, we use the following commands:

$$\begin{array}{l} >> A = [3\ \ -1\ 0; -1\ 2\ \ -1; 0\ \ -1\ 3] \\ >> [X, D] = eig(sym(A)) \\ X = \\ \quad [1, -1, 1] \\ \quad [-1, 0, 2] \\ \quad [1, 1, 1] \\ \\ D = \\ \quad [4, 0, 0] \\ \quad [0, 3, 0] \\ \quad [0, 0, 1] \end{array},$$

where the eigenvector in the first column of vector X corresponds to the eigenvalue in the first column of D, and so on.

Plotting Symbolic Expressions

We can easily plot a symbolic expression using the *ezplot* command. To plot a symbolic expression Z that contains one or two variables, the *ezplot* command is:

$$>> ezplot(Z)$$

or

$$>> ezplot(Z, [min, max])$$

or

$$>> ezplot(Z, [xmin, xmax, ymin, ymax]).$$

For example, we can plot a graph of symbolic expression $Z = (2x^2+2)/(x^2-6)$, using the following commands:

$$\begin{array}{l} >> syms\ x \\ >> Z = (2 * x\hat{\ } 2{+}2)/(x\hat{\ } 2{-}64) \\ >> ezplot(Z) \end{array}.$$

we obtained Figure D.1.
Note that *ezplot* can also be used to plot a function that is given in a parametric form. For example, when $x = \cos 2t$ and $y = \sin 4t$, we use the following commands:

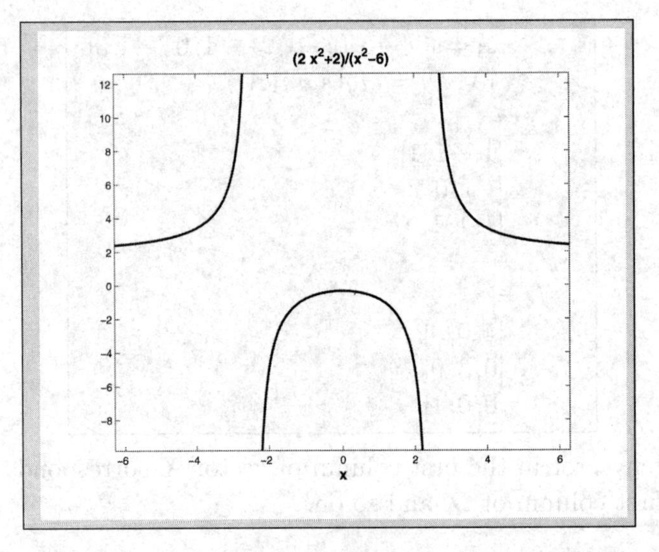

Figure D.1: Graph of $Z = (2x^2 + 2)/(x^2 - 6)$.

```
>> syms t
>> x = cos(2 * t)
>> x = sin(4 * t)
>> ezplot(x, y)
```

we obtained Figure D.2.

Symbolic Math Toolbox Functions

Listed below are some of the Symbolic Math Toolbox functions:

Symbolic Math Toolbox Functions	**Definitions**
diff	differentiate
int	integration
limit	limit of an expression
symsum	summation of series
taylor	Taylor series expansion
det	determinant
diag	create or extract diagonals
eig	eigenvalues and eigenvectors

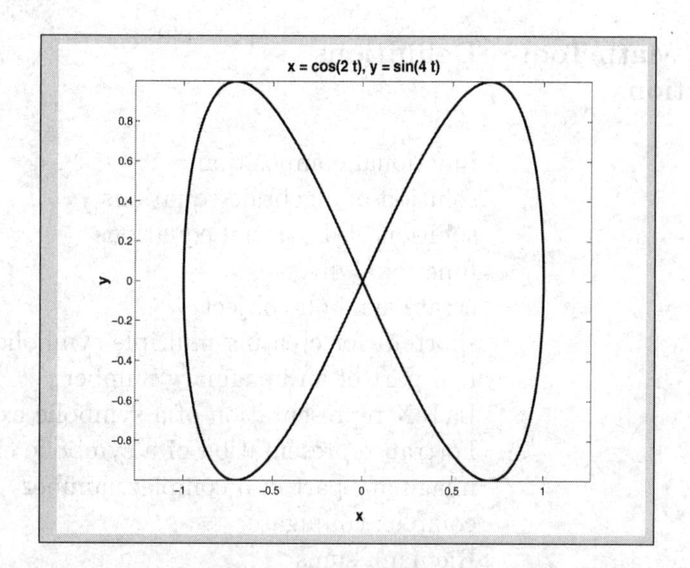

Figure D.2: Graph of function in a parametric form.

Symbolic Math Toolbox Functions	**Definitions**
inv	inverse of a matrix
expm	exponential of a matrix
rref	reduced row echelon form
null	basis for null space
svd	singular value decomposition
poly	characteristic polyonmial
rank	rank of a matrix
tril	lower triangle
triu	upper triangle
collect	collect common terms
expand	expand polynomials and elementary functions
factor	factor a expression
simplify	simplification
simple	search for shortest form
pretty	pretty print of symbolic expression
findsym	determine symbolic variables
subexpr	rewrite in terms of subexpresions
numden	numerator and denominator

Symbolic Math Toolbox Functions	**Definitions**
compose	functional composition
solve	solution of algebraic equations
desolve	solution of differetial equations
finverse	functional inverse
sym	create symbolic object
syms	shortcut for creating multiple symbolic objects
real	real part of an imaginary number
latex	LaTeX representation of a symbolic expression
fortran	Fortran representation of a symbolic expression
imag	imaginary part of a complex number
conj	complex conjugate
resums	Riemann sums
taylortool	Taylor seriec calculator
funtool	Tfunction calculator
digits	set variable precision accuracy
vpa	variable precision arithmetic
double	convert symbolic matrix to double
char	convert sym object to string
poly2sym	function calculator
sym2poly	symbolic polynomial to coefficient vector
fix	round toward zero
floor	round toward minus infinity
ceil	round toward plus infinity
int8	convert symbolic matrix to signed 8-bit integer
int16	convert symbolic matrix to signed 16-bit integer
int32	convert symbolic matrix to signed 32-bit integer
int64	convert symbolic matrix to signed 64-bit integer
uint8	convert symbolic matrix to unsigned 8-bit integer
uint16	convert symbolic matrix to unsigned 16-bit integer
dirac	dirac delta function
zeta	Riemann zeta function
cosint	cosine integral
sinint	sine integral
fourier	fourier transform
ifourier	inverse fourier transform
laplace	laplace transform

Symbolic Math Toolbox Functions	**Definitions**
ilaplace	inverse laplace transform
ztrans	z-transform
iztrans	inverse z-transform
ezplot	function plotter
ezplot3	3-D curve plotter
ezpolar	polar coordinate plotter
ezcontour	contour plotter
ezcontourf	filled contour plotter
ezmesh	mesh plotter
ezmeshc	combined mesh and contour plotter
ezsurf	surface plotter
ezsurfc	combine surface and contour plotter

Appendix E

Answers to Selected Exercises

Chapter 1

1. 10, 37, $\dfrac{99}{128}$

3. $(.00011)_2$, 0.00625, 0.0025

5. *Absolute error* $= 1/3 \times 10^{-4}$, *Relative error* $= 10^{-4}$

7. **(a)** *Absolute error* $= 7.346 \times 10^{-6}$, *Relative error* $= 2.3383 \times 10^{-6}$
(b) *Absolute error* $= 7.346 \times 10^{-4}$, *Relative error* $= 2.3383 \times 10^{-4}$

9. 0.132×10^2 $(m = 3, e = 2)$, -0.12532×10^2 $(m = 5, e = 2)$,
0.16×10^{-1} $(m = 2, e = -1)$

11. **(a)** Chopping: $e = 2.718 \times 10^0$, Rounding: $e = 2.718 \times 10^0$
(b) Chopping: $e = 2.718281 \times 10^0$, Rounding: $e = 2.718282 \times 10^0$
(c) Chopping: $e = 1.010110111 \times 2^1$,
Rounding: $e = 1.010111000 \times 2^1$

Chapter 2

1. $x_{10} = -1.84141$

3. **(a)** $x_{16} = 0.35173$ **(b)** $x_{15} = 0.97300$

5. **(a)** $x_8 = 1.53906$ **(b)** $|\alpha - x_8| \leq \frac{3-1}{2^8} = 0.00781$

7. $x_8 = 0.59431$

9. $x_9 = 3.73308$

11. **(a)** Since g is continuous in $[0,1]$ and $g(0) = 0.5 \in [0,1]$ and $g(1) = 0.82436 \in [0,1]$. Also, $g'(x) = \frac{1}{4}e^{.5x}$ and $g'(0) = 0.25$, $g'(1) = 0.41218$, so $|g'(x)| < 1$ for $x \in [0,1]$

(b) $x_3 = 0.58705$ **(c)** $|\alpha - x_3| \leq \frac{(0.41218)^3}{1 - 0.41218}|0.5 - 0| = 0.05957$

13. Since $g(x) = \frac{x^2 - 4kx + 7}{4}$, then

$$g(1) = 1 = \frac{1 - 4k + 7}{4}, \quad \text{gives} \quad k = 1.$$

Also,

$$g'(x) = \frac{2x - 4k}{4} \quad \text{gives} \quad g'(1) = -\frac{1}{2} \neq 0$$

a linear convergence.

15. **(a)** Since g is continuous in $[3,4]$, but $g(3) \notin [3,4]$, no fixed-point exists.

(b) Since $g(3) \in [3,4]$ and $g(4) \in [3,4]$ and also, $g'(x) = 12/x^3$, g is continuous and g' exists on $[3,4]$ and $|g'(x)| \leq< 1$ for $x \in (3,4)$. Then the Theorem 2.2 implies that a unique fixed-point exists in $[3,4]$. With $k = \max_{3 \leq 4} |g'(x)| = 4/9$ and $x_0 = 3$, we have $x_1 = 10/3$ and

$$|\alpha - x_n| \leq \frac{(4/9)^n}{(1 - 4/9)}|10/3 - 3| = \frac{3(4/9)^n}{5}.$$

For the bound to be less than 0.001, we need $n = 8$.

17. $x_3 = -1.84141$

19. $x_3 = 0.0666657$

21. $x_3 = 1.53201,\qquad$ quadratic convergence

23. Since Newton's iteration function is

$$\begin{aligned} g(x) &= x - \frac{f(x)}{f'(x)} \\ g'(x) &= 1 - \frac{[f'(x)]^2 - f(x)f''(x)}{[f'(x)]^2} = \frac{f(x)f''(x)}{[f'(x)]^2}. \end{aligned}$$

Since $f'(\alpha) = 0$, using the Hopi'tal rule, it gives us

$$\begin{aligned} g'(x) &= \frac{f'(x)f''(x) + f(x)f'''(x)}{2f'(x)f''(x)} \\ g'(x) &= \frac{[f''(x)]^2 + f(x)f'''(x) + f'(x)f'''(x) + f(x)f^{(4)}}{2[f'(x)]^2 + 2f'(x)f'''(x)} \\ g'(\alpha) &= \frac{1}{2} \neq 0. \end{aligned}$$

25. $x_3 = -1.84141$

27. $x_4 = 0.56714$

29. The root of multiplicity of the given function is 3 as $f(1) = 0$ and

$$\begin{aligned} f' &= 2(x-1)\ln x + \frac{(x-1)^2}{x}: & f'(1) &= 0 \\ f'' &= 2\ln x + 4\frac{(x-1)}{x} - \frac{(x-1)^2}{x^2}: & f''(1) &= 0 \\ f''' &= \frac{6}{x} - \frac{6(x-1)}{x^2} + \frac{(x-1)^2}{x^3}: & f'''(1) &= 6 \neq 0. \end{aligned}$$

31. **(a)** $x_{23} = 1.00016$, **(b)**$x_4 = 0.9999998$, **(c)**$x_4 = 1.000001$

33. Since

$$\begin{aligned} g(x) &= 1 + x - \frac{x^2}{2} \\ g(\sqrt{2}) &= 1 + \sqrt{2} - 1 = \sqrt{2} \end{aligned}$$

so $\sqrt{2}$ is a fixed-point of $g(x)$. Also,

$$g'(x) = 1 - \frac{2x}{2}, \quad \text{and} \quad g'(\sqrt{2}) = -0.414214 \neq 0.$$

So the given sequence is linear convergent.

35. $\hat{x}_0 = 0.73139,\ \hat{x}_1 = 0.73609,\ \hat{x}_2 = 0.73765,\ \hat{x}_3 = 0.73847,\ \hat{x}_4 = 0.73880$

37. The roots are

Fixed-point method	Aitken accelerated method
$x_1 = 0.20270$	
$x_2 = 0.09230$	$\hat{x}_2 = 0.02702$
$x_3 = 0.04414$	$\hat{x}_3 = 0.00691$
$x_4 = 0.02160$	$\hat{x}_4 = 0.00175$

39. $x^{(3)} = 1.08828, \qquad y^{(3)} = 0.84434$

41. $x^{(4)} = -1.53112, \qquad y^{(4)} = 0.72888$

43. **(a)** $p(-2) = -362$ **(c)** $p(4) = -6735$

45. **(a)** $x^{(1)} = -0.7596,\ x^{(2)} = -0.7025$ **(c)** $x^{(1)} = -1.3076,\ x^{(2)} = -1.2141$

47. **(a)** Starting values: $x_0 = -2, x_1 = -1, x_2 = -1.5, x_{new} = -0.2097$ **(c)** Starting values: $x_0 = -0.5, x_1 = 1, x_2 = 0.5, x_{new} = 2.2248E - 006$

49. **(a)** Starting values: $x_0 = -3.5, x_1 = -2, x_2 = -2.5, x_{new} = -2.6556$
(c) Starting values: $x_0 = 0, x_1 = 1.5, x_2 = 1, x_{new} = 0.4675$

51. $(u_1, v_1) = (1.3333, 0.6667)$ and $(u_2, v_2) = (1.3246, 0.7544)$

Chapter 3

1. $C = \begin{pmatrix} 1 & -5 & -1 \\ -2 & 1 & -3 \\ -2 & -1 & -8 \end{pmatrix}$

3. $|AB| = \begin{vmatrix} 1 & -5 & -1 \\ -2 & 1 & -3 \\ -2 & -1 & -8 \end{vmatrix} = 0$

5. $(AB)^{-1} = B^{-1}A^{-1} = \begin{pmatrix} 1 & -1 \\ -1 & 2 \end{pmatrix}$

7. $\det(A) = 72, \qquad \det(C) = 0$

9. (a) $Adj(A) = \begin{pmatrix} 4 & -2 \\ 3 & 1 \end{pmatrix}, \quad A^{-1} = \begin{pmatrix} 2/5 & -1/5 \\ 3/10 & 1/10 \end{pmatrix}$

(c) $Adj(A) = \begin{pmatrix} -1 & -1 & 1 \\ -1 & 1 & 1 \\ 1 & -1 & -1 \end{pmatrix}, \quad A^{-1} = \begin{pmatrix} 1/2 & 1/2 & -1/2 \\ 1/2 & -1/2 & 1/2 \\ -1/2 & 1/2 & 1/2 \end{pmatrix}$

11. (a) $\mathbf{x} = [1/5, 2/5]^T$, **(c)** $\mathbf{x} = [0, 1, 0]^T$

13. (a) $\det(A) = -35, \det(A_1) = -35, \det(A_2) = -70$

$\det(A_3) = -105, \quad \mathbf{x} = [1, 2, 3]^T.$

(c) $\det(A) = -60, \det(A_1) = -240, \det(A_2) = -150$

$\det(A_3) = -190, \quad \mathbf{x} = [4, 9/4, 19/6]^T.$

15. (a) $\mathbf{x} = [4/3, 4/3, 1/3]^T$, **(c)** $\mathbf{x} = [0, 4, 7]^T$, **(e)** $\mathbf{x} = [2, 0.2, 0, 0.8]^T$

17. $\text{rank}(A) = 3, \quad \text{rank}(B) = 3, \quad \text{rank}(C) = 2$

19. $\text{rank}(A) = 3, \quad \text{rank}(B) = 3, \quad \text{rank}(C) = 2$

21. (a) WP: $\mathbf{x} = [-500.001, 333.667]^T$ PP: $\mathbf{x} = [-499.977, 333.651]^T$

(c) WP: $\mathbf{x} = [1, 1]^T$ PP: $\mathbf{x} = [1, 0.999]^T$

23. (a) $\mathbf{x} = [8, 2.077, 1.308]^T$, **(c)** $\mathbf{x} = [-2.474, 3.579, -0.579, -0.316]^T$

25. (a) $A^{-1} = \begin{pmatrix} -0.147 & 0.934 & -0.557 \\ -0.164 & 0.229 & -0.492 \\ 0.082 & -0.147 & 0.246 \end{pmatrix}$

(c) $A^{-1} = \begin{pmatrix} 0.249 & 0.123 & 0.586 & 0.235 \\ 0.123 & 0.308 & 0.147 & 0.586 \\ 0.586 & 0.147 & 0.308 & 0.123 \\ 0.235 & 0.586 & 0.123 & 0.249 \end{pmatrix}$

27. **(a)** $LU = \begin{pmatrix} 1 & 0 & 0 \\ -1.5 & 1 & 0 \\ 0.5 & -0.2 & 1 \end{pmatrix} \begin{pmatrix} 2 & -1 & 1 \\ 0 & 2.5 & 0.5 \\ 0 & 0 & 0.6 \end{pmatrix}$,

$\mathbf{x} = [-2, 2.333, 10.333]^T$

(c) $LU = \begin{pmatrix} 1 & 0 & 0 \\ 0.5 & 1 & 0 \\ 1.5 & 0 & 1 \end{pmatrix} \begin{pmatrix} 2 & 2 & 2 \\ 0 & 1 & 0 \\ 0 & 0 & 1 \end{pmatrix}$, $\mathbf{x} = [3, -4, 1]^T$

(e) $LU = \begin{pmatrix} 1 & 0 & 0 \\ 2 & 1 & 0 \\ 2 & 0 & 1 \end{pmatrix} \begin{pmatrix} 1 & -1 & 0 \\ 0 & 1 & 1 \\ 0 & 0 & -1 \end{pmatrix}$, $\mathbf{x} = [1, -1, 1]^T$

29. **(a)** $L = \begin{pmatrix} 1.414 & 0 & 0 \\ -0.707 & 1.223 & 0 \\ 0.707 & -0.408 & 1.155 \end{pmatrix}$, $\mathbf{x} = [0.5, 2.5, 2.5]^T$

(c) $L = \begin{pmatrix} 2.236 & 0 & 0 \\ 0.894 & 1.789 & 0 \\ 1.342 & -0.112 & 2.046 \end{pmatrix}$, $\mathbf{x} = [0.597, 0.194, 0.104]^T$

31. **(a)** $\|\mathbf{x}\|_1 = 12, \quad \|\mathbf{x}\|_2 = 7.0711, \quad \|\mathbf{x}\|_\infty = 6$

(c) $\|\mathbf{x}\|_1 = 8.5, \quad \|\mathbf{x}\|_2 = 5.2202, \quad \|\mathbf{x}\|_\infty = 4$

33. **(a)** $\|\mathbf{A}^3\|_1 = 2996, \qquad \|\mathbf{A}^3\|_\infty = 23$

(c) $\|\mathbf{B}C\|_1 = 427, \qquad \|\mathbf{B}C\|_\infty = 531$

35. **(a)** $\mathbf{x}^{(13)} = [2, 4, 3]^T$

(c) $\mathbf{x}^{(12)} = [-0.102, 0.539, 0.328]^T$

(e) $\mathbf{x}^{(14)} = [-0.673, 0.096, 0.711]^T$

(g) $\mathbf{x}^{(12)} = [0.588, -0.147, 0.559]^T$

37. **(a)** $\mathbf{x}^{(8)} = [2, 4, 3]^T$

(c) $\mathbf{x}^{(6)} = [-0.102, 0.539, 0.328]^T$

(e) $\mathbf{x}^{(8)} = [-0.673, 0.096, 0.711]^T$

(g) $\mathbf{x}^{(7)} = [0.588, -0.147, 0.559]^T$

39. **(a)** $\rho(A) = 4 > 1$, divergent.

41. **(a)** $\mathbf{x}^{(8)} = [2, 4, 3]^T$

(c) $\mathbf{x}^{(6)} = [-0.102, 0.539, 0.328]^T$

(e) $\mathbf{x}^{(8)} = [-0.673, 0.096, 0.711]^T$

(g) $\mathbf{x}^{(8)} = [0.588, -0.147, 0.559]^T$

43. $\omega = 1.172, \quad \mathbf{x}^{(9)} = [0.375, 1.750, 2.375]^T$

Gauss-Seidel method needed 15 iterations.

Jacobi method needed 26 iterations.

45. $\rho_J = 0.809017, \quad \rho_{GS} = 0.654508, \quad \rho_{SOR} = 0.25962$

47. **(a)** $\mathbf{x}^{(2)} = [0.51814, -0.72359, -1.94301]^T$

$\mathbf{r}^{(2)} = [1.28497, -0.80311, 0.64249]^T$

(c) $\mathbf{x}^{(2)} = [0.90654, 0.46729, -0.33645, -0.57009]^T$

$\mathbf{r}^{(2)} = [-1.45794, -0.59813, -0.26168, -2.65421]^T$

49. **(a)** $K(A) = 3.969$, Well-conditioned

(c) $K(A) = 42.2384$, Ill-conditioned

51. $K(A) = 100$ (ill-conditioned),

$\mathbf{r} = [-0.02, 0.02]^T$, *relative error* $= 1$

53. $K(A) \geq \dfrac{6.06}{0.24} = 25.25$

55. $\delta\mathbf{x} = [-100, 101]^T$, $K(A) = 404.01$, Ill-conditioned

57. $\delta\mathbf{x} = [1.9976, 0.6675]^T$, $K(A) = 80002$, Ill-conditioned

59. $\delta\mathbf{x} = [-17, 20]^T$, $K(A) = 8004$, Ill-conditioned

61. Simple Gaussian Elimination $\mathbf{x}^{(1)} = [-10, 1.01]^T$
Residual Corrector Method $\mathbf{x}^{(2)} = [10, 1]^T$

Chapter 4

1. $p_2(x) = 1.7321 + 0.2887x + 0.1203x^2$, $p_2(1) = 2.1411$

3. $p_3(1.1) = 1.688211$, Actual Error $= 0.0000381$

5. $p_2(2.3) = 0.4548$

7. $p_2(0.5) = 2.3212$, $E_B = 0.03125$ and $p_2(2.8) = 5.9399$,
$E_B = 0.007111$

9. $p_3(1.1) = -3.6860$, $E_B = 3.73308$

11. **(a)** Divided difference table is:

2.0000	0	0	0
3.7200	1.7200	0	0
8.3900	4.6700	1.4750	0
21.0600	12.6700	4.0000	0.8417

(b) $p_2(2.2) = 10.2840$ and $p_3(2.2) = 10.1224$

13. **(a)** Divided difference table is:

2.0000	0	0	0	0
2.2361	0.2361	0	0	0
2.4495	0.2134	−0.0113	0	0
2.6458	0.1963	−0.0086	0.0009	0
2.8284	0.1826	−0.0069	0.0006	−0.0001

(b) $p_3(5.9) = 2.4290$ and$p_4(5.9) = 2.4290$
(c) $E_3 = 0.0005742$ and $E_4 = 0.00000004$

15. All three divided differences can be expanded as

$$\frac{(x_2 - x_1)f_0 - (x_2 - x_0)f_1 + (x_1 - x_0)f_2}{(x_2 - x_1)(x_2 - x_0)(x_1 - x_0)}$$

17. $f[0, 1, 0] = 0.7183$

19. $f(1.21) \approx 0.9347$

21. $h = 5^o,\ x = 52^o,\ s = 1.4,\ f(52^o) \approx 0.7880$

23. $h = 0.25,\ x = 0.6,\ s = 1.4,$
(a) $f(0.6) \approx p_2(0.6) = 0.6416$ **(b)** Error $= 0.00027$

25. $h = 2,\ x = 11.8,\ s = -1.1,\quad f(11.8) \approx 1727$

27. **(a)** $h = 0.1,\ x = 1.05,\ s = 0.5,\ f(1.05) \approx 0.0212$
(b) $h = 0.1,\ x = 1.55,\ s = -0.5,\ f(1.55) \approx 0.1903$

29. **(a)** $f_{i+4} - 4f_{i+3} + 6f_{i+2} - 4f_{i+1} + f_i$
(b) $f_i - 3f_{i-1} + 3f_{i-2} - f_{i-3}$
(c) $f_{i+1} + 2f_i + f_{i-1}$

31. $h = 5,\ x = 41,\ s = 0.2,\ f(41) \approx 2289.99$

33. $h = 10,\ x = 12516,\ s = -0.4,\ f(41) \approx 111.8749$

35. $h = 0.2,\ x = 1.62,\ s = 0.1,\ f(1.62) \approx 7.2429$

37. $h = 0.5,\ x = 1.22,\ s = 0.44,\ f(1.22) \approx 0.3887$

39.

$$\begin{aligned} s_0(x) &= 0.4(x - 0) + 3.5, \quad [0, 1] \\ s_1(x) &= 0.8(x - 1) + 3.9, \quad [1, 2] \\ s_2(x) &= 1.1(x - 2) + 4.7, \quad [2, 3] \end{aligned}$$

$s_0(0.55) = 3.7200,\quad s_1(1.15) = 4.0210,\quad s_2(2.5) = 5.2500$

41.

$$\begin{aligned} s_0(x) &= 0.9(x-0)+0.0, & [0,0.2] \\ s_1(x) &= 0.8(x-0.2)+0.18, & [0.2,0.3] \\ s_2(x) &= 0.75(x-0.3)+0.26, & [0.3,0.5] \end{aligned}$$

$s_0(0.15) = 0.1350,\quad s_1(0.25) = 0.2200,\quad s_2(0.45) = 0.3725$

43. $s_0(2.5) = 1.8333,\quad s_1(5.5) = 2.5000,\quad s_2(10.5) = 3.3571$
$E_0 = 0.0375,\quad E_1 = 0.0495,\quad E_2 = 0.0341$

45. Not cubic spline

47. (a)

$$\begin{aligned} s_0(x) &= 0.77(x+2)^3 + 0.0(x+2)^2 + 5.69(x+2) - 8.0, \\ & \quad [-2,-1.5] \\ s_1(x) &= -0.15(x+1.5)^3 - 1.15(x+1.5)^2 + 5.12(x+1.5) - 5.25, \\ & \quad [-1.5,-1] \\ s_2(x) &= 0.23(x+1)^3 - 1.39(x+1)^2 + 3.85(x+1) - 3.0, \\ & \quad [-1,1] \end{aligned}$$

(c)

$$\begin{aligned} s_0(x) &= 1.30(x-0)^3 + 0.0(x-0)^2 - 0.30(x-0) - 4.0, \\ & \quad [0,1] \\ s_1(x) &= -0.26(x-1)^3 + 3.91(x-1)^2 + 3.61(x-1) - 3.0, \\ & \quad [1,1.5] \\ s_2(x) &= -0.35(x+1.5)^3 + 3.52(x+1.5)^2 + 7.33(x+1.5) - 0.25, \\ & \quad [1.5,2] \end{aligned}$$

49. (a)

$$\begin{aligned} s_0(x) &= 0.11(x-1)^3 - 0.42(x-1)^2 + (x-1), \\ & \quad [1,2] \\ s_1(x) &= 0.01(x-2)^3 - 0.09(x-2)^2 + 0.49(x-2) + 0.69, \\ & \quad [2,3] \\ s_2(x) &= 0.01(x-3)^3 - 0.06(x-3)^2 + 0.34(x-3) + 1.10, \\ & \quad [3,4] \end{aligned}$$

(c)

$$\begin{aligned} s_0(x) &= -0.47(x-0)^3+1.47(x-0)^2+(x-0)+1.0, \\ &\quad [0,1] \\ s_1(x) &= 5.40(x-1)^3+0.07(x-1)^2+2.53(x-1)+3.0, \\ &\quad [1,2] \\ s_2(x) &= -15.13(x-2)^3+16.27(x-2)^2+18.87(x-2)+11.0, \\ &\quad [2,3] \end{aligned}$$

51. **(a)** $s(-1.7) = -6.3131,\ Error = 0.0231,$
$s'(-1.7) = 5.4821,\ Error = -0.0821$
(c) $s(1.1) = -2.6003,\ Error = 0.0303,$
$s'(1.1) = 4.3842,\ Error = -0.2158$

53. **(a)** $a = 1, b = 3,\ y = x + 3,\ E(a, b) = 0.0$
(c) $a = 5.4, b = -11,\ y = 5.4x - 11,\ E(a, b) = 1.2000$
(e) $a = 6.24, b = 1.45,\ y = 6.24x + 1.45,\ E(a, b) = 4.0120$

55. **(a)** $a = -1.782, b = 0.073, c = 1.818,\ y = -1.782x^2 + 0.073x + 1.818,$
$E(a, b) = 0.146$
(c) $a = 28.125, b = 0.3, c = -3.125,\ y = 28.125x^2 + 0.3x - 3.125,$
$E(a, b) = 0.2$
(e) $a = -0.6, b = -0.1, c = 2.5,\ y = -0.6x^2 - 0.1x + 2.5,$
$E(a, b) = 1.6$

57. **(a)** $a = 4.961, b = 0.037,\ y = 4.961e^{0.037x},\ E(a, b) = 0.452$
(c) $a = 0.980, b = 0.530,\ y = 0.980e^{0.530x},\ E(a, b) = 21.185$
(e) $a = 3.867, b = -0.508,\ y = 3.867e^{-0.508x},\ E(a, b) = 0.071$

59. $a = 1.11,\ b = -0.44,\ c = 2.11,\ Z = 1.11x - 0.44y + 2.11,$
$E(a, b, c) = 0.22$

61. **(a)** $x_1 = 0.1600, \quad x_2 = 1.1600$
(c) $x_1 = -1.7928, \quad x_2 = 0.3707, \quad x_3 = 1.6388$

63. $\hat{\mathbf{x}} = [x_1, x_2]^T = [-4, 9/5]^T$

65. $\hat{\mathbf{x}} = [x_1, x_2]^T = [5/3, -2]^T$

67. $\hat{\mathbf{x}} = [x_1, x_2]^T = [1, 0]^T$

69. $\hat{\mathbf{x}} = [x_1, x_2]^T = [1, -2]^T$

71. $\hat{\mathbf{x}} = [x_1, x_2, x_3]^T = [-0.3334, 0.3333, 0.6666]^T$

73. **(a)** $\hat{\mathbf{x}} = [x_1, x_2]^T = [0.4000, 0.2000]^T$
(c) $\hat{\mathbf{x}} = [x_1, x_2]^T = [0.4000, 0.2000]^T$

75. **(a)** $\hat{\mathbf{x}} = [x_1, x_2]^T = [0.9250, 0.0250]^T$
(c) $\hat{\mathbf{x}} = [x_1, x_2]^T = [0.8833, -0.0250]^T$

Chapter 5

1. **(a)** 23.10591068, - 0.16529103, 0.166731728
(b) 22.77675826, 0.16386139, 0.16457401

3. Forward difference approximately 30.00,
Backward difference approximately 20.00

5. 21.29455248, 0.000173356, 0.000175221

7. Forward difference approximately 21.29403551, 0.00034614, 0.00034667
Backward difference approximately 21.29402979, 0.00034933, 0.0003543

9. Central difference approximately 13.36970731, 0.01056639, 0.01059534
Forward difference approximately 13.3396478, 0.01949311, 0.02125174
Backward difference approximately 13.33804978, 0.02109113, 0.021132617

11. Forward difference approximately $f'(2.1) = 1.8836475$
Central difference approximately $f'(2.2) = 1.4765925$
Central difference approximation $f'(2.3) = 1.1445225$
Backward difference approximately $f'(2.4) = 1.774875$

13. Forward difference approximately $f'(8.1) = 3.09205$
Central difference approximately $f'(8.3) = 3.11615$
Central difference approximately $f'(8.5) = 3.139975$
Backward difference approximately $f'(2.4) = 3.163525$

15. 1.24999992, and the error bound 0.00000008,
exact error 0.00000010097

17. 3.6000000

19. 3.85160000, and the error bound 0.0002313

21. 0.23910000, and the error bound 0.0000067

23. 0.24990000, and the error bound 0.0010884

25. $n = 4$, $T_4(f) = 0.41817958$,
For $n = 6$, $T_6(f) = 0.42069568$

27. $T_5(f) = 0.49557207$

29. $T_4(f) = 0.84166667$ and the error bound $E_T(f) = 0.3333333$

31. $S_8(f) = 1.57078431$

33. $n = 4$, $S_4(f) = 0.42273587$,
For $n = 6$, $S_6(f) = 0.42269015$

35. $h = 0.0543$ and $n = 18$

37. $n = 3$, 0.3861, 0.0009, $n = 4$, 0.3863, 0.0001

39. 2.7974 0.0001

41. 2.0344 0.0172 0.0218

43. **(a)**

1	2.4750
2	2.4029
4	2.3850
8	2.3806
16	2.3795

(c)

1	0.8959
2	0.9061
4	0.9087
8	0.9093
16	0.9095

45. 0.9096 0.0002

47. (a)

2.5322			
2.5161	2.5107		
2.5121	2.5107	2.5107	
2.5111	2.5107	2.5107	2.5107

(c)

1.0000			
1.1000	1.1333		
1.1241	1.1322	1.1321	
1.1300	1.1320	1.1320	1.1320

49. (a)

1.1513			
1.1650	1.1695		
1.1682	1.1693	1.1692	
1.1690	1.1692	1.1692	1.1692

(c)

1.1605			
1.1633	1.1642		
1.1640	1.1642	1.1642	
1.1642	1.1642	1.1642	1.1642

51. 1.0980, 0.0006

53. $n = 2$, 0.9985, 0.0015, $n = 4$, 1.0010, 0.0010

55. (a) $n = 2$, 2.9985, 0.0015, $n = 4$, 1.0010, 0.0010
(c) $n = 2$, 2.2179, 0.0004, $n = 4$, 2.2180, 0.0003

Chapter 6

1. Using separation of variables: $y^2 - x^2 = c$

3.

(a)

x_i	y_i
0.2000	1.2000
0.4000	1.4480
0.6000	1.7696
0.8000	2.1955
1.0000	2.7626

(c)

x_i	y_i
0.2100	1.5170
0.2200	1.5342
0.2300	1.5518
0.2400	1.5696
0.2500	1.5876

5.

(a)

x_i	y_i
0.2000	−0.8200
0.4000	−0.6420
0.6000	−0.4368
0.8000	−0.1806
1.0000	0.1463

(c)

x_i	y_i
0.2000	1.5900
0.4000	1.2899
0.6000	1.0898
0.8000	0.9808
1.0000	0.9479

7.

(a)

x_i	y_i
1.2000	−0.8078
1.4000	−0.6385
1.6000	−0.5000
1.8000	−0.3911
2.0000	−0.3075

(c)

x_i	y_i
1.2000	0.9800
1.4000	0.9432
1.6000	0.9013
1.8000	0.8611
2.0000	0.8252

9. Heun's Method

(a)

x_i	y_i
0.7000	1.3690
0.9000	1.9472
1.1000	2.8765
1.3000	4.4120
1.5000	7.0236

(c)

x_i	y_i
1.2000	−0.6267
1.4000	−0.2464
1.6000	0.2073
1.8000	0.8343
2.0000	1.8688

Midpoint Method

(a)

x_i	y_i
0.7000	1.3680
0.9000	1.9442
1.1000	2.8696
1.3000	4.3974
1.5000	6.9937

(c)

x_i	y_i
1.2000	−0.6300
1.4000	−0.2522
1.6000	0.1983
1.8000	0.8183
2.0000	1.8328

11.

(a)

x_i	y_i
1.2000	1.4188
1.4000	1.8710
1.6000	2.3520
1.8000	2.8580
2.0000	3.3863

(c)

x_i	y_i
0.2000	0.9791
0.4000	0.9159
0.6000	0.8183
0.8000	0.7015
1.0000	0.5820

13.

(a)

x_i	y_i
1.0000	1.0000
1.2000	1.4188
1.4000	1.8711
1.6000	2.3520
1.8000	2.8580
2.0000	3.3863

(c)

x_i	y_i
0.0000	1.0000
0.2000	0.9790
0.4000	0.9159
0.6000	0.8182
0.8000	0.7015
1.0000	0.5820

15.

(a)

x_i	y_i	*Exact*
0.0000	0.0000	0.0000
0.2000	0.2040	0.2027
0.4000	0.4240	0.4228
0.6000	0.6747	0.6841
0.8000	0.9733	1.0296
1.0000	1.3360	1.5574

(c)

x_i	y_i	*Exact*
1.0000	0.0000	0.0000
1.2000	0.2700	0.2880
1.4000	0.7357	0.7840
1.6000	1.4601	1.5360
1.8000	2.4853	2.5920
2.0000	3.8545	4.0000

17.

(a)

x_i	*Pred*	*Corr*	*Exact*
0.0000	1.0000	1.0000	5.0000
0.2000	3.4600	3.4600	3.3456
0.4000	2.5668	2.5668	2.4377
0.6000	2.0487	2.0487	1.9394
0.8000	1.4394	1.7266	1.6660
1.0000	1.4296	1.5718	1.5159

(c)

x_i	*Pred*	*Corr*	*Exact*
2.0000	1.0000	1.0000	1.0000
2.2000	0.9800	0.9800	0.9802
2.4000	0.9228	0.9228	0.9231
2.6000	0.8349	0.8349	0.8353
2.8000	0.7267	0.7258	0.7261
3.0000	0.6070	0.6062	0.6065

19.

(a)

x_i	*Pred*	*Corr*	*Exact*
0.0000	1.0000	1.0000	5.0000
0.2000	3.4600	3.4600	3.3456
0.4000	2.5668	2.5668	2.4377
0.6000	2.0487	2.0487	1.9394
0.8000	1.7423	1.7179	1.6660
1.0000	1.5819	1.5379	1.5159

(c)

x_i	*Pred*	*Corr*	*Exact*
2.0000	1.0000	1.0000	1.0000
2.2000	0.9800	0.9800	0.9802
2.4000	0.9228	0.9228	0.9231
2.6000	0.8349	0.8349	0.8353
2.8000	0.7264	0.7258	0.7261
3.0000	0.6068	0.6061	0.6065

21.

(a)

x_i	y_i	z_i
1.0000	−1.0000	0.0000
1.2000	−0.4666	0.3646
1.4000	−0.0285	0.6729
1.6000	0.3501	0.9399
1.8000	0.6890	1.1755
2.0000	1.0001	1.3862

(c)

x_i	y_i	z_i
0.0000	1.0000	0.0000
0.2000	2.1511	−0.6593
0.4000	4.4097	−2.1845
0.6000	8.7655	−5.4460
0.8000	17.0598	−12.1079
1.0000	32.7051	−25.3183

23.

(a)

x_i	y_i	z_i
0.0000	3.0000	−6.0000
0.2000	1.9648	−4.4257
0.4000	1.2038	−3.2442
0.6000	0.6468	−2.3717
0.8000	0.2399	−1.7310
1.0000	−0.0563	−1.2532

(c)

x_i	y_i	z_i
1.0000	−0.5000	0.5000
1.2000	−0.4032	0.4098
1.4000	−0.3197	0.3378
1.6000	−0.2527	0.2808
1.8000	−0.2006	0.2357
2.0000	−0.1606	0.1998

25.

(a)

x_i	y_i
0.0000	0.0000
0.2000	−0.1611
0.4000	−0.2120
0.6000	−0.1012
0.8000	0.2564
1.0000	1.0000

(b)

x_i	y_i
1.0000	1.0000
1.2000	0.7191
1.4000	0.6630
1.6000	0.8974
1.8000	1.5718
2.0000	3.0001

Chapter 7

1. **(a)** $p(\lambda) = \lambda^3 - 7\lambda^2 + 4\lambda + 12,\ \lambda_1 = -0.3489, \lambda_2 = 0.6041, \lambda_3 = 4.7448$

$$\mathbf{x1} = \begin{pmatrix} 1 \\ 1 \\ 2 \end{pmatrix}, \quad \mathbf{x2} = \begin{pmatrix} 1 \\ 2 \\ 4 \end{pmatrix}, \quad \mathbf{x3} = \begin{pmatrix} 2 \\ 3 \\ 1 \end{pmatrix}$$

(c) $p(\lambda) = \lambda^3 - 4\lambda^2 - 5\lambda,\ \lambda_1 = 5, \lambda_2 = -1, \lambda_3 = 0$

$$\mathbf{x1} = \begin{pmatrix} -5 \\ -2 \\ 1 \end{pmatrix}, \quad \mathbf{x2} = \begin{pmatrix} 1 \\ -8 \\ -10 \end{pmatrix}, \quad \mathbf{x3} = \begin{pmatrix} 0 \\ 1/2 \\ 1 \end{pmatrix}$$

(e) $p(\lambda) = \lambda^3 - 2\lambda^2 - \lambda + 2,\ \lambda_1 = 2, \lambda_2 = 1, \lambda_3 = -1$

$$\mathbf{x1} = \begin{pmatrix} 0 \\ 1 \\ 1 \end{pmatrix},\quad \mathbf{x2} = \begin{pmatrix} 1 \\ 0 \\ 1 \end{pmatrix},\quad \mathbf{x3} = \begin{pmatrix} 1 \\ -1 \\ 1 \end{pmatrix}$$

3. (a) $k = 1$ (c) $k \neq 1/3$

5. (a) Diagonalizable, real distinct eigenvalues 2, -2, -1
(c) Not diagonalizable, real repeated eigenvalues -1, 2, 2
(e) Diagonalizable, real distinct eigenvalues -11, 7, -1

7. (a) $\begin{pmatrix} 1 & 1 & 2 \\ 1 & 2 & 3 \\ 2 & 4 & 1 \end{pmatrix}$, (c) $\begin{pmatrix} -5 & 1 & 0 \\ -2 & -8 & 1/2 \\ 1 & -10 & 1 \end{pmatrix}$, (e) $\begin{pmatrix} 0 & 1 & 1 \\ 1 & -1 & 0 \\ 1 & 1 & 1 \end{pmatrix}$

9. Similar matrices because they have same eigenvalues, $\lambda = 1, 1, 1$

11. (a) $\begin{pmatrix} 2 & 0 & 0 \\ 0 & 1 & 0 \\ 0 & 0 & -1 \end{pmatrix}$, (c) $\begin{pmatrix} 3 & 0 & 0 \\ 0 & -4 & 0 \\ 0 & 0 & -1 \end{pmatrix}$

13. (a) $Q = \begin{pmatrix} 1 & 0 & 0 \\ 0 & 1/\sqrt{2} & 1/\sqrt{2} \\ 0 & 1/\sqrt{2} & -1/\sqrt{2} \end{pmatrix}$, $D = \begin{pmatrix} -1 & 0 & 0 \\ 0 & 1 & 0 \\ 0 & 0 & 1 \end{pmatrix}$

(c) $Q = \begin{pmatrix} -2/3 & 1/3 & 2/3 \\ 2/3 & 2/3 & 1/3 \\ 1/3 & -2/3 & 2/3 \end{pmatrix}$, $D = \begin{pmatrix} 0 & 0 & 0 \\ 0 & 3 & 0 \\ 0 & 0 & 6 \end{pmatrix}$

(e) $Q = \begin{pmatrix} 2/3 & -2/3 & 1/3 \\ 1/3 & 2/3 & 2/3 \\ 2/3 & 1/3 & -2/3 \end{pmatrix}$, $D = \begin{pmatrix} 0 & 0 & 0 \\ 0 & 9 & 0 \\ 0 & 0 & 9 \end{pmatrix}$

15. (a) $\lambda^2 - 7\lambda + 22 = 0,\ A^3 = \begin{pmatrix} -100 & 81 \\ -108 & -19 \end{pmatrix},\quad A^4 = \begin{pmatrix} -524 & 105 \\ -140 & -419 \end{pmatrix}$

$$A^{-1} = \begin{pmatrix} 5/22 & -3/22 \\ 2/11 & 1/11 \end{pmatrix},\quad A^{-2} = \begin{pmatrix} 13/484 & -21/484 \\ 7/121 & -2/121 \end{pmatrix}$$

(c) $\lambda^3 - \lambda^2 + 3 = 0$, $A^3 = \begin{pmatrix} -3 & 1 & 1 \\ -3 & -3 & 0 \\ -3 & -2 & -2 \end{pmatrix}$, $A^4 = \begin{pmatrix} -6 & -2 & 1 \\ 0 & -3 & -3 \\ 3 & -5 & -2 \end{pmatrix}$

$$A^{-1} = \begin{pmatrix} 1/3 & 0 & -1/3 \\ 2/3 & 0 & 1/3 \\ 1/3 & 1 & -1/3 \end{pmatrix}, \quad A^{-2} = \begin{pmatrix} 0 & -1/3 & 0 \\ 1/3 & 1/3 & -1/3 \\ 2/3 & -1/3 & 1/3 \end{pmatrix}$$

17. **(a)** $\lambda^3 - 2\lambda^2 - \lambda + 2 = 0$, $A^{-1} = \begin{pmatrix} -5 & -1/2 & 5/2 \\ -4 & -1/2 & 3/2 \\ -12 & -1 & 6 \end{pmatrix}$

(c) $\lambda^3 + 2\lambda^2 - 11\lambda - 12 = 0$, $A^{-1} = \begin{pmatrix} -3/4 & -1/4 & -1 \\ -1/2 & -1/2 & 1 \\ 1/12 & -1/12 & 1/3 \end{pmatrix}$

19. **(a)** $\lambda^3 - 5\lambda^2 - 19\lambda + 65 = 0$, $\det(A) = -65$

$$adj(A) = \begin{pmatrix} -5 & -20 & 5 \\ -10 & 12 & -3 \\ 0 & 39 & -26 \end{pmatrix}$$

$$A^{-1} = \begin{pmatrix} 1/3 & 4/13 & -1/13 \\ 2/13 & -12/65 & 3/65 \\ 0 & -3/5 & 2/5 \end{pmatrix}$$

(c) $\lambda^4 + 6\lambda^3 - 3\lambda^2 - 1 = 0$, $\det(A) = -1$

$$adj(A) = \begin{pmatrix} 0 & -1 & 0 & -2 \\ -1 & 1 & 2 & -2 \\ 0 & -1 & -3 & 3 \\ 2 & -2 & -3 & 3 \end{pmatrix}$$

$$A^{-1} = \begin{pmatrix} 0 & 1 & 0 & 2 \\ 1 & -1 & -2 & 2 \\ 0 & 1 & 3 & -3 \\ -2 & 2 & 3 & -3 \end{pmatrix}$$

21. **(a)** $\lambda^{(4)}) = 5.7320$, $[0.8255, 1.000, 0.9225]^T$

(c) $\lambda^{(4)} = 3.3103, \quad \mathbf{X} = [0.4583, 0.0729, 1.0000]^T$

23. **(a)** $\lambda^{(4)} = 4.3731, \quad \mathbf{X} = [0.8897, -0.2962, 1.0000]^T$

(c) $\lambda^{(4)} = 4.4143, \quad \mathbf{X} = [0.3333, 0.1381, 1.0000]^T$

25. **(a)** The three real eigenvalues satisfies $0 \leq \lambda \leq 6$.

(c) The three real eigenvalues satisfies $-2 \leq \lambda \leq 4$.

27. $Q = \begin{pmatrix} 1 & 0 & 0 \\ 0 & 1 & 0 \\ 0 & 0 & 1 \end{pmatrix}, \qquad B = \begin{pmatrix} 2 & -3 & 6 \\ 0 & 3 & -4 \\ 0 & 2 & -3 \end{pmatrix}$

$C = \begin{pmatrix} 3 & -4 \\ 2 & -3 \end{pmatrix}$

C has eigenvalues 1, -1 with eigenvectors $[1, 1/2]^T$ and $[1, 1]^T$, respectively. The eigenvectors of B are $[0, 1, 1/2]^T$, $[-1, 1, 1]^T$, which are also the remaining eigenvectors of matrix A.

29. **(a)** $\lambda = 13.6138, -0.1635, -5.4502$

$$\mathbf{X} = \begin{pmatrix} 0.7000 & -0.6414 & -0.3140 \\ 0.4357 & 0.7319 & -0.5240 \\ 0.5659 & 0.2300 & 0.79118 \end{pmatrix}$$

(c) $\lambda = 11.5915, 6.6240, -7.2154$

$$\mathbf{X} = \begin{pmatrix} 0.7933 & -0.1710 & -0.5843 \\ 0.4934 & 0.7429 & 0.4525 \\ 0.3567 & -0.6472 & 0.6737 \end{pmatrix}$$

(e) $\lambda = -1.6073, 4.9988, 14.5592, 4.0494$

$$\mathbf{X} = \begin{pmatrix} 0.6068 & 0.4539 & 0.4156 & -0.5031 \\ -0.4738 & 0.5814 & 0.5323 & 0.3928 \\ -0.4513 & -0.4775 & 0.5215 & -0.5444 \\ 0.4513 & -0.4775 & 0.5215 & 0.5444 \end{pmatrix}$$

31. **(a)** $\lambda = 6.6569, -4.6569, 1.000$
(c) $\lambda = 3.8284, -1.8284, 1.0000$

33. **(a)** $T = \begin{pmatrix} 1 & 4 & 0 \\ 4 & 1 & 4 \\ 0 & 4 & 1 \end{pmatrix}, \quad \lambda = 6.6569, -4.6569, 1.000$

(c) $T = \begin{pmatrix} 1 & -2 & 0 \\ -2 & 1 & -2 \\ 0 & -2 & 1 \end{pmatrix}, \lambda = 3.8284, -1.8284, 1.0000$

35. **(a)** $T = \begin{pmatrix} 2.0000 & -5.0000 & 0.0000 \\ -5.0000 & 10.0800 & 0.4400 \\ 0.0000 & 0.4400 & -0.0800 \end{pmatrix}$

$\lambda = 12.4807, -0.4807, 0.0000$

(c) $T = \begin{pmatrix} 4.0000 & 4.5826 & 0.0000 & 0.0000 \\ 4.5826 & 5.0476 & -3.3873 & 0.0000 \\ 0.0000 & -3.3873 & 7.1666 & -1.1701 \\ 0.0000 & 0.0000 & -1.1701 & 5.7858 \end{pmatrix}$

$\lambda = 11.1117, 6.7240, 4.9673, -0.8030$

37. **(a)** $A^{(15)} = \begin{pmatrix} 5.4713 & -0.7373 & 2.1419 \\ -0.0097 & -3.4713 & 1.9705 \\ 0.0000 & 0.0000 & 1.0000 \end{pmatrix}$

$\lambda = 5.4713, -3.4713, 1.0000$

(c) $A^{(15)} = \begin{pmatrix} 3.8284 & -0.0001 & 0.0000 \\ -0.0001 & -1.8284 & -0.0002 \\ 0.0000 & -0.0002 & 1.0000 \end{pmatrix}$

$\lambda = 3.8284, -1.8284, 1.0000$

39. (a) $\lambda = 4.3028, 0.70, 0,$ Iterations $= 8$

(c) $\lambda = 3, 2, 1,$ Iterations $= 24$

41. (a) $\begin{pmatrix} 1 & 7.6 & 4 \\ 5 & 2.2 & 3 \\ 0 & 4.72 & 2 \end{pmatrix}$

(c) $\begin{pmatrix} 9 & 3 & -5 & -2 \\ 2 & 5 & -14 & -5 \\ 0 & 2 & -14 & -7 \\ 0 & 0 & 20 & 13 \end{pmatrix}$

43. (a) $\sigma_1 = 9, \sigma_2 = 8, \sigma_3 = 3$

(c) $\sigma_1 = 2, \sigma_2 = 6, \sigma_3 = 9$

45. Let A be an orthogonal matrix, then $A^T = A^{-1}$. Moreover, the singular values of A are eigenvalues of matrix $A^TA = A^{-1}A = I$, since the only eigenvalue of an identity matrix is 1.

47. (a)

$$A = UDV^T = \begin{pmatrix} 0 & 1 \\ 1 & 0 \end{pmatrix} \begin{pmatrix} 3 & 0 \\ 0 & 2 \end{pmatrix} \begin{pmatrix} -1 & 0 \\ 0 & -1 \end{pmatrix}$$

(c)

$$A = UDV^T = \begin{pmatrix} \frac{1}{\sqrt{3}} & 0 & \frac{2}{\sqrt{6}} \\ \frac{1}{\sqrt{3}} & \frac{1}{\sqrt{2}} & -\frac{1}{\sqrt{6}} \\ -\frac{1}{\sqrt{3}} & \frac{1}{\sqrt{2}} & \frac{1}{\sqrt{6}} \end{pmatrix} \begin{pmatrix} \sqrt{3} & 0 \\ 0 & \sqrt{2} \\ 0 & 0 \end{pmatrix} \begin{pmatrix} 0 & 1 \\ 1 & 0 \end{pmatrix}$$

49. **(a)** $\mathbf{x} = [x_1, x_2]^T = [0.25, -0.25]^T$
(c) $\mathbf{x} = [x_1, x_2, x_3]^T = [-0.6154, 2.6923, 1.3846]^T$

Appendix F

About the CD-ROM

- Included on the CD-ROM are the MATLAB ".m" files from the text, simulations, third party software, and other files related to topics in numerical analysis.
- See the "README" files for any specific information/system requirements related to each file folder, but most files will run on Windows 2000 or higher and Linux.

Bibliography

[1] Abramowitz M. and I. A. Stegum I. A.(eds): Handbook of Mathematical Functions, National Bureau of Standards, 1972.

[2] Achieser, N. I.: Theory of Approximation, Dover, New York, 1993.

[3] Ahlberg, J., E. Nilson, and J. Walsh: The Theory of Splines and their Application, Academic Press, New York (1967).

[4] Akai, T. J.: Applied Numerical Methods for Engineers, John Wiley & Sons, New York, 1993.

[5] Allgower, E. L., K. Glasshoff, and H. O. Peitgen (eds.): Numerical Solution of Nonlinear Equations, LNM**878**, Springer-Verlag, 1981.

[6] Atkinson, K. E. and W. Han: An Introduction to Numerical Analysis, 3rd ed., John Wiley, New York, 2004.

[7] Axelsson, O.: Iterative Solution Methods, Cambridge University press, New York, 1994.

[8] Ayyub, B. M. and R. H. McCuen: Numerical Methods for Engineers, Prentice Hall, Upper Saddle River, NJ, 1996.

[9] Bender, C. M. and S. A. Orszag: Advanced Mathematical Methods for Scientists and Engineers, McGraw-Hill, New York, 1978.

[10] Blum, E. K.: Numerical Analysis and Computation: Theory and Practice, Addison-Wesley, Reading, MA, 1972.

[11] Borse, G. H.: Numerical Methods with MATLAB, PWS, Boston, 1997.

[12] Bronson, R.: Matrix Methods - An Introduction, Academic Press, New York, 1969.

[13] Buchanan, J. L. and P. R. Turner: Numerical Methods and Analysis, McGraw-Hill, New York, 1992.

[14] Burden, R. L. and J. D. Faires: Numerical Analysis, 8th ed., Brooks/Cole Publishing Company, Boston, 2005.

[15] Butcher, J.: The Numerical Analysis of Ordinary Differential Equations, John Wiley, New York, 1987.

[16] Carnahan, B., A. H. Luther, and J. O. Wilkes: Applied Numerical Methods, John Wiley & Sons, New York, 1969.

[17] Chapra, S. C. and R. P. Canale: Numerical Methods for Engineers, 3rd ed., McGraw-Hill, New York, 1998.

[18] Cheney, E. W.: Introduction to Approximation Theory, McGraw-Hill, New York, 1982.

[19] Ciarlet, P. G.: Introduction to Numerical Linear Algebra and Optimization, Cambridge University Press, Cambridge, 1989.

[20] Coleman, T. F. and C. Van Loan: Handbook for Matrix Computations, SAIM, Philadelphia, 1988.

[21] Conte, S. D. and C. de Boor: Elementary Numerical Analysis, 3rd ed., McGraw-Hill, New York, 1980.

[22] Dahlquist, G. and A. Bjorck: Numerical Methods, Prentice Hall, Englewood Cliffs, NJ, 1974.

[23] Daniels, R. W.: An Introduction to Numerical Methods and Optimization Techniques, North-Holland, New York, 1978.

[24] Datta, B. N.: Numerical Linear Algebra and Application, Brook/Cole, Pacific Grove, CA, 1995.

[25] Davis, P. J.: Interpolation and Approximation, Dover, New York, 1975.

[26] Davis, P. J. and P. Rabinowitz: Methods of Numerical Integration, 2nd ed., Academic Press, 1984.

[27] Epperson, J. F.: An Introduction to Numerical Methods and Analysis, John Wiley & Sons, Chichester, 2001.

[28] Etchells, T. and J. Berry: Learning Numerical Analysis Through Derive, Chartwell- Bratt, Kent, 1997.

[29] Etter, D. M. and D. C. Kuncicky: Introduction to MATLAB, Prentice Hall, Englewood Cliffs, NJ, 1999.

[30] Evans, G.: Practical Numerical Analysis, John Wiley & Sons, Chichester, England, 1995.

[31] Fatunla, S. O.: Numerical Methods for Initial-Value Problems in Ordinary Differential Equations, Academic Press, New York, 1988.

[32] Fausett, L. V.: Numerical MethodsAlgorithms and Applications, Prentice-Hall, Englewood Cliffs, NJ, 2003.

[33] Ferziger, J. H.: Numerical Methods for Engineering Application, John Wiley & Sons, New York. 1981.

[34] Forsythe, G. E. and C. B. Moler: Computer Solution of Linear Algebraic Systems,Prentice Hall, Englewood Cliffs, NJ, 1967.

[35] Fox, L.: Numerical Solution of Two-Point Boundary-Value Problems in Ordinary Differential equations, Dover, New York, 1990.

[36] Fox, L.: An Introduction to Numerical Linear Algebra, Oxford University Press, New York, 1965.

[37] Fröberg, C. E.: Introduction to Numerical Analysis, 2nd ed., Addison-Wesley, Reading, Massachusetts, 1969.

[38] Fröberg, C. E.: Numerical Mathematics: Theory and Computer Application, Benjamin/Cummnings, Menlo Park, CA, 1985.

[39] Gerald, C. F. and P. O. Wheatley: Applied Numerical Analysis, 7th ed., Addison-Wesley, Reading, Massachusetts, 2004.

[40] Gilat A.: MATLAB- An Introduction with Applications, John Wiley & Sons, New York, 2005.

[41] Goldstine, H. H.: A History of Numerical Analysis from the 16th Through the 19th Century, Springer-Verlag, New York, 1977.

[42] Golub, G. H.: Studies in Numerical Analysis, Washington, DC, MAA, 1984.

[43] Golub, G. H. and J. M. Ortega: Scientific Computing and Differential Equations, Academic Press, New York, 1992.

[44] Golub, G. H. and C. F. van Loan: Matrix Computation, 3rd ed., Johns Hopkins University Press, Baltimore, MD, 1996.

[45] Goldstine, H. H.: A History of Numerical Analysis from the 16th Through the 19th Century, Springer-Verlag, New York, 1977.

[46] Greenspan, D. and V. Casulli: Numerical Analysis for Applied Mathematics, Science and Engineering, Addison-Wesley, New York, 1988.

[47] Greeville, T. N. E.: Theory and Application of Spline Functions, Academic Press, New York, 1969.

[48] Griffiths, D. V. and I. M. Smith: Numerical Methods for Engineers, CRC Press, Boca Raton, FL, 1991.

[49] Hageman, L. A. and D. M. Young: Applied Iterative Methods, Academic Press, New York, 1981.

[50] Hager, W. W.: Applied Numerical Linear Algebra, Prentice Hall, Englewood Cliffs, NJ, 1988.

[51] Hamming, R. W.: Introduction to Applied Numerical Analysis, McGraw-Hill book company, New York, 1971.

[52] Henrici, P. K.: Elements of Numerical Analysis, John Wiley & Sons, New York, 1964.

[53] Hildebrand, F. B.: Introduction to Numerical Analysis, 2nd edition, McGraw-Hill, New York, 1974.

[54] Hoffman, Joe. D.: Numerical Methods for Engineers and Scientists,McGraw-Hill, New York, 1993.

[55] Hohn, F. E.: Elementary Matrix Algebra, 3rd ed., Macmillan, New York, 1973.

[56] Horn, R. A. and C. R. Johnson: Matrix Analysis, Cambridge University Press, Cambridge, 1985.

[57] Hornbeck, R. W. Numerical Methods, Prentice-Hall, Englewood Cliffs, NJ, 1975.

[58] Householder, A. S.: The Numerical Treatment of a Single Non-linear Equation, McGraw-Hill, New York, 1970.

[59] Hultquist, P. E.: Numerical Methods for Engineers and Computer Scientists, Benjamin/Cummnings, Menlo Park, CA, 1988.

[60] Hunt, B. R., R. L. Lipsman, and J. M. Rosenberg: A Guide to MATLAB for Beginners and Experienced Users, Cambridge University Press, Cambridge, 2001.

[61] Isaacson, E. and H. B. Keller: Analysis of Numerical Methods, John Wiley & Sons, New York, 1966.

[62] Jacques, I. and C. Judd: Numerical Analysis, Chapman and Hall, New York, 1987.

[63] Jennings, A.: A Matrix Computation for Engineers and Scientists, John Wiley & Sons, London, 1977.

[64] Johnson, L. W. and R. D. Riess: Numerical Analysis, 2nd ed., Addison-Wesley, Reading, MA, 1982.

[65] Johnston, R. L.: Numerical MethodsA Software Approach, John Wiley & Sons, New York, 1982.

[66] Kahanger, D., C. Moler, and S. Nash: Numerical Methods and Software, Prentice Hall, Englewood Cliffs, NJ, 1989.

[67] Kharab, A. and R. B. Guenther: An Introduction to Numerical MethodsA MATLAB Approach, Chapman & Hall/CRC, New York, 2000.

[68] Kincaid, D. and W. Cheney: Numerical AnalysisMathematics of Scientific Computing, 3rd ed., Brooks/Cole Publishing Company, Boston, 2002.

[69] King, J.T.: Introduction to Numerical Computation, McGraw-Hill, New York, 1984.

[70] Knuth, D. E.: Seminumerical Algorithms, 2nd ed., Vol. 2 of The Art of Computer Programming, Addison-Wesley, Reading, MA, 1981.

[71] Lastman, G. J. and N. K. Sinha: Microcomputer-Based Numerical Methods for Science and Engineering, Saunders, New York, 1989.

[72] Lawson, C. L. and R. J. Hanson: Solving Least Squares Problems, SIAM, Philadelphia, 1995.

[73] Leader, J. J.: Numerical Analysis and Scientific Computation, Addison-Wesley, Reading, MA, 2004.

[74] Linear, P.: Theoretical Numerical Analysis, Wiley, New York, 1979.

[75] Linz, P. and R. L. C. Wang: Exploring Numerical MethodsAn Introduction to Scientific Computing Using MATLAB, Jones and Bartlett Publishers, Boston, 2002.

[76] Marcus, M.: Matrices and Matlab, Prentice Hall, Englewood Cliffs, NJ, 1993.

[77] Maron, M. J. and R. J. Lopez: Numerical Analysis: A Practical Approach, 3rd ed., Wadsworth, Belmont, CA, 1991.

[78] Mathews, J. H.: Numerical Methods for Mathematics, Science and Engineering, 2nd ed., Prentice-Hall Englewood Cliffs, NJ, 1987.

[79] Meyer, C. D.: Matrix Analysis and Applied Linear Algebra, SIAM, Philadelphia, 2000.

[80] Moore, R. E.: Mathematical Elements of Scientific Computing, Holt, Reinhart & Winston, New York, 1975.

[81] Mori, M. and R. Piessens (eds.): Numerical Quadrature, North Holland, New York, 1987.

[82] Morris, J. L.: Computational Methods in Elementary Theory and Application of Numerical Analysis, John Wiley & Sons, New York, 1983.

[83] Nakamura, S.: Applied Numerical Methods in C, Prentice-Hall, Englewood Cliffs, NJ, 1993.

[84] Nakos, G. and D. Joyner: Linear Algebra With Applications, Brooks/Cole Publishing Company, Boston, 1998.

[85] Neumaier, A.: Introduction to Numerical Analysis, Cambridge University Press, Cambridge, 2001.

[86] Nicholson, W. K.: Linear Algebra With Applications, 4th ed., McGraw-Hill Ryerson, New York, 2002.

[87] Noble, B. and J. W. Daniel: Applied Linear Algebra, 2nd ed., Prentice-Hall, Englewood Cliffs, NJ, 1977.

[88] Ortega, J. M.: Numerical Analysis-A Second Course, Academic Press, New York; 1972.

[89] Quarteroni, Alfio: Scientific Computing with MATLAB, Springer-Verlag, Berlin Heidelberg, 2003.

[90] Ralston, A. and P. Rabinowitz: A first course in Numerical Analysis, 2nd Edition, McGraw-Hill, New York, 1978.

[91] Recktenwald, G.: Numerical Methods with MATLABImplementation and Application, Prentice Hall, Englewood Cliffs, NJ, 2000.

[92] Rice, J. R.: Numerical methods, Software and Analysis, McGraw-Hill, New York, 1983.

[93] Rorrer, C. and H. Anton: Applications of Linear Algebra, John Wiley & Sons, New York, 1977.

[94] Schatzman, M.: Numerical AnalysisA Mathematical Introduction, Oxford University Press, New York, 2002.

[95] Scheid, F.: Numerical Analysis, McGraw-Hill, New York, 1988.

[96] Schilling, R. J. and S. L. Harris: Applied Numerical Methods for Engineers using MATLAB and C, Brooks/Cole Publishing Company, Boston, 2000.

[97] Steward, B. W.:Introduction to Matrix Computations, Academic Press, New York, 1973.

[98] Stewart, G. W.: Afternotes on Numerical Analysis, SIAM, Philadelphia, 1996.

[99] Store, J. and R. Bulirsch: Introduction to Numerical Analysis, Springer-Verlag, New York, 1980.

[100] Strang, G.: Linear Algebra and Its Applications 3rd ed., Brooks/Cole Publishing Company, Boston, 1988.

[101] Suli, E. and D. Mayers: An Introduction to Numerical Analysis, Cambridge University Press, Cambridge, 2003.

[102] The Mathworks, Inc.: Using MATLAB. The Mathworks, Inc., Natick, MA, 1996.

[103] The Mathworks, Inc.: Using MATLAB Graphics. The Mathworks, Inc., Natick, MA, 1996.

[104] The Mathworks, Inc.: MATLAB Language Reference Manual. The Mathworks, Inc., Natick, MA, 1996.

[105] Trefethen, L. N. and D. Bau III: Numerical Linear Algebra, SIAM, Philadelphia, 1997.

[106] Traub, J. F.: Iterative Methods for the Solution of Equations, Prentice-Hall, Englewood Cliffs, NJ, 1964.

[107] Turner, P. R.: Guide to Scientific Computing, 2nd ed., Macmillan Press, Basingstoke, 2000.

[108] Usmani, R. A.: Numerical Analysis for Beginners, D and R Texts Publications, Manitoba, 1992.

[109] Vandergraft, J. S.: Introduction to Numerical Computations, Academic Press, New York, 1978.

[110] Verga, R. S.: Matrix Iterative Analysis, Prentice-Hall, Englewood Cliffs, NJ. 1962.

[111] Wilkinson, J. H.: The Algebraic Eigenvalue Problem, Clarendon Press, Oxford, England, 1965.

[112] Wilkinson, J. H. and C. Reinsch: Linear Algebra, vol. II of Handbook for Automatic Computation, Springer-Verlag, New York, pp. 418, 1971.

[113] Wood, A.: Introduction to Numerical Analysis, Addison-Wesley, Reading, MA, 1999.

Index

Q

R

S